Fundamentals of Ocean Climate Models

Fundamentals of Ocean Climate Models

Stephen M. Griffies

PRINCETON UNIVERSITY PRESS
PRINCETON AND OXFORD

Published by Princeton University Press, 41 William Street, Princeton, New Jersey 08540.
In the United Kingdom: Princeton University Press, 3 Market Place, Woodstock, Oxfordshire OX20 1SY.

Library of Congress Cataloguing-in-Publication Data

Griffies, Stephen M., 1962-
Fundamentals of ocean climate models / Stephen M. Griffies.
p. cm.
Includes bibliographical references and index.
ISBN 0-691-11892-2 (acid-free paper)
1. Ocean circulation–Mathematical models. 2. Ocean circulation–Computer simulation. 3. Atmospheric physics–Mathematical models. 4. Atmospheric physics–Computer simulation. I. Title.

GC228.5.G75 2004
551.46′2′015118–dc22 2004043409

British Library Cataloging-in-Publication Data is available

This book has been composed in LaTex.

The publisher would like to acknowledge the author of this volume for providing the camera-ready copy from which this book was printed.

Printed on acid-free paper.

pup.princeton.edu

Printed in the United States of America

10 9 8 7 6 5 4 3 2 1

Para Adi. Muchas gracias por tu amor, cariño. Sat Nam.

Contents

FOREWORD

Ocean models embody a wide range of oceanic knowledge. There are issues concerned with the numerical representation of processes and of conservation equations on different finite spatial grids. There is the need to respect the smallness of diapycnal mixing and particularly not to allow horizontal diffusion to cause false diapycnal density fluxes. Do the model variables represent the Eulerian mean quantities, or are they the result of some other type of averaging process? Is the Boussinesq approximation really made in a model or is this simply a matter of interpretation? Should the diffusion tensor be symmetric, and if not, what is the physical justification for the skew component of diffusion? Where this book really shines is that rather than simply presenting workable recipes to each of these issues (and many more), the underlying physics is expertly described, usually first in differential equations and then in the discrete representation on the numerical grid. Because of this clarity and attention to detail, the book will appeal not only to ocean modellers but to a much broader spectrum of oceanographers.

As a branch of fluid dynamics, distinguishing features of the ocean are its rather strong stratification and the smallness of its diapycnal mixing processes. For climate purposes the seemingly small diapycnal mixing processes are very important yet their faithful representation in models has been a formidable obstacle. While we have known the importance of diapycnal processes in controlling the thermohaline ocean circulation for some decades, it is only in the past decade that we have learned how to control the amount of diapycnal mixing in z-coordinate ocean models, and then only at coarse resolution. This goal has been achieved by clear thinking about the conservation of properties in ocean models, particularly the way in which diffusion is imposed. With this book we now have the relevant averaged model equations derived carefully from first principles, and the subtleties associated with interpreting averaged quantities should no longer need to be glossed over.

The present book can be described as providing comprehensive treatment of the following subjects (i) deriving the ocean's conservation equations from first principles, (ii) carefully considering the issues that arise when these equations are averaged, (iii) describing the many numerical procedures that are used to integrate the averaged equations forward in time, and (iv) providing the tensorial underpinning so that the equations can be transformed consistently onto different grids on the spherical earth. This book clearly fills a void in the oceanographic literature. The usual development skips the first two topics and treats the averaged equations as given (which begs the question of what the model variables might represent), while the subject of tensor analysis is familiar to too few oceanographers, especially given the range of grids that are now in common use in ocean models.

Two key advances in ocean modelling over the past decade are the rotation of the (symmetric) diffusion tensor to be aligned with respect to the local neutrally buoyant directions, and the discovery and implementation of the so-called Gent-McWilliams mixing scheme for mesoscale eddies. These two topics are expertly described in this book and together they occupy one third of its pages. This is entirely appropriate since both these advances have been crucial for controlling the false diapycnal mixing that otherwise occurs across sloping density surfaces in low-resolution z-coordinate models. Both advances have been hard-won by oceanographers. The solution of the first involved the subtle thermobaric nature of the equation of state of seawater, while the second relies on the action of mesoscale eddies to release potential energy, with the mixing scheme being best viewed as either an extra quasi-Stokes advection or alternatively as an additional amount of diffusion which happens to be skew diffusion. These issues are subtle and are not covered in existing oceanographic texts, and yet they have been at the root of the biggest improvements in ocean climate modelling in more than twenty years.

This book truly lives up to its title. By delving into the physical basis for the choices made in present ocean models, the book does indeed establish the fundamental basis of ocean models. Where the issues are not yet agreed on by the oceanographic community, the book takes on the character of a review of these research issues. The book will prove invaluable to ocean modellers and to those concerned with how fundamental ocean physics is represented in ocean models.

Trevor J. McDougall
Hobart, Australia

PREFACE

The central purpose of this book is to contribute to a rationalization of the physical, mathematical, and numerical foundations of computer models used to understand and predict the global ocean circulation. The presentation is geared toward students and researchers in ocean and climate science who aim to understand the physical content of the equations used in ocean models and to become exposed to methods for solving these equations. Much of the formulation is general, and so applicable to any ocean model. However, for purposes of presenting concrete examples, attention is focused on z-coordinate models and generalizations based on isomorphisms with the z-model structure. In their traditional construction, z-models represent the most common model class used to simulate the global ocean climate system. Future ocean model research and development certainly will evolve beyond the limitations of z-models (see Chapter 6). Nonetheless, their maturity and popularity, and the ease of generalizations beyond them, warrant a thorough discussion. Furthermore, their relative simplicity aids in the pedagogical treatment of the subject.

It is hoped that this book can partially fill a niche in the literature whereby a (mostly) first principles presentation of ocean climate models is given, with discussions extending from fundamental ocean fluid mechanics to detailed parameterization and discretization issues. This aim is to be distinguished from a book on ocean *modeling* or a book on ocean *physics*. To do justice to ocean modeling, a book must provide extensive discussions of simulations and their relation to Nature's ocean. To do justice to ocean physics, we would need to remove much of what remains fundamental to the model builder yet which can be characterized as "ocean model engineering." Further discussions of theoretical solutions and phenomena would also be required of such a book.

To the student, some topics in ocean model fundamentals can appear complex and esoteric. This point has contributed to what is arguably a growing distance between scientists who analyze numerical ocean model simulations and scientists who understand details of the model's inner workings. The two groups ideally should be working hand-in-hand, so that model users who compare simulations to observations can provide feedback to educate model builders about what is realistic, and not so realistc, regarding the simulations. Likewise, model builders must communicate to the user certain of the model's fundamental aspects, and its limitations, so as to lend perspective and experience to the user.

When the model user knows little of the model fundamentals, and the model builder is detached from observations of Nature's ocean, models stand the chance of losing credibility, and large-scale observations can lack a rational mechanistic description. This situation is unhealthy for the science.* It is hoped that this book provides the climate science student and/or researcher with a digestible presentation of ocean models so as to assist in removing their mystery and opacity. In the end, this book will be worth the effort (to read and to write) if it helps to reduce the distance between model builder and model user.

WHAT THIS BOOK IS AND IS NOT

The scientist who builds an ocean model aims to provide a tool facilitating the understanding of ocean physics. Ideally, the numerical tool is congruent with physical reality, and so the model builder must be well versed in ocean physics. By exposing a selection of fundamental physical issues that arise when building an ocean model, this book may allow the student to garner a better appreciation, if not an improved understanding, of the underlying notions and methods forming the foundation of models.

To assist in providing a readable book, presentations are self-contained and generally start from a basic physical perspective. To garner the most from the discussions, some patience and persistence is asked of the reader since many of the discussions are dense. This is perhaps the most important prerequisite for making it through this book from cover to cover. As with most treatments of continuum mechanics, some level of mathematical sophistication, fully explained in the text, proves useful to present physical and mathematical results cleanly and in a generality necessary for developing global ocean models. To help with such material, a working knowledge of classical mechanics, vector calculus, and basic fluid mechanics, each realized from undergraduate physics courses and/or graduate studies in the geosciences, is useful.

This book is not a text on oceanography, geophysical fluid dynamics, or numerical methods. Hence, there are no problem sets given, no computer simulations described, and only very few theoretical solutions presented. Furthermore, little mention is made of ocean observations, and knowledge of basic numerical methods is assumed. For more complete treatments of these subjects, the reader is referred to one or more of the many excellent texts and monographs in the literature. A selection of these, familiar to the author, includes the following.

- Pickard and Emery (1990) and Tomczak and Godfrey (1994) present observational descriptions of the ocean and so introduce the reader to the phenomenology that a computer model aims to simulate.

- Gill (1982), Apel (1987), Pedlosky (1987), Cushman-Roisin (1994), Pedlosky (1996), Mellor (1996), Salmon (1998), and Pedlosky (2003) discuss various aspects of geophysical fluid dynamics and theoretical physical oceanography.

- Washington and Parkinson (1986), Trenberth (1992), Peixoto and Oort (1992), Chassignet and Verron (1998), Haidvogel and Beckmann (1999), Kantha and Clayson (2000a), and Kantha and Clayson (2000b) introduce the reader to various aspects of climate physics and climate modeling.

*In an ideal world, those who focus on the elements of ocean model design and construction should spend time at sea, and those who regularly go to sea should learn more about models and use them to help rationalize datasets.

- The books and reviews by Haltiner and Williams (1980), O'Brien (1986), Bryan (1989), Durran (1999), and Randall (2000) describe numerical methods of use for geophysical fluid dynamics.

- The book edited by Siedler et al. (2001) is noteworthy for its many lucid contributions by leading experts. This book affords the reader a clear view at the state-of-the-art in ocean observations, theory, and modeling.

Given the above caveats about what is not contained in this book, it would be remiss not to admit that there are numerous further topics missing that arguably should be present in a monograph on ocean models. Their absence is due to the author's physical and mental limitations, rather than any reflection on their lack of importance. It is hoped that this book nonetheless serves as a useful complement to other presentations, and that it provides a stepping-stone for future expositions which may fill the many holes contained here.

It is worth commenting a bit here on the level of technical sophistication employed in this book. As discussed above, the aim is to rationally and systematically formulate the equations to be used by numerical ocean climate models. As these models are posed in a spherical geometry, it is necessary to make some use of notions from non-Euclidean geometry. This is facilitated by general tensor analysis. Hence, this book generally does not shy away from the use of tensor analysis where it is needed, yet a conscious effort is made to use more familiar vector calculus notation when sufficient. Overall, it is hoped that the mathematical treatment is self-contained enough to allow the interested reader to become proficient in the tools of tensor analysis. However, those uninterested in such details can readily jump across places where tensor gymnastics are employed, without too much loss of continuity.

This book maintains the premise that *physical, mathematical, and numerical details matter when formulating the algorithms of ocean climate models*. This perspective prompted the author to exploit many of the tensor analysis tools discussed above, as well as ideas from functional methods to discretize the elliptic transport operators appearing in certain subgrid scale (SGS) schemes. However, the approach requires defense since there remains a large degree of uncertainty in the SGS schemes appropriate for ocean climate simulations. The neutral physics and horizontal friction operators discussed in Parts 4 and 5 are prime examples.

A large degree of physical uncertainty may argue for the use of a somewhat "impressionistic" approach regarding the level of detail warranted to formulate the mathematical and numerical aspects of ocean models. However, the perspective taken in this book is that it is precisely because of the large physical uncertainty that modelers should pay a great deal of attention to mathematical and numerical details. Doing so gives a more solid technical foundation to the algorithms, thus allowing the scheme's physical ideas to more clearly translate into rational behavior from the simulation. Importantly, mathematical and numerical rigor does not translate into physical rigor (nor is the complement true). Nor should it be construed as providing weak physical ideas with a sheen of legitimacy. Instead, such attention to detail attempts to provide a sound venue from which to better represent any physics, or lack thereof, available from the scheme. Doing so provides the physical modeler with a better tool for understanding.

Quite generally, what is argued for in this book is an increased level of *honest algorithm* development in ocean climate models. By this is meant that physical, mathematical, and numerical details should be exposed, and a sincere attempt should be made to rationally address and document them. Furthermore, assumptions and uncertainties should be clearly articulated rather than obscured. This may prove no more successful in improving the simulations than an impressionistic approach. However, there are cases where less rigorous methods have proven less successful. Additionally, more rigor and honesty will promote an improved facility to reason why simulations from various models agree, or differ, from one another, and in so doing assist in reducing uncertainty in model predictions.

ORGANIZATION

This book is organized into parts, each of which have multiple chapters. It is assumed that the book will *not* be read cover-to-cover. Consequently, the contents of each part are written so that they can largely be read independently. Chapters likewise aim for such independence, though less so. This organizational style allows the book to be useful for those reading it in a more or less arbitrary manner, so long as some of the basics are appreciated. It is hoped that this approach enhances the book's readability, accessibility, and utility as both a monograph and reference to students and researchers.

Chapter 1 starts the book with an introduction to ocean climate models. It is here that we are exposed to some reasons why ocean models are of primary importance for climate science. This chapter also helps to motivate the more fundamental, and by necessity mathematical, development presented in subsequent chapters.

Part 1 begins the main part of the book by focusing on the equations describing stratified fluid dynamics on a rotating sphere. Chapter 2 introduces some physical ideas underlying a continuum description of the ocean fluid. It also highlights common approximations made in ocean climate models. Chapters 3 and 4 derive the hydrodynamical equations of the ocean fluid. Chapter 5 presents aspects of ocean energetics and thermo-hydrodynamics. Chapter 6 presents some mathematical results applicable to a generalized vertical coordinate description of the ocean. This chapter also highlights some deficiencies inherent in the different model classes now in use for simulating the ocean.

The equations described in Part 1 offer a precise mathematical description of a particular realization of the ocean fluid. To make use of this description requires an infinite level of knowledge about the ocean's state, i.e., the positions and velocities of each fluid parcel, and full information about forces acting on the parcels. In practice, we are always faced with less than infinite information in both space and time. Hence, it is necessary to derive alternative equations via a coarse-graining procedure. That is, we need to determine approximate, averaged, or mean field equations for the coarsened ocean fluid.

Part 2 presents three chapters illustrating various issues with averaged descriptions of the ocean. Chapter 7 begins with a general discussion of subgrid scale processes, with a focus on those contributing to dianeutral transport. Such transport is spread non-uniformly throughout the World Ocean, with most occurring in regions near ocean boundaries. Dianeutral transport plays an important role in affecting the ocean's stratification and vertical distribution of tracers, and this chapter provides an introduction to some of the issues. Chapter 8 is the first of two chapters discussing some formal issues related to averaging the equations. Here, we accept the fact that the precise state of the turbulent ocean is not physically measurable with finite instruments, thus necessitating a statistical approach. In particular, this chapter provides a conceptual and mathematical interpretation of the fields discretized by a z-model. It is argued that these equations represent an ensemble mean, as measured at a fixed point in space-time, of individual realizations of the ocean fluid. The subgrid scales considered in this average are associated with the small-scale dianeutral mixing arising from three-dimensional turbulence. In Chapter 9, we focus on the transport of volume and tracers by mesoscale eddies. This transport, which is mostly two-dimensional and oriented according to *neutral directions*, is the dominant means whereby tracers are stirred within the ocean interior, with tracer gradients ultimately dissipated by small scale mixing processes. The discussion in this chapter motivates a particular interpretation of model variables for those cases where the simulations do not explicitly resolve mesoscale eddies. Furthermore, in this context we see how the choice of vertical coordinate strongly affects the simplicity of the mathematical description of adiabatic transport.

Part 3 presents three chapters on the semi-discrete equations of an ocean model, with focus on those equations used in a z-model. Semi-discrete refers to equations that have some parts discretized, some parts continuous. Chapter 10 begins the discussion with an introduction to issues arising when discretizing the primitive equations. This material is quite terse and is presented mostly to anticipate later discussions given in other chapters. The aim for the remainder of this part is to describe in general terms the equations of a discrete z-model, and to outline methods used to time step these equations. In particular, Chapter 11 discusses the semi-discrete version of the mass and tracer budgets, and highlights the importance of maintaining compatibility between these budgets. Chapter 12 discusses various methods used to march the primitive equations forward in time.

Chapter 9 introduces a framework for thinking about transport in the ocean interior. The parameterization of this transport in ocean models constitutes the *neutral physics* part of the models. Notably, the integrity of z-model simulations has greatly improved since modelers started employing such schemes (for a review, see Griffies et al., 2000a). Part 4 of this book aims to rationalize the neutral physics schemes commonly used in z-models. This aim is far from realized, since so many aspects of subgrid scale closure applied to ocean models are not deductive. Nonetheless, we give it our best shot.

In the first two neutral physics chapters, Chapters 13 and 14, we explore physical and mathematical aspects of neutral physics transport operators. Our focus in these discussions is on neutral physics in the ocean interior, as this has been the emphasis thus far in the literature. In Chapter 15, we consider how modelers handle the neutral physics schemes next to boundaries. This material is relatively fresh and is the subject of intense research. Chapter 16 finishes this part with details of a method for discretizing neutral physics in z-models.

Horizontal friction used in ocean models plays an important role in determining the integrity of the simulations. Additionally, the state of the art in ocean climate model horizontal friction remains at an engineering stage, with modelers encouraged to use horizontal friction largely for purposes of maintaining numerical stability. The more scientifically satisfying situation is to consider horizontal friction as a parameterization of unresolved physical processes, such as in the kinetic theory of gases. However, there is no theory for how to close the momentum equation at scales relevant to ocean climate studies. Hence, horizontal friction used in ocean climate models is *ad hoc*. Indeed, some argue that dissipation of kinetic energy, which all the model implementations of friction aim to do, is an egregiously incorrect basis on which to design the schemes (e.g., Holloway, 1999). Either of these points would remain purely academic, were it not for the importance that friction plays in the simulations. Its details are crucial especially in boundary and/or equatorial regions, where geostrophy breaks down. Hence, it is critical that modelers understand the rationale underlying some of the commonly used schemes, if only to motivate approaches that lead to improvements.

Given the importance to simulations of horizontal friction, Part 5 is offered as a tutorial for these issues as they are commonly practiced in ocean climate modeling. Chapter 17 focuses on the friction force in the continuum, with mathematics developed in Part 6 of some utility. Chapter 18 then considers how to choose a viscosity to set the friction force's magnitude in an ocean climate model. The issues discussed in these chapters are generic, whereas Chapter 19 finishes this part of the book with a discretization of the horizontal friction operator on a B-grid.

Tensor analysis is not commonly taught in oceanographic or atmospheric science curricula. However, an understanding of general tensor analysis is necessary when formulating the equations of rotating fluid dynamics in a spherical geometry. Hence, tensor analysis is needed by those actually building ocean climate models, and by those aiming to fully understand details of the model's equations. Two chapters are included at the end of the book that aim to develop tensor analysis from a relatively fundamental level, with examples drawn from ocean fluid dynamics. Chapter 20 introduces the basic notions, and Chapter 21 expands on these notions while deriving some key results from calculus on curved manifolds. The presentation is aimed at the reader who has a solid foundation in undergraduate vector calculus. No exposure to Cartesian or general tensors is assumed. The pace may be a bit slow for those who have some exposure to tensor analysis, yet it is hoped that both experienced and inexperienced readers will find something of use here. For those aiming to garner a "just-the-facts" treatment, some attempt has been made to allow one to only briefly read or to skip much of this material. For this purpose, salient points are summarized as needed in the appropriate places throughout the main text, with Section 21.12 offered as a more general summary.

CONCERNING AN EMPHASIS ON SGS PHYSICS

Some comments are offered here to anticipate how the reader may feel after reading the large parts of this book addressing SGS physics. For those who may feel unsatisfied, consider three possible reasons. First, parameterizations of the ocean SGS are based on partially deductive arguments, and a great number of inspired guesses. Describing the subgrid scale, and turbulence in particular, is highly nontrivial and incomplete, both formally and phenomenologically. There are few general principles and few unambiguous observations. Second, as mentioned in many places throughout this book, simulations are dependent on details of the closure schemes. The reader will need to take this point on faith, since no simulations are presented in this book. Griffies et al. (2000a) reviews numerous studies illustrating this point. It is unsettling that simulation integrity depends to a large degree on aspects of the models that are uncertain. Unfortunately, this is the state of the art in ocean modeling. Third, sensible implementations of many closure schemes in a global model require a relatively sophisticated level of mathematical and numerical tools. The reader may feel little motivation for learning the fancy tools, if the physical ideas underlying the tools are only partially complete or even wrong. However, experience has shown that cavalier implementations of closure schemes can produce spurious results, thus rendering a poor or incorrect understanding of what the scheme is physically supposed to be doing in the simulation. Hence, fleshing out whether a scheme makes sense physically for global ocean simulations often requires a nontrivial investment in methods and techniques. Notably, the use of sophisticated mathematical and numerical tools, many of which are described in this book, in no way should be mistaken for sanctioning a physical theory as being sound, robust, or the final word on the subject.

Given the incomplete status of many ideas in ocean subgrid scale closure, one may argue that there is little reason to write a book, or a large part of a book, that places so much energy on documenting schemes now in use. Indeed, with the title to this book using the word "fundamentals," one may feel this should require it to present notions that are in universal agreement within the ocean science community. Such is not the case for *ocean model* fundamentals, nor is it likely that the community will reach such a state. Ocean scientists thrive on debate. But then why not at least wait until ideas are clarified a bit more? How much more?

Attempts to tie together various strands of understanding can help to reveal those parts of the story that are incomplete or wrong. There are indeed many incomplete stories in this book, and elsewhere in ocean modeling. Some are likely very wrong. Nonetheless, science makes progress when both what is well established, and what is not so well established yet is done in practice, are pedagogically articulated. One goal of this book is to do just that. If this goal is realized, even partially, then some success will have been achieved. Fundamental methods, techniques, formulations, assumptions, and presumptions must be exposed to the point that researchers and especially students can better understand what sorts of problems are in need of being solved and where there are holes in understanding. Oceanography requires such for its evolution into a mature science.

OCEAN CLIMATE MODEL DEVELOPMENT

Let us close this preface with some personal reflections on what is necessary for the climate science community to produce the highest quality numerical models. Start by noting the obvious: ocean climate models are not conceived one year, to be then publicly released and supported the next. Instead, they take years, indeed decades, of creative passion and obsession from many scientists and engineers. It is only via patience and persistence that an ocean climate model is successfully realized.

There are many phases of development that an ocean climate model sees. The first phase can be considered a *vision* phase, where model equations are written down and then debated by theorists and modelers alike. Material in Parts 1 and 2 of this book address this phase. Upon arriving at a set of suitable equations, one moves onto the prototype phase, where the continuum equations are discretized and coded into one's favorite computer language (Matlab, Fortran, C, etc.) using various numerical methods. Simplified tests are conducted and analyzed using the prototype model, and modifications made to address problems and limitations. In Parts 3, 4, and 5 of this book, we encounter examples of how to take the continuum equations and cast them into a form appropriate for certain ocean models, and methods are presented to solve these equations. No numerical tests are presented, as this would add much to an already long book.

Assuming that the developer of the prototype model is confident that the model produces a physically relevant simulation, one then moves onto the third and perhaps most taxing phase of ocean climate model development. It is taxing since it requires a significant level of interaction and compromise with other model developers and model users, possessing their own needs and desires. This phase sees the prototype model, which is basically a raw skeleton, become penetrated and surrounded by a tremendous amount of infrastructure and superstructure. These added layers serve many purposes, not the least of which are (A) to allow the ocean model be be readily coupled to other component models, such as models of the atmosphere, sea ice, biogeochemistry, etc.; (B) to facilitate the model's use on various computer platforms, each of which possesses idiosyncrasies understood by few but experienced by many; (C) to provide diagnostic capabilities that render the model accessible to scientific probing; and (D) to remain flexible enough so that its algorithms can be modified, tested, and further refined. Only by successfully completing the third phase does an ocean model become appropriate for serious ocean climate studies.

Model documentation is essential during the third phase, and afterward, since multiple users make contact with the model and require clear, complete, and pedagogical treatments of the algorithms. Without this, the model may tend toward becoming a black box, with only a few developers privy to its underlying "knobs and handles." In contrast, models that are well documented allow for easier usage and so present fewer barriers to the developer and nondeveloper alike. Furthermore, honest documentation allows for educated scrutiny from those scientists who may not have the time or ability to penetrate the computer code, but who can provide valuable comments on the methods and fundamentals. This situation clearly advances the state of ocean climate model integrity, and thus advances the climate science field in general.

The final stage of model development comes about when limitations of a particular model become so apparent that new efforts focus on building a new model prototype, thus starting the process over again. The motivation for starting over may come from new paradigms in computer architecture, novel ideas regarding ocean model algorithms, or simply the desire to understand and have control over the details of the model. The magnitude of the redevelopment effort should not be underestimated by those aiming to start over.

Ocean model development as a process flourishes only when there is an intimate marriage between research and development, with each phase of model building requiring an unpredictable amount of time to debate and explore various research avenues. Allowing adequate time requires dedication and support from funding agencies and managers. Absent such, ocean climate model development is handicapped, the integrity of simulations compromised, and the depth of scientific understanding shallow. For this reason, it is critical to maintain a research and development environment that fosters a healthy balance between addressing the exigencies of the moment with visions extending out years and decades. Doing so will help climate modeling to mature into a hard science. Short of this goal, the massive problems of climate science, including anthropogenic climate change, will remain elusive. It is the profound responsibility of leaders in the international climate science field to foster a balanced research and development environment, especially now that the questions of climate change are at the forefront of society's concern, and the signals appear to be rising above ambient levels of natural variability.

DISCLAIMER

Early versions of this book grew from various notes and publications written over many years to document the *Modular Ocean Model* from the Geophysical Fluid Dynamics Laboratory. Hence, elements of this book originate from scientific papers published while the author was a GFDL employee. Reference to these papers is provided at the appropriate place in the text. However, no sentences in this book are taken verbatim from published papers. The material has been extensively reworked, refined, and digested to enhance pedagogical value well beyond that appropriate for technical papers or reports. Additionally, the bulk of this book's writing occupied personal time, and was not part of any official assigned government duty.

ACKNOWLEDGMENTS

During my years in climate research, the Geophysical Fluid Dynamics Laboratory (GFDL) has provided a friendly and exciting environment for studying the earth's climate. GFDL has nurtured honest and rich scientific relationships through the active exchange of knowledge, experience, and ideas, and it encourages scientists to attack critical research problems that may take decades to resolve. These are perhaps the greatest legacies of GFDL's founder Joseph Smagorinsky, whose prescient and uncompromising vision for climate science from the 1950's to 1980's is still unfolding in the twenty-first century.

This book draws its inspiration from the many creative researchers within the oceanography and climate science communities. At an increasing rate, these communities share a sense that scientifically based models of sound physical, mathematical, and numerical integrity are effective tools to further understanding of the ocean and the earth's climate. I also owe a debt to the mathematical engineers and physicists encountered during my undergraduate and graduate studies in engineering, mathematics, and physics. Their lessons regarding how clear and general mathematical statements lead to deep physical understanding are hopefully reflected in these pages.

I have had valuable input from readers of various drafts of this book over the years. Each of their comments helped me to clarify or correct many discussions. There are too many readers to thank in full. Nonetheless, I owe Greg Holloway and Trevor McDougall special thanks for their extra level of detailed, and often critical, input.

About the Cover

This book deals with rather abstract elements of computer models used to simulate the global ocean circulation. To balance this focus, the cover reflects the reality that the computer models aim to emulate. The cover's spectacular image depicts how the atmospheric winds, oceanic waves, and icebergs interact within the Southern Ocean (as well as many other parts of the high latitude oceans).

This photograph was taken by Gerhard Dieckmann of the Alfred Wegener Institute for Polar and Marine Research in Bremerhaven, Germany. He took this photograph on 16 November 1988 (middle of the Southern Hemisphere spring) from the oceanographic research vessel RV Polarstern when it was located within the Drake Passage (57°W longitude between 58°S and 59°S latitude). The rough seas shown in the image are typical for this part of the world, which experiences the strongest climatological winds on the planet. According to the Polarstern's metereologist, "A gale cyclone west of the Drake Passage moved to the Bellingshausen Sea and its frontal system crossed the ship's track during the following night with snowfall and winds reaching force 8[†] from northeasterly, later northwesterly directions." The photo was taken just as the sun peered through a dark sky.

[†]The Beaufort Wind Scale categorizes wind speeds according to readily visible signatures. The scale is named after its creator, Admiral Sir Francis Beaufort of the British Navy, who developed it in the early 1800's. A Beaufort wind of force 8 is also known as a "fresh gale," and it corresponds to winds between 62 and 74 km hr^{-1}, or 39 and 46 miles hr^{-1}. This wind is evidenced by long streaks of foam that appear on the ocean surface.

List of Symbols

As with any book on a topic in mathematical physics, there are a plethora of symbols. This situation often provokes frustration from the reader who does not fully appreciate the meaning of a symbol, perhaps since it was defined an unknown number of chapters earlier, or perhaps because it was poorly defined from the start. Authors, in contrast, often lament the limitations of commonly used symbols for communicating their ideas in a precise and unambiguous manner. Furthermore, in a book such as this one, it is inevitable that symbols must be used to represent more than one object due to limitations of the common alphabets employed.

An attempt is made throughout this book to minimize the symbols used without sacrificing clarity. A balance has hopefully been achieved whereby symbols are defined where they are used, and redefined later when revisited in another context, or used for another meaning, or when a reminder is useful. Even so, it may prove useful to have a list of common symbols used in this book as well as the page where the symbol's definition can be found. For this purpose, symbols are organized here according to whether they are Latin, Greek, or mathematical operators. Note that many symbols used only for a particular section are not listed here, as their local definition should be sufficient.

Latin symbols

A = horizontal viscosity used in ocean models (page 391).

C = tracer mass per mass of seawater, i.e., the tracer mass fraction, or tracer *concentration* (page 88).

C_p = specific heat at constant pressure (page 94).

$\langle C \rangle^\rho$ = density weighted ensemble average tracer concentration (page 180).

c_s = speed of sound (page 116).

D = depth of a vertical column of ocean fluid extending from the free surface at $z = \eta$ to the solid earth at $z = -H$ (page 33).

$\mathrm{d}A = \mathrm{d}x\,\mathrm{d}y$ = time-independent horizontal cross-sectional area of a vertical column of ocean fluid (page 33).

$\mathrm{d}A_{(\eta)}$ = Area element on the generally curved ocean free surface (page 30).

$\mathrm{d}A_{(-H)}$ = Area element on the generally curved solid earth boundary (page 31).

$\mathrm{d}M$ = infinitesimal mass of a fluid parcel (page 13).

$\mathrm{d}V$ = infinitesimal volume of a fluid parcel (page 14).

$\mathbf{e}_{\overline{a}}$ = local orthonormal basis directions, with $\overline{a} = 1, 2, 3$ representing three orthogonal basis directions (page 130).

e_{mn} = deformation or rate of strain tensor (page 381).

e_{S} = shearing rate of strain deformation (page 391).

e_{T} = tension rate of strain deformation (page 391).

$\mathbf{F}$ = subgrid scale flux of tracer concentration (page 87).

$\mathbf{F}^{(\mathbf{v})}$ = three-dimensional force per unit mass (or acceleration) due to subgrid scale frictional effects (page 54).

$\mathbf{F}^{(\mathbf{u})}$ = horizontal force per unit mass due to subgrid scale frictional effects (page 54).

f = Coriolis parameter or the inertial frequency (page 46).

g = effective gravitational acceleration (page 45).

g_{mn} = metric tensor (page 481).

$\mathcal{G}$ = the determinant of the metric tensor (page 30).

H = solid earth boundary for an ocean basin at $z = -H$ (page 26).

$\mathcal{H}$ = enthalpy per unit mass (page 113).

$\mathcal{H}^{o}$ = potential enthalpy per unit mass (page 113).

$h = z_{,\rho}$ = specific thickness of an isopycnal layer (page 194).

$\overline{h}^{\rho}$ = isentropic ensemble mean specific thickness of an isopycnal layer (page 195).

δh = thickness of a generalized layer (page 129).

$\mathcal{I}$ = internal energy per mass (page 93).

J = Jacobian of transformation between two sets of coordinates (page 14).

$\mathbf{J}$ = subgrid scale flux of a tracer (page 89).

$\mathbf{J}$ = tracer mixing tensor (page 204).

$\mathbf{J}_{q}$ = heat flux (page 104).

$\mathcal{K}_{(\mathrm{nh})}$ = kinetic energy per unit mass of a nonhydrostatic fluid parcel (page 97).

$\mathcal{K}$ = kinetic energy per unit mass of a hydrostatic fluid parcel (page 97).

$\mathcal{L}$ = angular momentum per unit mass (page 82).

$\mathcal{M}$ = advective metric frequency (page 53).

$\hat{\mathbf{N}}$ = orientation direction used for the ocean surface or ocean bottom (page 31).

$\hat{\mathbf{n}}$ = unit normal direction (page 31).

p = pressure at a point in the ocean fluid, including the pressure from the atmosphere and sea ice living above the ocean (page 50).

q = diabatic heating (page 91).

q_{w} = surface fresh water flux, which is the volume of moisture entering the ocean, per unit time, per unit horizontal cross-sectional area $\mathrm{d}A = \mathrm{d}x\,\mathrm{d}y$ (page 34).

$\tilde{q}_w^*$ = modified mean surface fresh water flux (page 178).

R = radial coordinate corresponding to the ellipsoid of best fit to the sea-level geopotential (page 25).

$\mathbf{S}$ = slope of a generalized layer with respect to the horizontal (page 129).

$S = |\mathbf{S}|$ = magnitude of the slope of a generalized layer with respect to the horizontal (page 129).

S_x and S_y = two components to the slope, $\mathbf{S}$, of a generalized layer with respect to the horizontal (page 129).

$\mathcal{S}$ = tracer concentration source (page 87).

s = salinity in grams of salt per kilogram of seawater (page 87).

s = generalized vertical coordinate (page 128).

T = *in situ* temperature (page 90).

T = tracer concentration when used in the discussion of neutral physics discretization (page 351).

t = time as measured by an Eulerian observer (page 12).

$\mathbf{U} = (U, V)$ = two-dimensional vertically integrated horizontal velocity field (page 34).

$\mathcal{U}$ = arbitrary divergence-free transport velocity introduced in the discussion of skewsion (page 191).

$\overline{\mathbf{U}}^{\#}(\overline{z}^{\rho})$ = ensemble mean horizontal transport of fluid beneath the modified mean potential density surface (page 199).

$\mathbf{U}^{qs}$ = quasi-Stokes transport (page 201).

$\mathbf{u} = (u, v)$ = two-dimensional horizontal components of the physical velocity vector for a fluid parcel (page 28).

$\mathbf{u}^{\text{bolus}}$ = two-dimensional horizontal bolus velocity (page 196).

$\hat{\mathbf{u}}$ = effective two-dimensional horizontal transport velocity (page 195).

$\hat{\mathbf{u}}$ = Horizontal baroclinic velocity used for splitting the ocean dynamics into fast and slow modes (page 247).

$\overline{\mathbf{u}}^{z}$ = Barotropic velocity used for splitting the ocean dynamics into fast and slow modes (page 247).

$\overline{\mathbf{u}}^{\#}$ = transformed residual mean (TRM) horizontal velocity (page 199).

$\mathbf{v} = (u, v, w)$ = three-dimensional physical velocity vector for a fluid parcel (page 28).

$\mathcal{V}$ = global tracer variance (page 288).

$\dot{\mathbf{v}}$ = material acceleration of a fluid parcel (page 13).

$\tilde{\mathbf{v}} = \rho \, \mathbf{v} / \rho_o$ = normalized linear momentum density (momentum per unit volume) carried by a fluid parcel (page 37).

$\langle \mathbf{v} \rangle^{\rho}$ = density weighted velocity (page 180).

$\overline{\mathbf{v}}^{\#}$ = transformed residual mean (TRM) three-dimensional velocity (page 199).

$\mathbf{v}^*$ = three-dimensional non-divergent velocity suggested by Gent and McWilliams (1990) (page 209).

w = vertical component to the velocity vector (page 28).

w = work done on a fluid parcel (page 91).

$w^{(s)}$ = dia-surface velocity component (page 139).

$\mathbf{x} = (x^1, x^2, x^3)$ = three-dimensional Cartesian position vector of a fluid parcel (page 11). Components to $\mathbf{x}$ are written x^m with the tensor label $m = 1, 2, 3$.

$\dot{\mathbf{x}} = \mathrm{d}\mathbf{x}/\mathrm{d}t = \mathbf{v}$ = velocity of a fluid parcel (page 13).

z = vertical position of a fluid parcel taken from the sea-level geopotential (page 25).

$\overline{z}^{\rho}$ = mean vertical position of a potential density surface ρ (page 197).

$z_{,s}$ = specific thickness of the generalized s layer (page 129).

GREEK SYMBOLS

α = thermal expansion coefficient as determined by changes in potential temperature (page 116).

α_T = thermal expansion coefficient as determined by changes in *in situ* temperature (page 95).

β = saline contraction coefficient with potential temperature held fixed (page 116).

$\beta = f_{,y}$ = planetary vorticity gradient (page 66).

Γ = adiabatic lapse rate as a function of changes in pressure: $\mathrm{d}T = \Gamma\,\mathrm{d}p$ (page 95).

$\widehat{\Gamma}$ = adiabatic lapse rate as a function of changes in depth $\mathrm{d}T = -\widehat{\Gamma}\,\mathrm{d}z$ (page 95).

$\Delta\tau$ = discrete time interval used in a numerical ocean model. Specifically, the time step used for time cycling the slower dynamics (page 244).

Δt = discrete time interval used in a numerical ocean model. Specifically, the time step used for time cycling the fast dynamics (page 244).

δ_{mn} = Kronecker symbol, also known as the unit tensor (page 447).

ϵ = frictional dissipation associated with molecular viscosity (page 99).

ϵ_{mnp} = Levi-Civita symbol, or alternating symbol, for Cartesian coordinates (page 460).

ε_{mnp} = covariant Levi-Civita symbol (page 460).

ζ = spatial coordinates of a fluid parcel when described in material space-time (page 10). Components of ζ are denoted ζ^a, with the tensor label $a = 1, 2, 3$.

ζ = entropy of a fluid parcel (page 93).

ζ = vertical component of the fluid vorticity (page 73).

η = ocean surface height or sea level, which is the displacement of the ocean surface from its resting position at $z = 0$ (page 26).

η^* = modified mean ocean surface height (page 176).

θ = potential temperature of a fluid parcel (page 112).

Θ = conservative temperature of a fluid parcel (page 114).

$\iota = \sqrt{-1}$ (page 267).

κ = frictional viscosity (page 100).

κ = tracer diffusivity (page 306).

$\Lambda^{a}{}_{\bar{a}}$ = transformation matrix between two sets of coordinates (pages 133 and 447).

λ = longitudinal position on the earth (page 25).

λ_1 = first baroclinic Rossby radius (page 311).

μ = chemical potential of a constituent of seawater (page 91).

ν = molecular kinematic viscosity of water (page 98).

$\boldsymbol{\xi}$ = three-dimensional position vector for a parcel as written in terms of an arbitrary coordinate system (page 446). Components of $\boldsymbol{\xi}$ are written ξ^a, with the tensor label $a = 1, 2, 3$.

Π = potential vorticity (page 72)

$\boldsymbol{\Pi}$ = canonical momentum for a point particle (page 77)

ϖ = mass of a fluid parcel per volume in material space (page 13).

ρ = *in situ* mass per unit volume of a fluid parcel, where volume is that in $\mathbf{x}$-space (pages 14 and 90).

$\widetilde{\rho}$ = modified mean potential density (page 198).

ρ_o = constant reference density used for the Boussinesq approximation (page 37).

ρ_{pot} = potential density (page 117).

ρ_{w} = *in situ* density of fresh water (page 37).

Σ = tracer source (page 87).

σ = terrain-following vertical coordinate (page 124).

σ = entropy source (page 111).

τ = discrete time step for an ocean model (page 244).

τ = time coordinate of a fluid parcel when measured by a material or Lagrangian observer (page 10).

τ^{mn} = frictional stress tensor (page 384).

$\boldsymbol{\tau}^{\text{bottom}}$ = stress on the ocean fluid imparted by the solid earth boundary (page 107).

$\boldsymbol{\tau}^{\text{wind}}$ = stress on the ocean fluid imparted by atmospheric winds and/or sea ice (page 107).

$\Phi = g\,z$ = gravitational potential energy per unit mass (page 44).

ϕ = latitudinal position on a sphere (page 25).

$\overline{\phi}$ = half the mean density variance (page 203).

$\overline{\boldsymbol{\Psi}}^{\#}$ = vector streamfunction for the TRM velocity $\overline{\mathbf{v}}^{\#}$ (page 199).

Ω = Earth's angular velocity directed upward from the North Pole (page 43).

OPERATORS

d/dt = material time derivative for a fluid parcel, equivalent to ∂_τ (page 13).

δ = infinitesimal path dependent increment of a thermodynamic variable (page 91).

δ = infinitesimal variation in the form of a function used when computing a functional derivative (page 346).

$\nabla = (\partial_x, \partial_y, \partial_z)$ = three-dimensional Eulerian gradient operator (page 28).

∇_s = two-dimensional gradient operator taken on a constant s surface (page 134).

∇_z = two-dimensional gradient operator taken on a constant z surface (page 28).

∇_ρ = two-dimensional gradient operator taken on a constant ρ surface (page 140).

∇_σ = two-dimensional gradient operator taken on a constant σ surface (page 126).

∂_τ = material time derivative for a fluid parcel, equivalent to d/dt (page 12).

$\wedge$ = vector cross-product, which is a specific form of the more general *wedge* product (page 448). The wedge symbol is preferred over the alternative x, since x can be confused with the spatial coordinate.

$\overline{(\ )}^{\rho}$ = mean over an ensemble of fluid parcels, each having the same potential density ρ, the same horizontal position (x, y), and the same time t (page 194).

$\overline{(\ )}^{z}$ = an Eulerian average obtained by an observer at a fixed point (x, y, z, t) (page 196).

Partial derivative: $u^m_{,n} = \partial_n u^m$ = partial derivative in the n direction of the vector field u^m (page 29 as well as page 483). The comma notation helps to distinguish the partial derivative operator from the zth component of a tensor.

Covariant derivative: $u^m_{;n} = n^{th}$ covariant derivative component of the vector field u^m (page 483). The semicolon notation generalizes the comma notation, with the two agreeing when the space is flat and Cartesian coordinates are used. In general, the two differ, as they do in a spherical geometry.

Chapter One

OCEAN CLIMATE MODELS

The purpose of this chapter is to introduce ocean climate models and their use in climate science.

1.1 OCEAN MODELS AS TOOLS FOR OCEAN SCIENCE

A column of ocean water only 3m thick contains as much heat capacity as the full atmospheric column above (Gill, 1982). Hence, the oceans, which cover roughly 70% of the earth's surface, provide a large reservoir for heat and other constituents of the earth's climate system, such as the increasing amounts of anthropogenic carbon dioxide. Through its buffering abilities and relatively slow time scales, the ocean represents the flywheel of the earth's climate system. That is, as goes the ocean, so goes the climate system.

A scientific understanding of the ocean's time mean state, as well as its variability about this mean and its stability to various forms of perturbations, represents a key goal of physical oceanography and climate science. Due to our inability to perform controlled experiments on large-scale systems studied in the geosciences, such as the earth's climate and its component subsystems, computer models represent a critical tool for rationalizing climate phenomena. Indeed, computer models are becoming the primary tools used to study and predict physical, chemical, and biological characteristics of the ocean fluid, reflecting the growing power of computers, improved knowledge and observations of the ocean, and enhancements in the realism of ocean model simulations.

That ocean models are increasingly being used by all sorts of scientists, including those without direct experience developing models, is a sign that the models have enhanced their physical integrity over the past decades to a level deserving a general respect within the broader climate science community. Correspondingly, as model usage increases, model developers have a growing responsibility to ensure that their codes are physically based, numerically sound, and well documented. Given this mandate, one aim of this book is to establish a level of ocean model documentation that goes beyond the usual technical discussion that assumes the model user is familiar with the fundamentals and understands the physical meanings of the mathematical symbols. Instead, we develop the equations from a (mostly) first principles perspective and take some care to nurture a physical understanding of the mathematics.

1.2 OCEAN CLIMATE MODELS

Models of the ocean range in complexity from idealized theoretical models whose solutions can be summarized by a few lines of mathematics, to realistic global ocean circulation models encompassing many equations and requiring thousands or millions of lines of computer code to solve. The main focus of this monograph concerns the realistic models, and in particular the fundamentals of their formulation. It is important to note that a scientific understanding of ocean climate phenomena is often realized most profoundly by the creative and judicious use of a hierarchy of models. The most realistic models play an important part in this hierarchy, but they are not the full story.

We use the term *ocean climate model* as a means to distinguish models that simulate the World Ocean over climatologically relevant time scales from those that simulate, say, coastal, regional, or basin scale dynamics. Distinctions between these ocean modeling subfields is decreasing, largely due to the steady growth in computer power that allows modelers to dispense with some of the simplifications required only a few years ago. Distinctions are also becoming fewer due to realizations by practitioners that elements of the ocean strongly interact across many spatial and temporal scales. That is, ocean modeling subfields overlap in crucial manners. Nonetheless, the finite nature of both scientists and their tools introduces a difference in focus, with choices made by practitioners in one subfield often unacceptable to those in another.

Simulations of the World Ocean over time scales appropriate for climate (e.g., decades to millennia) involve extremely rich and complex arrays of flow regimes and interactions between components of the climate system. Additionally, the ocean is largely forced at its upper and lower boundaries, with interior flow relatively ideal. In particular, high latitude oceanography involves strong interactions between the ocean with sea ice and rivers, and intense air-sea heat fluxes induce deep convection and the associated formation of deep water masses. Tropical oceanography involves intense equatorial current systems with rapid adjustments to wind forcing associated with equatorial Kelvin, Rossby, and instability waves, and a powerful interannual mode of air-sea variability known as El Niño in the Pacific. Oceanography in the subtropical and subpolar latitudes is dominated by large-scale gyres with meandering and eddying boundary currents forming their western margins. Furthermore, solid earth boundaries provide a leading order influence on the ocean circulation. For example, meridional boundaries block otherwise zonal flow except within certain parts of the Southern Ocean, variations in topography cause flows to feel the bottom throughout many crucial parts of the World Ocean, and straits and sills funnel water from marginal seas, such as the Mediterranean and Greenland, into the larger ocean basins. A primary goal of ocean climate modeling is to simulate the global ocean circulation over these various regimes, given just the boundary forcing. This is a highly nontrivial goal.

As discussed in Section 6.2, there are three general model classes that have been used for ocean modeling. These classes are distinguished by the manner used to discretize the vertical direction. Indeed, as argued in Griffies et al. (2000a), the choice of vertical coordinate represents the most fundamental choice that can be made when designing an ocean model.

Since the 1960's, *z-coordinate* ocean models, or simply *z*-models, have been the dominant class for global ocean climate simulations. This is the model class with which the author is most familiar, and which forms the focus of some of the latter parts of this book. Characteristics of z-models, as well as the other two classes (*isopycnal* and *terrain following*), are described in Section 6.2. Each model class has advantages and disadvantages when simulating various flow regimes encountered in ocean climate modeling. Only two of the three (isopycnal and z) have routinely been used for global circulation studies. One reason that z-models presently dominate ocean climate modeling is that their relative simplicity has allowed them to be used for many decades, going back to the work of Bryan (1969), Bryan and Lewis (1979), and Cox (1984). In contrast, the isopycnal model class requires more sophisticated numerical schemes, whose development did not mature until the 1980's. The third model class, the terrain following sigma models, remain the model of choice for coastal oceanographers, but they have largely remained absent from simulations of global ocean climate.

Ocean climate models continue to evolve. For example, many egregious problems identified with early representatives of the different model classes are now remedied by more mature numerical treatments. Nonetheless, as argued in Section 6.2, each class has basic limitations that warrant developing models with generalized vertical coordinates. The hope is that by generalizing the treatment of the vertical, a well-designed model with this capability can reduce or remove many of the egregious problems of the less flexible models based on a single vertical coordinate choice. This remains a topic of intense research in the ocean model development community.

Research into ocean model fundamentals and algorithm development can take many years to penetrate into the common practice of major climate modeling centers. The reason is largely related to the extreme complexity involved with building coupled global climate models. It takes teams of researchers years to build and refine a coupled model, with significant feedback and compromise necessary in order to successfully mesh the needs of various component modelers. Notably, an ocean model suitable for coupled climate simulations is far more than a dynamical kernal. In addition, it must consist of a full suite of physical parameterizations of unresolved processes, diagnostics allowing its simulations to be readily analyzed, infrastructure providing a means to *talk* to the computer that is running the model, a superstructure with appropriate handles for interacting with other component models (e.g., sea ice and atmosphere models) necessary for climate simulations, and computational sophistication rendering it efficient on the many computer platforms employed by research laboratories and universities.

1.3 CHALLENGES OF CLIMATE CHANGE

Since the 1990's, thousands of scientists worldwide have been contributing to the development of extensive reports on climate change science, with the latest being Houghton et al. (2001). This work is in response to the increasing scientific evidence that industrial society represents a nontrivial geophysical force. Common questions that arise are: What should we expect? How much of the observed climate change is due to humans? Providing sound scientifically based answers to these and other questions is profoundly difficult. Indeed, as lucidly described in the book by Philander (1998), unequivocal answers are not forthcoming from climate science. Instead, as with weather prediction, probabilistic statements are the best the science can provide.

As discussed in Houghton et al. (2001), we are at a stage in climate science where the wide variety of climate models yield a general consensus regarding the large-scale effects of increased greenhouse gases. Quite simply, the planet is warming and will likely continue to do so, with higher latitudes feeling the relative effects more than lower latitudes. However, when quantitative questions are posed, models provide varying projections. Part of the spread is related to the chaotic nature of the climate system. Part is due to large uncertainties in future greenhouse gas emission scenarios. Yet some is due to differing details of the model formulations and their parameterizations. It is on this latter issue that climate scientists can make further progress through research and development.

Given the critical importance of models for understanding climate and predicting its future behavior, it is incumbent on model developers to impose the highest standards on model integrity. In particular, ocean climate models should incorporate realistic parameterizations and sound numerical formulations (for reviews, see Chassignet and Verron, 1998; Griffies et al., 2000a). Yet they must do so at a level of computational expense that does not overly handicap the abilities of the earth system modeler to incorporate other components of the climate system, and to fully investigate various scenarios. Within the ocean science community, this mandate to improve the models entrains hundreds of researchers such as process oriented physicists, chemists, biologists, observational oceanographers, numerical algorithm developers, software engineers, ocean climate modelers, and others. It is anticipated that the questions of climate change will continue to strongly influence and motivate all areas of climate science for many years.

PART 1

Fundamental Ocean Equations

This part of the book presents the fundamental equations describing the ocean fluid. These equations are based on classical continuum physics using both Newtonian mechanics and quasi-equilibrium thermodynamics. The material is for the most part independent of the niceties of ocean model numerics. Instead, we focus on the basics. Hence, the following chapters will be of interest to those aiming to understand the physical content of the mathematical equations underlying ocean climate models.

The tensor analysis described in Part 6 is necessary to arrive at certain results presented in this part of the book. However, requirements of the reader to penetrate Part 6 are actually quite minimal. Additionally, little previous knowledge of continuum mechanics is assumed. Instead, the presentation is aimed at the reader who has a solid foundation in undergraduate physics and vector calculus, but not a formal course on continuum mechanics.

Chapter Two

BASICS OF OCEAN FLUID MECHANICS

This chapter introduces some of the principles and assumptions forming the foundations for a mathematical description of oceanic fluid motion. It is from this foundation, as well as that provided by ocean thermodynamics, that numerical models are then formulated.

2.1 SOME FUNDAMENTAL OCEAN PROCESSES

As an introduction to a study of ocean hydrodynamics, consider some physical processes that are part of the "zoo" of oceanic dynamics. Representing or parameterizing these processes in realistic ocean circulation models is essential in order for the models to be of use as a tool for climate science. The following list is illustrative, not comprehensive, and is given as motivation for the interested reader to pursue a deeper understanding by referring to one or more of the texts listed in the Preface.

Physical processes in the ocean span space-time scales from millimeters and seconds to global and millennial. These processes can be classified into large-scale currents with time-dependent fluctuations manifesting as waves, turbulence, large-scale fluctuations of the mean circulation, and fundamental changes in the overall ocean circulation (e.g., transitions of the overturning circulation). Important examples in the wave-like regime include surface gravity waves set up when the atmospheric winds perturb the ocean surface. Indeed, these surface gravity waves are crucial for transmitting atmospheric momentum into the ocean's upper planetary boundary layer since they provide a frictional element to the otherwise smooth ocean surface. In the ocean interior, undulations of the stratified density surfaces are generated through a variety of forcing mechanisms, such as tidal motion over gradients in bottom topography. The breaking of these waves is thought to contribute to internal mixing of density classes, and so helps determine stratification of the ocean interior. At the ocean's boundaries and continental shelves, waves, such as coastally trapped waves, help to transmit information around ocean basins, with similar waves set up at the equator where a vanishing Coriolis force acts as an effective boundary allowing for the existence of equatorially trapped Kelvin waves. Finally, at planetary scales, Rossby waves owe their existence to the differential Coriolis force (the β-effect) experienced by large-scale motion on the rotating spherical earth. These waves are important for setting up the ocean's general circulation.

As with any hydrodynamical system, some fluctuations can feed off dynamical instabilities of the time mean flow and grow over time. For example, a heavy fluid parcel overlying a lighter parcel rapidly moves vertically to its level of neutral buoyancy as it releases its potential energy. This effect leads to convective overturning, which is an important process for the formation of large masses of ocean water possessing similar properties. An additional process involved with deep water formation is the cascade of heavy water down topographic slopes in certain parts of the World Ocean. These unstable flows are quite turbulent and entrain much of the ambient fluid. Another unstable process, which is ubiquitous in the ocean, arises when gravitationally stable density fronts, common in a rotating stratified fluid, are perturbed so that the sloped front breaks down into unstable undulations that transfer potential energy into kinetic energy. This process, known as *baroclinic instability*, forms one of the most important instability mechanisms in geophysical fluids.

In general, after an initial rapid growth phase, unstable fluctuations saturate via nonlinear wave breaking and tend to evolve into a highly chaotic or turbulent state. Such turbulent states occur on all scales from micro to planetary, and they efficiently stir and mix ocean properties across these scales.

At the largest scales, turbulence associated with baroclinic instability, known as geostrophic turbulence, represents the large-scale "storms" of the ocean, analogous to the synoptic storms comprising the atmosphere's weather. Such eddies represent a very large part of the ocean's energetic motions, and so are important for understanding many aspects of the ocean structure, such as the distribution of tracers. Due to differences in vertical stratification between the ocean and atmosphere, the scale of oceanic geostrophic eddies, set roughly according to the first internal radius of deformation (e.g., Pedlosky, 1987; Stammer, 1997; Smith and Vallis, 2002), is some ten times smaller than the first internal radius of deformation in the atmosphere. Hence, an explicit representation of geostrophic eddies in numerical ocean models requires about 1000 times more refined space-time grid resolution than in atmospheric models.* This nontrivial increase in computational cost makes the explicit representation of the ocean's geostrophic eddies in numerical models far more difficult than the representation of the atmosphere's geostrophic eddies. This difference in scales has kept the problem of parameterizing eddies at the forefront of theoretical physical oceanography.

The oceanic processes mentioned above and numerous others are important for understanding, modeling, and predicting the ocean circulation. Such issues are of fundamental interest scientifically, and they have driven the fields of physical oceanography and ocean modeling for much of their history. They have furthermore become increasingly important to society within the past few decades due to the nontrivial human effects on the earth's climate system. Numerical models are thus being relied upon to help answer questions of profound importance to life on this planet. Physically sound and rational numerical models are critical as we seek an understanding of the climate's response to the greenhouse gas experiment now being conducted.

*A factor of ten comes from each of the two horizontal directions. Also, the model time step must be correspondingly smaller to maintain numerical stability, thus increasing the model cost by another factor of ten.

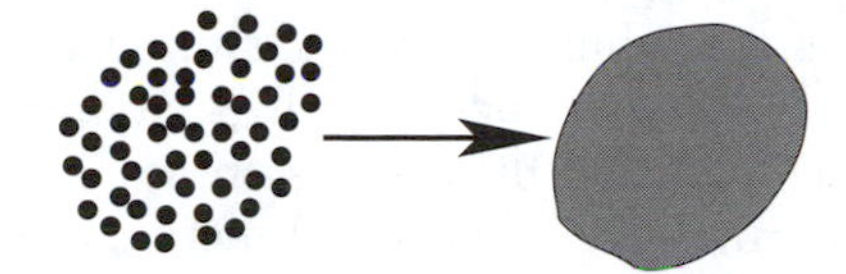

Figure 2.1 A small parcel of water, say on the order 10^{-9}cm^3, is comprised of a huge number of interacting molecules (roughly 3×10^{19}), schematically drawn in the left figure. For purposes of geophysical fluid dynamics, it is safe to ignore details of the individual molecules and approximate the collection of molecules as a continuum, as drawn in the right figure. This constitutes the *continuum hypothesis*.

2.2 THE CONTINUUM HYPOTHESIS

There are approximately 18 grams of pure water per mole, where a mole has roughly 6×10^{23} molecules (Avogadro's number). A tiny parcel of pure water with dimensions on the order of 10^{-3}cm has a volume of 10^{-9}cm^3. With a density of $1\,\text{g}\,\text{cm}^{-3}$, this parcel has about 3×10^{19} molecules. As noted in Section 1.2 of Batchelor (1967), 10^{-9}cm^3 is extremely tiny from a macroscopic perspective, since gross properties such as density, tracer concentration, and temperature are effectively uniform over this volume. Yet the 3×10^{19} molecules in this volume mean that it is gigantic from the perspective of molecular dynamics. The same-sized parcel of air, with approximately 29 grams per mole, at the earth's surface and $0°$C, has a density of $1.3 \times 10^{-3}\text{g}\,\text{cm}^{-3}$. Therefore, a 10^{-9}cm^3 parcel of air at the earth's surface has roughly 3×10^{10} molecules, which is again a huge number.

These numbers suggest that for the purposes of geophysical fluid dynamics, and for many other areas of fluid mechanics, it is extremely accurate to characterize a fluid parcel with volume on the order 10^{-9}cm^3 as *macroscopically small yet microscopically large*. The macroscopically small characterization is warranted since such fluid parcels have a near uniform set of thermodynamic properties (temperature, salinity, tracer concentration, etc.). Hence, the first and second laws of thermodynamics can be applied to such parcels using a local thermodynamic equilibrium assumption, which constitutes the basis for classical irreversible or non-equilibrium thermodynamics (DeGroot and Mazur, 1984). That is, each fluid parcel, though generally far from equilibrium with other parcels, itself constitutes an open thermodynamic system in near-equilibrium.

The microscopically large characterization warrants treating the fluid as a *mathematical continuum*. That is, no concern is placed on the discrete nature of molecules when studying large-scale ocean dynamics. Consequently, limits can be taken as volumes go to zero when formulating the differential laws of fluid motion. Figure 2.1 illustrates this proposition, which formally is known as the *continuum hypothesis*. The physical processes mentioned in Section 2.1, as well as others of importance to large-scale ocean dynamics, can be described using the continuum hypothesis.

As the above discussion suggests, the continuum fluid equations are in principle related to more fundamental equations of motion, describing the multitude of molecular degrees of freedom, through a series of averaging operations. Hence, the laws of continuum fluid dynamics represent a *mean field theory*. A rigorous proof of this statement is available only for the simplest of gaseous systems, and can be found in statistical mechanics books such as Huang (1987). Also, the reader interested in these issues may find the discussion in Chapter 1 of Salmon (1998) enlightening. For the ocean, as for other liquid media, the continuum hypothesis has not been rigorously proven. Nonetheless, its validity is evidenced by its high degree of success in describing the bulk of liquid and gas motions, including those relevant for geophysical fluid dynamics. Hence, the continuum hypothesis serves the purposes of this book.

After making the continuum hypothesis, the differential laws of hydrodynamics and thermodynamics are derived by applying the basic ideas from Newtonian dynamics and thermodynamics to a collection of local thermodynamically equilibrated fluid parcels. The resulting field equations possess information regarding a broad spectrum of fluid motion, with spatial scales reaching from the parcel level to those relevant for global ocean dynamics, and temporal scales on the order of milliseconds to millennia. The continuum equations, often called the Navier-Stokes equations, have been shown through numerous experiments over the past two centuries to be very successful in describing fluid flow. That these equations are relevant over such a large space-time range represents a spectacular success in providing a rational mathematical description of natural phenomena.

2.3 KINEMATICS OF FLUID MOTION

Kinematics is the branch of mechanics concerned with the intrinsic aspects of motion, the geometry of space-time where motion occurs, and the conceptual and mathematical tools used to describe motion. It is not concerned with the forces or causes of motion, all of which are within the purview of *dynamics*. Kinematical results are therefore very generic and of great utility for a variety of dynamical situations.

The purpose of this section is to introduce some kinematic results useful when describing the motion of fluid parcels. In particular, relationships are established here between *Eulerian* and *Lagrangian* descriptions of that motion.

2.3.1 Material and fixed-space coordinates

In describing the dynamics of fluid parcels, one may take the perspective that the parcels form a continuum of deformable "particles." In this way, many of the ideas from point particle mechanics are transferable to fluid parcel mechanics. To proceed in this manner, it is necessary to distinguish between the infinite number of fluid parcels. Recall that in classical particle mechanics, particles are distinguished by an integer label. For a continuum of fluid parcels, the integer label becomes continuous. That is, the fluid parcels are distinguished by ascribing a value for a continuous label, ζ, to each parcel filling up the fluid continuum. Associated with this *material space coordinate* is the co-moving or *material time coordinate*, τ, measured by an observer riding along with the fluid parcels. The coordinates (ζ, τ) are coordinates in *material* space-time. Material coordinates are often called *Lagrangian* or *convected* coordinates.

For ideal parcel motions, dynamics in material space-time is trivial: $\partial_\tau \zeta =$

0. This equation simply means that parcels maintain their identity for all time. That is, once labeled, a parcel maintains the same value for the label as it moves throughout the fluid (see Figure 2.2). Such ideal fluid motion can be a good approximation to realistic motions in parts of the ocean's interior. For a non-ideal fluid system, parcel identity preservation is only an approximation, since a parcel's identity is lost over time whenever properties are interchanged or mixed between adjacent parcels. In this case, $\partial_\tau \zeta$ is no longer zero.

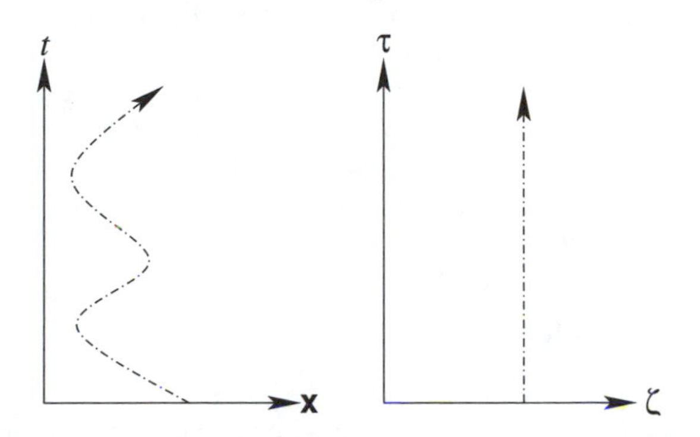

Figure 2.2 Illustration of the space-time paths or trajectories exhibited by two ideal fluid parcels. The left panel shows the trajectories as seen in a fixed-space or Eulerian perspective. The paths are typically curved, which represents motion due to waves, advection, etc. The right panel shows the same trajectories in the material space-time. These trajectories are trivial for ideal parcels since they maintain the same value for their spatial material coordinate ζ, hence exhibiting a linear path in material space-time. The two pictures are related through the time-dependent coordinate transformation $\mathbf{x} = \mathbf{x}(\zeta, \tau)$ and $t = \tau$.

Measurements of fluids discussed here are taken by observers living on a generally curved non-Euclidean surface or volume embedded in a background Euclidean space-time. In particular, for geophysical fluid dynamics, observers can be considered to be on the surface of a rotating sphere (a non-Euclidean space), approximating the rotating earth, where the sphere is embedded in a Euclidean space-time background (the space-time of Newtonian mechanics).

A physical description of the fluid, motivated by the particle mechanics analog, includes the trajectory of fluid parcels as they move in Euclidean space-time. A set of coordinates can be used to specify these trajectories (Cartesian, spherical, cylindrical, etc.), and these are sometimes called *physical* coordinates. This name is not preferred here since there is nothing "unphysical" about the material coordinates. Rather, the name *fixed* or *position* coordinates is preferred. In turn, it should be noted that the "fixed" coordinates for ocean modeling are actually rotating coordinates since they are fixed relative to a terrestrial observer's rotating frame of reference. Hence, the coordinates commonly used in geophysical fluid dynamics are strictly not fixed in the sense of inertial coordinates.

The continuum of fluid parcel trajectories is represented by the vector field

$$\mathbf{x} = \mathbf{x}(\zeta, \tau). \tag{2.1}$$

The vector field $\mathbf{x}(\zeta, \tau)$ measures the Cartesian position of a parcel with material space-time coordinates (ζ, τ). This position is taken relative to some fixed origin in

Euclidean space. For geophysical fluids, it is sufficient to assume that fluid parcels move at speeds well below the speed of light. Therefore, the co-moving time τ and the time t measured by observers at rest with respect to the origin $\mathbf{x} = 0$ can be taken to be the same,

$$t = \tau. \tag{2.2}$$

The relations (2.1) and (2.2) specify a time-dependent coordinate transformation, or mapping, between the material space and the fixed space. Since the time parameters are the same, it is sufficient to distinguish these two space-times simply by identifying material space-time as ζ-space and fixed space-time as $\mathbf{x}$-space. Note that the number of geometric dimensions needed to describe the material space is the same as the number of dimensions in the fixed space. For example, one may choose to define the material label as the initial position of the field of fluid parcels, $\zeta = \mathbf{x}(\zeta, \tau = 0)$.

2.3.2 Lagrangian and Eulerian descriptions

The *Lagrangian description* of a fluid is afforded by knowledge of the continuum of parcel trajectories $\mathbf{x}(\zeta, \tau)$. The Lagrangian equations of motion are written much as are the equations for a point particle, yet with forcing terms written generally as partial derivatives with respect to the material coordinate ζ and functional derivatives with respect to the parcel trajectory $\mathbf{x}(\zeta, \tau)$.

In contrast, an *Eulerian description* is afforded by knowledge of the velocity field

$$\mathbf{v}(\mathbf{x}(\zeta, \tau), t) = \partial_\tau \mathbf{x}(\zeta, \tau). \tag{2.3}$$

This relation defines the mapping between the Eulerian and Lagrangian descriptions of the fluid: knowledge of one description implies knowledge of the other as specified through this first order differential equation.

2.3.3 Transformations between ζ-space and x-space

As mentioned in Section 2.3.1, the relation between the space point $\mathbf{x}$ and the parcel trajectory $\mathbf{x}(\zeta, \tau)$ defines a time-dependent mapping between the fixed space-time with the material space-time. In component form, this mapping takes the form

$$x^m = x^m(\zeta^a, \tau) \tag{2.4}$$

$$t = \tau, \tag{2.5}$$

where $m = 1, 2, 3$ labels coordinates in $\mathbf{x}$-space and $a = 1, 2, 3$ labels coordinates in ζ-space. Although identification of the time parameters, $t = \tau$, has been made, it is important to use different symbols. The reason is that partial derivatives with respect to τ are taken with the material space coordinates ζ constant, whereas partial derivatives with respect to t are taken with the fixed space coordinates $\mathbf{x}$ held fixed. As such, the material space time derivative is related to the fixed space time derivative through the expression

$$\partial_\tau = \partial_t + \sum_{m=1}^{3} \partial_\tau x^m \, \partial_m \tag{2.6}$$

$$= \partial_t + \partial_\tau x^m \, \partial_m, \tag{2.7}$$

which follows from the rules of coordinate transformations (see Sections 20.4 and 20.5). The second equation introduced the Einstein summation convention, where repeated indices are summed (see Section 20.3). This notation is used throughout this book. This result is more often written using the *total* or *material time derivative*

$$\frac{\mathrm{d}}{\mathrm{d}t} = \partial_t + \mathbf{v} \cdot \nabla, \tag{2.8}$$

where $\mathbf{v}$ is the Eulerian velocity field defined by equation (2.3). In the following, an over-dot is sometimes used as shorthand for material time derivative. For example, the velocity of a fluid parcel can be written in one of the following equivalent manners:

$$\begin{aligned} \mathbf{v} &= \partial_\tau \mathbf{x} \\ &= \mathrm{d}\,\mathbf{x}/\mathrm{d}t \\ &= \dot{\mathbf{x}}. \end{aligned} \tag{2.9}$$

The over-dot notation brings many fluid dynamic equations into forms analogous to those encountered in particle dynamics, where similar notation is employed.

The nonlinear term

$$\mathbf{v} \cdot \nabla = \dot{x}^m \partial_m \tag{2.10}$$

appearing in the material time derivative (equation (2.8)) is known as the *transport* or *advection* operator, with advection the usual terminology in geophysical fluid dynamics. Advection arises since an Eulerian observer at a fixed position $\mathbf{x}$ is not a co-moving or material observer. The Lagrangian observer is co-moving with the fluid parcel, and so has no need to explicitly take note of advection.

The mass of a parcel of seawater includes the mass of fresh water plus the mass of dissolved salts and other chemical and/or biological species. The infinitesimal mass of a fluid parcel can be written

$$\mathrm{d}M = \varpi(\zeta, \tau)\,\mathrm{d}\zeta. \tag{2.11}$$

In this expression, $\varpi(\zeta, \tau)$ is the fluid mass density in material space, and $\mathrm{d}\zeta = \mathrm{d}\zeta^1\,\mathrm{d}\zeta^2\,\mathrm{d}\zeta^3$ is a useful shorthand for the volume element in material space.* If the parcel mass is conserved, then

$$\partial_\tau(\mathrm{d}M) = 0. \tag{2.12}$$

Consequently, the mass density in material space-time for mass conserving parcels must be independent of material time,

$$\varpi(\zeta, \tau) = \varpi(\zeta). \tag{2.13}$$

It is more common to consider the mass density in fixed space-time. For this purpose, the familiar rules of calculus† can be used to determine an expression for the mass element written using the $\mathbf{x}$-space coordinates

$$\mathrm{d}M = \varpi\,\mathrm{d}\zeta \tag{2.14}$$

$$= \varpi\,J^{-1}\,\mathrm{d}\mathbf{x}, \tag{2.15}$$

*Because the physical dimensions of the material coordinates $(\zeta^1, \zeta^2, \zeta^3)$ need not be length, the physical dimensions of the material volume element $\mathrm{d}\zeta$ need not be length^3.

†See Schutz (1980), or most books on differential forms, for a more elegant proof in terms of the volume 3-form.

where

$$
\begin{aligned}
J &= \det(x^m_a) \\
&= \frac{\partial \mathbf{x}}{\partial \zeta} \\
&= \frac{1}{3!} \epsilon_{mnp} \epsilon^{abc} \left(\frac{\partial x^m}{\partial \zeta^a} \frac{\partial x^n}{\partial \zeta^b} \frac{\partial x^p}{\partial \zeta^c} \right)
\end{aligned}
\tag{2.16}
$$

is the determinant of the transformation matrix between the position coordinates and material coordinates; that is, it is the Jacobian, and ϵ is the totally antisymmetric Levi-Civita symbol defined on page 460.* Therefore, the mass density of the fluid as expressed in the position space, ρ, is related to the mass density of the fluid as expressed in material space, ϖ, by the expression

$$
\rho = \varpi J^{-1}. \tag{2.17}
$$

Note that uniform mass density in material space ($\varpi =$ constant) does not imply uniform mass density in position space, and vice versa.

2.3.4 Mass conservation

It is important to carefully choose suitable domains within a fluid medium over which to formulate the equations of motion. As in other treatments, such as Batchelor (1967) and Gill (1982), we choose to focus on fluid parcels whose total mass, determined by summing the mass of the various constituents within the parcel, remains fixed regardless of the dynamical interactions.† We already encountered such parcels in the previous subsection. Notably, since mass is equal to density times volume, constant mass fluid parcels using an $\mathbf{x}$-space description can have both changing density and changing volume, yet the product of the two remains constant.

There are fluid regions where mass conservation is not respected. Such regions are termed *nonmaterial* in this context, whereas regions with constant mass are termed *material*. As mass for the fluid parcels considered here is constant, regardless of the dynamical interactions, mass conservation forms an area of fluid kinematics. Given this basis, it is possible to derive differential statements of mass conservation using an Eulerian description. The derivation proceeds in a manner useful for establishing other fluid kinematical results.

In ζ-space, conservation of parcel mass implies that $\partial_\tau \varpi = 0$, as discussed in Section 2.3.3. For $\mathbf{x}$-space, the consequences of mass conservation are found by setting $\partial_\tau(\rho \, dV) = 0$, where

$$
dV = d\mathbf{x} \tag{2.18}
$$

is a common notation used for the infinitesimal volume of a fluid parcel. Note that since mass is a scalar under the Galilean transformations relevant for geophysical fluid dynamics, the differential mass conservation law takes the same form in any frame of reference.

---

*Recall our use of the summation convention in the last expression of equation (2.16).

†As discussed at the beginning of Chapter II in DeGroot and Mazur (1984), for a multi-component fluid this choice constitutes an interpretation of the velocity of a fluid parcel as that of the center of mass of the various fluid constituents within the parcel.

With the mass of a parcel written

$$\mathrm{d}M(\zeta) = \rho(\mathbf{x}(\zeta, \tau), t)\, \mathrm{d}\mathbf{x}(\zeta, \tau), \tag{2.19}$$

mass conservation implies

$$\begin{aligned}
0 &= \frac{\mathrm{d}}{\mathrm{d}t} \ln(\mathrm{d}\, M) \\
&= \frac{\mathrm{d}}{\mathrm{d}t} \ln(\rho\, \mathrm{d}V) \\
&= \frac{\mathrm{d}}{\mathrm{d}t} \ln(\rho\, \mathrm{d}x^1\, \mathrm{d}x^2\, \mathrm{d}x^3) \\
&= \frac{\mathrm{d} \ln \rho}{\mathrm{d}t} + \frac{\mathrm{d}}{\mathrm{d}x^m} \dot{x}^m.
\end{aligned} \tag{2.20}$$

The parcel mass conservation equation (2.20) is satisfied by all fluid parcels in the continuum, which implies that it holds at each point of the fluid at all times. It is therefore convenient to introduce the Eulerian velocity field $\mathbf{v} = \dot{\mathbf{x}} = \mathrm{d}\mathbf{x}/\mathrm{d}t$, which leads to the statement of parcel mass conservation

$$\frac{\mathrm{d} \ln \rho}{\mathrm{d}t} = -\nabla \cdot \mathbf{v}. \tag{2.21}$$

Expanding the material time derivative allows this equation to be written in the form of an Eulerian conservation law

$$\rho_{,t} + \nabla \cdot (\mathbf{v}\, \rho) = 0, \tag{2.22}$$

with

$$\rho_{,t} = \partial \rho / \partial t \tag{2.23}$$

a useful shorthand commonly used in this book.

The mass balance equation (2.22) says that an observer at a fixed point in space measures time changes in the density of fluid parcels according to the convergence of $\mathbf{v}\, \rho$ onto that point. Note that $\mathbf{v}\, \rho$ is the linear momentum per unit volume. Hence, when the momentum per unit volume of fluid parcels converges onto a region, this convergence introduces a positive time tendency for the mass per volume in that region. This forms the physical content of the mass conservation law written from an Eulerian perspective.

2.3.5 Volume conservation

For many purposes, a seawater parcel can be approximated as incompressible in the sense that changes in the volume of a parcel are small,

$$\partial_\tau\, (\mathrm{d}V) \approx 0. \tag{2.24}$$

This is the case, in particular, when the time changes of a parcel's density are very small relative to the density

$$\frac{\mathrm{d} \ln \rho}{\mathrm{d}t} \approx 0. \tag{2.25}$$

Letting the approximation become an equality leads to a fluid whose parcels materially conserve their volume,

$$\partial_\tau\, (\mathrm{d}V) = 0 \Rightarrow \nabla \cdot \mathbf{v} = 0. \tag{2.26}$$

Materially conserved volume is part of the *Boussinesq* approximation, and this approximation has found much use in ocean modeling. Notably, volume conservation *does not* mean that the fluid maintains a materially constant density. Indeed, material changes in density affect changes in hydrostatic pressure, which in turn drive many of the ocean's current systems, especially those at depth.

2.4 KINEMATICAL AND DYNAMICAL APPROXIMATIONS

The very success of the Navier-Stokes equations makes them often cumbersome to apply in practice. That is, since they encompass such a huge spectrum of dynamical motions, the theoretician, modeler, and experimentalist typically find it difficult to focus on certain elements of these motions while using the full set of equations. Reducing the spectral range over which the equations are valid has motivated a multitude of methods to approximate the equations of fluid motion.

The purpose of this section is to summarize some common approximations made in ocean climate modeling. More precise descriptions of these approximations are provided in later chapters. For now, just the general ideas are presented.

2.4.1 Hydrostatic approximation

Pressure is a force per unit area acting on a fluid parcel. Pressure at a point within a fluid at rest in a gravitational field is given by the weight of fluid above the point per unit horizontal cross-sectional area. Hence, the vertical pressure gradient is given by the buoyancy. This situation constitutes the *hydrostatic balance*. When the fluid is in motion, vertical pressure gradients are also affected by vertical accelerations and friction. However, for many geophysically relevant fluid motions, the dominant balance in the vertical momentum equation remains the hydrostatic balance. The level to which the hydrostatic balance remains dominant is directly proportional to the ratio of vertical (H) to horizontal (L) length scales of motion. For many motions of direct interest in ocean climate modeling, the vertical length scale is on the order of a few meters to a few kilometers, whereas the horizontal length scale is on the order of tens to thousands of kilometers. Thus, H/L is typically far less than $1/10$.

In addition to simplifying the computation of pressure, the hydrostatic approximation filters out sound waves, which are three-dimensional pressure fluctuations (e.g., Gill, 1982; Apel, 1987). There remains, however, an acoustic mode known as the Lamb wave that is not filtered by the hydrostatic approximation. As noted by DeSzoeke and Samelson (2002), for the atmosphere the Lamb wave is a vertically evanescent wave trapped at the lower boundary. Hence, a non-Boussinesq (Section 2.4.3) atmospheric model will support this mode. In the ocean, the Lamb wave is subsumed into the external gravity mode due to the large difference between the vertical scale height based on compressibility (order 100-200 km) and the ocean depth (order 5 km) (see Section 4.8.2). Consequently, the Lamb wave in hydrostatic ocean models does not affect the model's stability relative to Boussinesq models where all acoustic modes are filtered.

For studies aiming to simulate motions with large vertical accelerations and relatively small horizontal scales, such as those occurring in convective regions, the hydrostatic approximation can be an unacceptable limitation. The paper by Marshall et al. (1997) provides a lucid discussion of nonhydrostatic Boussinesq ocean modeling.

2.4.2 Shallow ocean approximation

Relative to the earth's radius, the ocean is a shallow layer of fluid moving in an approximately spherical geometry. When measuring the distance between two points within the ocean, one must use a metric tensor (Chapter 20). The metric tensor components are generally functions of the latitudinal, longitudinal, and radial position within the ocean.

Recognizing the huge scale separation between the depth of the ocean fluid and the radius of the earth, the *shallow ocean approximation* drops the radial dependence of the metric tensor components (see Section 3.2). Upon doing so, it sets the radius to a constant. Dropping radial dependence of the metric leads to some implications for the angular momentum of a fluid parcel moving in this geometry. These issues are discussed in Section 4.10. Note that the shallow ocean approximation is sometimes called the *traditional approximation* in the literature.

The dynamical equations for a hydrostatic fluid parcel moving in a spherical geometry with this approximate metric constitute the *primitive equations*. The term "primitive" refers to the use of velocity in the momentum budget, rather than the vorticity and divergence commonly used in other formulations. The primitive equations are those most commonly used in global ocean climate models.

2.4.3 Boussinesq approximation

One common approximation employed in ocean climate models is the Boussinesq approximation. From a kinematic perspective, a Boussinesq fluid parcel maintains the same volume during its transport through the fluid, regardless of the dynamical interactions. Without volume sources, such as those from precipitation, evaporation, or rivers, a Boussinesq ocean maintains a constant total volume. This behavior is to be contrasted to the more fundamental mass conservation property of non-Boussinesq fluid parcels, whereby an ocean without mass sources maintains a fixed total mass. Figure 2.3 provides a schematic of the differences.

The Boussinesq approximation has been commonly used in ocean climate models due to the near incompressibility of ocean fluid parcels, and thus the near conservation of volume maintained by these parcels. Additionally, a Boussinesq fluid does not support acoustic fluctuations (e.g., Gill, 1982), thus filtering out the climatologically unimportant acoustic waves. By filtering out such fast and unimportant waves, an ocean model based on the Boussinesq approximation can support larger time steps without violating the Courant-Friedricks-Levy (CFL) condition (e.g., Haltiner and Williams, 1980; Durran, 1999). Larger time steps make the model more efficient and thus enhance its utility for climate studies.

Like any approximate description, a Boussinesq fluid has its limitations. Most notably for climate purposes, a Boussinesq fluid does not render an accurate computation of the sea level height. The reason is that it does not incorporate fluctuations in the depth averaged density field (Section 3.4.3, page 36). Such *steric effects* are not incorporated into volume conserving kinematics. The example of

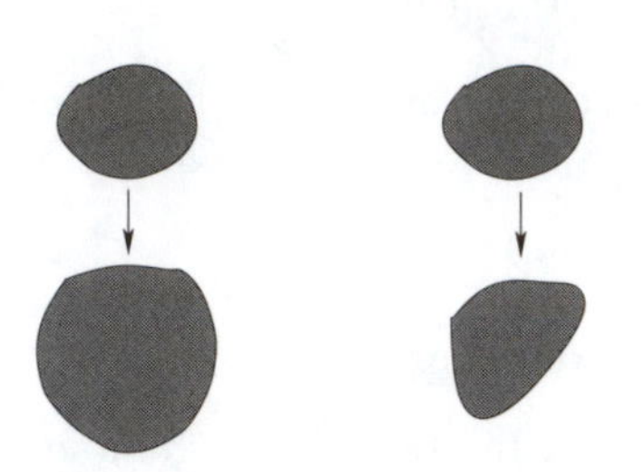

Figure 2.3 The left schematic illustrates a mass conserving fluid parcel, as in a non-Boussinesq ocean. As it is deformed by currents and affected by changes in temperature, salinity, and pressure, the parcel's volume and density generally change, yet its mass $\mathrm{d}M = \rho\,\mathrm{d}V$ remains constant. In contrast, a Boussinesq fluid parcel evolves without changing its volume $\mathrm{d}V = \mathrm{d}M/\rho$, but allows for changes in its density and mass. The right schematic is meant to illustrate such constant-volume deformations. The particular example of heating is revealing. Heating a mass conserving parcel reduces its density and so increases its volume, whereas heating a volume conserving parcel reduces its density and so reduces its mass via *virtual mass fluxes* removing mass from the parcel.

uniformly heating the ocean illustrates this point. As with the example given in Figure 2.3, uniformly heating a non-Boussinesq ocean causes its density to decrease and its volume to increase in order to maintain a constant mass. Hence, the non-Boussinesq ocean's sea level rises under uniform heating. In contrast, uniformly heating a Boussinesq ocean causes its density to decrease and its mass to decrease in order to maintain a constant volume. Hence, the Boussinesq ocean's sea level remains constant under uniform heating. The sea level height prognostically computed from Boussinesq ocean models is therefore insufficient for purposes of predicting sea level changes associated with global warming.

The pressure within a Boussinesq model is compromised due to the inaccurate computation of the ocean surface height. Additionally, as volume instead of mass is conserved, Boussinesq fluid parcels generally change their mass via unphysical *virtual mass fluxes*, thus influencing the weight of a fluid column, again affecting the pressure field.

Greatbatch (1994) and Mellor and Ezer (1995) showed that diagnostic adjustments can be made to account for inaccuracies in the Boussinesq fluid's surface height. However, these adjustments have their limitations, and they do not overcome problems associated with bottom pressure errors as discussed by Greatbatch et al. (2001). Nonetheless, it has been argued by Losch et al. (2004) that uncertainties in other parts of the model formulation, such as subgrid scale parameterizations, can cause as large, or larger, differences in practice between solutions from non-Boussinesq or Boussinesq models. Even so, for those aiming to reduce the *a priori* assumptions used to formulate a model, it may prove desirable to start from the more fundamental non-Boussinesq approach. As discussed in Chapter 6, pressure is a natural vertical coordinate for non-Boussinesq hydrostatic models (see also DeSzoeke and Samelson, 2002).

2.4.4 Rigid lid approximation

The rigid lid approximation assumes that the surface elevation of the ocean fluid is static. Doing so filters out high frequency fluctuations of the ocean associated with motions of the ocean surface. Most notably, *external* or *barotropic* gravity wave fluctuations cause undulations in the sea level height, and these waves can move at speeds some 100 times faster than the next fastest *internal* or *baroclinic* fluctuations. For many climate modeling purposes, external gravity waves are of only indirect interest, and so their absence from the model is of little physical consequence yet of great computational advantage due to the reduced CFL constraints. It is for these reasons that the methods of Bryan (1969) exploited the rigid lid approximation.

A physically unsatisfying consequence of assuming a static surface height is that direct simulations of ocean tides are precluded. Barotropic tides manifest as undulations of the sea level, and so require the use of a *free surface* ocean model. Additionally, as commonly implemented, a rigid lid ocean model does not allow fresh water to cross the ocean surface. That is, the typical form of a rigid lid ocean maintains a constant volume. This is a severe restriction on the utility of the ocean model, since fresh water fluxes are important in their effect on the ocean's buoyancy field. There are well-known indirect methods to include hydrological forcing on the rigid lid model, yet they are *ad hoc* and physically unsatisfying (e.g., Huang, 1993; Griffies et al., 2001). Ocean climate modelers have therefore been moving away from the rigid lid assumption, and towards less restrictive free surface methods that allow for tides and fresh water.

A computationally unsatisfying consequence of the rigid lid approximation is that it introduces an elliptic problem with attendant global boundary conditions. The World Ocean has a very complex geometry and topography, as well as rapidly changing surface forcing from the atmosphere. Solving elliptic equations under these circumstances is a computationally difficult problem. Additionally, when implementing an algorithm on a parallel computer, it is computationally advantageous to have the algorithm be local in the sense that the solution depends only on nearest neighbor information. For the rigid lid method and other elliptic methods, nonlocal boundary conditions often arise, such as the so-called *island integrals* derived by Bryan (1969) for the rigid lid. These boundary conditions are awkward to implement in a parallel computational environment and can lead to inefficiencies in computation.

The above discussion illustrates some consequences of the rigid lid approximation. Many other approximations commonly made in theoretical geophysical fluid dynamics, such as those associated with *balanced systems* reviewed by Salmon (1998), also lead to elliptic problems and attendant physical and computational limitations. Hence, although elliptic algorithms possess some advantages from a theoretical perspective, ocean climate modelers are moving away from them. The less approximated system is proving to be physically more accurate and more general, algorithmically more straightforward, and computationally more tractable on parallel computers.

2.4.5 Trends toward reduced approximations

As explained in the previous discussion, approximations may be useful for many purposes, yet they often come at the cost of reducing the physical integrity and/or accuracy of the solution. Indeed, as with the rigid lid approximation, they often have poor computational efficiency. Hence, there is a general trend in ocean climate modeling toward reducing many of the common approximations (for a review of such developments, see Griffies et al., 2000a).

There are three general reasons to reduce the number of approximations. First, measurement techniques are getting quite accurate and will improve in the future. For example, as discussed by Hughes et al. (2000), it is anticipated that remote methods will soon be able to measure ocean bottom pressures accurately enough to distinguish between a Boussinesq and non-Boussinesq fluid. Second, approximations such as the rigid lid approximation place restrictions on the utility of the model for realistic ocean climate simulations incorporating sea ice, rivers, hydrological cycles, and tides. Third, the motivation for making approximations in the early days of ocean modeling was largely based on insufficient computer power. Three to four decades of ocean modeling have taught us there will "never" be enough computer power to do all the experiments desired. Nonetheless, present-day computers do provide sufficient power to run global models with increasingly realistic situations exhibiting improved skill. Hence, to simulate the global ocean more accurately over its broad range of temporal and spatial scales, modelers are not interested in handicapping their abilities by making restrictive assumptions on the basic model equations. This motivation has led to the gradual elimination of the previously ubiquitous Boussinesq and rigid lid approximations in ocean climate models. In another decade or two, global modelers may also find it feasible and useful to dispense with the hydrostatic approximation.

2.5 AVERAGING OVER SCALES AND REALIZATIONS

Oceanic flow is generally nonlinear, chaotic, and turbulent. Correspondingly, the ocean's spatial and temporal scales are extremely broad, with fluctuations reaching from the parcel level and seconds to the global scale and millennia. Coupling occurs between the scales due to nonlinear cascades of flow properties, such as energy and tracer variance. Furthermore, it is not possible to measure every spatial point within a fluid, nor all time instances. These practical limitations restrict access to information about the turbulent fluid's spatial and temporal properties. Sensitive dependence on initial conditions in this turbulent flow then severely limits its predictive capabilities. This lack of information and limited predictability motivates a formulation of averaged or mean field fluid equations.

To derive the mean field ocean equations, one may envision a multitude of ocean flows, each differing by their initial and/or boundary conditions. Averaging over these realizations leads to *ensemble averaged* equations of motion. Formally, there are an infinite number of realizations in an ensemble. Alternatively, averages over an imagined infinite-sized ensemble are replaced by integrations over a probability distribution setting the likelihood of a particular phase space configuration. Ensemble averages allow for an unambiguous interchange between the averaging operator and integral/differential operators appearing in the unaveraged equations of motion, and so are preferred here. Additionally, space-time average operators formally lead to issues of spectral leakage and the attendant details of how to design the filter in a more sophisticated manner.

Although aiming to provide some rigor to the averaging applied in this book, the level of rigor is far from satisfying to those adept in statistical physics. In particular, no claim is made as to whether the ocean fluid satisfies the *ergodic hypothesis*.* Making some of these notions rigorous for a general fluid remains an unsolved area of statistical physics.

Conceptually, the treatment here considers a "distance" in phase space between members of an ensemble as being directly proportional to the spatial and temporal scales that remain unresolved by measurements and/or a discrete numerical model. For example, ensemble members may differ by the phase of a wave or eddying fluctuation occurring on a spatial scale of tens of kilometers. For measurements and/or models that resolve only scales greater than hundreds of kilometers, these *mesoscale* features remain unresolved and so constitute part of the subgrid scale (SGS) regime that is averaged over by the ensemble averaging process (see Chapter 9 for more discussion). Other features smaller than mesoscale fluctuations also constitute dynamics within the unresolved spectrum, and so likewise must be averaged when formulating the equations describing the large-scale features (see Chapters 7 and 8).

As discussed in Chapter 8, a practical and useful goal for the averaging methods is to produce averaged kinematical and dynamical equations maintaining the same mathematical form as the unaveraged equations. Doing so allows for mathematical properties of the unaveraged equations to be shared by the averaged equations. Although mathematically similar, terms appearing in the two sets of equations have distinct physical interpretations. In particular, the averaged equations contain terms unknown to the "averaged observer." These terms take a form dependent on the dynamics occupying the SGS.

Specifying SGS terms is a complicated problem in *turbulence closure theory*. Notably, most closures contain formally unjustified steps, thus prompting a multitude of different closures. Unfortunately, ocean model simulations have proven to be quite sensitive to details of the closures. This is perhaps the most unsatisfying aspect of ocean climate modeling. In turn, the formulation and testing of SGS closures are active and exciting areas of research with numerous areas available for advancing understanding and simulation integrity.

2.6 NUMERICAL DISCRETIZATION

A mathematical description of fluid motion starts from the continuum approximation, whereby molecular dynamical scales are averaged out to reveal a mean field set of equations. Then, depending on one's interest in dynamical regimes, selected approximations are introduced to reduce the huge scales of motion allowed by the full continuum set of equations. Next, due to incomplete information and sensi-

*For many purposes in ocean modeling, the ergodic hypothesis says that ensemble averages equal time averages.

tive dependence on initial conditions, averaging methods are required to form a set of *model equations* with SGS operators determined in terms of averaged fields in order to close the equations.

After these mathematical and physical steps are finalized, the ocean modeler must formulate the averaged equations of motion in a manner accessible to finite computational capabilities. That is, the approximated, averaged, closed, continuum partial differential ocean model equations are cast into a discrete form so that they are accessible to the numerical/algebraic methods interpretable by computers. In turn, algorithms are developed in order to evolve numerical simulations forward in time.

There are various methods for casting the continuum equations into discrete form. One method integrates the continuum equations over finite grid cells, defining cell-averaged variables, thus resulting in a set of discrete *finite volume* equations of motion (see Section 10.1). Most ocean climate models follow this approach to some degree.

Numerical discretization breaks the continuum symmetries present in the partial differential equations of fluid mechanics. Hence, solutions realized on the computer may not always correspond to those allowed by the continuum equations. Eliminating such spurious numerical solutions is a constant concern of the numerical modeler. In general, significant care is required so that the numerical fluid is a close analog to the continuum fluid.

Assuming that the continuum equations are successfully cast into a numerical algorithm, a numerically realized fluid provides the modeler with a powerful tool to probe the many dynamical regimes present in the equations of motion. That is, numerical models are powerful tools to perform investigative experiments and to extrapolate observations into the future via prediction. For example, the modeler is able to test ideas by altering subgrid scale parameters, changing boundary forcing or domain geometry, refining grid mesh sizes to allow the flow to become more nonlinear, simplifying the dynamical degrees of freedom via approximations, identifying dominant balances by measuring terms within the model equations, and projecting scenarios for future ocean climate change due to alterations in atmospheric radiative forcing. Such direct experimental methods are unavailable for studies of large spatial and temporal scales, due to our inability to perform reproducible and controllable experiments with Nature's climate system. Furthermore, the suite of solutions available numerically is far larger than available analytically. Hence, when combined with analytical methods of highly idealized configurations, laboratory experiments of small-scale phenomena, and *in situ*, remote, and paleo-ocean observations, numerical models round out the suite of tools for ocean climate science. Indeed, numerical models have become the main repository for observations and theories of ocean climate.

2.7 CHAPTER SUMMARY

This chapter introduced the reader to a variety of topics fundamental to a kinematic description of the ocean fluid. At the core of this description is the continuum hypothesis. This hypothesis allows one to ignore details of the huge numbers of molecular degrees of freedom possessed by a fluid. It thus facilitates the formulation of the kinematic, dynamic, and thermodynamic equations of the ocean fluid as a continuum set of partial differential equations. These equations, which are developed in Chapters 3, 4, and 5, have proven sufficient for the purposes of geophysical fluid dynamics, and hence for ocean modeling.

The continuum hypothesis then leads to a basis for the mathematical description of fluid motion. In particular, when fluid motion is formulated in terms of parcel trajectories, one obtains a Lagrangian form of fluid dynamics which has much in common with point particle dynamics. Hence, a great deal of fundamental understanding in fluid dynamics has arisen from the Lagrangian approach.

For purposes of formulating the equations of an ocean model, it has been found useful to combine Lagrangian and Eulerian methods, where the Eulerian observer is fixed with respect to a background reference frame. A basic theme of this book is that both Lagrangian and Eulerian methods have their place in analyzing oceanic flows and formulating the equations of an ocean model.

A number of approximations applied to the equations of ocean fluid dynamics have been made over the years by researchers formulating equations for particular regimes of flow. In many cases, these approximations were originally aimed at providing a simpler set of equations amenable to faster solution on a computer. However, some of the approximations, especially those leading to an elliptic problem, turn out to be physically cumbersome and computationally inefficient on the parallel computers commonly used in early twenty-first century modeling. Hence, there is a general movement toward the development of ocean climate models with fewer approximations, in hopes that the basic equations used by these models will allow for a more accurate simulation as well as a simpler computational algorithm.

The ocean fluid is turbulent. Hence, statistical methods are needed to formulate equations appropriate for ocean modeling. That is, ocean model equations must respect our ignorance of the precise trajectory of the ocean fluid in its phase space. In so doing, equations discretized by a numerical model describe the mean of an infinite ensemble of ocean fluid states. Statistical methods introduce subgrid scale (SGS) closure terms into the equations used by an "averaged observer." The SGS closure problem remains one of the most intractable problems in theoretical physical oceanography. Additionally, details of the SGS closure terms can have a nontrivial impact on the simulation of ocean climate. Consequently, SGS closure schemes are ripe for advance in both modeling and theory.

Upon reaching a point where approximations have been made, statistical averaging performed, and SGS operators prescribed, the numerical algorithm developer must consider how to cast the continuum partial differential equations onto a finite space-time lattice. This task is highly nontrivial. In particular, symmetries present on the lattice are distinct from those in the continuum. Indeed, it is common to find that the different symmetries allow for spurious unphysical modes to be present in the numerical fluid. The numericist must therefore exercise care to provide a sound discretization so that the numerical fluid resembles the continuum fluid. This task remains at the forefront of numerical oceanography, where sound and elegant algorithms are always sought. In this context, "sound and elegant" means an algorithm possessing robust physical integrity, clean and straightforward numerical methods, and efficient computational performance.

Chapter Three

KINEMATICS

The purpose of this chapter is to present some kinematics forming a part of ocean fluid mechanics.

3.1 INTRODUCTION

In this chapter, we explore certain aspects of ocean fluid kinematics, with the next chapter focusing on ocean fluid dynamics. Hence, these two chapters formulate mathematical equations of use for rationalizing the global ocean circulation. The presentations vary from fundamental physical notions to general results from tensor analysis. It is hoped that by the end of these two chapters that the reader will have gained a physical and mathematical understanding of ocean fluid mechanical equations.

Sections 3.2 and 3.3 provide some key mathematical results of use in formulating the equations of global ocean fluid mechanics. Much of this material is treated more fully in the tensor analysis Chapters 20 and 21. However, the basic ideas can be garnered here, so long as the reader has some experience with vector calculus. Section 3.4 then introduces some kinematic aspects of stratified fluids on a sphere. This section compares the distinct kinematics of mass conserving versus volume conserving fluid parcels. Section 3.5 finishes the chapter with a summary.

3.2 MATHEMATICAL PRELIMINARIES

This section summarizes some mathematical results from Part 6 of this book that prove to be of use for describing a fluid parcel in a rotating spherical geometry.

3.2.1 Coordinates for large-scale motion

Veronis (1973) and Gill (1982) discuss how the earth's geometry can be well approximated by an oblate spheroid, with the equatorial radius larger than the polar. With this geometry, surfaces of constant geopotential are represented by surfaces with a constant oblate spheroid radial coordinate, such as described on page 662 of Morse and Feshbach (1953). However, the metric functions, which determine how to measure distances between points on the spheroid, are less convenient to use than the more familiar spherical metric functions.

Veronis (1973) and Gill (1982) (see in particular page 91 of Gill) indicate that it is possible, within a high level of accuracy, to maintain the best of both situations. That is, the surfaces of constant "radius" r are interpreted as best fit oblate spheroidal geopotentials, yet the metric functions used to measure distance between points in the surface are approximated as spherical. As the metric functions determine the geometry of the surface, and hence the form of the equations of motion, the equations are exactly those which result when using spherical coordinates on a sphere. Hence, throughout this book, the geometry of the earth is spherical, yet the radial position r represents a surface of constant geopotential.

From a mathematical physics perspective, a key goal of large-scale ocean fluid mechanics boils down to formulating the equations of fluid motion on the surface of a rotating planet with spherical geometry. Notably, the derivations must carefully account for details of the sphere's non-Euclidean geometry (see Chapters 20 and 21 for more on differential geometry). This is a fundamental distinction from nongeophysical formulations of fluid mechanics, such as described in Landau and Lifshitz (1987), where it is often sufficient to consider Euclidean space. A rigorous and lucid treatment of fluid motion on arbitrary smooth surfaces is given by Aris (1962).

Figure 3.1 illustrates the relation between spherical and Cartesian coordinates of use for describing fluid dynamics on a rotating sphere. Mathematically, Cartesian coordinates are related to spherical coordinates via the transformation

$$\begin{aligned} x^1 &= r\cos\phi\cos\lambda \\ x^2 &= r\cos\phi\sin\lambda \\ x^3 &= r\sin\phi. \end{aligned} \tag{3.1}$$

Formally, r is the radial distance from the sphere's center. However, following the interpretation mentioned above, for ocean fluid dynamics r is interpreted as the radial coordinate of a geopotential surface. The angular coordinate ϕ is the latitude, with values $\phi = 0$ at the equator and $\phi = \pi/2(-\pi/2)$ at the north (south) poles. Finally, $0 \leq \lambda \leq 2\pi$ is the longitude, with positive values measured eastward from the prime meridian passing through Greenwich, England. The axis of rotation passes from the southern pole to the northern pole. Both sets of coordinates are fixed on the rotating planet (non-inertial coordinates).

Note that for many idealized geophysical fluid studies, Cartesian coordinates refer to those defined locally to a tangent plane at some point on the surface of the rotating planet. Such β-plane or f-plane coordinates (e.g., Gill, 1982; Pedlosky, 1987), are distinct from the Cartesian coordinates defined here. Throughout this book, Cartesian coordinates refer to coordinates whose origins are at the center of the rotating earth.

The radial position of a fluid parcel in the ocean can be written

$$r = R + z, \tag{3.2}$$

where z represents the deviation of the parcel's position from the sea level geopotential. As noted by Gill (1982) (see his page 91),

$$R = 6.367 \times 10^6\,\mathrm{m} \tag{3.3}$$

corresponds to the ellipsoid of best fit to the sea level geopotential. This is the appropriate value for the "earth's radius" of use in ocean climate models. Note that R in ocean climate models is often taken as the slightly larger value $R = 6.371 \times 10^6\,\mathrm{m}$. This value corresponds to the radius of a sphere with the same volume as the earth (Gill, 1982, page 597).

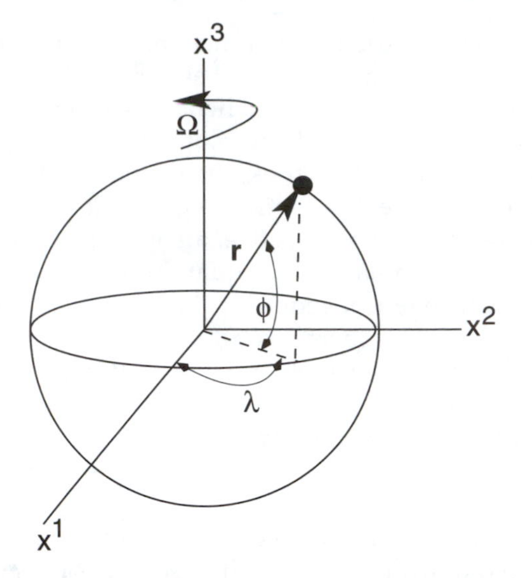

Figure 3.1 Schematic of spherical and Cartesian coordinates used for describing the motion of fluid parcels on a rotating spherical planet. The coordinates are defined by equation (3.1). The coordinate origin is at the center of a sphere, and the rotation axis is aligned through the sphere's north pole. The planet rotates with an angular velocity Ω in a direction counter clockwise when looking down from the north pole.

For an ocean fluid at rest, $z = 0$ represents the surface of the ocean, which is a geopotential surface. The surface of the solid earth living beneath the ocean is taken as

$$z = -H, \tag{3.4}$$

whereas

$$z = \eta \tag{3.5}$$

is the deviation of the ocean's free surface from a state of rest. The coordinate z is generally termed the "vertical" coordinate of the parcel, since it measures the position of the parcel in the direction of the effective gravity force, that is, perpendicular to the sea level geopotential. Models based on the use of z to measure the vertical position are termed *z-coordinate* or *geopotential coordinate* ocean models. For present purposes, the coordinate z is a useful vertical coordinate to formulate the equations of ocean fluid mechanics.

3.2.2 Lateral orthogonal coordinates

In geophysical fluid dynamics, coordinates are chosen so that the two *lateral* coordinate directions are perpendicular to the local vertical, where the vertical is defined by the effective gravitational force.* The term *horizontal* is sometimes used

*In Section 7.2, we generalize the notion of *lateral* to embody motions that occur along neutral directions. However, as noted in Section 6.4, coordinates that are oriented according to neutral directions are of limited use for geophysical fluid modeling.

synonymously with lateral, though it should be remembered that motion under consideration is actually in a spherical geometry rather than on a flat plane. The central reason that this coordinate specification is chosen is that the vertical force balance, dominated by the hydrostaic balance (Section 4.5) is cleanly separated by the much smaller force balance in the horizontal directions. The spherical angles (λ, ϕ) shown in Figure 3.1 represent the most familiar examples of orthogonal lateral coordinates used to locate positions on a sphere.

Global ocean climate models are typically written using angular coordinates that generalize the spherical angles (λ, ϕ). That is, they use general locally orthogonal coordinates to specify angular positions. The central reason for doing so is that the spherical coordinates possess an awkward coordinate singularity at $\phi = \pi/2$, which is within the Arctic Ocean.* This singularity creates numerous problems for realistic ocean simulations of the Arctic, since the grid area near the North Pole shrinks to zero, thus necessitating a prescription for how to handle this point. Methods for maintaining numerical stability of simulations using these coordinates have introduced unwanted side-effects, with solution integrity and/or computational efficiency greatly compromised. Hence, common practice in ocean climate models is to employ generalized orthogonal curvilinear coordinates.

As discussed in Chapter 21, the use of orthogonal coordinates (ξ^1, ξ^2) allows for the squared infinitesimal distance between two points in the ocean to be written as

$$(\mathrm{d}s)^2 = (h_1\,\mathrm{d}\xi^1)^2 + (h_2\,\mathrm{d}\xi^2)^2 + (\mathrm{d}z)^2, \tag{3.6}$$

where the metric, or stretching functions h_1 and h_2 are non-negative. The lack of cross-terms in this squared distance is a result of the coordinate orthogonality (indeed, it defines orthogonality).†

In terms of *physical* distances with dimensions of length, the infinitesimal length increments in the two lateral directions are

$$\begin{aligned} \mathrm{d}x &= h_1\,\mathrm{d}\xi^1 \\ \mathrm{d}y &= h_2\,\mathrm{d}\xi^2. \end{aligned} \tag{3.7}$$

Correspondingly, the line element takes the compact form

$$(\mathrm{d}s)^2 = (\mathrm{d}x)^2 + (\mathrm{d}y)^2 + (\mathrm{d}z)^2, \tag{3.8}$$

the volume of an infinitesimal fluid parcel is given by

$$\begin{aligned} \mathrm{d}V &= (h_1\,\mathrm{d}\xi^1)\,(h_2\,\mathrm{d}\xi^2)\,\mathrm{d}z \\ &= \mathrm{d}x\,\mathrm{d}y\,\mathrm{d}z, \end{aligned} \tag{3.9}$$

and the physical components of the horizontal partial derivatives are

$$\partial_x = h_1^{-1}\,\partial_1 \tag{3.10}$$

$$\partial_y = h_2^{-1}\,\partial_2. \tag{3.11}$$

Figure 3.2 illustrates these formulae.

---

*The coordinate singularity at $\phi = -\pi/2$ is of no consequence, so long as there is an Antarctic continent as in today's geological arrangment.

†The author knows of no global ocean climate model that uses non-orthogonal coordinates to describe lateral motion.

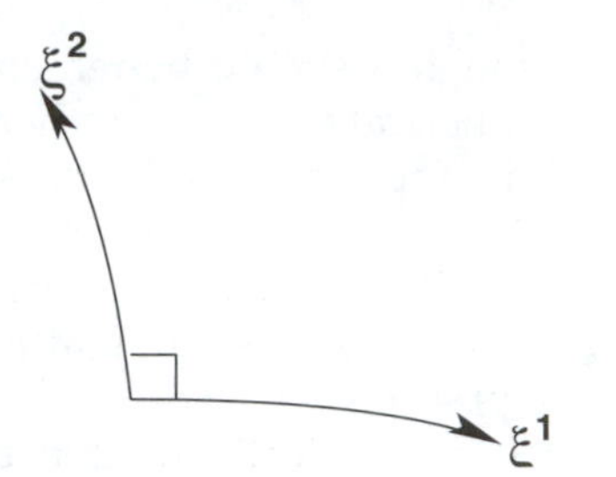

Figure 3.2 Generalized horizontal orthogonal coordinates of use for global ocean models. The coordinate lines intersect at right angles, but generally do not follow lines parallel to constant longitude or latitude. An infinitesimal horizontal region has area given by $\mathrm{d}A = (h_1\,\mathrm{d}\xi^1)\,(h_2\,\mathrm{d}\xi^2) = \mathrm{d}x\,\mathrm{d}y$. For spherical coordinates $(\xi^1, \xi^2) = (\lambda, \phi)$, the infinitesimal horizontal distances are $\mathrm{d}x = (r\cos\phi)\,\mathrm{d}\lambda$ and $\mathrm{d}y = r\,\mathrm{d}\phi$.

Although the introduction of physical displacements in equation (3.7) brings the metric tensor into a form analogous to that for three-dimensional Euclidean space, the non-Euclidean nature of the sphere manifests by the nonvanishing commutator

$$\begin{aligned}[\partial_x, \partial_y] &= \partial_x\,\partial_y - \partial_y\,\partial_x \\ &= (\partial_x \ln \mathrm{d}y)\,\partial_y - (\partial_y \ln \mathrm{d}x)\,\partial_x,\end{aligned} \tag{3.12}$$

which vanishes only when the geometry is flat instead of curved. In particular, use of spherical coordinates with $(\xi^1, \xi^2) = (\lambda, \phi)$ leads to

$$[\partial_x, \partial_y] = \left(\frac{\tan\phi}{R}\right)\partial_x. \tag{3.13}$$

3.2.3 Convention for the gradient operator

The familiar notation

$$\nabla = (\partial_x, \partial_y, \partial_z) \tag{3.14}$$

is used in this book for the three-dimensional gradient operator. When acting on a function that depends only on the horizontal spatial directions, such as the surface height η or bottom topography H defined in Section 3.4.2, ∇ reduces to a horizontal gradient. The notations ∇_h and ∇_z are also commonly used to signify a horizontal gradient on a constant depth surface. This notation is not needed, however, when acting on a depth-independent function, such as η and H. Likewise, when acting on the three-dimensional velocity vector

$$\mathbf{v} = (\mathbf{u}, w) = (u, v, w), \tag{3.15}$$

the divergence operator can be written

$$\nabla \cdot \mathbf{v} = \nabla \cdot \mathbf{u} + \partial_z w. \tag{3.16}$$

On the right-hand side, $\nabla \cdot \mathbf{u}$ is the same divergence as $\nabla_z \cdot \mathbf{u}$ since $\mathbf{u}$ is horizontal. The more streamlined notation $\nabla \cdot \mathbf{u}$ is used when applicable.

3.2.4 The comma notation for partial derivatives

As motivated by the tensor analysis discussions of Chapters 20 and 21, it is convenient to use the operator and/or comma notation for partial derivatives. For example, the local time derivative of density can be written in one of three equivalent ways

$$\partial\rho/\partial t = \partial_t \rho = \rho_{,t}. \tag{3.17}$$

Likewise, the vertical shear of the horizontal current can be written

$$\partial \mathbf{u}/\partial z = \partial_z \mathbf{u} = \mathbf{u}_{,z}. \tag{3.18}$$

The comma notation succinctly distinguishes the component of a tensor, written without the comma, from the partial derivative of a function.

3.2.5 The shallow ocean approximation

The *shallow ocean approximation** says that the metric functions h_1 and h_2 introduced in equation (3.6) are dependent *only* on the lateral coordinates (ξ^1, ξ^2). Radial dependence of the metric functions is reduced to the constant radial factor R discussed in Section 3.2.1. This approximation reflects the small vertical deviations of a fluid parcel relative to the earth's radius.

Distances used to compute partial derivatives, covariant derivatives, areas, and volumes are determined by a metric tensor whose components are functions only of the lateral position on the sphere. Additionally, assumptions regarding the metric function dependence, as well as assumptions about the smallness of vertical accelerations associated with the hydrostatic approximation, have implications toward the energy and angular momentum conservation laws. Some discussion of this point is provided in Section 4.10.

3.3 THE DIVERGENCE THEOREM AND BUDGET ANALYSES

In studying ocean fluid mechanics, one often asks questions about budgets over finite domains or regions in the ocean. To understand these budgets requires an understanding of fluid mechanics within the domain as well as boundary effects acting on the domain. When formulating a budget analysis, the differential equations governing fluid parcels are integrated over the region. Notably, the *divergence theorem* converts the volume integral of a divergence into a surface integral on the bounding area. The purpose of this section is to summarize some of the salient mathematical points useful in the formulation of such budgets.

A key assumption often made for ocean budget analyses is that the bottom and surface boundaries are smooth and have an outward normal with a nonzero projection in the vertical. A smooth version of the ocean surface (i.e., no breaking waves), and smooth solid earth boundary (i.e., no overhangs or caves) provide examples.

The integral theorems from Cartesian vector analysis transform in a straightforward manner to arbitrary coordinates in arbitrary smooth spaces. For example,

---

*This approximation is distinct from the *shallow water* approximation discussed in Section 4.6.3.

the divergence of a vector $\mathbf{F}$ over a region can be written*

$$\int dV\, \nabla \cdot \mathbf{F} = \int dA_{(\hat{\mathbf{n}})}\, \mathbf{F} \cdot \hat{\mathbf{n}}, \tag{3.19}$$

with dV the volume for infinitesimal regions inside the domain, $\hat{\mathbf{n}}$ the outward normal direction on the domain boundary, and $dA_{(\hat{\mathbf{n}})}$ the surface area for infinitesimal regions on the domain boundary. This equation is as expected from Cartesian vector analysis. However, it is important to note that with general coordinates $\boldsymbol{\xi}$, the volume element is given by

$$dV = \sqrt{\mathcal{G}}\, d\boldsymbol{\xi}, \tag{3.20}$$

where $\mathcal{G}$ is the determinant of the metric tensor used to measure distances within the volume, and $d\boldsymbol{\xi} = d\xi^1\, d\xi^2\, d\xi^3$. The area element on the surface is given likewise, but with one spatial dimension fewer and with generally different coordinates used to describe the surface. Within the volume, the divergence of a vector is given by

$$\nabla \cdot \mathbf{F} = \mathcal{G}^{-1/2}\, (\mathcal{G}^{1/2}\, F^m)_{,m} \tag{3.21}$$

thus leading to the result

$$dV\, \nabla \cdot \mathbf{F} = d\boldsymbol{\xi}\, (\mathcal{G}^{1/2}\, F^m)_{,m} \tag{3.22}$$

where repeated indices are summed.

Consider an explicit example relevant for ocean fluid mechanics: the divergence of a vector integrated over an ocean basin. In this case, the two bounding surfaces are the free surface at $z = \eta$ and the solid earth boundary at $z = -H$. We find it convenient to consider the *inward* normal at the solid earth boundary, and the *outward* normal at the ocean surface. For a flat bottom and flat surface ocean, both normals then reduce to the vertical direction $\hat{\mathbf{z}}$. In general, this convention leads to

$$\int dV\, \nabla \cdot \mathbf{F} = \int_{z=\eta} dA_{(\eta)}\, \hat{\mathbf{n}} \cdot \mathbf{F} - \int_{z=-H} dA_{(-H)}\, \hat{\mathbf{n}} \cdot \mathbf{F}, \tag{3.23}$$

where again $\hat{\mathbf{n}}(\eta)$ is the outward normal at the surface and $\hat{\mathbf{n}}(-H)$ is the inward normal at the solid earth boundary. For fluxes that have zero normal components at the solid earth boundaries, the last term vanishes. More generally, it is necessary to evaluate the surface integrals.

To evaluate surface integrals, various surface coordinates can be used. For example, one may use locally orthogonal coordinates defined on the generally undulating surface, such as those discussed in Section 6.4 on page 130. More commonly, it is convenient to use the familiar horizontal coordinates x, y, which are useful in cases when the surface and bottom normals have a nonzero projection in the vertical. Since the ocean surface and solid earth boundary are not generally horizontal, care is needed to write the area elements in terms of the usual horizontal coordinates. For this purpose, results from Section 20.13.2 on page 461 lead to the following expression for the area element at the ocean surface:

$$dA_{(\eta)} = |\nabla(-\eta + z)|\, dx\, dy, \tag{3.24}$$

---

*See Section 21.10 on page 474 for discussion.

where $\nabla(-\eta + z) = -\nabla\eta + \hat{\mathbf{z}}$ points upward in a direction normal to the ocean surface. Similar considerations render the area element at the solid earth boundary

$$\mathrm{d}A_{(-H)} = |\nabla(H+z)|\, \mathrm{d}x\, \mathrm{d}y, \tag{3.25}$$

with $\nabla(H+z) = \nabla H + \hat{\mathbf{z}}$ pointing upward in a direction normal to the solid earth boundary. Correspondingly, the outward normal at the ocean surface is

$$\begin{aligned} \hat{\mathbf{n}}(\eta) &= \frac{\nabla(-\eta+z)}{|\nabla(-\eta+z)|} \\ &= \frac{-\nabla\eta + \hat{\mathbf{z}}}{\sqrt{1+(\nabla\eta)^2}} \end{aligned} \tag{3.26}$$

and the inward normal at the solid earth boundary is

$$\begin{aligned} \hat{\mathbf{n}}(-H) &= \frac{\nabla(H+z)}{|\nabla(H+z)|} \\ &= \frac{\nabla H + \hat{\mathbf{z}}}{\sqrt{1+(\nabla H)^2}}. \end{aligned} \tag{3.27}$$

Because the numerator in the normals appear quite often, it is convenient to introduce the *outward orientation direction* at the surface

$$\hat{\mathbf{N}}(\eta) = \nabla(-\eta + z) \tag{3.28}$$

and the *inward orientation direction* at the bottom

$$\hat{\mathbf{N}}(-H) = \nabla(H+z). \tag{3.29}$$

Bringing these results together leads to

$$\int \mathrm{d}V\, \nabla \cdot \mathbf{F} = \int_{z=\eta} \mathrm{d}x\, \mathrm{d}y\, \hat{\mathbf{N}} \cdot \mathbf{F} - \int_{z=-H} \mathrm{d}x\, \mathrm{d}y\, \hat{\mathbf{N}} \cdot \mathbf{F}. \tag{3.30}$$

3.4 VOLUME AND MASS CONSERVING KINEMATICS

As discussed in Section 2.4.3 on page 17, Boussinesq fluid parcels conserve their volume, whereas non-Boussinesq fluid parcels conserve their mass. In this section, we highlight the kinematic issues related to volume and mass conservation. In turn, derivations are given for kinematic boundary conditions applicable to Boussinesq and non-Boussinesq fluids.

3.4.1 Mass and volume conservation for parcels

To get started, recall the discussion given in Section 2.3.4 on page 14. In that discussion, constraints were derived that the three-dimensional velocity field $\mathbf{v} = (\mathbf{u}, w)$ must satisfy for mass and volume conserving fluid parcels. With the mass of a parcel of seawater given by $\mathrm{d}M = \rho\, \mathrm{d}V$, mass conservation for the parcel leads to

$$\rho_{,t} + \nabla \cdot (\mathbf{u}\, \rho) + (w\, \rho)_{,z} = 0, \tag{3.31}$$

whereas volume conservation leads to

$$\nabla \cdot \mathbf{u} + w_{,z} = 0. \tag{3.32}$$

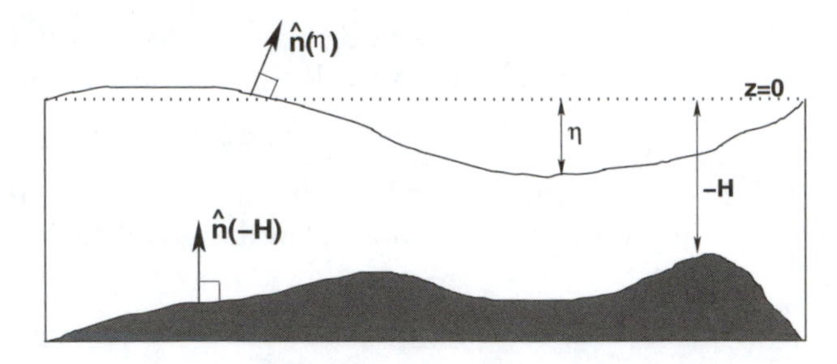

Figure 3.3 A control volume of fluid for use in discussing volume and mass conservation and kinematic boundary conditions. The vertical side boundaries are stationary and penetrable. The bottom topography at $z = -H$ is generally non-flat, impenetrable, and stationary. The free surface at $z = \eta$ is generally non-flat, penetrable, and time dependent. The surface outward normal is $\hat{\mathbf{n}}(\eta) = \nabla(-\eta + z)/|\nabla(-\eta + z)|$, and the surface area element is $\mathrm{d}A_{(\eta)} = |\nabla(-\eta + z)|\,\mathrm{d}x\,\mathrm{d}y$. The bottom inward normal is $\hat{\mathbf{n}}(-H) = \nabla(H + z)/|\nabla(H + z)|$, and the area element is $\mathrm{d}A_{(-H)} = |\nabla(H + z)|\,\mathrm{d}x\,\mathrm{d}y$. Undulations of the surface height are exaggerated here for illustrative purposes. In the real ocean, topographic variations can reach from the solid earth boundary to the ocean surface (thousands of meters), whereas surface height variations are on the order of a few meters. Furthermore, for mathematical simplicity, the bottom and surface boundaries are assumed never to overturn or break, which means the orientation vectors always have a positive vertical component.

Equations (3.31) and (3.32) are both referred to as *continuity equations*, with the relevant one implied by knowing whether the fluid is mass conserving (equation (3.31)) or volume conserving (equation (3.32)).

When not encountering the gravitational force, Boussinesq fluid parcels are assumed to have a mass given by $\mathrm{d}M = \rho_o\,\mathrm{d}V$ for purposes of developing the momentum and tracer budgets, where ρ_o is a constant reference density. When the mass is multiplied by the gravitational acceleration, the Boussinesq parcels have the non-Boussinesq mass $\mathrm{d}M = \rho\,\mathrm{d}V$. The ambivalent nature of this formulation translates into a corrupted gravitational potential energy budget, as discussed in Section 5.3.3 on page 102. In this case, volume conserving Boussinesq parcels can have potential energy budgets affected by unphysical changes in mass. We see this effect via

$$\begin{aligned} \mathrm{d}\,(\mathrm{d}M)/\mathrm{d}t &= \mathrm{d}(\rho\,\mathrm{d}V)/\mathrm{d}t \\ &= \mathrm{d}V\,(\mathrm{d}\rho/\mathrm{d}t) + \rho\,\mathrm{d}\,(\mathrm{d}V)/\mathrm{d}t \\ &= \mathrm{d}V\,(\mathrm{d}\rho/\mathrm{d}t) \\ &\neq 0, \end{aligned} \tag{3.33}$$

where we set $\mathrm{d}\,(\mathrm{d}V)/\mathrm{d}t = 0$ due to volume conservation. These mass changes are termed *virtual* due to their unphysical nature, although they are quite *real* insofar as the Boussinesq mechanics is affected by them.

3.4.2 Volume conservation for finite domains

The purpose of this section is to apply the constraint of volume conservation over a finite control volume. Expressions are given for the time tendency of the free ocean surface height as well as kinematic boundary conditions satisfied at the ocean surface and solid earth boundary. Additionally, some consideration is given to understanding how to quantify the flux of volume passing across boundaries of the finite domain.

To get started, consider the control volume of fluid shown in Figure 3.3. The top boundary is the ocean surface, with $z = \eta$ a time-dependent vertical deviation from a resting ocean state at $z = 0$. The lower boundary is the solid earth at $z = -H$. This boundary is assumed stationary for the purposes of ocean climate studies. Side boundaries are assumed vertical and penetrable. Considerations here thus focus on vertical columns of ocean fluid.

The volume of fluid within the column is given by

$$\begin{aligned} V &= \int\limits_{\text{column}} \mathrm{d}V \\ &= \int \mathrm{d}x\,\mathrm{d}y \int\limits_{-H}^{\eta} \mathrm{d}z \\ &= \int D\,\mathrm{d}A, \end{aligned} \tag{3.34}$$

where

$$D = H + \eta \tag{3.35}$$

is the vertical thickness of the fluid column as a function of horizontal position and time, and

$$\mathrm{d}A = \mathrm{d}x\,\mathrm{d}y \tag{3.36}$$

is the horizontal cross-sectional area element.

If the control volume is filled with volume conserving fluid parcels, conservation of volume over the finite domain implies that the volume's time tendency

$$V_{,t} = \int \mathrm{d}A\, D_{,t} \tag{3.37}$$

equals the sum of all volume fluxes passing across the domain boundaries. There are generally three types of boundary fluxes of volume.

The first boundary flux passes through the solid earth boundary and is associated with geological effects. These fluxes are typically ignored for ocean climate studies, in which case the normal component of the velocity vector vanishes at the solid earth

$$\hat{\mathbf{n}} \cdot \mathbf{v} = 0 \qquad \text{at } z = -H. \tag{3.38}$$

With the inward normal at the solid earth given by $\hat{\mathbf{n}}(-H) = \nabla(H+z)/|\nabla(H+z)|$, the no normal flow constraint takes the form

$$\mathbf{u} \cdot \nabla H + w = 0 \qquad \text{at } z = -H. \tag{3.39}$$

This boundary condition is termed *kinematic* since it is satisfied independent of the dynamics. It represents a constraint on the fluid flow set by the solid earth boundary.

The second boundary flux passes across the vertical sides of the domain. The convergence of these fluxes leads to time changes of the fluid volume in the domain

$$\left(\int_{-H}^{\eta} \mathrm{d}z\,\mathbf{u}\right)_{\text{left}} - \left(\int_{-H}^{\eta} \mathrm{d}z\,\mathbf{u}\right)_{\text{right}} = -\int \mathrm{d}A\,\nabla\cdot\int_{-H}^{\eta} \mathrm{d}z\,\mathbf{u}. \tag{3.40}$$

The third boundary flux is due to the volume passing across the ocean free surface. This flux is associated with precipitation P ($P > 0$ for moisture entering the ocean), evaporation E ($E > 0$ for moisture leaving the ocean), and river runoff R ($R > 0$ when river runoff enters the ocean). It is useful to associate a direction $\hat{\mathbf{n}}_w$ for the moisture that passes through the ocean free surface. In this way, the volume of moisture passing across the free surface, per time, per free surface area, can be represented as a vector $\hat{\mathbf{n}}_w\,(P - E + R)$.* The projection of $\hat{\mathbf{n}}_w\,(P - E + R)$ onto the normal at the ocean surface, $\hat{\mathbf{n}}(\eta)$, multiplied by the area on the free surface, $\mathrm{d}A_{(\eta)}$, yields the volume flux of moisture crossing the free surface. This moisture flux is written as†

$$\begin{aligned}\hat{\mathbf{n}}(\eta)\cdot\hat{\mathbf{n}}_w\,(P - E + R)\,\mathrm{d}A_{(\eta)} &= (-\nabla\eta + \hat{\mathbf{z}})\cdot\hat{\mathbf{n}}_w\,(P - E + R)\,\mathrm{d}A \\ &\equiv q_{\mathrm{w}}\,\mathrm{d}A.\end{aligned} \tag{3.41}$$

Equation (3.24) was used for the first equality to relate the area on the free surface, $\mathrm{d}A_{(\eta)}$, to the horizontal cross-sectional area, $\mathrm{d}A$. As defined, q_{w} is the volume of moisture passing across the ocean free surface, per unit time, per unit horizontal cross-sectional area $\mathrm{d}A = \mathrm{d}x\,\mathrm{d}y$.

Bringing these results together leads to the volume budget over the fluid column

$$\begin{aligned}V_{,t} &= \int \mathrm{d}A\,D_{,t} \\ &= \int \mathrm{d}A\left(q_{\mathrm{w}} - \nabla\cdot\int_{-H}^{\eta}\mathrm{d}z\,\mathbf{u}\right).\end{aligned} \tag{3.42}$$

Since the horizontal cross-sectional area is arbitrary, equality holds only if

$$\eta_{,t} + \nabla\cdot\left(\int_{-H}^{\eta}\mathrm{d}z\,\mathbf{u}\right) = q_{\mathrm{w}}. \tag{3.43}$$

Writing the vertically integrated horizontal velocity as

$$\mathbf{U} = \int_{-H}^{\eta}\mathrm{d}z\,\mathbf{u} \tag{3.44}$$

*The different components of the moisture flux generally have different normal directions. Generalizations from the present discussion to such cases are straightforward.

†The moisture flux q_{w} represents for the ocean surface what the dia-surface velocity component $w^{(s)}$ does for an arbitrary smooth surface $s = s(x, y, z, t)$ within the ocean interior (see Section 6.7).

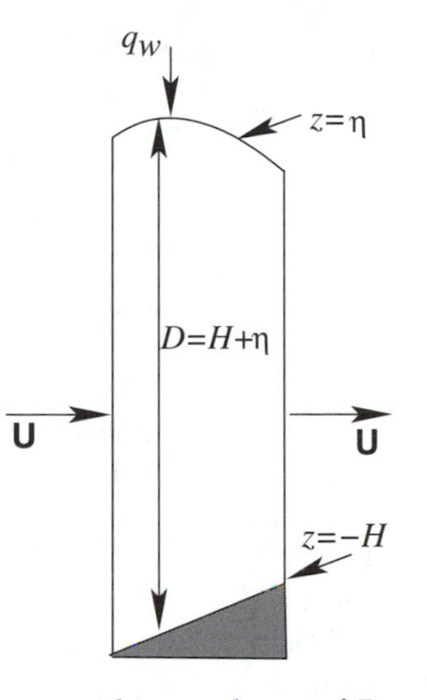

Figure 3.4 The balance of volume within a column of Boussinesq fluid extending from the free surface to the solid earth boundary is determined by three processes as given by equation (3.45): $\eta_{,t} = -\nabla \cdot \mathbf{U} + q_{\mathrm{w}}$. These effects are (1) time-dependent fluctuations of the ocean surface height, (2) the convergence of volume associated with the vertically integrated horizontal currents, and (3) the input of volume across the ocean surface from fresh water.

leads to the *surface height equation* for a volume conserving fluid

$$\eta_{,t} = -\nabla \cdot \mathbf{U} + q_{\mathrm{w}}. \tag{3.45}$$

The convergence term affects surface height via changes in the motion of fluid columns within the ocean domain, whereas the fresh water term affects surface height via changes in surface boundary forcing. Figure 3.4 illustrates this very important kinematic balance.

For the global ocean, again assuming no water flux through the solid earth boundary, the only means for the volume of seawater to change is through moisture fluxes at the ocean surface. Thus, a global balance of volume takes the form

$$\partial_t \left(\int \mathrm{d}A\, D \right) = \int \mathrm{d}A\, q_{\mathrm{w}}. \tag{3.46}$$

The solid earth boundary is assumed to be time-independent, and so this global balance reduces to

$$\partial_t \left(\int \mathrm{d}A\, \eta \right) = \int \mathrm{d}A\, q_{\mathrm{w}}. \tag{3.47}$$

This is an important relation to maintain in the numerical Boussinesq ocean model to ensure that total volume is indeed conserved.

There are various means to derive the kinematic boundary condition at the ocean free surface. One approach starts by applying the divergence operator onto the vertical integral in equation (3.43) to find

$$(\partial_t + \mathbf{u}(\eta) \cdot \nabla)\, \eta + \mathbf{u}(-H) \cdot \nabla H + \int_{-H}^{\eta} \mathrm{d}z\, \nabla \cdot \mathbf{u} = q_{\mathrm{w}}. \tag{3.48}$$

Recall that volume conserving fluid parcels satisfy $\nabla \cdot \mathbf{u} + w_{,z} = 0$ (equation (3.32)). Use of this divergence free condition, as well as the bottom kinematic boundary condition (3.39), yields the *surface kinematic* boundary condition for a volume conserving fluid

$$(\partial_t + \mathbf{u} \cdot \nabla)\, \eta = q_{\mathrm{w}} + w \qquad \text{at } z = \eta. \tag{3.49}$$

An alternative form of this condition is given by

$$\eta_{,t} = \hat{\mathbf{N}} \cdot \mathbf{v} + q_{\mathrm{w}} \qquad \text{at } z = \eta, \tag{3.50}$$

with the surface orientation direction $\hat{\mathbf{N}}$ given by equation (3.28). This balance says that local time tendencies of the surface height are driven by fresh water fluxes at the surface as well as the projection of the surface velocity onto the surface orientation direction. Just as for the bottom kinematic boundary condition (3.39), the surface boundary condition (3.49) is *kinematic* since it depends only on the geometry of the free surface and the fluxes of volume crossing it.

Comparing the boundary condition (3.50) with the column integrated budget (3.45) leads to the balance

$$\hat{\mathbf{N}} \cdot \mathbf{v} = -\nabla \cdot \mathbf{U} \qquad \text{at } z = \eta. \tag{3.51}$$

This expression says that convergence of the vertically integrated transport in a Boussinesq fluid is balanced by the projection of the surface velocity onto the surface normal. For the special case of a *rigid lid* Boussinesq ocean (see Section 2.4.4), $\nabla \cdot \mathbf{U} = 0$, which then leads to $\hat{\mathbf{N}} \cdot \mathbf{v} = 0$, and this is typically taken to mean $w = 0$.

3.4.3 Mass conservation for finite domains

We now proceed in a manner parallel to the previous section, but consider a control volume full of mass conserving fluid parcels. The total mass of fluid, M, inside the control volume is given by

$$M = \int \mathrm{d}A \int_{-H}^{\eta} \rho \, \mathrm{d}z, \tag{3.52}$$

where ρ is the mass density of seawater. Mass conservation implies that the time tendency of mass

$$M_{,t} = \int \mathrm{d}A \, \partial_t \left(\int_{-H}^{\eta} \mathrm{d}z \, \rho \right) \tag{3.53}$$

changes due to imbalances in the mass flux passing across the boundaries. Following the analysis given in Section 3.4.2 allows us to note that the only means of affecting the mass in the fluid column are through fluxes crossing the ocean free surface, and convergences within the ocean domain. Hence, generalizing the result (3.42) to a mass conserving fluid leads to

$$M_{,t} = \int \mathrm{d}A \left(q_{\mathrm{w}} \, \rho_{\mathrm{w}} - \nabla \cdot \int_{-H}^{\eta} \mathrm{d}z \, \rho \, \mathbf{u} \right). \tag{3.54}$$

The term

$$q_{\mathrm{w}}\,\rho_{\mathrm{w}}\,\mathrm{d}A = \hat{\mathbf{n}}(\eta)\cdot\hat{\mathbf{n}}_{w}\,(P - E + R)\,\rho_{\mathrm{w}}\,\mathrm{d}A_{(\eta)} \tag{3.55}$$

represents the mass flux of fresh water (mass per unit time per unit area) crossing the free surface, where ρ_{w} is the *in situ* density of the fresh water. Equivalently, $q_{\mathrm{w}}\,\rho_{\mathrm{w}}\,\mathrm{d}A$ is the momentum density of the fresh water in the direction normal to the ocean surface. Equating the time tendencies given by equations (3.53) and (3.54) leads to a mass balance within each vertical column of fluid in the control volume

$$\partial_t \left(\int_{-H}^{\eta} \mathrm{d}z\,\rho \right) + \nabla\cdot\left(\int_{-H}^{\eta} \mathrm{d}z\,\rho\,\mathbf{u} \right) = q_{\mathrm{w}}\,\rho_{\mathrm{w}}. \tag{3.56}$$

Taking the control volume to be the full ocean domain leads to the mass balance

$$\partial_t \left(\int \mathrm{d}A \int_{-H}^{\eta} \mathrm{d}z\,\rho \right) = \int_{z=\eta} \mathrm{d}A\,q_{\mathrm{w}}\,\rho_{\mathrm{w}}. \tag{3.57}$$

Hence, the ocean's mass changes only due to the passage of fresh water across the free surface. This is the non-Boussinesq analog to the volume conservation statement given by equation (3.46).

3.4.3.1 *Tilde notation for density weighted velocity*

In deriving the free surface equation for a mass conserving fluid, it is useful to follow the discussion in McDougall et al. (2003a) by introducing the density weighted velocity field

$$\rho_o\,\tilde{\mathbf{v}} = \rho\,\mathbf{v} \tag{3.58}$$

and the vertical integral of its horizontal components

$$\tilde{\mathbf{U}} = \int_{-H}^{\eta} \mathrm{d}z\,\tilde{\mathbf{u}}. \tag{3.59}$$

Notably, $\rho\,\mathbf{v} = \rho_o\,\tilde{\mathbf{v}}$ represents a mass flux (mass per unit time per unit area), and so $\tilde{\mathbf{v}}$ is the mass flux normalized by the constant Boussinesq reference density*

$$\rho_o = 1035\,\mathrm{kg/m^3}. \tag{3.60}$$

Equivalently, $\rho_o\,\tilde{\mathbf{v}}$ is the linear momentum density (momentum per unit volume) carried by a fluid parcel. Likewise, $\rho_o\,\tilde{\mathbf{U}}$ is the vertically integrated horizontal mass flux, or vertically integrated horizontal momentum density.

With these definitions, the mass balance (3.56) takes the form

$$\partial_t \left(\int_{-H}^{\eta} \mathrm{d}z\,\rho \right) = -\rho_o\,\nabla\cdot\tilde{\mathbf{U}} + q_{\mathrm{w}}\,\rho_{\mathrm{w}}. \tag{3.61}$$

This equation should be compared to the analogous result (3.45) valid for Boussinesq fluids.

---

*Note that $\rho_o = 1035\,\mathrm{kg/m^3}$ is preferred to the commonly used $\rho_o = 1000\,\mathrm{kg/m^3}$. The reason is that, as stated on page 47 of Gill (1982), the ocean density varies less than 2% from $1035\,\mathrm{kg/m^3}$, whereas it deviates further from $1000\,\mathrm{kg/m^3}$.

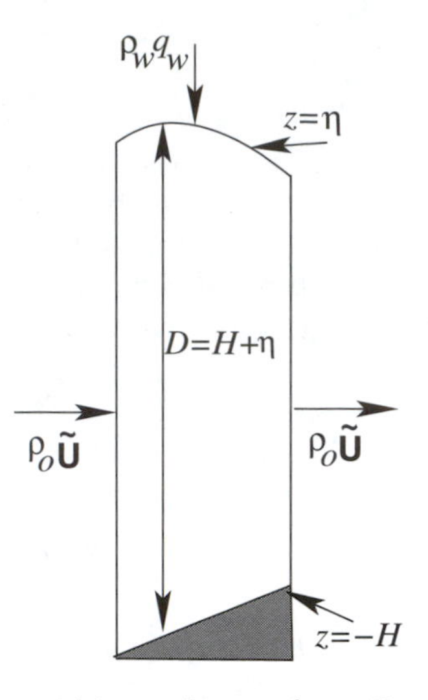

Figure 3.5 The balance of mass within a column of non-Boussinesq fluid extending from the free surface to the solid earth boundary is determined by four processes as given by equation (3.64): $\overline{\rho}^z\,\eta_{,t} = -\rho_o\,\nabla\cdot\tilde{\mathbf{U}} + \rho_{\mathrm{w}}\,q_{\mathrm{w}} - D\,\partial_t\,\overline{\rho}^z$. These effects are (1) time-dependent fluctuations of the ocean surface height, (2) the convergence of mass associated with the vertically integrated horizontal linear momentum, (3) the input of mass across the ocean surface from fresh water, and (4) time tendencies in the vertically averaged density.

3.4.3.2 Steric effects

To isolate the surface height tendency, and thus to highlight the main differences between the mass budget given by equation (3.61) and the volume budget in equation (3.45), two approaches are apparent. One approach applies the time derivative in equation (3.61) to give

$$\rho(\eta)\,\eta_{,t} = -\rho_o\,\nabla\cdot\tilde{\mathbf{U}} + q_{\mathrm{w}}\,\rho_{\mathrm{w}} - \int_{-H}^{\eta} \mathrm{d}z\,\rho_{,t}. \tag{3.62}$$

An alternative arises by introducing the vertically averaged density

$$\overline{\rho}^z = \frac{1}{H+\eta}\int_{-H}^{\eta} \mathrm{d}z\,\rho, \tag{3.63}$$

thus leading to the surface height equation for a mass conserving fluid

$$\overline{\rho}^z\,\eta_{,t} = -\rho_o\,\nabla\cdot\tilde{\mathbf{U}} + \rho_{\mathrm{w}}\,q_{\mathrm{w}} - D\,\partial_t\,\overline{\rho}^z, \tag{3.64}$$

where $D = H + \eta$ is the total thickness of the fluid column. Figure 3.5 illustrates the terms in this equation.

To understand the differences between how surface height evolves for a Boussinesq and non-Boussinesq fluid, one can compare either equation (3.62) or (3.64) to the Boussinesq equation (3.45). For the non-Boussinesq fluid, a density ratio weights the convergence $-\nabla\cdot\tilde{\mathbf{U}}$ as well as the fresh water contribution, whereas the Boussinesq fluid has no density ratio weighting. Although the density ratios indicate a difference between the budgets, the convergence and fresh water are directly analogous for both fluids. Again, the convergence arises from changes in the vertically integrated fluid column motion, and the fresh water term arises from surface boundary forcing.

The term $-D\,\partial_t \ln \overline{\rho}^z$, which is derived by dividing $\overline{\rho}^z$ from both sides of equation (3.64), or the analogous term in equation (3.62), represent a fundamentally new process that provides a change in free surface height upon changing the vertically averaged density. If the vertically averaged density decreases, the surface height increases, and vice versa. This *steric effect* is absent from the Boussinesq fluid's surface height equation, thus necessitating a diagnostic correction to compute the absolute surface height for sea level studies with Boussinesq models, as discussed by Greatbatch (1994) and Mellor and Ezer (1995). However, as noted by Greatbatch et al. (2001), the diagnostic correction misses a potentially significant level of spatial variability in the surface height that is captured by the non-Boussinesq surface height. Such differences may be important for sea level studies, especially those associated with climate change and under situations where ageostrophic effects are important. Losch et al. (2004) provide further discussion within the context of other model uncertainties.

3.4.3.3 *Surface kinematic boundary condition*

To derive the surface kinematic boundary condition, apply the time and space derivatives on the left-hand side of equation (3.56) to yield

$$\rho(\eta)\,(\partial_t + \mathbf{u}(\eta)\cdot\nabla)\,\eta + \mathbf{u}(-H)\cdot\nabla H + \int_{-H}^{\eta} \mathrm{d}z\,(\rho_{,t} + \nabla\cdot(\mathbf{u}\,\rho)) = q_{\mathrm{w}}\,\rho_{\mathrm{w}}. \quad (3.65)$$

Recall from Section 3.4.1 that a mass conserving fluid parcel satisfies $\rho_{,t} + \nabla\cdot(\mathbf{u}\,\rho) + (w\,\rho)_{,z} = 0$ (equation (3.31)). Using this result with the bottom kinematic boundary condition (3.39) leads to the surface kinematic boundary condition applicable for a mass conserving fluid

$$(\partial_t + \mathbf{u}\cdot\nabla)\,\eta = (\rho_{\mathrm{w}}/\rho)\,q_{\mathrm{w}} + w \qquad \text{at } z = \eta. \quad (3.66)$$

Comparison with the analogous result (3.49) applicable to a volume conserving fluid reveals that the only difference is the density ratio (ρ_{w}/ρ) weighting the fresh water flux. Introducing the linear momentum density $\rho_o\,\tilde{\mathbf{v}} = \rho\,\mathbf{v}$ leads to the equivalent form

$$(\rho\,\partial_t + \rho_o\,\tilde{\mathbf{u}}\cdot\nabla)\,\eta = \rho_{\mathrm{w}}\,q_{\mathrm{w}} + \rho_o\,\tilde{w} \qquad \text{at } z = \eta. \quad (3.67)$$

Finally, introducing the surface orientation direction $\hat{\mathbf{N}} = \nabla\,(-\eta + z)$ leads to

$$\rho\,\eta_{,t} = \rho_o\,\hat{\mathbf{N}}\cdot\tilde{\mathbf{v}} + \rho_{\mathrm{w}}\,q_{\mathrm{w}} \qquad \text{at } z = \eta. \quad (3.68)$$

Comparing the column integrated mass budget (3.62) with the surface kinematic boundary condition (3.68) leads to the balance

$$\rho_o\,\hat{\mathbf{N}}\cdot\tilde{\mathbf{v}} = -\rho_o\,\nabla\cdot\tilde{\mathbf{U}} - \int_{-H}^{\eta} \mathrm{d}z\,\rho_{,t} \qquad \text{at } z = \eta. \quad (3.69)$$

This expression says that projection of the surface velocity onto the surface orientation direction is balanced by convergence of vertically integrated momentum per volume and the depth integrated time tendencies in the seawater density. The second term on the right-hand side, as noted already, accounts for the steric effect of density changes within a fluid column, and this term is absent in the Boussinesq fluid.

3.5 CHAPTER SUMMARY

We focused in this chapter on fluid kinematics as applied to the ocean. Section 3.2 started by introducing some mathematical notation and results of use to describe the motion of a fluid parcel on a rotating planet with spherical geometry. In particular, Figure 3.1 is worth a periodic revisit as it provides a useful orientation for the various formulations, especially those in Chapter 4. Additionally, the "comma-notation" of Section 3.2.4 should be highlighted here, whereby a useful shorthand for the partial derivative is written

$$\partial\rho/\partial t = \rho_{,t}. \tag{3.70}$$

This notation and its generalization to other partial derivatives are used throughout this book.

Section 3.2.2 noted that use of orthogonal coordinates to describe the lateral position of a fluid parcel allows for the introduction of the *physical* displacements $\mathrm{d}x = h_1\,\mathrm{d}\xi^1$ and $\mathrm{d}y = h_2\,\mathrm{d}\xi^2$, with h_1 and h_2 the stretching functions mapping coordinate displacements $\mathrm{d}\xi^1$ and $\mathrm{d}\xi^2$ into physical displacements $\mathrm{d}x$ and $\mathrm{d}y$. Use of physical displacements for writing distances and derivative operators helps to render formulae describing fluid motion in a spherical geometry, where the metric functions h_1 and h_2 are nontrivial (see equation (3.6)), to a form reminscient of Cartesian coordinates on a plane. This is especially useful when deriving the equations of ocean fluid dynamics in Chapter 4. Important distinctions are present, however, due to the noncommutativity of the lateral partial differential operators on the sphere: $\partial_x = h_1^{-1}\,\partial_1$ and $\partial_y = h_2^{-1}\,\partial_2$ (see equation (3.12)).

The shallow ocean approximation (Section 3.2.5) follows from noting that large-scale ocean fluid motion occurs with vertical deviations very small relative to the earth's radius. Hence, it is quite accurate to reduce the spatial dependence of the metric functions h_1 and h_2 to just the lateral position, with their radial dependence set to the constant $R = 6.367 \times 10^6$ m, with this value for R corresponding to the best fit oblate spheroid to the sea level geopotential (Section 3.2.1).

The complicated geometry of the World Ocean and the use of arbitrary coordinates to describe the motion of fluid parcels necessitate a bit of sophistication in how the divergence theorem is applied when deriving budgets over finite domains. Section 3.3 spends some time explaining the necessary mathematics. When the budget includes a full column of ocean fluid from the surface to the solid earth boundary, then

$$\int \mathrm{d}V\,\nabla\cdot\mathbf{F} = \int\limits_{z=\eta} \mathrm{d}x\,\mathrm{d}y\,\hat{\mathbf{N}}\cdot\mathbf{F} - \int\limits_{z=-H} \mathrm{d}x\,\mathrm{d}y\,\hat{\mathbf{N}}\cdot\mathbf{F}, \tag{3.71}$$

where $\hat{\mathbf{N}}(\eta) = \nabla(-\eta + z)$ is the outward orientation direction at the ocean surface, and $\hat{\mathbf{N}}(-H) = \nabla(H + z)$ is the inward orientation direction at the solid earth bottom. This result is used throughout this book for arbitrary domains and general coordinates.

Section 3.4 compared and contrasted the kinematics of fluid parcels and fluid columns that maintain constant volume (Boussinesq parcels) and constant mass (non-Boussinesq parcels). The starting point for this comparison is the continuity equations for volume conserving fluid parcels

$$\nabla \cdot \mathbf{u} + w_{,z} = 0 \tag{3.72}$$

and mass conserving fluid parcels

$$\rho_{,t} + \nabla \cdot (\mathbf{u}\,\rho) + (w\,\rho)_{,z} = 0, \tag{3.73}$$

where

$$\mathbf{v} = (\mathbf{u}, w) \tag{3.74}$$

is the three-dimensional velocity field. Integration over the depth of the ocean transforms these fluid parcel relations into volume and mass conservation budgets for fluid columns. In particular, such relations provide expressions for the time tendency of the surface height in terms of the convergence of depth integrated flow onto the column, surface forcing, and steric effects due to time tendencies of density within the column. Explicitly, volume conserving fluids have a surface height tendency given by

$$\eta_{,t} = -\nabla \cdot \mathbf{U} + q_{\mathrm{w}}, \tag{3.75}$$

whereas for mass conserving fluids, the free surface is deduced from

$$\overline{\rho}^{z}\,\eta_{,t} = -\rho_o\,\nabla \cdot \tilde{\mathbf{U}} + \rho_{\mathrm{w}}\,q_{\mathrm{w}} - D\,\partial_t\,\overline{\rho}^{z}. \tag{3.76}$$

Notably, the *steric* term, $-D\,\partial_t\,\overline{\rho}^{z}$, expressing changes in surface height due to tendencies in the depth averaged density, is absent in the volume conserving fluid. This precludes the volume conserving fluid from directly prognosing the full surface height of the ocean.

As noted by Losch et al. (2004), differences between a Boussinesq and a non-Boussinesq simulation can be small for coarsely resolved steady flow, relative to uncertainties in other terms within the model equations. Even so, one may prefer the formulation of an ocean model in terms of a non-Boussinesq fluid in order to reduce the level of approximations, and to have more confidence in the simulation's accuracy when moving to fine resolution and realistic time-dependent boundary forcing. Furthermore, diagnosis of the surface height from a Boussinesq model, which can be reasonably straightforward as described by Greatbatch (1994), still represents an added step beyond that needed for running the prognostic Boussinesq model. A more satisfying approach is to employ a non-Boussinesq algorithm so that the sea level computed in the numerical model is influenced by steric effects, and so corresponds directly to that measured by observations. Huang et al. (2001), DeSzoeke and Samelson (2002) and Marshall et al. (2003) discuss how the equations for a Boussinesq fluid written in z-coordinates are isomorphic to a non-Boussinesq fluid written in pressure coordinates. This isomorphism facilitates a clean and efficient mapping between discrete representations of these two fluids using the same numerical algorithm. Further discussion on these matters is provided in Chapter 6.

Chapter Four

DYNAMICS

The purpose of this chapter is to present various aspects of ocean fluid mechanics, with a focus on the dynamical equations.

4.1 INTRODUCTION

Mechanical equations describing ocean circulation are based on Newton's laws applied to a stratified fluid continuum moving on a rotating planet. Discussion of ocean fluid dynamics begins in Section 4.2, which provides basic elements for understanding motion in a rotating reference frame. This discussion relies on ideas from point particle mechanics, and so should be familiar to those who have taken a classical mechanics course. Section 4.3 makes a brief excursion into theoretical continuum mechanics in order to provide a rationale for applying Newton's laws to a continuous medium such as the ocean fluid. This discussion lends some confidence to the use of a Lagrangian perspective in Section 4.4 to derive the equations for fluid parcels moving in a rotating spherical geometry.

The full content of ocean fluid dynamics is contained in Newton's laws, which provide a powerful summary of our understanding of ocean motions. However, an understanding of how the dynamics manifests itself is assisted by exploring various implications of Newton's laws as realized in the ocean. This is the focus of the remaining sections in this chapter. We start in Section 4.5 by highlighting implications of the hydrostatic balance, which is a balance maintained by fluids satisfying the *primitive equations*. In particular, some discussion is provided of the differences in hydrostatic pressure felt at the ocean bottom when the fluid domain is occupied by a Boussinesq fluid versus a non-Boussinesq fluid. Section 4.6 then investigates the dynamics of a vertical column of fluid, where the column extends from the free surface to the ocean bottom. The dynamics of such columns corresponds to that of the *external* or *barotropic* mode of the linearized primitive equations. Sections 4.7 and 4.8 illustrate some of the dominant scales of motion found in a rotating and stratified fluid. This discussion aims to render an intuitive sense for the flow of large-scale fluid motion on the earth.

The next three sections provide a formal understanding of the dynamical equations. For this purpose, Section 4.9 derives the evolution equations for vorticity and potential vorticity. These fields are central to theoretical geophysical fluid dynamics. The discussion here is brief and mathematically focused. Whole texts and review articles are devoted to these important topics, and the purpose here is to introduce just a few basic results. Section 4.10 then derives the dynamical equation for a particle moving on a rotating sphere, with derivations proceeding from Hamilton's principle. This discussion aims to clarify notions of energy and angu-

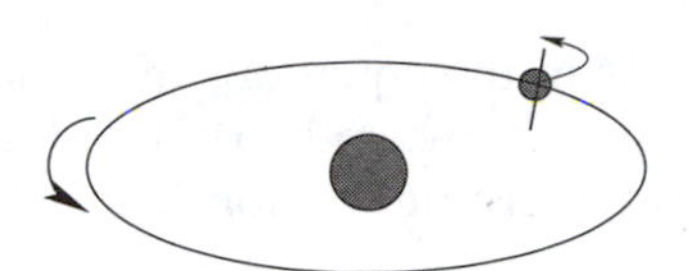

Figure 4.1 The earth-sun system, whereby the angular velocity of the earth is associated both with rotation about the earth's axis and with its orbit about the sun. This angular velocity determines the strength of the Coriolis force.

lar momentum for an ideal fluid parcel described by primitive equation dynamics. Finally, angular momentum is discussed in Section 4.11, where its conservation on a smooth sphere is contrasted to the balance of linear momentum, which is conserved on a smooth plane. This discussion introduces some intuitive notions of symmetries and conservation laws. A formalism for exploiting this intuition provides theoretical physicists with powerful tools to understand the relationship between kinematics as it deals with geometry and dynamics as it deals with conservation laws.

4.2 MOTION ON A ROTATING SPHERE

As discussed in Section 3.2, it is quite accurate for ocean climate modeling to approximate the earth's geometry as spherical, and rotating with angular velocity $\boldsymbol{\Omega}$ about an axis through the North Pole (see Figure 3.1 on page 26). Here we present some notions that are useful when exploring dynamics in a reference frame fixed on a rotating planet.

4.2.1 Earth's angular velocity

The earth's angular velocity is comprised of two main contributions: the spin of the earth about its axis and the orbit of the earth about the sun (see Figure 4.1). Other astronomical motions can be neglected for geophysical fluid dynamics. Therefore, in the course of a single period of 24 hours, or $24 \times 3600 = 86400$ seconds, the earth experiences an angular rotation of $(2\pi + 2\pi/365.24)$ radians. As such, the angular velocity of the earth is given by

$$\begin{aligned}\Omega &= \left(\frac{2\pi + 2\pi/365.24}{86400\text{s}}\right)\\ &= \left(\frac{\pi}{43082}\right)\text{s}^{-1}\\ &= 7.2921 \times 10^{-5}\text{s}^{-1}.\end{aligned} \tag{4.1}$$

To an extremely high degree of accuracy, this angular velocity is assumed constant in time for purposes of geophysical fluid dynamics.

4.2.2 Inertial and non-inertial reference frames

The preferred frame of reference for describing geophysical fluids is that of a terrestrial observer fixed on the rotating earth. A rotating frame is not inertial, and so there are non-inertial terms in the equations of motion that must be taken into account. To derive the relations between inertial and non-inertial observers, it is useful to employ the general operator relation (e.g., Marion and Thornton, 1988; Gill, 1982).

$$\left(\frac{\mathrm{d}}{\mathrm{d}t}\right)_{\text{inertial}} = \left(\frac{\mathrm{d}}{\mathrm{d}t}\right)_{\text{rotating}} + \mathbf{\Omega} \wedge . \tag{4.2}$$

Applying this relation to the position vector $\mathbf{x}$ of a fluid parcel leads to

$$\mathbf{v}_{\text{inertial}} = \mathbf{v} + \mathbf{\Omega} \wedge \mathbf{x}, \tag{4.3}$$

where $\mathbf{v} = \mathbf{v}_{\text{rotating}}$ is the parcel's velocity in the rotating frame. Note that because $\mathbf{x}$ is the position vector for a fluid parcel, the time derivative $\mathrm{d}/\mathrm{d}t$ is the material time derivative. Applying the relation twice leads to

$$\begin{aligned} \left(\frac{\mathrm{d}^2\,\mathbf{r}}{\mathrm{d}t^2}\right)_{\text{inertial}} &= \left(\left(\frac{\mathrm{d}}{\mathrm{d}t}\right)_{\text{rotating}} + \mathbf{\Omega} \wedge\right)\left(\left(\frac{\mathrm{d}\mathbf{x}}{\mathrm{d}t}\right)_{\text{rotating}} + \mathbf{\Omega} \wedge \mathbf{x}\right) \\ &= \frac{\mathrm{d}^2\mathbf{x}}{\mathrm{d}t^2} + 2\,\mathbf{\Omega} \wedge \mathbf{v} + \mathbf{\Omega} \wedge (\mathbf{\Omega} \wedge \mathbf{x}) \\ &= \dot{\mathbf{v}} + 2\,\mathbf{\Omega} \wedge \mathbf{v} - \nabla\,(\Omega\, r\, \cos\phi)^2/2 \end{aligned} \tag{4.4}$$

where $\dot{\mathbf{v}}$ is the parcel's acceleration in the rotating frame, $\mathbf{\Omega}$ is assumed to be constant in time, and the spherical coordinates are defined in Figure 3.1. The contribution $-2\,\mathbf{\Omega} \wedge \mathbf{v}$ is called the *Coriolis acceleration*, and $\nabla\,(\Omega\, r\, \cos\phi)^2/2$ is the *centrifugal acceleration*. Note that the factor of 2 in the Coriolis acceleration is a straightforward result of applying the operator relation (4.2) twice to the parcel's position vector.

4.2.3 Effective gravitational force

Excursions by a fluid parcel in the radial direction away from the center of the earth introduce changes to the particle's gravitational and centrifugal potential energies. It is useful to combine these two accelerations into an *effective gravitational force*.

If the earth is assumed to be a sphere, then the gravitational potential energy of a point mass at a radius $r = z + R$ from the earth's center is given by

$$\begin{aligned} m\,\Phi &= -\frac{G\,M_e\,m}{z + R} \\ &\approx -\frac{G\,M_e\,m}{R} + \left(\frac{G\,M_e}{R^2}\right) m\,z \\ &= \text{constant} + g_{\mathrm{e}}\,m\,z, \end{aligned} \tag{4.5}$$

where m is the mass of the point particle, Φ is the gravitational potential energy per unit mass, R is the earth's radius, and

$$g_{\mathrm{e}} = G\,M_{\mathrm{e}}/R^2 \tag{4.6}$$

is the gravitational acceleration. In this equation, $G = 6.67 \times 10^{-11}\mathrm{N\,m^2\,kg^{-2}}$ is Newton's gravitational constant and $M_{\mathrm{e}} = 5.98 \times 10^{24}\,\mathrm{kg}$ is the mass of the earth.

The restriction $z << R$ is relevant for fluid parcel motion in the ocean, where this inequality is maintained to a high degree of accuracy for all motions. For this reason, it is safe to drop terms going as $(z/R)^2$ or higher order in the gravitational potential energy.

Combining the gravitational potential for a point particle at rest with that from the centrifugal force leads to the effective gravitational force acting on the particle on a rotating spherical earth (e.g., pages 40 and 48 of Fetter and Walecka, 1980)

$$\begin{aligned} m\,\mathbf{g} &= -m\,\nabla(r\,g_e - (\Omega\,r\cos\phi)^2/2) \\ &= -m(g_e - r\,\Omega^2\cos^2\phi)\,\hat{\mathbf{r}} - (m\,r\,\Omega^2\cos\phi\,\sin\phi)\,\hat{\boldsymbol{\phi}}\,. \end{aligned} \tag{4.7}$$

The effective gravitational force is noncentral due to the effects of rotation. Hence, if the earth were an ideal fluid, matter would flow from the poles toward the equator, thus ensuring that the earth's surface would everywhere be perpendicular to the effective gravitational acceleration, $\mathbf{g}$. Indeed, the earth does exhibit a slight equatorial bulge. Furthermore, inhomogeneities in the earth's composition and surface loading by continents, glaciers, and seawater make its shape differ from the ideal case.

Because of its equatorial bulge and polar flattening, the earth is not spherical. In Section 3.2.1, it was noted that a more accurate description of the earth's shape is oblate spheroidal. In these coordinates, geopotentials are approximated by surfaces of constant oblate spheroidal radial coordinate. For example, $R = 6.367 \times 10^6$ m (equation (3.3)) represents the oblate spheriod that best fits the sea level geopotential. Even though oblate spheriodal coordinates are better than spherical ones for describing geopotentials, it is possible, to a high degree of accuracy, to describe the earth's geometry as spherical. That is, the metric functions used to measure distances are the same as those used with spherical coordinates.

There are two implications of this interpretation of the coordinates and geometry. First, agreeing to measure the zero of gravitational potential energy on the sea level geopotential (i.e., the surface of a resting ocean at $z = 0$) allows one to ignore the dynamically irrelevant constant in equation (4.5). Second, the nonradial nature of the gravitational force in equation (4.7) is now absent. This is the most important implication of the coordinate interpretation, since a nonradial component to the gravitational force would complicate the description of geophysical motion. Absorbing the centrifugal term into an effective gravitational potential then leads to the effective gravitational acceleration vector

$$\begin{aligned} \mathbf{g} &= -\nabla\Phi \\ &= -g\,\hat{\mathbf{z}}. \end{aligned} \tag{4.8}$$

Ocean modelers could use a gravitational acceleration which is a local function of space, as determined by inhomogeneities in the earth's composition. However, to a high level of accuracy, it is sufficient to use a gravitational acceleration determined by a mean over the ocean surface. As reported on page 132 of Moritz (2000), the gravitational acceleration at the earth's surface as a function of latitude is given by

$$g = 9.780327\,[1 + 0.0053024\,\sin^2\phi - 0.0000058\,(\sin 2\phi)^2]. \tag{4.9}$$

At a latitude of 45°, $g = 9.806\,\mathrm{m\,s^{-2}}$, which is the value used by ocean models based on Cox (1984). The areal average of g over the earth's surface is $9.7976\,\mathrm{m\,s^{-2}}$

(e.g., Gill, 1982, page 597). Alternatively, it may be more appropriate to take an areal average over the ocean surface, which yields*

$$g = 9.7963\,\mathrm{m\,s^{-2}}. \tag{4.10}$$

The sensitivity of simulated large-scale ocean circulation to which of these three values is used will likely be negligible.

4.2.4 Coriolis acceleration

Besides the centrifugal acceleration, motion in a rotating frame introduces the Coriolis acceleration

$$\mathbf{a}_c = -2\,\boldsymbol{\Omega} \wedge \mathbf{v}. \tag{4.11}$$

The Coriolis acceleration plays a crucial role in the large-scale motion of fluids on the earth. Notably, it is a function of the velocity and is directed perpendicular to the velocity direction. Consequently, it does not perform work on the fluid parcel. In this way, the Coriolis force

$$\mathbf{F}_c = -2\,(\rho\,\mathrm{d}V)\,\boldsymbol{\Omega} \wedge \mathbf{v} \tag{4.12}$$

acting on a fluid parcel is akin to the Lorentz force acting on a charged particle in an electromagnetic field (see, e.g., Jackson, 1975).

As discussed in Section 3.2, the shallow ocean sets the metric tensor components equal to their values at the sea level geopotential $r = R$. This assumption influences the angular momentum associated with motion in this geometry. Namely, angular momentum about the earth's center is computed with a moment-arm that has a fixed radius $r = R$. Hence, motion in the vertical direction does not alter angular momentum in the shallow ocean approximation. Correspondingly, it is necessary to drop the nonradial component of the earth's angular rotation vector when computing the Coriolis force, since this component alters angular momentum when moving in the vertical direction. Thus, with the shallow ocean approximation, the Coriolis force is given by

$$\mathbf{F}_c = -f\,(\rho\,\mathrm{d}V)\,\hat{\mathbf{z}} \wedge \mathbf{v}. \tag{4.13}$$

The Coriolis parameter

$$f = 2\,\Omega\,\sin\phi \tag{4.14}$$

is of fundamental importance for geophysical fluid motion.

4.2.5 Inertial oscillations

To garner some experience with how the Coriolis force affects parcel motion, consider the case of a fluid parcel affected *only* by the Coriolis force. In this case, the equation of motion is given by

$$(\mathrm{d}/\mathrm{d}t + f\,\hat{\mathbf{z}} \wedge)\,\mathbf{u} = 0, \tag{4.15}$$

*This alternative was suggested by Trevor McDougall and David Jackett (2002, personal communication).

which in component form is

$$\dot{u} - f\,v = 0 \tag{4.16}$$
$$\dot{v} + f\,u = 0, \tag{4.17}$$

where $\dot{\mathbf{u}} = \mathrm{d}\mathbf{u}/\mathrm{d}t$ is the parcel's horizontal acceleration. Taking the time derivative of the first equation and using the second equation lead to

$$\begin{aligned} \ddot{u} - f\,\dot{v} &= \ddot{u} + f^2\,u \\ &= 0. \end{aligned} \tag{4.18}$$

Similar manipulations for the meridional velocity equation render the first order equation (4.15) equivalent to the second order free oscillator equation

$$(\mathrm{d}^2/\mathrm{d}t^2 + f^2)\,\mathbf{u} = 0. \tag{4.19}$$

Motions that satisfy this equation are termed *inertial oscillations* (e.g., Gill, 1982), and they have period given by

$$\begin{aligned} T_{\text{inertial}} &= \frac{2\,\pi}{f} \\ &= \frac{11.97}{|\sin\phi|}\,\text{hour}, \end{aligned} \tag{4.20}$$

where $\Omega = 7.292 \times 10^{-5}\mathrm{s}^{-1}$. The period of inertial oscillations is smallest at the poles, where $\phi = \pm\pi/2$ and $T_{\text{smallest}} \approx 12$ hour. Inertial oscillations are commonly seen by current meters in the ocean, especially in higher latitude regions where diurnal (day-night) variations in the wind forcing have a strong projection onto the inertial period.

4.3 PRINCIPLES OF CONTINUUM DYNAMICS

The purpose of this section is to summarize some general dynamical principles forming the basis of classical continuum mechanics. The book by Aris (1962) is recommended for a more thorough discussion. Establishing these ideas helps to support the application of Newton's laws of motion to infinitesimal fluid parcels in Section 4.4.

4.3.1 Position, velocity, and acceleration

The position of a parcel of a continuous medium is described by a general coordinate $\vec{\xi} = (\xi^1, \xi^2, \xi^3)$, with $\xi^3 = z$ the choice when using geopotential coordinates, as in this chapter. As discussed in Chapter 20, coordinates carry contravariant labels; this is signaled by placing indices "upstairs" on ξ^a. In other words, the position of a parcel is given by the *vector* position* $\vec{\xi} = \xi^a\,\vec{e}_a$ with respect to an appropriate set of basis vectors $\vec{e}_a$. Note that the arrow over an object, thus symbolizing a vector, is typically replaced in the following by boldface for brevity in notation.

*Recall the summation convention whereby repeated indices are summed.

The time rate of change of the position defines the parcel's velocity. As time does not carry tensorial properties in nonrelativistic dynamics, the velocity $\mathbf{v} = \mathrm{d}\boldsymbol{\xi}/\mathrm{d}t = \mathrm{d}(\xi^a \vec{e}_a)/\mathrm{d}t$ is also a vector. Likewise, the parcel's acceleration $\mathbf{a} = \mathrm{d}\mathbf{v}/\mathrm{d}t$ is a vector. In order for Newton's second law of motion to be consistent from a tensorial perspective, it follows that the forces per unit mass acting on a parcel, which drive the acceleration, are vector quantities. The tensorial properties of the equations for fluid flow are based on these very basic statements about position, velocity, and acceleration.

4.3.2 External/body and internal/contact forces

The force vectors acting on an arbitrary volume of a continuous medium are of two kinds. The first are *external or body* forces, such as gravitation (including tidal forces), Coriolis, and electromagnetic forces. These forces act throughout the extent of the medium. The total body force acting on the volume is given by the integral of the external force per unit mass $\mathbf{f}$ multiplied by the mass of the medium

$$\mathbf{F}_{\text{external}} = \int \mathbf{f}\,\rho\,\mathrm{d}V. \tag{4.21}$$

For example, the gravitational force acting on a volume of fluid is given by

$$\mathbf{F}_{\text{gravity}} = \int \mathbf{g}\,\rho\,\mathrm{d}V, \tag{4.22}$$

where $\mathbf{g}$ is the acceleration of gravity, or equivalently the gravitational force per unit mass.

The second kind of forces are *internal or contact* forces, such as pressure forces and frictional forces. These forces act on a volume of continuous media by affecting the volume's boundary. Note that point particles, having no finite extent, do not experience internal or contact forces. Cauchy's stress principle (e.g., Aris, 1962, Chapter 5) states that the force per unit area exerted by the material outside the volume is a function of the position within the volume, the time, and the orientation of the boundaries of the surface. It is not necessary here to justify this principle, since its consequences are familiar and intuitive. For internal boundaries, the total contact force exerted on the volume V through its boundaries is given by

$$\mathbf{F}_{\text{internal}} = \int \mathbf{t}_{(\hat{\mathbf{n}})}\,\mathrm{d}A_{(\hat{\mathbf{n}})}, \tag{4.23}$$

where $\hat{\mathbf{n}}$ is the outward normal direction* to the domain boundary,

$$\mathbf{t}_{(\hat{\mathbf{n}})} = \mathbf{t}_{(\hat{\mathbf{n}})}(\boldsymbol{\xi}, t, \hat{\mathbf{n}}) \tag{4.24}$$

is the force per unit area, or *contact stress*, exerted on the boundary by the surrounding material, and $\mathrm{d}A_{(\hat{\mathbf{n}})}$ is the area element on the boundary. The functional dependence of the contact stress on the normal direction can be solved through invoking Newton's second law, as now described.

---

*Mathematically, the normal to a surface is a *one-form*, as discussed in Chapter 20.

4.3.3 Local equilibrium of contact stresses

The conservation of linear momentum (Newton's second law) states that the sum of the external and internal forces acting on a fluid volume are equal to the acceleration

$$\frac{\mathrm{d}}{\mathrm{d}t}\int \mathbf{v}\,\rho\,\mathrm{d}V = \int \mathbf{f}\,\rho\,\mathrm{d}V + \int \mathbf{t}_{(\hat{\mathbf{n}})}\,\mathrm{d}A_{(\hat{\mathbf{n}})}. \tag{4.25}$$

An important consequence of this relation follows from allowing the volume to go to zero smoothly without changing the shape. Assuming the integrands are well behaved, the volume integrals vanish as L^3, where L is a characteristic length for the volume. The surface integral, however, vanishes only as L^2. Consequently, dividing equation (4.25) by L^2 yields the relation

$$\lim_{L\to 0}\frac{1}{L^2}\int \mathbf{t}_{(\hat{\mathbf{n}})}\,\mathrm{d}A_{(\hat{\mathbf{n}})} = 0. \tag{4.26}$$

This result implies that the *stresses acting on the volume are locally in equilibrium*. In other words, the sum of the internal stresses acting on all sides of an infinitesimal volume vanish. In particular,

$$\mathbf{t}_{(\hat{\mathbf{n}})} = -\mathbf{t}_{(-\hat{\mathbf{n}})}, \tag{4.27}$$

which is a statement of Newton's third law.

4.3.4 The stress tensor

Now apply the local equilibrium property to an infinitesimal tetrahedron with four flat sides, three of which are aligned along the negative coordinate axes directions, and the fourth directed in an arbitrarily slanted direction (see Aris, 1962, figure 5.1) defined by the normal $\hat{\mathbf{n}} = n_a\,\tilde{e}^a$, with $\tilde{e}^a$ for $a = 1,2,3$ representing a basis of one-forms.* Local equilibrium implies

$$\mathbf{t}_{(\hat{\mathbf{n}})}\,\mathrm{d}A_{(\hat{\mathbf{n}})} - \mathbf{t}_{(1)}\,\mathrm{d}A_1 - \mathbf{t}_{(2)}\,\mathrm{d}A_2 - \mathbf{t}_{(3)}\,\mathrm{d}A_3 = 0, \tag{4.28}$$

where again $\mathrm{d}A_{(\hat{\mathbf{n}})}$ is the area of the slanted face, $\mathrm{d}A_a$ are the areas of the other three faces, and each face has a corresponding force per unit area directed normal to the respective face. The negative signs arise from the projection of the areas onto the negative coordinate directions. Now the projection of the areas $\mathrm{d}A_a$ onto the direction defined by $\hat{\mathbf{n}}$ is given by $\mathrm{d}A_a = n_a\,\mathrm{d}A_{(\hat{\mathbf{n}})}$. Hence, equation (4.28) takes the form

$$\mathbf{t}_{(\hat{\mathbf{n}})} = \mathbf{t}_{(1)}\,n_1 + \mathbf{t}_{(2)}\,n_2 + \mathbf{t}_{(3)}\,n_3. \tag{4.29}$$

Consequently, the stress in an arbitrary direction defined by $\hat{\mathbf{n}}$ can be written as the linear sum of the three stresses projected along the coordinate axes, each weighted by the projection of the normal along those axes.

Equation (4.29) can be summarized by introducing T^{ba} as the ath component of the stress, $\mathbf{t}_{(b)}$, and $t^b_{(\hat{\mathbf{n}})}$ as the bth component of the stress, $\mathbf{t}_{(\hat{\mathbf{n}})}$. With this notation, local equilibrium implies that the stress along a direction defined by $\hat{\mathbf{n}}$ is linearly related to $\hat{\mathbf{n}}$ through

$$t^b_{(\hat{\mathbf{n}})} = T^{ba}\,n_a. \tag{4.30}$$

*See Section 20.6 starting on page 452 for discussion of one-forms.

Since $t^b_{(\hat{\mathbf{n}})}$ are components to a vector, and n_a are components to a one-form, the linear proportionality coefficients T^{ba} form the components to a second order *stress tensor*.

The reasoning leading to the stress tensor implies that the internal stresses $\mathbf{t}_{(\hat{\mathbf{n}})}$, which are generally functions $\mathbf{t}_{(\hat{\mathbf{n}})}(\boldsymbol{\xi}, t, \hat{\mathbf{n}})$ of space, time, and the normals, are actually quite simple in that they can be summarized by nine quantities $T^{ab}(\boldsymbol{\xi}, t)$ that provide the linear proportionality between stresses and the normals. Additionally, since the result of this reasoning leads to a proper tensorial or covariant expression (4.30), the result remains form-invariant under an arbitrary change in coordinates.

4.3.5 Friction and pressure

When parcels exchange momentum with other parcels and/or boundaries, this exchange can be represented via the components of a symmetric stress tensor, whose covariant divergence leads to a force acting on the boundaries of the parcels. The forces arise from molecular viscosity and generally act to reduce the kinetic energy of parcels. When averaging over turbulent realizations of the ocean fluid, subgrid scale transfers of momentum are generally far larger than those associated with molecular viscosity. More discussion of this point is given in Chapter 8 when presenting the ensemble averaged equations. Further details of friction used in ocean models are given in Chapters 17–19.

There are two types of stress of concern in this book: diagonal stresses associated with pressure, and stresses associated with friction. In this case, the stress tensor components are given by

$$T^{ab} = \tilde{\tau}^{ab} - p\, g^{ab}. \tag{4.31}$$

In this equation, p is the pressure, which is a force per unit area, g^{ab} is the metric tensor which summarizes the local geometry, and $\tilde{\tau}^{ab}$ is the frictional stress tensor. The frictional stress tensor is also known as the *deviatoric* stress tensor (e.g., Aris, 1962; Batchelor, 1967), as it represents *deviations* from the static case when stress is due solely to pressure.

Given the expression (4.31) for the stress tensor, it is useful to recall the expression (4.23) for internal forces, and thus to write

$$\mathbf{F}_{\text{internal}} = \int (\tilde{\boldsymbol{\tau}} \cdot \hat{\mathbf{n}} - p\, \hat{\mathbf{n}})\, \mathrm{d}A_{(\hat{\mathbf{n}})}, \tag{4.32}$$

where $\tilde{\boldsymbol{\tau}} \cdot \hat{\mathbf{n}}$ is shorthand for $\tilde{\tau}^{ab}\, n_b$, and the integral is again taken over the bounding surface of the domain whose outward normal is $\hat{\mathbf{n}}$. It is notable that the pressure considered here is the same as the pressure used for equilibrium and non-equlibrium thermodynamical considerations in Chapter 5. Given this expression for internal forces acting on the boundary of a fluid domain, it is seen that positive pressure acts in the direction opposite to the surface's outward normal. That is, it acts in a compressive manner. Deviatoric stresses create more general forces on the surface, which can have compressive, expansive, and/or shearing characteristics. More details of frictional stresses are discussed in Chapter 17.

4.3.6 Equations of motion

The relation (4.30) between the contact stresses and the stress tensor brings the equation of motion (4.25) to the form

$$\frac{\mathrm{d}}{\mathrm{d}t}\int \rho v^a \,\mathrm{d}V = \int \rho f^a \,\mathrm{d}V + \int T^{ab}\, n_b \,\mathrm{d}A. \tag{4.33}$$

The covariant form of the divergence theorem, or Gauss's law, can be applied to the area integral to yield

$$\frac{\mathrm{d}}{\mathrm{d}t}\int \rho v^a \,\mathrm{d}V = \int \rho f^a \,\mathrm{d}V + \int T^{ab}_{\;;b} \,\mathrm{d}V, \tag{4.34}$$

where the semicolon denotes a covariant derivative (see Section 21.2.2, page 468). Since the volume under consideration is arbitrary, this integral relation is satisfied only if it is satisfied as a differential balance for infinitesimal fluid parcels. This provides a theoretical basis for the derivations provided in Section 4.4.

4.4 DYNAMICS OF FLUID PARCELS

Given the theory developed in Section 4.3, Newton's laws can be applied with impunity to infinitesimal fluid parcels. As discussed in Section 2.3.4, the fluid parcels of interest are those that conserve their mass as they move throughout the fluid. Hence, at issue here is a derivation of the dynamical equations for a mass conserving fluid parcel traveling on a general trajectory $\mathbf{x}(\zeta, \tau)$ with velocity $\mathbf{v} = \partial_\tau \mathbf{x}(\zeta, \tau)$, where (ζ, τ) are the material space-time coordinates of the parcel (see discussion in Section 2.3). Newton's second law is used, whereby the time tendency of the fluid parcel's linear momentum $(\rho\,\mathrm{d}V)\,\mathbf{v}$ is equated to forces acting on the parcel. Although this approach is Lagrangian in character, Eulerian equations are derived which consider time tendencies for the linear momentum density $\rho\,\mathbf{v} = \rho(\mathbf{x}, t)\,\mathbf{v}(\mathbf{x}, t)$ instead of the parcel's trajectory $\mathbf{x}(\zeta, \tau)$.

Many details in this section rely on the tensor analysis presented in Chapters 20 and 21. For those interested in verifying details of the derivations, reference is given to the relevant sections in these chapters. For those not interested, portions of the discussion relying on these results can be easily accepted without sacrificing continuity in the overall presentation.

4.4.1 Time tendency of linear momentum

The fluid parcel's linear momentum has a time tendency given by

$$\begin{aligned} \partial_\tau(\mathbf{v}\,\rho\,\mathrm{d}V) &= (\rho\,\mathrm{d}V)\,\partial_\tau \mathbf{v} \\ &= (\rho\,\mathrm{d}V)\,\frac{\mathrm{d}\mathbf{v}}{\mathrm{d}t}, \end{aligned} \tag{4.35}$$

where $\partial_\tau = \mathrm{d}/\mathrm{d}t$ is the material time derivative following the fluid parcel, and $\partial_\tau(\rho\,\mathrm{d}V) = 0$ since the parcel conserves its mass. For a parcel moving in a spherical geometry, it is necessary to use the tensor analysis tools presented in Chapters 20 and 21 to express the parcel's acceleration, $\mathrm{d}\mathbf{v}/\mathrm{d}t$, in general orthogonal horizontal coordinates.

Start by noting that a tensorial component of the parcel's acceleration is given by

$$\begin{aligned}\frac{\mathrm{d}v^a}{\mathrm{d}t} &= v^a_{,t} + v^b\, v^a_{;b} \\ &= v^a_{,t} + (v^b\, v^a)_{;b} - v^a\, v^b_{;b}\end{aligned} \tag{4.36}$$

where $a = 1, 2, 3$ is a label for the three spatial coordinates, and the semicolon notation signifies the covariant derivative (Section 21.2.2). The second step introduced the covariant divergence of the velocity, written as $v^b_{;b}$. It can be replaced through use of mass conservation (3.31)

$$\rho\, v^b_{;b} = -(\partial_t + v^a\, \partial_a)\, \rho, \tag{4.37}$$

to yield

$$\rho\, \frac{\mathrm{d}v^a}{\mathrm{d}t} = (\rho\, v^a)_{,t} + (\rho v^a\, v^b)_{;b}. \tag{4.38}$$

The product $\rho v^a\, v^b$ comprises the elements of a second order symmetric tensor. As shown in Section 21.6, the covariant divergence of such a tensor is given by

$$(\rho v^a\, v^b)_{;b} = \frac{1}{\sqrt{\mathcal{G}}} \left(\sqrt{\mathcal{G}}\, v^a\, v^b\right)_{,b} + \Gamma^a_{bc}\, v^b\, v^c. \tag{4.39}$$

In this expression, $\sqrt{\mathcal{G}} = h_1\, h_2\, h_3$ is the square root of the metric tensor's determinant (Section 3.2) and Γ^a_{bc} are the Christoffel symbols (Section 21.4). In the geophysical fluid applications considered in this monograph, the metric functions h_a have time dependence limited to dependence on position and not explicitly on time. For this case, multiplication by the metric functions h_a, without an implied sum over a, leads to the physical components of the density weighted acceleration*

$$\rho\, \frac{\mathrm{d}v^{(a)}}{\mathrm{d}t} = (\rho\, v^{(a)})_{,t} + \nabla \cdot (\rho\, v^{(a)}\, \mathbf{v}) + \rho\, v^b\, h_a\, (v^c\, \Gamma^a_{bc} - v^a \partial_b\, \ln h_a), \tag{4.40}$$

where

$$v^{(a)} = (\mathbf{u}, w) = (u, v, w) \tag{4.41}$$

are the physical velocity components, each with dimensions length per time. Use of the Christoffel symbols for orthogonal coordinates leads to

$$\rho\, \frac{\mathrm{d}u}{\mathrm{d}t} = (\rho\, u)_{,t} + \nabla \cdot (\rho\, u\, \mathbf{v}) - \rho\, v\, (v\, \partial_x \ln h_2 - u\, \partial_y \ln h_1) \tag{4.42}$$

$$\rho\, \frac{\mathrm{d}v}{\mathrm{d}t} = (\rho v)_{,t} + \nabla \cdot (\rho\, v\, \mathbf{v}) + \rho\, u\, (v\, \partial_x \ln h_2 - u\, \partial_y \ln h_1) \tag{4.43}$$

$$\rho\, \frac{\mathrm{d}w}{\mathrm{d}t} = (\rho\, w)_{,t} + \nabla \cdot (\rho\, w\, \mathbf{v}). \tag{4.44}$$

*Physical components of a general tensor have the same dimensions of the components to Cartesian tensors. For example, the physical components of velocity, $v^{(a)}$, have units of length per time. The tensor components v^a depend on the coordinates and so can have arbitrary dimensions. Components to physical tensors are denoted with a parentheses surrounding the index. Details of physical tensor components are summarized in Section 21.12.

As in Section 3.2, now introduce the physical displacements $\mathrm{d}x = h_1\,\mathrm{d}\xi^1$, $\mathrm{d}y = h_2\,\mathrm{d}\xi^2$ and use $\partial_1(\mathrm{d}\xi^2) = \partial_2(\mathrm{d}\xi^1) = 0$. Doing so eliminates the stretching functions in favor of the physical displacements

$$\rho\,\frac{\mathrm{d}u}{\mathrm{d}t} = (\rho\,u)_{,t} + \nabla\cdot(\rho\,u\,\mathbf{v}) - \rho\,v\,\left(v\,\partial_x \ln \mathrm{d}y - u\,\partial_y \ln \mathrm{d}x\right) \tag{4.45}$$

$$\rho\,\frac{\mathrm{d}v}{\mathrm{d}t} = (\rho\,v)_{,t} + \nabla\cdot(\rho\,v\,\mathbf{v}) + \rho\,u\,\left(v\,\partial_x \ln \mathrm{d}y - u\,\partial_y \ln \mathrm{d}x\right) \tag{4.46}$$

$$\rho\,\frac{\mathrm{d}w}{\mathrm{d}t} = (\rho\,w)_{,t} + \nabla\cdot(\rho\,w\,\mathbf{v}). \tag{4.47}$$

This equation can be written in the shorthand form

$$\rho\,\frac{\mathrm{d}\mathbf{v}}{\mathrm{d}t} = (\rho\,\mathbf{v})_{,t} + \nabla\cdot(\rho\,\mathbf{v}\,\mathbf{v}) + \mathcal{M}\,(\hat{\mathbf{z}}\wedge\rho\,\mathbf{v}), \tag{4.48}$$

where

$$\mathcal{M} = v\,\partial_x \ln \mathrm{d}y - u\,\partial_y \ln \mathrm{d}x \tag{4.49}$$

defines an *advective metric frequency*. In spherical coordinates where

$$\begin{aligned} \mathrm{d}x &= (r\,\cos\phi)\,d\lambda \\ \mathrm{d}y &= r\,d\phi \\ \partial_x &= (r\,\cos\phi)^{-1}\partial_\lambda \\ \partial_y &= R^{-1}\partial_\phi, \end{aligned} \tag{4.50}$$

the advective metric frequency is given by

$$\mathcal{M} = (u/r)\,\tan\phi \tag{4.51}$$

thus leading to

$$\rho\,\frac{\mathrm{d}\mathbf{v}}{\mathrm{d}t} = (\rho\,\mathbf{v})_{,t} + \nabla\cdot(\rho\,\mathbf{v}\,\mathbf{v}) + (u/r)\,\tan\phi\,(\hat{\mathbf{z}}\wedge\rho\,\mathbf{v}). \tag{4.52}$$

The advective metric frequency is typically quite small due to the relatively large size of the earth, where $R \approx 6.37\times10^6\,\mathrm{m}$. A zonal current of $0.1\,\mathrm{m\,s^{-1}}$ at 45°N latitude has an advective metric frequency $\mathcal{M} \approx 1.6\times10^{-8}\mathrm{s}^{-1}$, or roughly 10^{-4} times smaller than the Coriolis or inertial frequency $f = 1.03\times10^{-4}\mathrm{s}^{-1}$.

The advective metric force density $\mathcal{M}\,(\hat{\mathbf{z}}\wedge\rho\,\mathbf{v})$ arises from the nontrivial metric functions h_1 and h_2, and it appears as an added rotational force analogous to the Coriolis force. It arises from the coordinate frame of reference, whereas the Coriolis force appears in a non-inertial rotating reference frame. As with the Coriolis force, the advective metric force does no work on a fluid parcel since it is perpendicular to the parcel's velocity: $\mathbf{v}\cdot(\hat{\mathbf{z}}\wedge\mathbf{v}) = 0$. Hence, it does not affect the parcel's kinetic energy. This result provides a good check on the manipulations, since work and kinetic energy are scalars and so cannot be influenced by coordinate-dependent metric components. Additionally, the frequency $\mathcal{M}$ vanishes for some coordinate choices and remains nonzero for others. Hence, $\mathcal{M}$ is not a tensor.*

*See Section 21.1, page 466 for a discussion of properties satisfied by a tensor.

4.4.2 Internal forces

As discussed in Section 4.3.5, there are two internal forces of interest when formulating the equations for an ocean model: pressure and friction. These forces can be represented as the divergence of the stress tensor, as discussed in Section 4.3.4. That is, the general expression for the stress tensor components is

$$T^{ab} = \tilde{\tau}^{ab} - p\, g^{ab} \tag{4.53}$$

with the internal force per unit volume given by the divergence

$$\nabla \cdot \mathbf{T} = \rho\, \mathbf{F}^{(\mathbf{v})} + \mathbf{F}_{(p)} \tag{4.54}$$

with $\nabla \cdot \mathbf{T} = T^{ab}_{;b}$ a shorthand for the covariant divergence of the stress tensor. The force due to pressure acting on the boundaries of a fluid parcel is given by the volume of the parcel times the gradient of pressure

$$\mathbf{F}_{(p)} = -\mathrm{d}V\, \nabla p, \tag{4.55}$$

and the frictional force is given by

$$(\rho\, \mathrm{d}V)\, \mathbf{F}^{(\mathbf{v})} = (\rho\, \mathrm{d}V)\, (F^{(u)}, F^{(v)}, F^{(w)}). \tag{4.56}$$

More details of frictional stresses employed by ocean models are discussed in Chapters 8 and 17–19.

4.4.3 Evolution of linear momentum density

Setting the time tendency of a fluid parcel's linear momentum equal to the forces acting on the parcel yields the equations of motion

$$\mathrm{d}V\, [(\rho\, \mathbf{v})_{,t} + \nabla \cdot (\rho\, \mathbf{v}\, \mathbf{v}) + \mathcal{M}\, (\hat{\mathbf{z}} \wedge \rho\, \mathbf{v})] = \mathrm{d}V\, [-\rho\, g\, \hat{\mathbf{z}} - f\, \hat{\mathbf{z}} \wedge \rho\, \mathbf{v} - \nabla p + \rho\, \mathbf{F}^{(\mathbf{v})}]. \tag{4.57}$$

Dividing by the parcel's volume leads to the time tendency for the linear momentum density

$$(\rho\, \mathbf{v})_{,t} + \nabla \cdot (\rho\, \mathbf{v}\, \mathbf{v}) + \mathcal{M}\, (\hat{\mathbf{z}} \wedge \rho\, \mathbf{v}) = -\rho\, g\, \hat{\mathbf{z}} - f\, \hat{\mathbf{z}} \wedge \rho\, \mathbf{v} - \nabla p + \rho\, \mathbf{F}^{(\mathbf{v})}. \tag{4.58}$$

This equation forms the basis for averaging the equations of motion over an ensemble of realizations (Chapter 8) as well as for discretizing the dynamics (Chapter 12).

4.4.4 Vector invariant velocity equation

Another form of the equation describing fluid parcel dynamics appears in many contexts, such as when deriving the vorticity equation discussed in Section 4.9. It also serves as an alternative starting point for numerical discretization. This equation goes by the name *vector invariant velocity equation* since it is written in a coordinate invariant form. Notably, the advection frequency $\mathcal{M}$ in equation (4.58) vanishes for some coordinate choices, thus indicating the nontensorial character of this object. In contrast, all terms in the vector invariant velocity equation remain intact regardless of the coordinates.

To derive the vector invariant velocity equation, return to the derivation in Section 4.4.1. Instead of focusing on the linear momentum density, consider a parcel's velocity, whose time tendency is given by

$$\frac{dv^a}{dt} = v^a_{,t} + v^b\, v^a_{;b} \tag{4.59}$$

where again v^a is a tensorial component of the velocity field and all tensor labels run over the spatial range 1, 2, 3. The nonlinear advection term can be written

$$\begin{aligned} v^b\, v^a_{;b} &= v^b\, g^{ac}\, v_{c;b} \\ &= v^b\, g^{ac}\,(v_{c;b} - v_{b;c}) + v^b\, g^{ac}\, v_{b;c} \end{aligned} \tag{4.60}$$

where $g^{ac}_{;b} = 0$ (see Section 21.2.4, page 470). The second term in equation (4.60) can be written as the covariant derivative of the kinetic energy per unit mass

$$\begin{aligned} g^{ac}\, v^b\, v_{b\,;c} &= (g^{ac}/2)\,(v^b\, v_b)_{;c} \\ &= (g^{ac}/2)\,(v^b\, v_b)_{,c} \end{aligned} \tag{4.61}$$

where the last step notes that the covariant derivative of a scalar equals to its partial derivative (see equation (21.11), page 469). Use of expression (21.14) for the covariant derivative of a one-form component leads to the first term taking the form

$$\begin{aligned} v^b\, g^{ac}\,(v_{c;b} - v_{b;c}) &= v^b\, g^{ac}\,(v_{c,b} - \Gamma^p_{cb}\, v_p - v_{b,c} + \Gamma^p_{bc}\, v_p) \\ &= g^{ac}\, v^b\,(v_{c,b} - v_{b,c}) \end{aligned} \tag{4.62}$$

where the last step used the symmetry property, $\Gamma^p_{bc} = \Gamma^p_{cb}$, of the Christoffel symbols (Section 21.4). The antisymmetric term $v_{c,b} - v_{b,c}$ is reminiscent of vorticity. Indeed, introducing the covariant curl given in Section 21.8 leads to

$$\begin{aligned} (\vec{\omega} \wedge \vec{v})_a &= (\mathrm{curl}\,\vec{v} \wedge \vec{v})_a \\ &= \varepsilon_{abc}\,(\mathrm{curl}\,\vec{v})^b\, v^c \\ &= \varepsilon_{abc}\,\varepsilon^{bpq}\, v_{q,p}\, v^c \\ &= (\delta^p_c\,\delta^q_a - \delta^p_a\,\delta^q_c)\, v_{q,p}\, v^c \\ &= v^c\,(v_{a,c} - v_{c,a}) \end{aligned} \tag{4.63}$$

where use was made of properties satisfied by the covariant Levi-Civita symbol ε_{abc} (Section 20.12), and

$$(v^1, v^2, v^3) = \vec{v} \tag{4.64}$$

form the three tensorial components of the velocity vector. This result gives

$$\frac{dv^a}{dt} = v^a_{,t} + (g^{ac}/2)\,(v^b\, v_b)_{,c} + g^{ac}\,(\vec{\omega} \wedge \vec{v})_c. \tag{4.65}$$

As in Section 4.4.1, now specialize the result (4.65) to the locally orthogonal coordinates commonly used in ocean models, in which case the metric tensor components can be written $g_{ab} = h_a\, h_b\, \delta_{ab}$ with no implied summation, and where the covariant Levi-Civita symbol ε_{abc} is given by $\varepsilon_{abc} = h_1\, h_2\, h_3\, \epsilon_{abc}$, with ϵ_{abc} the flat-space Levi-Civita symbol (Section 20.12). Additionally, time dependence of the metric is at most implicit, as may arise through dependence on position on the sphere. It is then straightforward to derive an expression for the time tendency of a physical velocity component for a parcel

$$\frac{dv^{(a)}}{dt} = v^{(a)}_{,t} + (1/2)\,(\mathbf{v}\cdot\mathbf{v})_{,(a)} + (\boldsymbol{\omega}\wedge\mathbf{v})_{(a)}, \tag{4.66}$$

where again the physical components of the velocity are given by $v^{(a)} = h_a\, v^a$ with no implied summation. A boldface symbol characterizes the physical components to a vector such as the velocity, $\mathbf{v} = (v^{(1)}, v^{(2)}, v^{(3)})$.

The expression (4.66) is more compactly written as

$$\rho\,\frac{d\mathbf{v}}{dt} = \rho\,(\partial_t + \boldsymbol{\omega}\wedge\,)\,\mathbf{v} + (\rho/2)\,\nabla(\mathbf{v}\cdot\mathbf{v}). \tag{4.67}$$

Hence, bringing together all the forces acting on a fluid parcel leads to the vector-invariant form of the velocity equation

$$[\partial_t + (\boldsymbol{\omega} + \mathbf{f}\,)\wedge\,]\,\mathbf{v} = \nabla\,(\Phi - \mathbf{v}\cdot\mathbf{v}/2) - \rho^{-1}\,\nabla p + \mathbf{F}^{(\mathbf{v})}, \tag{4.68}$$

where $\mathbf{g} = -\nabla\,\Phi = g\,\hat{\mathbf{z}}$ is the gravitational acceleration (equation (4.8)). Comparison of this equation with that for the linear momentum density (4.58) highlights the different treatment of nonlinear transport. The vector invariant velocity equation (4.68) exposes vorticity and kinetic energy per unit mass, whereas the linear momentum equation (4.58) focuses on nonlinear self-advection along with the coordinate dependent advection metric.

As stated earlier, the vector invariant velocity equation is useful for certain theoretical developments, such as deriving the vorticity equation. Additionally, some modelers use it as their starting point for numerical discretization, with those building Boussinesq C-grid models (see Section 10.2) generally preferring this form (for a review, see Griffies et al., 2000a). In Chapter 8 we focus on the linear momentum equation (4.58) as the starting point for obtaining the ensemble averaged dynamical equations, as this renders a straightforward discretization of a non-Boussinesq fluid in z-models following the methods of McDougall et al. (2003a) and Greatbatch and McDougall (2003).

4.5 HYDROSTATIC PRESSURE

The hydrostatic balance is a fundamental balance maintained by fluids described by the primitive equations. The purpose of this section is to discuss features of Boussinesq and non-Boussinesq fluids that maintain hydrostatic balance.

4.5.1 Hydrostatic balance

A fluid in static equilibrium maintains a balance between the vertical pressure gradient and the gravitational force per volume

$$p_{,z} = -\rho\,g. \tag{4.69}$$

This equation is known as the *hydrostatic balance*. Scaling analysis (e.g., Gill, 1982) shows that the hydrostatic balance remains the dominant balance within the vertical momentum equation, so long as the vertical length scales of motion are much smaller than the horizontal length scales. As such scales are relevant for large-scale ocean climate modeling, global ocean models typically assume a hydrostatic balance, and this constitutes a basic assumption of the primitive equations. Integrating the hydrostatic balance vertically from the ocean surface determines the pressure at a point in the ocean

$$p(z) = p_{\mathrm{a}} + g \int_{z}^{\eta} \mathrm{d}z' \, \rho(z'), \tag{4.70}$$

where p_{a} is the pressure applied at the sea surface by media outside the ocean (e.g., loading by the atmosphere and/or sea ice).

4.5.2 Bottom pressure and the Boussinesq approximation

Assuming hydrostatic balance in a constant gravitational field, pressure at the solid earth boundary (the ocean bottom) is a function of the density within the overlying vertical fluid column, the thickness of the column, and the pressure applied onto the ocean surface by other media such as the atmosphere and sea ice. Hence, there are two reasons that the bottom pressure differs for a Boussinesq and non-Boussinesq fluid: (1) differences in sea level, and (2) differences in density within the column.

To further understand these differences, ignore applied loading from other media, and consider two fluid columns of equal depth and equal density, thus producing equal bottom pressures

$$p_{\mathrm{bottom}} = g \int_{-H}^{\eta} \mathrm{d}z \, \rho. \tag{4.71}$$

Let one column be filled with a volume conserving fluid, the other with a mass conserving fluid, and examine the different time evolutions of bottom pressure for these two columns.

The time tendency for bottom pressure in the mass conserving fluid is given by

$$\begin{aligned} \partial_t \, p_{\mathrm{bottom}} &= g \, \partial_t \left(\int_{-H}^{\eta} \mathrm{d}z \, \rho \right) \\ &= g \, (-\rho_o \, \nabla \cdot \tilde{\mathbf{U}} + \rho_{\mathrm{w}} \, q_{\mathrm{w}}), \end{aligned} \tag{4.72}$$

where the second equality follows from the mass balance equation (3.61). Equation (4.72) says that bottom pressure changes according to changes in the mass per unit area of fluid within the vertical column. In particular, for a mass conserving fluid at rest with zero currents, uniform heating of the fluid, which maintains the fluid at rest, does not change the bottom pressure, since the mass of fluid in the column remains the same, and only the thickness of the column changes. Integration of the bottom pressure tendency over the full ocean bottom area leaves only a contribution from surface mass fluxes, regardless of the buoyancy fluxes within the ocean interior. This result makes sense because bottom pressure integrated over the full area of ocean bottom represents the total gravitational force applied to the ocean bottom from the mass of fluid. With a constant gravitational acceleration, this force changes in time only when the ocean fluid mass changes.

Now consider the same issues for a volume conserving fluid. In this case the time tendency of bottom pressure is

$$\begin{aligned}\partial_t\, p_{\text{bottom}} &= g\left(\rho(\eta)\,\eta_{,t} + \int\limits_{-H}^{\eta} \mathrm{d}z\,\rho_{,t}\right)\\ &= g\left(\rho(\eta)\,(-\nabla\cdot\mathbf{U} + q_{\text{w}}) + \int\limits_{-H}^{\eta} \mathrm{d}z\,\rho_{,t}\right),\end{aligned} \tag{4.73}$$

where the volume budget given by equation (3.45) was used. Comparison to equation (4.72) for the mass conserving fluid reveals some differences in details of the first and second terms, plus an extra term $\int_{-H}^{\eta} \mathrm{d}z\,\rho_{,t}$ associated with the depth integrated density time tendency. When uniformly heating a volume conserving fluid at rest, the extra density tendency term leads to a reduction in bottom pressure. This result can be understood in the following way. Recall from Section 2.4.3 that uniformly heating a column of volume conserving fluid reduces its density while keeping its surface height fixed in order to maintain the same volume. Therefore, the mass within the column is reduced. Reducing the mass within the column without changing its thickness leads to a reduction in bottom pressure, as revealed by equation (4.73).

These examples illustrate some unphysical effects associated with a Boussinesq fluid. If one can precisely and accurately measure the ocean's bottom pressure using satellites, as suggested by Hughes et al. (2000), then comparison of these data with a numerical simulation is handicapped if the model discretizes a Boussinesq fluid.

4.6 DYNAMICS OF HYDROSTATIC FLUID COLUMNS

Section 3.4 considered the kinematics of infinitesimal fluid parcels as well as finite domains of fluid, including columns of ocean fluid. The present section follows up by deriving the dynamical equations of a vertically integrated column of hydrostatically balanced ocean fluid. Particular attention is given to a column extending from the ocean bottom, where it interacts with the solid earth, to its free surface, where it interacts with the atmosphere, sea ice, rivers, and so on.

The motion of fluid columns is described by equations in two spatial dimensions, instead of the three dimensions available for parcels. These equations are derived by vertically integrating the dynamical equations for a parcel from the ocean bottom to its surface. The two-dimensional vertically integrated dynamics, when combined with two-dimensional fluid column kinematics, constitutes the *vertically integrated mode*. This mode is not a "mode" in the mathematical sense of eigenmode. However, use of the term is not without relevance. Namely, the vertically integrated mode exhibits similar dynamics to the *external* mode, also known as the *barotropic* mode. The external mode, rigorously defined only for a flat bottom ocean, is the zeroth vertical eigenmode for the linearized primitive equations (e.g., Gill, 1982; Killworth et al., 1991).

The external mode is strictly two-dimensional only when applying a *rigid lid* upper boundary condition.* For a free surface ocean, the external mode has a weak vertical dependence, thus breaking the exact analogy with the vertically integrated mode (Gill, 1982; Killworth et al., 1991). Nonetheless, because of the overall similarity between the external mode and the vertically integrated mode, the two are often considered synonymous in general discussions. However, for some special considerations, such as when designing algorithms to numerically solve the primitive equations, it is important to be mindful of the distinction.

Dynamics of the vertically integrated mode have much in common with dynamics of a single homogeneous layer of shallow fluid. Indeed, linearizing the vertically integrated primitive equations by dropping nonlinear interactions between the external mode and other modes, the external mode equations reduce to the linear shallow water equations. Such fluid layers are well studied in geophysical fluid dynamics since they allow one to focus on the effects of rotation as isolated from the effects of stratification (e.g., Gill, 1982; Pedlosky, 1987; Cushman-Roisin, 1994).

For the World Ocean, the speed of external mode disturbances is roughly 100 times faster than the speed of the next fastest disturbances. These *external mode gravity waves* propagate such physical effects as tidal fluctuations and tsunamis. Indeed, such disturbances are often described quite well by a single layer of shallow fluid, rather than the full baroclinic system. To explicitly resolve external waves in the three-dimensional numerical model would place a severe restriction on the model's efficiency, since the model time step would need to be small enough to resolve these relatively fast waves. This motivates one to formulate methods for time stepping the ocean primitive equations which extract or *split* the fast external mode dynamics from the slower *baroclinic* or *internal* mode dynamics. Details of this splitting are given in Chapter 12.

4.6.1 An approximate hydrostatic pressure

Recall that the hydrostatic pressure is given by equation (4.70): $p(z) = p_{\mathrm{a}} + g\int_z^\eta \mathrm{d}z'\,\rho$. Fluctuations in this pressure are associated with fluctuations in the pressure p_{a} applied at the ocean surface, in the ocean surface height, and in the ocean density. Fluctuations in the surface height generally occur on the faster external mode time scale, whereas fluctuations in density generally occur on the slower internal mode time scale. Fluctuations in the applied pressure can also be fast, when they arise from atmospheric fluctuations. In order to incorporate the fast dynamics into the vertically integrated mode, it is useful to separate the fast and slow contributions to the pressure field prior to vertically integrating the equations of motion.

For this purpose, consider the identity

$$\int_z^\eta \rho\,\mathrm{d}z' = \int_z^0 \rho\,\mathrm{d}z' + \int_0^\eta \rho\,\mathrm{d}z'. \tag{4.74}$$

The region of fluid between $z = 0$ and $z = \eta$ is only a few meters for typical flows considered in large-scale ocean modeling. For present purposes, assume the

*See Sections 2.4.4 and 12.10 for discussions of the rigid lid approximation.

fluid in this region to be vertically well mixed (i.e., vertically homogeneous) due to strong mechanical and/or buoyant forcing. In this case

$$\int_0^\eta \rho \,\mathrm{d}z \approx \eta\, \rho(z=0). \tag{4.75}$$

This approximation brings the hydrostatic pressure to the form

$$p = p_\mathrm{a} + p_\mathrm{s} + p_\mathrm{b}, \tag{4.76}$$

where

$$p_\mathrm{b} = g \int_z^0 \rho \,\mathrm{d}z \tag{4.77}$$

is known as the *baroclinic* or *internal* pressure field, and

$$p_\mathrm{s} = g\, \eta\, \rho(z=0) \tag{4.78}$$

is known as the *surface pressure* field. Figure 4.2 illustrates this partitioning of the pressure field.

The baroclinic pressure is the hydrostatic pressure associated with density in the vertical column between some depth, z, and a resting ocean surface, $z = 0$. In turn, the horizontal gradient of this field, $\nabla_z p_\mathrm{b} = g \int_z^0 \nabla_z \rho \,\mathrm{d}z'$, arises from baroclinic effects (i.e., horizontal density gradients) in that part of the ocean between the resting ocean surface and the depth z.

Care should be taken to distinguish the surface pressure p_s, which is the pressure due to ocean fluid between $z = 0$ and $z = \eta$, with the applied pressure p_a, which is the pressure felt at the ocean surface, $z = \eta$. Both the applied and surface pressures are independent of depth, whereas the baroclinic pressure is depth dependent. Notably, when $\eta < 0$, the surface pressure is negative. Many Boussinesq models approximate the surface pressure with $p_\mathrm{s} \approx g\, \eta\, \rho_o$, where ρ_o is the constant Boussinesq reference density. This approximation only saves a small amount of computer time, and does not simplify the equations in a fundamental manner. Hence, even when making the Boussinesq approximation, modelers may choose to maintain the more accurate hydrostatic result $p_\mathrm{s} = g\, \eta\, \rho(z=0)$, as proposed by Griffies et al. (2001).

4.6.2 Vertically integrated momentum

Vertically integrating the momentum equation given in Section 4.4.3 and using the surface and bottom kinematic boundary conditions (Section 3.4) lead to

$$(\partial_t + f\, \hat{\mathbf{z}} \wedge)\, \tilde{\mathbf{U}} = -(H + \eta)\, \nabla\, (p_\mathrm{a} + p_\mathrm{s})/\rho_o + \mathbf{G}, \tag{4.79}$$

where

$$\mathbf{G} = \mathbf{u}(\eta)\, \rho_\mathrm{w}\, q_\mathrm{w} + \int_{-H}^{\eta} \mathrm{d}z\, [-\nabla\, (p_\mathrm{b}/\rho_o) - (\nabla \cdot (\rho\, \mathbf{v}\, \mathbf{u}) + \mathcal{M}\, \hat{\mathbf{z}} \wedge \rho\, \mathbf{u}) + \rho\, \mathbf{F}^\mathbf{v}], \tag{4.80}$$

and $\tilde{\mathbf{U}}$ is the vertical integral of the horizontal momentum density

$$\rho_o\, \tilde{\mathbf{U}} = \int_{-H}^{\eta} \mathrm{d}z\, \rho\, \mathbf{u} \tag{4.81}$$

introduced in equation (3.59). The nonlinear forcing imparted by $\mathbf{G} \neq 0$ couples the vertically integrated dynamics to the depth-dependent dynamics.

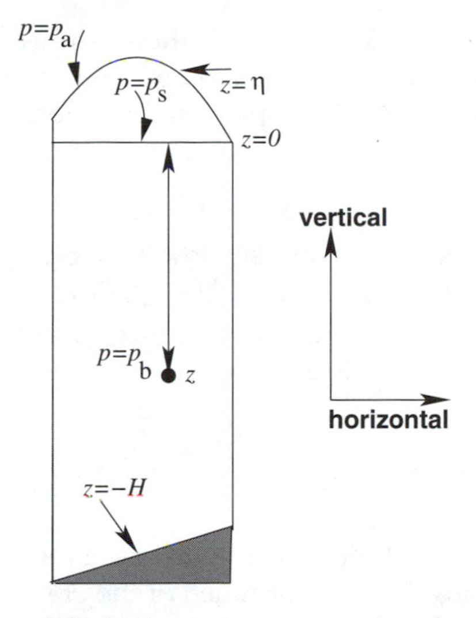

Figure 4.2 The partitioning of the hydrostatic pressure given by equation (4.76). The pressure at a depth z is given by the sum of: (1) the baroclinic pressure, p_{b}, which is the pressure associated with the density between the depth z and the resting ocean surface, $z = 0$; (2) the surface pressure p_{s}, which is the pressure associated with the vertically uniform ocean density between the resting ocean surface $z = 0$ and the free surface, $z = \eta$; (3) the applied pressure p_{a} arising from loading by the overlying atmosphere and/or sea ice.

4.6.3 Linear shallow water waves

It is useful to bring these results into a context familiar from the study of *shallow water* waves (Gill, 1982). For this purpose, drop the nonlinear coupling term **G**, assume the fluid to be Boussinesq, ignore moisture fluxes across the free surface, and drop the applied atmospheric plus sea ice pressure. These restrictions lead to the momentum and volume balances

$$\begin{aligned} \rho_o\,(\partial_t + f\,\hat{\mathbf{z}}\wedge)\,\mathbf{U} &= -(H+\eta)\,\nabla\,p_{\mathrm{s}} \\ \eta_{,t} &= -\nabla\cdot\mathbf{U}. \end{aligned} \tag{4.82}$$

Furthermore, assume that the surface pressure gradient is dominated by the surface height gradient

$$\nabla\,p_{\mathrm{s}} \approx \rho_o\,g\,\nabla\,\eta, \tag{4.83}$$

thus leading to

$$\begin{aligned} (\partial_t + f\,\hat{\mathbf{z}}\wedge)\,\mathbf{U} &= -g\,(H+\eta)\,\nabla\,\eta \\ \eta_{,t} &= -\nabla\cdot\mathbf{U}. \end{aligned} \tag{4.84}$$

The $\eta\,\nabla\eta$ term is the only nonlinearity in this system, and it is dropped now in order to focus on linear disturbances.

A time derivative of the volume budget substituted into the momentum budget reveals a linear wave equation for surface height

$$(\partial_{tt} - g\,H\,\nabla^2)\,\eta = -\nabla\wedge(f\,\mathbf{U}). \tag{4.85}$$

In the absence of the Coriolis force ($f = 0$), these surface height disturbances are unforced linear *shallow water waves*. These are the external gravity waves mentioned at the start of this section. The speed of these nondispersive waves is given by

$$c_{\text{shallow}} = \sqrt{g\,H}. \tag{4.86}$$

For an ocean of depth $H \approx 5000$m, shallow water waves travel at more than $220\,\text{m}\,\text{s}^{-1}$, or nearly $800\,\text{km}\,\text{hr}^{-1}$. Since the disturbances travel faster through deeper water, the waves pile up as they approach a coast. Eventually, the nonlinear term $\eta\,\nabla\eta$ is non-negligible and the waves break onto the coast.

4.7 FLUID MOTION IN A RAPIDLY ROTATING SYSTEM

Rotation plays a fundamental role in planetary scale motions, such as atmosphere and ocean fluid dynamics. Indeed, for much of the observed motion of the ocean and atmosphere, the earth can be considered a rapidly rotating planet. This fact may come as a surprise, since our everyday existence does not appear to be affected by rotation (e.g., trees are not horizontally bending over!). However, large-scale atmospheric wind patterns oriented parallel, not perpendicular, to lines of constant pressure reflect on the importance of rotation. A similar orientation occurs for the large-scale ocean current systems. Both behaviors are a direct result of the Coriolis force.

The purpose of this section is to present some dynamical implications of rotation in geophysical fluids. For this purpose, a few scales are introduced to the equations of motion. We examine how these scales compare with one another in order to understand the dominant balances. The discussion here is quite brief, with texts on geophysical fluid dynamics recommended for a more thorough treatment (see recommended texts in the Preface).

4.7.1 The Rossby number

The *Rossby number* is a dimensionless number measuring the ratio of inertial (or advective) horizontal accelerations to the acceleration associated with the Coriolis force. That is, let the horizontal advective acceleration be given by the order of magnitude

$$(\mathbf{u} \cdot \nabla)\,\mathbf{u} \sim U^2/L, \tag{4.87}$$

where U and L are typical velocity and length scales characterizing the horizontal motion under consideration. Correspondingly, let the magnitude of the Coriolis acceleration be given by

$$f\,\hat{\mathbf{z}} \wedge \mathbf{u} \sim |f|\,U. \tag{4.88}$$

Their ratio defines the Rossby number

$$\begin{aligned} \text{Ro} &= \frac{U^2/L}{|f|\,U} \\ &= \frac{U}{|f|\,L}. \end{aligned} \tag{4.89}$$

With fixed length and velocity scales, the Rossby number is smallest near the poles, is positive away from the poles, and goes to infinity at the equator. In general, the length and velocity scales are not constant, so the Rossby number is a function of more than just the latitude.

A complementary way to define the Rossby number is as a ratio of time scales. Let the horizontal material acceleration of a parcel be given by the order of magnitude

$$\frac{D\mathbf{u}}{Dt} \sim \frac{U}{T}, \tag{4.90}$$

where T is a time scale determined by the material motion of the parcel (i.e., an advective time scale). Now take the ratio of the material acceleration and the Coriolis force per mass to write the Rossby number

$$\begin{aligned} \mathrm{Ro} &= \frac{1}{|f|\,T} \\ &= \frac{T^{-1}}{|f|}. \end{aligned} \tag{4.91}$$

Thus, for motions which have a low frequency T^{-1} compared to the *rotational inertial frequency* $|f|^{-1}$, the Rossby number is small. In both ways of writing the Rossby number, a small Ro is associated with regimes of flow where the earth's rotation plays a crucial role on the dynamics.

Two examples are useful. First consider a kitchen sink. Here, the length scale is $L = 1\text{m}$ (typical sink size), the velocity scale is $U = .01 - 0.1\,\text{m}\,\text{s}^{-1}$ (typical velocity), thus giving a typical time scale for sink motion of $L/U \approx 10\,\text{s} - 100\,\text{s}$. At 30° latitude, where $f = 2\Omega \sin\phi = \Omega$, the Rossby number for fluid motion in a sink is

$$\mathrm{Ro} \approx 10^2 - 10^3. \tag{4.92}$$

This is a large number, and it indicates that the Coriolis force is entirely negligible for kitchen sink fluid dynamics. This result explains the difficulty of experimentally correlating the preferred rotational direction of water leaving a sink's drain to the particular hemisphere.

Next consider motion of a Gulf Stream ring. Here, the typical length scale is $L = 10^5\text{m}$ and typical velocity scale is $U = 0.1 - 1.0\text{m}\,\text{s}^{-1}$, thus leading to a time scale $L/U \approx 10^5\,\text{s} - 10^6\,\text{s}$. For this motion, again assumed to be at 30° latitude, the Rossby number is

$$\mathrm{Ro} \approx 10^{-2} - 10^{-1}. \tag{4.93}$$

This is a small number, and it indicates the importance of Coriolis acceleration on dynamics of Gulf Stream rings.

4.7.2 Geostrophy

Under the influence of horizontal pressure forces, a parcel is accelerated down the pressure gradient (movement from higher pressure to lower pressure). When gaining a nonzero velocity as it moves down the pressure gradient, the parcel's horizontal velocity $\mathbf{u}$ couples to the Coriolis parameter f, thus giving rise to a nonzero horizontally oriented Coriolis force per unit mass $\mathbf{F}_c = -f\,\hat{\mathbf{z}} \wedge \mathbf{u}$ (recall discussion in Section 4.2.4). Hence, for large-scale dynamics, the Coriolis force, in a manner directly analogous to the Lorentz force in electrodynamics, acts perpendicular to the parcel's motion. In the northern hemisphere, where $f > 0$, the Coriolis force acts to the right of the parcel's motion, thus causing counterclockwise motion around low pressure centers and clockwise motion around high pressure centers (Figure 4.3). In the southern hemisphere, where $f < 0$, it acts in the opposite direction.

When pressure and Coriolis forces balance, parcel motion is *along* lines of constant pressure. The importance of this *geostrophic* balance is evidenced by observing that large-scale winds in the atmosphere are roughly aligned parallel to isobars. Circulation around oceanic mesoscale eddies (the ocean's "weather") similarly respects geostrophy.

Geostrophic balance becomes important when the Rossby number is small. In this case, the leading order dynamical balance in the momentum equation is between the Coriolis force and the pressure force

$$f\,\hat{\mathbf{z}} \wedge \mathbf{u} = -\rho_0^{-1}\,\nabla_z p. \tag{4.94}$$

Note that the equator is special since the Coriolis parameter $f = 2\,\Omega\,\sin\phi$ vanishes. Hence, geostrophy is not maintained at the equator.

When oriented in the same sense as the earth's rotation (i.e., same sign of the Coriolis parameter), rotational motion of a parcel is said to be in a *cyclonic* sense. Oppositely oriented motion is *anti-cyclonic*. For example, as mentioned above, geostrophic motion around a low pressure center in the northern hemisphere is counterclockwise. Using the right-hand rule, this represents a positively oriented rotation. Hence, with $f > 0$ in the north, counterclockwise motion is cyclonic. Similarly, in the south, geostrophic motion around a low pressure center is clockwise, which is a negatively oriented rotational motion (again, recall the right-hand rule). In the south, where $f < 0$, clockwise motion around a low pressure center is also cyclonic (see Figure 4.3).

4.7.3 Taylor-Proudman theorem and vertical stiffening

Consider again a Boussinesq fluid maintaining the geostrophic balance (4.94). Taking $\hat{\mathbf{z}} \wedge$ on this relation leads to the expression for the geostrophic velocity

$$\rho_o\,f\,\mathbf{u} = \hat{\mathbf{z}} \wedge \nabla\,p. \tag{4.95}$$

Operating on this expression with the divergence operator leads to

$$\nabla \cdot (f\,\mathbf{u}) = 0, \tag{4.96}$$

where $\nabla \cdot (\hat{\mathbf{z}} \wedge \nabla_z\,p) = 0$. For motion on an f-plane, where the Coriolis parameter is a constant, the flow is two-dimensionally nondivergent: $\nabla \cdot \mathbf{u} = 0$. Volume conservation with $\nabla \cdot \mathbf{v} = 0$, along with the two-dimensional nondivergence condition $\nabla \cdot \mathbf{u} = 0$, can only be satisfied if $w_{,z} = 0$. If the vertical velocity vanishes on the bottom, say for a flat bottom domain, then $w = 0$ throughout a vertical column. The flow is therefore two-dimensional. This result is known as the *Taylor-Proudman theorem*. Laboratory experiments can be performed where a dye is inserted into a rapidly rotating unstratified fluid. After a few rotation periods, the dye forms vertical sheets, or "Taylor curtains," aligned with the axis of

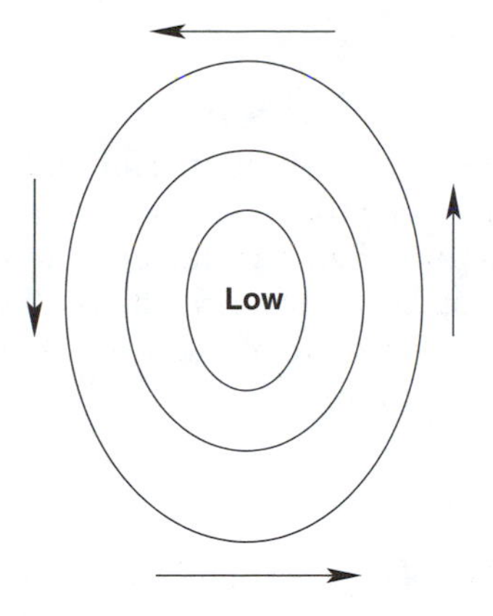

Figure 4.3 Geostrophic motion around a low pressure center in the northern hemisphere. This motion is termed *cyclonic* since it occurs in the same sense (as defined by the right-hand rule) relative to the earth's rotation ($f = 2\Omega \sin\phi > 0$ in the north). Cyclonic motion around a low pressure center occurs in the opposite sense in the southern hemisphere, where $f < 0$.

rotation. The fluid is said to have obtained a "vertical stiffness" due to the effects of rotation.

In the ocean, the idealized assumptions leading to the Taylor-Proudman theorem are rarely satisfied due to the presence of stratification (i.e., vertical density variations; see Section 4.8). Nonetheless, there is a tendency for vertical velocities to be quite small due to the effects of rotation, even smaller than the incompressible scaling $W/H \sim U/L$ would indicate.*

Additionally, for unstratified or linearly stratified fluids, there is a tendency for large-scale turbulent motions that maintain a nearly geostrophic balance (i.e., *geostrophic turbulence*) to cascade energy into the gravest (i.e., the largest scale) vertical mode. This largest vertical scale mode is termed the *barotropic* mode (see Section 4.6). Smaller vertical scales of variation are captured by an infinite hierarchy of *baroclinic* modes. Salmon (1998) provides a lucid discussion of geostrophic turbulence.

4.7.4 Sverdrup balance

The Coriolis parameter f is a function of latitude. Hence, the nondivergence condition $\nabla \cdot (f\,\mathbf{u}) = 0$ satisfied by the geostrophically balanced flow leads to

$$\beta\, v = f\, w_{,z}. \tag{4.97}$$

This relation is known as the *Sverdrup balance*. It is a very powerful result that has been used frequently in theories of the ocean circulation (e.g., Rhines, 1986; Pedlosky, 1987, 1996). It says that vertical shear in the vertical velocity balances a meridional current, with the Coriolis parameter f and the planetary vorticity

*Incompressibility $\partial_x u + \partial_y v + \partial_z w = 0$ leads to the relation $U/L \sim W/H$, where W is a typical vertical velocity scale, H is a typical vertical length scale, and U and L are the corresponding horizontal scales.

gradient

$$\beta = f_{,y} \tag{4.98}$$

determining the sense and strength of the meridional flow. A common source of vertical velocity shear arises when there is a nonzero curl in the wind stress acting on the ocean surface. Vorticity is then transferred to the ocean via frictional effects that cause *Ekman pumping* or *Ekman suction* (depending on the sign of the curl). These effects alter the vertical structure of the vertical velocity, and so, through the Sverdrup balance, induce a meridional flow.

4.7.5 Thermal wind balance

Taking the vertical derivative of a geostrophically balanced Boussinesq fluid leads to

$$\rho_o \, f \, \hat{\mathbf{z}} \wedge \mathbf{u}_{,z} = -\nabla_z \, p_{,z}. \tag{4.99}$$

Use of the hydrostatic relation $p_{,z} = -g\,\rho$ leads to the following balance between vertical shears in the horizontal velocity field, and horizontal density gradients,

$$\hat{\mathbf{z}} \wedge \mathbf{u}_{,z} = \left(\frac{g}{f\,\rho_o}\right) \nabla\rho. \tag{4.100}$$

Taking $\hat{\mathbf{z}} \wedge$ leads to the more conventional form

$$\mathbf{u}_{,z} = -\left(\frac{g}{f\,\rho_o}\right) \hat{\mathbf{z}} \wedge \nabla\rho. \tag{4.101}$$

This diagnostic relation is termed *thermal wind balance*, which is a name originating from its atmospheric analog. Thermal wind balance says that the geostrophic velocity has a vertical *thermal wind shear** in cases where density has a horizontal gradient.

A density field with a horizontal gradient is termed a *baroclinic* density field. Baroclinicity in the density field leads to a nontrivial thermal wind shear. Correspondingly, it is only the baroclinic piece of the velocity field which is related to horizontal gradients in density through the thermal wind balance; barotropic velocities have nearly zero vertical shear (see Section 4.6).

Due to the increased solar radiation reaching the equator relative to the poles, the zonally averaged ocean temperature decreases poleward. Neglecting salinity effects on density, this poleward reduction in temperature corresponds to a poleward increase in density. Also, for a stably stratified fluid, density increases with depth. Figure 4.4 illustrates this situation. The zonal average removes all zonal variations, thus putting $\rho_{,x} = 0$ and so reducing the zonally averaged thermal wind relation to the following statement about the vertical shear of the zonal velocity,

$$u_{,z} = \left(\frac{g}{f\,\rho_o}\right) \rho_{,y}. \tag{4.102}$$

Hence, with a zonally averaged density increasing to the north, $\rho_{,y} > 0$, the zonally averaged thermal wind velocity is eastward. It also increases when moving

---

**Shear* as used in fluid mechanics is synonymous with "derivative." It is most often used when referring to derivatives in the velocity field, as in the present context.

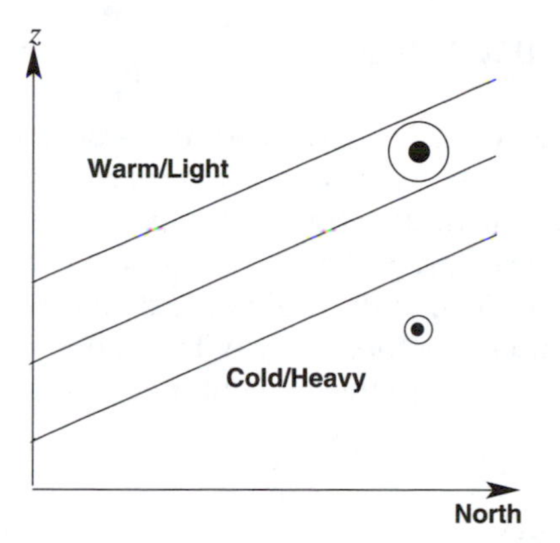

Figure 4.4 Schematic of thermal wind balance in the northern hemisphere, with north to the right and east out of the page. Shown here are surfaces of constant density (termed *isopycnals* in oceanography), which in the absence of salinity are parallel to constant temperature surfaces. Density increases with depth and when moving poleward. The zonal average removes zonal variations, thus leading to $\rho_{,x} = 0$. The thermal wind velocity associated with this density field is eastward (toward the reader), and it increases when moving upward in the ocean (surface intensified). The same eastward thermal wind velocity results in the southern hemisphere, where $\rho_{,y} < 0$ and $f < 0$.

upward in the ocean: $u_{,z} > 0$, which yields a surface intensified zonal velocity field. In the southern hemisphere, poleward increasing density leads to $\rho_{,y} < 0$. With $f < 0$ south of the equator, the southern hemisphere also has an eastward thermal wind velocity. Note that movement toward the poles, where $|f|$ increases, leads to a smaller thermal wind shear.

Thermal wind is a diagnostic relation, just as are geostrophy and hydrostasy. Hence, it cannot be used to predict future evolution of the fluid. However, it does represent a valid steady state balance of a frictionless rotating fluid. That is, in the presence of rotation, a fluid can exist in steady state with nonflat isopycnals. This would not occur in a nonrotating fluid, where steady unforced motion is reached only when isopycnals are flat. The dynamical stability of a sloped isopycnal solution, however, remains to be determined, and forms the subject of baroclinic instability studies (e.g., Gill, 1982; Pedlosky, 1987).

Vertical integration of the thermal wind relation, along with knowledge of the geostrophic velocity at a point along the integration path, allows for determination of the full geostrophic velocity in terms of density. This determination, however, is valid only up to an arbitrary depth-independent constant. Without direct velocity measurements, the specification of this unknown barotropic velocity is unavailable just from density measurements. This is the origin of the "depth of no motion" problem in observational oceanography. Gill (1982) provides further discussion.

4.8 VERTICAL STRATIFICATION

Gravitationally induced convective overturning occurs when density decreases with depth, thus making deep water parcels lighter than shallow parcels. After the overturning has settled down, the stratification is stable with dense water living beneath light water. The result is that the large-scale ocean is for the most part stably stratified in the vertical, with small-scale convective overturning acting to rapidly remove intermittent regions of gravitational instability. The purpose of this section is to introduce some of the elements needed to describe stably stratified fluids.

4.8.1 Buoyancy frequency

When the vertical density profile is gravitationally stable, yet there is a small vertical adiabatic displacement of a parcel so that it moves into a region of slightly heavier fluid, a restoring force brings the parcel back toward its original position. The strength of the restoring force is determined by the vertical stratification in the water column. Generally, as the displaced parcel returns to its original position, it overshoots, thus setting up an oscillation known as a *buoyancy oscillation*. For a Boussinesq fluid, the squared frequency of this oscillation is given by

$$N^2 = -\left(\frac{g}{\rho_o}\right)\rho_{,z} \tag{4.103}$$

where $\rho_{,z} < 0$ for a stably stratified column of water. The stronger the stratification, corresponding to a larger vertical density gradient $|\rho_{,z}|$, the higher the buoyancy frequency N (in radians per second). That is, in highly stratified fluids, buoyancy oscillations are fast; in weakly stratified fluids, buoyancy oscillations are slow.

4.8.2 Vertical depth scale

For much of the World Ocean, seawater density is very close to a static stably stratified background vertical density profile $\overline{\rho}(z)$. In particular, this is the case assumed when studying quasi-geostrophic (QG) dynamics (e.g., Pedlosky, 1987).

The background profile is generally not uniform, yet it is monotonic. For example, in many parts of the ocean, density increases rather abruptly when moving down through the upper few hundred meters of the water column (large values of the buoyancy frequency $\overline{N}(z)$). This strong transition region is known as the *pycnocline*. If the temperature is also changing, which it usually will since density is a function of temperature, then the region of strong temperature transition is known as the *thermocline*. For salinity, strong vertical gradients occur in the *halocline*.

The depth scale depth over which density varies most rapidly is on the order of 1 km. This is the depth scale that is most important for the subsequent discussions. However, there is another depth scale that reflects the near incompressibility of the ocean fluid. This *scale depth* H_{scale} is defined via

$$H_{\text{scale}} \sim \rho_o \, |\partial_z \overline{\rho}|^{-1}. \tag{4.104}$$

In most of the ocean, density varies by roughly 1-2% from top to bottom. With $\rho_o = 1035\,\text{kg}\,\text{m}^{-3}$, the ocean's scale depth is then on the order of 100-200 km. This scale is far larger than the typical ocean depth of 5 km, and it reflects the large degree to which the ocean can be approximated as an incompressible fluid for many purposes.

4.8.3 The Rossby radius and Froude number

Combined effects of buoyancy and rotation yield the richness of quasi-geostrophic motions. Hence, the buoyancy frequency and the Coriolis parameter play central roles in QG theory. The ratio of these two frequencies N/f in regions of nontrivial vertical stratification is typically around 100. Hence, rotational inertial oscillations (usually just called *inertial oscillations*) have period $T_f = 2\pi/f$ about 100 times longer than buoyancy oscillations with period $T_b = 2\pi/N$.

Letting the squared buoyancy frequency N^2 refer to a value typical of a particular flow regime, one can define a length scale

$$\begin{aligned}\lambda &= H\left(\frac{N}{f}\right)\\ &= \frac{\sqrt{gH}}{f}\end{aligned} \tag{4.105}$$

known as the *Rossby radius*. With $H \approx 1$ km the depth over which density varies most, and $N/f \approx 100$, the Rossby radius is roughly 100 km. In general, the Rossby radius is a crucial scale in geophysical fluids. First, it sets the scale for the most unstable baroclinic waves leading to baroclinically unstable flow (e.g., Pedlosky, 1987). Second, although it is not obvious that the scale of linearly unstable waves also sets the scale for equilibrium geostrophic turbulence, there is strong empirical and theoretical evidence to suggest that this scale is indeed the one where most of the baroclinic energy in ocean mesoscale eddies resides.*

Given the importance of the Rossby radius, the dimensionless ratio between the horizontal scales of interest and the Rossby radius

$$\mathrm{Fr} = \left(\frac{L}{\lambda}\right)^2 \tag{4.106}$$

is also important in determining the spatial regime of the flow. This ratio is known as the *Froude number*. Depending on the taste of the writer, often the inverse Froude number will be used instead; it is known as the *Burger number*

$$\mathrm{Bu} = \mathrm{Fr}^{-1}. \tag{4.107}$$

Quasi-geostrophic theory focuses on regimes where the Froude number is order one or a bit smaller (equivalently, where the Burger number is order one or a bit larger).

---

*The reader is encouraged to read the GFD book by Salmon (1998) for an accounting of theoretical issues of geostrophic turbulence. The paper by Smith and Vallis (2002) provides further elaboration and reference to observational data supporting the importance of the first baroclinic Rossby radius of deformation for setting scales of ocean eddies.

4.8.4 The Richardson number

The *Richardson number* is given by the ratio of the squared buoyancy frequency to the squared vertical shear of the horizontal velocity field,

$$\mathrm{Ri} = \frac{N^2}{|\mathbf{u}_{,z}|^2}. \tag{4.108}$$

In regions where Ri << 1, the vertical shear is strong and the flow tends to be unstable to an instability known as *Kelvin-Helmholtz instability* (e.g., Cushman-Roisin 1994). In these regions, there is enough kinetic energy in the vertical shear to extract potential energy from the stratification, and this extraction process occurs via a dynamical instability. In contrast, for large-scale highly stratified flow, the Richardson number is quite large, with $\mathrm{Ri} \sim 100$ common. This is the regime where quasi-geostrophy is relevant. Even though the Richardson number is large for QG dynamics, the flow can still go unstable due to baroclinic instability (Gill, 1982; Pedlosky, 1987; Cushman-Roisin, 1994).

For many discussions of quasi-geostrophy, it is useful to associate a typical value for the Richardson number to the various scales of the flow. Setting the vertical scale equal to H, the horizontal velocity scale to U, and the squared buoyancy frequency to a scale N^2 leads to

$$\mathrm{Ri} = \frac{N^2 H^2}{U^2}. \tag{4.109}$$

Consequently, the Rossby, Richardson, and Froude numbers are related by

$$\begin{aligned}
\mathrm{Ro}^2\,\mathrm{Ri} &= \left(\frac{U}{f L}\right)^2 \left(\frac{N H}{U}\right)^2 \\
&= \left(\frac{N H}{f L}\right)^2 \\
&= \left(\frac{\lambda}{L}\right)^2 \\
&= \mathrm{Fr}^{-1}.
\end{aligned} \tag{4.110}$$

4.9 VORTICITY AND POTENTIAL VORTICITY

Vorticity is one of the most important dynamical variables in fluid mechanics. Furthermore, the associated potential vorticity scalar is key to understanding and predicting aspects of geophysical fluid flows. This section introduces these vorticities by deriving their evolution equations. More complete discussions, with physical interpretations, are available in many places, with Gill (1982), Pedlosky (1987), Müller (1995), and Salmon (1998) recommended for a geophysical fluid focus. Cartesian tensors are used to simplify the mathematics. The resulting equations are easily translated into general coordinates by writing them in a covariant manner as detailed in Chapter 21.

4.9.1 Vorticity evolution

To derive the vorticity equation, start from the vector-invariant form of the velocity equation derived in Section 4.4.4, rewritten here for completeness,

$$(\partial_t + \underline{\omega} \wedge)\, \mathbf{v} = \nabla\, (\Phi - \mathbf{v} \cdot \mathbf{v}/2) - (1/\rho)\, \nabla\, p + \mathbf{F}^{(\mathbf{v})}. \tag{4.111}$$

In this equation,

$$\underline{\omega} = \omega + \mathbf{f} \tag{4.112}$$

is the absolute vorticity, which is the sum of the *relative* vorticity ω and *planetary* vorticity $\mathbf{f}$. Acting on the velocity equation with the curl eliminates the conservative forcing from gravity and the kinetic energy density. To tidy up the resulting equation, use the identity

$$\begin{aligned} [\nabla \wedge (\underline{\omega} \wedge \mathbf{v})]_p &= \epsilon_{pqr}\, \epsilon_{rst}\, (\underline{\omega}_s\, v_t)_{,q} \\ &= (\delta_{ps}\, \delta_{qt} - \delta_{pt}\, \delta_{qs})\, (\underline{\omega}_s\, v_t)_{,q} \\ &= (\underline{\omega}_p\, v_q)_{,q} - (\underline{\omega}_q\, v_p)_{,q} \\ &= (\mathbf{v} \cdot \nabla)\underline{\omega}_p + \underline{\omega}_p\, \nabla \cdot \mathbf{v} - (\underline{\omega} \cdot \nabla) v_p, \end{aligned} \tag{4.113}$$

where use was made of the vanishing divergence of the absolute vorticity: $\nabla \cdot \underline{\omega} = 0$. Additionally, properties of the Levi-Civita symbol ϵ_{pqr} (Section 20.12) were used. Bringing these results together yields the material evolution of absolute vorticity

$$\frac{\mathrm{d}\,\underline{\omega}}{\mathrm{d}t} = \underbrace{-\underline{\omega}\, (\nabla \cdot \mathbf{v})}_{\text{vortex stretching}} + \underbrace{(\underline{\omega} \cdot \nabla)\, \mathbf{v}}_{\text{vortex tilting}} + \underbrace{\rho^{-2}\, (\nabla \rho \wedge \nabla p)}_{\text{baroclinicity}} + \underbrace{\nabla \wedge \mathbf{F}^{(v)}}_{\text{friction}}. \tag{4.114}$$

The four terms on the right-hand side represent various manners whereby a parcel's absolute vorticity is modified. The names associated with these terms represent the mechanisms under which vorticity is affected. A discussion of the physics of these mechanisms is outside the scope of the present considerations. Instead, Chapter 2 of Pedlosky (1987) is highly recommended to garner a physical understanding.

4.9.2 Potential vorticity evolution

Now introduce a scalar function that is materially conserved, $\mathrm{d}\psi/\mathrm{d}t = 0$, and consider evolution of vorticity projected onto a direction normal to surfaces of constant ψ. For this purpose, note that

$$\frac{\mathrm{d}\, \psi_{,a}}{\mathrm{d}t} = -\mathbf{v}_{,a} \cdot \nabla \psi. \tag{4.115}$$

This result then leads to

$$\nabla \psi \cdot \frac{\mathrm{d}\, \underline{\omega}}{\mathrm{d}t} = \frac{\mathrm{d}\, (\nabla \psi \cdot \underline{\omega})}{\mathrm{d}t} + \nabla \psi \cdot (\underline{\omega} \cdot \nabla) \mathbf{v}. \tag{4.116}$$

More than an arbitrary conserved scalar, consider those satisfying $\nabla \psi \cdot (\nabla \rho \wedge \nabla p) = 0$, that is, where the *baroclinicity vector* $\nabla \rho \wedge \nabla p$ is parallel to surfaces of constant ψ. This is the case for functions $\psi = \psi(\rho, p)$, where $\nabla\psi = \psi_{,\rho}\nabla\rho + \psi_{,p}\nabla p$. For such scalars, the vorticity equation leads to

$$\frac{\mathrm{d}\,(\nabla\psi \cdot \underline{\omega})}{\mathrm{d}t} = -\nabla\psi \cdot \underline{\omega}\,(\nabla \cdot \mathbf{v}) + \nabla\psi \cdot \nabla \wedge \mathbf{F}^{(v)}. \tag{4.117}$$

Mass conservation $(\partial_t + \mathbf{v}\cdot\nabla)\,\rho = -\rho\,\nabla\cdot\mathbf{v}$ eliminates the divergence $\nabla\cdot\mathbf{v}$ to give

$$\frac{\mathrm{d}\Pi}{\mathrm{d}t} = \rho^{-1}\,\nabla\psi \cdot \nabla \wedge \mathbf{F}^{(v)}, \tag{4.118}$$

where

$$\Pi = \rho^{-1}\,\underline{\omega}\cdot\nabla\psi \tag{4.119}$$

is the *potential vorticity*. By definition, $\rho\,\Pi$ is the projection of the absolute vorticity onto the direction normal to surfaces of constant ψ. Equation (4.119) thus says that inviscid fluid motion materially conserves the potential vorticity Π. This is an especially important and useful conservation property of ideal ocean fluid dynamics.

The density weighted potential vorticity $\rho\,\Pi$ can be written as a total divergence

$$\rho\,\Pi = \nabla\cdot(\underline{\omega}\,\psi) \tag{4.120}$$

since $\nabla\cdot\underline{\omega} = 0$. Haynes and McIntyre (1987, 1990) show that this form of $\rho\,\Pi$ implies that for general fluid flow, even non-ideal and diabatic flow, sources of $\rho\,\Pi$ are present *only* at domain boundaries. This result is reasonable, given the total divergence form of $\rho\,\Pi$. This property distinguishes $\rho\,\Pi$, which can be considered a "potential vorticity substance," from other tracer substances whose sources can generally be anywhere in the fluid.

4.9.3 Hydrostatic-Boussinesq potential vorticity

Just as there are many approximate forms of the ocean fluid dynamical equations, there are corresponding forms of the potential vorticity field. To illustrate these forms, consider here the potential vorticity equation appropriate for an adiabatic, inviscid, Boussinesq, hydrostatic fluid. Many of the following steps have their analog in the more general discussion just given, but they are presented here for pedagogical purposes.

The dynamical and thermodynamical equations satisfied by the fluid are

$$\rho_o\,(\mathrm{d}/\mathrm{d}t + f\,\hat{\mathbf{z}}\wedge)\,\mathbf{u} = -\nabla_z p \tag{4.121}$$

$$p_{,z} = -g\,\rho \tag{4.122}$$

$$\nabla\cdot\mathbf{v} = 0 \tag{4.123}$$

$$\frac{\mathrm{d}\,\rho}{\mathrm{d}t} = 0, \tag{4.124}$$

where $\mathbf{u} = (u, v, 0)$ is the horizontal velocity, $\nabla_z = (\partial_x, \partial_y, 0)$ is the horizontal gradient (taken on constant a geopotential surface), equation (4.122) is the hydrostatic balance, and equation (4.124) expresses the conservation of potential density for the adiabatic fluid. Note that potential density is here equated to *in situ* density, thus allowing the same symbol, ρ, to be used in the hydrostatic equation as in the potential density equation.*

*See Section 5.7 for a discussion of ocean density.

The vector invariant form of the velocity equation is derived by noting that the advection of horizontal momentum can be written

$$(\mathbf{v} \cdot \nabla)\, \mathbf{u} = (1/2)\, \nabla_z\, (\mathbf{u} \cdot \mathbf{u}) + \zeta\, \hat{\mathbf{z}} \wedge \mathbf{u} + w\, \partial_z\, \mathbf{u}, \tag{4.125}$$

which leads to

$$(\partial_t + w\, \partial_z + \underline{\zeta}\, \hat{\mathbf{z}} \wedge)\, \mathbf{u} = -\nabla_z\, (\mathbf{u} \cdot \mathbf{u}/2 + p/\rho_o), \tag{4.126}$$

where

$$\begin{aligned} \underline{\zeta} &= \zeta + f \\ &= \hat{\mathbf{z}} \cdot \nabla \wedge \mathbf{v} + f \end{aligned} \tag{4.127}$$

is the vertical component of the absolute vorticity. Operating on equation (4.126) with $\hat{\mathbf{z}} \cdot \nabla \wedge$ leads to the material evolution

$$\frac{\mathrm{d}\underline{\zeta}}{\mathrm{d}t} = w_{,z}\, \underline{\zeta} + w_{,y}\, u_{,z} - w_{,x}\, v_{,z}. \tag{4.128}$$

This equation can be put in a more concise form by noting that the nonhydrostatic relative vorticity

$$\boldsymbol{\omega} = \hat{\mathbf{x}}\, (w_{,y} - v_{,z}) + \hat{\mathbf{y}}\, (u_{,z} - w_{,x}) + \hat{\mathbf{z}}\, (v_{,x} - u_{,y}) \tag{4.129}$$

becomes

$$\boldsymbol{\omega} = -\hat{\mathbf{x}}\, v_{,z} + \hat{\mathbf{y}}\, u_{,z} + \hat{\mathbf{z}}\, (v_{,x} - u_{,y}) \tag{4.130}$$

upon moving to the hydrostatic primitive equations. This case leads to

$$\frac{\mathrm{d}\underline{\zeta}}{\mathrm{d}t} = (\boldsymbol{\omega} \cdot \nabla)\, w, \tag{4.131}$$

which can be deduced from the general vorticity equation (4.114).

For an adiabatic Boussinesq fluid, potential density is a conserved scalar whose gradient can be projected onto the absolute vorticity to define potential vorticity. Hence, one is led to consider

$$\rho_o\, \Pi = \underline{\boldsymbol{\omega}} \cdot \nabla \rho \tag{4.132}$$

with the relative vorticity given by the hydrostatic form (4.130) and the ρ^{-1} factor present in the potential vorticity (4.119) reduced to ρ_o^{-1} for the Boussinesq fluid. As defined, Π is proportional to the projection of the absolute vorticity onto the three-dimensional potential density stratification, and its units are those of vorticity. Note that by assuming a Boussinesq fluid, the potential vorticity Π is itself given by a total divergence, rather than $\rho\,\Pi$ as for the non-Boussinesq fluid.

To verify that Π given by equation (4.132) is materially conserved by the ideal flow, one may proceed in a direct manner and thus write

$$\rho_o\, \Pi = f\, \rho_{,z} + (v_{,x}\, \rho_{,z} - v_{,z}\, \rho_{,x}) + (u_{,z}\, \rho_{,y} - u_{,y}\, \rho_{,z}) \tag{4.133}$$

and then compute the material evolution of each term on the right-hand side. Some intermediate steps are noted here for completeness

$$\begin{aligned}
\frac{\mathrm{d}}{\mathrm{d}t}\, (f\, \rho_{,z}) &= v\, \beta\, \rho_{,z} - \mathbf{v}_{,z} \cdot \nabla \rho \\
\frac{\mathrm{d}}{\mathrm{d}t}\, (v_{,x}\, \rho_{,z}) &= \rho_{,z}\, (-f\, u_{,x} - p_{,xy}/\rho_o - \mathbf{v}_{,x} \cdot \nabla v) - v_{,x}\, \mathbf{v}_{,z} \cdot \nabla \rho \\
\frac{\mathrm{d}}{\mathrm{d}t}\, (v_{,z}\, \rho_{,x}) &= \rho_{,x}\, (-f\, u_{,z} + g\, \rho_{,y}/\rho_o - \mathbf{v}_{,z} \cdot \nabla v) - v_{,z}\, \mathbf{v}_{,x} \cdot \nabla \rho \\
\frac{\mathrm{d}}{\mathrm{d}t}\, (u_{,z}\, \rho_{,y}) &= \rho_{,y}\, (f\, v_{,z} + g\, \rho_{,x}/\rho_o - \mathbf{v}_{,z} \cdot \nabla u) - u_{,z}\, \mathbf{v}_{,y} \cdot \nabla \rho \\
\frac{\mathrm{d}}{\mathrm{d}t}\, (u_{,y}\, \rho_{,z}) &= \rho_{,z}\, (\beta\, v + f\, v_{,y} - p_{,xy}/\rho_o - \mathbf{v}_{,y} \cdot \nabla u) - u_{,y}\, \mathbf{v}_{,z} \cdot \nabla \rho
\end{aligned}$$

where $\beta = f_{,y}$ is the planetary vorticity gradient. Expansion of these terms reveals that indeed $\mathrm{d}\Pi/\mathrm{d}t = 0$.

4.9.4 Planetary geostrophic potential vorticity

To finish this excursion into various forms of potential vorticity, consider potential vorticity for the inviscid planetary geostrophic equations

$$\rho_o\,(f\,\hat{\mathbf{z}}\wedge\mathbf{u}) = -\nabla_z p \tag{4.134}$$

$$p_{,z} = -g\,\rho \tag{4.135}$$

$$\nabla\cdot\mathbf{v} = 0 \tag{4.136}$$

$$\frac{\mathrm{d}\,\rho}{\mathrm{d}t} = 0. \tag{4.137}$$

The momentum equation is a statement of geostrophic balance between the Coriolis force and pressure force (Section 4.7.2). It neglects relative vorticity and the material evolution of velocity. The remaining equations are the same as considered in the ideal hydrostatic Boussinesq system in Section 4.9.3. The equations of planetary geostrophy have been studied extensively in oceanography, especially when considering thermocline theory (e.g., Pedlosky, 1987, 1996).

Potential vorticity for the planetary geostrophic system is given by the planetary vorticity times the vertical stratification

$$\begin{aligned}\rho_o\Pi &= f\,\rho_{,z}\\ &= -(f\,\rho_o/g)\,N^2\end{aligned} \tag{4.138}$$

with $N^2 = -(g/\rho_o)\,\rho_{,z}$ the squared buoyancy frequency for a Boussinesq fluid. That is, all contributions to the hydrostatic Boussinesq potential vorticity $\rho_o\,\Pi = \underline{\omega}\cdot\nabla\rho$ that arise from relative vorticity are neglected. All that remains is the contribution from the Coriolis term $\hat{\mathbf{z}}\,f\cdot\nabla\rho = f\,\rho_{,z}$.

Proof that Π is materially conserved is straightforward. The main steps are included here for completeness. First, note that

$$\rho_o\frac{\mathrm{d}\Pi}{\mathrm{d}t} = \beta\,v\,\rho_{,z} - f\,\mathbf{u}_{,z}\cdot\nabla\rho - f\,w_{,z}\,\rho_{,z} \tag{4.139}$$

where $\mathrm{d}\rho/\mathrm{d}t = 0$. Next take the vertical derivative of the geostrophic balance to derive the thermal wind equations (Section 4.7.5)

$$f\,\mathbf{u}_{,z} = (g/\rho_o)\,\hat{\mathbf{z}}\wedge\nabla\rho. \tag{4.140}$$

This expression implies that the vertical shear of the geostrophic velocity is orthogonal to the horizontal density gradient

$$\begin{aligned}\mathbf{u}_{,z}\cdot\nabla\rho &= (g/f\,\rho_o)\,(\hat{\mathbf{z}}\wedge\nabla\rho)\cdot\nabla\rho\\ &= 0.\end{aligned} \tag{4.141}$$

Now use the continuity equation to replace $w_{,z}$ with the horizontal convergence $w_{,z} = -\nabla\cdot\mathbf{u}$, and use geostrophy to relate this convergence to the meridional advection of planetary vorticity

$$\begin{aligned}\nabla\cdot(f\,\mathbf{u}) &= \beta\,v + f\,\nabla\cdot\mathbf{u}\\ &= 0.\end{aligned} \tag{4.142}$$

This result substituted into equation (4.139) leads to the desired material conservation property $\mathrm{d}\Pi/\mathrm{d}t = 0$.

4.10 PARTICLE DYNAMICS ON A ROTATING SPHERE

The purpose of this section is to highlight some dynamical issues related to the shallow ocean approximation introduced in Section 3.2, as well as the hydrostatic approximation. Namely, what is the form for the energy and angular momentum of a fluid parcel moving in the geometry defined by the shallow ocean approximation and where vertical motions are sub-dominant to horizontal? Both of these questions can be addressed with manipulations of the hydrostatic primitive equations derived in the previous sections. However, it is interesting to provide a complementary discussion that is not commonly presented in the geophysical fluids literature. For this purpose, the focus here turns to the dynamics of a point particle freely moving around a rotating massive planet with spherical geometry. As mentioned in Section 4.4, a Lagrangian perspective on fluid parcel mechanics has much in common with point particle mechanics. Hence, many of the results for the point particle apply to the fluid parcel. However, the fundamental distinction, pointed out in Section 4.3.2, is that the boundary of a continuous fluid parcel feels the effects of other parcels through internal or contact forces such as pressure and friction. Correspondingly, a continuous fluid has internal energy, whereas an ideal point particle does not.

To derive the equations for point particle motion, we employ Hamilton's principle instead of Newton's second law. Hamilton's principle leads to the same dynamical equations as Newton's second law. The key difference is that it formulates dynamics from a least action variational principle. The action principle readily highlights conservation laws and their connection to symmetries, and this represents the fundamental connection between dynamics and geometry. These issues are well known from analytical mechanics (e.g., Fetter and Walecka, 1980; Marion and Thornton, 1988) and have their direct analog in fluid dynamics (Salmon, 1988, 1998; Müller, 1995). In the following, some experience with these topics is needed at the level of an undergraduate mechanics course.

4.10.1 A fluid parcel and a point particle

Application of Newton's second law in Section 4.4 led to the equation of motion for a fluid parcel

$$(\rho \, \mathrm{d}V) \left(\frac{\mathrm{d}}{\mathrm{d}t} + f \, \hat{\mathbf{z}} \wedge \right) \mathbf{v} = -g \, (\rho \, \mathrm{d}V) \, \hat{\mathbf{z}} - \mathrm{d}V \, \nabla p + (\rho \, \mathrm{d}V) \, \mathbf{F}^{(\mathbf{v})}. \tag{4.143}$$

Now consider a classical point particle with mass

$$m = \rho \, \mathrm{d}V \tag{4.144}$$

moving on a rotating sphere. Notably, this particle obeys nearly the same equation of motion. The key differences are (1) internal forces (i.e., pressure and friction) are absent for the point particle since these forces arise from the finite extent of the continuous fluid parcel; and (2) the mass times acceleration $(\rho \, \mathrm{d}V) \, \mathrm{d}\mathbf{v}/\mathrm{d}t$ of the parcel becomes $m \, \dot{\mathbf{v}} = m \, \ddot{\mathbf{x}}$ for the particle, where the dot represents a time derivative, with time measured in the frame of the moving particle. Hence, the point particle satisfies

$$m \left(\frac{\mathrm{d}}{\mathrm{d}t} + f \, \hat{\mathbf{z}} \wedge \right) \dot{\mathbf{x}} = -g \, m \, \hat{\mathbf{z}}. \tag{4.145}$$

In the following, this relation is rederived using alternative methods for the purpose of clarifying how the shallow ocean and hydrostatic approximations affect the energy and angular momentum of the particle, and hence of the fluid parcel.

4.10.2 Lagrangian function

To use Hamilton's principle, one requires expressions for the kinetic and potential energies of the particle. The gravitational potential energy has been discussed in Section 4.2.3. Focus is now placed on the kinetic energy, which then provides an expression for the particle's Lagrangian function.

If the particle is at a fixed position with respect to a reference frame rotating with the earth, it experiences a constant *solid body* rotation represented by the velocity

$$\mathbf{U} = \boldsymbol{\Omega} \wedge \mathbf{x}, \tag{4.146}$$

where $\boldsymbol{\Omega}$ is the angular velocity directed upwards from the north pole, and $\mathbf{x}$ is the Cartesian position vector of the particle with respect to the earth's center (see Figure 3.1 for an illustration). Allowing the particle to move with respect to the rotating sphere yields the particle's inertial velocity

$$\dot{\mathbf{x}}_{\text{inertial}} = \dot{\mathbf{x}} + \mathbf{U}, \tag{4.147}$$

where $\dot{\mathbf{x}}$ is the particle's velocity as measured in the rotating reference frame. Correspondingly, the particle's inertial kinetic energy is given by

$$K = (m/2)\,(\dot{\mathbf{x}} + \mathbf{U})^2, \tag{4.148}$$

which leads to the Lagrangian function (defined as the difference between kinetic and potential energies)

$$\begin{aligned} L &= K - P \\ &= (m/2)\,(\dot{\mathbf{x}} + \mathbf{U})^2 - m\, g_e\, r. \end{aligned} \tag{4.149}$$

Use of the spherical coordinates defined in Figure 3.1 leads to the solid body velocity

$$\mathbf{U} = \Omega\, r\, \cos\phi\, (-\sin\lambda, \cos\lambda, 0), \tag{4.150}$$

the velocity measured in the rotating frame

$$\begin{aligned} \dot{\mathbf{x}} &= \frac{\mathrm{d}}{\mathrm{d}t}\,(x, y, z) \\ &= \frac{\mathrm{d}}{\mathrm{d}t}\,(r\,\cos\phi\cos\lambda, r\,\cos\phi\sin\lambda, r\,\sin\phi), \end{aligned} \tag{4.151}$$

and the kinetic energy

$$K = (m/2)\left[(\dot{r}^2 + r^2\dot{\phi}^2 \; + r^2\cos^2\phi\,\dot{\lambda}^2) + (\Omega^2\, r^2\,\cos^2\phi) + (2\,\Omega\, r^2\dot{\lambda}\,\cos^2\phi)\right]. \tag{4.152}$$

4.10.3 Hamiltonian dynamics

The Hamiltonian function for the particle is generally given by

$$H = \boldsymbol{\Pi} \cdot \dot{\mathbf{x}} - L, \tag{4.153}$$

where

$$\boldsymbol{\Pi} = \frac{\partial L}{\partial \dot{\mathbf{x}}} \tag{4.154}$$

is the *canonical momentum*.* For the point particle in the rotating reference frame

$$\boldsymbol{\Pi} = m\,(\dot{\mathbf{x}} + \mathbf{U}), \tag{4.155}$$

thus leading to the Hamiltonian

$$\begin{aligned} H &= m\,(\dot{\mathbf{x}} + \mathbf{U}) \cdot \dot{\mathbf{x}} - (m/2)\,(\dot{\mathbf{x}} + \mathbf{U})^2 + m\,g_e\,r \\ &= (m/2)\,\dot{\mathbf{x}}^2 + m\,r\,(g_e - \mathbf{U}^2/(2r)). \end{aligned} \tag{4.156}$$

Hence, the Hamiltonian is the sum of the kinetic energy of the particle as measured in the rotating reference frame, plus an effective gravitational potential energy arising from the earth's gravitational attraction and the centrifugal force

$$P_{\text{eff}} = m\,r\,[g_e - \mathbf{U}^2/(2r)]. \tag{4.157}$$

The gradient of P_{eff} defines the local vertical direction, $\hat{\mathbf{z}}$, perpendicular to geopotential surfaces (see Section 4.2.3).

Because the physical system exhibits symmetry with respect to constant translations in time, general principles of analytic mechanics show that the Hamiltonian remains unchanged during the particle's motion. Additionally, the physical system remains invariant under rotations about the north pole. This symmetry leads to yet another conserved quantity, the angular momentum, as shown in Section 4.10.4.

To employ the formalism of Hamiltonian dynamics for deriving the equations of motion, it is necessary to write the Hamiltonian function in terms of the canonical momentum

$$H = \frac{(\boldsymbol{\Pi} - m\,\mathbf{U})^2}{2\,m} + P_{\text{eff}}. \tag{4.158}$$

This form for the Hamiltonian suggests an interpretation of the physical system motivated by its analog to a charged particle moving in an electromagnetic field (e.g., Fetter and Walecka, 1980, section 33). The two systems have the same mathematical form for the kinetic energy, where $\mathbf{U}$ is the analog of the electromagnetic vector potential and $\nabla \wedge \mathbf{U}$ is the analog of the magnetic field. This correspondence is to be expected since the Coriolis force is mathematically analogous to the Lorentz force acting on a moving charged particle. It suggests an interpretation of the particle moving around the rotating sphere as a particle moving within the *field* associated with the solid body velocity $\mathbf{U}$ as well as the gravitational field.

Hamilton's canonical equations of motion are given by

$$\frac{\partial H}{\partial \boldsymbol{\Pi}} = \dot{\mathbf{x}} \tag{4.159}$$

$$\frac{\partial H}{\partial \mathbf{x}} = -\dot{\boldsymbol{\Pi}}. \tag{4.160}$$

For the particle, the first equation leads to an identity

$$\frac{\partial H}{\partial \boldsymbol{\Pi}} = (1/m)\,(\boldsymbol{\Pi} - m\,\mathbf{U}) = \dot{\mathbf{x}}, \tag{4.161}$$

*Do not confuse $\boldsymbol{\Pi}$ with the unrelated potential vorticity scalar defined by equation (4.118).

whereas the second equation leads to

$$\begin{aligned}\frac{\partial H}{\partial x^a} &= -m\,\dot{\mathbf{x}}\cdot\frac{\partial \mathbf{U}}{\partial x^a}+\frac{\partial P_{\text{eff}}}{\partial x^a} \\ &= -\dot{\Pi}_a.\end{aligned} \tag{4.162}$$

Taking the time derivative of equation (4.161) and substituting equation (4.162) leads to the second order equation of motion

$$m\,\ddot{\mathbf{x}} = m\,\dot{\mathbf{x}}\wedge(\nabla\wedge\mathbf{U}) - \nabla P_{\text{eff}}. \tag{4.163}$$

To reach this equation, note that the time derivative of the rotating frame's velocity is required from the frame of the moving particle. Hence, it is given by the total derivative

$$\dot{\mathbf{U}} = (\partial_t + \dot{\mathbf{x}}\cdot\nabla)\,\mathbf{U}, \tag{4.164}$$

where $\partial_t\mathbf{U} = 0$ since the sphere is rotating at a constant rate. Standard vector identities lead to

$$\nabla\wedge\mathbf{U} = 2\,\boldsymbol{\Omega}, \tag{4.165}$$

where $\boldsymbol{\Omega}$ is assumed to be spatially constant. Thus, the equation of motion takes the form

$$m\left(\frac{\mathrm{d}}{\mathrm{d}t}+2\,\boldsymbol{\Omega}\wedge\right)\dot{\mathbf{x}} = -\nabla P_{\text{eff}}, \tag{4.166}$$

as expected from a derivation of equation (4.145).

4.10.4 Conservation laws

The system's spatial symmetry is reflected by choosing spherical coordinates, in which case the Lagrangian L takes the form

$$L + P_{\text{eff}} = (m/2)\left[(\dot{r}^2 + r^2\dot{\phi}^2 + r^2\cos^2\phi\,\dot{\lambda}^2) + (2\,\Omega\,r^2\dot{\lambda}\,\cos^2\phi)\right], \tag{4.167}$$

with

$$P_{\text{eff}} = m\,g_{\text{e}}\,r - (m/2)\,(\Omega\,r\cos\phi)^2. \tag{4.168}$$

Closer connection to the fluid case is realized by introducing the velocity vector's physical components

$$(u,v,w) = ((r\cos\phi)\,\dot{\lambda}, r\,\dot{\phi}, \dot{r}), \tag{4.169}$$

which leads to

$$L = (m/2)(u^2+v^2+w^2+2\,\Omega\,u\,r\cos\phi) - P_{\text{eff}}. \tag{4.170}$$

Spherical components to the canonical momentum

$$\begin{aligned}\boldsymbol{\Pi} &= (\Pi_r, \Pi_\lambda, \Pi_\phi) \\ &= \left(\frac{\delta L}{\delta\dot{r}}, \frac{\delta L}{\delta\dot{\lambda}}, \frac{\delta L}{\delta\dot{\phi}}\right)\end{aligned} \tag{4.171}$$

are given by

$$\Pi_r = m\dot{r} \tag{4.172}$$

$$\Pi_\lambda = m\,(r\cos\phi)^2\,(\dot{\lambda}+\Omega) \tag{4.173}$$

$$\Pi_\phi = m\,r^2\,\dot{\phi}, \tag{4.174}$$

which can be brought into a more familiar form by introducing the velocity components (u, v, w)

$$\Pi_r = m\,w \tag{4.175}$$

$$\Pi_\lambda = m\,(r\,\cos\phi)\,(u + \Omega\,r\,\cos\phi) \tag{4.176}$$

$$\Pi_\phi = m\,r\,v. \tag{4.177}$$

Note the different physical dimensions of the momentum components, reflecting the different dimensions of the spherical coordinates (r, λ, ϕ).

The zonal canonical momentum Π_λ represents the angular momentum of the particle about the rotation axis extending from the south to north poles of the sphere, with $r\,\cos\phi$ the distance to the rotation axis, and $u + \Omega\,r\,\cos\phi$ the particle's total zonal velocity. Because the Lagrangian is independent of the zonal angle, λ, reflecting the system's symmetry about the north pole, Π_λ is a constant of the motion

$$\frac{\mathrm{d}\,\Pi_\lambda}{\mathrm{d}t} = 0. \tag{4.178}$$

The corresponding Hamiltonian takes the form

$$\begin{aligned} H &= (m/2)\,(\dot{r}^2 + r^2\dot{\lambda}^2\cos^2\phi + r^2\dot{\phi}^2) + P_{\text{eff}} \\ &= (m/2)\,(u^2 + v^2 + w^2) + P_{\text{eff}} \\ &= \frac{\Pi_r^2}{2\,m} + \frac{[\Pi_\lambda - m\,\Omega\,(r\,\cos\phi)^2]^2}{2\,m\,(r\,\cos\phi)^2} + \frac{\Pi_\phi^2}{2\,m\,r^2} + P_{\text{eff}}, \end{aligned} \tag{4.179}$$

which is also conserved because of time translation symmetry

$$\frac{\mathrm{d}\,H}{\mathrm{d}t} = 0. \tag{4.180}$$

Following Salmon (1998), approximations made within the Hamiltonian enable one to directly track the approximations' effects on conservation laws. Consider now two approximations: (1) the particle's motion is dominated by the horizontal, and (2) vertical excursions are much smaller than the earth's radius. The first assumption corresponds to the hydrostatic approximation, and the second to the shallow ocean approximation. Dropping the radial component to the canonical momentum leads to the Hamiltonian

$$H = (m/2)\,(u^2 + v^2) + P_{\text{eff}}. \tag{4.181}$$

Additionally, the conserved angular momentum Π_λ takes the form

$$\Pi_\lambda = m\,(R\,\cos\phi)\,(u + \Omega\,R\,\cos\phi), \tag{4.182}$$

where again $r = R + z \approx R$.

This work establishes the conserved energy and angular momentum for a point particle whose motion corresponds to a hydrostatic fluid within the shallow ocean geometry. As noted in Section 4.11, the corresponding results for the fluid parcel are modified by the following. First, the conservation laws apply over the extent of the finite fluid domain and only in the absence of friction. Second, angular momentum is conserved in the absence of zonal boundaries, since in the presence of such boundaries there is no longer rotational symmetry. It is because of zonal land/sea boundaries throughout much of the World Ocean that angular momentum is not commonly discussed in oceanography (see, however, Holloway and Rhines, 1991). It does play a more prominent role in meteorology (e.g., Holton, 1992).

4.11 SYMMETRY AND CONSERVATION LAWS

The mathematical formulation of physical processes often leads to the time tendencies associated with these processes being written as the convergence of a flux. A finite volume integration of this convergence over a grid cell leads, through use of the divergence theorem discussed in Section 3.3, to a finite difference of the fluxes across the grid cell boundaries. This formulation provides a very convenient starting point for numerical discretization since it allows one to readily build into the numerics certain conservation properties essential for physical integrity of the model.

When considering motion in a spherical geometry and the transport of non-scalar fields such as the linear momentum, one encounters tendency terms that cannot be written solely as a convergence. Instead, they acquire added terms taking the form of a source. The advective metric frequency in the budget for linear momentum (Section 4.4.1) provides one such example. As seen in Section 19.4, horizontal friction acting on linear momentum on the sphere also cannot be written solely as the convergence of a "viscous" flux across grid cells. Notably, both of these processes can be written solely as a convergence when considering fluid motion on a flat plane using Cartesian coordinates. Hence, these extra terms are associated with the underlying spherical geometry. The purpose of this section is to describe the physical and geometric reasons for these extra source-like terms.

4.11.1 Motion on an infinite plane

To get started, consider fluid motion on an infinite flat plane. In the absence of external forces that act to make a particular horizontal direction special, the environment maintains translational symmetry in either of the horizontal directions. Consequently, the horizontal momentum in either direction is conserved.* Mathematically, this result means that the linear momentum of a fluid parcel satisfies a conservation equation. That is, the forces affecting the time tendency of this momentum are represented as a total divergence. These statements take their mathematical form as the time tendency for the momentum density

$$(\rho\, u^m)_{,t} = (T^{mn} - \rho\, u^m\, u^n)_{,n} + \rho\, f^m, \tag{4.183}$$

where the tensor labels extend over the horizontal directions, Cartesian coordinates x, y are used for convenience, and conservation of mass has been used in the form

$$\rho_{,t} + (\rho\, u^n)_{,n} = 0. \tag{4.184}$$

Additionally, the stress tensor has been written

$$T^{mn} = \tau^{mn} - \delta^{mn}\, p, \tag{4.185}$$

thus representing the sum of the symmetric frictional stress tensor and a diagonal pressure stress tensor. In the absence of external forces f^m, or in the case when these forces can be derived as the divergence of a scalar, the total horizontal momentum $\int \rho\, u^m \, \mathrm{d}V$ is a constant in time.

*See Chapter II of Landau and Lifshitz (1976) for a lucid discussion of symmetries and their connection to conservation laws.

In addition to momentum in a particular direction, the discussion in Section 17.3.3 shows that, so long as the stress tensor is symmetric and there are no external forces, there is an angular momentum conservation law. For motion on the plane, this conservation law arises from symmetry of the unforced motion under rotations about the vertical axis. That is, angular momentum about the vertical direction is conserved in the absence of external forces or boundary effects. Mathematically, the conservation of angular momentum can be derived from the momentum equation in a similar manner to that used in Section 17.3.3. For completeness, the derivation is summarized here.

The angular momentum per unit volume has components given by

$$\rho \, \mathcal{L}_m = \rho \, \epsilon_{mnp} \, x^n \, u^p, \tag{4.186}$$

where ϵ_{mnp} is the Cartesian representation of the totally antisymmetric Levi-Civita tensor defined in Section 20.12. As the motion is restricted to a horizontal plane, the labels n, p are here limited to $1, 2$. Hence, $m = 3$ represents the angular momentum about the vertical axis. Using conservation of mass, the momentum equations, and symmetry of the stress tensor, it is straightforward to determine the conservation law

$$(\rho \, \mathcal{L}_m)_{,t} + (\rho \, \mathcal{L}_m \, u^p)_{,p} = \epsilon_{mnp} \, [(x^n \, T^{pq})_{,q} + x^n \, \rho \, f^p]. \tag{4.187}$$

The first term on the right-hand side takes the form of a total divergence, and the second term represents external torques. In the absence of external torques and boundary effects (i.e., when the geometry and external forcing are rotationally symmetric), then $\int \rho \, \mathcal{L}_m \, \mathrm{d}V$ is a constant in time.

4.11.2 Angular momentum about the North Pole

In general, for unforced ideal motion on a manifold containing a translational symmetry, momentum in the direction of this symmetry is conserved. Likewise, if the manifold contains an axis of symmetry, the angular momentum about this axis is conserved. In either case, the form of the equation describing the evolution of the density of the conserved quantity takes the form of a conservation law. To be more specific, for unforced motion on a sphere, the three components of angular momentum about three independent axes of the sphere are conserved. For unforced motion on a rotating sphere, angular momentum about the axis of rotation is conserved. We saw this in Section 4.10 for the motion of a particle around a sphere.

For either the rotating or the nonrotating sphere, linear momentum in any direction along the surface of the sphere is *not* conserved. The reason is that the manifold is not flat, thus removing the translational symmetry required for linear momentum conservation. This is the reason that linear momentum evolution cannot be written in a purely conservative form for fluid motion in a spherical geometry.

To mathematically illustrate these points, recall the spherical coordinate form of the budget for the zonal and meridional linear momentum density (Section 4.4)

$$(\rho \, \mathbf{u})_{,t} + \nabla \cdot (\rho \, \mathbf{v} \, \mathbf{u}) + (\mathcal{M} + f) \, (\hat{\mathbf{z}} \wedge \rho \, \mathbf{u}) = -\nabla p + \rho \, \mathbf{F^u}, \tag{4.188}$$

with $\mathcal{M} = (u/R) \, \tan \phi$ the spherical coordinate advection metric frequency. Use of mass conservation in the form $\rho_{,t} + \nabla \cdot (\rho \, \mathbf{v}) = 0$ leads to the budget for zonal velocity

$$(\partial_t + \mathbf{v} \cdot \nabla)\, u - v\, (\mathcal{M} + f) = -\rho^{-1}\, p_{,x} + F^u. \tag{4.189}$$

Multiplication by the distance to the polar axis, $R \cos \phi$, yields

$$(\partial_t + \mathbf{v} \cdot \nabla)\, (u\, R\, \cos \phi) = -(R\, \cos \phi / \rho)\, p_{,x} + v\, f\, R\, \cos \phi + R\, \cos \phi\, F^u. \tag{4.190}$$

Use of the identity

$$(\partial_t + \mathbf{v} \cdot \nabla)\, (\Omega\, R^2 \cos^2 \phi) = -f\, v\, R\, \cos \phi \tag{4.191}$$

yields

$$(\partial_t + \mathbf{v} \cdot \nabla)[\, R\, \cos \phi\, (u + R\, \Omega\, \cos \phi)] = (R\, \cos \phi)\, (-\rho^{-1}\, p_{,x} + F^u). \tag{4.192}$$

Alternatively, reuse of mass conservation gives

$$(\rho\, \mathcal{L})_{,t} + \nabla \cdot (\rho\, \mathcal{L}\, \mathbf{v}) = (R\, \cos \phi)\, (-p_{,x} + \rho\, F^u), \tag{4.193}$$

where

$$\mathcal{L} = R \cos \phi\, (u + R\, \Omega\, \cos \phi) \tag{4.194}$$

is the angular momentum per unit mass about the north polar axis for a fluid parcel subject to the shallow ocean approximation. Substituting

$$(R \cos \phi)\, p_{,x} = p_{,\lambda}, \tag{4.195}$$

and using the explicit form for the zonal friction force per volume, $\rho\, F^u$, as presented in Section 19.4, one finds that $\int (\rho\, \mathrm{d}V)\, \mathcal{L} = \int (\rho\, R^2\, \cos \phi\, d\lambda\, d\phi)\, \mathcal{L}$ is a constant for latitudes and depths where a full zonal circuit about the sphere is possible. Such geometry is common in the atmosphere, yet it holds only for the Antarctic Circumpolar Current in the ocean.

4.11.3 Advective and frictional metric terms

It is useful to provide a mathematical statement concerning the origin of the advective and frictional metric terms. For this purpose, consider the linear momentum of a fluid parcel moving on an arbitrary manifold. To describe such motion, it is useful to employ the tensor analysis considered in Section 4.4 and Chapters 20 and 21.* This material gives

$$\rho \frac{\mathrm{d}u^m}{\mathrm{d}t} = T^{mn}_{;n} + \rho\, f^m. \tag{4.196}$$

Acceleration of the parcel takes the form

$$\begin{aligned} \rho\, \frac{\mathrm{d}u^m}{\mathrm{d}t} &= \rho(u^m_{,t} + u^n\, u^m_{;n}) \\ &= (\rho\, u^m)_{,t} + (\rho\, u^m\, u^n)_{;n} \end{aligned} \tag{4.197}$$

where conservation of mass on a curved manifold has been used: $\rho_{,t} + (\rho\, u^n)_{;n} = 0$. Hence, the time tendency for the momentum density is given by

$$(\rho\, u^m)_{,t} = (T^{mn} - \rho\, u^m\, u^n)_{;n} + \rho\, f^m. \tag{4.198}$$

*See especially the summary of tensor analysis given in Section 21.12.

This equation is written in the same form as the Cartesian equivalent (4.183), except that now the derivatives are covariant and so contain information about the generally curved manifold. Expanding the covariant divergence using equation (17.173) yields

$$(\rho\, u^m)_{,t} = (\sqrt{\mathcal{G}})^{-1}\, [\sqrt{\mathcal{G}}\, (T^{mn} - \rho\, u^m\, u^n)]_{,n} + (T^{ab} - \rho\, u^a\, u^b)\, \Gamma^m_{ab} + \rho\, f^m. \tag{4.199}$$

The Christoffel symbol Γ^m_{np} accounts for the spatial dependence of the basis vectors which define the directions on the manifold. Recalling the form of the stress tensor T^{mn} given by equation (4.185), we identify the term $\Gamma^m_{ab}\, \tau^{ab}$ as the "frictional metric term" and the $\rho\, u^a\, u^b\, \Gamma^m_{ab}$ term as the "advective metric term."

As discussed in Section 21.10, integration of a quantity over the volume of a finite grid cell in arbitrary coordinates means performing an integral of the form

$$\int \psi\, \mathrm{d}V = \int \sqrt{\mathcal{G}}\, \mathrm{d}\xi^1\, \mathrm{d}\xi^2\, \mathrm{d}\xi^3\, \psi, \tag{4.200}$$

where

$$\mathrm{d}V = \sqrt{\mathcal{G}}\, \mathrm{d}\xi^1\, \mathrm{d}\xi^2\, \mathrm{d}\xi^3 \tag{4.201}$$

is the coordinate invariant volume element. For example, in spherical coordinates

$$\mathrm{d}V = R^2 \cos\phi\, d\lambda\, d\phi\, \mathrm{d}z. \tag{4.202}$$

Hence, integration of the metric terms in equation (4.199) over a finite volume does not lead to them being transformed into the difference of terms across domain boundaries. However, the first term in equation (4.199) can be written in such a way. For flat space using Cartesian coordinates, the metric terms drop out, thus recovering the results discussed in Section 4.11.1.

4.12 CHAPTER SUMMARY

The main purpose of this chapter was to derive dynamical equations describing the motion of fluid parcels on a rotating spherical earth and to introduce various results following from these equations. Building experience with the dynamical equations is important for those who are developing ocean model algorithms and for those aiming to understand simulations.

Section 4.2 began by describing particle motion observed from a reference frame on a rotating sphere, with the Coriolis force

$$\mathbf{F}_{\mathrm{c}} = -2\, (\rho\, \mathrm{d}V)\, \boldsymbol{\Omega} \wedge \mathbf{v} \tag{4.203}$$

playing a critical role in geophysical fluid dynamics. After a brief introduction in Section 4.3 to some formal ideas in continuum mechanics, Section 4.4 derived the evolution equation for linear momentum density of a fluid parcel. Without the shallow ocean approximation discussed in Section 3.2, this evolution equation is given by

$$(\rho\, \mathbf{v})_{,t} + \nabla \cdot (\rho\, \mathbf{v}\, \mathbf{v}) + \mathcal{M}\, (\hat{\mathbf{z}} \wedge \rho\, \mathbf{v}) = -\rho\, g\, \hat{\mathbf{z}} - \boldsymbol{\Omega} \wedge \rho\, \mathbf{v} - \nabla p + \rho\, \mathbf{F}^{(\mathbf{v})}. \tag{4.204}$$

In this equation,

$$\mathbf{v} = (\mathbf{u}, w) \tag{4.205}$$

is the three-dimensional velocity of a fluid parcel, and $\rho\,\mathbf{v}$ is its linear momentum per unit volume. $\boldsymbol{\Omega}$ is the earth's angular rotation rate with direction upward from the North Pole (see Figure 3.1). Its presence arises from taking a noninertial reference frame fixed on the rotating spherical earth. The advection metric frequency

$$\mathcal{M} = v\,\partial_x \ln \mathrm{d}y - u\,\partial_y \ln \mathrm{d}x \tag{4.206}$$

is a source accounting for the nonconservation of linear momentum on the sphere. Notably, neither the Coriolis force nor that associated with the advection metric frequency does work on the parcel, and so they do not alter its energy. Internal forces arise from the nontrivial extent of fluid parcels, thus allowing them to experience deformations such as shears and strains. The stress tensor embodies these interactions, and its divergence gives rise to a pressure gradient force per unit volume $-\nabla p$ as well as a frictional force per volume $\rho\,\mathbf{F}^{(\mathbf{v})}$. Pressure is a force per unit area directed normal to a surface, and friction results from the irreversible viscous exchange of momentum. Finally, $-\hat{\mathbf{z}}\,\rho\,g$ is the gravitational force per unit volume acting on the parcel.

Section 4.4.4 focused on the velocity vector $\mathbf{v}$ rather than the linear momentum density $\rho\,\mathbf{v}$, which led to the *vector invariant* form of the velocity equation (again written here without the shallow ocean approximation)

$$[\partial_t + (\boldsymbol{\omega} + \boldsymbol{\Omega})\wedge]\,\mathbf{v} = \nabla[\Phi - \mathbf{v}\cdot\mathbf{v}/2] - \rho^{-1}\,\nabla p + \mathbf{F}^{(\mathbf{v})}, \tag{4.207}$$

where $\boldsymbol{\omega} = \nabla \wedge \mathbf{v}$ is the relative vorticity. This form of the velocity equation is the starting point for many ocean models, especially those formulated on a C-grid.* In contrast, most B-grid models employ the *advective* form which follows from equation (4.204). The vector-invariant velocity equation also forms the starting point when deriving results related to vorticity and potential vorticity, as seen in Section 4.9.

Nonradial components of the Coriolis force are dropped when making the shallow ocean approximation (Section 3.2.5), so that

$$\boldsymbol{\Omega} \wedge (\rho\,\mathbf{v}) \rightarrow \hat{\mathbf{z}}\,f \wedge (\rho\,\mathbf{v}) \tag{4.208}$$

with

$$f = 2\,\Omega\,\sin\phi \tag{4.209}$$

the Coriolis parameter measuring the strength of the Coriolis force. The momentum budget with the shallow ocean approximation is given by equation (4.58)

$$(\rho\,\mathbf{v})_{,t} + \nabla\cdot(\rho\,\mathbf{v}\,\mathbf{v}) + \mathcal{M}\,(\hat{\mathbf{z}} \wedge \rho\,\mathbf{v}) = -\rho\,g\,\hat{\mathbf{z}} - f\,\hat{\mathbf{z}} \wedge \rho\,\mathbf{v} - \nabla p + \rho\,\mathbf{F}^{(\mathbf{v})}. \tag{4.210}$$

In addition to simplifying the Coriolis force, it is common in large-scale dynamics to assume that the vertical momentum equation is dominated by the hydrostatic balance

$$p_{,z} = -\rho\,g. \tag{4.211}$$

The shallow ocean and hydrostatic balance lead to the evolution of horizontal linear momentum density

$$(\rho\,\mathbf{u})_{,t} + \nabla\cdot(\rho\,\mathbf{v}\,\mathbf{u}) + (f + \mathcal{M})\,(\hat{\mathbf{z}} \wedge \rho\,\mathbf{u}) = -\rho\,g\,\hat{\mathbf{z}} - \nabla_z p + \rho\,\mathbf{F}^{(\mathbf{u})}. \tag{4.212}$$

---

*See Section 10.2 for a discussion of B-grid and C-grid models.

This equation is the momentum equation for the ocean's primitive equations.* It forms the basis for the discretized ocean model dynamics discussed in Chapter 12.

The remaining sections focused on developing some theoretical and practical experience with the basic equations. To get started, Section 4.5 explored properties of hydrostatic pressure. As stated above, fluids satisfying the primitive equations maintain the hydrostatic balance, and so one often places the adjective "hydrostatic" in front of "primitive equations." In this section, some attention was given to understanding the different behavior of hydrostatic pressure at the ocean bottom for Boussinesq (volume conserving) versus non-Boussinesq (mass conserving) fluids. A Boussinesq fluid possesses some unphysical properties associated with *virtual mass fluxes* set up in response to the constraint that parcels conserve volume instead of mass.

From an algorithmic perspective, the hydrostatic balance prompts one to distinguish a vertically integrated flow from one depending on depth. The evolution equations for these hydrostatic fluid columns, colloquially known as the *external* or *barotropic* modes, were derived in Section 4.6, with the result given by equation (4.79)

$$(\partial_t + f\,\hat{\mathbf{z}} \wedge)\,\tilde{\mathbf{U}} = -(H+\eta)\,\nabla\,(p_{\mathrm{a}} + p_{\mathrm{s}})/\rho_o + \mathbf{G}. \tag{4.213}$$

In this equation, $\mathbf{G}$ represents the vertical integral of the nonlinear forcing, as given by equation (4.80), and $\tilde{\mathbf{U}}$ is the vertical integral of the horizontal momentum density

$$\rho_o\,\tilde{\mathbf{U}} = \int_{-H}^{\eta} \mathrm{d}z\,\rho\,\mathbf{u}. \tag{4.214}$$

The pressure p_{a} is that applied to the ocean surface by sea ice and/or atmospheric loading, and p_{s} is the hydrostatic pressure at $z = 0$ due to seawater between $z = 0$ and the free surface at $z = \eta$. The linearized form of these equations reduces to the momentum equation for a shallow water fluid.

Sections 4.7 and 4.8 presented some phenomenology relevant for large-scale ocean fluid dynamics. Here, we encountered the Rossby number, which measures the relative strength of advective accelerations to Coriolis accelerations, and the Richardson number, which measures the relative strength of stratification to vertical shear in the horizontal velocity. Some of the scaling relations that form the basis for quasi-geostrophy were more generally discussed. These sections were provided to give the reader a taste for various scales and notions critical for understanding ocean dynamics.

Section 4.9 considered various forms of potential vorticity corresponding to different approximate versions of the momentum equations. Potential vorticity is materially conserved for ideal fluid motions. It has proven to be an extremely useful scalar field for understanding fundamental aspects of geophysical fluid dynamics. Although it can be thought of as a tracer due to its being materially conserved for ideal flow, its form is very special:

$$\begin{aligned} \Pi &= \rho^{-1}\,\underline{\boldsymbol{\omega}} \cdot \nabla\psi \\ &= \rho^{-1}\,\nabla \cdot (\underline{\boldsymbol{\omega}}\,\psi), \end{aligned} \tag{4.215}$$

*The evolution of tracers, discussed in Section 5.1, also forms part of the ocean primitive equations.

where the second expression follows because $\nabla \cdot \underline{\omega} = 0$. Hence, $\rho \Pi$ is a total divergence. This property leads to some rather novel, and non-tracer-like, properties, as highlighted in the papers by Haynes and McIntyre (1987) and Haynes and McIntyre (1990).

Section 4.10 derived the evolution equation for a particle moving on the surface of a sphere. The derivation used Hamilton's principle, rather than Newton's laws, as Hamilton's principle allows us to focus on the connection between symmetries and conservation laws. The main point of this diversion into classical particle mechanics was to clarify the form for the energy and angular momentum of a particle whose vertical accelerations are far smaller than its horizontal ones, and whose vertical excursions are far smaller than the earth's radius. This motion corresponds to a hydrostatic fluid with the shallow ocean approximation. Section 4.11 followed up on this discussion by illustrating why linear momentum is not conserved on a sphere, even for ideal fluid motions, whereas it is conserved on a plane. The reason is that linear translations do not represent a symmetry of the sphere, and so total linear momentum is not conserved, even in the absence of friction or boundaries. Instead, total angular momentum is the appropriate conserved momentum on the sphere. The lack of conservation for total linear momentum means that the evolution equation for linear momentum density contains a source term arising from the spherical geometry. This is the *advective metric frequency* given by equation (4.206).

Chapter Five

THERMO-HYDRODYNAMICS

The mechanical equations from Chapters 3 and 4 are combined here with the notions of thermodynamics to formulate a *thermo-hydrodynamical* description of the ocean fluid. Along the way, we present ocean energetics as well as the evolution equation for oceanic material constituents, such as salts and other chemical species as well as biological matter. Upon completion of this and the previous two chapters, the reader should have a firm understanding of both the mechanical and thermodynamical aspects of the equations used to describe the ocean.

5.1 GENERAL TYPES OF OCEAN TRACERS

For considerations of non-equilibrium thermodynamics later in this chapter, we introduce a temperature-like scalar field that measures the heat content within a parcel of seawater. This field evolves in a manner similar to other ocean *tracers*. Therefore, it is useful to introduce here the various forms of ocean tracers and to derive their evolution equation.

There are three general types of ocean tracers: (1) tracers representing the concentration of material constituents in the fluid (e.g., passive tracers and salinity); (2) tracers representing the thermodynamic properties of the fluid (e.g., entropy, enthalpy, and temperature); and (3) tracers embodying dynamical properties of the fluid (e.g., potential vorticity discussed in Section 4.9).

5.1.1 Material tracers

Fluid parcels generally contain material constituents that are transported as the parcel moves through the ocean. These constituents are commonly termed *tracers* since they are useful for *tracing* pathways of fluid transport.*

Important *passive tracers* include the biological species of phytoplankton and zooplankton, which form the base of the ocean's food chain; radioactive isotopes input to the ocean due to nuclear bomb testing in the 1950's and 1960's; and purposely released tracers, such as sulfur hexafluoride (SF_6), used to quantify rates of mixing (Ledwell et al., 1993). *Passive* here refers to the near inability of these tracers to alter ocean density. Hence, they have negligible effects on ocean dynamics.

Salinity is a generic term referring to the large number of ionic constituents in seawater. The dominant ions are sodium Na^{+}, chloride Cl^{-}, sulfate SO_4^{--}, and magnesium Mg^{++}. At the common salinity of 35 parts "salt" per thousand parts solution, Table 4.1 of Apel (1987) indicates there are roughly 19.345 grams of chlo-

*A similar terminology is commonly used in the medical sciences when probing human physiology using chemical or radioactive tracers.

ride ion, 10.752 grams of sodium ion, 1.295 gram of magnesium ion, and 2.701 grams of sulfate ion per kilogram of water. These, and the other trace ions, maintain a near constant ratio throughout the World Ocean, hence allowing for the single term "salinity" to have near global validity. Since salinity affects the ocean density field, it is termed an *active* tracer, along with temperature. It is notable that ocean biologists often reserve the term active for biologically active constituents. The terminology used in this book is consistent with that used in the physical oceanography community.

Material tracers have a total mass within a fluid parcel written

$$\mathrm{d}M_C = (\mathrm{d}M)\, C, \tag{5.1}$$

where $\mathrm{d}M = \rho\,\mathrm{d}V$ is the total mass of the parcel of seawater (including fresh water and all material constituents), and

$$C = \frac{\mathrm{d}M_C}{\mathrm{d}M} \tag{5.2}$$

$$= \frac{\rho_C}{\rho} \tag{5.3}$$

is the mass of tracer per mass of seawater. The second equality introduced the density of tracer C per volume. C is often termed the *tracer concentration* or *tracer mass fraction*. Notably, since the seawater mass and tracer mass are both scalar quantities, the tracer concentration is likewise a scalar field. Hence, tracers are often called *scalars* in the fluid dynamics literature.*

Conserving tracer mass within a parcel of seawater leads to the conservation statement†

$$\begin{aligned} \frac{\mathrm{d}\,[(\rho\,\mathrm{d}V)\,C]}{\mathrm{d}t} &= (\rho\,\mathrm{d}V)\,\frac{\mathrm{d}C}{\mathrm{d}t} \\ &= 0, \end{aligned} \tag{5.4}$$

where the total parcel mass is also assumed to be constant: $\mathrm{d}(\rho\,\mathrm{d}V)/\mathrm{d}t = 0$. Combining the Eulerian form of material tracer conservation $\mathrm{d}C/\mathrm{d}t = (\partial_t + \mathbf{v}\cdot\nabla)\,C = 0$ with mass conservation $\rho_{,t} + \nabla\cdot(\rho\,\mathbf{v}) = 0$ leads to the Eulerian conservation law for tracer mass per volume $\rho\,C$ in a seawater parcel

$$(\rho\,C)_{,t} + \nabla\cdot(\rho\,C\,\mathbf{v}) = 0. \tag{5.5}$$

As discussed below, additional terms appear on the right-hand side when sources or sinks of tracer are found within the fluid or at the boundaries.

In most numerical ocean models, tracer concentration is time-stepped by discretizing the Eulerian equation (5.5) for the tracer mass per volume, and separately time-stepping the conservation for ocean mass per volume $\rho_{,t} + \nabla\cdot(\rho\,\mathbf{v}) = 0$. Compatibility between the two discretized conservation laws is maintained by having the discrete equation for tracer mass per volume reduce to the discrete

*See Section 20.5 starting on page 451 for a mathematical definition of scalar fields.

†As noted by DeGroot and Mazur (1984) (see their Section II.2), the velocity of a constituent within a parcel need not be the same as the velocity of the center of mass of the parcel. In this more general case, tracer conservation becomes $\mathrm{d}C/\mathrm{d}t = -\nabla\cdot\mathbf{J}_C$, where $\mathbf{J}_C = \rho_C\,(\mathbf{v}_C - \mathbf{v})$. In our treatment, it is assumed that constituents of an ideal fluid parcel move with the velocity of the parcel's center of mass. Nonideal behavior associated with mixing introduces diffusive fluxes that generally depend on properties of the constituent (equation (5.7)).

mass per volume equation upon setting tracer concentration to unity: $C = 1$. Such compatibility is very important to maintain numerically to eliminate spurious sources and sinks of mass and tracer.

For purposes of establishing the momentum and tracer budgets, the mass of a Boussinesq seawater parcel is given by $\mathrm{d}M = \rho_o \,\mathrm{d}V$. Material conservation of tracer mass $(\rho_o \,\mathrm{d}V)\, C$ still leads to material conservation of tracer concentration: $\mathrm{d}C/\mathrm{d}t = 0$, since now the parcel's volume is conserved. Combining material conservation of tracer concentration with parcel volume conservation, $\nabla \cdot \mathbf{v} = 0$, leads to the Eulerian conservation law

$$C_{,t} + \nabla \cdot (\mathbf{v}\, C) = 0. \tag{5.6}$$

As for the mass conserving fluid, this equation is especially useful as a numerical starting point since it manifests the need to maintain compatibility between volume and tracer conservation.

There are two general ways that the mass of tracer within a seawater parcel can change. First, there can be tracer-dependent sources or sinks of tracer mass within the fluid domain or at the boundaries. Second, there can be irreversible molecular or turbulent mixing effects that move tracer mass between parcels. Many of these mixing effects can be mathematically represented by the convergence of a tracer flux $\mathbf{J}$ whose details generally depend on properties of the tracer. Such fluxes are often called *non-advective* fluxes to distinguish them from the advective flux $\rho\, C\, \mathbf{v}$. For a non-Boussinesq fluid, the time tendency for density weighted tracer concentration is written

$$(\rho\, C)_{,t} = -\nabla \cdot (\rho\, C\, \mathbf{v} + \mathbf{J}) + \Sigma, \tag{5.7}$$

where Σ is the tracer source. For a Boussinesq fluid, this equation takes the form

$$C_{,t} = -\nabla \cdot (C\, \mathbf{v} + \mathbf{J}/\rho_o) + \Sigma/\rho_o. \tag{5.8}$$

Because the ρ_o factor is awkward to carry around, it is often convenient to introduce the tracer concentration flux $\mathbf{F}$ via

$$\mathbf{J} = \rho\, \mathbf{F}. \tag{5.9}$$

For material tracers, the tracer flux $\mathbf{J}$ has dimensions of tracer mass per area per time, whereas the tracer concentration flux $\mathbf{F}$ has dimensions of tracer concentration times velocity. In a similar fashion, it is useful to introduce a tracer concentration source $\mathcal{S}$

$$\Sigma = \rho\, \mathcal{S}. \tag{5.10}$$

With these new terms, the tracer equation for a non-Boussinesq fluid takes the form

$$(\rho\, C)_{,t} = -\nabla \cdot (\rho\, C\, \mathbf{v} + \rho\, \mathbf{F}) + \rho\, \mathcal{S}, \tag{5.11}$$

whereas for a Boussinesq fluid it is

$$C_{,t} = -\nabla \cdot (C\, \mathbf{v} + \mathbf{F}) + \mathcal{S}. \tag{5.12}$$

5.1.2 Thermodynamical tracers

The density of a seawater parcel is a function of temperature along with salinity and pressure. The functional relation

$$\rho = \rho(T, s, p) \tag{5.13}$$

is known as the *equation of state* for seawater (Section 5.7). In this relation, T is the *in situ* temperature, which is the temperature that a thermometer measures at the pressure and salinity of the fluid parcel maintained at its local environment. Since density affects pressure via the hydrostatic balance, variations in temperature and salinity yield important *thermo-haline* forces for ocean currents. Therefore, it is crucial to provide an accurate and precise description of the temperature and salinity distributions to understand and simulate ocean currents.

Developing an understanding of heat in the ocean requires some of the tools of thermodynamics to be discussed in subsequent sections. Thermodynamics is a phenomenological discipline whose fundamentals lie in statistical mechanics (e.g., Huang, 1987; Reichl, 1987). As discussed in Section 2.2, geophysical fluid dynamical systems of interest for ocean climate modeling involve parcels of seawater in quasi-thermodynamical equilibrium. Hence, although these fluid parcels evolve in time, the laws of thermodynamics can be used to establish balances of heat, work, internal energy, and entropy within the fluid. Use of equilibrium thermodynamics for time-dependent phenomena falls under the area of quasi-equilibrium thermodynamics (or linear irreversible thermodynamics). The methods of irreversible thermodynamics, as articulated by such textbooks as DeGroot and Mazur (1984) and Landau and Lifshitz (1987), are commonly used in oceanography to derive equations expressing the evolution of heat within a parcel of seawater.

The traditional manner for describing heat within the ocean is via *potential temperature* (Section 5.6.1), and its evolution is typically simulated by ocean models. The reason for preferring potential temperature is that it is conserved under adiabatic and isentropic motion, whereas *in situ* temperature changes due to pressure effects even without the transfer of heat. Conservative equations are generally more amendable to straightforward physical interpretations in terms of flux divergences leading to local temporal changes. They also lend themselves to standard numerical discretizations of the transport operators. Hence, preference is given to a conservative quantity for describing heat transport in the ocean, and potential temperature has served this purpose for sometime.

Even though potential temperature is standard in the oceanographic literature, work by McDougall (2003) highlights the fact that potential temperature is less conservative in the ocean than is *potential enthalpy*. Therefore, a more accurate description of ocean heat content, and hence the density of a fluid parcel in an ocean model, is obtained by using potential enthalpy rather than potential temperature. This point is explored further in Section 5.6.

5.1.3 Dynamical tracers

Dynamical tracers form the third tracer class. These tracers are built from generally nonlinear combinations of the dynamical fields. Potential vorticity, discussed in Section 4.9, is an important dynamical tracer for geophysical fluid dynamics.

5.2 BASIC EQUILIBRIUM THERMODYNAMICS

When discussing the continuum hypothesis in Section 2.2, it was noted that a fluid parcel represents a macroscopically small yet microscopically large system in quasi-thermodynamical equilibrium. Hence, there are innumerable molecular degrees of freedom that are averaged over when describing a fluid as a continuum. Internal energy embodies the energy of these internal degrees of freedom, such as arise from thermal agitation and molecular interactions. The total energy of a macroscopic system in quasi-thermodynamical equilibrium is thus the sum of the system's kinetic, potential, *and* internal energies.

The methods of equilibrium thermodynamics allow one to relate small changes in work and heat applied to a parcel of seawater to changes in its internal energy. For many oceanographic purposes, seawater can be considered a two-component fluid consisting of fresh water and salt, where salt is a generic term representing a suite of dissolved ions in seawater whose ratio is nearly constant over the ocean (Section 5.1.1). Therefore, following Fofonoff (1962) as well as Davis (1994a), it is useful to consider thermodynamical relations for a multi-component fluid, and then to specialize to the two-component case.

5.2.1 First Law of thermodynamics

The First Law of thermodynamics establishes a relationship between infinitesimal changes of a system's internal energy to changes in the system's heat, work, and composition

$$d\mathcal{I}^{e} = \delta q + \delta w + \sum_{k} \mu_k \, dM_k. \tag{5.14}$$

In this equation, $\mathcal{I}^{e}$ is the system's internal energy, q is the heat applied to the system, w is the work applied to the system, μ_k is the chemical potential associated with species component k, and M_k is the mass of component k. As discussed in Section 5.2.3, the internal energy is proportional to the size of the system, with systems having more volume and mass having more energy. Quantities of this sort are termed *extensive* and are labeled with the superscript e in the following.

Changes in heat and work applied to a system, δq and δw, depend on details of the path taken to realize the changes. This path dependence motivates the use of the δ symbol to signify an *inexact* differential. Their sum is path-independent, thus identifying $d\mathcal{I}^{e}$ as an *exact* differential just like differential changes in mass of a species, dM_k, or differential changes in the mechanical energy, $d\,[M\,(\mathbf{v}^2/2 + \Phi)]$. Correspondingly, internal energy is known as a thermodynamic *state function* since its value depends only on properties of the equilibrium state, not the path used to reach that state.

There are many ways that forces can do work to a parcel of seawater. However, focus here is on work associated with quasi-static* changes to the parcel's volume† V^{e}

$$\delta w = -p \, dV^{e}, \tag{5.15}$$

**Quasi-static* refers to an idealized situation whereby a thermodynamic system moves from one state to another via an infinite number of intermediate equilibrium states. See Callen (1985) for details.

†The symbol V^{e} for the parcel volume is used in this section, instead of dV, in order to maintain consistency with the thermodynamic literature.

where p is the pressure applied to the boundaries of the parcel. The negative sign in the pressure-work relation (5.15) arises since compressing the fluid parcel into a smaller volume ($\mathrm{d}V^{\mathrm{e}} < 0$) requires positive work be applied to the parcel ($\delta w > 0$). Pressure is termed an intensive variable, measuring the *intensity* of a force conjugate to the extensive variable V^{e}. Pressure also provides an *integrating factor* that relates the inexact differential δw to the exact differential $\mathrm{d}V^{\mathrm{e}}$. Note that for a Boussinesq seawater parcel, its incompressibility means that no pressure-work can be applied to this parcel, since in this case $\mathrm{d}V^{\mathrm{e}} = 0$ by definition.

Just as for work, heating applied to the parcel is assumed to occur in a quasi-static manner. For such processes, heating is related to changes in entropy via

$$\delta q = T\,\mathrm{d}\zeta^{\mathrm{e}}, \tag{5.16}$$

with T the absolute *in situ* temperature (an intensive variable as well as an integrating factor) and ζ^{e} the entropy of the system (an extensive variable and a state function).

5.2.2 Fundamental thermodynamic relation

With work limited to quasi-static pressure-work associated with volume changes, and heat applied quasi-statically, substitution of relations (5.15) and (5.16) into the First Law (5.14) leads to

$$\mathrm{d}\mathcal{I}^{\mathrm{e}} = T\,\mathrm{d}\zeta^{\mathrm{e}} - p\,\mathrm{d}V^{\mathrm{e}} + \sum_{k} \mu_k\,\mathrm{d}M_k. \tag{5.17}$$

This relation holds between infinitesimal changes in thermodynamical state functions. Hence, although derived for quasi-static processes from the First Law using connections to work and heat, equation (5.17) holds for arbitrary infinitesimal changes; its connection to the First Law holds only for quasi-static processes.

Equation (5.17) represents an important *fundamental thermodynamic* relation, often called the *first Gibbs relation*. It is the starting point for many manipulations in thermodynamics. In particular, it leads to the following relations between intensive variables and the partial derivatives of internal energy with respect to extensive quantities

$$\left(\frac{\partial \mathcal{I}^{\mathrm{e}}}{\partial \zeta^{\mathrm{e}}}\right)_{V^{\mathrm{e}},M_k} = T \tag{5.18}$$

$$\left(\frac{\partial \mathcal{I}^{\mathrm{e}}}{\partial V^{\mathrm{e}}}\right)_{\zeta^{\mathrm{e}},M_k} = -p \tag{5.19}$$

$$\left(\frac{\partial \mathcal{I}^{\mathrm{e}}}{\partial M_k}\right)_{\zeta^{\mathrm{e}},V^{\mathrm{e}}} = \mu_k. \tag{5.20}$$

In these expressions, partial derivatives are taken with the noted variables held constant.

5.2.3 Internal energy and homogeneous functions

The fundamental thermodynamic relation (5.17) indicates that internal energy is naturally considered a function of entropy, volume, and mass

$$\mathcal{I}^{\mathrm{e}} = \mathcal{I}^{\mathrm{e}}(\zeta^{\mathrm{e}}, V^{\mathrm{e}}, M_k). \tag{5.21}$$

Now scale the system by an arbitrary parameter λ. Under this operation, the extensive variables entropy, volume, and mass scale by the same scale factor. Through the fundamental relation (5.17), the internal energy scales likewise, giving

$$\mathcal{I}^{e}(\lambda\,\zeta^{e},\lambda\,V^{e},\lambda\,M_{k}) = \lambda\,\mathcal{I}^{e}(\zeta^{e},V^{e},M_{k}). \tag{5.22}$$

A function that scales in this way is termed a *homogeneous function of degree one*. Differentiating both sides of this identity with respect to λ, setting λ to unity, and using the partial derivative identities (5.18)–(5.20) yields

$$\mathcal{I}^{e} = T\,\zeta^{e} - p\,V^{e} + \sum_{k} \mu_{k}\,M_{k}. \tag{5.23}$$

This result represents a special case of Euler's theorem of homogeneous functions. Taking the differential of this equation, and using the fundamental relation (5.17) leads to the *Gibbs-Duhem* relation

$$\zeta^{e}\,\mathrm{d}T - V^{e}\,\mathrm{d}p + \sum_{k} M_{k}\,\mathrm{d}\mu_{k} = 0. \tag{5.24}$$

5.2.4 Fundamental relation and specific quantities

For many purposes in fluid dynamics, it proves convenient to consider fundamental thermodynamic relations for a system of unit mass. For this purpose, one scales away the mass of the system by setting the scale factor $\lambda = M^{-1}$ and introducing the *specific* quantities

$$\mathcal{I}^{e} = M\,\mathcal{I} \tag{5.25}$$

$$\zeta^{e} = M\,\zeta \tag{5.26}$$

$$V^{e} = M\,\rho^{-1} \tag{5.27}$$

$$M_{k} = M\,C_{k}. \tag{5.28}$$

In the last equality, C_k is the mass fraction or concentration of species k in the seawater parcel that was introduced in equation (5.2). Substituting the specific quantities into the fundamental relation (5.17) and using expression (5.23) for the internal energy leads to the fundamental relation in terms of intensive state variables

$$\mathrm{d}\mathcal{I} = T\,\mathrm{d}\zeta - p\,\mathrm{d}\rho^{-1} + \sum_{k} \mu_{k}\,\mathrm{d}C_{k}. \tag{5.29}$$

As mentioned earlier, seawater can be considered a binary system of salt and fresh water. In this case,

$$C_{\text{salt}} + C_{\text{water}} = 1. \tag{5.30}$$

Introducing this constraint into (5.29) leads to

$$\mathrm{d}\mathcal{I} = T\,\mathrm{d}\zeta - p\,\mathrm{d}\rho^{-1} + \mu\,\mathrm{d}C, \tag{5.31}$$

where $C = C_{\text{salt}}$ is the concentration of salt, and

$$\mu = \mu_{\text{salt}} - \mu_{\text{water}} \tag{5.32}$$

is the relative chemical potential. The *salinity* s, with units parts per thousand, is related to C_{salt} via

$$s = 1000\,C_{\text{salt}}. \tag{5.33}$$

The range of salinity in the ocean (roughly, $0 \leq s \leq 40$) is more convenient than the range of C_{salt}, and so salinity is more commonly used in oceanography. Note that in some treatments of the equation of state for seawater, the "salinity" is taken to be $1004.867\, C_{\text{salt}}$ (Section 5.7.1). Care should be taken to ensure that the units assumed in an ocean model for salinity are consistent with the form of the model's equation of state.

5.2.5 Adiabatic lapse rate

Consider a finite mass of water with uniform and fixed salinity (e.g., a fresh water lake). Let this water be static and rest in a very thin horizontal layer so each water parcel has approximately the same gravitational potential energy. Consequently, all fluid parcels experience the same pressure and hence the same temperature. Now isentropically (i.e., adiabatically and reversibly) rearrange the fluid into a raised vertical column. Doing so requires work to be done against the gravitational field since the center of mass has increased in height. The result is a column of fluid with increased gravitational potential energy. Gravity imparts a hydrostatic pressure (Section 4.5) that makes pressure at the bottom of the fluid column greater than at the top. One may then ask how this difference in pressure affects temperature in the fluid. To quantify this effect, it is necessary to perform some standard thermodynamic manipulations valid for a single component fluid, such as fresh water (all manipulations here are for C =constant).

To start, introduce the specific heat at constant pressure

$$\begin{aligned} C_p &\equiv (\delta q / \mathrm{d}T)_p \\ &= T\, (\partial \zeta / \partial T)_p \end{aligned} \tag{5.34}$$

where the p subscript indicates that pressure is held fixed when varying temperature. Specific heat measures the change in heat associated with a change in temperature, here taken with the fluid pressure held fixed. The second form of this expression arises by assuming that changes in heat occur quasi-statically, hence allowing for use of the relation between heating and entropy $\delta q = T\, d\zeta$. With entropy a function of temperature and pressure, infinitesimal changes in entropy are given by

$$\mathrm{d}\zeta = (\partial \zeta / \partial T)_p\, \mathrm{d}T + (\partial \zeta / \partial p)_T\, \mathrm{d}p. \tag{5.35}$$

Substituting the definition of heat capacity leads to

$$T\, \mathrm{d}\zeta = C_p\, \mathrm{d}T + T\, (\partial \zeta / \partial p)_T\, \mathrm{d}p. \tag{5.36}$$

It is useful to eliminate $(\partial \zeta / \partial p)_T$ in favor of a more easily measurable quantity. For this purpose, note that use of the fundamental thermodynamic relation (5.31) (with $\mathrm{d}C = 0$) leads to

$$T\, (\partial \zeta / \partial T)_p = (\partial \mathcal{I} / \partial T)_p + p\, (\partial \rho^{-1} / \partial T)_p. \tag{5.37}$$

Likewise, equation (5.31) implies

$$T\, (\partial \zeta / \partial p)_T = (\partial \mathcal{I} / \partial p)_T + p\, (\partial \rho^{-1} / \partial p)_T. \tag{5.38}$$

Applying $(\partial / \partial p)_T$ to equation (5.37) and $(\partial / \partial T)_p$ to equation (5.38), and then subtracting, leads to the identity

$$(\partial \zeta / \partial p)_T = -(\partial \rho^{-1} / \partial T)_p. \tag{5.39}$$

Introducing the thermal expansion coefficient

$$\begin{aligned} \alpha_T &= -\rho^{-1} (\partial \rho / \partial T)_p \\ &= \rho (\partial \rho^{-1} / \partial T)_p \end{aligned} \tag{5.40}$$

yields an expression for changes in entropy in terms of changes in temperature and pressure

$$\begin{aligned} T \, \mathrm{d}\zeta &= C_p \, \mathrm{d}T - T (\partial \rho^{-1} / \partial T)_p \, \mathrm{d}p \\ &= C_p \, \mathrm{d}T - (T \, \alpha_T / \rho) \, \mathrm{d}p. \end{aligned} \tag{5.41}$$

C_p and α_T are readily measurable *response functions*, thus making equation (5.41) a very useful expression for infinitesimal entropy changes. Note that a subscript T to α distinguishes this expansion coefficient from that defined according to potential temperature (Section 5.7.1).

Equation (5.41) indicates that a change in temperature associated with isentropic motion through a pressure field is given by

$$\mathrm{d}T = (T \, \alpha_T / \rho \, C_p) \, \mathrm{d}p, \tag{5.42}$$

where $\mathrm{d}\zeta = 0$ for isentropic changes. Hence, temperature indeed changes when pressure changes, even though there has been no heat exchanged with the environment. The combination

$$\Gamma = \frac{T \, \alpha_T}{\rho \, C_p} \tag{5.43}$$

is termed the adiabatic lapse rate for temperature as a function of pressure. Alternatively, note that a static fluid in a gravity field maintains hydrostatic balance, thus allowing the total pressure differential to be directly related to the vertical differential

$$\mathrm{d}p = -g \, \rho \, \mathrm{d}z, \tag{5.44}$$

which leads to the temperature change

$$\mathrm{d}T = -(g \, T \, \alpha_T / C_p) \, \mathrm{d}z. \tag{5.45}$$

In this case, introduce the adiabatic lapse rate

$$\widehat{\Gamma} = \frac{g \, T \, \alpha_T}{\rho \, C_p} \tag{5.46}$$

for temperature as a function of depth. The lapse rates determine the change in temperature (the *lapse*) a fluid parcel experiences as it is isentropically moved vertically through a hydrostatically balanced pressure field.

McDougall and Feistel (2003) provide a discussion of the lapse rate in terms of molecular dynamics. In particular, they note that the lapse rate, being proportional to the thermal expansion coefficient α_T, can be negative when α_T is negative. This occurs in cool fresh water. Hence, although work is done on the parcel under increasing pressure, the parcel temperature can decrease.

5.3 ENERGY OF A FLUID PARCEL

In this section we derive energy balances for fluid parcels. Energetics for non-Boussinesq (mass conserving) and Boussinesq (volume conserving) parcels are compared and contrasted.

5.3.1 Kinematics and dynamics

Before getting started, it is useful to summarize the kinematical and dynamical equations discussed in Chapters 3 and 4. In particular, mass conservation for a non-Boussinesq fluid parcel in its Eulerian form is given by (Section 3.4)

$$\rho_{,t} + \nabla \cdot (\rho \mathbf{v}) = 0, \tag{5.47}$$

where

$$\mathbf{v} = (\mathbf{u}, w) = (u, v, w) \tag{5.48}$$

is the three-dimensional velocity with $\mathbf{u} = (u, v)$ the horizontal components. Mass conservation in its Lagrangian form

$$\begin{aligned} \frac{1}{dM}\frac{d(dM)}{dt} &= \frac{1}{\rho}\frac{d\rho}{dt} + \frac{1}{dV}\frac{d(dV)}{dt} \\ &= 0 \end{aligned} \tag{5.49}$$

gives

$$\begin{aligned} \frac{1}{dV}\frac{d(dV)}{dt} &= \rho\frac{d\rho^{-1}}{dt} \\ &= \nabla \cdot \mathbf{v}, \end{aligned} \tag{5.50}$$

where the second equality follows from the Eulerian form of mass conservation (5.47). Equation (5.50) says that the relative change in volume of a fluid parcel equals the divergence of the three-dimensional velocity field. This relation then leads to the volume conservation relation appropriate for incompressible Boussinesq fluids

$$\frac{1}{dV}\frac{d(dV)}{dt} = \nabla \cdot \mathbf{v} = 0. \tag{5.51}$$

Mass conservation can be used to relate the Lagrangian and Eulerian forms of time evolution for a scalar via

$$\rho\frac{d\psi}{dt} = (\rho\,\psi)_{,t} + \nabla \cdot (\mathbf{v}\,\rho\,\psi). \tag{5.52}$$

This relation will be used frequently in the following. Its Boussinesq analog is given by

$$\frac{d\psi}{dt} = \psi_{,t} + \nabla \cdot (\mathbf{v}\,\psi). \tag{5.53}$$

In addition to constraints imposed by mass and volume conservation for parcels within the ocean interior, the fluid maintains the following kinematic constraints at the solid earth boundary and the surface

$$\mathbf{v} \cdot \hat{\mathbf{N}} = 0 \qquad \text{at } z = -H \tag{5.54}$$

$$\mathbf{v} \cdot \hat{\mathbf{N}} = \eta_{,t} - (\rho_w/\rho)\, q_w \qquad \text{at } z = \eta, \tag{5.55}$$

where the orientation directions are given by

$$\hat{\mathbf{N}} = \nabla\,(H + z) \qquad \text{at } z = -H \tag{5.56}$$

$$\hat{\mathbf{N}} = \nabla\,(-\eta + z) \qquad \text{at } z = \eta. \tag{5.57}$$

For a Boussinesq fluid, the $z = \eta$ boundary condition is recovered by setting $\rho_w/\rho = 1$. These boundary conditions are relevant when considering energy budgets over the full ocean domain.

Energetics for the hydrostatic and shallow ocean fluid (i.e., primitive equations) as well as the unapproximated nonhydrostatic fluid are discussed here. The hydrostatic approximation (Section 4.5) reduces the vertical momentum equation to the inviscid statement of hydrostatic balance

$$p_{,z} = -\rho g. \tag{5.58}$$

The horizontal momentum equation (Section 4.4) for the hydrostatic and shallow ocean fluid takes the form

$$(\rho \mathbf{u})_{,t} + \nabla \cdot (\rho \mathbf{v}\, \mathbf{u}) + (f + \mathcal{M})\, \hat{\mathbf{z}} \wedge (\rho \mathbf{v}) = -\nabla p + \rho \mathbf{F}^{(\mathbf{u})}. \tag{5.59}$$

The nonhydrostatic case which has not made the shallow ocean approximation has a similar momentum equation for all three velocity components

$$(\rho \mathbf{v})_{,t} + \nabla \cdot (\rho \mathbf{v}\, \mathbf{v}) + (\mathbf{\Omega} + \hat{\mathbf{z}}\, \mathcal{M}) \wedge (\rho \mathbf{v}) = -\nabla p - \hat{\mathbf{z}}\, \rho g + \rho \mathbf{F}^{(\mathbf{v})}, \tag{5.60}$$

where $\mathbf{\Omega}$ is the angular rotation of the earth about the North Pole. The momentum balance for a Boussinesq fluid in either case results by setting $\rho = \rho_o$, except when multiplying the gravitational acceleration.

5.3.2 Kinetic energy

The kinetic energy of a fluid parcel is given by

$$(\rho\, dV)\, \mathcal{K}_{(nh)} = (\rho\, dV)\, \mathbf{v}^2/2, \tag{5.61}$$

where $\mathbf{v}^2 = \mathbf{v} \cdot \mathbf{v}$ and

$$\mathcal{K}_{(nh)} = \mathbf{v}^2/2 \tag{5.62}$$

is the kinetic energy per unit mass. This is the appropriate form of kinetic energy for a nonhydrostatic fluid parcel, hence the "nh" subscript. For hydrostatic fluids, kinetic energy associated with vertical motions is many orders smaller than that associated with horizontal motions. Hence, in this case

$$\mathcal{K} = \mathbf{u}^2/2 \tag{5.63}$$

is the appropriate kinetic energy per mass. Note that Bokhove (2000) showed that the Hamiltonian function* for a hydrostatic fluid has a kinetic energy associated just with horizontal motions, thus formalizing the relevance of $\mathbf{u}^2/2$ for the hydrostatic fluid's kinetic energy per unit mass.

5.3.2.1 *Kinetic energy evolution*

Use of the momentum equation (5.59) for the hydrostatic parcel leads to

$$(\rho \mathcal{K})_{,t} + \nabla \cdot (\rho \mathbf{v}\, \mathcal{K}) = \mathbf{u} \cdot (-\nabla p + \rho \mathbf{F}^{(\mathbf{u})}), \tag{5.64}$$

*See Section 4.10 starting on page 75 for an example of a Hamiltonian function for particle dynamics.

and similar manipulations for a nonhydrostatic parcel render

$$(\rho \mathcal{K}_{(\mathrm{nh})})_{,t} + \nabla \cdot (\rho \mathbf{v} \mathcal{K}_{(\mathrm{nh})}) = \mathbf{v} \cdot (-\nabla p - \hat{\mathbf{z}} \rho g + \rho \mathbf{F}^{(\mathbf{v})}). \tag{5.65}$$

Equation (5.53) was used here to relate Eulerian to Lagrangian forms for the evolution of a scalar. Notice how the rotational terms, from both the Coriolis force and advective metric contribution, drop out from the kinetic energy budgets. This result should be expected, since kinetic energy is a scalar under rotations.

The pressure gradient term in both kinetic energy budgets can be interpreted as work by currents against pressure gradients. For example, if currents are directed down the pressure gradient ($-\mathbf{v} \cdot \nabla p > 0$ for the nonhydrostatic fluid, or $-\mathbf{u} \cdot \nabla p > 0$ for hydrostatic), then kinetic energy increases as the flow speed increases. The opposite occurs for flow directed up the pressure gradient. Flow directed parallel to isobars, such as geostrophically balanced currents ($\mathbf{u} \cdot \nabla p = 0$), does not alter kinetic energy.

5.3.2.2 Dissipation from molecular viscosity

Frictional dissipation of kinetic energy in the ocean occurs due to viscous effects at the Kolmogorov scale, which is the scale where the Reynolds number associated with molecular viscosity is on the order of unity. We consider here some mathematical results appropriate for analyzing these effects. For convenience, Cartesian coordinates are used as these are sufficient for describing many small-scale processes.

The friction force per volume appearing in the momentum equation (Section 5.3.1) at the Kolmogorov scale is associated with molecular viscosity. The mathematical form of this force is typically written as a three-dimensional isotropic Laplacian

$$\rho \mathbf{F}^{(\mathbf{v})} = \nabla \cdot (\gamma \nabla \mathbf{v}). \tag{5.66}$$

The Laplacian form is suggested by arguments taken from the kinetic theory of gases, where one can show rigorously that the Laplacian is indeed appropriate (e.g., Reif, 1965; Huang, 1987). There is, however, no analogous proof for liquids. This form for friction must therefore be consider an ansätz (i.e., an inspired guess), whose relevance is verified *a posteriori* by empirical measurements.

In equation (5.66), the strength of friction is scaled by the non-negative number

$$\gamma = \nu \rho, \tag{5.67}$$

which is the molecular *dynamic viscosity* for water. Typical values are around $\gamma \approx 10^{-3}\,\mathrm{kg\,m^{-1}\,s^{-1}}$ (Gill, 1982). More commonly considered in applications is the *kinematic viscosity*

$$\nu = \gamma / \rho, \tag{5.68}$$

whose values for water are around $\nu \approx 10^{-6}\,\mathrm{m^2\,s^{-1}}$. Oceanographers typically mean "kinematic viscosity" when they say "viscosity." The dimensions clarify which viscosity is actually being considered. With this value for the kinematic viscosity, a unit Reynolds number

$$\mathrm{Re} = U \nu / L, \tag{5.69}$$

with a velocity of $U = 10^{-1}\,\mathrm{m\,s^{-1}}$ taken as typical for the ocean, leads to a Kolmogorov length $L = \nu/U \approx 10^{-3}\,\mathrm{m}$, or a millimeter. This is the scale where the momentum transfer due to molecular viscous exchanges is as important as advective transport. This is also the scale discussed in Section 2.2 when considering the continuum hypothesis fundamental to fluid dynamics.

The form (5.66) for the friction vector leads to

$$\begin{aligned} \rho\,\mathbf{v}\cdot\mathbf{F}^{(\mathbf{v})} &= v_m\,\partial_n\,(\gamma\,\partial_n\,v_m) \\ &= (1/2)\,\nabla\cdot(\gamma\,\nabla\mathbf{v}^2) - \gamma\,\partial_m\,\mathbf{v}\cdot\partial_m\,\mathbf{v} \\ &= \underbrace{\nabla\cdot(\gamma\,\nabla\mathcal{K}_{(\mathrm{nh})})}_{\text{viscous flux divergence}} - \underbrace{\gamma\,\partial_m\,\mathbf{v}\cdot\partial_m\,\mathbf{v}}_{\text{viscous dissipation}}, \end{aligned} \tag{5.70}$$

where the tensor labels $m, n = 1, 2, 3$ are summed when repeated.* The divergence term redistributes kinetic energy within the fluid via the effects of viscous interaction between fluid parcels. When integrated globally, effects from this redistribution reduce to boundary effects via the divergence theorem (Section 3.3). The negative semidefinite term represents a kinetic energy sink associated with local viscous dissipation. It is commonly written as

$$\epsilon = \nu\,\partial_m\,\mathbf{v}\cdot\partial_m\,\mathbf{v} \geq 0. \tag{5.71}$$

As noted by McDougall (2003), frictional dissipation in the ocean interior associated with molecular viscosity is on the order†

$$\epsilon \approx 10^{-9}\,\mathrm{W\,kg^{-1}}. \tag{5.72}$$

The number $10^{-9}\,\mathrm{W\,kg^{-1}}$ sounds small, as indeed it is. To put it into perspective, divide by the heat capacity of seawater $C_p \approx 3989\,\mathrm{Joules\,kg^{-1}\,K^{-1}}$ (see equation (5.142)). Doing so reveals that frictional dissipation via molecular viscosity warms seawater at a rate less than 10^{-3} °K per hundred years. This is a negligible amount of heating from a large-scale ocean circulation perspective.

Using these results in the nonhydrostatic kinetic energy equation (5.65) leads to the material evolution of kinetic energy

$$\rho\,\mathrm{d}\mathcal{K}_{(\mathrm{nh})}/\mathrm{d}t = -\nabla\cdot[p\,\mathbf{v} - \rho\,\nu\,\nabla\,\mathcal{K}_{(\mathrm{nh})}] - w\,\rho\,g - (p/\rho)\,\mathrm{d}\rho/\mathrm{d}t - \rho\,\epsilon. \tag{5.73}$$

The following steps were used to reach this result:

$$\begin{aligned} \mathbf{v}\cdot\nabla p &= \nabla\cdot(p\,\mathbf{v}) - p\,\nabla\cdot\mathbf{v} \\ &= \nabla\cdot(p\,\mathbf{v}) + (p/\rho)\,\mathrm{d}\rho/\mathrm{d}t, \end{aligned} \tag{5.74}$$

with the last step employing equation (5.50) to relate the relative change in specific volume to the velocity divergence. The first term on the right-hand side of equation (5.73) is the divergence of a kinetic energy flux arising from pressure and viscosity. The second term represents a conversion between kinetic and gravitational potential energy, as revealed when considering the potential energy budget discussed in Section 5.3.3. The third term is a source arising from pressure-work acting to alter the volume of a fluid parcel. This term vanishes for Boussinesq fluid parcels since they maintain a constant volume. The final term is the sink from viscous dissipation.

---

*Cartesian tensor notation does not distinguish between a raised or lowered tensor label. This convention is sufficient for this section where coordinates are assumed to be Cartesian.

†Note that $\mathrm{W\,kg^{-1}} = \mathrm{m^2\,s^{-3}}$.

5.3.2.3 Dissipation in hydrostatic ocean models

Although the discussion in this chapter focuses on ocean fundamentals, this is an opportune time to expose some points regarding frictional dissipation in hydrostatic ocean models. For this purpose, we need to anticipate results from Chapter 17, where it is noted that frictional dissipation in the interior of ocean models occurs near the grid scale. The reason is that the model's friction operator is designed to have scale-selective behavior similar to that arising from the Laplacian operator that parameterizes molecular friction. With such operators, the dissipation strength is increased as the wavelength of fluctuations decreases. That is, the dissipation preferentially selects the smaller scales rather than the larger scales.

Even though the mathematical operators have much in common, there is a major distinction between molecular dissipation appropriate for the ocean and frictional dissipation employed in ocean climate models. Namely, the level of model dissipation is typically many orders larger than that in the real ocean. The reason is that the model grid scale (order tens to hundreds of kilometers) is many orders larger than the Kolmogorov scale (order millimeters). Hence, discretizing the ocean into finite sized grid cells necessitates an increase in the model viscosity so that the Reynolds number at the model's grid scale is roughly unity. Absent this level of model friction, the simulation is exposed to possible spurious numerical behavior, such as a physically large level of power at the grid scale, and potentials for numerical instability. These numerical issues are detailed in Section 18.1.

In the real ocean, viscous dissipation is converted to local heating. However, it is not appropriate to do the analogous conversion in ocean models since the model's huge levels of frictional dissipation would translate into an unphysically large level of heating. Among other spurious effects, such large levels of heating would create an unphysical diapycnal mass flux altering simulated water mass properties. Hence, ocean models do not consider the transfer of kinetic energy dissipation to heating. This is distinct from the practice common in atmospheric models, where it is important to consider this local heating (e.g., Becker, 2003). More is said on this point when considering the internal energy budget in Section 5.3.5.

We present now the mathematical form of frictional dissipation occuring in hydrostatic ocean models. For this purpose, it is important to return to the use of generalized horizontal coordinates since they are appropriate for global ocean models. Results from Chapter 17 are also required, where details are provided of the frictional stress employed by the models. For the present discussion, it is sufficient to quote the salient results from Chapter 17.

Primitive equation global ocean models only concern themselves with the horizontal friction force, since the vertical momentum equation is reduced to the inviscid hydrostatic balance. The horizontal friction force in hydrostatic models can be written

$$\rho F^{m} = \tau^{mn}_{;n} + (\rho \kappa u^{m}_{,z})_{,z}. \tag{5.75}$$

In this equation, τ^{mn} are components of the symmetric 2×2 frictional stress tensor $\boldsymbol{\tau}$, the semicolon denotes a covariant derivative (see Section 21.2, page 468), and the summation convention is employed where repeated indices are summed over their horizontal range with $m, n = 1, 2$. The covariant divergence of the horizontal stress tensor accounts for friction arising from horizontal deformations in the simulated fluid. It is called *horizontal friction* for this reason. The second term arises from vertical deformations, and is called *vertical friction*. Consistent with the model grid being coarser than the Kolmogorov length scale, the vertical viscosity $\kappa > 0$ is typically much larger than the molecular viscosity.

To expose the effects of model friction on the kinetic energy budget, proceed as for the molecular case. The contribution from vertical friction takes on the form

$$\mathbf{u} \cdot (\rho \kappa \mathbf{u}_{,z})_{,z} = -\rho \kappa (\mathbf{u}_{,z})^2 + (\rho \kappa \mathcal{K}_{,z})_{,z} \tag{5.76}$$

where $(\mathbf{u}_{,z})^2 = \mathbf{u}_{,z} \cdot \mathbf{u}_{,z}$ is the squared vertical shear of the horizontal velocity field. This equation is of the same form as that occurring with molecular friction, except that here the derivatives are vertical rather than three-dimensional. The contribution from horizontal friction leads to

$$\begin{aligned} g_{mn}\, u^m\, \tau^{np}_{;p} &= (g_{mn}\, u^m\, \tau^{np})_{;p} - g_{mn}\, u^m_{;p}\, \tau^{np} \\ &= (g_{mn}\, u^m\, \tau^{np})_{;p} - e_{np}\, \tau^{np} \end{aligned} \tag{5.77}$$

where

$$2\, e_{np} = u_{n\,;p} + u_{p\,;n} \tag{5.78}$$

are components to the strain tensor that embodies information about strain within the fluid. Properties of this tensor are described in Section 17.3.1. It is also noted that the covariant derivative of the metric tensor components vanish, $g_{mn\,;p} = 0$ (Section 21.2.4). This property allows the metric tensor to commute with the covariant derivative.

Just as for the molecular case and the vertical friction case, equation (5.77) exposes two terms: a divergence and a sink. The divergence term, written in the suggestive *dyadic* form

$$\nabla \cdot (\mathbf{u} \cdot \boldsymbol{\tau}) \equiv (g_{mn}\, u^m\, \tau^{np})_{;p} \tag{5.79}$$

locally redistributes momentum via horizontal frictional stresses. As described in Section 17.4, the $e_{np}\, \tau^{np}$ scalar is non-negative in ocean models,

$$\mathbf{e} \cdot \boldsymbol{\tau} \equiv e_{np}\, \tau^{np} \geq 0, \tag{5.80}$$

with zero the result only in the absence of horizontal deformations. Hence, this term is a local sink of kinetic energy arising from the horizontal friction stresses in the hydrostatic model.

In summary, the effects of friction on kinetic energy in hydrostatic ocean models take the form

$$\begin{aligned} \rho\, \mathbf{u} \cdot \mathbf{F}^{(\mathbf{u})} &= \nabla \cdot (\mathbf{u} \cdot \boldsymbol{\tau}) + (\rho \kappa \mathcal{K}_{,z})_{,z} - \mathbf{e} \cdot \boldsymbol{\tau} - \rho \kappa (\mathbf{u}_{,z})^2 \\ &= \underbrace{\nabla \cdot [\mathbf{u} \cdot \boldsymbol{\tau} + \hat{\mathbf{z}}\, \rho \kappa \mathcal{K}_{,z}]}_{\text{viscous flux divergence}} - \underbrace{[\mathbf{e} \cdot \boldsymbol{\tau} + \rho \kappa (\mathbf{u}_{,z})^2]}_{\text{viscous dissipation}}. \end{aligned} \tag{5.81}$$

As expected, there is a direct correspondence between this form and that realized from molecular friction acting in the nonhydrostatic case (equation (5.70)). The present case is distinguished by anisotropy between vertical and horizontal friction, as well as a more general form of the horizontal friction stress tensor.

The result (5.81) is now used in the kinetic energy budget (5.64) to determine the material evolution of kinetic energy in a hydrostatic fluid

$$\begin{aligned} \rho\, \mathrm{d}\mathcal{K}/\mathrm{d}t = &-\nabla \cdot [p\, \mathbf{v} - \mathbf{u} \cdot \boldsymbol{\tau} - \hat{\mathbf{z}}\, \rho \kappa \mathcal{K}_{,z}] \\ &- w\, \rho\, g - (p/\rho)\, \mathrm{d}\rho/\mathrm{d}t - [\mathbf{e} \cdot \boldsymbol{\tau} + \rho \kappa (\mathbf{u}_{,z})^2]. \end{aligned} \tag{5.82}$$

The following steps were used to reach this result:

$$\begin{aligned} \mathbf{u} \cdot \nabla p &= \nabla \cdot (p\,\mathbf{v}) - w\,p_{,z} - p\,\nabla \cdot \mathbf{v} \\ &= \nabla \cdot (p\,\mathbf{v}) + w\,\rho\,g + (p/\rho)\,\mathrm{d}\rho/\mathrm{d}t, \end{aligned} \tag{5.83}$$

where the hydrostatic balance $p_{,z} = -\rho g$ was used, as well as equation (5.50) to relate the relative change in parcel specific volume to the velocity divergence. Comparison of equation (5.82) to the nonhydrostatic kinetic energy budget (5.73) reveals the only distinction is the form that friction contributes to the budget.

5.3.3 Gravitational potential energy

Now consider the gravitational potential energy of a fluid parcel, taken with respect to the reference level $z = 0$ at the surface of a resting ocean.

5.3.3.1 *Non-Boussinesq parcels*

The gravitational potential energy of a non-Boussinesq fluid parcel is

$$(\rho\,\mathrm{d}V)\,\Phi = (\rho\,\mathrm{d}V)\,g\,z, \tag{5.84}$$

where $\Phi = g\,z$ is the potential energy per unit mass, and the gravitational acceleration g is assumed to be constant over the ocean fluid. The material evolution of potential energy is thus given by

$$(\rho\,\mathrm{d}V)\,\frac{\mathrm{d}\Phi}{\mathrm{d}t} = (\rho\,\mathrm{d}V)\,g\,w. \tag{5.85}$$

That is, only motion in the vertical direction, parallel to the gravitational force, alters the parcel's gravitational potential energy. Mixing, surface forcing, and subgrid scale (SGS) processes affect potential energy only insofar as they affect the vertical velocity. Use of mass conservation in the form (5.47) leads to the Eulerian conservation law

$$(\rho\,\Phi)_{,t} + \nabla \cdot (\rho\,\mathbf{v}\,\Phi) = w\,\rho\,g. \tag{5.86}$$

5.3.3.2 *Boussinesq parcels*

When developing budgets for a Boussinesq fluid parcel, the parcel is assumed to have mass $\rho_o\,\mathrm{d}V$, *except* when the parcel feels the gravitational force, where its mass is that of a non-Boussinesq parcel, $\rho\,\mathrm{d}V$. Hence, for purposes of deriving the potential energy budget, proceed as for the non-Boussinesq case, yet constrain the parcel to maintain a constant volume instead of a constant mass

$$\begin{aligned} \frac{\mathrm{d}\,(\rho\,\mathrm{d}V\,\Phi)}{\mathrm{d}t} &= g\,\mathrm{d}V\,\frac{\mathrm{d}\,(\rho\,z)}{\mathrm{d}t} \\ &= \rho\,\mathrm{d}V\left(g\,w + g\,z\,\frac{\mathrm{d}\ln\rho}{\mathrm{d}t}\right). \end{aligned} \tag{5.87}$$

The $(\rho\,\mathrm{d}V)\,g\,w$ term is the same as for the non-Boussinesq parcel, whereas the $g\,z\,\mathrm{d}\ln\rho/\mathrm{d}t$ term is new. Again, this new term arises from the assumption that the Boussinesq parcel's volume remains constant, not its mass. Hence, if the parcel changes its density, as through fluxes of buoyancy, the parcel's mass must change to maintain its fixed volume. The mass changes via the presence of *virtual mass fluxes*. Since Boussinesq fluids are so common in ocean modeling, it is worth discussing these issues a bit further.

5.3.3.3 Distinctions between budgets

The $\mathrm{d}\ln\rho/\mathrm{d}t$ term in equation (5.87) means that the gravitational potential energy of a Boussinesq fluid parcel is directly affected by buoyancy fluxes, including those at the subgrid scale. A particular example is the SGS flux arising from the Gent and McWilliams (1990) closure, which provides a sink for potential energy, so long as there is baroclinicity (horizontal density gradients) in the fluid domain.* In contrast, horizontal fluxes do not alter the gravitational potential energy, since they are directed along surfaces of constant geopotential.

To further discuss the different potential energy budgets, consider a parcel that is heated within a stratified ocean. Heating reduces the parcel's density, which increases its buoyancy and so causes it to develop an upward velocity until it reaches a new level of neutral buoyancy. If the parcel is non-Boussinesq, then its mass $(\rho\,\mathrm{d}V)$ remains fixed, so its gravitational potential energy increases upon moving upward: $g\,(\rho\,\mathrm{d}V)\,\mathrm{d}z/\mathrm{d}t > 0$. If the parcel is Boussinesq, its volume is fixed, so the potential energy change is given by $g\,\mathrm{d}V\,\mathrm{d}(z\,\rho)/\mathrm{d}t$. Notably, for a parcel at the ocean surface with the rigid lid assumption, heating the Boussinesq parcel leads to a reduction in potential energy, whereas heating the non-Boussinesq parcel leads to no change. In general, it is unclear how the potential energy will change for the Boussinesq parcel. Although the potential energy budget for Boussinesq fluids is unsatisfying from a fundamental perspective, it has yet to be determined how important it is to properly represent the potential energy budget in global ocean models.

5.3.4 Total mechanical energy

To derive a budget for the total mechanical energy (kinetic plus potential) for a nonhydrostatic non-Boussinesq seawater parcel affected by molecular friction, one adds the kinetic energy budget (5.73) to the gravitational potential energy budget (5.86), to get

$$\rho\,\frac{\mathrm{d}(\mathcal{K}_{(\mathrm{nh})}+\Phi)}{\mathrm{d}t} = -\nabla\cdot[p\,\mathbf{v}-\rho\,\nu\,\nabla\,\mathcal{K}_{(\mathrm{nh})}] - (p/\rho)\,\mathrm{d}\rho/\mathrm{d}t - \rho\,\epsilon. \tag{5.88}$$

Notice how the *conversion* term $w\,\rho\,g$ drops out. The reason is that when it contributes to an increase in kinetic energy, it does so at the price of a reduction in potential energy, and vice versa. The conversion between potential and kinetic energy is familiar from elementary mechanics (e.g., the simple pendulum). For a hydrostatic non-Boussinesq parcel, add the kinetic energy budget (5.82) to the gravitational potential energy budget (5.86), to give

$$\rho\,\frac{\mathrm{d}(\mathcal{K}+\Phi)}{\mathrm{d}t} = -\nabla\cdot[p\,\mathbf{v}-\mathbf{u}\cdot\boldsymbol{\tau}-\hat{\mathbf{z}}\,\rho\,\kappa\,\mathcal{K}_{,z}] - (p/\rho)\,\mathrm{d}\rho/\mathrm{d}t - [\mathbf{e}\cdot\boldsymbol{\tau}+\rho\,\kappa\,(\mathbf{u}_{,z})^2]. \tag{5.89}$$

Both the nonhydrostatic and hydrostatic budgets indicate that mechanical energy for a parcel is affected by the divergence of pressure and viscous fluxes, as well as a sink from frictional dissipation and a source from pressure-work. Closure for the

*More is said of the effects that the Gent and McWilliams (1990) scheme has on potential energy in Section 14.2.1.

total energy budget comes about by noting that modifications to the mechanical energy by frictional dissipation and pressure-work are compensated by opposite signed terms in the internal energy budget, thus maintaining a conserved total mechanical energy plus internal energy.

5.3.5 Mechanical plus internal energies

As discussed in Section 5.2, internal energy represents the energy of the molecular degrees of freedom that are averaged out when formulating a continuum description of fluid motion. Consequently, the total energy per mass $\mathcal{E}$ (specific energy) of a fluid parcel is written

$$\mathcal{E} = \mathbf{v}^2/2 + \Phi + \mathcal{I}, \tag{5.90}$$

with $\mathbf{v}^2/2$ the kinetic energy per mass, Φ the gravitational potential energy per mass, and $\mathcal{I}$ the internal energy per mass. Macroscopically, this relation *defines* the specific internal energy $\mathcal{I}$ as the total energy minus the mechanical energy. Microscopically, $\mathcal{I}$ embodies the energy of molecular thermal agitation and molecular interactions. For a hydrostatic fluid,

$$\mathcal{E} = \mathbf{u}^2/2 + \Phi + \mathcal{I} \tag{5.91}$$

since horizontal motions are many orders larger than vertical ones.

Energy conservation for a closed fluid system means that total energy per mass of a fluid parcel evolves according to

$$\rho \frac{\mathrm{d}\mathcal{E}}{\mathrm{d}t} = -\nabla \cdot \mathbf{J}_{\mathcal{E}} \tag{5.92}$$

for some flux of energy $\mathbf{J}_{\mathcal{E}}$. Closure of the full fluid system means that the normal component to the energy flux vanishes at system boundaries. Nonzero normal flux components arise for open fluid systems. The World Ocean is an open system, since it is in contact with fluxes from other components in the climate system: the atmosphere, sea ice, rivers, and so on.

Based on considerations of mechanical energy flux for a nonhydrostatic parcel affected by molecular viscosity, define the flux of total energy as

$$\mathbf{J}_{\mathcal{E}} = \mathbf{v}\, p - \rho\, \nu\, \nabla\, \mathcal{K}_{(\mathrm{nh})} + \mathbf{J}_q. \tag{5.93}$$

The *heat flux* $\mathbf{J}_q$ is generally a function of temperature as well as tracer concentration (for discussions, see Fofonoff, 1962; Gregg, 1984; Davis, 1994a; Landau and Lifshitz, 1987). Subtracting equation (5.88) for mechanical energy evolution from equation (5.92) leads to the expression for internal energy evolution

$$\rho \frac{\mathrm{d}\mathcal{I}}{\mathrm{d}t} = -\nabla \cdot \mathbf{J}_q + (p/\rho) \frac{\mathrm{d}\,\rho}{\mathrm{d}t} + \rho\, \epsilon. \tag{5.94}$$

Internal energy of a parcel is thus affected by the convergence of heat fluxes, and sources due to pressure-work and frictional dissipation. Notice how internal energy is increased as the parcel's volume is decreased (and density increased), as expected from the thermodynamic considerations in Section 5.2.1. In the absence of irreversible effects due to heat transport and momentum friction, internal energy is affected only by pressure-work. Consequently, Boussinesq parcels, which are incompressible and so cannot receive pressure-work, maintain a constant internal energy when undergoing reversible transport.

For a hydrostatic parcel, the flux of total energy is defined as

$$\mathbf{J}_{\mathcal{E}} = \mathbf{v}\, p - \mathbf{u} \cdot \boldsymbol{\tau} - \hat{\mathbf{z}}\, \rho\, \kappa\, \mathcal{K}_{,z} + \mathbf{J}_q. \tag{5.95}$$

Subtracting equation (5.89) for mechanical energy evolution from equation (5.92) leads to the expression for internal energy evolution in the hydrostatic fluid

$$\rho \frac{\mathrm{d}\mathcal{I}}{\mathrm{d}t} = -\nabla \cdot \mathbf{J}_q + (p/\rho) \frac{\mathrm{d}\rho}{\mathrm{d}t} + [\mathbf{e} \cdot \boldsymbol{\tau} + \rho\, \kappa\, (\mathbf{u}_{,z})^2]. \tag{5.96}$$

The interpretation of this budget is analogous to that for the nonhydrostatic fluid. Note, however, an important caveat insofar as this budget is concerned for ocean model simulations. As mentioned in Section 5.3.2.3, ocean model frictional dissipation is huge relative to molecular viscous dissipation. Hence, insisting that the model's internal energy budget closes by converting its frictional dissipation to heating, and thus into internal energy as in equation (5.96), would lead to severe and spurious degradation of water mass properties. It is for this reason that frictional heating is ignored in ocean climate models. Consequently, the internal energy budget is not closed in the models.

5.4 GLOBAL MECHANICAL ENERGY BALANCE

When developing ocean models, it is useful to consider the global properties of the discrete equations, such as their energy balances. Often, one aims to have the discrete equations maintain balances analogous to those of the continuum equations. To guide in this development, we examine here the mechanical energy balances over the full ocean domain for a hydrostatic fluid.

5.4.1 Generic budget

We are generally interested in quantities of the form

$$\begin{aligned} \Psi &= \int (\rho\, \mathrm{d}V)\, \psi \\ &= \int \mathrm{d}A \int_{-H}^{\eta} \mathrm{d}z\, \rho\, \psi, \end{aligned} \tag{5.97}$$

where $\mathrm{d}A = \mathrm{d}x\, \mathrm{d}y$ is the time-independent horizontal area element and ψ is an arbitrary scalar fluid property (e.g., energy per mass, tracer concentration).* Changes in Ψ are determined by the passage of the property ψ across the domain boundaries (e.g. wind stress, fresh water, heating, etc.), and changes in the amount of ψ within the ocean domain (e.g., source/sink terms, pressure-work, dissipation, etc.).

*Analogous budgets are often considered for material volumes moving with the fluid. These budgets make use of Reynold's transport theorem (e.g., Aris, 1962). Considerations here are not Lagrangian in nature, and so the discussion is different.

To determine time changes in Ψ, first consider the integral of mass weighted material transport of ψ over the ocean domain

$$\int (\rho \mathrm{d}V) \frac{\mathrm{d}\psi}{\mathrm{d}t} = \int \mathrm{d}V[(\rho\,\psi)_{,t} + \nabla \cdot (\rho\,\mathbf{v}\,\psi)]. \tag{5.98}$$

To bring this result into a more convenient form, use the divergence theorem (Section 3.3) to handle the divergence term $\nabla \cdot (\rho\,\mathbf{v}\,\psi)$. For the time derivative, use Leibnitz's rule, which specifies how a derivative and an integral commute when the limits of integration are nonconstant:

$$\frac{\partial}{\partial y}\left(\int\limits_{\phi_1(y)}^{\phi_2(y)} F(x,y)\,\mathrm{d}x\right) = F(\phi_2,y)\,\partial_y\,\phi_2 - F(\phi_1,y)\,\partial_y\,\phi_1 + \int\limits_{\phi_1(y)}^{\phi_2(y)} \partial_y\,F(x,y)\,\mathrm{d}x. \tag{5.99}$$

This rule is standard from differential calculus.

Leibnitz's Rule brings the local time derivative in equation (5.98) to the form

$$\begin{aligned}\int \mathrm{d}A \int\limits_{-H}^{\eta} \mathrm{d}z\,(\rho\,\psi)_{,t} &= \partial_t\left(\int \mathrm{d}A \int\limits_{-H}^{\eta} \mathrm{d}z\,\rho\,\psi\right) - \int\limits_{z=\eta} \mathrm{d}A\,\eta_{,t}\,(\rho\,\psi)\\ &= \Psi_{,t} - \int\limits_{z=-\eta} \mathrm{d}A\,\eta_{,t}\,\rho\,\psi,\end{aligned} \tag{5.100}$$

where the last step identifies Ψ from equation (5.97), and the horizontal cross-sectional area element $\mathrm{d}A$ is time-independent. Notice how passage of the time derivative from inside to outside the vertical integral picked up a surface term associated with the time-dependent surface height. Use of the divergence theorem gives

$$\int \mathrm{d}V\,\nabla \cdot (\rho\,\mathbf{v}\,\psi) = \int\limits_{z=\eta} \mathrm{d}x\,\mathrm{d}y\,\hat{\mathbf{N}} \cdot \rho\,\mathbf{v}\,\psi \tag{5.101}$$

with $\hat{\mathbf{N}}(\eta) = \nabla\,(-\eta + z)$ the surface orientation direction, and we assumed zero material flux through the ocean bottom at $z = -H$. Use of the surface kinematic boundary condition (5.55) leads to

$$\int \mathrm{d}V\,\nabla \cdot (\rho\,\mathbf{v}\,\psi) = \int\limits_{z=\eta} \mathrm{d}x\,\mathrm{d}y\,(\eta_{,t} - q_{\mathrm{w}}\,\rho_{\mathrm{w}}/\rho)\,\rho\,\psi. \tag{5.102}$$

Combining this result with equation (5.100) gives the global budget

$$\Psi_{,t} = \int\limits_{z=\eta} \mathrm{d}A\,\rho_{\mathrm{w}}\,q_{\mathrm{w}}\,\psi + \int (\rho \mathrm{d}V)\,\frac{\mathrm{d}\psi}{\mathrm{d}t}. \tag{5.103}$$

The left-hand side represents the time tendency of a domain integrated fluid property, such as the kinetic energy or the tracer mass. In the first term on the right-hand side, $\mathrm{d}A\,\rho_{\mathrm{w}}\,q_{\mathrm{w}}$, is the mass per time of fresh water crossing the ocean surface, and so $(\mathrm{d}A\,\rho_{\mathrm{w}}\,q_{\mathrm{w}})\,\psi$ represents the transport of the ψ property across the surface with the fresh water. The final term represents material transport of the fluid property as integrated over the ocean domain. It is notable that with $\psi = 1$, the balance reduces to mass conservation discussed in Section 3.4.3. For Boussinesq fluids, $\rho_{\mathrm{w}} \rightarrow \rho$ for the $z = \eta$ surface term, and $\rho \rightarrow \rho_o$ for the volume integrated term, except for the potential energy budget to be considered in Section 5.4.3.2.

5.4.2 Kinetic energy budget

The kinetic energy of a hydrostatic fluid is given by

$$K = \int (\rho \, \mathrm{d}V) \, \mathcal{K}, \tag{5.104}$$

where $\mathcal{K} = \mathbf{u}^2/2$ is the kinetic energy per mass of the parcel arising from horizontal motion. With $\psi = \mathcal{K}$, results from Section 5.4.1 yield

$$K_{,t} = \int\limits_{z=\eta} \mathrm{d}A \, \rho_{\mathrm{w}} \, q_{\mathrm{w}} \, \mathcal{K} + \int (\rho \, \mathrm{d}V) \, \mathbf{u} \cdot \frac{\mathrm{d}\mathbf{u}}{\mathrm{d}t}. \tag{5.105}$$

The first term represents the passage of kinetic energy through the ocean surface with fresh water fluxes. For the second term, the momentum equation (5.59) leads to $\mathbf{u} \cdot \rho \, \mathrm{d}\mathbf{u}/\mathrm{d}t = \mathbf{u} \cdot (-\nabla p + \rho \, \mathbf{F}^{\mathbf{u}})$, thus yielding

$$K_{,t} = \int\limits_{z=\eta} \mathrm{d}A \, \rho_{\mathrm{w}} \, q_{\mathrm{w}} \, \mathcal{K} + \int \mathrm{d}V \, \mathbf{u} \cdot (-\nabla p + \rho \, \mathbf{F}^{(\mathbf{u})}). \tag{5.106}$$

Massaging this result highlights the effects from boundary forcing. For this purpose, write the pressure term as

$$-\int \mathrm{d}V \, \mathbf{u} \cdot \nabla p = \int \mathrm{d}V \, [-\nabla \cdot (\mathbf{v} \, p) + p \, \nabla \cdot \mathbf{v} - w \, \rho \, g], \tag{5.107}$$

where hydrostatic balance was used. The divergence theorem then yields

$$-\int \mathrm{d}V \, \mathbf{u} \cdot \nabla p = \int\limits_{z=\eta} \mathrm{d}A \, p \, \hat{\mathbf{N}} \cdot \mathbf{v} + \int \mathrm{d}V \, (p \, \nabla \cdot \mathbf{v} - w \, \rho \, g). \tag{5.108}$$

In the first term, $p(\eta)$ is the applied pressue at the ocean surface from the overlying atmosphere or sea ice. Hence,

$$\int\limits_{z=\eta} \mathrm{d}A \, p \, \hat{\mathbf{N}} \cdot \mathbf{v} = \int\limits_{z=\eta} \mathrm{d}A_{(\eta)} \, p \, \hat{\mathbf{n}} \cdot \mathbf{v} \tag{5.109}$$

represents the integrated effects of pressure-work applied on the ocean at the ocean free surface.

For the friction term, $\mathbf{u} \cdot \mathbf{F}^{(\mathbf{u})}$, employ equation (5.81) for the hydrostatic fluid, thus yielding

$$\int \mathrm{d}V \, \rho \, \mathbf{u} \cdot \mathbf{F}^{(\mathbf{u})} = \int \mathrm{d}V \, \left(\nabla \cdot [\mathbf{u} \cdot \boldsymbol{\tau} + \hat{\mathbf{z}} \, \rho \, \kappa \, \mathbf{u} \cdot \mathbf{u}_{,z}] - [\mathbf{e} \cdot \boldsymbol{\tau} + \rho \, \kappa \, (\mathbf{u}_{,z})^2] \right). \tag{5.110}$$

The second term is a sink due to frictional dissipation in the ocean interior, and the divergence reduces to boundary contributions. To illustrate a common form in which the boundary terms are written when formulating ocean models, write

$$\rho \, \kappa \, \mathbf{u}_{,z} = \boldsymbol{\tau}^{\mathrm{wind}} \qquad \text{at } z = \eta \tag{5.111}$$

for the ocean surface, and

$$\rho \, \kappa \, \mathbf{u}_{,z} = \boldsymbol{\tau}^{\mathrm{bottom}} \qquad \text{at } z = -H \tag{5.112}$$

for the ocean bottom. These terms represent surface and bottom stresses* applied to the ocean boundaries from the atmosphere and sea ice on the surface, and the

*Recall that stress has the same dimensions as pressure, $\mathrm{kg \, m^{-1} \, s^{-2}}$.

solid earth at the ocean bottom. Setting $\mathbf{u} \cdot \boldsymbol{\tau}$ to zero applies when assuming a no-slip condition. These assumptions then lead to

$$\int \mathrm{d}V\, \rho\, \mathbf{u} \cdot \mathbf{F}^{(\mathbf{u})} = \int\limits_{z=\eta} \mathrm{d}A\, \mathbf{u} \cdot \boldsymbol{\tau}^{\text{wind}} - \int\limits_{z=-H} \mathrm{d}A\, \mathbf{u} \cdot \boldsymbol{\tau}^{\text{bottom}} - \int \mathrm{d}V \left[\mathbf{e} \cdot \boldsymbol{\tau} + \rho\, \kappa\, (\mathbf{u}_{,z})^2\right]. \tag{5.113}$$

Bringing the various contributions together leads to the global budget for kinetic energy in a hydrostatic fluid

$$K_{,t} = \int\limits_{z=\eta} \underbrace{\mathrm{d}A\, [\rho_{\mathrm{w}}\, q_{\mathrm{w}}\, \mathcal{K} + p\, \hat{\mathbf{N}} \cdot \mathbf{v} + \mathbf{u} \cdot \boldsymbol{\tau}^{\text{wind}}]}_{\text{surface forcing}} - \int\limits_{z=-H} \underbrace{\mathrm{d}A\, \mathbf{u} \cdot \boldsymbol{\tau}^{\text{bottom}}}_{\text{bottom forcing}} + \int \underbrace{\mathrm{d}V\, (p\, \nabla \cdot \mathbf{v} - w\, \rho\, g)}_{\text{inviscid effects}} - \int \underbrace{\mathrm{d}V\, [\mathbf{e} \cdot \boldsymbol{\tau} + \rho\, \kappa\, (\mathbf{u}_{,z})^2]}_{\text{viscous dissipation}}. \tag{5.114}$$

Note that the only difference between this budget and that appropriate for a non-hydrostatic fluid experiencing the effects of molecular viscosity is in the form of viscous dissipation (Section 5.3.2). The physical content of the individual terms appearing in equation (5.114) has been discussed in the formulation leading up to this budget. Nonetheless, having all pieces in one place helps to highlight the intimate interplay between boundary forcing, inviscid dynamics within the ocean interior, and viscous dissipation.

5.4.3 Gravitational potential energy budgets

We consider here the gravitational potential energy budgets for non-Boussinesq and Boussinesq fluid parcels.

5.4.3.1 Non-Boussinesq fluid

A finite domain of fluid has gravitational potential energy

$$P = \int (\rho\, \mathrm{d}V)\, \Phi, \tag{5.115}$$

where $\Phi = g\, z$ is the potential energy per mass of a fluid parcel. Results from Section 5.4.1, with $D\psi/\mathrm{d}t = g\, w$, lead to the global potential energy budget for a non-Boussinesq fluid

$$P_{,t} = \int\limits_{z=\eta} \mathrm{d}A\, (g\, z\, \rho_{\mathrm{w}}\, q_{\mathrm{w}}) + \int (\rho \mathrm{d}V)\, g\, w. \tag{5.116}$$

Hence, the potential energy increases when the height of the ocean surface increases upon the introduction of fresh water (first term) or when the vertical motion causes mass within the ocean domain to rise (second term).

5.4.3.2 Boussinesq fluid

The total gravitational potential energy of a Boussinesq fluid has time evolution

$$P_{,t} = g \int_{z=\eta} \mathrm{d}A\,(\rho\, z\, \eta_{,t}) + g \int \mathrm{d}V\,(z\, \rho_{,t}). \tag{5.117}$$

The mass conservation identity $\rho_{,t} = -\nabla \cdot (\mathbf{v}\,\rho)$ is not available here for the volume conserving Boussinesq fluid, where $\nabla \cdot \mathbf{v} = 0$. This result highlights the very different potential energy budget for Boussinesq and non-Boussinesq fluids.

Because the Boussinesq approximation is so commonly used by ocean models, it is useful to provide some extra discussion of its potential energy budget. To proceed requires a few basic results from subsequent sections.* Consider the special case of a Boussinesq fluid with density equal to potential density with a linear equation of state. For this case, density's time tendency can be written†

$$\rho_{,t} = -\nabla \cdot (\rho\,\mathbf{v} + \mathbf{F}). \tag{5.118}$$

Here, $\mathbf{F}$ represents any flux acting either to mix or to stir density. Without $\mathbf{F}$, the evolution equation (5.118) is mathematically the same as the mass conservation equation. Notably, however, for the present case, besides linearizing the equation of state, the velocity field is assumed nondivergent $\nabla \cdot \mathbf{v} = 0$ and so equation (5.118) can equivalently be written in the advective form

$$\rho_{,t} = -\mathbf{v} \cdot \nabla \rho - \nabla \cdot \mathbf{F}. \tag{5.119}$$

To proceed, make use of the divergence theorem to derive

$$\begin{aligned} \int \mathrm{d}V\, z\, \rho_{,t} &= -\int \mathrm{d}V\, z\, \nabla \cdot (\rho\,\mathbf{v} + \mathbf{F}) \\ &= -\int \mathrm{d}V\, \nabla \cdot (z\,\rho\,\mathbf{v} + z\,\mathbf{F}) + \int \mathrm{d}V\,(\rho\, w + F^{z}) \\ &= -\int_{z=\eta} \mathrm{d}A\, \hat{\mathbf{N}} \cdot (z\,\rho\,\mathbf{v} + z\,\mathbf{F}) + \int \mathrm{d}V\,(\rho\, w + F^{z}), \end{aligned} \tag{5.120}$$

where no fluxes cross the solid earth boundary. Use of the surface kinematic boundary condition (5.55) leads to

$$P_{,t} = g \int_{z=\eta} \mathrm{d}A\, \eta\,(\rho_{\mathrm{w}}\, q_{\mathrm{w}} - \hat{\mathbf{N}} \cdot \mathbf{F}) + g \int \mathrm{d}V\,(w\,\rho + F^{z}). \tag{5.121}$$

This budget affords a physical interpretation largely consistent with that discussed for the Boussinesq parcel in Section 5.3.3.2. Notably, $-\hat{\mathbf{N}} \cdot \mathbf{F}$ represents surface buoyancy fluxes that either increase or decrease the Boussinesq fluid's potential energy, depending on the sign of the flux and its height relative to $z = 0$. Additionally, the volume integrated vertical flux $\int \mathrm{d}V\, F^{z}$ acts in a similar manner to the vertical velocity, transporting density against or with the gravitational field.

---

*Those new to the subject may wish to return to this discussion only after reading sections up to and through Section 5.7.3 on page 117.

†With a nonlinear equation of state, the nonflux form processes of cabbeling, thermobaricity, and halobaricity discussed in Section 14.1.7 (see also McDougall, 1987a,b) must be included on the right-hand side of the density equation. Here, these terms are dropped, as are other source/sink terms, since the equation of state is linear.

5.5 BASIC NON-EQUILIBRIUM THERMODYNAMICS

The purpose of this section is to introduce some notions of non-equilibrium thermodynamics. In particular, we derive an equation for the evolution of entropy within a nonhydrostatic fluid parcel affected by molecular viscosity. For this purpose, start with the fundamental thermodynamic relation (5.31): $\mathrm{d}\mathcal{I} = T\,\mathrm{d}\zeta - p\,\mathrm{d}v_{(s)} + \mu\,\mathrm{d}C$. This relation holds for a system infinitesimally close to thermodynamic equilibrium. Now assume that each fluid parcel is in local thermodynamic equilibrium, yet allow the full ocean system to be out of equilibrium. These assumptions yield the following internal energy time evolution

$$\rho\,\frac{\mathrm{d}\mathcal{I}}{\mathrm{d}t} = \rho\,T\,\frac{\mathrm{d}\zeta}{\mathrm{d}t} - \frac{p}{v_{(s)}}\,\frac{\mathrm{d}v_{(s)}}{\mathrm{d}t} + \mu\,\rho\,\frac{\mathrm{d}C}{\mathrm{d}t}. \tag{5.122}$$

This result allows one to transfer the methods of equilibrium thermodynamics to the linear irreversible thermodynamics of moving fluid parcels. Note that the term *linear* here refers to an assumption that the system is close to thermodynamic equilibrium. In this case, the dissipative thermodynamic fluxes are linear functions of the gradients of the thermodynamic state variables. Nonlinear effects are *not* absent, however, as there are nonlinear effects from advective transport, nonlinear source terms, a nonlinear equation of state, and nonlinear dependence of the transport coefficients. DeGroot and Mazur (1984) provide a full accounting of this subject, and Gregg (1984) and Davis (1994a) apply these methods to small-scale mixing in the ocean. Slightly different formulations can be found in Landau and Lifshitz (1987) and Batchelor (1967), and their approaches are preferred in the following.

Using equation (5.94) for the evolution of internal energy in equation (5.122) leads to the expression for evolution of entropy in a nonhydrostatic seawater parcel

$$T\,\rho\,\frac{\mathrm{d}\zeta}{\mathrm{d}t} = -\nabla\cdot\mathbf{J}_q + \rho\,\epsilon - \mu\,\rho\,\frac{\mathrm{d}C}{\mathrm{d}t}. \tag{5.123}$$

This equation implies that entropy of a fluid parcel evolves by three irreversible mixing processes: (1) convergence of heat fluxes; (2) mixing of momentum in a viscous fluid, thus creating frictional dissipation sources which ultimately increase a parcel's heat content; and (3) mass exchange resulting in salinity mixing. Correspondingly, a parcel generally maintains constant entropy if processes associated with its evolution are adiabatic, frictionless, and isohaline. Since the friction source is very small in the ocean, adiabatic isohaline transport is very nearly isentropic. Indeed, when ocean modelers refer to adiabatic and isohaline processes, they typically assume this to be synonymous with isentropic.*

It is important to note that the parcel's entropy may remain unchanged even in the event of heating and salinity diffusion. The reason is that these effects can compensate one another, thus canceling out. One may protest that the Second Law of thermodynamics implies that entropy must always increase when there are irreversible processes occuring. Indeed, the Second Law is respected. The loophole is that the fluid parcel under consideration here represents an open system.

*This meaning for isentropic ocean models is consistent with the models not including frictional heating (Section 5.3.2.3).

Hence, its entropy need not always increase when mixing events occur. However, the entropy of the parcel and its surroundings (thus representing a closed system) does increase. In summary, frictionless, adiabatic, isohaline evolution of a fluid parcel is isentropic. However, isentropic transport can involve non-adiabatic and non-isohaline effects, so long as these processes are fine-tuned so they cancel one another.

To identify local entropy sources and fluxes of entropy, assume salinity mixing occurs via the convergence of a salinity flux

$$\rho \frac{dC}{dt} = -\nabla \cdot \mathbf{J}_C, \tag{5.124}$$

thus leading to

$$\begin{aligned} \rho \frac{d\zeta}{dt} &= -\frac{1}{T} \nabla \cdot \mathbf{J}_q + \frac{\mu}{T} \nabla \cdot \mathbf{J}_C + \epsilon \frac{\rho}{T} \\ &= -\nabla \cdot (\mathbf{J_q}/T - \mu \mathbf{J}_C/T) \\ &\quad + \mathbf{J}_q \cdot \nabla(1/T) - \mathbf{J}_C \cdot \nabla(\mu/T) + (\rho/T)\,\epsilon. \end{aligned} \tag{5.125}$$

One can thus identify the entropy flux

$$\mathbf{J}_\zeta = \rho \mathbf{v}\, \zeta + \mathbf{J_q}/T - \mu \mathbf{J}_C/T \tag{5.126}$$

and entropy source

$$\sigma = \mathbf{J}_q \cdot \nabla(1/T) - \mathbf{J}_C \cdot \nabla(\mu/T) + (\rho/T)\,\epsilon. \tag{5.127}$$

The entropy source vanishes when all parcels are in thermodynamic equilibrium with one another, which is the case when the temperature and chemical potential are uniform throughout the ocean, and there is an absence of strain thus eliminating frictional dissipation. Gregg (1984) details the form of entropy sources in the ocean associated with small scale mixing processes.

5.6 THERMODYNAMICAL TRACERS

Heating and cooling of the ocean, as well as mass exchange, predominantly occur near the ocean surface. In contrast, transport in the interior is nearly adiabatic and isohaline (and so nearly isentropic, e.g., equation (5.123)). Hence, the surface ocean experiences irreversible processes that set characteristics of the water masses moving quasi-isentropically within the ocean interior. Useful labels for these water masses maintain their values when moving within the largely ideal ocean interior. Salinity is a good tracer for such purposes since it is altered predominantly by mixing between waters of varying concentrations. This constitutes a basic property of material tracers as considered in Section 5.1.1. We discuss here desirable properties of a thermodynamic tracer that tags the heat within a water parcel and evolves analogously to material tracers.

5.6.1 Potential temperature

Vertical adiabatic and isohaline motion in the ocean changes a fluid parcel's hydrostatic pressure, which thus causes its *in situ* temperature to change in proportion to the adiabatic lapse rate as given by equation (5.42), $\mathrm{d}T = \Gamma \mathrm{d}p$. Consequently, *in situ* temperature is not a useful thermodynamic variable to label water parcels of common origin. Instead, it is more useful to remove the adiabatic pressure effects.

Removing adiabatic pressure effects from *in situ* temperature leads to the concept of *potential temperature*. Potential temperature is the *in situ* temperature that a water parcel of fixed composition would have if it were isentropically transported from its *in situ* pressure to a reference pressure p_r, with the reference pressure typically taken at the ocean surface. Mathematically, the potential temperature θ is the reference temperature obtained via integration of $\mathrm{d}T = \Gamma \mathrm{d}p$ for an isentropic *in situ* temperature change with respect to pressure (e.g., Apel, 1987):

$$T = \theta(s, T, p_r) + \int_{p^r}^{p} \Gamma(s, \theta, p') \, \mathrm{d}p', \tag{5.128}$$

with Γ the lapse rate defined in terms of pressure changes (equation (5.43)). By definition, the *in situ* temperature T equals the potential temperature θ at the reference pressure $p = p_r$. Elsewhere, they differ by an amount determined by the adiabatic lapse rate. Beneath the diabatic surface mixed layer, a vertical profile of potential temperature is far more constant than *in situ* temperature.

The potential temperature of a parcel is constant when the parcel's entropy and material composition are constant. Mathematically, this result follows by noting that when entropy changes at a fixed pressure and composition, $p = p_r$ so that temperature equals potential temperature. Equation (5.41) then leads to

$$\mathrm{d}\zeta = C_p \, \mathrm{d} \ln \theta, \tag{5.129}$$

implying $\mathrm{d}\zeta = 0$ if and only if $\mathrm{d}\theta = 0$.

5.6.2 Potential enthalpy

Potential temperature has proven useful for many oceanographic purposes. However, we have yet to ask whether it is a convenient variable to mark the heat content in a parcel of seawater. Traditionally, it is the potential temperature multiplied by the heat capacity that is used for this purpose. Bacon and Fofonoff (1996) provide a review with suggestions for this approach. In contrast, McDougall (2003) argues that potential temperature multiplied by heat capacity is less precise, by some two orders of magnitude, than an alternative thermodynamic tracer called *potential enthalpy*.

To understand this issue from a mathematical perspective, consider the evolution equation for potential temperature

$$\rho \frac{\mathrm{d}\theta}{\mathrm{d}t} = -\nabla \cdot \mathbf{J}_\theta + \Sigma_\theta, \tag{5.130}$$

where $\mathbf{J}_\theta$ is a flux due to molecular diffusion, and Σ_θ is a source. That potential temperature evolves in this manner is ensured by its being a scalar field. Consider the mixing of two seawater parcels at the same pressure where the parcels have different potential temperature and salinity. In the absence of the source term, the equilibrated state consists of a single parcel with mass equal to the sum of the two separate masses, and potential temperature and salinity determined by their respective mass weighted means. Does this actually happen in the real ocean? That is, can source terms be ignored? Fofonoff (1962) and McDougall (2003) note that it is indeed the case for salinity (and any other material tracer due to conservation of matter), yet it is not the case for potential temperature. Instead, potential temperature contains source terms that alter the mass weighted average equilibrated state. In contrast, potential enthalpy (discussed below) maintains the desired conservative behavior when mixing at constant reference pressure, and nearly maintains this behavior if mixing parcels at a different pressure. Hence, ocean models which set the source term to zero upon mixing potential temperature are in error. McDougall (2003) quantifies this error.

Potential enthalpy is defined analogously to potential temperature. What motivates the use of potential enthalpy is the observation that the fundamental relation between thermodynamic state variables takes a nearly conservative form when written in terms of potential enthalpy. To see this point, return to the evolution of internal energy given by equation (5.94). Introducing the enthalpy per mass (specific enthalpy)

$$\mathcal{H} = \mathcal{I} + p/\rho \tag{5.131}$$

leads to

$$\rho\,\frac{\mathrm{d}\mathcal{H}}{\mathrm{d}t} = -\nabla\cdot\mathbf{J}_q + \frac{\mathrm{d}p}{\mathrm{d}t} + \rho\,\epsilon. \tag{5.132}$$

Dropping the frictional dissipation term arising from molecular friction leads to the approximate statement

$$\rho\frac{\mathrm{d}\mathcal{H}}{\mathrm{d}t} - \frac{\mathrm{d}p}{\mathrm{d}t} \approx -\nabla\cdot\mathbf{J}_q. \tag{5.133}$$

To proceed, note that the fundamental thermodynamic relation (5.31) becomes

$$\mathrm{d}\mathcal{H} = T\,\mathrm{d}\zeta + \rho^{-1}\,\mathrm{d}p + \mu\,\mathrm{d}C \tag{5.134}$$

in terms of enthalpy. Thus, enthalpy can be written as a function of entropy, salt concentration, and pressure,

$$\mathcal{H} = \mathcal{H}(\zeta, C, p). \tag{5.135}$$

Transport of a seawater parcel without changing heat, salt, or momentum occurs without change in entropy (equation (5.123)), thus rendering

$$\left(\frac{\partial\mathcal{H}}{\partial p}\right)_{\zeta,C} = \rho^{-1}. \tag{5.136}$$

Keeping salinity and entropy fixed (or equivalently fixed salinity and potential temperature) leads to

$$\mathcal{H}(\theta, s, p) = \mathcal{H}^{o}(\theta, s, p_r) + \int_{p^r}^{p} \rho^{-1}(\theta, s, p')\,\mathrm{d}p' \tag{5.137}$$

with $\mathcal{H}^{o}(\theta, s, p_r)$ defining the potential enthalpy of a parcel with potential temperature θ and salinity s. Taking the time derivative and using the approximate relation (5.133) yields

$$\frac{\mathrm{d}\mathcal{H}^{o}}{\mathrm{d}t} \approx -\rho^{-1}\,\nabla\cdot\mathbf{J}_q + \int_{p}^{p^r} \mathrm{d}p'\,\frac{\mathrm{d}\rho^{-1}(\theta, s, p')}{\mathrm{d}t}. \tag{5.138}$$

McDougall (2003) shows that for the ocean, the integral

$$\begin{aligned}\int_p^{p^r} \mathrm{d}p' \, \frac{\mathrm{d}\rho^{-1}(\theta, s, p')}{\mathrm{d}t} &= \int_p^{p^r} \mathrm{d}p' \left(\frac{\partial \rho^{-1}}{\partial \theta} \frac{\mathrm{d}\theta}{\mathrm{d}t} + \frac{\partial \rho^{-1}}{\partial s} \frac{\mathrm{d}s}{\mathrm{d}t} \right) \\ &= \frac{\mathrm{d}\theta}{\mathrm{d}t} \int_p^{p^r} \mathrm{d}p' \, \rho^{-1} \, \alpha - \frac{\mathrm{d}s}{\mathrm{d}t} \int_p^{p^r} \mathrm{d}p' \, \rho^{-1} \, \beta\end{aligned} \tag{5.139}$$

has magnitude on the order of the ocean's tiny levels of dissipation arising from molecular viscosity. These expressions introduced the thermal expansion coefficient $\alpha = -\partial \ln \rho / \partial \theta$ and saline contraction coefficient $\beta = \partial \ln \rho / \partial s$ (see Section 5.7). The time derivatives of potential temperature and salinity can be removed from the pressure integrals, since they are each independent of pressure. Given the smallness of $\int_p^{p^r} \mathrm{d}p' \, \mathrm{d}\rho^{-1}/\mathrm{d}t$, one can write the approximate potential enthalpy equation

$$\rho \, \frac{\mathrm{d}\mathcal{H}^o}{\mathrm{d}t} \approx -\nabla \cdot \mathbf{J}_q. \tag{5.140}$$

Hence, potential enthalpy is a state function that approximately specifies the heat in a parcel of seawater, and it evolves analogously to a material tracer such as salinity. See McDougall (2003) for a proof that $\mathcal{H}^o$ more accurately sets the heat for a parcel of seawater than does $C_p \, \theta$. Given that it does, McDougall suggests that *conservative temperature*

$$\Theta \equiv \frac{\mathcal{H}^o(\theta, s, p_r)}{C_p^o} \tag{5.141}$$

with $p_r = 0$ is more appropriate than potential temperature as a thermodynamic tracer for use in an ocean model, and generally for measuring heat in the ocean. In this equation

$$\begin{aligned}C_p^o &= \frac{\mathcal{H}(\theta = 25°\mathrm{C}, s = 35\mathrm{psu}, p_r = 0)}{25°\mathrm{C}} \\ &= 3989.245 \, \mathrm{J\,kg^{-1}\,°K^{-1}}\end{aligned} \tag{5.142}$$

is a heat capacity chosen to minimize the difference between $C_p^o \, \theta$ and potential enthalpy $\mathcal{H}^o(\theta, s, p_r)$ when averaged over the sea surface.* In the remainder of this book, we maintain the notation θ, recognizing the fact that Θ may instead be used. All formulas for density and thermodynamic fluxes can be generalized, as shown by McDougall (2003). From a fundamental perspective, McDougall (2003) provides a compelling case for the use of conservative temperature. Nonetheless, it remains a research topic to determine the significance to simulated ocean circulation of errors made in numerical models using potential temperature rather than conservative temperature.

5.7 OCEAN DENSITY

Density is an important variable to measure in the ocean and to compute in an ocean model. In particular, variations in the density field, via the hydrostatic bal-

*The value quoted by McDougall (2003) is $C_p^o = 3989.24495292815 \, \mathrm{J\,kg^{-1}\,°K^{-1}}$.

ance (equation (5.58)), provide one of the most important driving forces for large-scale circulation. In this section we summarize various forms of density as used in oceanography and ocean models.

5.7.1 *In situ* density

As mentioned earlier, density at a point in the ocean (often called the *in situ* density) is generally a function of temperature, salinity, and pressure,

$$\rho = \rho(T, s, p). \tag{5.143}$$

This equation is known as the *equation of state*, and its form is determined empirically. Recall that equation (5.128) provides a unique relation between potential temperature θ and temperature, given salinity and pressure. Hence, density can just as well be expressed as a function of potential temperature, salinity, and pressure

$$\rho = \rho(\theta, s, p). \tag{5.144}$$

Likewise, McDougall (2003) indicates that density can be written as a function of conservative temperature Θ defined in Section 5.6.2.

Writing the equation of state in terms of θ (or Θ) is more convenient for ocean models than in terms of *in situ* temperature, because ocean models prognostically solve for potential temperature (or conservative temperature). An accurate equation of state for use in ocean models using potential temperature has been given by McDougall et al. (2003b), and updated by Jackett et al. (2004). This work is based on that of Feistel (1993), Feistel and Hagen (1995), and Feistel (2003). Most ocean models are now switching to such accurate equations of state since the earlier approximate forms, such as Bryan and Cox (1972), maintain a relatively narrow range of salinity variations over which the equation is valid. With ocean models of refined grid resolution and realistic fresh water forcing, it is desirable to remove such limitations since model salinity can vary quite widely, especially near river mouths.

In formulations of the equation of state, the pressure variable is not the absolute pressure as measured at a point in the ocean fluid. Instead, it is conventional to use the *gauge* pressure (e.g., McDougall et al., 2003b). The gauge pressure is the absolute or *in situ* pressure at a point minus the standard atmosphere pressure

$$p_{\text{gauge}} = p - p_{\text{standard}}, \tag{5.145}$$

with

$$p_{\text{standard}} = 10.1325 \text{ dbars}. \tag{5.146}$$

As reviewed by Jackett et al. (2004), salinity measurements in the ocean are in units of *practical salinity units* (psu). These units, which are dimensionless, are close, but not equal, to the absolute salinity as measured as a concentration, say parts per thousand (ppt). The relation between salinity in psu and salinity in ppt depends on the precise ionic concentration of the seawater. However, for purposes of ocean climate models, it is quite accurate to set

$$s_{\text{ppt}} = 1.004867\, s_{\text{psu}}. \tag{5.147}$$

Most ocean models interpret their salinity as s_{psu}, since this facilitates direct comparison with ocean measurements of salinity. However, when diagnosing salt budgets (in terms of mass of salt), then it is important to note this conversion factor.

The functional relation $\rho = \rho(\theta, s, p)$ allows one to develop the material time derivative of density

$$\frac{\mathrm{d}\ln\rho}{\mathrm{d}t} = \left(\frac{\partial \ln\rho}{\partial p}\right)_{s,\theta}\frac{\mathrm{d}p}{\mathrm{d}t} + \left(\frac{\partial \ln\rho}{\partial s}\right)_{p,\theta}\frac{\mathrm{d}s}{\mathrm{d}t} + \left(\frac{\partial \ln\rho}{\partial \theta}\right)_{p,s}\frac{\mathrm{d}\theta}{\mathrm{d}t}. \tag{5.148}$$

Introducing the thermal expansion and saline contraction coefficients

$$\alpha = -\left(\frac{\partial \ln\rho}{\partial \theta}\right)_{p,s} \tag{5.149}$$

$$\beta = \left(\frac{\partial \ln\rho}{\partial s}\right)_{p,\theta} \tag{5.150}$$

and the squared speed of sound

$$c_s^2 = \left(\frac{\partial p}{\partial \rho}\right)_{s,\theta} \tag{5.151}$$

leads to

$$\begin{aligned}\frac{\mathrm{d}\ln\rho}{\mathrm{d}t} &= \frac{1}{\rho c_s^2}\frac{\mathrm{d}p}{\mathrm{d}t} + \beta\frac{\mathrm{d}s}{\mathrm{d}t} - \alpha\frac{\mathrm{d}\theta}{\mathrm{d}t} \\ &= -\nabla\cdot\mathbf{v}\end{aligned} \tag{5.152}$$

where the last step used mass conservation.

For fluid parcels undergoing isentropic and isohaline motion (frictionless motion at constant potential temperature and constant salinity; see equation (5.123)), pressure satisfies

$$\begin{aligned}\rho\,\mathrm{d}p/\mathrm{d}t &= (\rho\,p)_{,t} + \nabla\cdot(\rho\,\mathbf{v}\,p) \\ &= -(\rho\,c_s)^2\,\nabla\cdot\mathbf{v},\end{aligned} \tag{5.153}$$

whose linearized fluctuations are known as *acoustic modes*. Note that for the special case of a Boussinesq parcel with $\nabla\cdot\mathbf{v} = 0$, the speed of sound is effectively infinite. Hence, each term in equation (5.152) vanishes for isohaline, adiabatic, Boussinesq parcels.

In the more general case where parcels mix as they are materially transported, e.g., from molecular diffusion, the potential temperature and salinity terms in equation (5.152) become

$$\frac{1}{\rho c_s^2}\frac{\mathrm{d}p}{\mathrm{d}t} + \beta\,\nabla\cdot(\kappa_s\,\nabla s) - \alpha\,\nabla\cdot(\kappa_\theta\,\nabla\theta) = -\nabla\cdot\mathbf{v}, \tag{5.154}$$

with κ_s and κ_θ the molecular diffusivities for salinity and potential temperature, respectively. This equation indicates that in addition to material changes in pressure, the mixing of salinity and potential temperature act, via mass continuity, to balance changes in a parcel's volume. For example, absent salinity and pressure effects, raising the potential temperature of a mass conserving fluid parcel via molecular diffusion ($\nabla\cdot(\kappa_\theta\,\nabla\theta) > 0$) causes an increase in the parcel's volume ($\mathrm{d}\ln(\mathrm{d}V)/\mathrm{d}t = \nabla\cdot\mathbf{v} > 0$) when the thermal expansion coefficient α is positive.

5.7.2 Potential density

Since isentropic motion of a frictionless fluid parcel generally occurs at materially constant potential temperature and materially constant salinity (see equation (5.123)), it is convenient to combine the evolution of these two active tracers into the evolution of a single variable. *Potential density* is one such combination. By definition, potential density ρ_{pot} is the density a fluid parcel would have if isentropically moved to a reference pressure p_r, often taken as pressure at the ocean surface

$$\rho_{\text{pot}} = \rho(\theta, s, p_r). \tag{5.155}$$

Hence, the material evolution of potential density is given by

$$\frac{\mathrm{d} \ln \rho_{\text{pot}}}{\mathrm{d}t} = \beta \frac{\mathrm{d}s}{\mathrm{d}t} - \alpha \frac{\mathrm{d}\theta}{\mathrm{d}t}, \tag{5.156}$$

where α and β are evaluated at the reference pressure p_r. For general isentropic motion where potential temperature and salinity are materially conserved and frictional dissipation is negligible (see equation (5.123)), potential density is likewise materially conserved. This behavior is in contrast to *in situ* density, whose evolution is given by equation (5.152).

In many parts of the ocean, especially those close to the reference pressure, potential density is monotonically stacked in the vertical and thus forms a useful method for layering the ocean. Many physical processes related to tracer transport naturally occur within, rather than across, potential density layers. More precisely, these processes occur predominantly along *neutral directions* (e.g., McDougall, 1987a; Gent and McWilliams, 1990; Griffies et al., 1998), which are directions tangent to the locally defined (local value of p) potential density surface. We have more to say about *neutral physics* processes in Chapters 9 and 13–16.

5.7.3 Idealized equations of state

For certain purposes, it is useful to approximate the equation of state used in ocean models. One common idealization is to equate the *in situ* density to the potential density

$$\rho(\theta, s, p) = \rho_{\text{pot}}(\theta, s, p_r). \tag{5.157}$$

Models using this idealized thermodynamics remove pressure effects from the *in situ* density, thus becoming incompressible or Boussinesq. The use of potential density in such models to compute hydrostatic pressure causes inaccuracies in the horizontal pressure gradient. In particular, the thermal wind relations (Section 4.9.4), valid for geostrophic flows, are compromised. These inaccuracies are typically unimportant for idealized studies, such as dynamical explorations of adiabatic isopycnal layered models (e.g., Chapter 6). Yet the inaccuracies can be important for realistic simulations, thus necessitating a more accurate approach (e.g., Sun et al., 1999).

To further idealize the Boussinesq fluid's thermodynamics, it is convenient to compute potential density as a linear function of potential temperature and salinity

$$\rho_{\text{pot}} = \rho_o \left[1 - \alpha\,(\theta - \theta_o) + \beta\,(s - s_o)\right], \tag{5.158}$$

where α, β, θ_o, and s_o are constants. Indeed, some authors consider this expression for density to be part of the Boussinesq approximation (e.g., Chandrasekhar, 1961), although this is distinctly not the case in the oceanographic literature. Those wishing to avoid complications of two active tracers further simplify this expression by dropping the salinity dependence.

5.7.4 Quasi-non-Boussinesq approximation

An intermediate approximation for the equation of state, often used in early ocean climate models (e.g., Bryan and Cox, 1972), replaces the dependence on *in situ* pressure with the hydrostatic pressure felt by a fluid of density ρ_o

$$\rho(\theta, s, p) \approx \rho(\theta, s, p = g\,|z|\,\rho_o), \tag{5.159}$$

where $z < 0$ is the depth of the fluid parcel and ρ_o is the constant Boussinesq reference density. Notably, this expression for density removes local time-dependence from the pressure field within the non-Boussinesq continuity equation. Hence, a non-Boussinesq fluid using the equation of state (5.159) is transformed into a *quasi-non-Boussinesq fluid*, where the term "quasi" signals that the fluid admits no acoustic modes. Durran (1999) refers to this approximation as *pseudo-incompressibility*. Notably, this method for filtering acoustic modes does not introduce a new elliptic problem, in contrast to certain other wave filtering methods such as the rigid lid approximation of Bryan (1969) (see Sections 2.4.4 and 12.10).

Use of the approximate density (5.159) in the continuity equation must be accompanied by its use in the tracer and velocity budgets to maintain self consistency. However, one recovers significant accuracy in quasi-non-Boussinesq fluids by employing the proper pressure dependence within the equation of state for purposes of computing horizontal pressure gradients. The importance of such accuracy was emphasized by Dewar et al. (1998).

5.8 CHAPTER SUMMARY

Thermo-hydrodynamics is the study of fluid motion (hydrodynamics) along with the flow of heat, tracers, etc.. This field of study is fundamental to chemical engineering, where fluid flow with tracers, and chemical reactions, is ubiquitous. It is also fundamental to oceanography, and we touched on some of the basic issues in this chapter.

The notion of a scalar tracer field was formally introduced at the start of this chapter. Three sorts of tracers were identified: (1) material tracers such as salinity, chemical, and/or biological species; (2) thermodynamic tracers such as temperature, enthalpy, and entropy; and (3) dynamical tracers, such as vorticity, helicity, and potential vorticity. Section 5.1 presented a discussion of material tracers, whose concentration (mass fraction) in seawater is given by the ratio

$$C = \frac{\mathrm{d}M_C}{\rho\,\mathrm{d}V}, \tag{5.160}$$

with $\mathrm{d}M_C$ the mass of tracer constituent in the seawater parcel and $\rho\,\mathrm{d}V$ the total mass of the parcel (mass of water plus mass of material constituents). If the water parcel maintains a constant seawater mass as well as material constituent mass, then the parcel satisfies the conservation equation for tracer concentration

$$(\rho C)_{,t} + \nabla \cdot (\rho C \mathbf{v}) = 0. \tag{5.161}$$

The Boussinesq form of this equation is obtained by setting ρ to the constant Boussinesq reference density ρ_o. This discussion highlighted the importance of maintaining compatibility between the tracer and mass conservation equations within a numerical model.

A substantial part of this chapter focused on thermodynamical tracers, with various forms of temperature the key example of such tracers. To introduce these tracers, we first reviewed elements of equilibrium thermodynamics in Section 5.2. The First Law of thermodynamics relates the infinitesimal changes in a system's internal energy to changes in work and heat applied to the system. The internal energy represents the energy maintained by the huge numbers of internal molecular degrees of freedom which are averaged over in a continuum formulation of ocean fluid mechanics.

Section 5.3 turned to a discussion of the evolution of mechanical and internal energies within a fluid parcel. This discussion allowed for a comparison of the evolutions in nonhydrostatic and hydrostatic parcels, as well as Boussinesq and non-Boussinesq parcels. Particular notice was made of the somewhat unphysical terms appearing in the Boussinesq gravitational potential energy budget, arising from the constraint that the parcel maintains constant volume rather than constant mass. This introduces unphysical *virtual mass* fluxes to the Boussinesq fluid. Section 5.4 followed up the study of parcel energetics with energetics for a global ocean domain. Some of the effects of boundary forcing were exposed here, which provide the ultimate source of kinetic energy for the ocean via atmospheric winds.

The discussion in these two sections provided a useful venue to highlight the huge discrepency between the very tiny levels of frictional dissipation in the real ocean associated with molecular viscosity, and the relatively huge levels employed by global ocean climate models. Given the smallness of the molecular viscosity, and the correspondingly tiny Kolmogorov scale, it may be conjectured that some decades will pass before global ocean models reduce their level of numerical dissipation to the levels of molecular dissipation. Until then, the internal energy budget of ocean models will remain inaccurate, even for models maintaining the full nonhydrostatic and non-Boussinesq equations. The reason for this inaccuracy is that frictional dissipation is so large in the global models that it cannot be transferred into frictional heating, as required to close the internal energy budget in the model. If one insisted on closing the budget by admitting frictional heating, then the spuriously huge levels of this heating would soon degrade the simulation's water mass characteristics, thus making the simulation of little physical use.

Section 5.5 introduced elements of linear irreversible thermodynamics, which considers how thermodynamic notions are applied to systems near, but not at, equilibrium. This formalism has been applied to many aspects of the ocean, such as deriving the evolution of oceanic entropy and heat. This subject then led into a discussion of thermodynamical tracers in Section 5.6. This section discussed attributes of potential temperature, which is a superior temperature variable for ocean models than *in situ* temperature. The work of McDougall (2003), discussed in Section 5.6.2, suggests a related but distinct heat-like variable that has superior conservation properties relative to potential temperature. This temperature variable has yet to be systematically employed in ocean models.

Ocean density and its various forms finished the chapter in Section 5.7. The equation of state, which relates salinity, pressure, and temperature to density, is an important diagnostic equation used in ocean models. Computing an accurate density ensures that pressure gradients are likewise computed accurately. Doing so requires a realistic simulation of the temperature and salinity fields.

We are ready to document the basic set of ocean equations. For this purpose, refer back to the summaries given for Chapter 3 in Section 3.5, and for Chapter 4 in Section 4.12. First, the nonhydrostatic, non-Boussinesq equations are given by

$$(\rho \mathbf{v})_{,t} + \nabla \cdot (\rho \mathbf{v}\,\mathbf{v}) + (\mathbf{\Omega} + \hat{\mathbf{z}}\,\mathcal{M}) \wedge (\rho \mathbf{v}) = -\nabla p - \hat{\mathbf{z}}\,\rho g + \rho \mathbf{F}^{(\mathbf{v})} \tag{5.162}$$

$$(\rho \theta)_{,t} + \nabla \cdot (\rho \theta \mathbf{v} + \mathbf{J}_\theta) = \Sigma_\theta \tag{5.163}$$

$$(\rho s)_{,t} + \nabla \cdot (\rho s \mathbf{v} + \mathbf{J}_s) = \Sigma_s \tag{5.164}$$

$$\rho_{,t} + \nabla \cdot (\rho \mathbf{v}) = 0 \tag{5.165}$$

$$\rho = \rho(\theta, s, p). \tag{5.166}$$

The hydrostatic and shallow ocean approximations lead to the non-Boussinesq primitive equations

$$(\rho \mathbf{u})_{,t} + \nabla \cdot (\rho \mathbf{v}\,\mathbf{u}) + (f + \mathcal{M})\,\hat{\mathbf{z}} \wedge (\rho \mathbf{v}) = -\nabla_z p + \rho \mathbf{F}^{(\mathbf{u})} \tag{5.167}$$

$$p_{,z} = -\rho g \tag{5.168}$$

$$(\rho \theta)_{,t} + \nabla \cdot (\rho \theta \mathbf{v} + \mathbf{J}_\theta) = \Sigma_\theta \tag{5.169}$$

$$(\rho s)_{,t} + \nabla \cdot (\rho s \mathbf{v} + \mathbf{J}_s) = \Sigma_s \tag{5.170}$$

$$\rho_{,t} + \nabla \cdot (\rho \mathbf{v}) = 0 \tag{5.171}$$

$$\rho = \rho(\theta, s, p). \tag{5.172}$$

The Boussinesq equations are found by setting ρ to ρ_o in all places, except where it multiplies the gravitational acceleration.

Our work in this chapter, as well as in Chapters 3 and 4, utilized the geopotential coordinate z to measure distance in the vertical. This coordinate is orthogonal to the lateral coordinates, which are typically orthogonal between themselves. Such orthogonality allows one to focus on the essential dynamical laws without getting involved in difficulties arising from non-orthogonality. However, for geophysical fluid dynamics and ocean modeling, analytical and simulation capabilities are greatly enhanced by relaxing this orthogonality constraint, at least so far as the vertical coordinate is allowed to be non-orthogonal with the two orthogonal lateral coordinates. For this reason, in Chapter 6 we revisit many of the results from this chapter and Chapters 3 and 4 by allowing for such non-orthogonality. A discussion of these issues is essential for understanding many ongoing research topics at the forefront of ocean model design.

Chapter Six

GENERALIZED VERTICAL COORDINATES

So far in this book, we have employed three orthogonal coordinates to formulate the basic equations of ocean fluid mechanics and thermodynamics. Relaxing orthogonality between the vertical and horizontal greatly facilitates one's ability to analyze ocean phenomena and to develop interesting and very compelling sorts of ocean models. The purpose of this chapter is to present some of the salient issues related to these *generalized vertical coordinates*. Work with models formulated using such coordinates remains at the forefront of ocean model development. Upon completing this chapter, the reader should have a firm understanding of the physical motivation for generalizing the vertical coordinate and a working knowledge of the often subtle mathematics involved.

6.1 INTRODUCTION

The choice of vertical coordinate is the most important aspect of an ocean model's design. This choice strongly prejudices the model's representation of various resolved dynamical processes and determines the details of how to parameterize unresolved processes. Hence, there is a significant amount of physical, mathematical, and numerical research aimed at optimizing the choice of vertical coordinate. This research is taking place on many levels, from the fundamental formulation stage focusing on the most convenient form of the continuum equations to the practical numerical implementation stage where the pros and cons of various algorithms are debated. Currently, there is no clear single best choice for realistic global ocean climate simulations. Each vertical coordinate has certain advantages and disadvantages, many of which are complementary. This result argues for the use of *hybrid* or *generalized* vertical coordinates, where one tailors the choice based on the particular flow regime.

Generalized vertical coordinates as used in geophysical fluid models are often termed *non-orthogonal projected coordinates*, for reasons which will become clear in this chapter. As the issues of vertical coordinate are quite distinct from horizontal coordinates, simplicity of presentation warrants the use of Cartesian (x, y) horizontal coordinates. Transformations to generalized horizontal coordinates follow the techniques discussed in Chapters 20 and 21.

References for the mathematical formalism described in this chapter include the pioneering paper by Starr (1945). Starr was the first to systematically present the mathematical aspects of generalized vertical coordinates. He also argued for their utility for models in geophysical fluid dynamics (GFD). Bleck (1978) illustrates the geometric aspects of generalized vertical coordinates and shows how to construct an energetically consistent numerical model using them. Bleck is a pioneer in the use of isopycnal ocean models, isentropic atmospheric models, and generalized vertical coordinate ocean models (Bleck, 2002). An appendix in McDougall (1995) provides further words in support of Bleck's geometric interpretation of these coordinates.

6.2 CONCERNING THE CHOICE OF VERTICAL COORDINATE

To set the stage for mathematical investigations of generalized vertical coordinates, start by considering the stylized schematic of an ocean basin shown in Figure 6.1. Although cartoon-like in appearance, this figure illustrates three fundamental regimes of ocean dynamics key to understanding and simulating the ocean climate system.

The surface mixed layer, also known as the planetary boundary layer (PBL), is a region at and near the ocean surface. This region is strongly coupled to other components of the climate system, such as the atmosphere, sea ice, rivers, human activities, and so on. The PBL is typically very well mixed in the vertical, and the relevant physical processes are mostly three-dimensional boundary layer processes. A parameterization of these processes is necessary in a primitive equation ocean climate model, since explicit representation requires very high resolution (of order meters to centimeters in all three directions) as well as nonhydrostatic physics.

In contrast to the surface layer, the ocean interior is close to ideal (i.e., isentropic as defined in Section 5.5). In particular, most transport processes in this region, associated with mesoscale eddies, occur along surfaces of constant potential density, with very little mixing of properties between potential density layers. Hence, this region is often modeled by stacked immiscible fluid layers, each satisfying coupled shallow water equations. Coupling between layers occurs from stresses imparted by one layer on another. The text by Cushman-Roisin (1994) provides a lucid discussion of this physical system.

The solid earth boundary plays a very important role in directing ocean currents (i.e., land-sea boundaries, straights, throughflows) and interacting with overlying currents by affecting the density and hence the pressure. Additionally, there are many regions of the ocean where turbulent bottom boundary layer (BBL) processes act to strongly mix and transport water masses. These processes can be critical for determining the characteristics of many deep and intermediate water masses of the World Ocean.

As interesting and complex as physical processes are within a particular regime, oceanographers are also beginning to appreciate the importance of interactions between the regimes. For example, mesoscale eddies do interact with the mixed layer as well as the ocean topography. Likewise, mixed layer processes can reach very deep into the ocean in regions of deep convection. Garnering an understanding of such interactions, and then parameterizing them, constitutes an intense area of current research.

Each of the regimes indicated in Figure 6.1 suggests a corresponding vertical coordinate for use in simulating flow active in these regimes. We now compare basic features of these coordinates. The text by Haidvogel and Beckmann (1999) provides a more thorough treatment of the following discussion of ocean models.

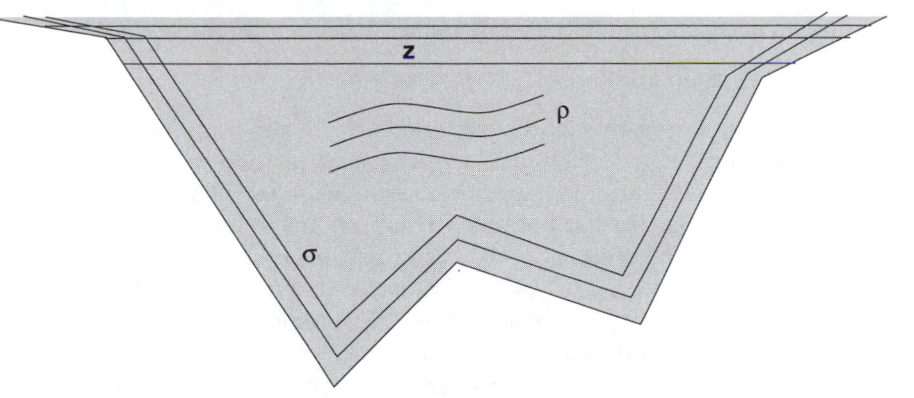

Figure 6.1 Stylized rendition of an ocean basin, illustrating three fundamental regimes of ocean dynamics critical to the ocean climate system. The surface mixed layer is naturally represented using z-coordinates (or even more naturally with pressure coordinates); the isentropic interior is naturally represented using isopycnal or ρ-coordinates; and the bottom topography is naturally represented using sigma or σ-coordinates.

6.2.1 Z-coordinate ocean models

The simplest and oldest choice of vertical coordinate is z, which is the vertical distance from a resting ocean surface* at $z = 0$, with z positive upward and $z = -H(x, y)$ the ocean bottom. This coordinate was chosen to formulate the equations of ocean mechanics and thermodynamics in Chapters 3 through 5. The vertical direction $\hat{\mathbf{z}}$ is orthogonal to the two horizontal directions. Note that z-models are also often referred to as *geopotential* coordinate models, in which z is the vertical displacement with respect to a local approximation to the sea level geopotential (Section 3.2.1). Z-models are presently those most widely used for the study of ocean climate.†

Z-models have been around for many decades, with the pioneering work of Bryan (1969), Semtner (1974), and Cox (1984) providing the community with the first examples of working ocean climate models. The author of this book has worked extensively with the Modular Ocean Model (MOM) (e.g., Pacanowski et al., 1991; Pacanowski, 1995; Pacanowski and Griffies, 1999; Griffies et al., 2004), which is a descendent of the Bryan-Cox-Semtner models. Indeed, much of the present book describes elements of ocean models motivated by an aim to document MOM.

The main advantages of z-models are the following.

- Simple (and often naive) numerical discretization methods have been used, with some success.
- The PBL is naturally parameterized.
- Thermodynamical effects are well represented, and in particular implement-

*A static ocean under hydrostatic balance.

†See Griffies et al. (2000a) for a tabulation of publicly supported ocean climate models.

ing an accurate equation of state for seawater is straightforward.

There are three main disadvantages of z-models.

- Representation of the solid earth boundary is generally cumbersome, as it is discretized with rectangular steps instead of piecewise linear fits to the topography. Rectangular steps impose an artificial distinction between vertical side walls and flat bottoms. There are means to overcome aspects of this awkward representation by using partial or shaved bottom cells, as illustrated in Figure 6.2. These methods have been successfully implemented in many popular z-models. However, there remains the issue of parameterizing the bottom boundary layer flow, which is quite important in the ocean. In general, even with more faithful representation of bottom topography, the z-coordinate framework is cumbersome for representing bottom intensified flows and transport.

- The representation of dynamics and physics in the ideal ocean interior, away from the side and bottom boundaries, requires great care in z-models. We discuss certain of the issues in Part 4 of this book. Suffice it to say that the smallness of the mixing between potential density layers places an extremely tough constraint on how "sloppy" z-model tracer transport numerics can be.

- Use of a free surface algorithm, which is desirable for physical reasons (see Chapter 12), imposes a limitation on deviations of the surface height to avoid a vanishing surface model grid cell.* As ocean climate models refine their vertical grid spacing, this limitation restricts one's ability to simulate large deviations in tidal fluctuations, or large displacements of the free surface due to sea ice.

6.2.2 Sigma-coordinate ocean models

The sigma-coordinate is given by

$$\sigma = \frac{z - \eta}{H + \eta}, \tag{6.1}$$

where $\eta(x, y, t)$ is the displacement of the ocean surface from its resting position $z = 0$, and $z = -H(x, y)$ is the ocean bottom. As defined, the dimensionless coordinate σ takes values of $\sigma = 0$ at the ocean surface and $\sigma = -1$ at the bottom. The σ-coordinate provides a monotonic mapping between the position of a parcel within a time-dependent vertical column of ocean fluid (where time dependence arises from fluctuations in the free surface), to a unit interval fixed in time. That is, the definition (6.1) defines a unique mapping between z and σ. Monotonicity is required in order to have σ be a valid vertical coordinate for uniquely describing a point in the vertical. Since the σ surfaces vary according to bottom topography, σ-models are also often called *terrain following* models.

Ocean σ-models are relatives of similar models used by atmospheric scientists starting in the late 1950's. As described in the text by Haltiner and Williams (1980),

---

*Figure 11.1, page 224 illustrates this issue.

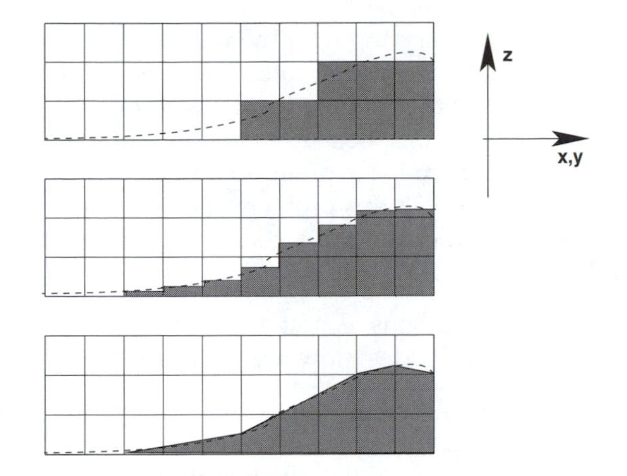

Figure 6.2 Comparing representations of the ocean bottom in z-models. Top: The old-fashioned "full cell" approach in which vertical thicknesses of all cells are independent of latitude and longitude. Middle: The "partial cell" approach discussed by Adcroft et al. (1997) and Pacanowski and Gnanadesikan (1998). Here, vertical thicknesses of the bottom cells can vary according to the topographic features. Bottom: The "shaved cell" approach of Adcroft et al. (1997), in which the bottom cell is a piecewise linear fit to the topography. Both full and partial cells have discontinuous representations of the bottom, whereas the shaved cell has a continuous depth but a discontinuous gradient of the depth. This figure is based on Figure 3 of Griffies et al. (2000a) and Figure 4 from Adcroft et al. (1997).

the terrain following coordinate was introduced by Phillips (1957) to better represent the atmospheric planetary boundary layer. It does so by aligning the "vertical" coordinate parallel to the undulating lower boundary. Blumberg and Mellor (1987) succeeded in building the first ocean σ-model. Various incarnations of this model, as well as related efforts such as those led by researchers at Rutgers University and the University of California at Los Angeles (for reviews of σ-models, see Greatbatch and Mellor, 1999; Ezer et al., 2002), are today used by thousands of scientists and engineers, most of whom focus on regional and/or coastal phenomena. In addition to its pioneering the use of a σ-coordinate for the ocean, the model of Blumberg and Mellor (1987) was the first primitive equation ocean model to employ a free surface. Among other advantages, this choice allowed their model to better represent tidal phenomena. We have more to say about free surface ocean models in Chapter 12.

The main advantages of σ-models are the following.

- They provide a smooth representation of the solid earth boundary, and so provide for a natural representation of bottom boundary layer physics and bottom intensified currents.

- Thermodynamic effects associated with the equation of state are well represented.

There are three main problems with the σ-coordinate models.

- In the presence of arbitrary topography, the PBL cannot be as well represented using σ as with the z-coordinate. The reason is that the vertical distance between grid points generally increases as one moves away from the side continental shelf regions, hence leaving the PBL with potentially coarse vertical resolution in the middle of an ocean basin. As discussed in Haidvogel and Beckmann (1999), this problem can be easily overcome by use of slightly modified definitions of the σ coordinate, and so is not a problem of fundamental importance.

- For many of the same reasons as noted for z-models, the representation of the ocean interior's nearly ideal flow regime is unnatural in σ-models.

- A second major problem with σ-models is that they have a difficult time accurately representing the horizontal pressure gradient. Because the surfaces of constant σ are not generally flat, the horizontal pressure gradient, which is perpendicular to the local vertical direction $\hat{\mathbf{z}}$ as defined by gravity, has a projection along and across σ surfaces. That is, σ-models employ a non-orthogonal set of coordinates. The result is a horizontal pressure gradient consisting of two terms
$$\begin{aligned} \nabla_z p &= (\nabla_\sigma - \nabla_\sigma z\, \partial_z)\, p \\ &= \nabla_\sigma p + \rho g\, \nabla_\sigma z. \end{aligned} \tag{6.2}$$
The coordinate transformation (6.33) on page 134 relating lateral gradients was used in the first equality, and the hydrostatic relation $p_{,z} = -\rho g$ was used in the second. In words, $\nabla_z p$ is the horizontal pressure gradient taken along surfaces of constant depth z, whereas $\nabla_\sigma p$ is the pressure gradient along surfaces of constant σ. The term $\nabla_\sigma z$ is the slope of the σ surface relative to constant depth surfaces. In the ocean, especially next to the continental shelves and certain seamounts, the slope of σ surfaces can reach above $1/100$. These slopes are nontrivial, and they tend to cause the "sigma coordinate correction term" $\rho g\, \nabla_\sigma z$ to be on the order of $\nabla_\sigma p$. Hence, maintaining a very accurate numerical representation of the two terms is quite important. Otherwise, errors in either term can contribute to spurious pressure gradients that drive unphysical currents.

There have been numerous efforts aimed at reducing the pressure gradient errors found in σ-models. The review by Ezer et al. (2002) cites cases where smarter schemes can make a nontrivial difference. Even so, the problems with pressure gradient errors, as well as the only recent development in σ-models of neutral physics schemes relevant for the ocean interior (see Part 4), have meant that σ-models have not been widely used by those focusing on global climate research. Indeed, as noted in the review of ocean climate models by Griffies et al. (2000a), there are presently no publications where a global ocean climate simulation* uses a σ-model or related terrain following coordinate models. It is hoped that the next few years will see more contributions to the global problem from this community, as these models can provide a very useful complement to other models used for climate.

*Defined here as a prognostic global model run for more than a century using realistic boundary forcing and a sea ice model.

6.2.3 Isopycnal-coordinate ocean models

Another choice for vertical coordinate is the potential density ρ_{pot}, which is density referenced to an arbitrary fixed pressure (Section 5.7.2). In most cases, the "pot" label is omitted for brevity, and we will do so in this book. This coordinate is a close analog to the atmosphere's entropy or potential temperature. In an idealized ocean consisting of a simplified equation of state, potential density defines a monotonic layering of the ocean fluid, and so it is a valid choice of vertical coordinate in this case. In general, physical processes, such as turbulent stirring and mixing from mesoscale eddies, have a strong tendency to occur along these surfaces, rather than across them. Indeed, the ocean interior can, to a very good approximation, be considered a fluid comprised of stacked immiscible potential density layers.

The main advantages of ρ-models are the following:

- They provide a suitable framework for simulating and analyzing the ocean's fluid dynamics occurring away from boundaries.
- So long as the ρ-model has enough density layers, it provides a natural enhancement of resolution in regions where density gradients are tight, such as in strong fronts (e.g., eddies, boundary currents, overflows). This tightening may be very helpful in capturing many dynamically important ocean regions using a minimal level of model grid resolution.
- The simulation of ocean climate in the presence of mesoscale eddies in z-models and σ-models is difficult because of the nontrivial levels of spurious diapycnal mixing that can be associated with the numerical transport operators (Section 13.1.1). Isopycnal models maintain all spurious mixing within density classes, and so do not allow for spurious diapycnal mixing. This is a distinct advantage of the ρ-models for long-term eddying ocean climate simulations.

The main problems with the isopycnal coordinate models are the following.

- The real ocean has a complicated equation of state. It turns out that there is accordingly no materially conserved density coordinate that is also monotonic with depth. For example, zero pressure was originally the common reference pressure used to define potential density in isopycnal ocean models. Yet this pressure is inadequate for the deep ocean. In particular, it leads to such unphysical effects as simulated Southern Ocean Deep Water living above North Atlantic Deep Water. Recently, 2000 decibars has been chosen as the reference pressure for realistic isopycnal models. This pressure leads to fewer regions with spurious water mass inversions (e.g., Sun et al., 1999). In general, the equation of state, as well as wide variations in ocean salinity such as may arise from river runoff, ice melt, and strong evaporation, adds complications to isopycnal coordinate models when used for realistic ocean climate simulations (see also Hallberg, 2003b).
- A purely isopycnal model is an inappropriate framework for implementing parameterizations of the surface PBL. The reason is that the PBL is approximately vertically unstratified in density, and so a pure isopycnal model will have only a single grid point to represent the mixed layer. This limitation

is nontrivial, since the mixed layer is where important interactions occur between the ocean, atmosphere, and sea ice. It is for this reason that bulk mixed layer models are appended to isopycnal models to simulate the surface mixed layer (Hallberg, 2003a). Alternative efforts aim towards a hybrid vertical coordinate in which a pressure-coordinate is used for the PBL, an isopycnal-like coordinate is used for the interior, and σ-coordinate is used near the bottom (Bleck, 2002).

6.2.4 Summary of the vertical coordinates

As mentioned at the beginning of this section, there appears to be no single vertical coordinate that is without problems when aiming to realistically simulate the World Ocean's circulation over time scales relevant for climate. Hence, there is presently a great deal of research focused on developing ocean models that allow for any continuously defined vertical coordinate. Given such freedom, researchers can examine the benefits of certain generalized, or hybrid choices of vertical coordinates.

For much of the remainder of this chapter, we leave behind particular examples of vertical coordinates. The focus will instead be on elements of the ocean equations as formulated with generalized vertical coordinates.

6.3 GENERALIZED SURFACES

Consider a partitioning of the ocean into stacked layers of smooth surfaces. These surfaces can be defined mathematically by giving the value of a *generalized vertical coordinate*, denoted in the following by the letter s. The value of this coordinate is generally a function of space-time, and so it maintains the functional dependence

$$s = s(x, y, z, t). \tag{6.3}$$

Three examples of generalized vertical coordinates were encountered in Section 6.2: the trivial example of z-models has $s = z$, terrain following σ-models have $s = \sigma$ defined according to equation (6.1), and isopycnal models have $s = \rho_{\text{pot}}$, with ρ_{pot} the potential density referenced to a chosen pressure. Note that s is a common notation used in the literature, and so it will be used in this chapter. Care should be taken, however, not to confuse this symbol with ocean salinity. The context should be sufficient to make the distinction.

6.3.1 Constraining the vertical coordinate

To make generalized surfaces useful for modeling geophysical fluid dynamical flows, they must be constrained never to have an undulation parallel to the local vertical direction. Mathematically, this means that

$$s_{,z} \neq 0 \tag{6.4}$$

throughout the fluid. That is, the surfaces must retain a nontrivial vertical stratification. Correspondingly, $s_{,z}$ must remain single-signed and so s is monotonic with depth. These basic assumptions are maintained by the generalized vertical coordinates discussed in this chapter.

The constraint (6.4) allows one to orient the surfaces of constant generalized vertical coordinate unambiguously within the ocean fluid. It is trivially satisfied for z-models, where $s_{,z} = 1$. It is also satisfied for terrain following sigma surfaces, so long there is no attempt to represent "overhangs" or "caves" in the bottom topography. In certain stratified fluids, surfaces of constant potential density are monotonic with depth and so also satisfy this constraint.

6.3.2 Specific thickness

The constraint (6.4) ensures that slopes of the surfaces never become infinite. If they become infinite, the transformation between depth z and generalized vertical coordinate s would become singular. Correspondingly, the Jacobian of transformation $\partial z/\partial s = z_{,s}$ (see Section 6.5.2) becomes infinite, and so there is no longer a one-to-one relation between the two coordinates. This situation is undesirable, as it leads to a non-unique specification of a point in the ocean.

The Jacobian of transformation $z_{,s}$ is often called the *specific thickness*, or *thickness* for short. This name can be motivated by considering a stacked set of s-layers. Invertibility of the transformation between z and s requires $z_{,s}$ to be single-signed. When multiplied by an infinitesimal increment of the generalized vertical coordinate, $\delta s > 0$, the product

$$\delta h = z_{,s}\, \delta s \tag{6.5}$$

represents the infinitesimal *thickness* separating the layers with generalized coordinate s and $s + \delta s$. The finite difference version of the thickness is appropriate for numerical models using generalized vertical coordinates.

6.3.3 Slope of a generalized layer

It is useful to characterize the orientation of surfaces with constant generalized vertical coordinate according to how they slope relative to surfaces of constant geopotential. For this purpose, the slope vector

$$\begin{aligned} \mathbf{S} &= \nabla_s z \\ &= -z_{,s}\, \nabla_z s \\ &= (S_x, S_y, 0) \end{aligned} \tag{6.6}$$

appears quite often. In this equation, $\nabla_s z$ is the horizontal gradient of the height of a fluid parcel as taken along surfaces of constant generalized vertical coordinate s. The second equality represents the horizontal gradient of the generalized surface function s as multiplied by the Jacobian of transformation between the s and z coordinate systems. This result follows from a coordinate transformation described in Section 6.5.

The two components, S_x and S_y, of the slope vector represent the projection of the slope of surfaces of constant generalized vertical coordinate onto the two horizontal coordinate directions. The squared magnitude of this vector is written

$$S = |\mathbf{S}|. \tag{6.7}$$

Components to the slope vector vanish when the generalized surface coincides with the horizontal (x, y) plane. The slope is infinite where the surface is unstratified in the vertical. That is, where $s_{,z} = 0$.

6.4 LOCAL ORTHONORMAL COORDINATES

Before delving into the mathematics of generalized vertical coordinates used in ocean modeling, it is useful to describe another set of coordinates. These coordinates are termed *local orthonormal coordinates*, and they are defined at each point on an arbitrary smooth surface by the locally orthonormal set of basis directions

$$\mathbf{e}_{\overline{1}} = \frac{\hat{\mathbf{y}} \wedge \nabla s}{|\hat{\mathbf{y}} \wedge \nabla s|} \tag{6.8}$$

$$\mathbf{e}_{\overline{2}} = \mathbf{e}_{\overline{3}} \wedge \mathbf{e}_{\overline{1}} \tag{6.9}$$

$$\mathbf{e}_{\overline{3}} = \frac{\nabla s}{|\nabla s|}. \tag{6.10}$$

Note that the basis directions are each well defined regardless of the orientation of the surface. However, for those special cases where $\mathbf{e}_{\overline{3}}$ has a single-signed projection onto the vertical direction $\hat{\mathbf{z}}$, which is the case with generalized vertical coordinates discussed in this chapter, then it is possible to introduce the slope vector $\mathbf{S} = (S_x, S_y)$ defined by equation (6.6) to yield

$$\mathbf{e}_{\overline{1}} = \frac{\hat{\mathbf{x}} + \hat{\mathbf{z}}\, S_x}{\sqrt{1 + S_x^2}} \tag{6.11}$$

$$\mathbf{e}_{\overline{2}} = \frac{-\hat{\mathbf{x}}\, S_x\, S_y + \hat{\mathbf{y}}\,(1 + S_x^2) + \hat{\mathbf{z}}\, S_y}{\sqrt{(1 + S^2)(1 + S_x^2)}} \tag{6.12}$$

$$\mathbf{e}_{\overline{3}} = \frac{(-\mathbf{S}, 1)}{\sqrt{1 + S^2}}. \tag{6.13}$$

Figure 6.3 shows a two-dimensional schematic of these basis vectors. "Lateral" distances are measured in the direction of the vectors $\mathbf{e}_{\overline{1}}$ and $\mathbf{e}_{\overline{2}}$, whereas "vertical" distances are measured parallel to the normal vector $\mathbf{e}_{\overline{3}}$. When the surfaces are horizontally aligned, the generalized basis reduces to the Cartesian basis $(\mathbf{e}_{\overline{1}}, \mathbf{e}_{\overline{2}}, \mathbf{e}_{\overline{3}}) = (\hat{\mathbf{x}}, \hat{\mathbf{y}}, \hat{\mathbf{z}})$. More generally, all three basis vectors change directions according to the dynamics and geometry of the surfaces.

The transformation matrix between the locally orthonormal coordinates and the fixed Cartesian basis $(\mathbf{e}_1, \mathbf{e}_2, \mathbf{e}_3) = (\hat{\mathbf{x}}, \hat{\mathbf{y}}, \hat{\mathbf{z}})$ is generally a space-time-dependent rotation, as can be deduced since both bases are orthonormal. Consequently, the transformation

$$\mathbf{e}_{\overline{a}} = \Lambda^{a}{}_{\overline{a}}\, \mathbf{e}_a \tag{6.14}$$

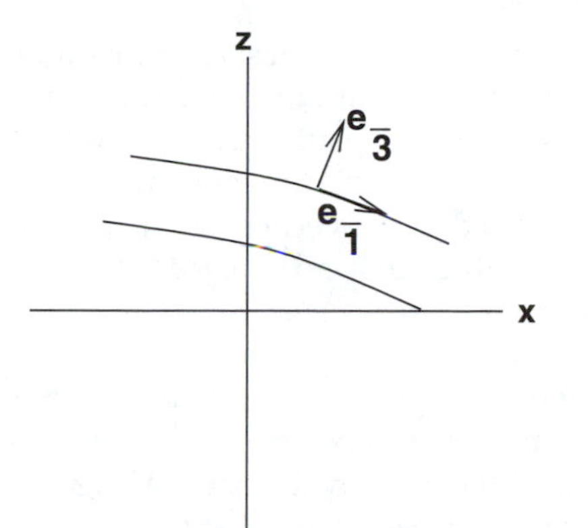

Figure 6.3 Schematic of local orthonormal coordinates. Two generalized surfaces are shown, with unit vectors indicated at a point on one of the layers. The unit vector $\mathbf{e}_{\overline{3}}$ is normal to the surface, and $\mathbf{e}_{\overline{1}}$ lies in the plane. The third direction $\mathbf{e}_{\overline{2}}$ is perpendicular to the page (pointing into the page). Lateral distances are measured in the plane of the surface, and vertical distances are measured normal to the surface.

is comprised of directional cosines, thus yielding

$$
\begin{aligned}
(\mathbf{e}_{\overline{1}}, \mathbf{e}_{\overline{2}}, \mathbf{e}_{\overline{3}}) &= (\hat{\mathbf{x}}, \hat{\mathbf{y}}, \hat{\mathbf{z}}) \begin{pmatrix} \hat{\mathbf{x}} \cdot \mathbf{e}_{\overline{1}} & \hat{\mathbf{x}} \cdot \mathbf{e}_{\overline{2}} & \hat{\mathbf{x}} \cdot \mathbf{e}_{\overline{3}} \\ \hat{\mathbf{y}} \cdot \mathbf{e}_{\overline{1}} & \hat{\mathbf{y}} \cdot \mathbf{e}_{\overline{2}} & \hat{\mathbf{y}} \cdot \mathbf{e}_{\overline{3}} \\ \hat{\mathbf{z}} \cdot \mathbf{e}_{\overline{1}} & \hat{\mathbf{z}} \cdot \mathbf{e}_{\overline{2}} & \hat{\mathbf{z}} \cdot \mathbf{e}_{\overline{3}} \end{pmatrix} \\
&= (\hat{\mathbf{x}}, \hat{\mathbf{y}}, \hat{\mathbf{z}}) \begin{pmatrix} \frac{1}{\sqrt{1+S_x^2}} & \frac{-S_x\, S_y}{\sqrt{(1+S^2)(1+S_x^2)}} & \frac{-S_x}{\sqrt{1+S^2}} \\ 0 & \frac{\sqrt{1+S_x^2}}{\sqrt{1+S^2}} & \frac{-S_y}{\sqrt{1+S^2}} \\ \frac{S_x}{\sqrt{1+S_x^2}} & \frac{S_y}{\sqrt{(1+S^2)(1+S_x^2)}} & \frac{1}{\sqrt{1+S^2}} \end{pmatrix}.
\end{aligned} \tag{6.15}
$$

As a check, note that the determinant of the transformation is unity and its inverse is given by its transpose, thus making it a rotation matrix.

The local orthonormal basis is used in formulating neutral physics operators of use in z-models as detailed in Section 14.1. However, the local orthonormal basis is not so useful to formulate the dynamical equations used in an ocean model. The problem is that each basis vector changes in time, and this proves inconvenient in practice. Additionally, the hydrostatic and geostrophic balances are very important for the large-scale ocean. As noted in Section 3.2.2, for theoretical convenience and numerical accuracy, it is crucial to have these balances described by separate diagnostic equations, and not to have any coupling within the equations describing these two balances. The locally orthonormal unit vectors do not satisfy this constraint in the general case.

6.5 MATHEMATICS OF GENERALIZED VERTICAL COORDINATES

When using generalized vertical coordinates in ocean and atmospheric models, the horizontal distance between two points is measured in the $(\hat{\mathbf{x}}, \hat{\mathbf{y}})$ directions, hence crossing through the vertical planes x =const and y =const. That is, the horizontal coordinates for a parcel are the *same* coordinates (x, y) also used in z-models. The vertical distance is measured parallel to the x =const and y =const planes; that is, it is measured in the $\hat{\mathbf{z}}$ direction, again just as for z-models. However, the crucial difference is that the vertical *position* of a parcel is not specified by giving the value of the depth z. Instead, it is specified by giving the value of the monotonic vertical coordinate $s = s(x, y, z, t)$.

It is notable that a similar set of generalized vertical coordinates have found use in other areas of theoretical physics. In particular, condensed matter physicists and biophysicists studying the dynamics of fluctuating membranes use these coordinates, where the coordinates go by the name *Monge gauge*. Their mathematical aspects are lucidly described in Section 10.4 of Chaikin and Lubensky (1995).

6.5.1 Projected aspect of vertical coordinates

As it is the generalized vertical coordinate s, not the vertical distance z, that is specified, lateral property gradients are taken along surfaces of constant s instead of surfaces of constant z. It is this property that prompts the adjective "projected" when referring to these coordinates (for further discussions, see Bleck, 1978; McDougall, 1995). Figure 6.4 further illustrates and describes this important, and subtle, point. Mathematically, these ideas imply that the transformation between geopotential and generalized vertical coordinates takes the form

$$t = t \tag{6.16}$$

$$x = x \tag{6.17}$$

$$y = y \tag{6.18}$$

$$s = s(x, y, z, t). \tag{6.19}$$

6.5.2 Transformation matrix

To develop equations describing the ocean fluid using generalized vertical coordinates, and to help understand more of their mathematical properties, it is useful to consider properties of transformations between geopotential coordinates

$$(\xi^0, \xi^1, \xi^2, \xi^3) = (t, x, y, z) \tag{6.20}$$

and generalized vertical coordinates

$$(\xi^{\overline{0}}, \xi^{\overline{1}}, \xi^{\overline{2}}, \xi^{\overline{3}}) = (t, x, y, s). \tag{6.21}$$

The abstract notation helps to organize the transformation properties according to the tensor calculus discussed in Chapters 20 and 21.

Again, it is only the manner in which one measures the vertical position that differs between these two sets of coordinates. That is, the coordinate transformation is just between the vertical coordinates. The horizontal coordinates and time are measured the same way in both coordinate systems. Hence, the vertical coordinate in one coordinate system is generally dependent on all the coordinates in the other system,

$$z = z(x, y, s, t) \tag{6.22}$$

$$s = s(x, y, z, t). \tag{6.23}$$

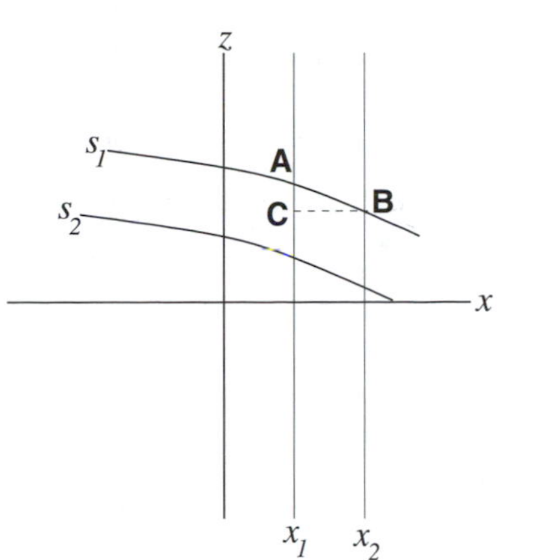

Figure 6.4 Schematic of projected coordinates as used in generalized vertical coordinate ocean models. The lateral distance between two points is measured in the $(\hat{\mathbf{x}}, \hat{\mathbf{y}})$ directions, hence crossing through the vertical planes $x =$ const and $y =$ const just as for z-coordinate ocean models. For example, the lateral distance between points A and B is not measured along the constant s surface. Instead, it is the same as the horizontal distance between C and B. The vertical distance is measured parallel to the intersection line of the $x =$ const and $y =$ const planes; that is, it is measured in the $\hat{\mathbf{z}}$ direction just as for geopotential coordinate models. However, the vertical position is specified not by giving the depth z. Rather, it is specified by giving the value of the generalized vertical coordinate $s = s(x, y, z, t)$. Lateral property gradients are thus taken along surfaces of constant s instead of surfaces of constant z. It is this property that prompts the adjective "projected" when referring to these coordinates. This figure is taken after Figure 1 of Bleck (1978).

This coordinate dependence leads to the transformation matrix between the "unbarred" geopotential coordinates and the "barred" generalized vertical coordinates

$$\Lambda^{a}{}_{\bar{a}} = \begin{pmatrix} 1 & 0 & 0 & 0 \\ 0 & 1 & 0 & 0 \\ 0 & 0 & 1 & 0 \\ z_{,t} & z_{,x} & z_{,y} & z_{,s} \end{pmatrix}. \tag{6.24}$$

The determinant of this transformation matrix, also known as the Jacobian, is given by

$$\begin{aligned} \det(\Lambda^{a}{}_{\bar{a}}) &= z_{,s} \\ &= (s_{,z})^{-1}. \end{aligned} \tag{6.25}$$

Recall the discussion in Section 6.3.2, which introduced the name *specific thickness* for the Jacobian of transformation. The inverse transformation is given by

$$\Lambda^{\bar{a}}{}_{a} = \begin{pmatrix} 1 & 0 & 0 & 0 \\ 0 & 1 & 0 & 0 \\ 0 & 0 & 1 & 0 \\ s_{,t} & s_{,x} & s_{,y} & s_{,z} \end{pmatrix}. \tag{6.26}$$

6.5.3 Transformation of partial derivatives

The rules established in Chapters 20 and 21 can be used to determine how tensors transform. First, we consider here how partial derivatives transform. As discussed in Section 20.10 (page 456), transformation of the partial derivative operator takes the form

$$\partial_{\overline{a}} = \Lambda^{a}{}_{\overline{a}}\, \partial_a. \tag{6.27}$$

In matrix-vector notation, with the matrix (6.24), this transformation can be written

$$(\partial_{\overline{0}}, \partial_{\overline{1}}, \partial_{\overline{2}}, \partial_{\overline{3}}) = (\partial_0, \partial_1, \partial_2, \partial_3) \begin{pmatrix} 1 & 0 & 0 & 0 \\ 0 & 1 & 0 & 0 \\ 0 & 0 & 1 & 0 \\ z_{,t} & z_{,x} & z_{,y} & z_{,s} \end{pmatrix}. \tag{6.28}$$

Note that when translating a tensor equation to a matrix-vector equation, care should be exercised to avoid interchanging the matrix with its transpose. Performing the multiplication in equation (6.28) yields

$$\left(\frac{\partial}{\partial t}\right)_s = \left(\frac{\partial}{\partial t}\right)_z + z_{,t}\, \partial_z \tag{6.29}$$

$$\left(\frac{\partial}{\partial x}\right)_s = \left(\frac{\partial}{\partial x}\right)_z + z_{,x}\, \partial_z \tag{6.30}$$

$$\left(\frac{\partial}{\partial y}\right)_s = \left(\frac{\partial}{\partial y}\right)_z + z_{,y}\, \partial_z \tag{6.31}$$

$$\partial_s = z_{,s}\, \partial_z. \tag{6.32}$$

In words, terms on the left-hand side are partial derivative operators taken in the generalized vertical coordinate frame, and those on the right-hand side are in the geopotential frame. The spatial portion of these transformations can be written in the more tidy form

$$\nabla_s + \hat{\mathbf{z}}\, \partial_s = (\nabla_z + \mathbf{S}\, \partial_z) + \hat{\mathbf{z}}\, z_{,s}\, \partial_z. \tag{6.33}$$

In this relation, ∇_z is the horizontal gradient taken on surfaces of constant geopotential, ∇_s is the horizontal gradient taken along surfaces of constant generalized vertical coordinate, and $\mathbf{S} = \nabla_s z$ is the slope vector introduced in Section 6.3.3 measuring the slope of the generalized surfaces relative to the horizontal.

6.5.4 Further useful identities

Further relations of use in subsequent manipulations are the triple products

$$\left(\frac{\partial s}{\partial z}\right)_t \left(\frac{\partial z}{\partial t}\right)_s \left(\frac{\partial t}{\partial s}\right)_z = -1 \tag{6.34}$$

$$\left(\frac{\partial s}{\partial z}\right)_x \left(\frac{\partial z}{\partial x}\right)_s \left(\frac{\partial x}{\partial s}\right)_z = -1 \tag{6.35}$$

$$\left(\frac{\partial s}{\partial z}\right)_y \left(\frac{\partial z}{\partial y}\right)_s \left(\frac{\partial y}{\partial s}\right)_z = -1. \tag{6.36}$$

These relations are familiar from manipulations in thermodynamics such as those used in deriving Maxwell's relations (e.g., Callen, 1985). Using the more compact notation with the implicit understanding of what coordinate frame the derivatives are acting in, the triple products imply

$$z_{,t} = -s_{,t}\, z_{,s} \tag{6.37}$$

$$z_{,x} = -s_{,x}\, z_{,s} \tag{6.38}$$

$$z_{,y} = -s_{,y}\, z_{,s} \tag{6.39}$$

where use was made of the identities

$$s_{,t} = (t_{,s})^{-1} \tag{6.40}$$

$$s_{,x} = (x_{,s})^{-1} \tag{6.41}$$

$$s_{,y} = (y_{,s})^{-1} \tag{6.42}$$

$$s_{,z} = (z_{,s})^{-1}. \tag{6.43}$$

6.5.5 Material time derivative and vertical velocity

When operating on a scalar field, the material time derivative $\mathrm{d}/\mathrm{d}t$ is itself a scalar under the usual nonrelativistic transformations relevant for GFD. Thus it takes the same form regardless of the coordinate system

$$\begin{aligned} \frac{\mathrm{d}}{\mathrm{d}t} &= \partial_0 + v^a \partial_a \\ &= \partial_{\bar{0}} + v^{\bar{a}} \partial_{\bar{a}}. \end{aligned} \tag{6.44}$$

Here, the velocity components v^a are those for the geopotential coordinate frame

$$\begin{aligned} (v^1, v^2, v^3) &= (\dot{x}, \dot{y}, \dot{z}) \\ &= (u, v, w) \end{aligned} \tag{6.45}$$

with $\dot{\psi} = \mathrm{d}\psi/\mathrm{d}t$ a shorthand for the material time derivative. The velocity components $v^{\bar{a}}$ are those for the generalized vertical frame

$$\begin{aligned} (v^{\bar{1}}, v^{\bar{2}}, v^{\bar{3}}) &= (\dot{x}, \dot{y}, \dot{s}) \\ &= (u, v, \dot{s}). \end{aligned} \tag{6.46}$$

In a nontensorial notation, the material time derivative is written

$$\begin{aligned} \frac{\mathrm{d}}{\mathrm{d}t} &= \left(\frac{\partial}{\partial t}\right)_z + \mathbf{u} \cdot \nabla_z + w \left(\frac{\partial}{\partial z}\right) \\ &= \left(\frac{\partial}{\partial t}\right)_s + \mathbf{u} \cdot \nabla_s + \dot{s} \left(\frac{\partial}{\partial s}\right). \end{aligned} \tag{6.47}$$

As discussed in Section 6.5.1, geopotential and generalized vertical models measure the same horizontal distances and the same time. Therefore, the horizontal velocity components are the same. This point is worth emphasizing, as in Bleck (1978). Indeed, often in the literature one can read phrases such as "the horizontal velocity along the isopycnal surface." Phrases such as this lend one to picture a velocity field that is oriented parallel to the isopycnal surface. However, as seen in

Section 6.7, the velocity generally has a nontrivial projection across the isopycnal surface, even when the flow is ideal (i.e., adiabatic and isohaline). The velocity vector is oriented parallel to isopycnals *only* in steady state ideal flow.

Although the horizontal velocity is the same in both coordinate systems, the vertical velocity component is distinct. The z-model vertical velocity component is related to the generalized vertical velocity component via a transformation using the material time derivative

$$\begin{aligned} w &= \frac{\mathrm{d}z}{\mathrm{d}t} \\ &= \left[\left(\frac{\partial}{\partial t}\right)_s + u\left(\frac{\partial}{\partial x}\right)_s + v\left(\frac{\partial}{\partial y}\right)_s + \dot{s}\left(\frac{\partial}{\partial s}\right) \right] z \\ &= z_{,t} + \mathbf{u}\cdot\mathbf{S} + \dot{s}\, z_{,s} \end{aligned} \tag{6.48}$$

where again $\mathbf{S} = \nabla_s z = -z_{,s}\, \nabla_z s$ is the slope vector introduced in Section 6.3.3. The complementary relation can be derived in a similar manner, where the vertical velocity component in the generalized vertical coordinate system is

$$\begin{aligned} \dot{s} &= \frac{\mathrm{d}s}{\mathrm{d}t} \\ &= \left[\left(\frac{\partial}{\partial t}\right)_z + u\left(\frac{\partial}{\partial x}\right)_z + v\left(\frac{\partial}{\partial y}\right)_z + w\left(\frac{\partial}{\partial z}\right) \right] s \\ &= s_{,t} + \mathbf{u}\cdot\nabla_z s + w\, s_{,z}. \end{aligned} \tag{6.49}$$

Use of the triple products (6.34)–(6.36) allows one to directly relate the vertical velocity components

$$s_{,z}\,\dot{z} = (\mathrm{d}/\mathrm{d}t - \mathbf{u}\cdot\nabla_z)\, s - s_{,t} \tag{6.50}$$

$$z_{,s}\dot{s} = (\mathrm{d}/\mathrm{d}t - \mathbf{u}\cdot\nabla_s)\, z - z_{,t} \tag{6.51}$$

where again $\dot{z} = w$. These formulae prove useful in the following discussions.

6.6 METRIC TENSORS

There are two metric tensors relevant when working with generalized coordinate surfaces.

6.6.1 Metric on the surface

A particular generalized surface can be defined by the algebraic equation

$$s(x, y, z, t) - s_{\mathrm{const}} = 0, \tag{6.52}$$

where s_{const} is a constant value for the generalized vertical coordinate. When performing spatial measurements on this surface, it is necessary to determine the metric tensor to be used to measure distances. For this purpose, note that if displacements are restricted to live on the particular surface $s = s_{\mathrm{const}}$, then by definition the differential of the generalized vertical coordinate vanishes,

$$\begin{aligned} \mathrm{d}s &= s_{,a}\, \mathrm{d}x^a \\ &= 0. \end{aligned} \tag{6.53}$$

Focusing on spatial measurements at a fixed time renders the vertical coordinate differential, on the surface, a function of the horizontal differentials through the relation

$$\mathrm{d}z = \mathbf{S} \cdot \mathrm{d}\mathbf{x}. \tag{6.54}$$

To measure the distance between two points on the surface, it is sufficient to specify their horizontal positions, since they are situated somewhere on the two-dimensional surface. In GFD, it is most common to focus on generalized coordinate surfaces embedded in three-dimensional Euclidean space. Hence, the squared distance between two infinitesimally close points is given by

$$\mathrm{d}\mathbf{x} \cdot \mathrm{d}\mathbf{x} = (1 + S_x^2)(\mathrm{d}x)^2 + (1 + S_y^2)(\mathrm{d}y)^2 + 2S_x\, S_y\, \mathrm{d}x\, \mathrm{d}y, \tag{6.55}$$

where equation (6.54) was used to eliminate $\mathrm{d}z$. This relation allows one to identify the metric tensor for the two-dimensional surface

$$g_{\bar{a}\bar{b}} = \begin{pmatrix} 1 + S_x^2 & S_x\, S_y \\ S_x\, S_y & 1 + S_y^2 \end{pmatrix}. \tag{6.56}$$

Likewise, components to the inverse metric are given by

$$g^{\bar{a}\bar{b}} = (1 + S^2)^{-1/2} \begin{pmatrix} 1 + S_y^2 & -S_x\, S_y \\ -S_x\, S_y & 1 + S_x^2 \end{pmatrix}. \tag{6.57}$$

Given the metric tensor, it is now possible to determine the area of an infinitesimal region on the surface. This proves useful when performing integrations over the surface. For this purpose, a result from the tensor analysis discussion in Section 20.13.2 (page 461) gives

$$\begin{aligned} \mathrm{d}A_{(s)} &= [\det(g_{\bar{a}\bar{b}})]^{1/2}\, \mathrm{d}x\, \mathrm{d}y \\ &= (1 + S^2)^{1/2}\, \mathrm{d}x\, \mathrm{d}y \\ &= |z_{,s}\, \nabla s|\, \mathrm{d}x\, \mathrm{d}y \end{aligned} \tag{6.58}$$

where $\det(g_{\bar{a}\bar{b}})$ is the determinant of the metric tensor. This result is used in Section 6.7 when discussing the dia-surface velocity component.

In many geophysical applications, the slope vector has components smaller than 10^{-2}. In this case, the generally nondiagonal metric tensor is well approximated by the unit (or Kronecker) tensor δ_{ab}, which is the metric for a flat manifold (see Section 20.4 starting on page 448). However, with some care it is not difficult to consider the more general case discussed here, whereby the slope can be an arbitrary finite value.

6.6.2 Metric for Euclidean space

Now consider the metric for three-dimensional Euclidean space, in which the vertical position of a point is measured by the generalized vertical coordinate s. A straightforward way to compute this metric is to transform the Cartesian metric δ_{ab} to generalized coordinates using

$$g_{\bar{a}\bar{b}} = \delta_{ab}\, \Lambda^{a}{}_{\bar{a}}\, \Lambda^{b}{}_{\bar{b}} \tag{6.59}$$

with just the spatial part of the transformation matrix equation (6.24). The result is the metric tensor

$$g_{\bar{a}\bar{b}} = \begin{pmatrix} 1+S_x^2 & S_x\,S_y & S_x\,z_{,s} \\ S_x\,S_y & 1+S_y^2 & S_y\,z_{,s} \\ S_x\,z_{,s} & S_y\,z_{,s} & z_{,s}^2 \end{pmatrix}. \tag{6.60}$$

The two-dimensional horizontal subspace of this metric tensor agrees with that given by equation (6.56). The metric (6.60) is that which would be used by someone measuring distances in three-dimensional Euclidean space using (x, y, s) as coordinates. As with the two-dimensional metric (6.56), the off-diagonal terms in (6.60) indicate the non-orthogonality of the (x, y, s) coordinates.

6.7 THE DIA-SURFACE VELOCITY COMPONENT

The following question arises frequently in geophysical fluid dynamics: What is the amount of fluid passing through a surface? The flux of fluid parcels crossing a surface is defined here as the *dia-surface* flux, with units volume per time per area (i.e., units of velocity). In determining this flux, one generally wishes to allow the surfaces to move in time through the fluid, and to have arbitrarily changing shapes. The purpose of this section is to describe the kinematics appropriate for determining the dia-surface flux for a slightly restricted class of surfaces, those defined by constant values of the generalized vertical coordinate. Some of the material in this section parallels Section 4 of Haynes and McIntyre (1990). See also Section 3.2 of Pedlosky (1996) for a complementary discussion.

At an arbitrary point on the surface of constant generalized vertical coordinate, the flux of fluid parcels in the direction normal to the surface is computed as

$$\text{seawater flux in direction } \hat{\mathbf{n}} = \mathbf{v}\cdot\hat{\mathbf{n}}, \tag{6.61}$$

with

$$\hat{\mathbf{n}} = \nabla s\,|\nabla s|^{-1} \tag{6.62}$$

the surface unit normal direction.* Introducing the material time derivative $\dot{s} = s_{,t} + \mathbf{v}\cdot\nabla s$ leads to the equivalent expression

$$\mathbf{v}\cdot\hat{\mathbf{n}} = (\dot{s} - s_{,t})|\nabla s|^{-1}. \tag{6.63}$$

Since the surface is moving, the net flux of seawater passing through the surface is obtained by subtracting the Eulerian velocity of the surface $\mathbf{v}^{(\text{ref})}$ in the $\hat{\mathbf{n}}$ direction from the Eulerian velocity component $\mathbf{v}\cdot\hat{\mathbf{n}}$ of the fluid parcels

$$\text{net flux of seawater through surface} = \hat{\mathbf{n}}\cdot(\mathbf{v} - \mathbf{v}^{(\text{ref})}). \tag{6.64}$$

The velocity $\mathbf{v}^{(\text{ref})}$ is the Eulerian velocity of a reference point fixed on the surface, and it is written

$$\mathbf{v}^{(\text{ref})} = \mathbf{u}^{(\text{ref})} + w^{(\text{ref})}\,\hat{\mathbf{z}}. \tag{6.65}$$

*Note that equation (6.13) used the symbol $\mathbf{e}_{\bar{3}}$ instead of $\hat{\mathbf{n}}$ for the normal direction. The more conventional notation $\hat{\mathbf{n}}$ is preferred here.

Since the reference point remains on the same $s = \text{const}$ surface, $\dot{s} = 0$ for the reference point. Consequently, the vertical velocity component $w^{(\text{ref})}$ is given by

$$w^{(\text{ref})} = -z_{,s}\,(s_{,t} + \mathbf{u}^{(\text{ref})} \cdot \nabla_z s), \tag{6.66}$$

where equation (6.49) was used with $\dot{s} = 0$. Hence, one can write

$$\begin{aligned} \hat{\mathbf{n}} \cdot \mathbf{v}^{(\text{ref})} &= \hat{\mathbf{n}} \cdot \mathbf{u}^{(\text{ref})} + \hat{\mathbf{n}} \cdot \hat{\mathbf{z}}\, w^{(\text{ref})} \\ &= |\nabla s|^{-1}(\nabla_z s \cdot \mathbf{u}^{(\text{ref})} - s_{,t} - \nabla_z s \cdot \mathbf{u}^{(\text{ref})}) \\ &= -s_{,t}\,|\nabla s|^{-1}, \end{aligned} \tag{6.67}$$

which then leads to the net flux of seawater crossing the surface

$$\begin{aligned} \hat{\mathbf{n}} \cdot (\mathbf{v} - \mathbf{v}^{(\text{ref})}) &= (\mathbf{v} \cdot \nabla s + s_{,t})\,|\nabla s|^{-1} \\ &= \dot{s}\,|\nabla s|^{-1}. \end{aligned} \tag{6.68}$$

The area normalizing the volume flux in the above discussion is the area $\mathrm{d}A_{(s)}$ of an infinitesimal region living on the surface of constant generalized vertical coordinate. This area is given by equation (6.58). Hence, the volume per time of fluid passing through the generalized surface is given by

$$\begin{aligned} \text{(volume/time) of fluid through surface} &= \hat{\mathbf{n}} \cdot (\mathbf{v} - \mathbf{v}^{(\text{ref})})\,\mathrm{d}A_{(s)} \\ &= |z_{,s}|\,\dot{s}\,\mathrm{d}x\,\mathrm{d}y, \end{aligned} \tag{6.69}$$

and the magnitude of this flux is

$$|\hat{\mathbf{n}} \cdot (\mathbf{v} - \mathbf{v}^{(\text{ref})})\,\mathrm{d}A_{(s)}| \equiv |w^{(s)}|\,\mathrm{d}x\,\mathrm{d}y, \tag{6.70}$$

where

$$w^{(s)} = z_{,s}\,\mathrm{d}s/\mathrm{d}t \tag{6.71}$$

measures the volume of fluid passing through the surface, per unit area $\mathrm{d}A = \mathrm{d}x\,\mathrm{d}y$ of the horizontal projection of the surface, per unit time. Since $w^{(s)}$ is not a vector, referring to it as the dia-surface *velocity* is a misnomer, although this is often done in the literature. Instead, the name dia-surface velocity *component* is more precise.*

The expression (6.71) for $w^{(s)}$ allows one to write the material time derivative (6.47) using the generalized vertical coordinate as

$$\frac{\mathrm{d}}{\mathrm{d}t} = \left(\frac{\partial}{\partial t}\right)_s + \mathbf{u} \cdot \nabla_s + w^{(s)}\left(\frac{\partial}{\partial z}\right), \tag{6.72}$$

where $\partial_s = z_{,s}\,\partial_z$. This form for the material time derivative motivates some to consider $w^{(s)}$ as a vertical velocity component that measures the rate at which fluid parcels penetrate the surface of constant generalized coordinate (see Appendix A to McDougall, 1995). One should be mindful, however, to distinguish $w^{(s)}$ from the generally different vertical velocity component $w = \mathrm{d}z/\mathrm{d}t$ used in z-models.

---

*In Section 3.4.2, we defined a similar velocity component, q_{w}, which is the volume of moisture passing across the ocean free surface, per unit time, per unit horizontal cross-sectional area $\mathrm{d}A = \mathrm{d}x\,\mathrm{d}y$. The kinematics in the two analyses is the same.

To make contact with the literature, it is useful to note the many equivalent forms of $w^{(s)}$,

$$\begin{aligned} w^{(s)} &= z_{,s}\, \mathrm{d}s/\mathrm{d}t \\ &= z_{,s}\, (s_{,t} + \mathbf{v} \cdot \nabla s) \\ &= -z_{,t} + (\mathbf{v} \cdot \hat{\mathbf{n}})\, z_{,s}\, |\nabla s| \\ &= -z_{,t} + (\mathbf{v} \cdot \hat{\mathbf{n}}) \sqrt{1 + S^2} \\ &= -z_{,t} + \mathbf{v} \cdot (-\mathbf{S}, 1) \\ &= w - (\partial_t + \mathbf{u} \cdot \nabla_s)\, z, \end{aligned} \tag{6.73}$$

where the identity (6.37) was used for the third equality, and $\mathbf{S} = \nabla_s z$ is the slope discussed in Section 6.3.3. The last two forms expose some special cases rendering useful experience with certain forms for $w^{(s)}$. For example, with horizontal surfaces, $\mathbf{S} = 0$ and so

$$w^{(s)} = -z_{,t} + w \qquad \text{when } \mathbf{S} = 0. \tag{6.74}$$

For the case when the horizontal surface is static ($z_{,t} = 0$), as in geopotential coordinate models, fluid parcels penetrate the surface at a rate set by the vertical velocity component w. For the case when horizontal surfaces change their depth at a rate $z_{,t}$ equal to that of the fluid parcels w, the surfaces retreat from fluid parcels at a rate ensuring that no parcels penetrate the surfaces. Now consider a static nonhorizontal surface of constant generalized vertical coordinate, in which case

$$w^{(s)} = w - \mathbf{u} \cdot \mathbf{S} \qquad \text{when } z_{,t} = 0. \tag{6.75}$$

Hence, even if fluid parcels are moving upward ($w > 0$), the dia-surface velocity component can be negative so long as the horizontal velocity projected onto the slope dominates ($\mathbf{u} \cdot \mathbf{S} > 0$ with $|\mathbf{u} \cdot \mathbf{S}| > |w|$). Indeed, if the slope becomes very steep, thus reducing the generalized surface's projection onto the horizontal plane, and the horizontal velocity remains nontrivial, then $w^{(s)}$ can become quite large.

When the generalized surface is a surface of constant potential density (i.e., an isopycnal), then $w^{(\rho)}$ is termed the *diapycnal* velocity component. For the more general case with neutral surfaces, $w^{(\rho)}$ is the *dianeutral* velocity component. In these cases,

$$w^{(\rho)} = z_{,\rho}\, \mathrm{d}\rho/\mathrm{d}t. \tag{6.76}$$

Hence, for flow materially conserving potential temperature and salinity so that potential density is materially conserved, $\mathrm{d}\rho/\mathrm{d}t = 0$, the diapycnal velocity component vanishes, so

$$w^{(\rho)} = 0 \qquad \text{when } \mathrm{d}\rho/\mathrm{d}t = 0. \tag{6.77}$$

That is, no fluid parcels penetrate isopycnal surfaces for the case of adiabatic and isohaline fluid flow. Additionally, the material time derivative simplifies for adiabatic and isohaline flow using isopycnal coordinates, since its spatial component reduces from three to two components

$$\frac{\mathrm{d}}{\mathrm{d}t} = \left(\frac{\partial}{\partial t}\right)_{\rho} + \mathbf{u} \cdot \nabla_{\rho} \qquad \text{when } \mathrm{d}\rho/\mathrm{d}t = 0. \tag{6.78}$$

These are very important results that motivate the use of an isopycnal coordinate description of the ocean interior, where flow is dominated by processess occurring along isopycnals. Bleck (1998) highlights this point in his review of isopycnal modeling. More advantages of isopycnal coordinates arise when discussing the kinematics of mesoscale eddy motion in Chapter 9.

In closing this discussion, note that only when the potential density surfaces are static does ideal flow lead to a three-dimensional velocity oriented parallel to potential density surfaces,

$$\mathbf{v} \cdot \nabla \rho = 0 \qquad \text{when } \mathrm{d}\rho/\mathrm{d}t = 0 \text{ and } \rho_{,t} = 0. \tag{6.79}$$

This sort of flow is often considered in studies of the thermocline (e.g., Pedlosky, 1996).

6.8 CONSERVATION OF MASS AND VOLUME FOR PARCELS

We have established a suite of mathematical results for the generalized vertical coordinate system. Now it is time to consider how a description of ocean fluid mechanics is formulated using these coordinates. Start by considering expressions for the conservation of mass and volume.

Recall from Chapter 3 that the mass of an infinitesimal parcel of fluid is written

$$\mathrm{d}M = \rho\, \mathrm{d}V, \tag{6.80}$$

where $\mathrm{d}V$ is the infinitesimal volume of the parcel and ρ is the *in situ* density. Focusing on fluid parcels that conserve mass, the discussion in Section 2.3.4 leads to the expression of mass conservation for each parcel,

$$\begin{aligned} \frac{\mathrm{d}}{\mathrm{d}t} \ln(\mathrm{d}M) &= \frac{\mathrm{d}}{\mathrm{d}t} \ln(\rho\, \mathrm{d}V) \\ &= 0. \end{aligned} \tag{6.81}$$

The Eulerian form of mass conservation, known as the *continuity equation*, depends on the coordinates used to describe the parcel's volume.

6.8.1 Z-coordinates

Using the geopotential z-coordinate system (x, y, z), the volume of a fluid element is $\mathrm{d}V = \mathrm{d}x\, \mathrm{d}y\, \mathrm{d}z$, thus leading to

$$\begin{aligned} 0 &= \frac{\mathrm{d}}{\mathrm{d}t} \ln(\rho\, \mathrm{d}V) \\ &= \frac{\mathrm{d}}{\mathrm{d}t} \ln(\rho\, \mathrm{d}x\, \mathrm{d}y\, \mathrm{d}z) \\ &= \frac{\mathrm{d} \ln \rho}{\mathrm{d}t} + \frac{\mathrm{d}}{\mathrm{d}x^a} \dot{x}^a. \end{aligned} \tag{6.82}$$

This expression for parcel mass conservation is satisfied by each of the parcels throughout the fluid, which implies that it holds at each point of the fluid at all times. It is therefore convenient to introduce the Eulerian velocity field $\mathbf{v} = \dot{\mathbf{x}}$ leading to the statement of parcel mass conservation

$$\frac{\mathrm{d} \ln \rho}{\mathrm{d}t} = -\nabla \cdot \mathbf{v}. \tag{6.83}$$

Expanding the material time derivative allows this equation to be written as an Eulerian conservation law

$$\rho_{,t} + \nabla \cdot (\mathbf{v}\,\rho) = 0. \tag{6.84}$$

Assuming the parcel's volume is conserved instead of its mass leads to the non-divergence condition on the velocity used in the Boussinesq z-coordinate ocean model

$$\frac{\mathrm{d}}{\mathrm{d}t}(\mathrm{d}V) = 0 \Rightarrow \nabla \cdot \mathbf{v} = 0. \tag{6.85}$$

6.8.2 Generalized vertical coordinates

With a generalized vertical coordinate, s, and Cartesian horizontal coordinates, (x, y), conservation of mass states that

$$\begin{aligned} \rho\,\mathrm{d}V &= \rho\,\mathrm{d}x\,\mathrm{d}y\,\mathrm{d}z \\ &= \rho\,\mathrm{d}x\,\mathrm{d}y\,z_{,s}\,\mathrm{d}s \end{aligned} \tag{6.86}$$

is materially constant. Therefore,

$$\begin{aligned} \frac{\mathrm{d}}{\mathrm{d}t}(\ln \rho\,\mathrm{d}V) &= \frac{\mathrm{d}(z_{,s})/\mathrm{d}t}{z_{,s}} + \frac{\dot{\rho}}{\rho} + \nabla_s \cdot \mathbf{u} + \partial_s\,\dot{s} \\ &= 0, \end{aligned} \tag{6.87}$$

where the material time derivative (Section 6.5.5) is given by

$$\frac{\mathrm{d}}{\mathrm{d}t} = \partial_t + \mathbf{u} \cdot \nabla_s + \dot{s}\,\partial_s \tag{6.88}$$

and horizontal derivatives are taken with constant s. Rearrangement leads to the mass continuity equation in generalized vertical coordinates

$$(\rho\,z_{,s})_{,t} + \nabla_s \cdot (\rho\,z_{,s}\mathbf{u}) + \partial_s(\rho\,z_{,s}\,\dot{s}) = 0. \tag{6.89}$$

6.8.3 Volume conservation in ρ-coordinates

Volume conservation states that $\mathrm{d}V = \mathrm{d}x\,\mathrm{d}y\,\mathrm{d}z = \mathrm{d}x\,\mathrm{d}y\,z_{,s}\,ds$ is a material constant, which leads to

$$(z_{,s})_{,t} + \nabla_s \cdot (z_{,s}\,\mathbf{u}) + \partial_s\,w^{(s)} = 0, \tag{6.90}$$

where again $w^{(s)}$ is the dia-surface velocity *component* given by equation (6.71). This equation is often called the thickness equation, since it provides a prognostic equation for the layer specific thickness $z_{,s}$ (Section 6.3.2). In the special case where $s = \rho_{\text{pot}}$ is the potential density, volume conservation takes the form

$$(z_{,\rho})_{,t} + \nabla_\rho \cdot (z_{,\rho}\,\mathbf{u}) + \partial_\rho\,w^{(\rho)} = 0, \tag{6.91}$$

where the "pot" label was dropped for brevity. For adiabatic and isohaline Boussinesq flow, $w^{(\rho)} = 0$, thus bringing the thickness equation to the form commonly used in studies of mesoscale eddies (e.g., Gent and McWilliams, 1990)

$$(z_{,\rho})_{,t} + \nabla_\rho \cdot (z_{,\rho}\,\mathbf{u}) = 0. \tag{6.92}$$

Hence, in isopycnal coordinates, the thickness and potential density equations are explicitly coupled. For adiabatic and isohaline flow, this coupling results in a simplification of the thickness equation since its transport operator reduces from three-dimensional to two-dimensional.

6.8.4 Mass conservation in hydrostatic fluids using pressure coordinates

In a hydrostatic fluid, pressure always increases with depth, and so pressure is a valid choice for vertical coordinate. In this case, it is appropriate to write the specific thickness weighted density in terms of pressure,

$$\begin{aligned} \rho z_{,s} &= \rho z_{,p} \, p_{,s} \\ &= -g^{-1} \, p_{,s} \end{aligned} \tag{6.93}$$

where the hydrostatic balance $p_{,z} = -\rho g$ was used for the second equality. The hydrostatic approximation therefore singles out pressure as being special, as can be seen by the absence of a time tendency term for mass conservation written in pressure coordinates

$$\nabla_p \cdot \mathbf{u} + \partial_p \dot{p} = 0. \tag{6.94}$$

To understand the simplicity of this result, note that the convergence of mass per horizontal area onto a column of hydrostatic fluid bounded above and below by surfaces of constant pressure is given by

$$g^{-1} \nabla_p \cdot \left(\int_{p_1}^{p_2} \mathbf{u} \, \mathrm{d}p \right) = -\nabla_p \cdot \left(\int_{z(p_1)}^{z(p_2)} \mathbf{u} \, \rho \, \mathrm{d}z \right). \tag{6.95}$$

As mass converges onto the column, there is a vertical divergence of the vertical velocity $\dot{p}$, thus causing fluid to cross surfaces of constant pressure. This is the physical content of the mass conservation equation (6.94).

Huang et al. (2001), DeSzoeke and Samelson (2002) and Marshall et al. (2003) exploit the simplicity of mass conservation in pressure coordinates, as well as a simple momentum balance, by mapping a Boussinesq z-model to a non-Boussinesq pressure coordinate model. Notably, atmospheric modelers have used pressure, or functions of pressure, for many years to simulate the compressible atmosphere (e.g., Haltiner and Williams, 1980).

6.9 KINEMATIC BOUNDARY CONDITIONS

The kinematic boundary conditions were derived in Section 3.4 for z-coordinates, and they are rewritten here in a material time derivative form

$$\frac{\mathrm{d}(z+H)}{\mathrm{d}t} = 0 \qquad \text{at } z = -H \tag{6.96}$$

$$\rho \left(\frac{\mathrm{d}(z-\eta)}{\mathrm{d}t} \right) = -\rho_{\mathrm{w}} \, q_{\mathrm{w}} \qquad \text{at } z = \eta. \tag{6.97}$$

Deriving the analogous results for s-coordinates requires us to relate the vertical velocity component $w = \dot{z}$ to the generalized vertical velocity component $\dot{s}$. Equation (6.50) provides the needed relation

$$s_{,z} \, \dot{z} = \dot{s} - \mathbf{u} \cdot \nabla_z s - s_{,t}. \tag{6.98}$$

Precisely, this relation holds on surfaces of constant depth. Extending it to the bottom and surface boundaries is achieved via coordinate transformations.

6.9.1 Kinematic boundary condition at the solid earth bottom

To apply equation (6.98) at the ocean bottom requires a transformation of the constant depth gradient ∇_z to a horizontal gradient taken along the bottom. We thus proceed as in Section 6.5 and consider a time-independent transformation

$$\begin{aligned} \overline{x} &= x \\ \overline{y} &= y \\ \overline{z} &= -H(x,y) \\ \overline{t} &= t. \end{aligned} \tag{6.99}$$

The horizontal gradient taken on constant depth surfaces, ∇_z, and the horizontal gradient along the bottom, $\nabla_{\overline{z}}$, are thus related by (see equation (6.33))

$$\nabla_{\overline{z}} = \nabla_z - (\nabla H)\,\partial_z. \tag{6.100}$$

Using this result in equation (6.98) yields

$$s_{,z}\,(w + \mathbf{u}\cdot\nabla H) = \dot{s} - (\partial_{\overline{t}} + \mathbf{u}\cdot\nabla_{\overline{z}})\,s \qquad \text{at} \quad z = -H. \tag{6.101}$$

The left-hand side vanishes due to the kinematic boundary condition (6.96), which then leads to

$$\dot{s} = (\partial_{\overline{t}} + \mathbf{u}\cdot\nabla_{\overline{z}})\,s \qquad \text{at} \quad s = s(x,y,z=-H,t). \tag{6.102}$$

The value of the generalized coordinate at the ocean bottom can be written in the shorthand form

$$s_{\text{bot}}(x,y,t) = s(x,y,z=-H,t) \tag{6.103}$$

which gives

$$\frac{\text{d}\,(s - s_{\text{bot}})}{\text{d}t} = 0 \qquad \text{at} \quad s = s_{\text{bot}}. \tag{6.104}$$

This relation is analogous to the z-coordinate result (6.96). Both follow since the bottom is an impenetrable material surface.

6.9.2 Kinematic boundary condition at the ocean surface

At the ocean surface, the coordinate transformation is time dependent

$$\begin{aligned} \overline{x} &= x \\ \overline{y} &= y \\ \overline{z} &= \eta(x,y,t) \\ \overline{t} &= t. \end{aligned} \tag{6.105}$$

The horizontal gradient operators are therefore related by

$$\nabla_{\overline{z}} = \nabla_z + (\nabla_z\,\eta)\,\partial_z \tag{6.106}$$

and the time derivatives satisfy an analogous equation

$$\partial_{\overline{t}} = \partial_t + \eta_{,t}\,\partial_z. \tag{6.107}$$

At the ocean surface, the relation (6.98) between vertical velocity components thus takes the form

$$s_{,z}\,w = \dot{s} - (\partial_{\overline{t}} + \mathbf{u}\cdot\nabla_{\overline{z}})\,s + s_{,z}\,(\partial_t + \mathbf{u}\cdot\nabla_z)\,\eta \qquad \text{at} \quad z = \eta. \tag{6.108}$$

Use of the z-coordinate kinematic boundary condition (6.97) leads to

$$\rho\, z_{,s}\, \dot{s} = -\rho_{\mathrm{w}}\, q_{\mathrm{w}} + \rho\, z_{,s}\, \dot{s}_{\mathrm{top}} \qquad \text{at} \quad s = s_{\mathrm{top}} \tag{6.109}$$

where

$$s_{\mathrm{top}} = s(x, y, z = \eta, t) \tag{6.110}$$

is the value of the generalized vertical coordinate at the ocean surface. Reorganizing the result (6.109) leads to

$$\rho\, z_{,s} \left(\frac{\mathrm{d}(s - s_{\mathrm{top}})}{\mathrm{d}t} \right) = -\rho_{\mathrm{w}}\, q_{\mathrm{w}} \qquad \text{at} \quad s = s_{\mathrm{top}}, \tag{6.111}$$

which is analogous to the z-coordinate result (6.97).

6.10 PRIMITIVE EQUATIONS

Generalized vertical coordinates are non-orthogonal, as exemplified by the non-diagonal metric tensor in Section 6.6. However, the "projected" nature of the coordinates (see Figure 6.4) allows one to formulate the momentum equations as a straightforward transformation from the geopotential form to the generalized vertical coordinate form. There are no nontrivial Christoffel symbols appearing in these equations, contrary to one's suspicion when noting the non-orthogonality of the coordinate system. Bleck (1978) emphasizes this point since there appeared to be some confusion in the literature prior to his paper, even though Starr (1945) had written the proper equations more than thirty years earlier.

6.10.1 Momentum balance

The horizontal momentum equations for a hydrostatic fluid employing the shallow ocean approximation are given by (Section 4.4)

$$(\rho\,\mathbf{u})_{,t} + \nabla \cdot (\rho\,\mathbf{v}\,\mathbf{u}) = -f\,\hat{\mathbf{z}} \wedge \rho\,\mathbf{u} - \nabla_z p, \tag{6.112}$$

or equivalently, by using mass conservation in the form (6.84), these equations are

$$\mathrm{d}\mathbf{u}/\mathrm{d}t = -f\,\hat{\mathbf{z}} \wedge \mathbf{u} - \rho^{-1}\,\nabla_z p. \tag{6.113}$$

Friction and sphericity have been dropped for simplicity. Use of the material time derivative in terms of the generalized vertical coordinate (Section 6.5.5) as well as mass conservation in the form (6.89) yields

$$(\rho\, z_{,s}\,\mathbf{u})_{,t} + \nabla_s \cdot (\rho\, z_{,s}\,\mathbf{u}\,\mathbf{u}) + (\rho\, z_{,s}\,\dot{s}\,\mathbf{u})_{,s} = -f\,\hat{\mathbf{z}} \wedge \rho\, z_{,s}\,\mathbf{u} - z_{,s}\,\nabla_z p. \tag{6.114}$$

The horizontal pressure gradient from Section 6.5.3 yields

$$\begin{aligned} \nabla_z p &= \nabla_s p - \nabla_s z\, p_{,z} \\ &= \nabla_s p + \rho\, g\, \nabla_s z, \end{aligned} \tag{6.115}$$

where the hydrostatic balance was used for the second equality. Using the hydrostatic balance once again in the form $g\,\rho\, z_{,s} = -p_{,s}$ leads to the pressure gradient expression

$$z_{,s}\,\nabla_z p = z_{,s}\,\nabla_s p - p_{,s}\,\nabla_s z \tag{6.116}$$

and the momentum equation

$$(\rho\, z_{,s}\,\mathbf{u})_{,t} + \nabla_s \cdot (\rho\, z_{,s}\,\mathbf{u}\,\mathbf{u}) + (\rho\, z_{,s}\,\dot{s}\,\mathbf{u})_{,s} = -f\,\hat{\mathbf{z}} \wedge \rho\, z_{,s}\,\mathbf{u} - (z_{,s}\,\nabla_s p - p_{,s}\,\nabla_s z). \tag{6.117}$$

The extra term $p_{,s}\,\nabla_s z$ was encountered in Section 6.2.2 when discussing the pressure gradient calculation in σ-coordinate models.

6.10.2 Summary of the primitive equations

It is a simple matter to derive the tracer balance in generalized vertical coordinates. Doing so yields the primitive equations

$$(\rho z_{,s}\, \mathbf{u})_{,t} + \nabla_s \cdot (\rho z_{,s}\, \mathbf{u}\, \mathbf{u}) + (\rho z_{,s}\, \dot{s}\, \mathbf{u})_{,s} = -f\, \hat{\mathbf{z}} \wedge \rho z_{,s}\, \mathbf{u} - z_{,s} \nabla_z p \tag{6.118}$$

$$(\rho z_{,s}\, C)_{,t} + \nabla_s \cdot (\rho z_{,s} \mathbf{u}\, C) + \partial_s(\rho z_{,s}\, \dot{s}\, C) = 0 \tag{6.119}$$

$$(\rho z_{,s})_{,t} + \nabla_s \cdot (\rho z_{,s} \mathbf{u}) + \partial_s(\rho z_{,s}\, \dot{s}) = 0 \tag{6.120}$$

$$p_{,s} = -z_{,s}\, \rho\, g, \tag{6.121}$$

where subgrid scale processes and sphericity are ignored for brevity, and where the horizontal pressure gradient is written

$$\nabla_z p = \nabla_s p + \rho\, g\, \nabla_s z. \tag{6.122}$$

When $s = z$, these equations reduce to the non-Boussinesq z-coordinate form considered in Chapters 3, 4 and 5. In the general case, there is an extra factor of the specific thickness $z_{,s}$ as well as the two terms needed to compute the horizontal pressure gradient.

6.10.3 Non-Boussinesq hydrostatic fluid in pressure coordinates

In the special case of pressure coordinates with $s = p$,

$$z_{,p}\, \nabla_z p = -\nabla_p z \tag{6.123}$$

and

$$\rho\, z_{,p} = -g^{-1}. \tag{6.124}$$

That is, the specific thickness weighted pressure gradient force is given by minus the slope of the isobaric surfaces. Hence, the non-Boussinesq hydrostatic primitive equations in pressure coordinates are given by

$$\mathbf{u}_{,t} + \nabla_p \cdot (\mathbf{u}\, \mathbf{u}) + (\dot{p}\, \mathbf{u})_{,p} = -f\, \hat{\mathbf{z}} \wedge \mathbf{u} - g\, \nabla_p z \tag{6.125}$$

$$C_{,t} + \nabla_p \cdot (\mathbf{u}\, C) + \partial_p(\dot{p}\, C) = 0 \tag{6.126}$$

$$\nabla_p \cdot \mathbf{u} + \partial_p \dot{p} = 0. \tag{6.127}$$

As mentioned earlier, Huang et al. (2001), DeSzoeke and Samelson (2002) and Marshall et al. (2003) exploit the isomorphism between these equations for a non-Boussinesq fluid using pressure coordinates, and those of a Boussinesq fluid using z-coordinates. Since most z-models have traditionally used the Boussinesq approximation, the isomorphism provides a ready means to remove the Boussinesq aproximation while maintaining the same underlying numerical algorithns. This isomorphism represents a very elegant result which will likely see more applications in the future as modelers aim to generalize their vertical coordinates systematically without starting completely from scratch. An alternative approach, which maintains use of the z-coordinate, is described by Greatbatch et al. (2001) and is pursued in Chapters 8 and 12. The advantage of the z-coordinate approach is that it is appropriate when developing nonhydrostatic models, for which use of a pressure coordinate requires some care.*

*The author knows of no pressure coordinate nonhydrostatic model algorithm.

6.10.4 Ideal Boussinesq fluid in ρ-coordinates

Isopyncal models are very useful for simulating an ideal Boussinesq fluid. Restricting attention to the case where potential density is equated to *in situ* density leads to the horizontal pressure gradient

$$\begin{aligned} \nabla_z p &= \nabla_\rho p + \rho g \nabla_\rho z \\ &= \nabla_\rho (p + \rho g z). \end{aligned} \tag{6.128}$$

Defining the Montgomery potential

$$M = p + \rho g z \tag{6.129}$$

leads to the ideal Boussinesq primitive equations written in isopycnal coordinates,

$$(z_{,\rho}\,\mathbf{u})_{,t} + \nabla_\rho \cdot (z_{,\rho}\,\mathbf{u}\,\mathbf{u}) + (z_{,\rho}\,\dot{\rho})_{,\rho} = -f\,\hat{\mathbf{z}} \wedge z_{,\rho}\,\mathbf{u} - \nabla_\rho M \tag{6.130}$$

$$(z_{,\rho}\,C)_{,t} + \nabla_\rho \cdot (z_{,\rho}\,\mathbf{u}\,C) + (z_{,\rho}\,\dot{\rho}\,C)_{,\rho} = 0 \tag{6.131}$$

$$(z_{,\rho})_{,t} + \nabla_\rho \cdot (z_{,\rho}\,\mathbf{u}) + (z_{,\rho}\,\dot{\rho})_{,\rho} = 0 \tag{6.132}$$

$$M_{,\rho} = \rho g. \tag{6.133}$$

Note the presence of only a single term for the horizontal pressure gradient force. Hence, for ideal flow where the equation of state is linear, isopycnal models do not suffer from difficulties numerically representing the horizontal pressure force. Some extra effort is required to compute the pressure gradient when allowing for a nonlinear equation of state, although the problems may be less difficult to handle than for σ- models (e.g., Sun et al., 1999; Hallberg, 2003b).

6.11 TRANSFORMATION OF SGS TRACER FLUX COMPONENTS

The parameterization of subgrid scale (SGS) processes is of fundamental importance to ocean models (see Part 2 of this book). Details of how these processes are parameterized depend on the choice of vertical coordinates. Here, we display the transformation properties of SGS tracer flux components under changes in vertical coordinates. We also comment on representing neutral physical processes with different vertical coordinates.

6.11.1 General formulation

The operator associated with SGS tracer transport is given by the convergence of a tracer flux

$$R = -\nabla \cdot \mathbf{J} \tag{6.134}$$

where $\mathbf{J}$ has dimensions of tracer mass per area per time. Note that it is often useful to identify a tracer concentration flux $\mathbf{F}$, related to the tracer flux via $\mathbf{J} = \rho\,\mathbf{F}$ (see equation (5.9)). For present purposes, it is sufficient to consider the tracer flux $\mathbf{J}$.

The SGS transport operator is a scalar, so it remains unchanged under a coordinate transformation. However, the flux components do change, as revealed by the following manipulations

$$
\begin{aligned}
-R &= \nabla_z \cdot \mathbf{J} + \partial_z J^z \\
&= (\nabla_s - \nabla_s z\, \partial_z) \cdot \mathbf{J} + s_{,z}\, \partial_s J^z \\
&= s_{,z} \left[z_{,s} \nabla_s \cdot \mathbf{J} + (\hat{\mathbf{z}}\, \partial_s - \nabla_s z\, \partial_s) \cdot \mathbf{J} \right] \\
&= s_{,z} \left[\nabla_s \cdot (z_{,s} \mathbf{J}) + \hat{\mathbf{z}}\, \partial_s J^z - \partial_s (\nabla_s z \cdot \mathbf{J}) \right] \\
&= s_{,z} \nabla_s \cdot (z_{,s} \mathbf{J}) + s_{,z}\, \partial_s \left[(\hat{\mathbf{z}} - \nabla_s z) \cdot \mathbf{J} \right] \\
&= s_{,z} \nabla_s \cdot (z_{,s} \mathbf{J}) + s_{,z}\, \partial_s \left(z_{,s} \nabla s \cdot \mathbf{J} \right),
\end{aligned}
\tag{6.135}
$$

where equation (6.33) was used to reach the second line, and

$$
z_{,s} \nabla s = \hat{\mathbf{z}} - \nabla_s z \tag{6.136}
$$

was used for the last (see Section 6.3.3). When integrating the tracer transport operator over a finite region of space, such as an ocean model grid cell, one considers the product of the volume of the region and the operator. The result (6.135) then leads to

$$
\begin{aligned}
\mathrm{d}x\, \mathrm{d}y\, \mathrm{d}z\, R &= \mathrm{d}x\, \mathrm{d}y\, \mathrm{d}s\, z_{,s}\, R \\
&= -\mathrm{d}x\, \mathrm{d}y\, \mathrm{d}s \left[\nabla_s \cdot (z_{,s} \mathbf{J}) + \partial_s (z_{,s} \nabla s \cdot \mathbf{J}) \right] \\
&= -\mathrm{d}x\, \mathrm{d}y\, \mathrm{d}s \left(\nabla_s \cdot \mathbf{J}^{\overline{h}} + \partial_s J^s \right),
\end{aligned}
\tag{6.137}
$$

which integrates to boundary contributions just as when using $s = z$. The last line identified the transformed tracer flux components

$$
\mathbf{J}^{\overline{h}} = z_{,s} \mathbf{J}^h \tag{6.138}
$$

$$
J^s = z_{,s} \nabla s \cdot \mathbf{J}. \tag{6.139}
$$

These equations relate the flux components $(\mathbf{J}^h, J^z)$ written in z-coordinates to those written in generalized vertical coordinates $(\mathbf{J}^{\overline{h}}, J^s)$.

6.11.2 Neutral physics in generalized coordinates

Part 4 of this book focuses on the issues of neutral physics as implemented and parameterized within a z-model. Key to this implementation is the slope of a neutral direction. For a neutral diffusion scheme, where the diffusion tensor is oriented relative to the neutral direction, the slope of the neutral direction relative to the surface of constant generalized vertical coordinate is required. For schemes requiring a measure of baroclinicity, such as the Gent and McWilliams (1990) scheme, the slope of the neutral direction relative to the geopotential surface is required, as this slope reflects on baroclinicity in the fluid and its available potential energy.

As discussed in Section 6.3.3, the neutral slope relative to a geopotential is given by

$$
\mathbf{S}_{(\rho/z)} = -z_{,\rho} \nabla_z \rho, \tag{6.140}
$$

with ρ the locally referenced potential density. The (ρ/z) subscript notation highlights that the neutral slope is computed relative to a geopotential. In generalized vertical coordinates, the horizontal gradient ∇_z is computed using the transformation (6.33), so that the neutral slope becomes

$$\begin{aligned}\mathbf{S}_{(\rho/z)} &= -z_{,\rho}\left(\nabla_s - \mathbf{S}_{(s/z)}\partial_z\right)\rho \\ &= -z_{,s}\left(\frac{\nabla_s\rho}{\rho_{,s}}\right) + \mathbf{S}_{(s/z)} \\ &= \mathbf{S}_{(\rho/s)} + \mathbf{S}_{(s/z)},\end{aligned} \tag{6.141}$$

where $\mathbf{S}_{(s/z)} = \nabla_s z$ is the slope of the generalized vertical coordinate surface relative to the geopotential, and $\mathbf{S}_{(\rho/s)}$ is the slope of the neutral direction relative to the generalized vertical coordinate surface. In words, this equation says that the slope of the neutral direction relative to the geopotential equals the slope of the neutral direction relative to the generalized vertical coordinate surface plus the slope of the generalized vertical coordinate surface relative to the geopotential. In isopycnal models, the slope $\mathbf{S}_{(\rho/s)}$ is very small for the most part, thus prompting isopycnal modelers to dispense with much of the numerical machinery described in Part 4 needed to implement neutral physical processes in z-models. For σ-models, $\mathbf{S}_{(s/z)}$ can be nontrivial in much of the model domain affected by topography, whereas in pressure coordinate models $\mathbf{S}_{(s/z)}$ is typically less than 10^{-4}.

Figure 6.5 illustrates the relation (6.141) between slopes. This figure shows a particular zonal-vertical slice, with slope given by the tangent of the indicated angle. That is, the x-component of the slope vectors are given by

$$\begin{aligned}S_{(s/z)} &= \tan\alpha_{(s/z)} \\ S_{(\rho/z)} &= \tan\alpha_{(\rho/z)} \\ S_{(\rho/s)} &= \tan\alpha_{(\rho/s)}.\end{aligned} \tag{6.142}$$

In this example, $S_{(s/z)} < 0$ whereas $S_{(\rho/z)} > 0$. Note that the angle between the generalized surface and the isopycnal surface, $S_{(\rho/s)}$, is larger in absolute value for this example than $S_{(\rho/z)}$. This case may be applicable to certain regions of σ-models, whereas for isopycnal models $S_{(\rho/s)}$ will generally be smaller than $S_{(\rho/z)}$.

6.12 CHAPTER SUMMARY

We touched on many issues in this chapter which are at the forefront of ocean model development, in particular, the issue of what vertical coordinate is appropriate for an ocean model. In the past, most models were designed around the choice of a single vertical coordinate. This choice greatly determines the form by which the dynamics is represented and unresolved physical processes are parameterized. Indeed, the choice of vertical coordinate is the single most important choice to be made when designing an ocean model.

The aim of many present-day ocean model research groups is to remove the need to make an *a priori* choice of vertical coordinate. Instead, the goal is to provide an algorithm whereby one may, within a single *model environment*, investigate any number of vertical coordinates. Arguably the most advanced model satisfying this goal is that designed by Bleck (2002), in which the near surface coordinate is

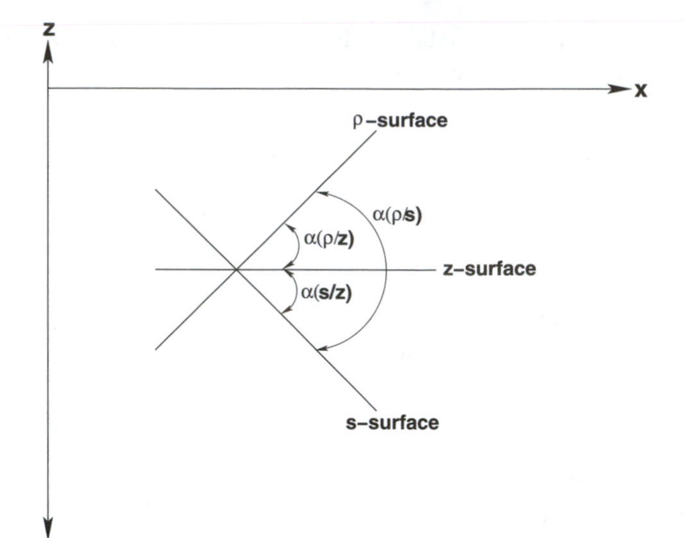

Figure 6.5 Relationship between the slopes of surfaces of constant depth, constant generalized vertical coordinate s, and potential density ρ. Shown here is a case where the slope is projected onto a single horizontal direction, so that the slope is given by the tangent of the indicated angle.

pressure, the interior coordinate is isopycnal, and the near bottom coordinate is sigma. This work represents yet another example of Bleck's pioneering efforts to push the envelope of ocean model design forward.

Motivations for various vertical coordinate choices presented in Section 6.2 were followed by mathematics required to derive and massage the ocean's kinematical and dynamical equations using a generalized vertical coordinate framework. Section 6.3 introduced some aspects of smooth generalized surfaces that fill the ocean domain. The monotonicity constraint, which is required for the vertical coordinate to be useful in practice, enables one to measure the slope of the surfaces with respect to the horizontal. The slope vector appears in many contexts when discussing generalized vertical coordinates, and its form is given by any one of the equivalent forms

$$\begin{aligned} \mathbf{S} &= \nabla_s z \\ &= -z_{,s} \, \nabla_z s \\ &= (S_x, S_y, 0). \end{aligned} \tag{6.143}$$

Again, because the surfaces are monotonically stacked in the vertical, the Jacobian of transform, $z_{,s}$, which is also known as the *specific thickness*, remains single-signed.

Sections 6.5 and 6.6 developed the calculus needed to derive the ocean's equations using generalized vertical coordinates. Additional discussions were provided of the geometric aspects of these coordinates. These aspects can be a bit subtle due to the non-orthogonality between the vertical and horizontal. For this reason, it is recommended that Figure 6.4 on page 133 be thoroughly digested by those aiming to employ generalized vertical coordinates in their studies.

Section 6.7 provided a discussion of the dia-surface velocity component

$$w^{(s)} = z_{,s}\, \mathrm{d}s/\mathrm{d}t. \tag{6.144}$$

As defined, $w^{(s)}$ measures the volume of fluid passing through a surface of constant generalized vertical coordinate, per unit area $\mathrm{d}A = \mathrm{d}x\,\mathrm{d}y$ of the horizontal projection of the surface, per unit time. Introducing the dia-surface velocity component allows for the material time derivative to be written

$$\frac{\mathrm{d}}{\mathrm{d}t} = \left(\frac{\partial}{\partial t}\right)_s + \mathbf{u} \cdot \nabla_s + w^{(s)} \left(\frac{\partial}{\partial z}\right), \tag{6.145}$$

thus motivating one to consider $w^{(s)}$ as a vertical velocity component that measures the rate at which fluid parcels penetrate the surface of constant generalized coordinate. Discussions of dia-surface velocity components appear frequently in the literature, such as when discussing the rate at which fluid passes across an isopycnal surface. The discussion in Section 6.7 should provide the reader with suitable geometric details to understand what this velocity component actually measures.

Sections 6.8 and 6.9 introduced some fluid kinematics encountered when using generalized vertical coordinates. In particular, the mass and volume continuity equations were presented. Attention was also placed on kinematic boundary conditions. Section 6.10 focused on the primitive equations in generalized vertical coordinates, as well as in a few special vertical coordinates. Section 6.11 finished the chapter with a discussion of how SGS tracer fluxes are transformed from one vertical coordinate system to another. We also exposed some issues of how to represent neutral physics in various vertical coordinate systems.

We close this chapter by noting that the goal of developing an ocean model *environment*, general enough to include all of the vertical coordinates in use today, remains a difficult task. Besides engineering details, which are innumerable, there is a fundamental issue related to algorithmic design, with Adcroft and Hallberg (2004) identifying two basic approaches that are presently incompatible.

The distinction between the two approaches can be summarized by how the two compute the dia-surface velocity component. First there is an *a posteriori* approach, which originates from algorithms based on the z-model and σ-model methods. Adcroft and Hallberg (2004) denote this as a *quasi-Eulerian vertical dynamics* or EVD. Here, the dia-surface velocity component is *diagnosed* via continuity *after* having computed the horizontal convergence and free surface evolution. Notably, all models solving the nonhydrostatic equations are based on this approach.

The second approach is an *a priori* approach, which originates from algorithms used in isopycnal models. Adcroft and Hallberg (2004) denote this as a *quasi-Lagrangian vertical dynamics* or LVD. Here, the dia-surface velocity component is *specified* rather than diagnosed. For example, in ideal fluid simulations using isopycnal models, the diapycnal velocity component is set to zero *a priori*, thus allowing the numerical model to retain a faithful isopycnal nature. For more general simulations, the diapycnal velocity component is set according to parameterizations of diapycnal processes, such as heating, mixing, thermobaricity, and so on. There is fundamentally no way that a model based on the *a posteriori* method, even if it used isopycnal coordinates, could guarantee that a diagnosed diapycnal velocity component would be identically zero for ideal flows. Truncation errors preclude it on the finite lattice of a numerical model.*

The generalized vertical coordinate model designed by Bleck (2002) orginates from an isopycnal model, and so it uses an *a priori* algorithm. In contrast, the generalized models built by Gerdes (1993), Mellor et al. (2002), and Marshall et al. (2003) originate from σ- and z-models, and so use an *a posteriori* algorithm. The author of the present book has more experience with *a posteriori* algorithms, and this is reflected in the discussion of primitive equation solution methods in Chapter 12. The two approaches represent nontrivial barriers that preclude one class of models based on a particular algorithm from running simulations in a manner identical to the other. Whether the two approaches can, or should, be unified, will likely remain at the forefront of algorithmic design questions, especially as we move into a research phase where general classes of model environments, rather than models, are the goal.

*See Section 7.5 (page 166) for issues of truncation errors and spurious mixing.

PART 2

Averaged Descriptions

The ocean is a turbulent fluid with dynamical scales over a huge range in space and time, and we are privy to very limited information about the fluid. Consequently, a practical description, such as that arising from an ocean model, must be based on a statistical-dynamical approach. This part of the book provides a very incomplete treatment of this problem, which arguably is the most intractable problem in classical physics, the problem of fluid turbulence. Indeed, the formalism here is naive relative to that employed in formal theories of turbulence and statistical physics. Nonetheless, the reader may garner from these pages some appreciation for issues that must be addressed when attempting to provide an averaged description of the ocean.

Subgrid scale (SGS) transport operators arise from averaging the equations of motion. These operators must be parameterized prior to integrating the mean field equations. Parameterizing SGS processes remains one of the most complex, rich, and critical areas of theoretical oceanography. There will always be regimes that an ocean model does not resolve, and how one parameterizes these regimes is a key element determining the physical integrity of simulations. Issues involved with SGS parameterizations are introduced in Chapter 7. Some attention is also given here to topics important for parameterizing small-scale dianeutral processes, which are processes that transfer properties across the smoothed neutral directions.

The two remaining chapters in this part focus on two different ensembles of ocean states. The dual ensemble approach presented here is largely motivated by the work of DeSzoeke and Bennett (1993). Corresponding to each of the two ensembles is an interpretation for the fields carried by an ocean model, with the interpretation depending on the physical regimes explicitly represented.

The first ensemble considers small scales of motion currently unapproachable by even the finest of global ocean model resolutions. Our viewpoint is fully Eulerian, in that we position ourselves at a fixed point in space-time and average over the ensemble. The result of averaging over small-scale (order meters) overturns and irreversible mixing processes, many of which are mentioned in Chapter 7, is a smooth averaged density profile.

The second ensemble focuses on the reversible stirring of fluid density in the nearly ideal ocean interior. This stirring is best viewed from the perspective of an isopycnal reference frame, where isopycnals are well defined stably stratified layers resulting from an average over small scales considered in the first ensemble. For the isopycnal reference frame, the horizontal position and time are the same as the Eulerian frame. The vertical position is Lagrangian for adiabatic and isohaline motion since it moves up and down based on undulations of a potential density surface. Mapping an isopycnal perspective back to the Eulerian perspective, a mapping needed to interpret coarse z-model simulations, requires some tools analogous to those used in the generalized Lagrangian mean formalism of Andrews and McIntyre (1978).

Focusing on two ensembles is traditional in the ocean literature, where small-scale mixing processes are first handled, and large scale stirring processes are handled separately. An ability to consider these ensembles separately is based on the hypothesis that the small and large scale processes only weakly interact. However, this is not the case for all flows, especially those near ocean boundaries. Theory has recently focused on the difficult problem of mesoscale eddies interacting with boundary layer processes. Ocean modelers, on the other hand, have been forced for some time to prescribe something for this part of the simulation, with many of the schemes quite *ad hoc*. Common practices are discussed in Chapter 15 for the surface boundary layer.

Determining a transparent and tidy set of averaged equations requires care in nomenclature, frames of reference, and interpretations. Some of the steps are subjective and based on convenience and desire. Indeed, how to average over an ensemble of fluid states is largely an art. One can certainly perform a traditional Reynolds decomposition, but will the resulting correlation terms be easily measurable and/or worth parameterizing? Mindfulness in choosing coordinates and weights can reduce the number of terms unavailable to the coarsened state, thus easing the burden on what needs to be parameterized. These kinematic issues become critical when attacking a statistical-dynamical problem for rotating-stratified flows. That is, care in setting up the problem goes a long way toward exposing the key physical results and rendering usable mean field equations.

Chapter Seven

CONCERNING UNRESOLVED PHYSICS

A fundamental problem in computational fluid dynamics is how to parameterize processes that live at space-time scales not explicitly represented by the computational grid. Due to nonlinear interactions between scales, details of these subgrid scale (SGS) processes are often critical to the structure of the larger scales explicitly represented. Hence, SGS processes cannot be ignored. Instead, they must be parameterized.

The purpose of this chapter is to provide an overview of the issues involved in SGS parameterizations, with a focus on issues related to small-scale processes that mix properties across neutral directions (i.e., *dianeutral* processes). Such processes affect the vertical stratification of properties in the ocean, and their parameterization in ocean models is critical for physically realistic simulations.

7.1 REPRESENTED DYNAMICS AND PARAMETERIZED PHYSICS

A process that is not resolved by a numerical model grid constitutes part of the model's *subgrid scale* (SGS). As we are concerned in this book with physics, unresolved processes are said to comprise the *SGS physics*. In contrast, space-time scales that are resolved by the model grid are termed *model dynamics*.

This terminology is somewhat loose. For example, resolved dynamical processes are often not fully resolved in a sense satisfactory to a rigorous numerical analyst. Likewise, unresolved processes are often not fully subgrid scale. Furthermore, even if there are plenty of grid points to resolve a dynamical process in principle, the model's discrete representation may not be faithful to the continuum dynamics due to inaccurate or inappropriate numerical techniques. Given these caveats, the loose distinction between *resolved dynamics* and *SGS physics* serves for the discussions in this chapter.

7.1.1 Resolved and SGS advective transport

An averaged description of nonlinear advective transport appearing in both the momentum and the tracer equations introduces SGS transport terms whose form is not prescribed by knowledge of just the resolved scales. Averaged advective transport serves as the canonical example of the split between resolved dynamics and SGS physics. Mathematically, this split can be seen by considering the average of advective transport for an arbitrary field ψ

$$\langle \mathbf{v}\,\psi \rangle = \langle \mathbf{v} \rangle \, \langle \psi \rangle + \langle \mathbf{v}'\,\psi' \rangle. \tag{7.1}$$

In this equation, the angle brackets denote an averaging operator, primes represent deviations from the average, and $\mathbf{v}$ is the velocity field. The first term, $\langle \mathbf{v} \rangle \, \langle \psi \rangle$, is explicitly represented in an ocean model via a choice for a numerical advection scheme. That is, this term is cast onto a discrete lattice and constitutes part of the model's resolved advective transport. The integrity of this representation of resolved advection is very dependent on the sophistication of the numerical advection scheme. Such matters are crucial to the model simulation, but are not part of the discussion here. The reader should refer to Durran (1999) for a thorough and pedagogical presentation of numerical advection.

Given knowledge only of the mean fields $\langle \mathbf{v} \rangle$ and $\langle \psi \rangle$, the correlation term $\langle \mathbf{v}' \, \psi' \rangle$ must be parameterized in order to account for its effects on the mean fields. Discussions in Chapters 8 and 9 provide specific examples of such SGS transport terms. The advective transport term $\langle \mathbf{v}' \, \psi' \rangle$ leads to a SGS transport whose effects on the mean flow is often nontrivial, especially in regions of the ocean where unresolved turbulent transport occurs. Hence, ignoring this term, or prescribing a closure in an *ad hoc* or cavalier manner, is generally not an acceptable option.

7.1.2 Information loss and the closure problem

Providing a systematic means to rationally parameterize SGS transport processes constitutes one of the fundamental unresolved problems in classical physics. This is the *turbulence closure problem*. Notably, there is not just one closure problem. Instead, there is a closure problem for each of the many unresolved physical processes. For example, the closure problem for unresolved mesoscale eddies is distinct from that for unresolved breaking internal waves. Details of the unresolved processes generally prompt the kinematic framework for posing the closure problem as well as the dynamical assumptions that lead to a choice for the form of the closure.

The closure problem results from a need to average over scales that are unresolved or unmeasured by the coarse-grained observer. Hence, the averaged equations contain less information than the unaveraged equations. This situation reflects the limits inherent in modeling and/or observing. Information loss also occurs when formulating a thermodynamic description of matter from statistical mechanics (e.g., Jaynes, 1957). From a statistical perspective, the two problems have much in common. Indeed, some effort has been focused on the use of information theory to motivate the form of turbulence closure in certain fluid systems (see, e.g., Kraichnan and Montgomery, 1980; Salmon, 1998; Holloway, 1999).

7.1.3 Analogies to Brownian motion

Theories of turbulence often start with analogies to the random walk motion of a Brownian particle (Reif, 1965). In this system, one considers the averaged motion of a particle within a fluid affected by random molecular bombardments. Well tested ideas from statistical physics lead to a diffusive form of the SGS operator in the mean field equations.

Following the Brownian particle example, fluid mechanical turbulence closures are often based on downgradient diffusion of quantities that are materially conserved in ideal flow, such as material and thermodynamic tracers. Analogously, it may also be appropriate to diffuse dynamical tracers that are materially conserved, such as potential vorticity (PV) (see Section 4.9). What a diffusive closure means is that SGS correlations between velocity and quasi-conserved tracers are parameterized as downgradient diffusion of the averaged tracer. Such closures are generally local, or semi-local, in that the fluxes are parameterized as a function of the state of the averaged flow field. The reader may wish to refer to Green (1970), Plumb (1979), Holland and Rhines (1980), and Rhines and Young (1982), where these issues are discussed, with particular attention placed on diffusion of potential vorticity.

Some processes, such as viscous transfer, can lead to local diffusive transfer of momentum. Additionally, pressure provides a mechanism for momentum to be radiated over large distances via wave motions. Hence, in contrast to quasi-conserved tracers, downgradient diffusion of linear momentum appears to be insufficient for parameterizing SGS momentum transport. Nonetheless, as discussed in Part 5 of this book, linear momentum diffusion is used in ocean climate models, in response to the need to satisfy fundamental numerical constraints.

It is commonly thought that tracer diffusion is both physically reasonable and numerically necessary. The main constraint is that tracer diffusion in ocean climate models must respect the huge separation in magnitude between neutral and dianeutral processes (see Section 7.2). Additionally, PV diffusion is the preferred approach for many theories of mesoscale eddy closures (Section 9.1). However, there are presently no global climate models whose closure schemes are fundamentally based on PV diffusion. Notably, arguments leading to Brownian motion diffusive closures of material tracers must be extended when considering a dynamically active tracer such as PV. For example, the rather unique properties of PV substance mentioned in Section 4.9, and described more thoroughly by Haynes and McIntyre (1987, 1990), must be respected when prescribing a PV-based closure.

7.2 LATERAL (NEUTRAL) AND VERTICAL PROCESSES

Turbulence in the ocean associated with small-scale processes tends to be three-dimensional. As an approximation, this mixing is typically assumed isotropic. At the larger scales (order many kilometers to hundreds of kilometers), transport exhibits a significant amount of anisotropy, with lateral processes far more efficient than vertical processes. In this section, we describe why this is the case, and make more precise what oceanographers mean by lateral.

7.2.1 Neutral directions

Consider a fluid stably stratified in the vertical yet without any horizontal stratification (i.e., flat isopycnals and so zero baroclinicity). Adiabatic and isohaline motion of a parcel in the vertical exposes the parcel to a restoring force due to its being surrounded by fluid with different buoyancy $-\rho g/\rho_o$. When displaced a small amount in the vertical, the parcel undergoes linear oscillations about its initial equilibrium position with a frequency proportional to the vertical density gradient (e.g., Gill, 1982). We discussed such buoyancy oscillations in Section 4.8 (page 68). In contrast, adiabatic and isohaline motion in the horizontal does not alter buoyancy, and so horizontal motion encounters no restorative force. That is, the horizontal defines a direction of neutral buoyancy, or *neutral direction*, for horizontal density surfaces.

For nonhorizontal isopycnals, neutral directions remain the directions in which a parcel can move in an adiabatic and isohaline manner, without altering its buoyancy. McDougall (1987a) formalized this definition and showed how to be mindful to define the potential density locally when the equation of state is nonlinear. The result of his analysis indicates that neutral directions are tangent to the locally referenced potential density surface. Therefore, the adiabatic and isohaline interchange of two fluid parcels along a neutral direction does not alter the buoyancy of either parcel, nor does it alter the gravitational potential energy of the combined two fluid parcel system. Locally referenced potential density can be considered a generalization of buoyancy (with a multiplicative factor of $-g/\rho_o$), and so the terms are used synonymously in the following.

This discussion indicates how stratification acts as an effective barrier to motion across neutral directions. A consequence of this barrier is that many transport processes tend to spread properties much more efficiently along, rather than across, neutral directions. Hence, neutral directions define what oceanographers mean by "lateral."

How "neutral" are processes in the ocean interior? This is fundamentally an empirical question.* Before seeking guidance from measurements, consider what would happen if a material displacement moved a parcel across, rather than along, a neutral direction. If this motion were to lead to an unstable stratification, as it generally could, then mixing processes would rapidly act to stabilize the parcels gravitationally. Hence, measuring the level of mixing between buoyancy classes provides a direct indication of how far fluid parcel trajectories deviate from neutral.

Direct measurements of tracer diffusivity in large regions of the ocean were pioneered with the purposeful passive tracer release experiments of Ledwell et al. (1993). These measurements indicate that on the large scales (order hundreds of kilometers), the associated neutral to dianeutral anisotropy in mixing can be as high as 10^8 in the ocean interior, with smaller anisotropies in regions of strong dianeutral mixing such as within boundary layers or above rough topography.

Another method for determining the dianeutral diffusivity uses the indirect approach suggested by Osborn (1980) and reviewed by Gregg (1987) and Davis (1994a). Here, momentum dissipation at small scales is directly measured, and dianeutral diffusivity is inferred based on a theoretical connection between buoyancy mixing and momentum dissipation. These *micro-structure* techniques indicate that the levels of interior dissipation are very small, and they are in general agreement with the direct tracer release measurements. They each indicate that the level of interior dianeutral mixing corresponds to a diffusivity on the order $10^{-5}\,\mathrm{m^2\,s^{-1}}$. This is a strong statement regarding the level to which the ocean interior respects the neutral orientation of transport. Ocean models, especially those used for purposes of climate simulations, must respect this anisotropy.

*A similar question is asked about numerical transport algorithms in Section 7.5.

7.2.2 Neutral and dianeutral physical processes

Because of the anisotropy in transport processes described above, oceanographers find it useful to classify SGS physical processes according to whether they predominantly act to transport properties in neutral directions or dianeutral directions. The processes associated with neutral directed SGS transport (both advective and diffusive) are dominated by mesoscale eddies, and their effects on the lateral distribution of oceanic properties require parameterization in present-day ocean climate models. As introduced in Chapter 9, kinematic and dynamic issues related to understanding and parameterizing these eddies constitute a large focus of current research. Parameterizations of such *neutral physics* processes, as implemented in z-models, are discussed in Part 4 of this book.

Even though the dominant transport of ocean properties in the interior is along, not across, neutral directions, those processes contributing to the material modification of buoyancy are crucial to establishing the oceanic stratification and vertical distribution of oceanic properties such as heat, salt, and nutrients. Certain of these dianeutral transport processes constitute the dianeutral SGS physics in an ocean model. Research on understanding and parameterizing processes affecting dianeutral transport is quite active. This book is remiss at not devoting a full set of chapters to this active field. To partially remedy this absence, we provide a terse discussion in Sections 7.3 and 7.4. A far more thorough compendium can be found in various contributions to Chassignet and Verron (1998), the rigorous treatment of oceanic microphysics by Davis (1994a,b), the review of small-scale mixing in the book by Kantha and Clayson (2000b), and the review articles by Holloway (1989) and Toole and McDougall (2001).

It is interesting to consider the analogous situation in atmospheric modeling. Here, SGS physics generally refers only to processes materially affecting buoyancy, such as convection and boundary layer processes. This situation is related to the larger Rossby radius in the atmosphere (some 10 times larger than in the ocean), which sets the scale for the quasi-geostrophic turbulent processes (i.e., synoptic weather) that dominate the processes associated with neutral physics. This larger spatial scale is reasonably well resolved by current global atmospheric models, thus alleviating the need to parameterize their effects.

7.3 BASIC MECHANISMS FOR DIANEUTRAL TRANSPORT

Small-scale (order millimeters to meters) turbulence leads to dianeutral mixing of fluid parcels. Such mixing is typically parameterized as Fickian diffusion of the material constituents and thermodynamic properties of the fluid (i.e., dianeutral diffusion of tracers as discussed in Section 7.4). Dianeutral tracer diffusion leads to dianeutral buoyancy diffusion. If the equation of state is linear, then dianeutral buoyancy diffusion is the only contributor to dianeutral buoyancy transport. In the presence of a nonlinear equation of state for seawater, processes parameterized as neutral diffusion of potential temperature and salinity lead to an additional contribution to dianeutral buoyancy transport which is itself nondiffusive in nature. The purpose of this section is to introduce some processes leading to dianeutral transport. Section 7.4 follows with a more mathematical discussion.

7.3.1 Mechanical and buoyant mixing

As noted in Section 1.2 of Kantha and Clayson (2000b), there are two basic methods whereby turbulent mixing is generated and maintained in the ocean: mixing via shear instabilities (mechanical mixing) and mixing via gravitational instabilities (buoyant mixing). Shear instabilities result from the differential motion of adjacent fluid parcels. An example includes the passage of internal waves through a quiescent background, in which wave breaking and/or shear instabilities prompt mixing (see, e.g., Polzin et al., 1997; Toole, 1998; Kantha and Clayson, 2000b). This mixing can affect the stratification of deep waters in the World Ocean. Another example is the wind induced mixing occurring in the surface ocean due to breaking surface waves. This mixing is critical for transferring momentum from the atmosphere into the ocean interior. A third example involves the strong vertical shears within the equatorial current system. Here, shears can induce unstable waves that break and thus mix properties vertically; they are thought to be essential for maintenance of the equatorial currents.

Gravitational instabilities occur when buoyancy forcing causes a fluid parcel to become gravitationally unstable. An important example occurs when buoyancy is lost to the atmosphere as a winter storm passes over the ocean at high latitudes (see, e.g., Send and Käse, 1998; Large, 1998; Marshall and Schott, 1999; Kantha and Clayson, 2000b). This process is associated with deep convection and is critical for the formation of water masses in the World Ocean. Another source of gravitational instability is the heating of deep waters by geothermal effects, thus inducing a rising column of fluid from the bottom to its level of neutral buoyancy.

If shear and/or buoyancy forcing are maintained, flow instabilities evolve into a fully developed three-dimensional turbulent state. Such turbulence is characterized by random behavior where fluid parcels readily mix their properties. Nonlinear interactions afforded by advection of momentum lead to a cascade of kinetic energy to ever smaller scales. Ultimately, kinetic energy introduced by forcing at the relatively large scales is dissipated by molecular viscosity at the Kolmogorov scale, which is the scale where the effects on kinetic energy due to molecular viscosity are comparable to advection (Section 5.3.2.2). That is, where the Reynolds number defined in terms of the molecular viscosity is on the order of unity. Notably, as three-dimensional turbulence leads to the dissipation of kinetic energy, the associated mixing leads to an increase in potential energy along with small levels of heating.* This behavior is in contrast to quasi-geostrophic turbulence, whereby potential energy contained in sloped isopycnal surfaces is transferred to kinetic energy via the mechanisms of baroclinic instability (Section 9.1).

7.3.2 The nonlinear equation of state

Besides mixing associated with mechanical and buoyant mechanisms, *cabbeling*, *thermobaricity*, and the weaker *halobaricity* lead to a nontrivial dianeutral transport when two fluid parcels mix with equal density yet different potential temperature and salinity. As seen in Sections 7.4.4 and 14.1.7 (page 302), these effects are associated with nonlinearities in the equation of state and lead to the material evolution

*As noted in Section 5.3.2, heating due to frictional dissipation by molecular viscosity is very small in the ocean.

of buoyancy that is distinctly nondiffusive. McDougall (1987b) provides a thorough discussion of these effects, whose importance in certain regions of the ocean can rival those from other mixing processes.

7.3.3 Double diffusion

Dianeutral buoyancy transport also arises when a multi-component fluid, such as seawater, is gravitationally stable yet possesses a nontrivial vertical gradient of its constituents. For seawater, this occurs when vertical temperature and salinity gradients have the same sign, and so contribute oppositely to the vertical density gradient. Due to the differing rates of molecular diffusion for temperature and salinity, mixing can ensue whereby potential energy is decreased. Notably, the resulting density flux is upgradient, although the temperature and salinity fluxes are both downgradient. Schmitt (1994), Schmitt (1998), Chapter 7 of Kantha and Clayson (2000b), and Toole and McDougall (2001) provide reviews of such *double diffusive* processes.

7.3.4 Summary of the processes

There is a plethora (a zoo!) of physical processes leading to dianeutral transport throughout the World Ocean. Broadly, large levels of mixing occur in the upper planetary boundary layer (the surface mixed layer) due to the effects of mechanical and buoyant forcing from the atmosphere, sea ice, and rivers. Large levels also occur in selected regions of the ocean bottom boundary layer, where cascading overflows of heavy fluid enter regions of lighter fluid. The strong flows can set up Kelvin-Helmholtz instabilities and/or hydraulic jumps that give rise to a significant level of entrainment with surrounding water as the overflow descends to a level of neutral buoyancy (see, e.g., Price and Baringer, 1994; Beckmann, 1998). Additionally, breaking internal waves scattered from rough topography, generated to a large degree by tidal motions (e.g., Munk and Wunsch, 1998), is now thought to provide nontrivial levels of mixing that extend hundreds of meters upward into the fluid column. Such mixing is also prevalent in shallow seas. Incorporating these processes, and others such as cabbeling, thermobaricity, double diffusion, breaking surface waves, and so on, into ocean climate models requires a synergistic relationship between ocean climate modelers, ocean process modelers, and ocean observationalists.

7.4 DIANEUTRAL TRANSPORT IN MODELS

The purpose of this section is to provide some mathematical statements in support the discussion in Sections 7.2 and 7.3. In so doing, we hope to clarify some of the main ideas.

7.4.1 Dianeutral velocity component

Recall the dianeutral velocity component from Section 6.7,

$$w^{(\rho)} = z_{,\rho} \frac{\mathrm{d}\rho}{\mathrm{d}t}. \tag{7.2}$$

In this equation, ρ symbolizes the locally referenced potential density, or buoyancy (sans the $-g/\rho_o$ factor). This equation defines the dianeutral velocity component $w^{(\rho)}$. In particular, there is no obligation at this point to average over the microscale fluctuations that lead to a "fuzziness" in surfaces of constant potential density. However, to make use of isopycnal coordinates (or more generally, neutral density coordinates), requires such an average, which necessarily leads to a coarsening of the field equations over small-scale mixing processes (see, e.g., DeSzoeke and Bennett, 1993; Davis, 1994a,b). This is assumed in the following, which allows one to write the material time derivative in isopycnal coordinates as

$$\frac{\mathrm{d}}{\mathrm{d}t} = \left(\frac{\partial}{\partial t}\right)_\rho + \mathbf{u} \cdot \nabla_\rho + w^{(\rho)} \left(\frac{\partial}{\partial z}\right). \tag{7.3}$$

This expression prompts one to interpret $w^{(\rho)}$ as a dianeutral velocity component measuring the rate at which fluid parcels penetrate a potential density surface.* When $w^{(\rho)}$ vanishes, the dynamics simplifies since the spatial component of the material time derivative reduces from three to two dimensions. Some kinematic aspects of such quasi-two-dimensional dynamics form the focus of Chapter 9. The concern here is with physical processes yielding a nontrivial $w^{(\rho)}$, thus leading to the material evolution of buoyancy.

7.4.2 Dianeutral advection of buoyancy

With the dianeutral velocity component $w^{(\rho)}$ nonzero, buoyancy is materially altered via

$$\frac{\mathrm{d}\rho}{\mathrm{d}t} = w^{(\rho)} \rho_{,z}. \tag{7.4}$$

This equation follows by definition from equation (7.2). It affords an interpretation of the material evolution of buoyancy according to dianeutral advection. That is, all forms of dianeutral transport of buoyancy are summarized by their effects on $w^{(\rho)}$, and so this perspective highlights the need to parameterize processes such as those mentioned in Section 7.3 that lead to a nonzero $w^{(\rho)}$, as these processes are key to providing a sound simulation of water mass transformations.

Expanding the material time derivative in equation (7.4) leads to the equivalent statement

$$(\partial_t + \mathbf{u} \cdot \nabla_z + w\,\partial_z)\rho = w^{(\rho)} \rho_{,z}. \tag{7.5}$$

Consequently, dianeutral advection, $w^{(\rho)}\rho_{,z}$, encompasses in one term what three terms describe using a z-coordinate description. This result highlights the fundamental distinction between the vertical velocity component w and the dianeutral velocity component $w^{(\rho)}$. This distinction was also noted from a kinematic perspective in Section 6.7. Hence, vertical advection is distinct from dianeutral advection, simply because (see equation (6.73))

$$\begin{aligned} w^{(\rho)} &= w - (\partial_t + \mathbf{u} \cdot \nabla_\rho)\, z \\ &\neq w. \end{aligned} \tag{7.6}$$

*See Section 6.7 for kinematic details leading to this interpretation.

7.4.3 The tracer equation with a diffusive closure

Ocean climate models traditionally time step salinity and potential temperature. The evolution of density is then diagnosed. The reason models time step tracers is because small-scale mixing in the ocean can typically be parameterized for tracers as downgradient diffusion. However, as seen here and mentioned in Section 7.3.2, dianeutral transport of buoyancy cannot generally be parameterized just as diffusion when the equation of state is nonlinear.

To expose these points it is sufficient to consider a Boussinesq fluid, in which case we can write the tracer concentration equation in the form

$$\frac{dC}{dt} = -\nabla \cdot \mathbf{F}, \tag{7.7}$$

where $\mathbf{F}$ is the tracer concentration flux. To arrive at this equation requires a coarsened description so that small-scale turbulent mixing processes have been averaged over. The formalism discussed in Chapter 8 is relevant for this coarsening. The convergence of $\mathbf{F}$ parameterizes the SGS mixing processes.* This coarsening renders, in particular, a smooth buoyancy field thus allowing one to exploit an isopycnal description, when convenient.

With small-scale mixing in the ocean associated with three-dimensional turbulence, it is common to assume that a diffusive parameterization is appropriate (Section 7.1.2). This assumption yields

$$\frac{dC}{dt} = \nabla \cdot (\mathbf{K} \cdot \nabla C), \tag{7.8}$$

where $\mathbf{K}$ is a second order diffusion tensor (positive semidefinite and symmetric). Mathematical properties of the diffusion tensor are discussed in Section 13.4, with further focus on the neutral diffusion tensor in Section 14.1. For present purposes, note that components to this tensor can generally be written in a projection operator form (see equation (14.13), page 298)

$$K^{mn} = A_I(\delta^{mn} - \hat{\rho}^m \hat{\rho}^n) + A_D \hat{\rho}^m \hat{\rho}^n. \tag{7.9}$$

In this equation, A_I and A_D are non-negative diffusivities associated with parameterized tracer mixing along and across neutral directions, and $\hat{\rho}$ is a unit direction perpendicular to a neutral direction. In the literature, processes occurring along neutral directions are often called *epineutral*, whereas those crossing neutral directions are *dianeutral*. For diffusion parameterizing three-dimensional isotropic turbulence, it makes sense to set $A_I = A_D$. However, it is useful to anticipate an anisotropy between along and across neutral directions for processes affected by buoyancy, which generally leads to $A_I >> A_D$.

Due to large inhomogeneities in unresolved dianeutral processes, the diffusivity A_D exhibits a widely varying space-time dependence. For example, in the quiescent ocean interior and pycnocline, measurements indicate that $A_D \approx 10^{-5}\,\text{m}^2\,\text{s}^{-1}$ (e.g., Ledwell et al., 1993; Toole et al., 1994; Kunze and Sanford, 1996), whereas A_D is enhanced one to three orders of magnitude above rough topography due to breaking internal waves (Polzin et al., 1997) or near rapid changes in bottom

*Large et al. (1994) introduced a nonlocal source term to parameterize nondiffusive dianeutral transport processes, especially those occurring in the ocean's upper planetary boundary layer. Such effects are ignored in the present discussion.

topographic slope (Toole et al., 1994, 1997; Polzin et al., 1996). The diffusivity is also very large in various boundary layers, such as the surface and bottom mixed layers.

Averaging over small-scale motions so that neutral directions are well defined leads to neutral slopes somewhat smaller than $1/100$ in the bulk of the ocean interior. Consequently, it is common to approximate the dianeutral diffusive contribution to the diffusion tensor (7.9) as

$$A_D \hat{\boldsymbol{\rho}}^m \hat{\boldsymbol{\rho}}^n \approx A_D \hat{\mathbf{z}}\, \hat{\mathbf{z}}. \tag{7.10}$$

As shown in Section 14.1.4 (page 299), these two expressions differ to within terms proportional to $A_D\, S^2$, which is quite small for neutral slopes $S \leq 1/100$. Hence, the use of vertical diffusion to represent dianeutral diffusion is accurate where neutral slopes are small, which includes most regions outside of boundary layers. A similar approximation for the along-slope diffusive fluxes, also detailed in Section 14.1.4, allows for

$$\mathbf{K} \cdot \nabla C \approx A_I \, \nabla_\rho C + \hat{\mathbf{z}}\, (A_I\, \mathbf{S} \cdot \nabla_\rho C + A_D\, C_{,z}), \tag{7.11}$$

where neglected neutral and dianeutral terms are proportional to $(A_I, A_D)\, S^2$, respectively, and

$$\nabla_\rho = \nabla_z + \mathbf{S}\, \partial_z \tag{7.12}$$

is the lateral gradient along a neutral direction (see equation (6.33), page 134), with $\mathbf{S}$ the slope of the neutral direction given by $\mathbf{S} = \nabla_\rho z$. Consequently, the tracer concentration equation under a diffusive closure takes the form

$$\frac{\mathrm{d}C}{\mathrm{d}t} = \nabla_z \cdot (A_I \, \nabla_\rho C) + \partial_z\, (A_I\, \mathbf{S} \cdot \nabla_\rho C + A_D\, C_{,z}). \tag{7.13}$$

7.4.4 Evolution of locally referenced potential density

With potential temperature and salinity satisfying evolution equations of the form (7.7) with a diffusive closure for small-scale mixing processes, it is straightforward to deduce the evolution equation for locally referenced potential density

$$\begin{aligned} \frac{\mathrm{d}\rho}{\mathrm{d}t} &= \rho_{,\theta} \frac{\mathrm{d}\theta}{\mathrm{d}t} + \rho_{,s} \frac{\mathrm{d}s}{\mathrm{d}t} \\ &= -\nabla \cdot (\rho_{,\theta}\, \mathbf{F}^{(\theta)} + \rho_{,s}\, \mathbf{F}^{(s)}) + \mathbf{F}^{(\theta)} \cdot \nabla \rho_{,\theta} + \mathbf{F}^{(s)} \cdot \nabla \rho_{,s} \end{aligned} \tag{7.14}$$

where $\rho_{,\theta} = \partial\rho/\partial\theta$ and $\rho_{,s} = \partial\rho/\partial s$ are the thermal and saline partial derivatives of the locally referenced potential density. The absence of a $\rho_{,p}\, \mathrm{d}p/\mathrm{d}t$ contribution to $\mathrm{d}\rho/\mathrm{d}t$ is due to a focus on locally referenced potential density, rather than *in situ* density. Even so, $\rho_{,\theta}$ and $\rho_{,s}$ are affected by pressure deviations when considering their spatial gradients

$$\nabla \rho_{,\theta} = \rho_{,\theta\theta} \nabla \theta + \rho_{,\theta s} \nabla s + \rho_{,\theta p} \nabla p \tag{7.15}$$

$$\nabla \rho_{,s} = \rho_{,s\theta} \nabla \theta + \rho_{,ss} \nabla s + \rho_{,sp} \nabla p. \tag{7.16}$$

The convergence $-\nabla \cdot (\rho_{,\theta}\, \mathbf{F}^{(\theta)} + \rho_{,s}\, \mathbf{F}^{(s)})$ in equation (7.14) represents the convergence of the diffusive flux of locally referenced potential density

$$\mathbf{F}^{(\rho)} = \rho_{,\theta}\, \mathbf{F}^{(\theta)} + \rho_{,s}\, \mathbf{F}^{(s)}. \tag{7.17}$$

Notably, the contribution to $\mathbf{F}^{(\rho)}$ from diffusion of potential temperature and salinity along neutral directions vanishes, as it must by definition. That is, one cannot diffuse buoyancy along a surface of constant buoyancy.* Hence, when considering small-scale mixing parameterized as diffusion, the flux of buoyancy reduces to just a vertical diffusive contribution

$$\mathbf{F}^{(\rho)} = -A_D\, \rho_{,z}\, \hat{\mathbf{z}}. \tag{7.18}$$

The term $\mathbf{F}^{(\theta)} \cdot \nabla \rho_{,\theta} + \mathbf{F}^{(s)} \cdot \nabla \rho_{,s}$ appearing in equation (7.14) does not take the form of a flux convergence. Hence, it is best thought of as a source term affecting buoyancy. It is nontrivial only when the second partial derivatives of density do not all vanish, and so they account for the nonlinear equation of state processes introduced in Section 7.3.2. Manipulations in Section 14.1.7 (page 302) illustrate how to identify the source terms arising from cabbeling, thermobaricity, and halobaricity, which are associated with temperature and salinity diffusion along neutral surfaces. A more complete presentation is given by McDougall (1987a,b).

In summary, small-scale mixing parameterized as tracer diffusion leads to the material evolution of buoyancy

$$\begin{aligned} \frac{d\rho}{dt} &= w^{(\rho)}\, \rho_{,z} \\ &= \partial_z\,(A_D\, \rho_{,z}) + \mathbf{F}^{(\theta)} \cdot \nabla \rho_{,\theta} + \mathbf{F}^{(s)} \cdot \nabla \rho_{,s}. \end{aligned} \tag{7.19}$$

The first term on the right-hand side arises from the dianeutral diffusion of potential temperature and salinity. The second and third terms arise from the nonlinear equation of state for seawater. These two terms can be the same order of magnitude in certain parts of the ocean (e.g., McDougall, 1987a,b). Nonetheless, many discussions in the literature ignore the nonlinear equation of state terms, and so the reader should be mindful of this limitation in those discussions.

Equation (7.19) provides a useful venue to highlight how certain ocean models provide a more or less faithful numerical representation of these processes. For this purpose, focus on the distinctions between isopycnal models and z-models, with σ-models sharing much with z-models in the present discussion.† Z-models trivially incorporate dianeutral diffusive processes via linear vertical diffusion of potential temperature and salinity, and they have no problem allowing for arbitrarily large vertical diffusivities, so long as the time stepping is handled implicitly. In contrast, diapycnal diffusion is a nonlinear process in isopycnal models, and details of how to handle large diffusivities stably have only recently been clarified by Hallberg (2000). Z-models incorporate effects from the nonlinear equation of state simply by using a realistic equation for seawater density, such as the one proposed by McDougall et al. (2003b). In contrast, these effects are nontrivial to incorporate into isopycnal models, although successful methods are now available, such as those from Sun et al. (1999) and Hallberg (2003b). Finally, the stirring effects arising from advective processes (appearing when the material time derivative is represented in its Eulerian form) are more cleanly handled by isopycnal models,

*This basic point is emphasized in Section 14.1.6 (page 301), where it is noted that a numerical implementation of neutral diffusion in a z-model must respect this property in order to remain numerically stable.

†See Section 6.2 starting on page 122 for a presentation of σ-models.

thus accounting for their use in many idealized studies of the adiabatic and isohaline ocean circulation. We have more to say in these regards in Sections 7.5 and 13.1.

7.5 NUMERICALLY INDUCED SPURIOUS DIANEUTRAL TRANSPORT

The huge anisotropy between neutral and dianeutral transport processes places a heavy burden on the numerical integrity of an ocean model. In particular, one cannot be too cavalier about how both resolved and parameterized transport processes are realized. Doing so risks not properly respecting the anisotropy. Until the past decade or so, physical levels of anisotropy in the ocean interior were not respected by the most common class of ocean climate models, the z-models which form the focus of later chapters in this book. However, recent advances documented in Part 4 have tremendously improved the integrity of coarse resolution z-models.

Problems remain for refined grid models that admit mesoscale eddies. As documented by Griffies et al. (2000b), mesoscale eddies transfer variance to the grid scale, and this variance must be dissipated. Otherwise, eddying simulations degenerate into a sea of grid noise. Unfortunately, common methods for absorbing variance introduce unphysically large levels of spurious dianeutral mixing, and this mixing can swamp the levels introduced by the physically based SGS parameterizations.

Although truncation errors generally get smaller as resolution is enhanced, refined resolution, which is usually associated with smaller numerical frictional dissipation, introduces new regimes of flow, that is, more energetic eddies. These eddies in turn pump variance to the grid scales, which further stresses the numerical integrity. This is a difficult and fundamental problem for the non-isopycnal class of ocean models. A possible means for resolving the spurious mixing problem is mentioned in Section 13.1 (page 283). There, it is conjectured that the neutral physics operators used in coarse resolution models, based on physical closure arguments, can also serve as useful numerical closure operators at the refined scales.

Additional problems with spurious mixing occur in z-models next to step-like topography (e.g., Winton et al., 1998). As noted by Beckmann (1998), much of the emphasis in simulating overflow processes in z-models is focused on providing a more suitable representational framework, rather than in parameterizing physical processes. That is, although there are physical reasons to expect enhanced mixing near the ocean bottom, z-models can spuriously provide far more than does the real ocean. Hence, consistent with our discussion in Section 6.2 (page 122), the z-model is cumbersome for the overflow problem, whereas the σ-model and the isopycnal model have been found to be far more suitable.

As noted in Section 6.2, isopycnal models using a linear equation of state trivially respect the large anisotropy between neutral and dianeutral processes. This point is highlighted in the review article by Bleck (1998). However, isopycnal models employing a realistic nonlinear equation of state must also contend with potentially unphysical levels of diapycnal advection. In this case, when isopycnal models advectively transport temperature and salinity along isopycnal surfaces, numerical isopycnal mixing associated with the advection schemes can introduce unphysically large amounts of cabbeling.* This issue is also relevant in z-models. It therefore appears that all model classes must contend with some level of spurious dianeutral processes. Algorithm developers must seek more refined methods which reduce the spurious transport to physically acceptable levels.

7.6 CHAPTER SUMMARY

In this chapter, we encountered some of the fundamental issues involved with parameterizing subgrid scale (SGS) processes in the ocean. Section 7.1 started the discussion by noting the distinction between represented dynamical processes and parameterized SGS processes. Although the distinction is not always clear from a rigorous perspective, it provides a useful point of reference.

Section 7.2 discussed the differences between lateral and vertical processes. In the ocean, lateral refers to neutral, where neutral directions are those directions in which a parcel can move without feeling a restoring force from buoyancy. Measurements in the ocean interior indicate a huge anisotropy between processes that transport tracers along neutral directions, and dianeutral mixing processes moving parcels across. Section 7.3 summarized some of the "zoo" of processes that lead to dianeutral transport. These processes tend to occur at the relatively small scales in the ocean, from centimeters to meters. They are thus largely unresolved by ocean climate models, and so must be parameterized. Section 7.4 then presented a formulation of dianeutral transport processes in models.

Dianeutral mixing processes are rather inhomogeneous in the ocean, since there are so many different processes that contribute to mixing. Parameterizing these processes occupies a large contingent of theoretical physical oceanography. Often after deriving a parameterization, the modeler wishes to test the scheme in an ocean model to garner a sense for how well the parameterization works, and how critical the parameterized process is for setting aspects of the large-scale circulation.

Section 7.5 exposed some of the difficulty to be expected when performing these parameterization tests in refined resolution (i.e., mesoscale eddying) simulations using non-isopycnal models. The problem is that these models can, in general, contain a level of spurious numerical mixing that in some cases overwhelms the levels of mixing induced by the physically based parameterization. This is not always the case when running coarsely resolved models without eddies, so long as the models employ reasonably accurate numerical transport methods and resolve the realized flows. But when mesoscale eddies are admitted, the eddies pump variance to the grid scale. The usual means for numerically dissipating variance (e.g., horizontal Laplacian or biharmonic mixing, or dissipative advection) can incur levels of spurious dianeutral mixing far larger than physically desired (Roberts and Marshall, 1998; Griffies et al., 2000b).

One may believe that upon grid refinement, truncation errors will get smaller. However, such is not necessarily the case when representing mesoscale eddies. The reason is that grid refinement is typically associated with a reduction in the strength of numerical dissipation operators in order to enhance the simulated

*Eric Chassignet, personal communication 2002.

eddy energy and so to exploit the resolution. Enhanced eddy energy also provides for an increased level of variance to pile up at the grid scales. Under such conditions, grid refinement results in a successively smaller length that must be resolved by the simulation. Under these circumstances, the relative truncation error can actually increase as the grid is refined, thus explaining how it is possible to have an increase in spurious mixing.

The spurious mixing problem is fundamental to the z-models and σ-models. It is a problem for active tracers such as temperature and salinity, and for passive tracers. This problem must be solved if these models are to remain physically viable for long-term ocean climate studies using resolutions that admit a vigorous mesoscale eddy field. But the question arises, how good is good enough? Or how bad is too bad? Certainly if the models can beat down levels of spurious mixing to well below physical levels of dianeutral mixing, then modelers can declare success. But what if the spurious mixing levels are no better than two or three times the measured levels? Are the solutions useless? Furthermore, how well do we know the measurements? Reducing spurious mixing by a factor of two or three is nontrivial when considering what is required algorithmically to refine a numerical transport scheme. But in some cases that is the level of accuracy of the measurements. These questions are important to answer in order to gauge the degree to which the models are faithful to the real ocean, and the degree to which further research should be focused on improving the algorithms.

Chapter Eight

EULERIAN AVERAGED EQUATIONS

The purpose of this chapter is to average the ocean equations over the small scales (order meters or less) where mixing processes dominate. Because these scales do not easily afford the specification of a smooth dynamical surface, such as an isopycnal or pressure surface, it is useful to take the average at a fixed point in space-time. That is, the average is Eulerian. The approach then leads to a particular interpretation of the variables used in numerical nonhydrostatic and hydrostatic z-models as a discrete representation of the ensemble averaged equations.

8.1 INTRODUCTION

The ocean is fundamentally turbulent. Given sensitive dependence on initial conditions and the limitations of ocean measurements, it is not possible to obtain complete knowledge of the ocean state. Therefore, when formulating the ocean's governing equations, it is necessary to recognize this limited access to information.

8.1.1 Ensemble averages

A common way to formulate a dynamical theory recognizing an incomplete level of information is to consider ensemble averages. As in statistical mechanics, ensemble averages in oceanography are obtained by formally considering an infinite number of ocean states, each of which is described by the kinematic and dynamic balances of Chapters 3 and 4, as well as the thermodynamic and tracer balances of Chapter 5. Our interpretation of ensemble averages follows that given in Section 2.5 (page 20), where the distance in phase space between members of an ensemble is directly related to the space-time scales that are not resolved by the "averaged observer." Notably, the presentation here is not rigorous. Instead, the discussion can be considered a first step towards a more rigorous yet presently intractable problem of closing the ocean equations using fully deductive methods.

The ensemble averages considered here are taken at a fixed point in space-time. These averages commute with space-time derivatives and integrals. As noted by DeSzoeke and Bennett (1993), such averages are relevant when coarsening a description over microstructure processes (order meters or less) that induce small-scale turbulent mixing between density classes (see Section 7.3). A particular result of averaging over the microscale is a smoothed mean density field suitable for use as a vertical coordinate in ocean climate models.

Even though averaging is done at a fixed space-time point, there remain ambiguities in details. Different methods reveal different aspects of the averaged or mean dynamics, as well as deviations from the mean. *Density weighted averag-*

ing plays a role in the method used here, with density referring to *in situ* density. Such averaging has recently become common in the oceanography literature (e.g., Smith, 1999; Lu, 2001; McDougall et al., 2003a). Additionally, it has been used in various other oceanographic studies such as Osborn and Cox (1972) and McDougall and Garrett (1992), as well as other areas of the compressible fluids literature, such as Hesselberg (1926) and Favre (1965) (see the footnote on pages 21-22 of Hinze, 1975).

8.1.2 Interpreting ocean model equations

A consistent interpretation of the ocean model equations is afforded by considering them to be discretizations of ensemble averaged continuum equations, where the ensemble members differ by details set at space-time scales smaller than the space-time scale of the model grid. Hence, besides their importance from a fundamental perspective, ensemble averaged ocean equations form the basis for the equations discretized in an ocean model.

The fixed space-time averages considered in this chapter are appropriate when averaging over small scales of motion where the dynamics is mostly isotropic in all three dimensions. For larger scales, such as those approaching the mesoscale, processes tend to occur along neutral directions, in which case the isopycnal averaging techniques discussed in Chapter 9 are far more suitable. For still larger scales, one may consider multiple realizations from numerical ocean models to explicitly form the ensemble members. An example includes the averaged response of the large-scale ocean circulation to a suite of atmospheric states (e.g., Griffies and Bryan, 1997).

8.1.3 Mean kinematics independent of dynamics

There are two forms of fluid parcel kinematics considered in this book: parcels conserving their volume and parcels conserving their mass. As described in Chapter 3, kinematic relations constrain the fluid motion, regardless of the dynamics. Key to the integrity and usability of the equations describing the ensemble averaged ocean, and consequently to the ocean model, is to have the kinematics of the averaged fluid remain independent of dynamical assumptions. In particular, closure terms, whose form depends on dynamical details, should not influence the kinematics of the averaged fluid or the ocean model.

A key motivation for using density weighted averaging is that it helps to keep the averaged parcel kinematics independent of dynamical closure assumptions. Providing a simple mapping between unaveraged and averaged kinematics generally does not require much thought when averaging the Boussinesq equations in z-coordinates, since volume conservation, $\nabla \cdot \mathbf{v} = 0$, is a linear constraint. Yet for non-Boussinesq fluids, mass conservation, $\rho_{,t} + \nabla \cdot (\mathbf{v}\,\rho) = 0$, is a nonlinear constraint, thus requiring extra consideration. Note that the converse is true when working in pressure coordinates, since mass conservation is a linear constraint for a non-Boussinesq hydrostatic fluid using pressure coordinates (see Section 6.8.4). This point illustrates the coordinate-dependent aspects of averaging, whereby one set of coordinates may be more convenient than another when averaging over certain scales.

As shown in Section 6.10.3, pressure coordinates are ideal for hydrostatic non-Boussinesq fluids given the simplicity of the ocean equations (Huang et al., 2001; DeSzoeke and Samelson, 2002; Marshall et al., 2003; Losch et al., 2004). Nonetheless, we focus here on *z*-coordinates based on their dominance in present-day ocean climate modeling. Additionally, it is useful to recognize that coordinates fixed in both space and time remain unambiguously defined regardless of the dynamical regime, whether it is hydrostatic or nonhydrostatic. In contrast, smooth surfaces formed by constant values of a dynamical coordinate, such as pressure, presume that an averaging has taken place over small scales where the coordinate can be discontinuous and/or highly corrugated. *Z*-coordinates become advantageous when aiming to resolve such small scales in the simulation,

Besides the mathematical utility of the density weighed approach for the non-Boussinesq system, McDougall et al. (2003a) argued for maintaining this average even when considering the averaged Boussinesq equations. Their reasoning is based on noting that the resulting Boussinesq system is far more accurate than the mean field equations resulting from non-density weighted averages. In this context, accuracy is based on comparing with the small levels of diapycnal mixing in the ocean interior.

8.2 THE NONHYDROSTATIC SHALLOW OCEAN EQUATIONS

This section starts the discussion by recalling pieces to the nonhydrostatic shallow ocean equations. Throughout this book, we focus on the dynamics of mass conserving fluid parcels. Mass conservation constrains the density and velocity to satisfy the continuity equation

$$\rho_{,t} + \nabla \cdot (\rho \, \mathbf{v}) = 0. \tag{8.1}$$

Conservation of material and/or thermodynamic tracers is reflected by the equation

$$(\rho C)_{,t} + \nabla \cdot (\rho C \, \mathbf{v}) = -\nabla \cdot (\rho \, \mathbf{F}) + \rho \, \mathcal{S}. \tag{8.2}$$

Note that the use of tracer flux in the form

$$\mathbf{J} = \rho \, \mathbf{F} \tag{8.3}$$

and tracer source

$$\Sigma = \rho \, \mathcal{S} \tag{8.4}$$

prove convenient in the following. Newton's second law applied to a mass conserving fluid parcel leads to the balance of linear momentum per volume

$$(\rho \, \mathbf{v})_{,t} + \nabla \cdot (\rho \, \mathbf{v} \, \mathbf{v}) + \mathcal{M} \, (\hat{\mathbf{z}} \wedge \rho \, \mathbf{v}) = -\rho \, g \, \hat{\mathbf{z}} - f \, \hat{\mathbf{z}} \wedge \rho \, \mathbf{v} - \nabla p + \rho \, \mathbf{F}^{(\mathbf{v})}. \tag{8.5}$$

The tracer flux $\mathbf{J}$, and its associated concentration flux $\mathbf{F}$, are interpreted as those arising from subgrid scale (SGS) molecular processes, such as molecular diffusion. Likewise, the frictional acceleration $\mathbf{F}^{(\mathbf{v})}$ is associated with momentum transport due to molecular viscosity. The averaging considered in this chapter introduces far more significant SGS processes associated with turbulent mixing and stirring.

The linear momentum density (momentum per volume)

$$\rho \, \mathbf{v} = \rho_o \, \tilde{\mathbf{v}} \tag{8.6}$$

plays a fundamental role in the dynamical balances. It proves to be especially useful when formulating the averaged equations. Additionally, it is useful to introduce an analogous form for the subgrid scale molecular fluxes of tracer concentration and momentum,

$$\rho\,\mathbf{F} = \rho_o\,\tilde{\mathbf{F}} \tag{8.7}$$

$$\rho\,\mathbf{F}^{(\mathbf{v})} = \rho_o\,\tilde{\mathbf{F}}^{(\mathbf{v})}. \tag{8.8}$$

Introducing these fields leads to the mass and tracer budgets

$$\rho_{,t} + \rho_o\,\nabla\cdot\tilde{\mathbf{v}} = 0 \tag{8.9}$$

$$(\rho\,C)_{,t} + \rho_o\,\nabla\cdot(C\,\tilde{\mathbf{v}}) = -\rho_o\,\nabla\cdot\tilde{\mathbf{F}} + \rho\,\mathcal{S}, \tag{8.10}$$

as well as the linear momentum budget

$$\begin{aligned}\tilde{\mathbf{v}}_{,t} + \nabla\cdot[(\rho_o/\rho)\,\tilde{\mathbf{v}}\,\tilde{\mathbf{v}}] + (\rho_o/\rho)\,\tilde{\mathcal{M}}\,\hat{\mathbf{z}}\wedge\tilde{\mathbf{v}}&\\ = -(\rho/\rho_o)\,g\,\hat{\mathbf{z}} - f\,\hat{\mathbf{z}}\wedge\tilde{\mathbf{v}} - \nabla(p/\rho_o) + \tilde{\mathbf{F}}^{(\mathbf{v})}.&\end{aligned} \tag{8.11}$$

In this equation, the advection metric frequency is written as

$$\begin{aligned}\mathcal{M} &= v\,\partial_x \ln \mathrm{d}y - u\,\partial_y \ln \mathrm{d}x\\ &= (\rho_o/\rho)\,(\tilde{v}\,\partial_x \ln \mathrm{d}y - \tilde{u}\,\partial_y \ln \mathrm{d}x)\\ &= (\rho_o/\rho)\,\tilde{\mathcal{M}}.\end{aligned} \tag{8.12}$$

Integrating the continuity equation over the full ocean column of thickness

$$D = H + \eta, \tag{8.13}$$

and assuming mass sources/sinks only at the ocean surface, lead to the balance of mass per unit area within an ocean column

$$(\overline{\rho}^{z}\,D)_{,t} = -\rho_o\,\nabla\cdot\tilde{\mathbf{U}} + \rho_{\mathrm{w}}\,q_{\mathrm{w}}, \tag{8.14}$$

as well as the surface and solid earth kinematic boundary conditions

$$\hat{\mathbf{N}}\cdot\mathbf{v} = \eta_{,t} - q_{\mathrm{w}}\,(\rho_{\mathrm{w}}/\rho) \qquad \text{at } z = \eta \tag{8.15}$$

$$\hat{\mathbf{N}}\cdot\mathbf{v} = 0 \qquad \text{at } z = -H. \tag{8.16}$$

Note that

$$\hat{\mathbf{N}} = \nabla\,(-\eta + z) \qquad \text{at } z = \eta \tag{8.17}$$

$$\hat{\mathbf{N}} = \nabla\,(H + z) \qquad \text{at } z = -H \tag{8.18}$$

define the surface and bottom orientation directions. Additionally,

$$\rho_o\,\tilde{\mathbf{U}} = \int_{-H}^{\eta} \mathrm{d}z\,\rho\,\mathbf{u} \tag{8.19}$$

is the depth integrated horizontal momentum density and

$$\overline{\rho}^{z} = D^{-1}\int_{-H}^{\eta} \mathrm{d}z\,\rho \tag{8.20}$$

is the depth averaged *in situ* density. The mass per unit time per unit horizontal area of fresh water crossing the ocean surface is given by $\rho_{\mathrm{w}}\,q_{\mathrm{w}}$, which is also a linear momentum per volume. As for the linear momentum per unit volume of ocean fluid $\rho\,\mathbf{v} = \rho_o\,\tilde{\mathbf{v}}$, it is useful to introduce the analogous quantity for the surface fresh water flux

$$\rho_\mathrm{w}\, q_\mathrm{w} \equiv \rho_o\, \tilde{q}_w, \tag{8.21}$$

which leads then to the mass balance

$$(\overline{\rho}^z\, D)_{,t} = -\rho_o\, \nabla\cdot\tilde{\mathbf{U}} + \rho_o\, \tilde{q}_w \tag{8.22}$$

and the surface kinematic boundary condition

$$\rho\,(\partial_t + \mathbf{u}\cdot\nabla)\,\eta = \rho_o\,\tilde{q}_w + \rho\, w \qquad \text{at } \; z = \eta. \tag{8.23}$$

The Boussinesq equations are recovered by setting $\rho \to \rho_o$ wherever it appears in the previous equations starting from equation (8.1), except when multiplying by the gravitational acceleration. The resulting continuity equation becomes a constraint on the three-dimensional velocity field

$$\nabla\cdot\mathbf{v} = 0. \tag{8.24}$$

The conservative form of the tracer concentration equation takes the form

$$C_{,t} + \nabla\cdot(C\,\mathbf{v}) = -\nabla\cdot\mathbf{F} + \mathcal{S}, \tag{8.25}$$

and linear momentum balance becomes

$$\mathbf{v}_{,t} + \nabla\cdot(\mathbf{v}\,\mathbf{v}) + \mathcal{M}\,\hat{\mathbf{z}}\wedge\mathbf{v} = -(\rho/\rho_o)\,g\,\hat{\mathbf{z}} - f\,\hat{\mathbf{z}}\wedge\mathbf{v} - \nabla(p/\rho_o) + \mathbf{F}^{(\mathbf{v})}. \tag{8.26}$$

Note that $\nabla\cdot\mathbf{v} = 0$ means the Boussinesq fluid parcels conserve their volume instead of their mass. Integrating the continuity equation over a full ocean column, and assuming volume sources/sinks only at the ocean surface, lead to the balance of volume per unit area within an ocean column

$$\eta_{,t} = -\nabla\cdot\mathbf{U} + q_\mathrm{w}, \tag{8.27}$$

as well as the surface and bottom kinematic boundary conditions

$$\hat{\mathbf{N}}\cdot\mathbf{v} = \eta_{,t} - q_\mathrm{w} \qquad \text{at } z = \eta \tag{8.28}$$

$$\hat{\mathbf{N}}\cdot\mathbf{v} = 0 \qquad \text{at } z = -H, \tag{8.29}$$

where $\mathbf{U} = \int_{-H}^{\eta} \mathrm{d}z\,\mathbf{u}$ is the depth integrated horizontal velocity field and q_w is the volume per unit time per unit horizontal area of fresh water crossing the ocean surface.

An important point to note is that the linear momentum density of a Boussinesq fluid is $\rho_o\,\mathbf{v}$, instead of the non-Boussinesq $\rho\,\mathbf{v}$. Hence, there is no difference between the velocity fields $\tilde{\mathbf{v}}$ and $\mathbf{v}$ for Boussinesq fluids. Ensemble averaging, however, breaks the symmetry between the averaged values of $\tilde{\mathbf{v}}$ and $\mathbf{v}$, as discussed next.

8.3 AVERAGED KINEMATICS

The averaged continuity equation using velocity (linear momentum per mass) as the fundamental transport field is given by

$$\rho_{,t} + \nabla\cdot(\rho\,\mathbf{v}) = 0 \rightarrow \langle\rho\rangle_{,t} + \nabla\cdot(\langle\mathbf{v}\rangle\,\langle\rho\rangle + \langle\mathbf{v}'\,\rho'\rangle) = 0, \tag{8.30}$$

where the angled brackets signify an Eulerian average, and primed variables are deviations from the average. The nonlinear correlation term in the averaged mass balance (8.30) is extremely cumbersome since to study the averaged kinematics it is necessary to specify the unknown correlation term $\nabla \cdot \langle \mathbf{v}' \rho' \rangle$. The form of this term is dependent on the space-time scales averaged over, and the associated SGS dynamics. In contrast, the averaged momentum per unit volume $\rho_o \tilde{\mathbf{v}}$ absorbs the correlation terms

$$\begin{aligned} \rho_o \langle \tilde{\mathbf{v}} \rangle &= \langle \rho \, \mathbf{v} \rangle \\ &= \langle \rho \rangle \langle \mathbf{v} \rangle + \langle \rho' \, \mathbf{v}' \rangle, \end{aligned} \tag{8.31}$$

thus leading to a continuity equation that remains form invariant upon averaging,

$$\rho_{,t} + \rho_o \nabla \cdot \tilde{\mathbf{v}} = 0 \rightarrow \langle \rho \rangle_{,t} + \rho_o \nabla \cdot \langle \tilde{\mathbf{v}} \rangle = 0. \tag{8.32}$$

It is useful to summarize these manipulations, which may appear to be a bit mysterious on first encounter. Notably, there has been no determination of the correlation $\langle \rho' \, \mathbf{v}' \rangle$. Rather, the Eulerian mean mass budget (8.32) *does not require* this correlation. Such is the case when focusing on the averaged linear momentum per unit volume, $\rho_o \langle \tilde{\mathbf{v}} \rangle$, instead of the averaged linear momentum per unit mass, $\langle \mathbf{v} \rangle$. There is nothing fundamental warranting the superiority of one averaged field versus the other. It is therefore valid to be guided by convenience in the resulting equations. These observations reflect the subjective nature of what turbulent fluctuations are needed for closing an averaged set of nonlinear equations. For purposes of working with the Eulerian averaged continuity equation, the single linear term $\rho_o \langle \tilde{\mathbf{v}} \rangle$ is far more convenient than the nonlinear alternative $\langle \rho \rangle \langle \mathbf{v} \rangle + \langle \rho' \, \mathbf{v}' \rangle$. This result promotes density weighted averaging for averaging the remaining kinematic and dynamic equations.

To further emphasize the utility of kinematics that does not contain closure terms, recall mass continuity for a hydrostatic fluid as written in pressure coordinates (Section 6.10.3),

$$\nabla_p \cdot \mathbf{u} + \partial_p \dot{p} = 0. \tag{8.33}$$

Choosing to perform an ensemble average while maintaining the same hydrostatic pressure and horizontal position leads to the ensemble mean continuity equation

$$\nabla_p \cdot \langle \mathbf{u} \rangle + \partial_p \langle \dot{p} \rangle = 0. \tag{8.34}$$

There are no correlations, since mass continuity is linear in pressure coordinates. As stated earlier, pressure coordinates are ideally suited for a non-Boussinesq hydrostatic fluid, since they are fundamentally mass based (pressure = mass of fluid per horizontal area). In contrast, depth is a volume based coordinate (depth = volume of fluid per horizontal area). Applying a density weight (mass per volume) within a z-coordinate Eulerian ensemble average introduces a mass weighting, and it results in a linear ensemble mean continuity equation, as occurs in pressure coordinates. That is, density weighting the fixed depth ensemble average for non-Boussinesq fluids recovers some of the advantages of unweighted ensemble averages taken at fixed hydrostatic pressure.

8.4 AVERAGED KINEMATICS OVER FINITE DOMAINS

Recall from Section 3.4.3 (page 36) that the kinematic boundary condition for the ocean surface was derived by considering the time tendency of mass within a finite ocean domain. We proceed here in a similar manner for the averaged non-Boussinesq fluid.

8.4.1 Averaged seawater mass and tracer mass

The average mass within the full ocean domain is a functional of the solid earth geometry of the ocean basins, the seawater density, and the ocean surface height. Assuming the solid earth geometry to be the same for all ensemble members leads to the functional relation

$$\begin{aligned} \langle M(\rho,\eta)\rangle &= \left\langle \int \mathrm{d}A \int_{-H}^{\eta} \mathrm{d}z\,\rho \right\rangle \\ &= \int \mathrm{d}A \left\langle \int_{-H}^{\eta} \mathrm{d}z\,\rho \right\rangle, \end{aligned} \tag{8.35}$$

where the second step follows since the horizontal boundaries of the ocean are fixed over the ensemble. Likewise, the total tracer mass within the ocean is given by

$$\begin{aligned} M_C &= \int (\rho\,\mathrm{d}V)\,C \\ &= \int \mathrm{d}A \int_{-H}^{\eta} \mathrm{d}z\,\rho\,C, \end{aligned} \tag{8.36}$$

and its average is

$$\begin{aligned} \langle M_C(\rho,\eta,C)\rangle &= \left\langle \int \mathrm{d}A \int_{-H}^{\eta} \mathrm{d}z\,\rho\,C \right\rangle \\ &= \int \mathrm{d}A \left\langle \int_{-H}^{\eta} \mathrm{d}z\,\rho\,C \right\rangle. \end{aligned} \tag{8.37}$$

When $C = 1$, the ocean tracer mass reduces to the total mass of water in the ocean.

In general, the average mass $\langle M\rangle$ of ocean fluid *does not* equal the mass of a fluid with average density $\langle\rho\rangle$ and average surface height $\langle\eta\rangle$. That is,

$$\langle M(\rho,\eta)\rangle \neq M(\langle\rho\rangle,\langle\eta\rangle). \tag{8.38}$$

Likewise for tracers,

$$\langle M_C(\rho,\eta,C)\rangle \neq M_C(\langle\rho\rangle,\langle\eta\rangle,\langle C\rangle). \tag{8.39}$$

Instead, it is necessary to introduce the average surface height $\langle\eta\rangle = \eta - \eta'$, average density $\langle\rho\rangle = \rho - \rho'$, and average tracer concentration $\langle C\rangle = C - C'$. For the ocean mass, these terms lead to

$$\begin{aligned} \left\langle \int_{-H}^{\eta} \mathrm{d}z\,\rho \right\rangle &= \left\langle \int_{-H}^{\langle\eta\rangle+\eta'} \mathrm{d}z\,(\langle\rho\rangle + \rho') \right\rangle \\ &= \left\langle \int_{-H}^{\langle\eta\rangle+\eta'} \mathrm{d}z\,\langle\rho\rangle \right\rangle + \left\langle \int_{\langle\eta\rangle}^{\langle\eta\rangle+\eta'} \mathrm{d}z\,\rho' \right\rangle, \end{aligned} \tag{8.40}$$

where $\langle \rho' \rangle = 0$ was used. The first term can be written

$$\left\langle \int_{-H}^{\langle \eta \rangle + \eta'} \mathrm{d}z \, \langle \rho \rangle \right\rangle = \int_{-H}^{\langle \eta \rangle} \mathrm{d}z \, \langle \rho \rangle + \left\langle \int_{\langle \eta \rangle}^{\langle \eta \rangle + \eta'} \mathrm{d}z \, \langle \rho \rangle \right\rangle$$
$$= \int_{-H}^{\langle \eta \rangle} \mathrm{d}z \, \langle \rho \rangle, \tag{8.41}$$

where $\langle \eta' \rangle = 0$ was used. The second term can be expanded in a Taylor series around $\eta' = 0$ to yield

$$\left\langle \int_{\langle \eta \rangle}^{\langle \eta \rangle + \eta'} \mathrm{d}z \, \rho' \right\rangle \approx \langle \eta' \, \rho' \rangle + (1/2) \, \langle (\eta')^2 \, \partial_z \, \rho' \rangle, \tag{8.42}$$

where ρ' on the right-hand side is evaluated at $z = 0$, and higher order correlations are dropped. Therefore, in order to formulate expressions for the averaged mass, it is necessary to introduce a closure assumption for the infinite number of correlation terms. Similar assumptions are needed for the averaged tracer mass.

8.4.2 Modified mean surface height

The above approach is unsatisfying since the averaged kinematics differ fundamentally from the unaveraged kinematics. In particular, the averaged kinematics depend on dynamical details determining the correlations. This result is symptomatic of working with the mean surface height field $\langle \eta \rangle$ and mean tracer $\langle C \rangle$, both of which are inconvenient.

An alternative approach considers the *modified mean* surface height η^*. This field is defined so that* the following global identity is satisfied:

$$\langle M_C \rangle = \int \mathrm{d}A \int_{-H}^{\eta^*} \mathrm{d}z \, \langle \rho C \rangle$$
$$= \int \mathrm{d}A \int_{-H}^{\eta^*} \mathrm{d}z \, \langle \rho \rangle \, \langle C \rangle^{\rho}, \tag{8.43}$$

where

$$\langle \rho \rangle \langle C \rangle^{\rho} = \langle \rho C \rangle \tag{8.44}$$

introduces the density weighted ensemble averaged tracer concentration $\langle C \rangle^{\rho}$. This average proves to be of further use when considering the average tracer within a fluid parcel in Section 8.5. In short, the introduction of η^* is defined so that

$$\langle M_C(\rho, \eta, C) \rangle \equiv M_C(\langle \rho \rangle, \eta^*, \langle C \rangle^{\rho}). \tag{8.45}$$

*It is notable that the modified surface height is analogous to the modified density field introduced in Section 9.3.4 (page 197), although from different motivation.

In words, the average tracer mass in the ocean (left-hand side) equals the mass of tracer in an ocean with average density $\langle\rho\rangle$, density weighted tracer $\langle C\rangle^{\rho}$, and surface height η^* (right-hand side). Setting $\langle C\rangle^{\rho}$ to unity reduces the average tracer mass to the average ocean fluid mass, as expected from compatibility between tracer and fluid masses.

It is possible, through pursuit of the Taylor series methods discussed previously, to express η^* locally in terms of $\langle\eta\rangle$, $\langle\rho\rangle$, $\langle C\rangle$, and correlations between η', C', and ρ'. Analogous derivations were provided by McIntosh and McDougall (1996). However, there is no need to pursue such algebra since η^* is sufficient to formulate the averaged kinematics of finite fluid domains. As seen in the following discussion, it satisfies local relations analogous to the unaveraged surface height η, and so it serves as a convenient mean field surface height (a similar approach is pursued by Greatbatch et al., 2001).

8.4.3 Kinematic boundary condition

Continuing to use the modified surface height leads to the time tendency for the averaged fluid mass

$$\langle M\rangle_{,t} = \int \mathrm{d}A \left(\langle\rho(\eta^*)\rangle\, \eta^*_{,t} + \int_{-H}^{\eta^*} \mathrm{d}z\, \langle\rho\rangle_{,t} \right). \tag{8.46}$$

Use of the average continuity equation (8.32) yields

$$\begin{aligned} \langle M\rangle_{,t} = & \int_{z=\eta^*} \mathrm{d}A \left(\langle\rho\rangle\, \eta^*_{,t} + \rho_o \langle\tilde{\mathbf{u}}\rangle \cdot \nabla\eta^* - \rho_o \langle\tilde{w}\rangle \right) \\ & + \int_{z=-H} \mathrm{d}A\, \rho_o \left(\langle\tilde{w}\rangle + \langle\tilde{\mathbf{u}}\rangle \cdot \nabla H \right) - \int \mathrm{d}A\, \nabla \cdot \int_{-H}^{\eta^*} \mathrm{d}z\, \rho_o \langle\tilde{\mathbf{u}}\rangle. \end{aligned} \tag{8.47}$$

Averaging the bottom kinematic boundary condition (3.39) leads to

$$\langle\tilde{w}\rangle + \langle\tilde{\mathbf{u}}\rangle \cdot \nabla H = 0, \tag{8.48}$$

where $\langle H\rangle = H$ since the solid earth boundary is the same for all ensemble members. Either no normal flow or periodic side boundary conditions allows one to drop the convergence term, thus leading to the balance

$$\langle M\rangle_{,t} = \int_{z=\eta^*} \mathrm{d}A \left(\langle\rho\rangle\, \eta^*_{,t} + \rho_o \langle\tilde{\mathbf{u}}\rangle \cdot \nabla\eta^* - \rho_o \langle\tilde{w}\rangle \right). \tag{8.49}$$

To proceed, recall that the ocean mass is altered only through passage of fresh water through the ocean surface. Averaging the budget for mass in the full ocean, equation (3.57), leads to

$$\begin{aligned} \langle M\rangle_{,t} &= \left\langle \int_{z=\eta} \mathrm{d}A\, \rho_{\mathrm{w}}\, q_{\mathrm{w}} \right\rangle \\ &= \rho_o \left\langle \int_{z=\eta} \mathrm{d}A\, \tilde{q}_w \right\rangle, \end{aligned} \tag{8.50}$$

where

$$\rho_w \, q_w = \rho_o \, \tilde{q}_w. \tag{8.51}$$

Defining a modified mean surface fresh water flux according to

$$\left\langle \int_{z=\eta} dA \, \tilde{q}_w \right\rangle \equiv \int_{z=\eta^*} dA \, \tilde{q}^*_w \tag{8.52}$$

leads to

$$\int_{z=\eta^*} dA \left(\langle\rho\rangle \, \eta^*_{,t} + \rho_o \, \langle\tilde{\mathbf{u}}\rangle \cdot \nabla\eta^* - \rho_o \, \langle\tilde{w}\rangle \right) = \rho_o \int_{z=\eta^*} dA \, \tilde{q}^*_w. \tag{8.53}$$

Since the horizontal area is arbitrary, the surface kinematic boundary condition is given by

$$\langle\rho\rangle \, \eta^*_{,t} + \rho_o \, \langle\tilde{\mathbf{u}}\rangle \cdot \nabla\eta^* = \rho_o \, \langle\tilde{w}\rangle + \rho_o \, \tilde{q}^*_w \quad \text{at } z = \eta^*. \tag{8.54}$$

In the Boussinesq limit this boundary condition becomes

$$\eta^*_{,t} + \langle\tilde{\mathbf{u}}\rangle \cdot \nabla\eta^* = \langle\tilde{w}\rangle + \tilde{q}^*_w \quad \text{at } z = \eta^*. \tag{8.55}$$

Note the absence of turbulence correlation terms in both surface boundary conditions. Therefore, both are of the same mathematical form as their respective unaveraged equations given in Section 8.2. Such form invariance resulted from introducing the modified mean fields η^* and $\tilde{q}^*_w$.

8.4.4 Mass within a column

Consider the budget for the mass per unit area of fluid contained within a single ocean column. For this case, maintain the convergence term in equation (8.47) and use the surface kinematic boundary condition (8.54) to find

$$\partial_t \left(\int_{-H}^{\eta^*} dz \, \langle\rho\rangle \right) + \rho_o \, \nabla \cdot \left(\int_{-H}^{\eta^*} dz \, \langle\tilde{\mathbf{u}}\rangle \right) = \rho_o \, \tilde{q}^*_w, \tag{8.56}$$

which is directly analogous to the unaveraged budget (3.56). Now introduce the vertically integrated averaged density

$$D^* \, \overline{\langle\rho\rangle}^z = \int_{-H}^{\eta^*} dz \, \langle\rho\rangle, \tag{8.57}$$

with

$$D^* = H + \eta^* \tag{8.58}$$

the total thickness of the modified mean fluid column, and the averaged horizontal momentum density of a column

$$\rho_o \, \langle\tilde{\mathbf{U}}\rangle = \rho_o \int_{-H}^{\eta^*} dz \, \langle\tilde{\mathbf{u}}\rangle. \tag{8.59}$$

These terms then lead to the mass balance for a column of fluid

$$\left(D^* \, \overline{\langle\rho\rangle}^z \right)_{,t} = -\rho_o \, \nabla \cdot \langle\tilde{\mathbf{U}}\rangle + \rho_o \, \tilde{q}^*_w. \tag{8.60}$$

The Boussinesq limit recovers the volume balance for the column

$$\eta^*_{,t} = -\nabla \cdot \langle\tilde{\mathbf{U}}\rangle + \tilde{q}^*_w. \tag{8.61}$$

Both balances are of the same mathematical form as their respective unaveraged balances given in Section 8.2.

8.5 AVERAGED TRACER

We now derive a mean field tracer equation compatible with the mean field continuity equation derived in Section 8.3.

8.5.1 Conventional approach

Consider a conventional Reynolds decomposition of the tracer equation for non-Boussinesq fluids. The Eulerian mean of the product of density and tracer is

$$\langle \rho\, C \rangle = \langle \rho \rangle \, \langle C \rangle + \langle \rho'\, C' \rangle, \tag{8.62}$$

and the mean triple product is

$$\langle \rho\, \mathbf{v}\, C \rangle = \langle \rho \rangle \, (\langle \mathbf{v} \rangle \langle C \rangle + \langle \mathbf{v}'\, C' \rangle) + \langle \mathbf{v} \rangle \langle \rho'\, C' \rangle + \langle C \rangle \langle \rho'\, \mathbf{v}' \rangle + \langle \rho'\, \mathbf{v}'\, C' \rangle. \tag{8.63}$$

These results lead to the mean tracer equation

$$\begin{aligned} &[\langle \rho \rangle \, \langle C \rangle + \langle \rho'\, C' \rangle]_{,t} \\ &\quad = -\nabla \cdot [\langle \rho \rangle \, (\langle \mathbf{v} \rangle \langle C \rangle + \langle \mathbf{v}'\, C' \rangle) + \langle \mathbf{v} \rangle \langle \rho'\, C' \rangle + \langle C \rangle \langle \rho'\, \mathbf{v}' \rangle + \langle \rho'\, \mathbf{v}'\, C' \rangle], \end{aligned} \tag{8.64}$$

where the source and molecular diffusion terms are dropped for brevity. This expression contains five turbulence correlation terms. A parameterization of these correlations is not generally known. Additionally, setting $\langle C \rangle = 1$ and $C' = 1$ recovers the mean continuity equation

$$\langle \rho \rangle_{,t} = -\nabla \cdot (\langle \rho \rangle \, \langle \mathbf{v} \rangle + \langle \mathbf{v}'\, \rho' \rangle) \,. \tag{8.65}$$

Recall that in Section 8.3 this form of the Eulerian mean continuity equation was seen to be inconvenient since it leads to a mean kinematics involving unknown turbulence correlation terms. Therefore, a conventional Reynolds decomposition of the tracer equation is unsatisfying since it leads to (1) multiple turbulence terms, and (2) an averaged tracer equation that is not consistent with the preferred form of the averaged continuity equation.

8.5.2 Density weighted approach

To motivate an alternative approach, start by recalling that in Section 5.1.1 (page 87), the material tracer concentration C is interpreted as the mass of tracer substance within a parcel of seawater per mass of seawater,

$$\begin{aligned} C &= \frac{\mathrm{d}M_C}{\mathrm{d}M} \\ &= \frac{\mathrm{d}M_C}{\rho\,\mathrm{d}V}. \end{aligned} \tag{8.66}$$

As defined, C is a scalar since the tracer mass and seawater parcel mass are both scalars. The mean equations derived in this chapter are based on an Eulerian average. From this perspective one considers the Eulerian averaged tracer mass within a *fixed* volume (such as a discrete model grid cell). In this case,

$$\begin{aligned} \langle \mathrm{d}M_C \rangle &= \langle \rho\, C \rangle \, \mathrm{d}V \\ &= \langle \rho \rangle \langle C \rangle^{\rho} \, \mathrm{d}V, \end{aligned} \tag{8.67}$$

where

$$\langle C \rangle^{\rho} \equiv \langle \rho\, C \rangle \, \langle \rho \rangle^{-1} \tag{8.68}$$

is the *density weighted* Eulerian averaged tracer concentration. Again, the average is taken at a fixed point in space-time, over a fixed infinitesimal volume. This averaged tracer was already encountered in Section 8.4.2, where it appears naturally when considering the averaged tracer mass in the full ocean domain. In addition to the density weighted tracer concentration, it proves useful to consider the density weighted velocity field

$$\begin{aligned} \langle \rho \rangle \, \langle \mathbf{v} \rangle^{\rho} &\equiv \langle \rho\, \mathbf{v} \rangle \\ &= \rho_o \langle \tilde{\mathbf{v}} \rangle \end{aligned} \tag{8.69}$$

for the purpose of seeking a mean field tracer equation.

Deviations from a density weighted average, defined according to

$$C = \langle C \rangle^{\rho} + C'_{\rho} \tag{8.70}$$

$$\mathbf{v} = \langle \mathbf{v} \rangle^{\rho} + \mathbf{v}'_{\rho} \tag{8.71}$$

satisfy

$$\langle \rho\, C'_{\rho} \rangle = 0 \tag{8.72}$$

$$\langle \rho\, \mathbf{v}'_{\rho} \rangle = 0. \tag{8.73}$$

Hence, the triple product term in the tracer equation takes the form

$$\begin{aligned} \rho\, C\, \mathbf{v} &= \rho\, (\langle C \rangle^{\rho} + C'_{\rho}) \, (\langle \mathbf{v} \rangle^{\rho} + \mathbf{v}'_{\rho}) \\ &= \rho\, \langle C \rangle^{\rho} \, \langle \mathbf{v} \rangle^{\rho} + \rho\, \langle C \rangle^{\rho}\, \mathbf{v}'_{\rho} + \rho\, C'_{\rho}\, \langle \mathbf{v} \rangle^{\rho} + \rho\, C'_{\rho}\, \mathbf{v}'_{\rho}, \end{aligned} \tag{8.74}$$

which has an average given by

$$\begin{aligned} \langle \rho\, C\, \mathbf{v} \rangle &= \langle C \rangle^{\rho} \, \langle \rho \rangle \, \langle \mathbf{v} \rangle^{\rho} + \langle \rho\, C'_{\rho}\, \mathbf{v}'_{\rho} \rangle \\ &= \rho_o \, \langle C \rangle^{\rho} \, \langle \tilde{\mathbf{v}} \rangle + \langle \rho\, C'_{\rho}\, \mathbf{v}'_{\rho} \rangle, \end{aligned} \tag{8.75}$$

where the identities (8.72) and (8.73) were used. Notably, the turbulence correlation terms from the conventional approach (equation (8.63)) have been reduced to a single unknown correlation term via use of density weighted averages. This is a very useful simplification.

8.5.3 Subgrid scale tracer fluxes

In addition to the molecular diffusive flux of tracer, the ensemble mean of

$$\rho\, \mathbf{F}_{\text{sgs}} = \rho\, C'_{\rho}\, \mathbf{v}'_{\rho} \tag{8.76}$$

provides a subgrid scale contribution to the evolution of the ensemble mean tracer. This term arises from the density weighted SGS fluctuations of tracer concentration and velocity. For ocean climate modeling, molecular diffusion is far smaller than the SGS contribution from $C'_{\rho}\, \mathbf{v}'_{\rho}$. Hence, molecular diffusion contributions will be formally incorporated into $\rho\, \mathbf{F}_{\text{sgs}}$ in the following development.

Just as it is useful to consider a density weighted velocity field $\tilde{\mathbf{v}}$ via the definition $\rho\, \mathbf{v} = \rho_o\, \tilde{\mathbf{v}}$, consider here the SGS tracer flux $\tilde{\mathbf{F}}_{\text{sgs}}$ defined by

$$\rho\, \mathbf{F}_{\text{sgs}} \equiv \rho_o\, \tilde{\mathbf{F}}_{\text{sgs}}. \tag{8.77}$$

With this definition, the ensemble mean tracer transport term takes the form

$$\langle \rho \, C \, \mathbf{v} \rangle = \rho_o \, \langle C \rangle^{\rho} \, \langle \tilde{\mathbf{v}} \rangle + \rho_o \, \langle \tilde{\mathbf{F}}_{\text{sgs}} \rangle, \tag{8.78}$$

thus yielding the time tendency for the ensemble averaged tracer concentration in a non-Boussinesq fluid

$$\partial_t \left(\langle \rho \rangle \langle C \rangle^{\rho} \right) + \rho_o \, \nabla \cdot \left(\langle \tilde{\mathbf{v}} \rangle \langle C \rangle^{\rho} \right) = -\rho_o \, \nabla \cdot \langle \tilde{\mathbf{F}}_{\text{sgs}} \rangle + \langle \rho \rangle \langle \mathcal{S} \rangle^{\rho}. \tag{8.79}$$

The Boussinesq results are recovered by setting $\langle \rho \rangle \rightarrow \rho_o$:

$$\partial_t \, \langle C \rangle^{\rho} + \nabla \cdot \left(\langle \tilde{\mathbf{v}} \rangle \langle C \rangle^{\rho} \right) = -\nabla \cdot \langle \tilde{\mathbf{F}}_{\text{sgs}} \rangle + \langle \mathcal{S} \rangle^{\rho}. \tag{8.80}$$

Note that the density weighted form of the averaged Boussinesq tracer equation (8.80) remains. According to McDougall et al. (2003a), doing so provides a set of equations compatible with the small levels of diapycnal mixing in the interior ocean. Also note that (1) setting the averaged tracer concentration to unity, $\langle C \rangle^{\rho} = 1$, (2) assuming SGS fluxes vanish, and (3) dropping source terms, we find that the tracer equation for both non-Boussinesq and Boussinesq systems recover their respective average continuity equations. Such *compatibility* between tracer and mass/volume conservation is important to maintain in the numerical model. If the sources are nontrivial when tracer is unity, that is, if sources introduce mass or volume, then compatibility requires sources to be part of the continuity equations as well.

8.5.4 Boussinesq limit in the averaged fluid

As shown in the previous discussion, the transition from Eulerian averaged non-Boussinesq to Eulerian averaged Boussinesq fluids arises by first averaging the non-Boussinesq equations and *then* setting $\langle \rho \rangle \rightarrow \rho_o$, except when multiplying by gravity. Correspondingly, averaged Boussinesq fluids satisfy $\langle \tilde{\mathbf{v}} \rangle = \langle \mathbf{v} \rangle^{\rho}$, whereas $\langle \tilde{\mathbf{v}} \rangle \neq \langle \mathbf{v} \rangle$. Hence, this approach breaks the symmetry between $\tilde{\mathbf{v}}$ and $\mathbf{v}$ present in the unaveraged Boussinesq equations.

8.5.5 A not-so-useful alternative

The density multiplier in the time tendency term for the non-Boussinesq tracer equation (8.79) may be eliminated by introducing the new tracer variable

$$\rho_o \, \tilde{C} = \rho \, C. \tag{8.81}$$

This variable is directly analogous to the velocity variable $\rho_o \, \tilde{\mathbf{v}} = \rho \, \mathbf{v}$. The ensemble mean of the new tracer variable is given by

$$\rho_o \, \langle \tilde{C} \rangle = \langle \rho \rangle \langle C \rangle^{\rho}, \tag{8.82}$$

thus leading the time tendency and advective portions of the tracer equation to take the form

$$\partial_t \left(\langle \rho \rangle \, \langle C \rangle^{\rho} \right) + \rho_o \, \nabla \cdot \left(\langle \tilde{\mathbf{v}} \rangle \, \langle C \rangle^{\rho} \right) = \langle \tilde{C} \rangle_{,t} + \nabla \cdot \left(\langle \mathbf{v} \rangle^{\rho} \langle \tilde{C} \rangle \right). \tag{8.83}$$

This form is mathematically similar to the Boussinesq tracer equation. However, this form does not manifest compatibility between tracer and mass conservation. Recall from the end of Section 8.5.2 that compatibility is manifest when setting $\langle C \rangle^{\rho} = 1$ in the tracer equation reduces the tracer budget to the mass budget. By setting $\langle \tilde{C} \rangle = 1$ throughout the fluid, equation (8.83) gives $\nabla \cdot \langle \mathbf{v} \rangle^{\rho} = 0$. However, $\nabla \cdot \langle \mathbf{v} \rangle^{\rho} = 0$ is not an expression of mass conservation. Rather, it is a solution generally realized only for the trivial case of uniform $\langle C \rangle^{\rho}$ and $\langle \rho \rangle$, as can be seen by the definition (8.82). Hence, for purposes of maintaining a manifest compatibility between tracer and mass conservation, it is more convenient to use $\langle C \rangle^{\rho}$ instead of $\langle \tilde{C} \rangle$ as the mean field tracer variable.

8.6 AVERAGED MOMENTUM BUDGET

When deriving an averaged momentum budget, the question arises whether we should work with the averaged form of the budget (8.5) written in terms of the linear momentum per mass $\mathbf{v}$, or the budget (8.11) written in terms of the linear momentum per volume $\rho_o \, \tilde{\mathbf{v}}$. As seen thus far, the linear momentum per volume $\rho_o \, \tilde{\mathbf{v}}$ is a very convenient transport variable for the Eulerian averaged conservation equations of mass and tracer. Hence, an averaged momentum budget is desired where time tendencies of $\langle \tilde{\mathbf{v}} \rangle$ can be determined. Since this is not the traditional method, we start by pointing out the form of the averaged momentum budget where the time tendency of $\langle \mathbf{v} \rangle$ is determined.

8.6.1 Time tendency for $\langle \mathbf{v} \rangle$

To isolate the time tendency of velocity, use mass conservation and divide by density in equation (8.5) to find

$$(\partial_t + \mathbf{v} \cdot \nabla + \mathcal{M} \, \hat{\mathbf{z}} \wedge) \mathbf{v} = -g \, \hat{\mathbf{z}} - f \, \hat{\mathbf{z}} \wedge \mathbf{v} - \nabla p / \rho + \mathbf{F}^{(\mathbf{v})}. \tag{8.84}$$

Taking an ensemble average leads to

$$(\partial_t + \langle \mathbf{v} \rangle \cdot \nabla + \langle \mathcal{M} \rangle \, \hat{\mathbf{z}} \wedge) \langle \mathbf{v} \rangle = -f \, \hat{\mathbf{z}} \wedge \langle \mathbf{v} \rangle - g \, \hat{\mathbf{z}} - \langle \nabla p / \rho \rangle + \langle \mathbf{F}^{(\mathbf{v})} \rangle. \tag{8.85}$$

This expression is achieved by assuming that nonlinear correlation terms arising from the advection terms are absorbed into the averaged friction vector

$$\langle \mathbf{F}^{(\mathbf{v})} \rangle = -\langle (\mathbf{v}' \cdot \nabla) \, \mathbf{v}' \rangle - \hat{\mathbf{z}} \wedge \langle \mathcal{M}' \, \mathbf{v}' \rangle, \tag{8.86}$$

where the molecular term is dropped since it is relatively small for averages over scales appropriate for ocean climate models. The pressure term can be written in terms of an average and correlations

$$\langle \nabla p / \rho \rangle \approx \frac{\nabla \langle p \rangle}{\langle \rho \rangle} + \frac{\langle \rho' \, \nabla p' \rangle}{\langle \rho \rangle^2} + \frac{\langle \rho' \, \rho' \rangle \, \nabla \langle p \rangle}{\langle \rho \rangle^3}, \tag{8.87}$$

where higher order correlations were dropped. The pressure-density turbulence correlation terms cannot be absorbed by the SGS friction vector $\langle \mathbf{F}^{(\mathbf{v})} \rangle$, since $\langle \mathbf{F}^{(\mathbf{v})} \rangle$ represents SGS transport terms, not pressure-density fluctuations. Hence, the pressure-density terms must be parameterized separately, where the parameterization depends on the dynamics at the SGS.

8.6.2 Time tendency for $\langle \tilde{\mathbf{v}} \rangle$

Now consider a density weighted average of the momentum equation (8.11) when written in terms of the momentum density $\rho_o \, \tilde{\mathbf{v}}$. As with the previous manipulations, averaging produces nonlinear correlations from the averaged nonlinear advection

$$\begin{aligned}\langle \rho \, \mathbf{v} \, \mathbf{v} \rangle &= \langle \rho \rangle \langle \mathbf{v} \rangle^{\rho} \langle \mathbf{v} \rangle^{\rho} + \langle \rho \, \mathbf{v}'_{\rho} \, \mathbf{v}'_{\rho} \rangle \\ &= \rho_o \, \langle \tilde{\mathbf{v}} \rangle \langle \mathbf{v} \rangle^{\rho} + \langle \rho \, \mathbf{v}'_{\rho} \, \mathbf{v}'_{\rho} \rangle\end{aligned} \tag{8.88}$$

and advection metric

$$\hat{\mathbf{z}} \wedge \langle \mathcal{M} \, \rho \, \mathbf{v} \rangle = \rho_o \, \hat{\mathbf{z}} \wedge \langle \tilde{\mathcal{M}} \rangle \langle \mathbf{v} \rangle^{\rho} + \hat{\mathbf{z}} \wedge \langle \rho \, \mathcal{M}'_{\rho} \, \mathbf{v}'_{\rho} \rangle, \tag{8.89}$$

where again $\rho_o \, \langle \tilde{\mathbf{v}} \rangle = \langle \rho \rangle \langle \mathbf{v} \rangle^{\rho}$.

To massage the SGS momentum terms, proceed in a manner similar to that done in Section 8.5.3 for the SGS tracer fluxes. That is, introduce the SGS momentum flux

$$\begin{aligned}\rho \, \mathbf{F}^{(\mathbf{v})}_{\text{sgs}} &= \nabla \cdot (\rho \, \mathbf{v}'_{\rho} \, \mathbf{v}'_{\rho}) + \hat{\mathbf{z}} \wedge \rho \, \mathcal{M}'_{\rho} \, \mathbf{v}'_{\rho} \\ &\equiv \rho_o \, \tilde{\mathbf{F}}^{(\mathbf{v})}_{\text{sgs}}\end{aligned} \tag{8.90}$$

and assume the much smaller molecular diffusion contribution to be absorbed by this flux. This form of the SGS momentum flux leads to the averaged non-Boussinesq momentum balance

$$\begin{aligned}&\langle \tilde{\mathbf{v}} \rangle_{,t} + \nabla \cdot (\langle \mathbf{v} \rangle^{\rho} \, \langle \tilde{\mathbf{v}} \rangle) + \langle \tilde{\mathcal{M}} \rangle \, \hat{\mathbf{z}} \wedge \langle \mathbf{v} \rangle^{\rho} \\ &\qquad = -(\langle \rho \rangle / \rho_o) \, g \, \hat{\mathbf{z}} - f \, \hat{\mathbf{z}} \wedge \langle \tilde{\mathbf{v}} \rangle - \nabla(\langle p \rangle / \rho_o) + \langle \tilde{\mathbf{F}}^{(\mathbf{v})}_{\text{sgs}} \rangle.\end{aligned} \tag{8.91}$$

Notably, there is no need to parameterize pressure-density correlations appearing in the pressure gradient force. In contrast, the discussion in Section 8.6.1 noted the need to parameterize $\langle \nabla p / \rho \rangle$ when formulating the equation for $\langle \mathbf{v} \rangle_{,t}$. The simplicity of the pressure gradient force is another reason for promoting $\langle \tilde{\mathbf{v}} \rangle$ as the fundamental transport field.

The momentum equation for a Boussinesq fluid arises by setting $\langle \rho \rangle \rightarrow \rho_o$, except when multiplying by gravity,

$$\begin{aligned}&\langle \tilde{\mathbf{v}} \rangle_{,t} + \nabla \cdot (\langle \tilde{\mathbf{v}} \rangle \, \langle \tilde{\mathbf{v}} \rangle) + \langle \tilde{\mathcal{M}} \rangle \, \hat{\mathbf{z}} \wedge \langle \tilde{\mathbf{v}} \rangle \\ &\qquad = -(\langle \rho \rangle / \rho_o) \, g \, \hat{\mathbf{z}} - f \, \hat{\mathbf{z}} \wedge \langle \tilde{\mathbf{v}} \rangle - \nabla(\langle p \rangle / \rho_o) + \langle \tilde{\mathbf{F}}^{(\mathbf{v})}_{\text{sgs}} \rangle.\end{aligned} \tag{8.92}$$

The pressure gradient force is identical in form to that appearing in the non-Boussinesq equation for $\langle \tilde{\mathbf{v}} \rangle_{,t}$. Consistent with the averaged tracer equation, the averaged velocity field used in the Boussinesq momentum equation is interpreted as $\langle \tilde{\mathbf{v}} \rangle$, which is distinct from $\langle \mathbf{v} \rangle$.

In summary, use of density weighted z-ensemble averages allows one to interpret the Boussinesq approximation as solely kinematic, where mass conservation is reduced to volume conservation. There is no associated dynamical approximation associated with the pressure gradient force, since it remains identical in form. This is also the case in the unaveraged equations, so long as the focus remains on the linear momentum per volume, $\rho \, \mathbf{v}$, instead of the velocity.

8.7 SUMMARY OF THE EULERIAN AVERAGED EQUATIONS

It is useful now to summarize the Eulerian averaged equations. The mass budgets for an infinitesimal fluid domain and for a vertical column of fluid are given by

$$\langle \rho \rangle_{,t} + \rho_o \, \nabla \cdot \langle \tilde{\mathbf{v}} \rangle = 0 \tag{8.93}$$

$$\left(D^* \, \overline{\langle \rho \rangle}^{z} \right)_{,t} = -\rho_o \, \nabla \cdot \langle \tilde{\mathbf{U}} \rangle + \rho_o \, \tilde{q}^{*}_{w}. \tag{8.94}$$

For the Boussinesq fluid, volume conservation for an infinitesimal domain and for a column are given by

$$\nabla \cdot \langle \tilde{\mathbf{v}} \rangle = 0 \tag{8.95}$$

$$\eta^{*}_{,t} = -\nabla \cdot \langle \tilde{\mathbf{U}} \rangle + \tilde{q}^{*}_{w}. \tag{8.96}$$

The surface and bottom kinematic boundary conditions for the non-Boussinesq fluid are

$$\langle \rho \rangle \, \eta^{*}_{,t} + \rho_o \, \langle \tilde{\mathbf{u}} \rangle \cdot \nabla \eta^{*} = \rho_o \, \langle \tilde{w} \rangle + \rho_o \, \tilde{q}^{*}_{w} \qquad \text{at } z = \eta^{*} \tag{8.97}$$

$$\langle \tilde{\mathbf{u}} \rangle \cdot \nabla H + \langle \tilde{w} \rangle = 0 \qquad \text{at } z = -H \tag{8.98}$$

and for the Boussinesq fluid

$$\eta^{*}_{,t} + \langle \tilde{\mathbf{u}} \rangle \cdot \nabla \eta^{*} = \langle \tilde{w} \rangle + \tilde{q}^{*}_{w} \qquad \text{at } z = \eta^{*} \tag{8.99}$$

$$\langle \tilde{\mathbf{u}} \rangle \cdot \nabla H + \langle \tilde{w} \rangle = 0 \qquad \text{at } z = -H. \tag{8.100}$$

The tracer budget for the non-Boussinesq fluid is

$$(\langle \rho \rangle \, \langle C \rangle^{\rho})_{,t} + \rho_o \, \nabla \cdot (\langle \tilde{\mathbf{v}} \rangle \, \langle C \rangle^{\rho}) = -\rho_o \, \nabla \cdot \langle \tilde{\mathbf{F}}_{\text{sgs}} \rangle + \langle \rho \rangle \langle \mathcal{S} \rangle^{\rho}, \tag{8.101}$$

and for the Boussinesq fluid

$$\partial_t \, \langle C \rangle^{\rho} + \nabla \cdot (\langle \tilde{\mathbf{v}} \rangle \langle C \rangle^{\rho}) = -\nabla \cdot \langle \tilde{\mathbf{F}}_{\text{sgs}} \rangle + \langle \mathcal{S} \rangle^{\rho}. \tag{8.102}$$

Finally, the non-Boussinesq momentum budget is

$$\begin{aligned} &\langle \tilde{\mathbf{v}} \rangle_{,t} + \nabla \cdot (\langle \mathbf{v} \rangle^{\rho} \, \langle \tilde{\mathbf{v}} \rangle) + \langle \tilde{\mathcal{M}} \rangle \, \hat{\mathbf{z}} \wedge \langle \mathbf{v} \rangle^{\rho} \\ &\qquad = -(\langle \rho \rangle / \rho_o) \, g \, \hat{\mathbf{z}} - f \, \hat{\mathbf{z}} \wedge \langle \tilde{\mathbf{v}} \rangle - \nabla(\langle p \rangle / \rho_o) + \langle \tilde{\mathbf{F}}^{(\mathbf{v})}_{\text{sgs}} \rangle, \end{aligned} \tag{8.103}$$

whereas the Boussinesq budget is

$$\begin{aligned} &\langle \tilde{\mathbf{v}} \rangle_{,t} + \nabla \cdot (\langle \tilde{\mathbf{v}} \rangle \, \langle \tilde{\mathbf{v}} \rangle) + \langle \tilde{\mathcal{M}} \rangle \, \hat{\mathbf{z}} \wedge \langle \tilde{\mathbf{v}} \rangle \\ &\qquad = -(\langle \rho \rangle / \rho_o) \, g \, \hat{\mathbf{z}} - f \, \hat{\mathbf{z}} \wedge \langle \tilde{\mathbf{v}} \rangle - \nabla(\langle p \rangle / \rho_o) + \langle \tilde{\mathbf{F}}^{(\mathbf{v})}_{\text{sgs}} \rangle. \end{aligned} \tag{8.104}$$

Recall that $\langle \tilde{\mathbf{v}} \rangle = \langle \mathbf{v} \rangle^{\rho}$ for a Boussinesq fluid, whereas $\rho_o \, \langle \tilde{\mathbf{v}} \rangle = \langle \rho \rangle \, \langle \mathbf{v} \rangle^{\rho}$ for the non-Boussinesq case. In the tracer equation, the SGS flux term is given by

$$\begin{aligned} \rho \, \mathbf{F}_{\text{sgs}} &= \rho \, C'_{\rho} \, \mathbf{v}'_{\rho} \\ &\equiv \rho_o \, \tilde{\mathbf{F}}_{\text{sgs}} \end{aligned} \tag{8.105}$$

which dominates the molecular diffusion flux. The friction vector likewise incorporates SGS turbulence terms

$$\begin{aligned} \rho \, \mathbf{F}^{(\mathbf{v})}_{\text{sgs}} &= \nabla \cdot (\rho \, \mathbf{v}'_{\rho} \, \mathbf{v}'_{\rho}) + \hat{\mathbf{z}} \wedge \rho \, \mathcal{M}'_{\rho} \, \mathbf{v}'_{\rho} \\ &\equiv \rho_o \, \tilde{\mathbf{F}}^{(\mathbf{v})}_{\text{sgs}}. \end{aligned} \tag{8.106}$$

The Eulerian averaged equations have the same mathematical form as the unaveraged equations given in Section 8.2. Precisely, the mapping between unaveraged and averaged fields is given by

$$\rho \rightarrow \langle \rho \rangle \tag{8.107}$$
$$p \rightarrow \langle p \rangle \tag{8.108}$$
$$\mathbf{v} \rightarrow \langle \mathbf{v} \rangle^{\rho} \tag{8.109}$$
$$\tilde{\mathbf{v}} \rightarrow \langle \tilde{\mathbf{v}} \rangle \tag{8.110}$$
$$C \rightarrow \langle C \rangle^{\rho} \tag{8.111}$$
$$\tilde{\mathbf{F}} \rightarrow \langle \tilde{\mathbf{F}}_{\text{sgs}} \rangle \tag{8.112}$$
$$\mathcal{S} \rightarrow \langle \mathcal{S} \rangle^{\rho} \tag{8.113}$$
$$\tilde{\mathbf{F}}^{(\mathbf{v})} \rightarrow \langle \tilde{\mathbf{F}}^{(\mathbf{v})}_{\text{sgs}} \rangle \tag{8.114}$$
$$\eta \rightarrow \eta^{*} \tag{8.115}$$
$$\tilde{q}_{w} \rightarrow \tilde{q}^{*}_{w}. \tag{8.116}$$

This mapping is very useful for purposes of extending properties of the unaveraged system to the averaged system, such as the energetic balances discussed in Chapter 5.

8.8 MAPPING TO OCEAN MODEL VARIABLES

Establishing a set of self-consistent equations describing the Eulerian averaged ocean is a necessary step toward writing down continuous equations to be discretized. As discussed in Section 8.1.2, discretization is applied to the averaged continuous equations, where averaging is taken over space-time scales omitted by the discrete grid. Discretization is *not* applied to the equations derived in Chapters 3–5 in which SGS processes are associated with molecular effects. The purpose of this section is to write the averaged equations in a form that affords their ready incorporation into an ocean model.

Maintaining a tidy form of the averaged continuity equation motivates discretizing $\langle \tilde{\mathbf{v}} \rangle$ instead of the conventional $\langle \mathbf{v} \rangle$. The distinction is nontrivial for both non-Boussinesq and Boussinesq ocean models. Hence, one has the correspondence*

$$\langle \tilde{\mathbf{v}} \rangle \rightarrow \mathbf{v}_{\text{model}}. \tag{8.117}$$

Analogously, it is appropriate to use $\langle C \rangle^{\rho}$ in the model instead of the conventional $\langle C \rangle$,

$$\langle C \rangle^{\rho} \rightarrow C_{\text{model}}. \tag{8.118}$$

Again, the distinction between $\langle C \rangle^{\rho}$ and $\langle C \rangle$ is nontrivial for both non-Boussinesq and Boussinesq ocean models. Finally, the surface height in the ocean model corresponds to the modified mean surface height η^{*} described in Section 8.4.2,

$$\eta^{*} \rightarrow \eta_{\text{model}}, \tag{8.119}$$

as will the modified mean surface fresh water flux

$$\tilde{q}^{*}_{w} \rightarrow (q_{\text{w}})_{\text{model}}. \tag{8.120}$$

*Note that the "model" suffix refers here to the continuous ocean model, since no discretization has yet occurred.

The density variable to be discretized by the ocean model is that which results from evaluating the nonlinear equation of state with the density weighted potential temperature and density weighted salinity

$$\rho_{\text{model}} = \rho(\langle\theta\rangle^{\rho}, \langle s\rangle^{\rho}, p_{\text{model}}), \tag{8.121}$$

with p_{model} the corresponding hydrostatic pressure. This density differs from the Eulerian averaged density $\langle\rho\rangle$ used in the mass continuity equation. McIntosh and McDougall (1996) and McDougall and McIntosh (2001) discuss the differences, which are small and so are ignored for the present discussion.* That is, we set

$$\rho_{\text{model}} \approx \langle\rho\rangle. \tag{8.122}$$

The hydrostatic approximation then leads to

$$\langle p\rangle \rightarrow p_{\text{model}}. \tag{8.123}$$

Use of these mappings, and dropping the "model" subscript for brevity, give the non-Boussinesq z-model equations

$$\rho_{,t} + \rho_o \nabla \cdot \mathbf{v} = 0 \tag{8.124}$$

$$(D\,\overline{\rho}^{z})_{,t} = -\rho_o \nabla \cdot \mathbf{U} + \rho_o\, q_{\text{w}} \tag{8.125}$$

$$\mathbf{v}_{,t} + \nabla \cdot [(\rho_o/\rho)\,\mathbf{v}\,\mathbf{v}] + (\rho_o/\rho)\,\mathcal{M}\,\hat{\mathbf{z}} \wedge \mathbf{v} = -(\rho/\rho_o)\,g\,\hat{\mathbf{z}} - f\,\hat{\mathbf{z}} \wedge \mathbf{v} - \nabla(p/\rho_o) + \mathbf{F}^{(\mathbf{v})} \tag{8.126}$$

$$(\rho\, C)_{,t} + \rho_o \nabla \cdot (\mathbf{v}\, C) = -\rho_o \nabla \cdot \mathbf{F} + \rho\, \mathcal{S}, \tag{8.127}$$

where use was also made of the correspondences

$$\langle \tilde{\mathbf{F}}_{\text{sgs}} \rangle \rightarrow \mathbf{F}_{\text{model}} \tag{8.128}$$

$$\langle \mathcal{S} \rangle^{\rho} \rightarrow \mathcal{S}_{\text{model}} \tag{8.129}$$

$$\langle \tilde{\mathbf{F}}^{(\mathbf{v})}_{\text{sgs}} \rangle \rightarrow \mathbf{F}^{(\mathbf{v})}_{\text{model}}. \tag{8.130}$$

Surface and bottom kinematic boundary conditions are set according to

$$\hat{\mathbf{N}} \cdot \mathbf{v} = (\rho/\rho_o)\,\eta_{,t} - q_{\text{w}} \qquad \text{at } z = \eta \tag{8.131}$$

$$\hat{\mathbf{N}} \cdot \mathbf{v} = 0 \qquad \text{at } z = -H \tag{8.132}$$

with

$$\hat{\mathbf{N}} = \nabla(-\eta + z) \qquad \text{at } z = \eta \tag{8.133}$$

$$\hat{\mathbf{N}} = \nabla(H + z) \qquad \text{at } z = -H \tag{8.134}$$

the surface and bottom orientation directions. The mapping from unaveraged to averaged fields, and then from averaged to model fields, is summarized in Table 8.1. This table is the key result from this chapter.

The continuous model equations presented above are nearly identical to the continuous unaveraged non-Boussinesq equations summarized in Section 8.2. Although in the end somewhat trivial (i.e., what a roundabout way to get back to the same equations!), the intermediate steps reveal a nontrivial interpretation of the fields discretized in a z-model.

*This point represents an awkward part of the formalism.

Unaveraged	Averaged	Model	Model Discrete
ρ	$\langle \rho \rangle$	ρ_{model}	ρ
p	$\langle p \rangle$	p_{model}	p
$\tilde{\mathbf{v}}$	$\langle \tilde{\mathbf{v}} \rangle$	$\mathbf{v}_{\text{model}}$	$\mathbf{v}$
$\mathbf{v}$	$\langle \mathbf{v} \rangle^{\rho}$	$\rho_{\text{model}}\, \mathbf{v}^{\rho}_{\text{model}} = \rho_o\, \mathbf{v}_{\text{model}}$	$\rho\, \mathbf{v}^{\rho} = \rho_o\, \mathbf{v}$
η	η^{*}	η_{model}	η
$\tilde{q}_w$	$\tilde{q}^{*}_w$	$(q_{\text{w}})_{\text{model}}$	q_{w}
C	$\langle C \rangle^{\rho}$	C_{model}	C
$\mathcal{S}$	$\langle \mathcal{S} \rangle^{\rho}$	$\mathcal{S}_{\text{model}}$	$\mathcal{S}$
$\tilde{\mathbf{F}}$	$\langle \tilde{\mathbf{F}}_{\text{sgs}} \rangle$	$\mathbf{F}_{\text{model}}$	$\mathbf{F}$
$\tilde{\mathbf{F}}^{(\mathbf{v})}$	$\langle \tilde{\mathbf{F}}^{(\mathbf{v})}_{\text{sgs}} \rangle$	$\mathbf{F}^{(\mathbf{v})}_{\text{model}}$	$\mathbf{F}^{(\mathbf{v})}$

Table 8.1 Correspondence between unaveraged continuous fields appropriate at the scale where the SGS terms arise from molecular effects, Eulerian averaged continuous fields, continuous model fields, and discrete model fields.

The Boussinesq model equations arise by setting $\rho_{\text{model}} \to \rho_o$, except when multiplying the gravitational acceleration

$$\nabla \cdot \mathbf{v} = 0 \tag{8.135}$$

$$\eta_{,t} = -\nabla \cdot \mathbf{U} + q_{\text{w}} \tag{8.136}$$

$$\mathbf{v}_{,t} + \nabla \cdot (\mathbf{v}\,\mathbf{v}) + \mathcal{M}\,\hat{\mathbf{z}} \wedge \mathbf{v} = -(\rho/\rho_o)\, g\, \hat{\mathbf{z}} - f\, \hat{\mathbf{z}} \wedge \mathbf{v} - \nabla(p/\rho_o) + \mathbf{F}^{(\mathbf{v})} \tag{8.137}$$

$$C_{,t} + \nabla \cdot (\mathbf{v}\, C) = -\nabla \cdot \mathbf{F} + \mathcal{S}, \tag{8.138}$$

with the surface and bottom kinematic boundary conditions

$$\hat{\mathbf{N}} \cdot \mathbf{v} = \eta_{,t} - q_{\text{w}} \qquad \text{at } z = \eta \tag{8.139}$$

$$\hat{\mathbf{N}} \cdot \mathbf{v} = 0 \qquad \text{at } z = -H. \tag{8.140}$$

As emphasized by Greatbatch et al. (2001) and McDougall et al. (2003a), upon making the hydrostatic approximation, these equations for the Boussinesq ocean model are identical to those integrated by traditional Boussinesq z-models, with the exception of details that have been absorbed by the turbulence tracer and momentum fluxes $\mathbf{F}$ and $\mathbf{F}^{(\mathbf{v})}$. Additionally, McDougall et al. (2003a) argue that the interpretation of model fields as proposed here allows for the Boussinesq equations to be far more accurate than the alternative interpretation. Hence, for this reason, and for reasons of mathematical elegance, one may prefer the interpretation summarized by Table 8.1 for the variables carried by the Boussinesq and non-Boussinesq versions of z-coordinate ocean climate models.

8.9 CHAPTER SUMMARY

The purpose of this chapter was to provide a systematic means for averaging the continuous equations over small-scale processe. These small-scale processes are typically associated with overturning or breaking waves on the centimeters to meters scale, and so they represent processes contributing to the dianeutral mixing discussed in Chapter 7.

Averaging the equations of motion requires a bit of formalism. In most cases, the formalism is guided by what one desires. That is, the averaging process is subjective. The averaged equations and the associated subgrid scale terms are a function of averaging. In this chapter, we chose to employ fixed space-time ensemble averages, that is, Eulerian averages. This perspective has its advantages pedagogically, as it maintains the z-coordinate perspective utilized in the early chapters (Chapters 2–5). Additionally, the Eulerian perspective is legitimate even for those situations where the use of density or pressure as vertical coordinate is less straightforward. That is, the Eulerian ensemble is well defined even in a highly turbulent nonhydrostatic fluid. Additionally, remaining within the z-coordinate frame provides an interpretation for the equations used in z-models, which remain the most common class of ocean climate models (Griffies et al., 2000a).

After motivating the use of density weighted averages, as well as introducing certain modified mean fields, the remainder of this chapter applied the formalism to the nonhydrostatic shallow ocean equations of Chapters 3–5 for the purpose of systematically deriving a set of ensemble mean equations. A summary of this chapter is best found by examining Table 8.1, which provides a correspondence between continuous fields whose SGS processes are associated with molecular effects, Eulerian ensemble averaged continuous fields, whose SGS processes are associated with small-scale ocean mixing, continuous model fields, which represent a relabeling of the averaged continuous fields, and discrete model fields, which are the fields to be discretized in subsequent parts of this book.

Chapter Nine

Kinematics of an Isentropic Ensemble

This chapter presents a set of kinematic tools applicable to the analysis of an ensemble of isentropic Boussinesq fluid parcels. For the purposes of this chapter, isentropic transport is considered to be frictionless, adiabatic, and isohaline. That is, isentropes are equated here to isopycnals. The tools introduced here are useful when formulating the mesoscale eddy problem in the ocean interior, since these eddies are well approximated by ideal fluid flow. Upon completion of this chapter, the reader should have a firm understanding of how to map results and ideas between z-coordinate and isopycnal coordinate frameworks. Because of their fundamental nature and utility, these and related techniques are becoming quite common in the theoretical geophysical fluid dynamics literature (e.g., Andrews and McIntyre, 1978; DeSzoeke and Bennett, 1993; McIntosh and McDougall, 1996; Kushner and Held, 1999; Held and Schneider, 1999; Killworth, 2001; McDougall and McIntosh, 2001; Schneider et al., 2003; Nurser and Lee, 2003a,b; Ferrari and Plumb, 2003).

9.1 PARAMETERIZING MESOSCALE EDDIES

Understanding and parameterizing mesoscale eddies in the ocean remains a topic of intense research within the physical oceanography community. The space scale characterizing the mesoscale eddies is largely determined by the first baroclinic Rossby radius (e.g., Stammer, 1997; Smith and Vallis, 2002), and the time scale is on the order of a few weeks to months. The space scale is smaller than the level of grid resolution in typical global ocean climate models, thus necessitating parameterizations aiming to capture eddy effects on the resolved scale flow.

The mesoscale eddy parameterization problem is important due to the nontrivial effects these eddies have on setting tracer distributions and water mass properties, especially within the Southern Ocean. Additionally, and quite uncomfortably, ocean climate models exhibit a high degree of sensitivity to their subgrid scale (SGS) operators. For these reasons, many research efforts have over the past two decades focused on improving the physical, mathematical, and numerical integrity of the SGS operators aiming to parameterize the eddies.

9.1.1 Posing the mesoscale eddy parameterization problem

The mesoscale eddy problem has typically been posed in the context of ideal Boussinesq fluid dynamics. This route is followed here. Further simplifications to the dynamical equations are rendered via the quasi-geostrophic approximation, and this has proved useful for understanding eddy mechanisms and scalings. Given a focus on ideal flow, we are not concerned with processes that alter the potential density of a fluid parcel, such as mixing or sources due to nonlinear effects in the equation of state for seawater density. Generalizations to the nonlinear equation of state are straightforward, as discussed by Gent et al. (1995), McIntosh and McDougall (1996) and McDougall and McIntosh (2001), whereby potential density and its associated specific thickness is translated into locally referenced potential density, or neutral density, and its associated specific thickness. Focusing on the linear equation of state case allows for the symbol ρ to refer in this chapter to potential density, as there is no distinction made here between potential density and *in situ* density.*

The parameterization problem involves many considerations, such as numerical constraints (e.g., dissipation operators are needed for numerical stability), physical observations (e.g., mesoscale eddies mix predominantly along neutral directions), physical/statistical postulates (e.g., eddies homogenize potential vorticity and/or eddies increase entropy in the sense of information theory). Different theoretical approaches have been tried with little consensus in the community as to what is optimal both theoretically and from the perspective of "good" modeling practice, largely defined as a practice that leads to unambiguous improvement in model realism. Griffies et al. (2000a) provide a review with numerous references.

The purpose of this chapter is modest in that no attempt is made to establish a dynamical theory leading to a parameterization scheme. Instead, the focus is on kinematic issues associated with volume and tracer budgets. These considerations provide a starting point for framing the mesoscale eddy parameterization problem, especially when posed in an Eulerian frame such as afforded by z-models. Yet they are not sufficient. More complete treatments can be found in references given in Griffies et al. (2000a).

9.1.2 Averaging at fixed depth versus at fixed neutral density

Chapter 8 showed how to systematically map between ocean model variables and *in situ* density weighted ensemble averaged fields (e.g., Section 8.8). The ensemble average from that chapter takes place at a fixed point in space-time. That is, it is an Eulerian average, and in particular, it occurs at a fixed depth.

A basic property of mesoscale eddies is that they stir and mix properties predominantly parallel to neutral directions. As discussed in Section 7.2.1, this property has been deduced from the small levels of dianeutral mixing measured in the ocean interior. It is important to build this empirical observation into the mathematical formulation of the mesoscale eddy closure problem. Consequently, when averaging over realizations of mesoscale eddies for the purpose of garnering a parameterization, we are prompted to consider a horizontal position that maintains a constant locally referenced potential density. Doing so leads to specific thickness weighted means, as discussed later in this chapter.

It is therefore convenient to distinguish two forms of ensemble averaging: this chapter focuses on averaging over the mesoscale whereas Chapter 8 averages over finer scales. An alternative is to average over the mesoscale and fine scales at once, thus combining the *in situ* density weighted averaging of Chapter 8 with

---

*See Section 5.7 for definitions of potential density and *in situ* density.

the specific thickness weighted means of this chapter. Greatbatch and McDougall (2003) pursue that approach. It turns out to be a straightforward exercise in the techniques introduced here and in Chapter 8.

9.1.3 Good modeling practice

A practical outcome of this chapter is an interpretation for SGS transport operators in the tracer equation integrated by models not resolving mesoscale eddies. In Chapter 13, it is argued that the SGS neutral transport operators are useful even for grids explicitly resolving mesoscale eddies. The main reason is that alternative transport operators that may be appropriate at resolutions admitting mesoscale eddies tend to introduce unphysically large levels of spurious mixing between water masses (Roberts and Marshall, 1998; Griffies et al., 2000b). From this perspective, the operators motivated from the mesoscale parameterization problem represent *good modeling practice* for all resolutions in ocean climate modeling.

9.1.4 Mesoscale eddies and diabatic processes

As stated above, focus is placed on kinematics appropriate for an ideal fluid. Generalizations which consider interactions between mesoscale eddies and irreversible mixing processes, such as those in boundary regions or via breaking internal waves in the ocean interior, remain an area of active research. The paper by Tanden and Garrett (1996) raised these issues by questioning the validity of a separation between adiabatic and diabatic processes. Related focus has largely been on deriving a general kinematic framework that describes how isentropes intersect boundaries where diabatic processes are vigorous, and generally how to specify boundary conditions for the mesoscale eddy parameterization schemes (e.g., Gent et al., 1995; McIntosh and McDougall, 1996; Kushner and Held, 1999; Held and Schneider, 1999; Killworth, 2001; McDougall and McIntosh, 2001; Schneider et al., 2003; Ferrari and Plumb, 2003). A discussion of how z-models handle the issues of boundaries is given in Chapter 15. That material is largely pragmatic in nature. A more theoretical discussion is beyond the scope of this book, and the above references should be consulted for more detail.

9.2 ADVECTION AND SKEWSION

There are two complementary ways of interpreting the transport operator used to reversibly stir a tracer. One can consider either the convergence of an advective flux, thus leading to *advection*, or the convergence of a skew flux, thus leading to *skewsion*. Notably, skewsion lends itself to a more streamlined theoretical development, and it provides a more robust numerical foundation (see Chapters 13–16 for numerical developments).

9.2.1 The vector streamfunction

Consider an arbitrary three-dimensional divergence-free velocity field

$$\nabla \cdot \mathcal{U} = 0. \tag{9.1}$$

The divergence-free condition represents a diagnostic relation, or constraint, that reduces the functional degrees of freedom for the velocity field from three to two. This constraint can be satisfied identically by introducing a vector streamfunction

$$\mathcal{U} = \nabla \wedge \boldsymbol{\Upsilon}. \tag{9.2}$$

The vector streamfunction $\boldsymbol{\Upsilon}$ is not completely specified by this relation, since the equally valid streamfunction

$$\boldsymbol{\Upsilon}' = \boldsymbol{\Upsilon} + \nabla\lambda \tag{9.3}$$

corresponds to the same velocity field $\mathcal{U}$. The arbitrary scalar function λ is known as a *gauge function*, and the freedom to modify the vector streamfunction through the addition of λ is known as *gauge freedom*. A similar symmetry is present in Maxwell's equations of electrodynamics (e.g., Jackson, 1975).

9.2.2 Advective and skew fluxes

The *advective tracer concentration flux*

$$\begin{aligned} \mathbf{F}_{(a)} &= \mathcal{U}\, C \\ &= (\nabla \wedge \boldsymbol{\Upsilon})\, C \end{aligned} \tag{9.4}$$

can be related to a *skew tracer concentration flux*

$$\begin{aligned} \mathbf{F}_{(s)} &= -\nabla C \wedge \boldsymbol{\Upsilon} \\ &= \mathbf{F}_{(a)} - \mathbf{F}_{(r)} \end{aligned} \tag{9.5}$$

through exploitation of the identity

$$\begin{aligned} \mathcal{U}\, C &= (\nabla \wedge \boldsymbol{\Upsilon})\, C \\ &= -(\nabla C \wedge \boldsymbol{\Upsilon}) + \nabla \wedge (C\, \boldsymbol{\Upsilon}). \end{aligned} \tag{9.6}$$

$\mathbf{F}_{(s)}$ represents a *skew* flux, which is a flux defined such that it is directed perpendicular to the tracer gradient $\mathbf{F}_{(s)} \cdot \nabla C = 0$. Since the *rotational flux*

$$\mathbf{F}_{(r)} = \nabla \wedge (C\, \boldsymbol{\Upsilon}) \tag{9.7}$$

has a vanishing divergence, the divergence of the skew flux and advective flux is identical,

$$\nabla \cdot \mathbf{F}_{(s)} = \nabla \cdot \mathbf{F}_{(a)}. \tag{9.8}$$

Hence, if these fluxes enter into the evolution of a tracer, as may occur if $\mathcal{U}$ represents a divergence-free transport velocity field, then one may equally well choose to use the skew flux or advective flux in describing the evolution. For example, splitting the velocity field into a resolved velocity $\mathbf{v}$ that is carried explicitly by a model, and an unresolved velocity $\mathcal{U}$ representing subgrid scale processes, allows one to write

$$\begin{aligned} C_{,t} + \nabla \cdot (\mathbf{v}\, C) &= -\nabla \cdot (\mathcal{U}\, C) \\ &= \nabla \cdot (\nabla C \wedge \boldsymbol{\Upsilon}). \end{aligned} \tag{9.9}$$

In various contexts, it is often convenient to introduce the antisymmetric transport tensor

$$\begin{aligned} F^{m}_{(s)} &= -\epsilon^{mnp}\, C_{,n}\, \Upsilon_{p} \\ &= -A^{mn}\, C_{,n} \end{aligned} \tag{9.10}$$

where A^{mn} represents a reorganization of the vector streamfunction

$$\begin{aligned} A^{mn} &= \epsilon^{mnp} \Upsilon_p \\ &= \begin{pmatrix} 0 & \Upsilon_3 & -\Upsilon_2 \\ -\Upsilon_3 & 0 & \Upsilon_1 \\ \Upsilon_2 & -\Upsilon_1 & 0 \end{pmatrix}, \end{aligned} \tag{9.11}$$

and ϵ^{mnp} is the totally antisymmetric Levi-Civita tensor defined in Section 20.12 (page 459). Hence, for any stirring process that is written in terms of an advection or skewsion, there is an associated antisymmetric stirring tensor. Furthermore, given the gauge freedom introduced in Section 9.2.1, it is always possible to specify a gauge so that

$$\Upsilon_3 \equiv 0. \tag{9.12}$$

This gauge is referred to as the *vertical* gauge and is found to be convenient in the following.

9.2.3 Complementary aspects of the fluxes

Consider now two aspects of the advective and skew tracer fluxes, which turn out to be complementary to one another. First, since the advective flux is directly proportional to the velocity, this flux vanishes when the velocity vanishes. In contrast, a skew flux vanishes when the tracer gradient vanishes. This property of the skew flux is also shared by a downgradient diffusive tracer flux.

Second, orientation or direction of the advective flux $\mathcal{U}\, C$ is determined by $\mathcal{U}$. For the case where $\mathcal{U}$ is a velocity, the orientation of this vector corresponds to fluid parcel trajectories. This orientation is the same for all tracers. The fluid parcels carry along with them a particular amount of tracer mass $(\rho \, dV)\, C$, hence leading to the notion of a flux representing the passage of a quantity of matter across an area per unit time. Furthermore, it is the component of the velocity (and hence the advective tracer flux) aligned parallel to gradients of the tracer that creates local changes in the tracer concentration. This property is seen from the advection equation

$$C_{,t} = -(\mathbf{v} + \mathcal{U}) \cdot \nabla C. \tag{9.13}$$

Hence, an advective tracer flux in an adiabatic flow *is not* generally oriented parallel to surfaces of constant tracer, although such is often indicated in the literature. Indeed, if it were, then the tracer field would be locally static since then $(\mathbf{v} + \mathcal{U}) \cdot \nabla C = 0$.

As a complement to the advective flux, the skew tracer flux is directed along lines of constant tracer. That is, the skew flux is neither upgradient nor downgradient since

$$\begin{aligned} \nabla C \cdot \mathbf{F}_{(s)} &= -\nabla C \cdot (\nabla C \wedge \boldsymbol{\Upsilon}) \\ &= 0. \end{aligned} \tag{9.14}$$

Hence, orientation of the skew flux is directly tied to the tracer field, with each tracer yielding a generally different orientation, all parallel to the respective tracer isolines. This property does not allow for the interpretation of a skew flux in terms of the passage of a quantity of matter across an area per unit time. This lack of familiarity in the interpretation does not preclude the utility of the skew flux.

9.3 VOLUME CONSERVATION

This section considers reversible stirring of volume in an ideal stratified Boussinesq fluid. *Stirring* is defined here as transport that can, in principle, be reversed, that is, where material properties of a fluid parcel remain unchanged. What does change is the arrangement of parcels within the fluid, and the shapes of parcels, which are generally stretched into finer scale filaments. Stirring by baroclinic eddies in the ocean is a good approximation to such ideal situations. Eventually, small-scale processes, such as those discussed in Chapter 7, mix properties irreversibly. The reader is encouraged to read Eckart (1948) for a pedagogical introduction to stirring and mixing.

Over long space and time scales, stirring by ocean mesoscale eddies can be considered chaotic, which motivates a stochastic perspective in which an ensemble of eddies is considered. The goal is to describe the ensemble mean properties of the fluid, with a focus in this section on the kinematics of parcel rearrangement.

An overbar with a density label, $\overline{(\ )}^{\rho}$, is used in the following to denote the ensemble mean over an ensemble of fluid parcels, each having the same potential density, ρ, the same horizontal position, (x, y), and the same time, t. Isopycnals undulate in space and time, which means that each ensemble member has a vertical position that is generally distinct from the fixed depth, z. When the context is clear, it is useful to drop the dependence on (x, y, t) in order to highlight the dependence on potential density and/or vertical position. Averaging over parcels maintaining the same potential density is often termed *isopycnal averaging* in the literature.

9.3.1 Two perspectives on volume conservation

Consider a stably stratified ideal Boussinesq fluid. Isopycnal coordinates introduced in Section 6.10.4 are ideally suited to describe the dynamics of this fluid. In these coordinates, the volume of a fluid parcel is written

$$\begin{aligned} dV &= dx\, dy\, dz \\ &= dx\, dy\, z_{,\rho}\, d\rho. \end{aligned} \tag{9.15}$$

Since the specific thickness $z_{,\rho}$ appears quite frequently, it is convenient to introduce the shorthand notation

$$h = z_{,\rho}. \tag{9.16}$$

Specific thickness is the Jacobian of transformation between geopotential coordinates, (x, y, z, t), and isopycnal coordinates, (x, y, ρ, t) (see Section 6.3.2). For stably stratified ideal fluids, h is one-signed, hence making the coordinate transformation well defined. Geometrically, the product $|h\, d\rho|$ represents the vertical distance, or *thickness*, between the two infinitesimally close density classes ρ and $\rho + d\rho$. For an ideal Boussinesq fluid parcel, material conservation of both volume and potential density implies conservation of the product of specific thickness and horizontal area $dx\, dy\, h$, which leads to the thickness equation

$$h_{,t} + \nabla_{\rho} \cdot (h\, \mathbf{u}) = 0, \tag{9.17}$$

with $\mathbf{u}$ the horizontal velocity field.

An Eulerian z-coordinate description of the reversible rearrangement of Boussinesq parcels is rendered via a combination of volume conservation, $\nabla \cdot \mathbf{v} = 0$, and material conservation of potential density, $\mathrm{d}\rho/\mathrm{d}t = 0$. When written as a skewsion process, the natural gauge is the vertical gauge $\boldsymbol{\Psi} = \mathbf{u} \wedge \hat{\mathbf{z}}$ introduced in Section 9.2.2 (equation (9.12)), since this gauge only requires the same horizontal velocity field $\mathbf{u}$ used with the isopycnal coordinate description. This gauge has an associated skew flux of potential density $\mathbf{F} = -\nabla\rho \wedge \boldsymbol{\Psi}$, which leads to the evolution

$$\rho_{,t} = \nabla \cdot (\nabla\rho \wedge \boldsymbol{\Psi}), \tag{9.18}$$

where all derivatives are here taken with fixed Eulerian coordinates, and the divergence operator is three-dimensional.

As an aside, note that Kushner and Held (1999) considered the case with potential vorticity replacing density and meridional stratification replacing vertical stratification. As pointed out by Held and Schneider (1999), this description might also be relevant for a surface layer where vertical density stratification can be weak yet meridional stratification nonzero. In this case, volume conservation using isopycnal coordinates is described by the alternate form of the thickness equation

$$\partial_t y_{,\rho} + \partial_x (y_{,\rho} u) + \partial_z (y_{,\rho} w) = 0, \tag{9.19}$$

where the partial derivatives are still taken along surfaces of constant density. Because the thickness in this case is only affected by the zonal and vertical velocities, the natural gauge for a skewsion description is a *meridional* gauge. Nurser and Lee (2003a,b) pursue these ideas for the ocean.

9.3.2 Exact ensemble mean kinematics in isopycnal coordinates

Consider an ensemble of ideal Boussinesq fluid parcels with the same infinitesimal volume, $\mathrm{d}V = \mathrm{d}x\,\mathrm{d}y\,\mathrm{d}z = \mathrm{d}x\,\mathrm{d}y\,h\,\mathrm{d}\rho$, and same potential density, ρ. Lacking any other marker, such as a tracer concentration, the ensemble members are distinguished from one another by values of their horizontal area, $\mathrm{d}A = \mathrm{d}x\,\mathrm{d}y$, and their specific thickness, h, that is, their geometric attributes. The ensemble members are assumed to be stirred by different stochastic realizations of the fluid flow. Since each flow realization alters the geometric properties of the parcels, a mean field description focuses on the mean of these geometric properties.

In isopycnal coordinates, (x, y, ρ, t), the thickness equation is satisfied by each ensemble member

$$h_{,t} + \nabla_\rho \cdot (h\,\mathbf{u}) = 0. \tag{9.20}$$

The ensemble mean over parcels with potential density ρ satisfies

$$\partial_t \overline{h}^\rho + \nabla_\rho \cdot \left(\overline{h}^\rho\,\overline{\mathbf{u}}^\rho + \overline{h'\,\mathbf{u}'}^\rho\right) = 0, \tag{9.21}$$

where primed variables represent deviations from the mean. The mean specific thickness $\overline{h}^\rho$ of parcels with potential density ρ therefore satisfies the conservation equation

$$\partial_t \overline{h}^\rho + \nabla_\rho \cdot (\overline{h}^\rho\,\widehat{\mathbf{u}}) = 0, \tag{9.22}$$

where

$$\begin{aligned}\widehat{\mathbf{u}} &= \frac{\overline{h\,\mathbf{u}}^{\rho}}{\overline{h}^{\rho}} \\ &= \overline{\mathbf{u}}^{\rho} + \frac{\overline{h'\,\mathbf{u}'}^{\rho}}{\overline{h}^{\rho}} \\ &= \overline{\mathbf{u}}^{\rho} + \mathbf{u}^{\text{bolus}}\end{aligned} \tag{9.23}$$

is an effective horizontal velocity. The first equality defines a *thickness weighted* averaging operator, and the last equality defines the horizontal *bolus velocity* $\mathbf{u}^{\text{bolus}}$ as introduced by Rhines (1982). Since each ensemble member is taken from a stably stratified fluid, the mean specific thickness $\overline{h}^{\rho}$ is single-signed and nonvanishing. The bolus velocity

$$\mathbf{u}^{\text{bolus}}(\rho) = \widehat{\mathbf{u}}(\rho) - \overline{\mathbf{u}}^{\rho}(\rho) \tag{9.24}$$

arises from the along-isopycnal correlations between thickness and horizontal velocity.

Quite conveniently, the mean conservation equation (9.22) takes the *same* mathematical form as the conservation equation (9.20) satisfied by each ensemble member. The key difference, of course, is that the ensemble mean thickness $\overline{h}^{\rho}$ is stirred by the effective horizontal velocity $\widehat{\mathbf{u}}(\rho)$, whereas the thickness of each ensemble member is stirred by a randomly different realization of the horizontal velocity $\mathbf{u}$. The simplicity of the resulting mean field description is afforded by use of the Lagrangian vertical coordinate ρ.

9.3.3 The non-utility of an Eulerian mean description

Although ρ-coordinates provide a natural description of the isentropic rearrangement of volume, it is useful to understand how to transform to z-coordinates for purposes such as interpreting results from z-coordinate ocean models.

It is useful to first illustrate mathematically why it is more convenient to employ fixed potential density averages (which are Lagrangian in the vertical) instead of fixed depth averages (which are Eulerian). For this purpose, it is sufficient to note that when sitting at a fixed Eulerian grid point, one measures a mean flux, $\overline{\mathbf{F}}^{z}$, that has a nonzero component across the mean potential density surface $\overline{\rho}^{z}$. Here, $\overline{(\ )}^{z}$ denotes an Eulerian average obtained by an observer at a fixed point (x, y, z, t) (see Figure 9.1). Precisely, although

$$\mathbf{F}(z) \cdot \nabla \rho(z) = 0 \tag{9.25}$$

for each realization of the isentropic flow, the Eulerian average satisfies

$$\overline{\mathbf{F}}^{z} \cdot \nabla \overline{\rho}^{z} = -\overline{\mathbf{F}' \cdot \nabla \rho'}^{z} \neq 0. \tag{9.26}$$

Herein lies the cumbersome nature of the Eulerian perspective: the behavior of the Eulerian mean flux differs fundamentally from that of each flow realization. This behavior of the ensemble mean is in contrast to the isopycnal description, where the ensemble mean thickness equation has the same mathematical form as the thickness equation satisfied by each ensemble member. The problem has already been mentioned. It is that an average taken at a fixed depth samples members from different ensembles defined by different potential densities. That is, the fixed depth average cannot be an isentropic average, and so it is of little use when aiming to describe a mean over an isentropic ensemble.

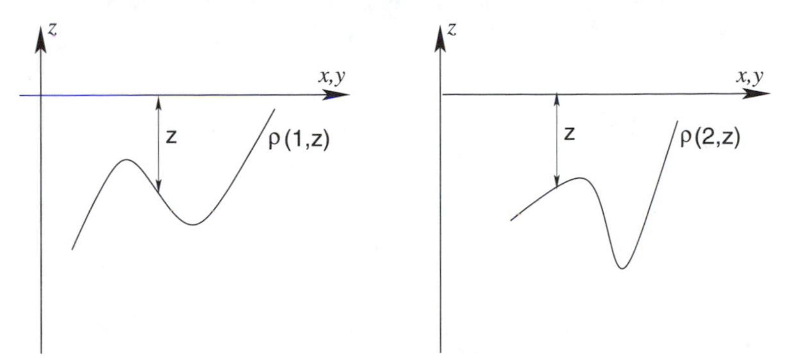

Figure 9.1 Schematic of the ensemble averaged potential density as measured by an observer at a fixed point (x, y, z, t) in space-time. In general, different members of the ensemble have potential density surfaces that live at different depths. That is, a fixed Eulerian space-time observer measures an ensemble mean potential density as the average over different potential density surfaces. For the case of a two-member ensemble as shown here, $2\,\overline{\rho}^{z}(z) = \rho(1,z) + \rho(2,z)$, where $\rho(1,z)$ is generally different from $\rho(2,z)$.

9.3.4 Transformed residual mean (TRM)

To find a more convenient z-coordinate formulation, one may choose to follow approaches used by DeSzoeke and Bennett (1993), McIntosh and McDougall (1996), Kushner and Held (1999), and McDougall and McIntosh (2001). Here, one introduces the displacement vector in the vertical

$$\boldsymbol{\xi} = \xi\,\hat{\mathbf{z}}. \tag{9.27}$$

When situated at a fixed point (x, y, t), this displacement vector identifies the vertical position of a parcel with a fixed potential density ρ. That is, $\xi(x, y, \rho, t)$ is the deviation of a potential density surface ρ from its mean vertical position

$$z(\rho) = \overline{z}^{\rho}(\rho) + \xi(\rho). \tag{9.28}$$

The ensemble mean of this deviation vanishes

$$\overline{\xi}^{\rho} = 0 \tag{9.29}$$

by definition. Figure 9.2 illustrates this situation for a two-member ensemble.

Define an isopycnally averaged version of a field Φ by fixing (x, y, ρ, t) and computing the ensemble mean. In symbols, one measures $\overline{\Phi(x, y, \overline{z}^{\rho} + \xi, t)}^{\rho}$. The question arises how does $\overline{\Phi}^{\rho}$ correspond to fields measured at the mean depth, $\overline{z}^{\rho}$, of the isopycnal? This question leads one to *define* a *modified mean* field satisfying*

$$\widetilde{\Phi}(x, y, \overline{z}^{\rho}, t) = \overline{\Phi(x, y, \overline{z}^{\rho} + \xi, t)}^{\rho}. \tag{9.30}$$

*A similar equality arises when considering generalized Lagrangian mean fields (Andrews and McIntyre, 1978). In this case, the displacement vector $\boldsymbol{\xi}$ has components in all three directions, instead of just the vertical considered here. Also, be careful not to confuse the tilde symbol used here for the modified mean field with the tilde used in Chapter 20 for one-forms. The meaning is completely distinct.

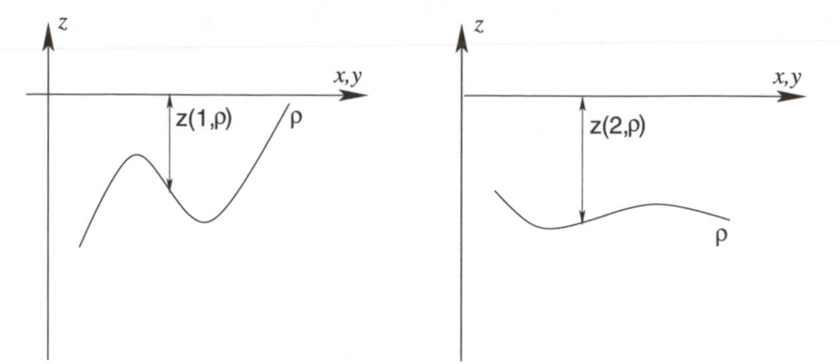

Figure 9.2 Schematic of the ensemble mean depth $\overline{z}^{\rho}(x, y, \rho, t)$ of a particular potential density surface ρ. In general, different members of an isentropic ensemble live at different depths. Therefore, when considering ensemble members with the same potential density, the ensemble mean depth is the average over the different members. For the case of a two-member ensemble, as shown here, $2\,\overline{z}^{\rho}(\rho) = z(1, \rho) + z(2, \rho)$, where the depth $z(1, \rho)$ is generally different from $z(2, \rho)$.

As defined, the modified mean field, $\widetilde{\Phi}$, when measured at the mean depth ,$\overline{z}^{\rho}$, is equal to the isopycnally averaged field, $\overline{\Phi}^{\rho}$. A particularly important modified mean field is the potential density

$$\widetilde{\rho}(x, y, \overline{z}^{\rho}, t) = \overline{\rho(x, y, \overline{z}^{\rho} + \xi, t)}^{\rho}. \tag{9.31}$$

Since the average is defined over constant potential density, ρ, the average operation can be removed to give the equality

$$\widetilde{\rho}(x, y, \overline{z}^{\rho}, t) = \rho(x, y, \overline{z}^{\rho} + \xi, t). \tag{9.32}$$

In words, this equality says that the modified mean density $\widetilde{\rho}$, when measured at the mean depth $\overline{z}^{\rho}$, is equal to the density ρ of the ensemble members, each measured at their respective depths $\overline{z}^{\rho} + \xi$. Generalizations of the modified mean potential density to a nonlinear equation of state lead to a modified mean neutral density (McDougall and McIntosh, 2001).

The relation $\widetilde{\rho}(\overline{z}^{\rho}) = \rho(\overline{z}^{\rho} + \xi)$ implies that $\widetilde{\rho}$ is the functional inverse of the mean depth, $\overline{z}^{\rho}$, of the ρ potential density surface (DeSzoeke and Bennett, 1993). From equation (9.30), it follows that

$$\begin{aligned} \widetilde{\Phi}(x, y, \overline{z}^{\rho}, t) &= \overline{\Phi(x, t, \overline{z}^{\rho} + \xi, t)}^{\rho} \\ &= \overline{\Phi}^{\rho}(x, y, \widetilde{\rho}, t). \end{aligned} \tag{9.33}$$

This is an important relation between an ensemble mean field defined at the modified mean potential density surface, $\overline{\Phi}^{\rho}(\widetilde{\rho})$, and the modified mean field defined at the mean depth of the isopycnal, $\widetilde{\Phi}(\overline{z}^{\rho})$. In an analogous fashion, *define* a relation between a thickness weighted field

$$\widehat{\Phi} = \frac{\overline{h\,\Phi}^{\rho}}{\overline{h}^{\rho}}, \tag{9.34}$$

which is defined as a function of density ρ, and a *transformed residual mean* (TRM) field $\overline{\Phi}^{\#}$ defined as a function of the mean depth (McDougall and McIntosh, 2001)

$$\overline{\Phi}^{\#}(x, y, \overline{z}^{\rho}, t) \equiv \widehat{\Phi}(x, y, \widetilde{\rho}, t). \tag{9.35}$$

Without thickness weighting, such as when h =constant for all ensemble members, the TRM field $\overline{\Phi}^{\#}(\overline{z}^{\rho})$ reduces to the modified mean field $\widetilde{\Phi}(\overline{z}^{\rho})$. Generally, these fields are distinct.

9.3.5 Exact ensemble mean kinematics in z-coordinates

The previous formalism is now applied to garner an exact z-coordinate description of the isentropic ensemble. For this purpose, interpret a vertical position, z, as the ensemble mean vertical position, $\overline{z}^{\rho}$, of a potential density surface. Mean fields defined at the fixed vertical position correspond to either modified mean fields when not thickness weighted (equation (9.33)), or TRM fields when thickness weighted (equation (9.35)). This interpretation of the vertical position, and fields defined at a vertical position, will be transferred to the discrete Eulerian lattice of a coarse resolution z-coordinate ocean model (see Section 9.5). Such an interpretation of coarse resolution z-models was originally proposed by McDougall and McIntosh (2001).

9.3.5.1 The total transport and the quasi-Stokes transport

As described using isopycnal coordinates, the mean specific thickness, $\overline{h}^{\rho}$, of the ρ isopycnal surface evolves via equation (9.22) through the stirring effects of the effective horizontal velocity $\widehat{\mathbf{u}}$. This velocity is thickness weighted, as defined by equation (9.23). According to the definition of a TRM field given by equation (9.35), an associated horizontal TRM velocity is given by

$$\widehat{\mathbf{u}}(x, y, \widetilde{\rho}, t) = \overline{\mathbf{u}}^{\#}(x, y, \overline{z}^{\rho}, t). \tag{9.36}$$

So far, it has been necessary to consider only the horizontal velocity components, from either the thickness weighted velocity or the TRM velocity. When working in z-models, it is useful to see how these velocities relate to a three-dimensional velocity. With a Boussinesq flow, a three-dimensional velocity used to transport properties is preferably nondivergent. Given the skewsion formalism of Section 9.2.1, one may associate the horizontal TRM velocity components with a three-dimensional nondivergent TRM velocity field

$$\overline{\mathbf{v}}^{\#} = \nabla \wedge \overline{\mathbf{\Psi}}^{\#}, \tag{9.37}$$

where $\overline{\mathbf{\Psi}}^{\#}$ is a vector streamfunction.

The definition of $\overline{\mathbf{v}}^{\#}$ is shown to be reasonable *a posteriori* based on the following interpretation of its streamfunction $\overline{\mathbf{\Psi}}^{\#}$. Choosing the vertical gauge (Section 9.2.2) allows

$$\begin{aligned} \overline{\mathbf{\Psi}}^{\#}(\overline{z}^{\rho}) &= \int_{-H}^{\overline{z}^{\rho}} (\overline{\mathbf{u}}^{\#}(z) \wedge \hat{\mathbf{z}})\, dz \\ &\equiv \overline{\mathbf{U}}^{\#}(\overline{z}^{\rho}) \wedge \hat{\mathbf{z}}, \end{aligned} \tag{9.38}$$

where $\overline{\mathbf{U}}^{\#}(\overline{z}^{\rho})$ is the mean horizontal transport. Vertical integration is over a column of the ocean from the bottom, $z = -H$, to the ensemble mean depth, $z = \overline{z}^{\rho}$, of the isopycnal ρ. $\overline{\mathbf{U}}^{\#}(\overline{z}^{\rho})$ is the ensemble mean transport of fluid beneath the potential density surface $\rho = \widetilde{\rho}$. Proving that such is the case is a useful exercise in the formalism.

For these purposes, start by noting that

$$\begin{aligned} \overline{\mathbf{U}}^{\#}(\overline{z}^{\rho}) &= \int_{-H}^{\overline{z}^{\rho}} \overline{\mathbf{u}}^{\#}(z)\,\mathrm{d}z \\ &= \int_{\rho(-H)}^{\widetilde{\rho}(\overline{z}^{\rho})} \widehat{\mathbf{u}}(\sigma)\,\overline{h}^{\sigma}\,\mathrm{d}\sigma. \end{aligned} \tag{9.39}$$

This relation follows from a change in variables from depth to density using $\mathrm{d}z = \overline{h}^{\sigma}\,\mathrm{d}\sigma$, and by recalling the inverse relationship between $\overline{z}^{\rho}$ and $\widetilde{\rho}(\overline{z}^{\rho})$ (end of Section 9.3.4). Equation (9.36) was also used to relate the horizontal thickness weighted velocity $\widehat{\mathbf{u}}(\sigma)$, which is a function of density, and the TRM velocity $\overline{\mathbf{u}}^{\#}(\overline{z}^{\rho})$, which is a function of depth. Now introduce the definition (9.23) for the thickness weighted velocity to yield

$$\begin{aligned} \overline{\mathbf{U}}^{\#}(\overline{z}^{\rho}) &= \int_{\rho(-H)}^{\widetilde{\rho}(\overline{z}^{\rho})} \overline{\mathbf{u}\,h}^{\sigma}\,\mathrm{d}\sigma \\ &= \int_{\rho(-H)}^{\rho(\overline{z}^{\rho}+\xi)} \overline{\mathbf{u}\,h}^{\sigma}\,\mathrm{d}\sigma, \end{aligned} \tag{9.40}$$

where the second step follows from the conservation property (9.32) satisfied by the modified mean potential density. The integrand $\overline{\mathbf{u}\,h}^{\sigma}\,\mathrm{d}\sigma$ is the ensemble mean of the horizontal transport passing within the infinitesimal density layer between σ and $\sigma + \mathrm{d}\sigma$. It is vertically integrated over each of the continuum of potential densities $\rho \leq \sigma \leq \rho(-H)$, thus yielding the ensemble mean transport of fluid beneath the potential density surface $\rho = \widetilde{\rho}$. This is the desired result.

Transferring the expression (9.40) to depth coordinates leads to the expression

$$\overline{\mathbf{U}}^{\#}(\overline{z}^{\rho}) = \overline{\int_{-H}^{\overline{z}^{\rho}+\xi} \mathbf{u}\,\mathrm{d}z}. \tag{9.41}$$

In words, without the average operation, this expression represents a vertical integral of the horizontal transport, where the vertical integral extends from the ocean bottom $z = -H$ to a depth $z = \overline{z}^{\rho} + \xi$ determined by one of the members of an ensemble whose members all have potential density $\rho = \widetilde{\rho}$. With the average operation, the mean of this vertical integral is computed over all the members of the ρ-ensemble.

To help interpret the transport, $\overline{\mathbf{U}}^{\#}(\overline{z}^{\rho})$, split the vertical integral into two parts

$$\overline{\mathbf{U}}^{\#}(\overline{z}^{\rho}) = \overline{\int_{-H}^{\overline{z}^{\rho}} \mathbf{u}\,\mathrm{d}z} + \overline{\int_{\overline{z}^{\rho}}^{\overline{z}^{\rho}+\xi} \mathbf{u}\,\mathrm{d}z}. \tag{9.42}$$

The first expression represents the ensemble mean horizontal transport between the bottom, $z = -H$, and the fixed depth, $z = \overline{z}^{\rho}$. Hence, it can be interpreted in the usual Eulerian average manner. The second term is the ensemble mean transport between the mean depth, $z = \overline{z}^{\rho}$, of the ρ isopycnal, and the depth, $z = \overline{z}^{\rho} + \xi$, of the various ensemble members. As a shorthand, write this split as

$$\overline{\mathbf{U}}^{\#}(\overline{z}^{\rho}) = \overline{\mathbf{U}}(\overline{z}^{\rho}) + \mathbf{U}^{\mathrm{qs}}(\overline{z}^{\rho}). \tag{9.43}$$

Following McDougall and McIntosh (2001), call

$$\mathbf{U}^{\mathrm{qs}}(\overline{z}^{\rho}) \equiv \overline{\int_{\overline{z}^{\rho}}^{\overline{z}^{\rho}+\xi} \mathbf{u}\,\mathrm{d}z} \tag{9.44}$$

the *quasi-Stokes transport* for the ρ isopycnal. This transport arises from eddy effects accounting for differences between the Eulerian mean transport $\overline{\mathbf{U}}(\overline{z}^{\rho})$ and the TRM transport $\overline{\mathbf{U}}^{\#}(\overline{z}^{\rho})$. This is characteristic of other forms of Stokes transport, which are defined between Eulerian and Lagrangian means (e.g., Andrews and McIntyre, 1978; Plumb, 1979). Note that, as the surface or bottom of the ocean is approached, the quasi-Stokes transport vanishes since there is no more fluid to be transported at such boundaries. McDougall and McIntosh (2001) emphasize this point (see their Section 8) as it provides a statement regarding the boundary conditions to be used for parameterizations of the quasi-Stokes transport (see Section 9.5).

9.3.5.2 *Evolution of modified mean density*

Following the skewsion formulation from Section 9.2, at the mean depth $z = \overline{z}^{\rho}$, the streamfunction $\overline{\mathbf{\Psi}}^{\#}$ defines an effective skew flux of potential density $\rho = \widetilde{\rho}$ given by

$$\overline{\mathbf{F}}^{\#} = -\nabla\widetilde{\rho} \wedge \overline{\mathbf{\Psi}}^{\#}. \tag{9.45}$$

Using the identity $\overline{\mathbf{\Psi}}^{\#} = \overline{\mathbf{U}}^{\#} \wedge \hat{\mathbf{z}}$, it is sometimes useful to write this expression in one of the forms

$$\begin{aligned} \overline{\mathbf{F}}^{\#} &= -\overline{\mathbf{U}}^{\#}\,\partial_z\widetilde{\rho} + \hat{\mathbf{z}}\,\overline{\mathbf{U}}^{\#}\cdot\nabla_z\widetilde{\rho} \\ &= -(\overline{\mathbf{U}}^{\#} + \hat{\mathbf{z}}\,\mathbf{S}\cdot\overline{\mathbf{U}}^{\#})\,\partial_z\widetilde{\rho}, \end{aligned} \tag{9.46}$$

where

$$\mathbf{S} = -\left(\frac{\nabla_z\widetilde{\rho}}{\partial_z\widetilde{\rho}}\right) \tag{9.47}$$

is the slope of the modified mean density field and $\nabla_z = (\partial_x, \partial_y, 0)$ is the horizontal gradient operator taken with constant depth $z = \overline{z}^{\rho}$. The convergence of the effective skew flux leads to a stirring of the modified mean density $\widetilde{\rho}$ at the mean depth $z = \overline{z}^{\rho}$,

$$\widetilde{\rho}_{,t} = \nabla\cdot(\nabla\widetilde{\rho}\wedge\overline{\mathbf{\Psi}}^{\#}). \tag{9.48}$$

This equation represents an exact z-coordinate specification of the evolution of the modified mean density due to stirring by the mean eddies. It corresponds directly to the evolution equation (9.18) satisfied at depth z by a single member of the ensemble. Contrast this elegant correspondence between unaveraged and averaged stirring processes with the lack of correspondence seen in the traditional Eulerian description of Section 9.3.3.

9.3.6 Approximate ensemble mean kinematics in z-coordinates

Equation (9.48) represents an exact z-coordinate description of the stirring of modified mean potential density. However, when working in a discrete z-model, all that is available is Eulerian information. The Lagrangian information used to realize this exact description must be approximated.

As mentioned at the start of Section 9.3.5, values of potential density carried by a coarsely resolved z-model, at fixed model grid depths z, are interpreted as the modified mean potential density $\widetilde{\rho}(z)$. Correspondingly, model depth levels are associated with the mean depth of the potential density surface, $z = \overline{z}^{\rho}$. This interpretation is in contrast to the usual association of the coarse model's Eulerian mean potential density $\overline{\rho}^{z}$ with the model potential density. The latter interpretation is not useful for coarse resolution z-models, as indicated by the discussion surrounding equation (9.26).

The approximation problem for z-models is a problem of how to compute the quasi-Stokes transport $\mathbf{U}^{qs}(z)$, where again $z = \overline{z}^{\rho}$ is assumed throughout this section. To do so in a z-model, the TRM transport is expanded in a Taylor series about the fixed model depth z via

$$\begin{aligned}\overline{\mathbf{U}}^{\#}(z) &= \overline{\int_{-H}^{z+\xi} \mathbf{u}(s)\,\mathrm{d}s} \\ &\approx \overline{\mathbf{U}}(z) + \overline{\mathbf{u}\,\xi}^{z} + \left(\overline{\partial_z \mathbf{u}\,\xi^2}^{z}\right)/2,\end{aligned} \tag{9.49}$$

where neglected terms are third order in deviation quantities. Note that the averages are taken at fixed vertical position, which accords with taking a Taylor series about the depth $z = \overline{z}^{\rho}$.

The averages in equation (9.49) are interpreted as follows. The first term is the usual Eulerian mean horizontal transport passing beneath the depth z. The second term, $\overline{\mathbf{u}(z)\,\xi}$ is the horizontal velocity evaluated at the mean depth, $z = \overline{z}^{\rho}$, multiplied by the deviation, ξ, of the potential density surface from its mean depth, all averaged at fixed depth. An Eulerian split of the horizontal velocity $\mathbf{u}(z)$ into its Eulerian mean $\overline{\mathbf{u}}^{z}$ and deviation $\mathbf{u}'(z)$ leads to the correlation

$$\overline{\mathbf{u}\,\xi}^{z} = \overline{\mathbf{u}'\,\xi}^{z}. \tag{9.50}$$

For the second order term, similar considerations lead to

$$\overline{\partial_z \mathbf{u}\,\xi^2}^{z} \approx \partial_z \overline{\mathbf{u}}^{z}\,\overline{\xi^2}^{z}, \tag{9.51}$$

where neglected terms are third order and higher. Combining these relations leads to the second order accurate expression

$$\overline{\mathbf{U}}^{\#}(z) \approx \overline{\mathbf{U}}(z) + \overline{\mathbf{u}'\,\xi}^{z} + (1/2)\,\overline{\xi^2}^{z}\,\partial_z \overline{\mathbf{u}}^{z}, \tag{9.52}$$

which can also be written in the suggestive form

$$\overline{\mathbf{U}}^{\#}(z) \approx \overline{\mathbf{U}}(z) + \overline{\xi\,(1 + (1/2)\,\xi\,\partial_z)\,\mathbf{u}}^{z}. \tag{9.53}$$

What remains is to determine the deviation ξ of the isopycnal in terms of fields at constant depth. For this purpose, use the identity (9.32) to give

$$\begin{aligned}\widetilde{\rho}(z) &= \rho(z+\xi) \\ &= \rho(z) + \partial_z \rho(z)\,\xi + \partial_{zz}\rho(z)\,\xi^2/2,\end{aligned} \tag{9.54}$$

where terms of third and higher order were neglected. Subtracting the Eulerian mean of equation (9.54) from the unaveraged equation (9.54), and noting that $\widetilde{\rho}$ is already a mean field, leads to the second order accurate expression for the deviation

$$\xi = -\rho'(z)/\partial_z \overline{\rho}^z, \tag{9.55}$$

where

$$\rho(z) = \overline{\rho}^z + \rho'(z). \tag{9.56}$$

To within the same order, the deviation can be written

$$\xi = -\rho'(z)/\partial_z \widetilde{\rho}(z), \tag{9.57}$$

which is more useful for applications in z-models since coarse versions of such models carry $\widetilde{\rho}$ rather than $\overline{\rho}$. Substituting the deviation (9.57) into the approximate expression (9.49) for the effective transport yields an approximate expression for the quasi-Stokes transport

$$\mathbf{U}^{\mathrm{qs}} \approx -\frac{\overline{\mathbf{u}'\,\rho'}}{\partial_z \widetilde{\rho}} + \frac{\overline{\phi}\,\partial_z \overline{\mathbf{u}}}{(\partial_z \widetilde{\rho})^2}, \tag{9.58}$$

where

$$2\,\overline{\phi}(z) = \overline{(\rho'(z))^2} \tag{9.59}$$

is the mean density variance, and recall that all depths are set to $z = \overline{z}^{\rho}$.

Substituting the deviation (9.57) into the approximate expression (9.54) yields, to within terms of third order, the relation

$$\widetilde{\rho}(z) \approx \overline{\rho}(z) - \partial_z \left(\frac{\overline{\phi}}{\partial_z \overline{\rho}} \right). \tag{9.60}$$

As for the quasi-Stokes transport, the modified mean density and Eulerian mean density, when evaluated at the same depth, differ by terms that are second order in eddy amplitude.

9.4 ENSEMBLE MEAN TRACER EQUATION

We now attach a tracer quantity to the ideal Boussinesq parcel and determine a mean field description for the tracer.* The transport of tracer by eddies has both a reversible stirring component and an irreversible mixing component.

9.4.1 Thickness weighted means

Equation (9.23) introduced a specific thickness weighted mean operator, which will prove to be quite useful when considering the mean tracer equation. In general, for any field Φ associated with a potential density layer ρ, define the decomposition

$$\begin{aligned} \Phi(\rho) &= \widehat{\Phi}(\rho) + \Phi''(\rho) \\ &= \frac{\overline{h\,\Phi}^{\rho}}{\overline{h}^{\rho}} + \Phi''. \end{aligned} \tag{9.61}$$

*Much in this section follows from Smith (1999) and McDougall and McIntosh (2001).

The quantity $\widehat{\Phi}$ is referred to as the mean thickness weighted field. It follows by definition that

$$\overline{h\,\Phi''}^{\rho} = 0. \tag{9.62}$$

9.4.2 Mean thickness weighted tracer equation

When attaching a tracer to fluid parcels, each member of the ensemble satisfies the isopycnal tracer equation

$$(\partial_t + \mathbf{u} \cdot \nabla_\rho)\, C = 0. \tag{9.63}$$

Combining the tracer and thickness equations leads to the thickness weighted tracer equation

$$(h\, C)_{,t} + \nabla_\rho \cdot (h\, \mathbf{u}\, C) = 0. \tag{9.64}$$

Hence, in isopycnal coordinates and in the absence of non-advective transport, the evolution of thickness weighted tracer occurs via the isopycnally oriented convergence of the two-dimensional thickness weighted horizontal advective flux, $h\,\mathbf{u}\,C$. Importantly, it is only the two-dimensional flux that appears here, whereas the analogous equation in z-coordinates involves the three-dimensional advective flux.

To address the problem of describing the ensemble mean tracer equation in isopycnal coordinates, decompose the tracer and velocity field into their thickness weighted average and deviation to give

$$[h\,(\widehat{C} + C'')]_{,t} + \nabla \cdot [h\,(\widehat{\mathbf{u}} + \mathbf{u}'')\,(\widehat{C} + C'')] = 0. \tag{9.65}$$

Taking an ensemble average over parcels with the same potential density, and using equation (9.62), yield the mean thickness weighted tracer equation

$$(\overline{h}^{\rho}\, \widehat{C})_{,t} + \nabla_\rho \cdot (\overline{h}^{\rho}\, \widehat{C}\, \widehat{\mathbf{u}}) = -\nabla_\rho \cdot (\overline{h\, C''\, \mathbf{u}''}^{\rho}). \tag{9.66}$$

Now introduce the correlation

$$\overline{h\, C''\, \mathbf{u}''}^{\rho} = \overline{h}^{\rho}\, \widehat{C''\, \mathbf{u}''} \tag{9.67}$$

(see equation (9.61)), and recall that the mean thickness $\overline{h}^{\rho}$ satisfies the mean thickness equation, $\partial_t \overline{h}^{\rho} + \nabla_\rho \cdot (\overline{h}^{\rho}\, \widehat{\mathbf{u}}) = 0$. These two points lead to the evolution equation for the mean thickness weighted tracer concentration

$$(\partial_t + \widehat{\mathbf{u}} \cdot \nabla_\rho)\, \widehat{C} = -\frac{1}{\overline{h}^{\rho}}\, \nabla_\rho \cdot (\overline{h}^{\rho}\, \widehat{C''\, \mathbf{u}''}). \tag{9.68}$$

9.4.3 Tracer mixing tensor

The correlation between tracer and velocity found on the right-hand side of the mean thickness weighted tracer equation (9.68) is typically written in terms of a tracer mixing tensor

$$\widehat{C''\, \mathbf{u}''} = -\mathbf{J} \cdot \nabla_\rho \widehat{C}. \tag{9.69}$$

This *definition* leads to the evolution equation

$$(\partial_t + \widehat{\mathbf{u}} \cdot \nabla_\rho)\, \widehat{C} = \frac{1}{\overline{h}^{\rho}}\, \nabla_\rho \cdot (\overline{h}^{\rho}\, \mathbf{J} \cdot \nabla_\rho \widehat{C}). \tag{9.70}$$

The ocean modeling community has traditionally assumed that the mixing tensor $\mathbf{J}$ is a positive-definite and symmetric *diffusion tensor*, which is also assumed isotropic in the isopycnal plane. It corresponds to the isopycnal coordinate version of the small angle diffusion tensor of Redi (1982), as first written down by Gent and McWilliams (1990). However, one can generally expect $\mathbf{J}$ to have both a symmetric and an antisymmetric component (e.g., Plumb, 1979; Plumb and Mahlman, 1987; Middleton and Loder, 1989). As in Section 9.2.2, the antisymmetric component stirs the tracer via skew diffusion. Importantly, such stirring by subgrid scale correlations appears *in addition* to that associated with the bolus transport arising from $\mathbf{u}^{\text{bolus}}$, where again $\mathbf{u}^{\text{bolus}}$ arises from correlations between velocity and thickness (equation (9.23)).

9.4.4 Mean tracer transport beneath a density surface

It is useful to further elucidate the relevance of mean thickness weighted fields. For this purpose, proceed as in Section 9.3.5.1 to consider the mean horizontal tracer transport occurring beneath a particular potential density surface $\rho = \widetilde{\rho}$,

$$\overline{\mathbf{C}}^{\#}(\overline{z}^{\rho}) = \overline{\int_{-H}^{\overline{z}^{\rho}+\xi} C\,\mathbf{u}\,\mathrm{d}z}. \tag{9.71}$$

Setting tracer concentration to unity recovers the expression (9.41) for the TRM transport. As in Section 9.3.5.1, which considered the mean horizontal transport of fluid beneath $\widetilde{\rho}$, one now has

$$\begin{aligned}\overline{\mathbf{C}}^{\#}(\overline{z}^{\rho}) &= \int_{\rho(-H)}^{\widetilde{\rho}(\overline{z}^{\rho})} \overline{C\,\mathbf{u}\,h}^{\sigma}\,\mathrm{d}\sigma \\ &= \int_{\rho(-H)}^{\widetilde{\rho}(\overline{z}^{\rho})} \overline{h}^{\sigma}\,\mathrm{d}\sigma\,(\widehat{C}\,\widehat{\mathbf{u}} + \widehat{C''\,\mathbf{u}''}) \\ &= \int_{\rho(-H)}^{\widetilde{\rho}(\overline{z}^{\rho})} \overline{h}^{\sigma}\,\mathrm{d}\sigma\,(\widehat{C}\,\widehat{\mathbf{u}} - \mathbf{J}\cdot\nabla_{\rho}\widehat{C}) \\ &= \int_{-H}^{\overline{z}^{\rho}} \mathrm{d}z\,(\widehat{C}\,\widehat{\mathbf{u}} - \mathbf{J}\cdot\nabla_{\rho}\widehat{C}).\end{aligned} \tag{9.72}$$

Hence, the mean thickness weighted fields naturally appear when considering such physically interesting quantities as the mean horizontal transport of a tracer beneath the modified mean potential density surface.

9.4.5 Summary of the mean field tracer equation

In summary, the parameterization problem for the mean thickness weighted tracer in isopycnal coordinates reduces to a parameterization of the bolus velocity $\mathbf{u}^{\text{bolus}}$, which again is related to the effective horizontal velocity field

$$\begin{aligned}\widehat{\mathbf{u}}(\rho) &= \frac{\overline{h\,\mathbf{u}}^{\rho}}{\overline{h}^{\rho}} \\ &= \overline{\mathbf{u}}^{\rho} + \frac{\overline{h'\,\mathbf{u}'}^{\rho}}{\overline{h}^{\rho}} \\ &= \overline{\mathbf{u}}^{\rho} + \mathbf{u}^{\text{bolus}}.\end{aligned} \tag{9.73}$$

It is necessary to also parameterize the tracer mixing tensor

$$\widehat{C''\,\mathbf{u}''} = -\mathbf{J}\cdot\nabla_{\rho}\widehat{C}, \tag{9.74}$$

which generally has symmetric (diffusive) and antisymmetric (stirring) components.

For a mean z-coordinate description, equation (9.35) is used to relate thickness weighted mean fields, defined as a function of ρ, and TRM fields, defined as a function of the mean depth of ρ, to write for the tracer field

$$\widehat{C}(x, y, \widetilde{\rho}, t) = \overline{C}^{\#}(x, y, \overline{z}^{\rho}, t). \tag{9.75}$$

The TRM tracer concentration $\overline{C}^{\#}$ is discretized by coarse resolution z-models. Equation (9.75), and the developed formalism, leads to the exact mean field tracer equation in z-coordinates

$$\partial_t \overline{C}^{\#} = \nabla\cdot(\nabla\overline{C}^{\#}\wedge\overline{\mathbf{\Psi}}^{\#}) + R(\overline{C}^{\#}), \tag{9.76}$$

where $R(\overline{C}^{\#})$ is the z-coordinate form of the mixing/stirring operator on the right-hand side of equation (9.70). In z-coordinates, this operator is generally comprised of the Redi (1982) isopycnal diffusion tensor, which produces a diffusive flux of $\overline{C}^{\#}$ oriented along surfaces of constant modified mean potential density $\widetilde{\rho}$, and an antisymmetric piece which has yet to be studied in the ocean modeling literature. $\overline{\mathbf{\Psi}}^{\#}$ is the TRM transport streamfunction defined by equation (9.38). Its parameterization is discussed in Section 9.5.

9.5 QUASI-STOKES TRANSPORT IN Z-MODELS

An eddy closure parameterization generally represents eddy correlation terms by expressions dependent only on the mean fields. Two correlation terms have been identified in the previous sections: the quasi-Stokes transport $\mathbf{U}^{\text{qs}}$ and the tracer mixing/stirring operator $R(\overline{C}^{\#})$. We focus in this section on the quasi-Stokes transport and its parameterization in z-models. With a z-coordinate focus, all averages refer here to Eulerian averages, and so the z label on the overbar is dropped for brevity.

As stated earlier, it is the modified mean potential density $\widetilde{\rho}$ that is carried by a coarse resolution Boussinesq z-model. Hence, the skew flux from equation (9.46)

$$\overline{\mathbf{F}}^{\#} = -\overline{\mathbf{U}}^{\#}\,\partial_z\widetilde{\rho} + \hat{\mathbf{z}}\,\overline{\mathbf{U}}^{\#}\cdot\nabla_z\widetilde{\rho} \tag{9.77}$$

must be parameterized. Again, for the isentropic ensemble with kinematic rearrangement of parcels, the horizontal TRM transport is written

$$\overline{\mathbf{U}}^{\#}(z) = \overline{\mathbf{U}}(z) + \mathbf{U}^{\text{qs}}(z), \tag{9.78}$$

with $z = \overline{z}^{\rho}$ the mean depth of the potential density, $\overline{\mathbf{U}}(z)$ the Eulerian mean horizontal transport explicitly carried by the z-model that passes beneath the depth z, and

$$\begin{aligned} \mathbf{U}^{\mathrm{qs}}(z) &= \overline{\int_{\overline{z}^{\rho}}^{\overline{z}^{\rho}+\xi} \mathbf{u}\, dz} \\ &\approx -\frac{\overline{\mathbf{u}'\,\rho'}}{\partial_z \widetilde{\rho}} + \frac{\overline{\phi}\, \partial_z \overline{\mathbf{u}}}{(\partial_z \widetilde{\rho})^2} \end{aligned} \tag{9.79}$$

is the horizontal eddy-induced, or quasi-Stokes transport which must be parameterized. The second, approximate expression is accurate to third order in eddy amplitude. Notably, one advantage of working with the skewsion or streamfunction approach is that it is only a horizontal transport $\mathbf{U}^{\mathrm{qs}}$ that must be parameterized, in contrast to the conventional Eulerian approach which aims to parameterize a three-dimensional eddy correlation $\overline{\mathbf{v}'\,\rho'}$.

9.5.1 Rotational and divergent components

The analysis so far has found the vertical gauge (Section 9.2.2) to be a natural means to represent both the resolved and unresolved potential density flux. However, as it is the convergence of the flux that determines the evolution of the modified mean potential density field $\widetilde{\rho}$, one can add various curls, or rotational terms, to the flux to put it into a more convenient form. For example, when representing the flux due to the resolved scale flow, it is possible to add the total rotational term $\nabla \wedge (\widetilde{\rho}\, \overline{\mathbf{U}} \wedge \hat{\mathbf{z}})$ in order to change the skew flux to an advection flux.

For the unresolved part of the flux, proportional to the quasi-Stokes transport $\mathbf{U}^{\mathrm{qs}}$, it is useful to remain within the skewsion framework as it is more convenient. However, there remains a gauge ambiguity that allows for the addition of an arbitrary rotational term to the horizontal skew flux $-\partial_z \widetilde{\rho}\, \mathbf{U}^{\mathrm{qs}}$. Indeed, when diagnosing eddy correlations in fine resolution models, there is generally a nontrivial rotational term (e.g., Lau and Wallace, 1979; Marshall and Shutts, 1981; Bryan et al., 1999) which is associated with the advection of potential density variance. This term is not easily parameterized, nor does it require parameterization since its convergence vanishes.

9.5.1.1 Flux divergence

Various approaches present themselves for handling the unresolved rotational term. The first approach is not to concern oneself with details of the diagnosed fluxes, but instead to focus on their convergence. This approach formed part of the analysis of Bryan et al. (1999), and it provides unambiguous information about what drives the mean flow.

9.5.1.2 Zonal symmetry and zonal means

The second approach is to work in a zonally symmetric channel and to employ zonal averages instead of, or in addition to, time and/or ensemble averages. All averaged fields then become functions of just the meridional and vertical spatial directions, and the effective transport reduces to just the meridional component.

More precisely, consider a zonally averaged form of the thickness equation using isopycnal coordinates

$$\partial_t \overline{h}^\rho + \partial_y (\overline{h}^\rho \, \widehat{v}) = 0, \tag{9.80}$$

with

$$\widehat{v} = \overline{v}^\rho + \frac{\overline{v' \, h'}^\rho}{\overline{h}^\rho} \tag{9.81}$$

the effective meridional velocity field. The $\partial_x (\overline{u \, h}^\rho)$ term present in the general three-dimensional case vanishes due to zonal symmetry and use of zonal averages. The corresponding exact z-coordinate description thence considers the two-dimensional effective skew flux

$$\overline{\mathbf{F}}^\# = \overline{V}^\# (-\hat{\mathbf{y}} \, \partial_z \widetilde{\rho} + \hat{\mathbf{z}} \, \partial_y \widetilde{\rho}) \tag{9.82}$$

whose convergence drives the evolution of the modified mean potential density field, $\widetilde{\rho}(y, z, t)$. The same approximations used to reach the approximate quasi-Stokes streamfunction (9.79) can be applied here to yield the third order accurate form for the effective transport passing beneath the averaged depth of the modified mean potential density field:

$$\int_{-H}^{\overline{z}^\rho} \widehat{v} \, \mathrm{d}z \approx \overline{V}(z) - \frac{\overline{v' \, \rho'}}{\partial_z \widetilde{\rho}} + \frac{\overline{\phi} \, \partial_z \overline{v}}{(\partial_z \widetilde{\rho})^2}. \tag{9.83}$$

The only ambiguity that remains in the meridional skew flux $-\overline{V}^\# \, \partial_z \widetilde{\rho}$ arises from the ability to add an arbitrary function of zonal position. Yet since all averaged fields are only functions of meridional and depth positions, this ambiguity is not a problem. The zonally symmetric case is therefore quite simple to consider.

9.5.1.3 *The rotational component*

The third approach is to actually take the effort to compute the rotational component. Doing so requires the specification of boundary conditions, and their form is not obvious, except for the case of periodicity in which both rotational and divergent components are periodic.

To expose the main issue, consider the case when $-\mathbf{U}^{\mathrm{qs}} \, \partial_z \widetilde{\rho}$ is diagnosed. For brevity, at each depth level, write

$$\mathbf{A}(x, y, t) = -\mathbf{U}^{\mathrm{qs}} \, \partial_z \widetilde{\rho}. \tag{9.84}$$

There is a self-consistent method for splitting $\mathbf{A}$ into its rotational and divergent components, and this method is standard in the fluid mechanics literature. For example, see Section 1.1 of Saffman (1992) and Section 17.2 of Panton (1996). Since the vector $\mathbf{A}$ is two-dimensional, the decomposition generally takes the form

$$\begin{aligned} \mathbf{A} &= \mathbf{A}_{\mathrm{div}} + \mathbf{A}_{\mathrm{rot}} \\ &= \nabla \Phi + \hat{\mathbf{z}} \wedge \nabla \chi, \end{aligned} \tag{9.85}$$

where the scalar fields Φ and χ satisfy

$$\nabla^2 \Phi = \nabla \cdot \mathbf{A} \tag{9.86}$$

$$\nabla^2 \chi = \hat{\mathbf{z}} \cdot \nabla \wedge \mathbf{A}. \tag{9.87}$$

Because of the no-normal flow boundary condition on vertical side walls

$$\mathbf{A} \cdot \hat{\mathbf{n}} = 0, \tag{9.88}$$

where $\hat{\mathbf{n}}$ is the outward normal at the boundary. When decomposing $\mathbf{A}$ into its rotational and divergent components, the boundary condition becomes

$$\hat{\mathbf{n}} \cdot \nabla \Phi + \hat{\mathbf{n}} \cdot (\hat{\mathbf{z}} \wedge \nabla \chi) = 0. \tag{9.89}$$

Notably, each component in general does *not* separately satisfy the no-normal flow condition.

To proceed, the scalar field χ is assumed to maintain regularity at infinity, periodicity on the sphere, but with no specified boundary conditions on the side walls. Upon solving for χ via a numerical elliptic solver or a Green's function approach, its value is known everywhere. In particular, it is known on the side boundaries, which in turn allows for the scalar field Ψ to be found by solving the boundary value problem

$$\nabla^2 \Psi = \nabla \cdot \mathbf{A} \tag{9.90}$$

$$\hat{\mathbf{n}} \cdot \nabla \Psi = -\hat{\mathbf{n}} \cdot (\hat{\mathbf{z}} \wedge \nabla \chi), \tag{9.91}$$

again using a numerical elliptic solver or a Green's function approach. Thus specifies the divergent and rotational components.

Although the above algorithm is self-consistent, it is not unique. Fox-Kemper et al. (2003) provide a summary of this issue. They suggest some integral diagnostics which are unique and so of use for the eddy-closure problem.

9.5.2 Skewsion according to Gent and McWilliams

A popular parameterization of the quasi-Stokes streamfunction is that proposed by Gent and McWilliams (1990) and Gent et al. (1995) (commonly referred to as "GM"). We discuss this parameterization more thoroughly in Section 14.2.1 (page 305). For now, note that this scheme closes the divergent part of the quasi-Stokes transport by setting

$$\mathbf{U}^{\mathrm{qs}} = -\kappa \mathbf{S}. \tag{9.92}$$

In this expression, $\mathbf{S}$ is the slope of the modified mean potential density surfaces (equation (9.47)), and $\kappa > 0$ is a diffusivity. The corresponding three-dimensional nondivergent velocity is given by

$$\mathbf{v}^* = -\partial_z (\kappa \mathbf{S}) + \hat{\mathbf{z}} \, \nabla_z \cdot (\kappa \mathbf{S}), \tag{9.93}$$

and the antisymmetric stirring tensor (Griffies, 1998) is

$$A^{mn} = \begin{pmatrix} 0 & 0 & -\kappa S_x \\ 0 & 0 & -\kappa S_y \\ \kappa S_x & \kappa S_y & 0 \end{pmatrix}. \tag{9.94}$$

The parameterized skew flux of modified mean potential density due to the quasi-Stokes transport is given by

$$\begin{aligned}\overline{\mathbf{F}}^{\text{qs}} &= -\mathbf{U}^{\text{qs}}\,\partial_z\widetilde{\rho} + \hat{\mathbf{z}}\,\mathbf{U}^{\text{qs}}\cdot\nabla_z\widetilde{\rho} \\ &= -\kappa\left(\nabla_z\widetilde{\rho} - \hat{\mathbf{z}}\,S^2\,\partial_z\widetilde{\rho}\right).\end{aligned} \tag{9.95}$$

This parameterization yields horizontal downgradient diffusion of modified mean potential density, combined with a vertical upgradient diffusion. Additionally, Gent et al. (1995) prescribe a diffusivity that vanishes on all boundaries, including the ocean surface. McIntosh and McDougall (1996) and McDougall and McIntosh (2001) present more discussion of vertical boundary conditions, which can be understood by considering the exact form of the quasi-Stokes transport defined by equation (9.44).

To see what the Gent et al. (1995) parameterization corresponds to in terms of subgrid scale correlations, expand the quasi-Stokes transport to second order (equation (9.79)) and thus equate to the parameterization (9.92)

$$\left(-\overline{\mathbf{u}'\,\rho'} + \frac{\overline{\phi}\,\partial_z\overline{\mathbf{u}}}{\partial_z\widetilde{\rho}}\right)_{\text{div}} \approx \kappa\,\nabla_z\widetilde{\rho}, \tag{9.96}$$

where only the divergent part of the eddy correlation is considered. The importance of the $\overline{\mathbf{u}}_z\overline{\phi}/\partial_z\widetilde{\rho}$ term has not been examined in the literature. For example, Treguier et al. (1997) and Roberts and Marshall (2000) ignore this term, thence equating the Gent et al. (1995) parameterization with the traditional horizontal downgradient diffusion of potential density $(\overline{\mathbf{u}'\,\rho'})_{\text{div}} = -\kappa\,\nabla_z\widetilde{\rho}$. Again, this is *not correct* when the variance term $\overline{\mathbf{u}}_z\overline{\phi}/\partial_z\widetilde{\rho}$ is important. Careful examination of eddying z-models is necessary to clarify the importance of the variance term.

9.5.3 Effective meridional volume transport

It is often of interest to compute the net transport of volume across a portion of the ocean. In particular, meridional-overturning streamfunctions allow one to visualize and quantify the zonally averaged circulation occurring in a closed basin or over the full globe. The quasi-Stokes transport provides a transport in addition to that from the resolved scale Eulerian mean transport. The parameterization of Gent et al. (1995) leads to a straightforward computation of the quasi-Stokes contribution. To see this, write the net meridional transport across a basin at a particular depth in the form

$$\begin{aligned}\mathcal{T}(y,z,t) &= \int \mathrm{d}x\,\overline{V}^{\#}(y,z,t) \\ &= \int \mathrm{d}x\,(\overline{V} - \kappa\,S_y).\end{aligned} \tag{9.97}$$

The parameterized quasi-Stokes transport adds a contribution that scales linearly with basin size, isopyncal slope, and diffusivity,

$$\mathcal{T}^{\text{qs}} \sim L\,S\,\kappa. \tag{9.98}$$

As an example, let $\kappa = 10^3\,\text{m}^2\,\text{s}^{-1}$, $S = 10^{-3}$, and $L = 10^7$ m, which yields $\mathcal{T} \approx 10 \times 10^6\,\text{m}^3\,\text{s}^{-1} = 10\,\text{Sv}$. Such transport can represent a nontrivial addition to that from the resolved scale velocity field.

9.5.4 The relation between isopycnal thickness diffusion and GM

Recall the ensemble mean thickness equation (9.22) derived in Section 9.3.2

$$\partial_t \overline{h}^{\rho} + \nabla_{\rho} \cdot (\overline{h}^{\rho}\, \widehat{\mathbf{u}}) = 0, \tag{9.99}$$

where $\widehat{\mathbf{u}} = \overline{\mathbf{u}}^{\rho} + \mathbf{u}^{\text{bolus}}$ provides an effective transport velocity for the ensemble mean thickness $\overline{h}^{\rho}$. Isopycnal correlations of horizontal velocity and thickness define the bolus velocity via $\overline{h}^{\rho}\, \mathbf{u}^{\text{bolus}} = \overline{h'\, \mathbf{u}'}^{\rho}$. Now consider a downgradient diffusive closure for this correlation

$$\begin{aligned} \overline{h}^{\rho}\, \mathbf{u}^{\text{bolus}} &= \overline{h'\, \mathbf{u}'}^{\rho} \\ &= -\mathbf{K} \cdot \nabla_{\rho} \overline{h}^{\rho} \end{aligned} \tag{9.100}$$

with $\mathbf{K}$ a symmetric and positive-definite 2×2 diffusion tensor. The mean thickness equation thus takes the form of an advection-diffusion equation

$$\partial_t \overline{h}^{\rho} + \nabla_{\rho} \cdot (\overline{h}^{\rho} \overline{\mathbf{u}}^{\rho}) = \nabla_{\rho} \cdot (\mathbf{K} \cdot \nabla_{\rho} \overline{h}^{\rho}). \tag{9.101}$$

To make a connection between the thickness diffusion closure (9.100) and the Gent and McWilliams (1990) closure discussed in Section 9.5.2, note that the mean thickness is related to the vertical gradient of the modified mean density

$$\overline{h}^{\rho} = (\partial_z \widetilde{\rho})^{-1}, \tag{9.102}$$

where the right-hand side is evaluated at a depth given by the mean of the isopycnal $z = \overline{z}^{\rho}$. Correspondingly, using $\nabla_{\rho} = \nabla_z + \mathbf{S}\, \partial_z$, where again $z = \overline{z}^{\rho}$ and $\mathbf{S}$ is the slope of the modified mean density field $\widetilde{\rho}$ (see equation (9.47)), gives

$$\begin{aligned} \nabla_{\rho} \ln \overline{h}^{\rho} &= -\overline{h}^{\rho}\, \nabla_{\rho}\, (\overline{h}^{\rho})^{-1} \\ &= -(\partial_z \widetilde{\rho})^{-1} (\nabla_z + \mathbf{S}\, \partial_z)\, \partial_z \widetilde{\rho} \\ &= -\frac{\partial_z\, (\nabla_z \widetilde{\rho})}{\partial_z \widetilde{\rho}} + \frac{\partial_{zz} \widetilde{\rho}\, \nabla_z \widetilde{\rho}}{(\partial_z \widetilde{\rho})^2} \\ &= \partial_z \mathbf{S}. \end{aligned} \tag{9.103}$$

Consequently, the bolus velocity takes the form

$$\begin{aligned} \mathbf{u}^{\text{bolus}} &= -\mathbf{K} \cdot \nabla_{\rho} \ln \overline{h}^{\rho} \\ &= -\mathbf{K} \cdot \partial_z \mathbf{S}. \end{aligned} \tag{9.104}$$

For the special case where $\mathbf{K}$ is independent of depth and proportional to the 2×2 identity matrix, then

$$\begin{aligned} \mathbf{u}^{\text{bolus}} &= -\partial_z\, (\kappa\, \mathbf{S}) \\ &= \mathbf{u}^{*}, \end{aligned} \tag{9.105}$$

where the horizontal component of the Gent and McWilliams (1990) velocity $\mathbf{u}^{*}$ was identified from equation (9.93). Again, this identity holds only for the special case of a vertically independent diffusivity tensor proportional to the identity. The relevance of a depth-independent diffusivity has been questioned by Killworth (1997) and Smith and Vallis (2002). For these cases, where one places the vertical derivative is crucial.

There has been no identification of the vertical velocity component $w^* = \nabla_z \cdot (\kappa \mathbf{S})$ from Gent and McWilliams (1990) with the vertical component of the bolus velocity. Indeed, there is no need to explicitly compute the vertical component of the bolus velocity since the thickness equation (9.99) is only concerned with the horizontal transport. However, for z-models, the full three-dimensional velocity is needed. Details of these points are provided by McDougall and McIntosh (2001).

Besides differences when the vertical diffusivity has depth dependence, identification of thickness diffusion with the Gent and McWilliams (1990) parameterization breaks down near boundaries. Indeed, this breakdown largely arises since the GM diffusivity cannot be considered depth-independent next to boundaries. Additionally, thickness diffusion next to solid earth boundaries leads to an increase in potential energy, with isopycnals creeping up the topographic slope. Such unphysical behavior motivates isopycnal modelers instead to use *interfacial height* diffusion to dissipate noise in the thickness field.* Given these nontrivial caveats, one often reads in the literature that the parameterization of Gent and McWilliams (1990) is the same as thickness diffusion. It is therefore pedagogically useful to see from what context such statements are motivated.

9.6 CHAPTER SUMMARY

At the start of this chapter, we introduced the mesoscale eddy parameterization problem and argued for the relevance of considering the adiabatic effects of these eddies distinct from other diabatic processes. This assumption is not valid everywhere, especially near boundaries. However, it does provide a useful theoretical starting point, and it serves the present pedagogical purposes. After this introductory discussion, Section 9.2 considered two complementary methods for describing the transport of a tracer via a nondivergent velocity field. The advective description is more common, but the skew flux description has some properties of use for subsequent material. As shown in Part 4 of this book, the skew approach also provides a useful numerical framework to implement various eddy closure schemes.

Section 9.3 entered into the main part of this chapter by considering the dual descriptions of volume conservation for an isentropic fluid parcel as provided by isopyncal and depth coordinates. The description provided for a single ensemble member was compared to that for an ensemble averaged flow. The isopycnal coordinate description is more natural, since the isentropic ensemble members each have the same potential density. However, an exact z-coordinate description can be found by introducing smarter methods of averaging. It is from this discussion that the two-dimensional *bolus velocity* was introduced, which arises from correlations between fluctuations in the horizontal velocity and specific thickness on an isopycnal. The transformed residual mean was also introduced in Section 9.3.4, and this mean provides the appropriate mean field in a z-coordinate description that is mindful of the isentropic nature of the ensemble. Section 9.4 extended the discussion of volume conservation to tracer conservation. This discussion introduced a mixing tensor proportional to correlations between tracer and velocity.

*This point was raised by Holloway (1997) and Griffies et al. (2000a).

This tensor generally contains symmetric and antisymmetric components, with the symmetric piece commonly implemented in coarse resolution ocean models. The antisymmetric piece is generally ignored.

We finished the chapter in Section 9.5 with a discussion of how to parameterize quasi-Stokes transport in z-models. This transport, identified in Section 9.3, represents the difference between the Eulerian mean and Lagrangian mean transport associated with isopycnals undulating in the presence of a mesoscale eddy field. Exact expressions for this transport are generally approximated in z-models, and the approximate forms then parameterized. A popular manner to parameterize this transport is via the scheme of Gent and McWilliams (1990) and Gent et al. (1995). We have more to say about how this scheme is implemented in ocean models in Part 4 of this book.

This was a difficult chapter to write. It likewise may have been difficult to digest, especially on first exposure. However, the topics discussed here are fundamental in many ways, and so the payoff to the persistent reader will be good. First, they address important questions about how to transfer isopycnal based Lagrangian kinematics, and thinking, to Eulerian z-coordinates kinematics. This provides a necessary step in attempts to analyze mesoscale eddying simulations using z-models, and to translate isopycnal based adiabatic eddy closures into z-coordinates. Second, they introduce the reader to issues which go beyond pragmatic questions of z-model analysis and interpretation. These more fundamental questions regard the nontrivial, and often very confusing topics of how to describe, understand, and parameterize mesoscale eddies as they encounter solid earth and upper mixed layer boundaries, and interact with physical processes active in these regions. This chapter did little to shed light on boundary issues. Nonetheless, much of the technology employed for these discussions follows from that given here.

PART 3
Semi-discrete equations and algorithms

The purpose of this part of the book is to cast the continuum mean field ocean equations onto a discrete lattice and then to present various solution algorithms for time stepping these equations on a computer. What lattice should be chosen? How should one discretize the various terms appearing in the equations, in both space and time? What vertical coordinate is best? How does one ensure that the numerical fluid looks like the ocean? These questions are fundamental to ocean models. There are few unambiguous answers.

In this part of the book, we use a fixed Eulerian lattice appropriate for a z-model. Making this choice introduces many egregious problems. The next few years will undoubtedly see research into alternatives, and promising ones exist today (see Chapter 6). Nonetheless, z-models are the most mature and popular choice for those aiming to simulate the global ocean climate. This in itself warrants a careful rationalization of their basic equations. Additionally, from a pedagogical perspective, z-model algorithms expose many generic issues, while at the same time describing elements of present-day working ocean climate models, such as the Modular Ocean Model of Griffies et al. (2004). That is, although focusing on z-model equations, many of the issues that must be addressed have their generalizations to other vertical coordinates. Hence, the discussion is in many ways much more general than details of the z-model equations may suggest.

This part of the book aims to perform the first steps toward obtaining a fully discretized set of equations appropriate for a z-model. The method used is to integrate the continuum equations over a discrete model grid cell. Formally, this approach is termed a *finite volume* approach. This approach is an effective and rational means to derive the semi-discrete finite differenced equations, including important points about boundary conditions.

After a short chapter introducing discrete methods, the remainder of this part focuses on the semi-discrete equations for mass, tracers, and momentum. Upon establishing the equations, various solution algorithms are presented along with some of their numerical characteristics.

Chapter Ten

DISCRETIZATION BASICS

The purpose of this chapter is to introduce some of the basic notions of discrete partial differential equations of use for solving the ocean equations on a lattice. This treatment is rather terse, with more complete treatments provided by one or more of the texts discussed in the Preface or references given in this chapter.

10.1 DISCRETIZATION METHODS

So far in this book, we have developed the continuum equations for the unaveraged fluid state (Part 1), and the continuum equations for various averaged states (Part 2). We now consider how to discretize the continuous model equations onto a discrete z-coordinate lattice. This book is short on pedagogical discussions of general discretization techniques, since the computational fluids literature has many examples (e.g., Haltiner and Williams, 1980; Durran, 1999). However, for completeness, it is useful to introduce here some of the issues, reserving detail for later chapters.

Discretization of continuous equations often starts by integrating the equations over the finite space-time lattice onto which the equations are to be numerically integrated. For example, introduce a space-time averaged field via the averaging operator

$$\overline{\psi}^{\mathbf{x},t} = \frac{1}{\Delta V}\frac{1}{\Delta t}\int_{-\Delta V/2}^{\Delta V/2} \mathrm{d}V \int_{-\Delta t/2}^{\Delta t/2} \mathrm{d}t\, \psi(x,y,z,t), \tag{10.1}$$

where $\Delta V = \Delta x\,\Delta y\,\Delta z$ is the volume of the lattice grid cell, and $\mathrm{d}\mathbf{x} = \mathrm{d}x\,\mathrm{d}y\,\mathrm{d}z$ is the infinitesimal volume element. If the cell volume is constant in space and time, then such averaging leads to the discrete expression for a time derivative operator

$$\overline{\psi_{,t}}^{\mathbf{x},t} = \frac{\overline{\psi}^{\mathbf{x}}(t+\Delta t/2) - \overline{\psi}^{\mathbf{x}}(t-\Delta t/2)}{\Delta t}, \tag{10.2}$$

and similar expressions for spatial derivatives. Other averaging operators exist whereby a kernel is used to weight the averaged field, with the present example employing a "box-car" kernel. In either approach, the numerical model fields are interpreted as discrete representations of continuous averages of the continuous fields. This is the *finite volume* approach. Another approach interprets the numerical model fields as pointwise discrete representations of the continuous fields, thus prompting Taylor series methods for approximating derivative operators.

Beside the finite difference or finite volume approaches, there are many other methods in the computational fluid dynamics literature that may be suitable for ocean climate models. One approach expands the continuous fluid in terms of basis functions evaluated at discrete points of a lattice. Such *spectral* methods are popular in atmospheric models, yet have not found widespread use in ocean climate models, largely due to the increased complexity of the ocean domain relative to the atmosphere. Another approach introduces an unstructured lattice of grid points with the number of nearest neighbors to a grid cell not necessarily known *a priori*. These methods, common in the engineering literature, allow for great flexibility in the design of the computational grid for purposes of representing flow in and around complex geometry. As with spectral approaches, the unstructured approaches have yet to become commonplace in ocean climate modeling. Notably, these methods appear to have a difficult time solving both the rotating and stratified problem in an economical and numerically stable manner. The reader is referred to the text by Haidvogel and Beckmann (1999) for a more complete survey of the various numerical methods used in ocean models.

Whatever method is used, discretization breaks many of the symmetries maintained by the continuous partial differential equations. Consequently, the discrete equations can have solutions whose qualitative and/or quantitative behavior is not reflective of the continuous equations. It is the task of the numerical algorithm developer to ensure that any misrepresentation of the continuous system is of negligible physical consequence. Otherwise, the numerical model is of little use for physical modeling.

10.2 AN INTRODUCTION TO ARAKAWA GRIDS

A first course in finite difference methods for partial differential equations typically examines such equations as the Poisson equation, the heat equation, and the wave equation. Each is relevant for the ocean equations. One element that is quite distinct, however, is that the ocean equations contain more than one prognostic variable. In particular, the Boussinesq system contains two horizontal velocity components, a multitude of tracers, and the free surface height. Vertical velocity, density, and pressure are diagnosed. The non-Boussinesq system in z-models also has a prognostic equation for density via the mass continuity equation.*

Given the many variables to be time stepped or diagnosed, one may ask how to arrange the fields spatially. This question was asked early in the development of numerical weather forecasting, for both the vertical and horizontal grids. There is more than one answer, with the development given by Arakawa (1966) and Arakawa and Lamb (1977) forming the basis for horizontal finite difference grids in use today for the ocean climate modeling. Factors considered when determining these grids concern the fidelity by which linear geophysical fluid waves are represented. For example, the inertia-gravity and Rossby waves, each critical for transmitting information in the ocean, have different finite difference representations depending on the grid choice.

Arakawa and co-workers characterized a set of horizontal grids, commonly known as *Arakawa grids*. The staggered Arakawa B- and C-grids shown in Fig-

*As discussed in Section 11.3.4, this equation is actually used in a diagnostic manner in z-models.

ure 10.1 are most popular in ocean climate models.* The key difference between these two grids arises from the placement of the velocity field components relative to the tracer or density points. For the B-grid, each velocity component lives at the same position, thus simplifying the computation of the Coriolis force. For the C-grid, the velocity components are placed on their corresponding side of the tracer grid. This arrangement simplifies the representation of gravity waves,† as well as transport operators such as advection and friction (because the transport velocity is identical to the prognostic velocity). However, treatment of the Coriolis force is complicated on the C-grid, with common discretizations introducing a computational null mode.

The B- and C-grids have somewhat complementary properties, with some arguing for the relevance of the C-grid as the model grid resolution is refined, and the B-grid for coarser resolutions.‡ The resolutions available for most global models were quite coarse until the late 1980's. Hence, it made sense for Bryan (1969) to choose the B-grid. All models following in the Bryan-Semtner-Cox lineage, even those run at refined resolution, continue to use the B-grid (e.g., Bryan, 1969; Semtner, 1974; Cox, 1984; Pacanowski et al., 1991; Dukowicz and Smith, 1994; Pacanowski, 1995; Webb et al., 1998a; Pacanowski and Griffies, 1999; Griffies et al., 2004). It is for this reason that we employ the B-grid in our discussion of the semi-discrete primitive equations in Chapter 12.

Whether the B- or C-grid is preferable at higher resolutions remains an open question for ocean climate simulations. Likewise, there are choices to be made regarding the placement of variables on the vertical grid. What is the best arrangement? This also remains largely an open question for realistic global ocean climate simulations. What would be valuable is a realistic global ocean climate study using the same model code and identical forcing and land-sea mask, separately testing algorithms based on the B- and C-grids, and various alternative vertical grid placements. One difficulty in realizing this study is that most existing ocean climate models are written with just one grid choice. Hence, they require a significant amount of new code to enable both B- and C-grid capabilities, not to mention variations in vertical grids. Nonetheless, this is a desirable goal for those building new models, and it would greatly facilitate a clean systematic study of these important issues.

10.3 TIME STEPPING

Just as one must consider where to place fields discretely in space, it is also necessary to consider what time to evaluate the fields. For example, consider the equation $\phi_{,t} = F$. In the continuum, both sides are thought of as situated at the same time. However, when discretizing time, a finite difference approximation of $\phi_{,t}$ requires more than one discrete time value on the left-hand side. Where to evaluate the forcing function F to improve accuracy and numerical stability depends on details of the time stepping scheme. Additionally, as with spatial grid staggering,

---

*Griffies et al. (2000a) tabulates the grids used in many ocean climate models.

†For example, there are no null modes for gravity waves on the C-grid, whereas there are on the B-grid (see Section 18.1.4, page 413).

‡Coarseness in this regard refers to the grid scale relative to a baroclinic Rossby radius.

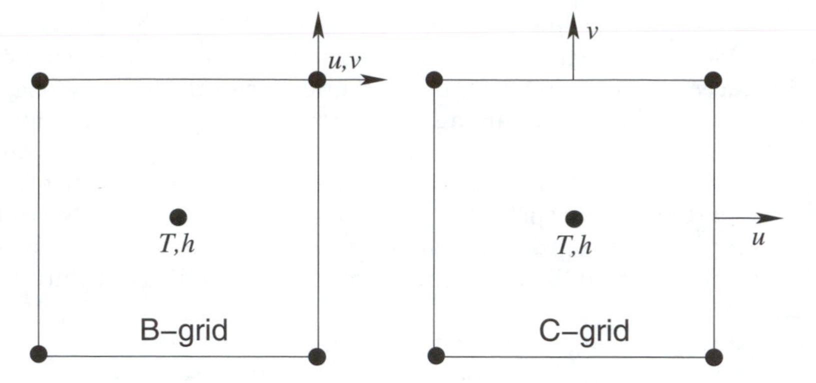

Figure 10.1 Schematic of the placement of model variables on the staggered horizontal Arakawa B- and C-grids. These are the two grids most commonly used in ocean climate models. T refers to tracer and density, u refers to the zonal velocity component, v refers to the meridional velocity componet, and h refers to layer thickness (as appears in isopycnal models) as well as the free surface height η.

it is possible to consider temporal grid staggering, where the prognostic variables are not all co-located in time (see Section 12.6).

In general, there are many time stepping schemes used in ocean climate models (Griffies et al., 2000a), with some models mixing schemes depending on what part of the equations is being considered. For example, the inviscid dynamics is often time stepped using the *leap-frog* scheme, whereby the time tendency $\phi_{,t}$ is approximated as $2\Delta t\, \phi_{,t} \approx \phi(t+\Delta t) - \phi(t-\Delta t)$ and the inviscid forcing terms are evaluated at the intermediate time t. This approach is accurate to second order in the discrete time step Δt. We have much more to say about time stepping schemes in Chapter 12, where various methods are presented to time step the ocean primitive equations.

The dissipative parts of the ocean equations, such as friction and diffusion, are unstable using a leap-frog scheme (e.g., Haltiner and Williams, 1980). Consequently, alternatives must be considered, with a *two-time level* scheme common. For the case of leap-frog inviscid dynamics, dissipative forcing terms are evaluated at the lagged time step $t - \Delta t$ rather than the intermediate step t.

These basic notions of time stepping are fundamental to many ocean models in the Bryan-Semtner-Cox lineage. Although it is still relatively popular, there are well known problems with the leap-frog approach to the inviscid dynamics which necessitate the introduction of time filters to remove a spurious mode that can cause numerical instability. We discuss one such filter in Section 12.4. Other time stepping schemes are becoming popular with newer ocean models. The book by Durran (1999) discusses many choices, and we mention one in Section 12.6.

10.4 CHAPTER SUMMARY

It is hoped that this chapter suffices to whet the appetite of the interested numericist about the various issues arising when developing a numerical algorithm for an ocean model. There is much more to come, with Chapters 11 and 12 following in this part of the book, and Chapters 16 and 19 documenting methods to discretize elliptic transport operators.

Our review of common practices highlighted the dominance of finite difference methods over the alternatives. Additionally, in ocean modeling, most horizontal spatial algorithms are based on either the Arakawa B- or C-grid, with the essential difference being the placement of the velocity components (Figure 10.1). Similar issues arise on the vertical grid, and no discussion is given here of those important questions. Spatial grid placement is paralleled by questions of where in time to evaluate fields. The common leap-frog and forward time stepping schemes were mentioned, with these and alternative schemes further considered in Chapters 11 and 12.

Much work in ocean model development consists of improving the flexibility of the models so that they can readily implement various space and time schemes. This is quite valuable since there are currently many viable schemes "on the market." A flexible model will therefore be able to exploit the scheme most suited for the application.

Chapter Eleven

MASS AND TRACER BUDGETS

The purpose of this chapter is to discuss the semi-discrete mass and tracer budgets appropriate for z-models using a free surface. Particular attention is paid to details of the surface grid cell, where the undulating free surface allows the volume of these cells to fluctuate in time. The reader of this chapter will learn some rudiments of how to take the continuum model equations and cast them into a form nearly ready for discretization in a numerical model. Additionally, this discussion details many elements of the Modular Ocean Model of Griffies et al. (2004).

11.1 SUMMARY OF THE CONTINUOUS MODEL EQUATIONS

The equations considered in this chapter represent discrete realizations of the continuum model equations for conservation of tracer and mass presented in Section 8.8 (page 185). The mass and tracer budgets are summarized here for completeness. The continuous non-Boussinesq mass and tracer equations are

$$\rho_{,t} + \rho_o \nabla \cdot \mathbf{v} = 0 \tag{11.1}$$

$$(D\,\overline{\rho}^{z})_{,t} = -\rho_o \nabla \cdot \mathbf{U} + \rho_o\, q_{\mathrm{w}} \tag{11.2}$$

$$(\rho\, C)_{,t} + \rho_o \nabla \cdot (\mathbf{v}\, C) = -\rho_o \nabla \cdot \mathbf{F} + \rho\, \mathcal{S}^{(C)}. \tag{11.3}$$

Note that mass sources can be present, but are omitted here for brevity. Finite domains have the associated surface and bottom kinematic boundary conditions

$$\hat{\mathbf{N}} \cdot \mathbf{v} = (\rho/\rho_o)\, \eta_{,t} - q_{\mathrm{w}} \qquad \text{at } z = \eta \tag{11.4}$$

$$\hat{\mathbf{N}} \cdot \mathbf{v} = 0 \qquad \text{at } z = -H, \tag{11.5}$$

with $\hat{\mathbf{N}} = -\nabla\,\eta + \hat{\mathbf{z}}$ the orientation direction at the ocean surface $z = \eta$, and $\hat{\mathbf{N}} = \nabla\, H + \hat{\mathbf{z}}$ the orientation direction at the solid earth bottom $z = -H(x, y)$. Additionally,

$$\rho\, \mathbf{v}^{\rho} = \rho_o\, \mathbf{v} \tag{11.6}$$

is the linear momentum per volume, and

$$D = H + \eta \tag{11.7}$$

is the thickness of a column of fluid extending from the surface to the solid earth bottom boundary. The Boussinesq equations are recovered by setting $\rho \rightarrow \rho_o$,

$$\nabla \cdot \mathbf{v} = 0 \tag{11.8}$$

$$\eta_{,t} = -\nabla \cdot \mathbf{U} + q_{\mathrm{w}} \tag{11.9}$$

$$C_{,t} + \nabla \cdot (\mathbf{v}\, C) = -\nabla \cdot \mathbf{F} + \mathcal{S}^{(C)} \tag{11.10}$$

with surface and bottom kinematic boundary conditions

$$\hat{\mathbf{N}} \cdot \mathbf{v} = \eta_{,t} - q_{\mathrm{w}} \qquad \text{at } z = \eta \tag{11.11}$$

$$\hat{\mathbf{N}} \cdot \mathbf{v} = 0 \qquad \text{at } z = -H. \tag{11.12}$$

Interpretation of the model fields as Eulerian ensemble means is given by Table 8.1. For models not resolving the mesoscale, the subgrid scale closure flux **F**, the model fields, and the model depth are interpreted according to the isentropic ensemble analysis given in Chapter 9.

11.2 TRACER AND MASS/VOLUME COMPATIBILITY

The tracer equation (11.3) represents a combination of mass and tracer conservation. In the special case where the tracer is set to a uniform constant and sources are set to zero, the tracer equation reduces to the mass continuity equation (11.1). Similarly, for a Boussinesq fluid, the tracer equation (11.10) reduces to volume continuity (11.8). Such compatibility is also maintained over vertical columns of fluid, where the column integrated mass conservation equation (11.2) is compatible with the column integrated non-Boussinesq tracer budget. Likewise, for a Boussinesq fluid, the column integrated volume conservation equation (11.9) is compatible with the column integrated Boussinesq tracer budget.

The importance of maintaining compatibility conditions in a numerical algorithm was emphasized by Griffies et al. (2001) when describing a free surface method for a Boussinesq fluid in a z-model. Time-dependent thicknesses for the surface cells called for extra care in the discretization relative to the rigid lid methods of Bryan (1969). Compatibility between the tracer and mass/volume budgets means that, for example, an initially uniform tracer concentration will not change in the absence of sources, even if the surface height changes. This compatibility condition is referred to as a *local conservation* property. It is distinguished from global conservation, which provides for constancy of the globally integrated tracer content in the absence of external sources.

Compatibility between tracer and mass/volume budgets means that the time stepping scheme used for the tracer and surface height must be the same. In the case of the Modular Ocean Model (Griffies et al., 2004), this means that either a leap-frog or two-time level forward scheme for the discrete tracer equation leads to the same scheme for the discrete surface height equation (Sections 12.4 and 12.6). It also means that mass continuity must be updated using the same scheme, as discussed in Section 11.3.

11.3 MASS BUDGET FOR A GRID CELL

To start the development, we vertically integrate volume and mass conservation over the thickness of a grid cell. The same approach is used in Section 11.5 for developing the discrete tracer budget and Section 12.2 for the discrete momentum budget. Mass conservation, although involving a time tendency, is implemented in a diagnostic manner to determine vertical advective velocities. For the Boussinesq fluid, similar diagnostic relations determine the vertical velocity, although

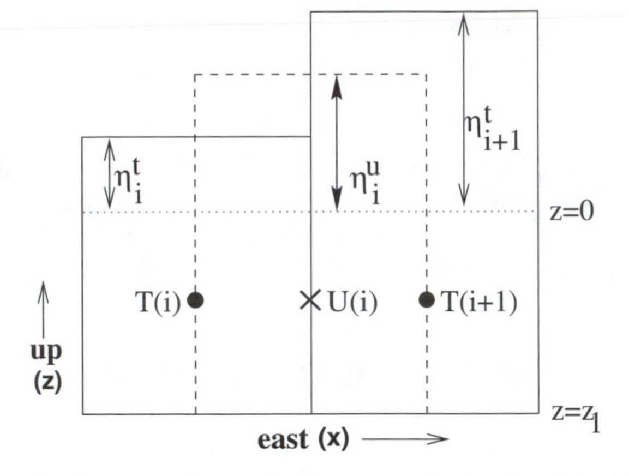

Figure 11.1 Schematic of rectangular surface tracer and velocity cells with a free surface z-model. This figure is modified slightly from the free surface paper of Griffies et al. (2001). Tracer points (T-points) are denoted by a solid dot and tracer cells are enclosed by solid lines; a horizontal velocity point (U-point) is denoted by an "x" and is enclosed by dashed lines. For a z-model, these points have fixed vertical position $z = z_1/2 < 0$, *regardless* of the value for the free surface height. Thickness of a tracer cell is $h^t = -z_1 + \eta^t$, where η^t is the prognostic surface height determined through volume conservation. Thickness of a velocity cell is $h^u = -z_1 + \eta^u$, where η^u is set to the minimum of the surrounding η^t values. Cell thicknesses are kept positive (i.e., no vanishing levels). In models without an explicit representation of tides, this constraint in practice means that $|z_1|$ must be greater than roughly 2 m. For models with tides, $|z_1|$ may need to be somewhat larger, depending on the tidal ranges considered. Additionally, when coupling to sea ice, the weight of the sea ice will depress the ocean surface by an amount roughly equivalent to the ice thickness. This process places another constraint on the thickness of the top model grid cell, or on the thickness of the sea ice. In either of these cases (tides or sea ice), limitations of the model cell thickness place an awkward constraint on the model's ability to run with fine vertical resolution. This represents a fundamental limitation of the z-models (see Section 11.8 for a discussion of "z-like" coordinates that overcome this limitation).

the density time tendency term is absent. Focusing on the non-Boussinesq mass and tracer budgets allows for the Boussinesq budget to be trivially recovered by setting density factors to the constant reference value ρ_o.

11.3.1 Notation for the vertical grid cells

In this chapter and in Chapter 12, we refer frequently to the schematic of discrete surface grid cells given in Figure 11.1. When considering momentum in Chapter 12, velocity variables are oriented horizontally relative to the tracer variables according to the B-grid shown in Figure 10.1 (page 220). In the vertical, the discrete tracer and horizontal velocity fields live at the same depth z_k, with the discrete integer label $k \in [1, N_k]$. The label $k = 0$ is used to denote the time-dependent ocean free surface $z_{k=0} = \eta$. All grid cells with $k > 1$ have volumes independent of time. Since the top cell with $k = 1$ is bounded by the free surface, its volume is time dependent. Hence, special care must be exercised in the following when discretizing this cell.

11.3.2 Diagnosing vertical velocity components

For Boussinesq fluids, the vertical velocity component is diagnosed from vertically integrating the nondivergence condition $\partial_z w = -\nabla \cdot \mathbf{u}$. As the continuity equation is first order, it requires a single boundary condition for its integration. There are two equivalent approaches: integrating from the top downward to the level of interest, and integrating from the bottom upward. Integration downward has been the common approach in many B-grid z-models. The reason is that it provides a useful diagnostic whereby the diagnosed bottom velocity on a tracer cell must vanish (for details, see Pacanowski and Griffies, 1999). Nonzero values help identify coding errors.

To diagnose the vertical velocity component in a mass conserving ocean model, it is necessary to integrate mass conservation $\rho_{,t} + \rho_o \nabla \cdot \mathbf{u} + \rho_o \, \partial_z w = 0$ vertically over a grid cell. Again there are two approaches: integrating downward or upward, with the downward approach also preferred.

Integration of mass conservation over a cell interior to the ocean ($k > 1$), whose volume is constant in time, yields

$$\begin{aligned}
0 &= \int_{z_k}^{z_{k-1}} \mathrm{d}z \, (\rho_{,t} + \rho_o \, \nabla \cdot \mathbf{v}) \\
&= \partial_t \int_{z_k}^{z_{k-1}} \mathrm{d}z \, \rho + \rho_o \, \nabla \cdot \int_{z_k}^{z_{k-1}} \mathrm{d}z \, \mathbf{u} + \rho_o \, (w_{z_{k-1}} - w_{z_k}) \\
&\rightarrow \partial_t \, (h \, \rho)_{z_k} + \rho_o \, \nabla \cdot (h \, \mathbf{u})_{z_k} + \rho_o \, (w_{z_{k-1}} - w_{z_k}),
\end{aligned} \tag{11.13}$$

where $h_k = z_{k-1} - z_k$ is the thickness of the tracer cell (the cell on which density is naturally defined). The continuous to discrete correspondence relation in the last step can be made by interpreting the discrete model variables as finite volume averages of the continuous variables over the grid cell. This interpretation will be implicit in the discussion, with added notation eschewed to minimize clutter. Solving for w_{z_k} leads to a relation for vertical advective velocity passing across the lower boundary of the kth interior grid cell

$$\rho_o \, w_{z_k} = \rho_o \, w_{z_{k-1}} + h_{z_k} \, \partial_t \, \rho_{z_k} + \rho_o \, \nabla \cdot (h \, \mathbf{u})_{z_k}. \tag{11.14}$$

The time tendency term is absent for the Boussinesq fluid, where mass conservation over the cell reduces to volume conservation.

Vertical integration over a surface ocean grid cell ($k = 1$) leads to

$$\begin{aligned}
\int_{z_1}^{\eta} \mathrm{d}z \, (\rho_{,t} + \rho_o \, \nabla \cdot \mathbf{v}) &= \partial_t \int_{z_1}^{\eta} \mathrm{d}z \, \rho + \rho_o \, \nabla \cdot \int_{z_1}^{\eta} \mathrm{d}z \, \mathbf{u} - \rho_o \, w_{z_1} - (\rho \, \eta_{,t} - \rho_o \, \hat{\mathbf{N}} \cdot \mathbf{v})_{z_0} \\
&\rightarrow \partial_t \, (h \, \rho)_{z_1} + \rho_o \, \nabla \cdot (h \, \mathbf{u})_{z_1} - \rho_o \, w_{z_1} - \rho_o \, q_{\mathrm{w}},
\end{aligned} \tag{11.15}$$

where $z_0 = \eta$ and the surface kinematic boundary condition (11.4) was used.* This result allows us to interpret the vertical advective velocity at the ocean surface as that arising from surface water fluxes

$$w_{z_0} = -q_{\mathrm{w}}, \tag{11.16}$$

where the minus sign arises from the sign convention applied to the fresh water flux q_{w}.

For some approaches, maintaining self-consistency with the discrete mass budget requires us to time step density instead of thickness weighted density.† This decision leads to the vertical advective velocity passing upward into the surface grid cell

$$\rho_o\, w_{z_1} = -\rho_o\, q_{\mathrm{w}} + h_{z_1}\, \partial_t\, \rho_{z_1} + \rho_{z_1}\, \eta_{,t} + \rho_o\, \nabla \cdot (h\, \mathbf{u})_{z_1}. \tag{11.17}$$

In this case, the recursive relation (11.14) for vertical advective velocity component can be used for all vertical cells if we define an "effective" vertical velocity $w^{\mathrm{eff}}_{z_0}$

$$\rho_o\, w^{\mathrm{eff}}_{z_0} = -\rho_o\, q_{\mathrm{w}} + \rho_{z_1}\, \eta_{,t} \tag{11.18}$$

for the non-Boussinesq case, and

$$w^{\mathrm{eff}}_{z_0} = -\nabla \cdot \mathbf{U} \tag{11.19}$$

for the Boussinesq case.

11.3.3 Advection velocity components

Equations (11.14) and (11.17) are semi-discrete. A full discretization requires averaging over the grid cell. Doing so results in all velocity components being averaged onto the sides and top faces of the grid cell. One then has a full specification of the convergence of mass onto the cell, which drives time tendencies in the grid cell's density.

When averaged onto cell faces, the velocity components are known as *advective velocities*, since they are used for advecting tracers (Section 11.5) or momenta. On the B-grid, a detailed form of the advective velocity components for tracers can be prescribed according to the needs of energetic consistency as detailed in Griffies et al. (2004). Furthermore, the relation between advective velocities for tracers and momentum can be specified according to a linear mapping, as discussed in Pacanowski and Griffies (1999) and Griffies et al. (2004). Notably, the same discretization of the advective velocities can be used whether the discrete fluid is Boussinesq or non-Boussinesq.

---

*As noted in Griffies et al. (2001), there is no need to linearize the surface kinematic boundary condition by assuming $\hat{\mathbf{N}} \approx \hat{\mathbf{z}}$. This assumption provides no simplification over the general case presented here.

†Approaches based on generalizing the older rigid lid methods time step the non-thickness weighted quantities (Killworth et al., 1991; Dukowicz and Smith, 1994; Griffies et al., 2001). Notably, there is nothing fundamental warranting one approach versus another, although thickness weighted tendencies lead to simpler statements of tracer conservation (Griffies et al., 2001, 2004). We have more to say about thickness weighted updates in Section 12.5.2.

11.3.4 Specifying the density time tendency

As the mass continuity equation is implemented in a diagnostic manner, it is necessary to specify the time tendency for the density. A prescription successfully used in the model of Griffies et al. (2004) is given by the leap-frog scheme

$$\partial_t \rho \approx \frac{\rho^{(e)}(\tau + \Delta\tau) - \rho^{(e)}(\tau - \Delta\tau)}{2\,\Delta\tau}, \tag{11.20}$$

where $\rho^{(e)}$ is a density field set according to linear extrapolation (Greatbatch et al., 2001)

$$\begin{aligned}\rho^{(e)}(\tau + \Delta\tau) &\equiv \rho(\tau) + [\rho(\tau) - \rho(\tau - \Delta\tau)] \\ &= 2\,\rho(\tau) - \rho(\tau - \Delta\tau).\end{aligned} \tag{11.21}$$

Although it adds to the model memory requirements, it is useful to carry the density variable $\rho^{(e)}$ explicitly in the model since it is used very often with the non-Boussinesq formulation. Note that a two-time level update

$$\partial_t \rho \approx \frac{\rho^{(e)}(\tau + \Delta\tau) - \rho^{(e)}(\tau)}{\Delta\tau} \tag{11.22}$$

is needed for models that discretize their tracer and velocity time tendencies in this manner (Section 12.6).

11.4 MASS BUDGET FOR A DISCRETE FLUID COLUMN

As seen in Section 11.3.2, the discrete fluid conserves mass by construction, since the vertical velocity components are derived by integrating mass conservation over a grid cell. Additionally, it is important to maintain mass conservation over a column of fluid extending from the ocean surface to the bottom. As in the continuum case discussed in Section 3.4.3 (page 36), this constraint leads to an expression for the time evolution of the surface height. A similar analysis is presented here for the discrete ocean fluid.

For this purpose, recall from Section 11.3.2 that the discrete vertical velocity components take the form

$$\rho_o\, w_{z_k} = \rho_o\, w_{z_{k-1}} + h_{z_k}\, \partial_t\, \rho_{z_k} + \rho_o\, \nabla \cdot (h\, \mathbf{u})_{z_k} \qquad k > 1 \tag{11.23}$$

$$\rho_o\, w_{z_1} = -\rho_o\, q_{\mathrm{w}} + h_{z_1}\, \partial_t\, \rho_{z_1} + \rho_{z_1}\, \eta_{,t} + \rho_o\, \nabla \cdot (h\, \mathbf{u})_{z_1} \qquad k = 1. \tag{11.24}$$

A vertical sum of these velocities over the depth of a discrete column with N_k vertical grid cells leads to an expression for the vertical velocity at the lower face of a bottom grid cell

$$\rho_o\, w_{z_{N_k}} = -\rho_o\, q_{\mathrm{w}} + \rho_o\, \nabla \cdot \mathbf{U} + \rho_{z_1}\, \eta_{,t} + \sum_{k=1}^{N_k} h_{z_k}\, \partial_t\, \rho_{z_k} \tag{11.25}$$

where $\mathbf{U} = \sum_{k=1}^{N_k} h_{z_k}\, \mathbf{u}_{z_k}$ is the vertically integrated horizontal velocity. For the B-grid, the land-sea boundary on the bottom of a column of tracer cells is assumed flat.* Hence, the bottom kinematic boundary condition leads to

$$w_{z_{N_k}} = 0 \qquad \text{at the lower face of a bottom tracer cell.} \tag{11.26}$$

*Section 22.3.3 of Pacanowski and Griffies (1999) provides further discussion of this point, as well as the bottom of the ocean on a velocity cell.

This boundary condition is maintained for a discrete mass conserving fluid so long as the surface height at the top of the tracer cell column satisfies

$$\rho_{z_1}\,\eta_{,t} = -\rho_o\,\nabla\cdot\mathbf{U} + \rho_o\,q_{\mathrm{w}} - \sum_{k=1}^{N_k} h_{z_k}\,\partial_t\,\rho_{z_k}. \tag{11.27}$$

This expression for mass balance of a fluid column is directly analogous to the continuum result given by equation (3.62) (page 38). Additionally, noting that interior vertical cell thicknesses are constant in time gives

$$\partial_t\left(\sum_{k=1}^{N_k} h_{z_k}\,\rho_{z_k}\right) = -\rho_o\,\nabla\cdot\mathbf{U} + \rho_o\,q_{\mathrm{w}}, \tag{11.28}$$

which is directly analogous to the continuum mass balance given by equation (3.61). For a Boussinesq fluid we have

$$\eta_{,t} = -\nabla\cdot\mathbf{U} + q_{\mathrm{w}}, \tag{11.29}$$

which is the same form as the continuum volume balance given by equation (3.45). The discrete form of these budgets provides the means to time step the surface height. Details of this time stepping are pursued in Sections 12.4 and 12.6. Finally, note that the expression (11.27) for $\eta_{,t}$ leads the relation (11.18) for the effective vertical velocity component to take the form

$$w^{\mathrm{eff}}_{z_0} = -\nabla\cdot\mathbf{U} - \rho_o^{-1}\sum_{k=1}^{N_k} h_{z_k}\,\partial_t\,\rho_{z_k}, \tag{11.30}$$

which reduces to

$$w^{\mathrm{eff}}_{z_0} = -\nabla\cdot\mathbf{U} \tag{11.31}$$

in the Boussinesq limit.

11.5 TRACER BUDGET FOR A GRID CELL

As for the discrete mass conservation budget, the discrete tracer budget is formulated by vertically integrating the continuous tracer budget $(\rho\,C)_{,t} + \rho_o\,\nabla\cdot(\mathbf{v}\,C) = -\rho_o\,\nabla\cdot\mathbf{F} + \rho\,\mathcal{S}$ over the depth of a model grid cell. Special care is taken when considering surface boundary conditions.

11.5.1 Vertical integration over a grid cell

Upon integrating over a grid cell, the tracer source term is interpreted in a finite volume sense

$$\int_{z_k}^{z_{k-1}} \mathrm{d}z\,\rho\,\mathcal{S} \rightarrow (h\,\rho\,\mathcal{S})_{z_k}. \tag{11.32}$$

For an interior grid cell, the material transport term integrates to

$$\int_{z_k}^{z_{k-1}} \mathrm{d}z\,[\,(\rho\,C)_{,t} + \rho_o\,\nabla\cdot(\mathbf{v}\,C)]$$

$$\rightarrow \partial_t(h\,\rho\,C)_{z_k} + \rho_o\,\nabla\cdot(h\,\mathbf{u}\,C)_{z_k} + \rho_o\,(w\,C)_{z_{k-1}} - \rho_o\,(w\,C)_{z_k}. \tag{11.33}$$

The advective tracer fluxes are computed on the sides of the cell, with their convergence contributing to the time tendency of the tracer value within the cell.

Equation (11.33) is semi-discrete. Full discretization of the thickness weighted advective fluxes depends on details of the chosen algorithm. Additionally, thickness weighting of the horizontal fluxes can be implemented according to methods of Adcroft et al. (1997) or Pacanowski and Gnanadesikan (1998) to account for the generally different adjacent cell thicknesses.

Vertical integration of the subgrid scale (SGS) flux term over an interior grid cell leads to

$$\int_{z_k}^{z_{k-1}} \mathrm{d}z \, \nabla \cdot \mathbf{F} \rightarrow \nabla \cdot (h \, \mathbf{F}^{\mathrm{h}})_{z_k} + (F^z_{z_{k-1}} - F^z_{z_k}), \tag{11.34}$$

where the three-dimensional SGS tracer flux is given by $\mathbf{F} = (\mathbf{F}^{\mathrm{h}}, F^z)$. Thickness weighting of these fluxes in a discrete model is also described by Pacanowski and Gnanadesikan (1998).

For a surface model grid cell, vertical integration leads to

$$\begin{aligned}
&\int_{z_1}^{\eta} \mathrm{d}z \, [\, (\rho \, C)_{,t} + \rho_o \, \nabla \cdot (\mathbf{v} \, C)] \\
&= \partial_t \int_{z_1}^{\eta} \mathrm{d}z \, (\rho \, C) + \rho_o \nabla \cdot \int_{z_1}^{\eta} \mathrm{d}z \, (\mathbf{u} \, C) - \rho_o \, (w \, C)_{z_1} - C_{z_0} \, (\rho \, \eta_{,t} - \rho_o \, \hat{\mathbf{N}} \cdot \mathbf{v})_{z_0} \\
&\rightarrow \partial_t (h \, \rho \, C)_{z_1} + \rho_o \, \nabla \cdot (h \, \mathbf{u} \, C)_{z_1} - \rho_o \, (w \, C)_{z_1} - \rho_o \, q_{\mathrm{w}} \, C_{z_0},
\end{aligned} \tag{11.35}$$

where the surface kinematic boundary condition (11.4) was used. Similar considerations for the divergence of the SGS flux lead to

$$\int_{z_1}^{z_0} \mathrm{d}z \, \nabla \cdot \mathbf{F} \rightarrow \nabla \cdot (h \, \mathbf{F}^{\mathrm{h}})_{z_1} - F^z_{z_1} + (\mathbf{F} \cdot \hat{\mathbf{N}})_{z_0}. \tag{11.36}$$

11.5.2 Vertical tracer flux at the ocean surface

The tracer flux is generally comprised of an advective and a SGS turbulent contribution

$$\mathbf{F}_C = \mathbf{v} \, C + \mathbf{F}. \tag{11.37}$$

In particular, the tracer flux crossing the ocean surface is given by

$$(\mathbf{F}_C \cdot \hat{\mathbf{N}})_{z_0} = -q_{\mathrm{w}} \, C_{z_0} + (\mathbf{F} \cdot \hat{\mathbf{N}})_{z_0}. \tag{11.38}$$

This flux has a positive sign when the tracer leaves the ocean surface.

In contrast to an interior vertical flux, ocean-only information is not sufficient to estimate the individual terms in the surface tracer flux (11.38). Instead, as for the surface momentum flux discussed in Section 12.2.4, the system is closed by prescribing a total tracer flux crossing the ocean surface via a boundary layer model or parameterization. Boundary layer models present the ocean with a total tracer flux Q_C, which is the tracer flux crossing the ocean surface from other component models, such as atmosphere, river, and sea ice models. Q_C can also be specified from data, or determined via damping to some specified tracer concentration.

The flux Q_C generally has a contribution from parameterized turbulence as well as a tracer flux with fresh water,

$$Q_C = -q_w\, C_w + Q_C^{(\text{turb})}, \tag{11.39}$$

where C_w is the tracer concentration in the fresh water. Hence, assuming a continuous flux at the ocean surface $Q_C = (\mathbf{F}_C \cdot \hat{\mathbf{N}})_{z_0}$ leads to

$$-q_w\, C_{z_0} + (\mathbf{F} \cdot \hat{\mathbf{N}})_{z_0} = -q_w\, C_w + Q_C^{(\text{turb})}. \tag{11.40}$$

In the absence of fresh water forcing,

$$(\mathbf{F} \cdot \hat{\mathbf{N}})_{z_0} = Q_C^{(\text{turb})} \qquad \text{if } q_w = 0. \tag{11.41}$$

However, in general it is not always appropriate to equate the individual terms in equation (11.40). In particular, C_w and C_{z_0} are not always the same. Salt provides the most striking example, where salinity of fresh water, s_w, is near zero yet the salinity on the ocean side of the ocean surface, s_{z_0}, is not generally zero. However, as seen below, other tracers, termed *neutral tracers*, do allow for equating the individual terms in equation (11.40).

11.5.3 Tracer budgets for surface grid cells

Given the above prescription for the ocean surface tracer flux, the tracer budget within a surface grid cell takes the form

$$\begin{aligned} \partial_t (h\,\rho\, C)_{z_1} = {} & -\rho_o\, \nabla \cdot (h\, \mathbf{u}\, C + h\, \mathbf{F}^{\text{h}})_{z_1} + \rho_o\, (w\, C + F^z)_{z_1} \\ & + \rho_o\, (q_w\, C_w - Q_C^{(\text{turb})}) + (h\, \rho\, \mathcal{S})_{z_1}. \end{aligned} \tag{11.42}$$

Note that boundary condition information which closes the surface cell tracer budget is knowledge of the tracer concentration in fresh water C_w, as well as the parameterized turbulent tracer flux $Q_C^{(\text{turb})}$. Neither the tracer concentration at the ocean surface C_{z_0} nor the turbulent flux $(\mathbf{F} \cdot \hat{\mathbf{N}})_{z_0}$ is explicitly needed.

The budget for a Boussinesq fluid simplifies by canceling density factors. Additionally, if we time step the tracer concentration instead of thickness weighted tracer concentration, and use the surface height equation $\eta_{,t} = -\nabla \cdot \mathbf{U} + q_w$, then

$$\begin{aligned} h_{z_1}\, \partial_t C_{z_1} = {} & -\nabla \cdot (h\, \mathbf{u}\, C + h\, \mathbf{F}^{\text{h}})_{z_1} + (w\, C + F^z)_{z_1} \\ & + q_w\, (C_w - C_{z_1}) + C_{z_1}\, \nabla \cdot \mathbf{U} - Q_C^{(\text{turb})} + (h\, \mathcal{S})_{z_1}. \end{aligned} \tag{11.43}$$

Hence, if the tracer concentration in the fresh water equals that in the surface ocean grid cell, the fresh water contribution to the time tendency drops out. Note, however, the influence of fresh water is still felt indirectly through its influence on the divergence $\nabla \cdot \mathbf{U}$.

11.5.4 Tracer concentration in fresh water

To compute the surface tracer budget, it is necessary to specify the tracer concentration C_w in the fresh water. Consider two general classes of tracers.

11.5.4.1 Neutral tracers

Most climate models do not carry information regarding the tracer concentration in their river or atmospheric component models. Furthermore, the tracer concentration C_{w} is often quite close to the tracer concentration within the adjacent ocean cell. Hence, for many purposes, it is appropriate to assume

$$C_{\mathrm{w}} \approx C_{z_0} \approx C_{z_1}. \tag{11.44}$$

Such tracers are called here *neutral tracers* since they do not add or subtract from the tracer concentration already present in the ocean cell, although they do alter the total tracer content. Notably, neutral tracers allow for the identification of individual terms in the surface flux equation (11.40)

$$(\mathbf{F} \cdot \hat{\mathbf{N}})_{z_0} = Q_C^{(\mathrm{turb})}, \tag{11.45}$$

even when there is fresh water forcing.

11.5.4.2 Salt

Due to a large hydration energy for salt, the air-sea interface effectively acts as an impenetrable barrier to salt transport. Therefore, neglecting the formation of sea spray and the interchange of salt with sea ice, and in the absence of salt entering through the solid earth boundaries, total ocean salt remains constant. Note that transfer of salt between the ocean and sea ice is commonly accounted for in ocean climate models. Yet without this transfer, the salt mass per unit volume $\rho\, s$ within a grid cell, where s is salinity (a dimensionless concentration), changes only through advective and turbulent fluxes from the interior ocean, as well as time tendencies in the volume of the grid cell.

Under the assumption that salt does not pass the air-sea interface, the total flux of salt passing across the ocean surface vanishes,

$$\begin{aligned} Q_s &= (\mathbf{F}_s \cdot \hat{\mathbf{N}})_{z_0} \\ &= 0, \end{aligned} \tag{11.46}$$

thus leading to the following salt budget in a surface grid cell

$$\partial_t (h\, \rho\, s)_{z_1} = -\rho_o\, \nabla \cdot (h\, \mathbf{u}\, s + h\, \mathbf{F}^{\mathrm{h}})_{z_1} + \rho_o\, (w\, s + F^z)_{z_1} + (h\, \rho\, \mathcal{S})_{z_1}. \tag{11.47}$$

As the salt flux is comprised of turbulent and advective terms, a zero surface salt flux implies a balance between the two contributions. On the ocean side of the interface, this balance leads to

$$\begin{aligned} (\mathbf{F}_s \cdot \hat{\mathbf{N}})_{z_0} &= -q_{\mathrm{w}}\, s_{z_0} + (\mathbf{F} \cdot \hat{\mathbf{N}})_{z_0} \\ &= 0. \end{aligned} \tag{11.48}$$

That is, the turbulence flux of salt on the ocean side of the ocean surface is proportional to the flux of salt associated with fresh water. It is this turbulence salt flux which is to be used for computing buoyancy fluxes in ocean mixed layer schemes (Section 11.6). Further discussion of this balance for the Boussinesq fluid can be found in Huang (1993), where the turbulence flux was called an *anti-advective flux*.

On the atmosphere/river/sea ice side, a zero total salt flux leads to the balance

$$\begin{aligned} Q_s &= -q_{\mathrm{w}}\, s_{\mathrm{w}} + Q_s^{(\mathrm{turb})} \\ &= 0. \end{aligned} \tag{11.49}$$

Assuming the salinity of fresh water vanishes then leads to

$$Q_s^{(\mathrm{turb})} = 0. \tag{11.50}$$

This result makes sense when recalling that a fluid with uniformly zero salinity, such as a fresh water river, trivially has no net transport of salinity, whether turbulent or otherwise.

These considerations for salt flux across the air-sea interface are modified when considering salt flux between the ocean and sea ice, since sea ice generally has a nonzero salinity $s_{\mathrm{w}} \neq 0$. Many modern sea ice models carry salinity content of the ice. For this case, ocean salt content changes when sea ice melts or forms.

11.6 FLUXES FOR TURBULENCE MIXED LAYER SCHEMES

Ocean mixed layer schemes such as that of Large et al. (1994) require the computation of the oceanic turbulence buoyancy flux at the ocean surface. Hence, it is useful to summarize expressions for the turbulence flux of potential temperature and salinity $(\mathbf{F} \cdot \hat{\mathbf{N}})_{z_0}$. Continuity of the tracer flux across the air-sea interface leads to (equation (11.40))

$$-q_{\mathrm{w}}\, C_{z_0} + (\mathbf{F} \cdot \hat{\mathbf{N}})_{z_0} = -q_{\mathrm{w}}\, C_{\mathrm{w}} + Q_C^{(\mathrm{turb})}, \tag{11.51}$$

which gives

$$(\mathbf{F} \cdot \hat{\mathbf{N}})_{z_0} = Q_C^{(\mathrm{turb})} + q_{\mathrm{w}}\,(C_{z_0} - C_{\mathrm{w}}). \tag{11.52}$$

Some special cases are now noted.

Recall from Section 11.5.4.1 that neutral tracers have $C_{z_0} = C_{\mathrm{w}}$. Potential temperature is a reasonable approximation to a neutral tracer. In this case,

$$(\mathbf{F} \cdot \hat{\mathbf{N}})_{z_0} = Q_\theta^{(\mathrm{turb})} \qquad \text{for neutral tracers with } C_{z_0} = C_{\mathrm{w}}, \tag{11.53}$$

as already noted in Section 11.5.4.1. This result always holds for rigid lid models since $q_{\mathrm{w}} = 0$ must be maintained in this case (Section 12.10 on page 278).

For ocean regions free of sea ice, recall from Section 11.5.4.2 that the total salt flux vanishes at the ocean surface. Hence,

$$(\mathbf{F} \cdot \hat{\mathbf{N}})_{z_0} = q_{\mathrm{w}}\, s_{z_0} \qquad \text{for tracers with zero air-sea flux such as salt.} \tag{11.54}$$

In the absence of fresh water flux, the turbulence flux of salt vanishes

$$(\mathbf{F} \cdot \hat{\mathbf{N}})_{z_0} = 0 \qquad \text{for salt if } q_{\mathrm{w}} = 0. \tag{11.55}$$

With nonzero fresh water fluxes, it may be appropriate to set salinity at the ocean surface using a centered approximation $s_{z_0} \approx (s_{z_1} + s_{\mathrm{w}})/2 = s_{z_1}/2$. However, it is arguable that a better estimate is $s_{z_0} \approx s_{z_1}$ for regions where fresh water fluxes are associated just with evaporation and precipitation. In this case, the turbulence salt flux is given by

$$(\mathbf{F} \cdot \hat{\mathbf{N}})_{z_0} = q_{\mathrm{w}}\, s_{z_1} \qquad \text{for salt if } s_{z_0} \approx s_{z_1}. \tag{11.56}$$

11.7 FLUX PLUS RESTORE BOUNDARY CONDITIONS

It is common to run ocean models with a flux provided from data or another component model, plus restoring to ocean data. The form for such a flux is given by

$$Q_C = (\gamma_C\, h_{z_1})\,(C_{z_1} - C^{(\text{data})}) + Q_C^{\text{turb},0} - q_w\, C_w, \tag{11.57}$$

where $Q_C^{\text{turb},0}$ is the turbulence flux from data or another model.* The coefficient γ_C is an inverse restoring time. The product $\gamma_C\, h_{z_1}$ is therefore a velocity scale. This velocity scale is often called the *piston velocity*

$$V_{\text{piston}} = \gamma_C\, h_{z_1} \tag{11.58}$$

with a larger V_{piston} resulting in a stronger restoring. Note that with a free surface, the piston velocity changes according to the thickness of the time-dependent top model grid cell h_{z_1} (see Figure 11.1). For less ambiguity in interpretation, the modeler may prefer to use the time-independent top cell thickness (defined for $\eta = 0$) instead, so to have a time-independent piston velocity.

The flux $V_{\text{piston}}\,(C_{z_1} - C^{(\text{data})})$ is positive when $C_{z_1} > C^{(\text{data})}$, indicating a transfer of tracer out of the model through the top, thus acting to damp the ocean surface tracer back toward $C^{(\text{data})}$. This restoring flux is considered part of the turbulent flux, as opposed to part of the advective fresh water flux. That is, the total turbulent flux is given by

$$Q_C^{(\text{turb})} = V_{\text{piston}}\,(C_{z_1} - C^{(\text{data})}) + Q_C^{\text{turb},0}. \tag{11.59}$$

The case of salinity restoring is somewhat distinct. There are two approaches. First, the restoring flux $\rho_o\, V_{\text{piston}}\,(s_{z_1} - s^{(\text{data})})$ can be considered an actual flux of salt. Yet, as mentioned in Section 11.5.4.2, salt does not generally cross the ocean surface. Instead, the ocean salt content is reasonably close to constant, at least for purposes of climate simulations. Hence, to maintain a constant salt content and to provide a local restoring for the salinity field, one can eliminate the salt flux in favor of an implied local fresh water flux via the equality

$$V_{\text{piston}}\,(s_{z_1} - s^{(\text{data})}) = q_w\, s_{z_1}, \tag{11.60}$$

which leads to the fresh water flux

$$q_w = V_{\text{piston}}\,(1 - s^{(\text{data})}/s_{z_1}). \tag{11.61}$$

As a check, note that if the ocean surface salinity s_{z_1} is larger than the salinity from the data, there is a positive input of fresh water to the ocean, thus damping the surface salinity back toward that of the data. There is no *a priori* reason that this fresh water flux will integrate to zero over the globe. Hence, its use will generally lead to a drift in ocean volume, unless a global adjustment is made at each model time step.

---

*Note that for heat, there are both *turbulent* fluxes, such as latent and sensible heat fluxes, as well as *radiative* fluxes, such as long and shortwave heat fluxes. For present purposes, both of these types of heat flux are assumed to be carried by $Q_C^{\text{turb},0}$.

11.8 Z-LIKE VERTICAL COORDINATE MODELS

Although straightforward to incorporate, boxes with time-dependent thickness add complication to the mass and tracer budgets. Most critically, the possibility of a vanishing surface grid cell places a constraint on the vertical grid spacing. This constraint becomes onerous when aiming to employ refined vertical resolution models in regions of strong tidal forcing and/or thick sea ice. The purpose of this section is to mention some alternatives which are presently being considered in the community.

To remove complications with a time-dependent vertical domain, and to avoid problems with vanishing top cells, the σ-coordinate given by (see equation (6.1), page 124)

$$\sigma = \frac{z - \eta}{H + \eta} \tag{11.62}$$

is a viable alternative. However, problems with pressure gradient errors and difficulties implementing neutral physical processes have prejudiced global ocean climate modelers away from these models (see Section 6.2.2, page 124).

Another alternative is a variant on the atmospheric "eta"-coordinate, given by*

$$\begin{aligned} z^* &= H\sigma \\ &= H\left(\frac{z-\eta}{H+\eta}\right). \end{aligned} \tag{11.63}$$

This *coastal coordinate* was introduced by Stacey et al. (1995) for studies with high vertical resolution and large tidal fluctuations. That is, just as for σ-coordinates, z^* can handle a large variation in the surface height, since the vertical coordinate squeezes together in those regions where η undulations are large. No grid cells are lost. Additionally, as for σ, z^* has a time-independent range, given here by $-H \le z^* \le 0$, the same range as encountered in the rigid lid z-models (Section 12.10, page 278). This reduces complications associated with the time-dependent vertical depth range $-H \le z \le \eta$ encountered with free surface z-models. Volume transport across z^* surfaces (Section 6.7) arising from tidal undulations is absent, since z^* moves with fluctuations in surface height. This may prove useful for reducing the levels of spurious diapycnal transport associated with numerical truncation errors with advection schemes (Section 7.5). Additionally, as illustrated for atmospheric modeling by Black (1994), the z^*-coordinate is quasi-horizontal, as seen trivially for the case when $\eta = 0$, where $z^* = z$. This property precludes it from being as useful for representing bottom boundary layer processes as the sigma coordinate. However, it reduces the spurious pressure gradient error to a level far smaller than the sigma models. It also facilitates the implementation of neutral physics schemes in a manner nearly identical to z-models.† Although it does not satisfy all the needs of ocean climate modelers, research into the utility of the z^*-coordinate, or its pressure-based analog

$$p^* = p_{\mathrm{bottom}}\left(\frac{p - p_{\mathrm{a}}}{p_{\mathrm{bottom}} + p_{\mathrm{a}}}\right), \tag{11.64}$$

*The symbol η is used in the atmospheric literature, but is not convenient here as we reserve this symbol for the ocean free surface height. The use of z^* for the ocean was suggested to the author by Alistair Adcroft, personal communication, 2002.

†See Section 6.11.2, page 148 for comments on neutral physics in generalized vertical coordinate models.

for the World Ocean holds some promise to resolve some of the cumbersome aspects of the free surface z-models, without losing the conveniences of this model class.*

11.9 CHAPTER SUMMARY

This chapter provided the reader with many details for how to go about discretizing the mass/volume and tracer budgets in an ocean model. Focus was placed on a discrete z-model using a free surface, with the Modular Ocean Model of Griffies et al. (2004) a realization of the formulation. However, much of the material generalizes to arbitrary vertical coordinates, such as the ones discussed in Chapter 6.

After a summary of the continuous model equations in Section 11.1, we spent time in Section 11.2 to illustrate the needs of compatibility between tracer and mass/volume conservation. Compatibility requires that the tracer equation reduce to the mass/volume equation when the tracer is set to a constant. This follows trivially by the definition of the tracer concentration as the mass of tracer per mass of seawater. However, the implications for algorithm design are nontrivial, as it links the space-time discretization used for the tracer equation to that of the mass/volume budget.

In Section 11.3 we considered the mass budget for a discrete model grid cell, with the volume budget for the Boussinesq fluid following trivially from setting ρ to the constant ρ_0. All cells interior to the ocean have time-independent volumes, which results in straightforward budget equations. At the ocean surface, however, the time-dependent free surface height complicates matters a bit, and so extra care was exercised when formulating budgets over this cell (see Figure 11.1).

The time tendency for density in the non-Boussinesq system is computed via an extrapolation technique. Doing so allows the vertical velocity to be diagnosed in a manner similar to the approach followed in the Boussinesq models. To diagnose vertical velocity within a column, it may be preferable to integrate down from the ocean surface to the bottom. This approach provides a useful check on the numerical integrity of the code, since the vertical velocity diagnosed at the lower face of the bottom tracer cell should vanish. Errors are readily found if this property is not manifest in the code. Considerations in Section 11.4 focused on establishing the surface height equation, which is obtained from a vertical integration of the mass/volume budget over a fluid column. The analysis in Section 11.5 for tracer mass followed similarly to that applicable for seawater mass. Here, details were discussed of the surface boundary conditions, which differ somewhat depending on details of the tracer and its content in the adjacent fluids. Sections 11.6 and 11.7 detailed how to specify the fluxes of tracer passing through the surface boundary. Finally, Section 11.8 mentioned some alternatives to the traditional z-coordinate. These alternatives aim to remove limitations of the free surface z-models related to the vanishing top model grid cell. Current research is testing the utility of these

*In equation (11.64), p_{bottom} = bottom pressure and p_{a} = pressure applied at the surface due to atmospheric and/or sea ice loading. Note that when the applied pressure p_{a} vanishes, p^* reduces to the pressure coordinate. In general, the use of pressure-based vertical coordinates, such as p^*, facilitates the introduction of non-Boussinesq dynamics, as discussed in Sections 6.8.4 and 6.10.3.

models for ocean climate simulations.

Chapter Twelve

ALGORITHMS FOR HYDROSTATIC OCEAN MODELS

The purpose of this chapter is to illustrate methods for time stepping the equations discretized in a hydrostatic z-model. Much of this material is relevant for arbitrary vertical coordinates, such as the ones discussed in Chapter 6. There are many ways to solve the ocean primitive equations, with various approaches commonly used in the literature. Our treatment is not comprehensive. Instead, the aim is to expose some of the more popular methods along with their pros and cons.

Splitting between the relatively slow dynamics and fast dynamics is motivated and detailed. Use of the hydrostatic approximation allows for the fast dynamics to be approximated by the vertically integrated equations, whereas the equations for the slow dynamics are three-dimensional. A popular solution algorithm for hydrostatic models is known as a *split-explicit* free surface method (e.g., Blumberg and Mellor, 1987; Killworth et al., 1991; Griffies et al., 2001). This method is illustrated using a leap-frog time stepping scheme. We also discuss a predictor-corrector scheme similar to that used by Hallberg (1997). An alternative approach solves the fast dynamical system implicitly in time (e.g., Dukowicz and Smith, 1994; Marshall et al., 1997; Campin et al., 2004), yet we do not discuss this method here.

We document two methods used in the Modular Ocean Model (Griffies et al., 2004) for updating the model equations. One method is based on a leap-frog time tendency, and this is currently the most popular method for ocean climate models (Griffies et al., 2000a). The second method can be interpreted as a *time staggered* approach, and is motivated from methods used by Hallberg (1997) and Campin et al. (2004). Time staggering resolves some fundamental problems with leap-frog based schemes, and so will likely prove to be more common in the future. After presenting the various time stepping schemes, we discuss their stability properties. This analysis further motivates us to jettison the leap-frog method, as it proves to be less stable than schemes discretizing the time tendency with two time levels. Finally, for completeness, we provide an overview of the older rigid lid method for Boussinesq fluids.

12.1 SUMMARY OF THE CONTINUOUS MODEL EQUATIONS

The equations considered in this chapter represent discrete realizations of the continuum hydrostatic model momentum, tracer, and kinematic equations presented in Section 8.8. These are the ocean *primitive equations*, and their form is repeated here for completeness:

$$\rho_{,t} + \rho_o \nabla \cdot \mathbf{v} = 0 \tag{12.1}$$
$$(\rho C)_{,t} + \rho_o \nabla \cdot (\mathbf{v} C) = -\rho_o \nabla \cdot \mathbf{F} + \rho \mathcal{S}^{(C)} \tag{12.2}$$
$$(D \overline{\rho}^z)_{,t} = -\rho_o \nabla \cdot \mathbf{U} + \rho_o q_w \tag{12.3}$$
$$\mathbf{u}_{,t} + \nabla \cdot (\mathbf{v}\, \mathbf{u}^\rho) + \mathcal{M}\, \hat{\mathbf{z}} \wedge \mathbf{v}^\rho = -f\, \hat{\mathbf{z}} \wedge \mathbf{v} - \nabla(p/\rho_o) + \mathbf{F}^{(\mathbf{u})} \tag{12.4}$$
$$p_{,z} = -\rho g. \tag{12.5}$$

The surface and bottom kinematic boundary conditions were summarized in Section 11.1 (page 222). In these expressions,

$$\rho\, \mathbf{v}^\rho = \rho_o\, \mathbf{v} \tag{12.6}$$

is the linear momentum density. The hydrostatic-Boussinesq equations are recovered by setting $\rho \longrightarrow \rho_o$, except when multiplying the gravitational acceleration:

$$\nabla \cdot \mathbf{v} = 0 \tag{12.7}$$
$$C_{,t} + \nabla \cdot (\mathbf{v}\, C) = -\nabla \cdot \mathbf{F} + \mathcal{S}^{(C)}. \tag{12.8}$$
$$\eta_{,t} = -\nabla \cdot \mathbf{U} + q_w \tag{12.9}$$
$$\mathbf{u}_{,t} + \nabla \cdot (\mathbf{v}\, \mathbf{u}) + \mathcal{M}\, \hat{\mathbf{z}} \wedge \mathbf{v} = -f\, \hat{\mathbf{z}} \wedge \mathbf{v} - \nabla(p/\rho_o) + \mathbf{F}^{(\mathbf{u})} \tag{12.10}$$
$$p_{,z} = -\rho g. \tag{12.11}$$

The corresponding kinematic boundary conditions were also summarized in Section 11.1. An interpretation of the model fields is summarized in Table 8.1.

12.2 BUDGET OF LINEAR MOMENTUM FOR A GRID CELL

Before developing methods for time stepping the ocean model equations, consider here a semi-discrete form of the equation for linear momentum. The techniques follow those applied to the mass and tracer budgets in Chapter 11, and the notation follows that in Section 11.3.1. For this purpose, the continuum model equations are integrated over the volume of a grid cell. The focus is on vertical integration since there are important points to highlight regarding the surface and bottom boundary conditions. Horizontal integration follows similarly, with boundary conditions on the sides simpler than on the surface and bottom. Where a choice is needed about horizontal placement of discrete fields, the B-grid is used (Figure 10.1). The result of these manipulations is a consistent discretization of the continuous equations. Some details are left unspecified, such as the precise form of advective and diffusive fluxes.

12.2.1 Material time derivative

To start, integrate the material time derivative

$$\begin{aligned} \rho \frac{D\mathbf{u}^\rho}{Dt} &= (\rho\, \mathbf{u}^\rho)_{,t} + \nabla \cdot (\mathbf{v}^\rho\, \rho\, \mathbf{u}^\rho) + \rho\, \mathcal{M}\, \hat{\mathbf{z}} \wedge \mathbf{u}^\rho \\ &= \rho_o\, (\mathbf{u}_{,t} + \nabla \cdot (\mathbf{v}\, \mathbf{u}^\rho) + \mathcal{M}\, \hat{\mathbf{z}} \wedge \mathbf{u}^\rho) \end{aligned} \tag{12.12}$$

over the depth of a grid cell. For the metric term, prescribe the finite volume correspondence

$$\int_{z_k}^{z_{k-1}} \mathrm{d}z\, \mathcal{M}\, \hat{\mathbf{z}} \wedge \mathbf{u}^{\rho} \rightarrow h_k\, (\mathcal{M}\, \hat{\mathbf{z}} \wedge \mathbf{u}^{\rho})_k, \tag{12.13}$$

where $h_k = z_{k-1} - z_k$ is the thickness of the velocity cell and k labels the vertical cell number. The correspondence (12.13) is afforded by assuming that correlation terms, which arise from integrating the nonlinear product over the grid cell, are absorbed into the subgrid scale (SGS) friction operator $\mathbf{F}^{(\mathbf{u})}$.

Vertical integration of the time and space derivative operators over the surface model grid cell (Figure 12.1) leads to

$$\begin{aligned}
\int_{z_1}^{\eta} \mathrm{d}z\, [\mathbf{u}_{,t} + \nabla \cdot (\mathbf{v}\, \mathbf{u}^{\rho})] &= \partial_t \left(\int_{z_1}^{\eta} \mathrm{d}z\, \mathbf{u} \right) + \nabla \cdot \left(\int_{z_1}^{\eta} \mathrm{d}z\, \mathbf{u}\, \mathbf{u}^{\rho} \right) \\
&\quad - (w\, \mathbf{u}^{\rho})_{z_1} + [\mathbf{u}^{\rho}\, (w - \mathbf{u} \cdot \nabla \eta) - \mathbf{u}\, \eta_{,t}]_{z=\eta} \\
&= \partial_t \left(\int_{z_1}^{\eta} \mathrm{d}z\, \mathbf{u} \right) + \nabla \cdot \left(\int_{z_1}^{\eta} \mathrm{d}z\, \mathbf{u}\, \mathbf{u}^{\rho} \right) \\
&\quad - (w\, \mathbf{u}^{\rho})_{z_1} - q_{\mathrm{w}}\, (\mathbf{u}^{\rho})_{z=\eta},
\end{aligned} \tag{12.14}$$

where $z_0 = \eta$, and use was made of the surface kinematic boundary condition (11.4) (page 222). In the discrete model, we make the finite volume correspondence

$$\partial_t \left(\int_{z_1}^{\eta} \mathrm{d}z\, \mathbf{u} \right) + \nabla \cdot \left(\int_{z_1}^{\eta} \mathrm{d}z\, \mathbf{u}\, \mathbf{u}^{\rho} \right) \rightarrow \partial_t (h\, \mathbf{u})_{z_1} + \nabla \cdot (h\, \mathbf{u}\, \mathbf{u}^{\rho})_{z_1}, \tag{12.15}$$

thus leading to

$$\begin{aligned}
&\int_{z_1}^{\eta} \mathrm{d}z\, [\mathbf{u}_{,t} + \nabla \cdot (\mathbf{v}\, \mathbf{u}^{\rho})] \\
&\quad \rightarrow \partial_t (h\, \mathbf{u})_{z_1} + \nabla \cdot (h\, \mathbf{u}\, \mathbf{u}^{\rho})_{z_1} - (w\, \mathbf{u}^{\rho})_{z_1} - q_{\mathrm{w}}\, (\mathbf{u}^{\rho})_{z_0},
\end{aligned} \tag{12.16}$$

where

$$h_1 = -z_1 + \eta > 0 \tag{12.17}$$

is the time-dependent thickness of the surface velocity cell. The $k > 1$ interior grid cells, which have time-independent thicknesses, have the finite volume correspondence

$$\int_{z_k}^{z_{k-1}} \mathrm{d}z\, [\mathbf{u}_{,t} + \nabla \cdot (\mathbf{v}\, \mathbf{u}^{\rho})] \rightarrow \partial_t (h\, \mathbf{u})_{z_k} + \nabla \cdot (h\, \mathbf{u}\, \mathbf{u}^{\rho})_{z_k} - (w\, \mathbf{u}^{\rho})_{z_k} + (w\, \mathbf{u}^{\rho})_{z_{k-1}}. \tag{12.18}$$

The thickness weighted horizontal fluxes can be computed as in Adcroft et al. (1997) or Pacanowski and Gnanadesikan (1998) to account for the generally different adjacent cell thicknesses.

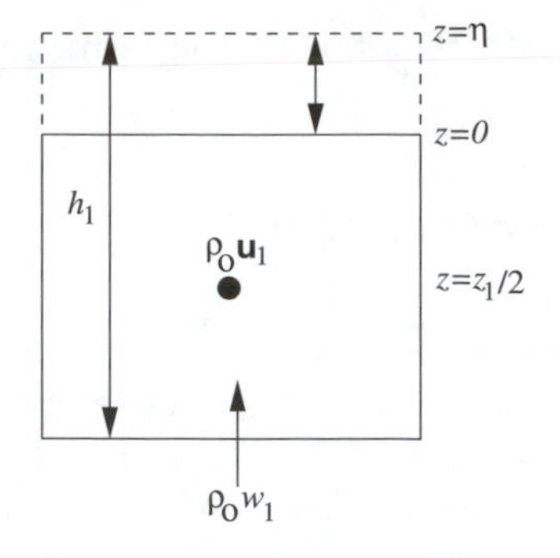

Figure 12.1 A grid cell used for computing the surface cell momentum budget. The cell's upper and lower boundaries are set by $z_0 = \eta$ and $z_1 < 0$, and the position of the grid point is $z_1/2$. See also Figure 11.1 on page 224 for more details.

Equations (12.16) and (12.18) represent semi-discrete forms of the continuous expression. A complete spatial discretization requires the computation of *advective velocities* $\mathbf{v}_{\text{adv}}$ on the respective sides of the velocity cell. The advective flux $\mathbf{v}_{\text{adv}}\,\mathbf{u}^{\rho}$ is then constructed through multiplying the advective velocity by an approximate value of $\mathbf{u}^{\rho}$ at the cell faces. The momentum advective fluxes commonly used in B-grid z-models are given in Griffies et al. (2004), where discrete energetic balances motivate the precise form. They are constructed using second order centered differences, as in Bryan (1969). For present purposes, schematic semi-discrete expressions, such as equations (12.16) and (12.18), are sufficient.

12.2.2 Horizontal pressure gradient

Vertical integration of the horizontal pressure gradient over the depth of a grid cell leads to the correspondence

$$\int_{z_k}^{z_{k-1}} \mathrm{d}z\, \nabla p \rightarrow h_k\, \nabla p_k, \tag{12.19}$$

where p_k is the hydrostatic pressure at the depth of the velocity point. Lin (1997) provides a detailed finite volume treatment of the pressure gradient for arbitrary vertical coordinates. For the special case of z-coordinates, his result reduces to that given here.

12.2.3 Friction due to vertical shears

The form of friction appropriate for an ocean model with spherical geometry, using generalized horizontal coordinates, is derived in Chapter 17. It is comprised of two terms,

$$\mathbf{F}^{(\mathbf{u})} = \mathbf{F}^{(\mathbf{u})}_{(\text{horz})} + \mathbf{F}^{(\mathbf{u})}_{(\text{vert})}, \tag{12.20}$$

where $\mathbf{F}^{(\mathbf{u})}_{(\text{horz})}$ arises from horizontal shears in the fluid and $\mathbf{F}^{\mathbf{u}}_{(\text{vert})}$ arises from vertical shears. To develop the full discrete equations, it is useful to detail here the contribution from vertical shears when integrated over a model cell. In general, $\mathbf{F}^{(\mathbf{u})}_{(\text{vert})}$ takes the form

$$\mathbf{F}^{(\mathbf{u})}_{(\text{vert})} = \partial_z \, (\kappa \, \mathbf{u}^{\rho}_{,z}), \tag{12.21}$$

where κ is a non-negative vertical viscosity arising from unresolved small-scale turbulence. As indicated here, it is important to compute friction using the velocity $\mathbf{u}^{\rho}$ instead of $\mathbf{u}$, where again $\rho \, \mathbf{u}^{\rho} = \rho_o \, \mathbf{u}$. The reason is that it is $\mathbf{u}^{\rho}$ that sets the kinetic energy, and friction is defined so that it directly affects kinetic energy. Likewise, it is $\mathbf{u}^{\rho}$ which should be used to determine the Richardson number for use in turbulence closure schemes. The distinction between $\mathbf{u}^{\rho}$ and $\mathbf{u}$ is not made for the Boussinesq fluids, where $\mathbf{u}^{\rho} = \mathbf{u}$, and the factors of density are replaced by ρ_o, which then cancel out from the friction vector.

Vertical integration of $\mathbf{F}^{(\mathbf{u})}_{(\text{vert})}$ over the depth of a model grid cell leads to

$$\int_{z_k}^{z_{k-1}} \mathrm{d}z \, \mathbf{F}^{(\mathbf{u})}_{(\text{vert})} = (\kappa \, \mathbf{u}^{\rho}_{,z})_{z=z_{k-1}} - (\kappa \, \mathbf{u}^{\rho}_{,z})_{z=z_k}. \tag{12.22}$$

Differences in the diffusive fluxes across a cell lead to a net vertical transfer of horizontal turbulent momentum into or out of the cell.

12.2.4 Special considerations for boundary cells

Written as a prognostic equation for thickness weighted horizontal velocity, the momentum budget vertically integrated over a model grid cell becomes

$$\begin{aligned}(\partial_t + f \, \hat{\mathbf{z}} \wedge) \, (h \, \mathbf{u})_{z_k} = & - (h \, \mathcal{M} \, \hat{\mathbf{z}} \wedge \mathbf{u}^{\rho})_{z_k} - \nabla \cdot (h \, \mathbf{u} \, \mathbf{u}^{\rho})_{z_k} \\ & - h_k \, \nabla p_k / \rho_o + (h \, \mathbf{F}^{(\mathbf{u})}_{(\text{horz})})_{z_k} \\ & - [\, w \, \mathbf{u}^{\rho} - \kappa \, \mathbf{u}^{\rho}_{,z} \,]_{z_{k-1}} + [\, w \, \mathbf{u}^{\rho} - \kappa \, \mathbf{u}^{\rho}_{,z} \,]_{z_k},\end{aligned} \tag{12.23}$$

where the surface vertical advection velocity is due to fresh water

$$w_{z_0} = -q_{\text{w}}, \tag{12.24}$$

(see equation (11.16)). The surface and bottom grid cells require some added consideration, and that is the focus of this subsection.

12.2.4.1 Vertical flux of horizontal momentum through the surface

Based on integrating the model equations over a grid cell, the vertical flux of horizontal momentum passing through the bottom of grid cell k is given by

$$\mathbf{F}^{z}_{z_k} = [\, w \, \mathbf{u}^{\rho} - \kappa \, \mathbf{u}_{,z} \,]_{z_k}. \tag{12.25}$$

The first term represents the vertical advective flux of horizontal momentum, and the second term is the vertical diffusive flux of horizontal momentum that parameterizes SGS processes. The signs are such that horizontal momentum of the k grid cell is gained if $w > 0$, or if horizontal velocity decreases upward. The vertical flux of horizontal momentum at the surface $z_{k=0} = \eta$ takes the form

$$\mathbf{F}^{z}_{z_0} = -[\, q_{\text{w}} \, \mathbf{u}^{\rho} + \kappa \, \mathbf{u}^{\rho}_{,z} \,]_{z_0}. \tag{12.26}$$

As with the ocean interior, the surface flux is comprised of two contributions: an advective flux, in this case due to fresh water crossing the ocean surface, and a turbulent flux parameterizing unresolved momentum mixing processes, such as those associated with atmospheric winds. The opposite sign on the fresh water term accounts for the convention that $q_{\text{w}} > 0$ for fresh water entering the ocean (moving downward), whereas $w > 0$ for parcels moving upward.

At the ocean surface, the vertical flux of horizontal momentum is given in terms of the generally unknown surface values for the horizontal velocity $(\mathbf{u}^{\rho})_{z_0}$ and its vertical shear $(\mathbf{u}^{\rho}_{,z})_{z_0}$. Unlike interfaces in the ocean interior, it is unclear how to approximate these terms since there is no ocean above the surface at $z_0 = \eta$. An alternative approach does not seek to approximate the terms individually. Rather, it assumes a continuous total momentum flux across the ocean surface. This approach now places the responsibility of garnering the momentum flux onto a boundary layer model. Such models depend on the multifaceted interactions between wind, waves, sea ice, rivers, etc. Similar considerations apply when formulating the surface boundary conditions for tracer fluxes in Section 11.5.2.

To proceed, assume this boundary layer model has produced a value for the total vertical flux of horizontal momentum, and this flux is written $-(\mathbf{u}^{\rho}_{\text{w}}\,\rho_o\,q_{\text{w}} + \boldsymbol{\tau}^{\text{wind}})$, with $\boldsymbol{\tau}^{\text{wind}}$ a stress imparted by wind (and sea ice) on the ocean surface. Continuity across the ocean surface renders

$$\begin{aligned} -\mathbf{F}^{z}_{z_0} &= [\,q_{\text{w}}\,\mathbf{u}^{\rho} + \kappa\,\mathbf{u}^{\rho}_{,z}\,]_{z=z_0} \\ &\equiv q_{\text{w}}\,\mathbf{u}^{\rho}_{\text{w}} + \rho_o^{-1}\,\boldsymbol{\tau}^{\text{wind}}. \end{aligned} \tag{12.27}$$

When there is no surface fresh water flux, there is a one-to-one relation between the stress $\boldsymbol{\tau}^{\text{wind}}$ and the diffusive momentum flux at the surface

$$\rho_o\,(\kappa\,\mathbf{u}^{\rho}_{,z})_{z_0} = \boldsymbol{\tau}^{\text{wind}} \qquad \text{if } q_{\text{w}} = 0. \tag{12.28}$$

In general, although the total momentum flux is continuous at the ocean surface, the two components of the flux in equation (12.27) need not be. For example, the fresh water velocity $\mathbf{u}^{\rho}_{\text{w}}$ is generally different from the ocean surface velocity $\mathbf{u}^{\rho}_{z_0}$. This may be important when coupling an ocean model to a detailed river model. Nonetheless, most present-day climate models (i.e., models circa 2004) assume the fresh water velocity is equal to the horizontal velocity $\mathbf{u}_{k=1}$ in the top model grid cell.

For some realizations of the model equations, it has been found to be convenient to time step the velocity $\mathbf{u}$ rather than the thickness weighted velocity $h\,\mathbf{u}$ (e.g., Killworth et al., 1991; Dukowicz and Smith, 1994; Griffies et al., 2001). For this case, the surface cell $k = 1$ has a velocity equation given by

$$\begin{aligned} h_{z_1}\,(\partial_t + f\,\hat{\mathbf{z}}\wedge)\,\mathbf{u}_{z_1} =& -\mathbf{u}_{z_1}\,\eta_{,t} - (h\,\mathcal{M}\,\hat{\mathbf{z}}\wedge\mathbf{u}^{\rho})_{z_1} - \nabla\cdot(h\,\mathbf{u}\,\mathbf{u}^{\rho})_{z_1} \\ & - (h\,\nabla p)_{z_1}/\rho_o + (h\,\mathbf{F}^{(\mathbf{u})}_{(\text{horz})})_{z_1} \\ & - [\,w\,\mathbf{u}^{\rho} - \kappa\,\mathbf{u}^{\rho}_{,z}\,]_{z_0} + [\,w\,\mathbf{u}^{\rho} - \kappa\,\mathbf{u}^{\rho}_{,z}\,]_{z_1}. \end{aligned} \tag{12.29}$$

In particular, combining the $\mathbf{u}_{z_1}\,\eta_{,t}$ term with the vertical flux of horizontal momentum at the ocean surface leads to

$$\begin{aligned} -\mathbf{u}_{z_1}\,\eta_{,t} - \mathbf{F}^{z}_{z_0} &= -\mathbf{u}_{z_1}\,\eta_{,t} + \rho_o^{-1}\,(\rho_o\,q_{\text{w}}\,\mathbf{u}^{\rho}_{\text{w}} + \boldsymbol{\tau}^{\text{wind}}) \\ &= \rho_o^{-1}\left(-\mathbf{u}^{\rho}_{z_1}\,\rho_{z_1}\,\eta_{,t} + \rho_o\,q_{\text{w}}\,\mathbf{u}^{\rho}_{\text{w}} + \boldsymbol{\tau}^{\text{wind}}\right), \end{aligned} \tag{12.30}$$

where $\rho_o\,\mathbf{u}_{z_1} = \rho_{z_1}\,\mathbf{u}^{\rho}_{z_1}$. For a Boussinesq fluid, density is set to ρ_o and volume conservation $\eta_{,t} = -\nabla\cdot\mathbf{U} + q_{\mathrm{w}}$ is used to find

$$-\mathbf{u}_{z_1}\,\eta_{,t} + \rho_o^{-1}\,(\rho_o\,q_{\mathrm{w}}\,\mathbf{u}^{\rho}_{\mathrm{w}} + \boldsymbol{\tau}^{\mathrm{wind}}) = \mathbf{u}_{z_1}\,\nabla\cdot\mathbf{U} + q_{\mathrm{w}}\,(\mathbf{u}_{\mathrm{w}} - \mathbf{u}_{z_1}) + \boldsymbol{\tau}^{\mathrm{wind}}/\rho_o. \quad (12.31)$$

Hence, for the commonly assumed case of equal fresh water and surface ocean velocities, $\mathbf{u}^{\rho}_{\mathrm{w}} = \mathbf{u}_{z_1}$, the fresh water flux drops out as an explicit contribution to the Boussinesq momentum budget. It nonetheless remains involved in the budget through its effects on the convergence $-\nabla\cdot\mathbf{U}$ within the surface height equation.

12.2.4.2 *Vertical flux of horizontal momentum through the bottom*

Considerations analogous to those relevant at the ocean surface also apply at the ocean bottom. Yet at the bottom there is no vertical advective momentum flux from the solid earth to the ocean, whereas there was a nonzero flux at the top due to fresh water. Furthermore, the bottom grid cell is typically assumed to have time-independent thickness. So the only contribution to the vertical flux of horizontal momentum arises from the SGS friction. For $k = N_k$ (the deepest grid cell in a column), this friction is commonly written

$$\begin{aligned} -\rho_o\,\mathbf{F}^{z}_{z_{Nk}} &= \rho_o\,(\kappa\,\mathbf{u}^{\rho}_{,z})_{z_k} \\ &= \boldsymbol{\tau}^{\mathrm{bottom}}, \end{aligned} \quad (12.32)$$

where $\boldsymbol{\tau}^{\mathrm{bottom}}$ is the bottom stress vector (with units of pressure). The bottom stress is often associated with currents or tides causing fluid to move over rough small scale topography. However, some SGS interactions between topography and mesoscale eddies act to increase the kinetic energy, as described by Holloway (1992).

12.2.4.3 *Hydrostatic pressure in the surface cell*

Section 4.6.1 (page 59) considered an approximate expression for the continuum hydrostatic pressure at an arbitrary depth

$$p = p_{\mathrm{a}} + p_{\mathrm{s}} + p_{\mathrm{b}}, \quad (12.33)$$

where

$$p_{\mathrm{b}} = g\int_{z}^{0}\rho\,\mathrm{d}z \quad (12.34)$$

is known as the *baroclinic* or *internal* pressure field,

$$p_{\mathrm{s}} = g\,\eta\,\rho(z=0) \quad (12.35)$$

is known as the *surface pressure* field, and p_{a} is the pressure at $z = \eta$ applied by the overlying atmosphere and/or sea ice. The baroclinic pressure is that pressure at depth $z < 0$ arising from ocean fluid contained between $z = 0$ and $z < 0$. The surface pressure is that pressure at $z = 0$ arising from the ocean fluid between $z = 0$ and $z = \eta$, assuming density in this layer is vertically uniform with value $\rho(z=0)$. The surface pressure is negative if $\eta < 0$.

Given the linear relation between pressure and density, it is natural to define the discrete pressure field on the density/tracer points. For a B-grid arrangement of variables, the offset in the horizontal between velocity and tracer/density points means that the hydrostatic pressure at the face of a velocity cell must be computed via a spatial average. The gradient of this pressure across the velocity cell determines the horizontal pressure gradient force acting on the fluid in the cell.

The particular case of a surface cell is worth highlighting. Here, pressure at the western face of a surface velocity cell and at the depth of the velocity point (see Figure 11.1, page 224) is given by

$$p = g\,\overline{\rho\,(|z_1|/2 + \eta)}^{y} + p_{\mathrm{a}}, \tag{12.36}$$

where $\overline{(\ )}^{y}$ is a meridional grid average and η is the surface height on the tracer point. A similar expression is used to estimate pressure on the meridional face of the velocity cell. The first piece of this pressure,

$$p_{\mathrm{b}} \equiv (g\,|z_1|/2)\,\overline{\rho}^{y}, \tag{12.37}$$

is a discretization of the baroclinic pressure in the surface cell. This is the pressure due to fluid between the resting surface $z = 0$ and the velocity point at $z = |z_1|/2$. Note that z_1 has been assumed independent of horizontal position, hence its removal from the meridional average operator. Such is the case for surface grid cells in z-models. The second contribution

$$p_{\mathrm{s}} \equiv g\,\overline{\rho\,\eta}^{y} \tag{12.38}$$

is a discretization of the surface pressure onto the western face of the velocity cell. Again, this is the hydrostatic pressure at $z = 0$ associated with fluid between $z = 0$ and $z = \eta$. Gradients in the surface pressure arise from those in the density weighted free surface height.

For Boussinesq models, it is common to approximate the surface pressure on the western face of the velocity cell as $p_{\mathrm{s}} \approx g\,\rho_o\,\overline{\eta}^{y}$, rather than use the hydrostatic form $g\,\overline{\rho\,\eta}^{y}$. With this approximation, surface pressure gradients arise *solely* from gradients in the free surface height. To maintain consistency with the hydrostatic baroclinic pressure field, while incurring only trivial computational expense, the hydrostatic form $p_{\mathrm{s}} = g\,\overline{\rho\,\eta}^{y}$ can be used even with the Boussinesq equations (Griffies et al., 2001).

12.3 STRATEGIES FOR TIME STEPPING MOMENTUM

The purpose of this section is to develop strategies for time stepping the momentum budget over the course of a discrete time step of length $\Delta\tau$. This goal is complicated by the presence of acoustic waves in non-Boussinesq fluids, and external or barotropic gravity waves in either non-Boussinesq or Boussinesq fluids. Resolving these fast modes requires $\Delta\tau$ to be too small for use in climate studies, where simulations are desired on the order of centuries to millennia. Hence, approximate methods are needed.

Additional complications arise in practice when allowing for the grid cell thicknesses to evolve. Thickness evolution represents a fundamentally new element to the traditional algorithms relevant for constant cells in z-models (e.g., Bryan, 1969; Semtner, 1974; Cox, 1984; Killworth et al., 1991; Dukowicz and Smith, 1994). However, imposing positive cell thicknesses for all cells, including the surface, means that modifications to the constant cell approach should be relatively modest.

12.3.1 Acoustic modes and the quasi-non-Boussinesq approximation

Acoustic waves are three-dimensional fluctuations in the pressure field (e.g., Gill, 1982; Apel, 1987). They travel at roughly $1500\,\mathrm{m\,s^{-1}}$. There is no evidence that resolving acoustic waves is essential for the physical integrity of ocean climate models. Hence, there is little motivation to resolve these waves for ocean climate purposes.

As mentioned in Section 2.4.1 (page 16), use of the hydrostatic approximation in ocean climate models acts to filter all acoustic modes except the Lamb wave. One can filter the Lamb wave either by taking the Boussinesq approximation or by assuming that *in situ* density appearing in mass continuity is a function of a time-independent pressure (Section 5.7.4, page 118). The latter method is described in Durran (1999), where he calls the resulting fluid *pseudo-incompressible*, as well as Greatbatch et al. (2001). An alternative name used this book is *quasi-non-Boussinesq*. This approximation, however, has *not* been found necessary in the non-Boussinesq form of the hydrostatic code described by Griffies et al. (2004).*

12.3.2 Computing *in situ* density

Recall from Section 5.7 (page 114) that density can be written as a function of potential temperature, salinity, and pressure

$$\rho = \rho(\theta, s, p). \tag{12.39}$$

Potential temperature and salinity evolve according to their respective prognostic tracer equations (Section 12.1). Pressure, however, is determined diagnostically in a z-model by vertically integrating the hydrostatic balance, and so pressure is itself a function of density.

As discussed in Griffies et al. (2001), a way to break the pressure-density loop is to approximate density at a model time step τ according to

$$\rho(\tau) = \rho(\theta(\tau), s(\tau), p(\tau - \Delta\tau)). \tag{12.40}$$

A slightly more accurate approximation is to estimate the pressure $p(\tau)$ according to a linear extrapolation using $p(\tau - \Delta\tau)$ and $p(\tau - 2\Delta\tau)$

$$p(\tau) \approx [p(\tau - \Delta\tau) - p(\tau - 2\Delta\tau)] + p(\tau - \Delta\tau). \tag{12.41}$$

Alternatively, one may use an iterative method, as in Dewar et al. (1998). But this approach can be time-consuming. Whatever method is used, the resulting density is a function of a time-dependent pressure field, as it should be. This density is then used for computing pressure $p(\tau)$ whose gradient sets the horizontal pressure gradient.

For purposes of computing mass continuity, a quasi-non-Boussinesq fluid uses the alternative density field

$$\rho(\tau) = \rho(\theta(\tau), s(\tau), -\rho_o\, g\, z), \tag{12.42}$$

where $-\rho_o\, g\, z$ is the hydrostatic pressure associated with a fluid of constant density ρ_o. Note that this is the form for density suggested by Bryan and Cox (1972)

*See Section 2.4.1 for further discussion of the Lamb wave and why it is unimportant for hydrostatic ocean modeling.

and used by many z-models until the middle 1990's (e.g., Cox, 1984; Pacanowski et al., 1991; Pacanowski, 1995). As desired, time tendencies of ρ do not have contributions from $\partial p/\partial t$, and so all acoustic modes are eliminated. As stated above, this approximation has not been found necessary in practice for the non-Boussinesq hydrostatic model of Griffies et al. (2004). Hence, the pressure-dependence of density as computed in equation (12.40) may be suitable.

12.3.3 Barotropic or external gravity waves

In both non-Boussinesq and Boussinesq fluids, external waves are roughly 100 times the speed of the next fastest internal wave or advective signal (see discussion in Section 4.6, page 58). Their propagation causes fluctuations in the ocean surface height as the waves propagate through the fluid columns. Fortunately, these external or barotropic gravity waves are nearly two-dimensional in structure, and so they are largely represented by the simpler dynamics of the vertically integrated fluid column. Consequently, if one can separate or split the fast vertically integrated dynamics and kinematics from the slow and more complicated vertically dependent dynamics and kinematics, there is a chance of substantially increasing the model's efficiency. Providing details for this split forms the bulk of the remainder of this chapter.

12.3.4 General form for the momentum equation

Recall that we wish to time step the semi-discrete momentum equation (12.23), written here as

$$\begin{aligned}(\partial_t + f\,\hat{\mathbf{z}}\wedge)\,(h\,\mathbf{u})_{z_k} = &-(h\,\mathcal{M}\,\hat{\mathbf{z}}\wedge\mathbf{u}^\rho)_{z_k} - \nabla\cdot(h\,\mathbf{u}\,\mathbf{u}^\rho)_{z_k} \\ &-(h\,\nabla p)_{z_k}/\rho_o + (h\,\mathbf{F}^{(\mathbf{u})}_{(\mathrm{horz})})_{z_k} \\ &-[\,w\,\mathbf{u}^\rho - \kappa\,\mathbf{u}^\rho_{,z}\,]_{z_{k-1}} + [\,w\,\mathbf{u}^\rho - \kappa\,\mathbf{u}^\rho_{,z}\,]_{z_k},\end{aligned} \tag{12.43}$$

where $w_{z_0} = -q_{\mathrm{w}}$ (equation (11.16)). To help develop a strategy for time stepping, write this equation in the shorthand form

$$(\partial_t + f\,\hat{\mathbf{z}}\wedge)\,(h\,\mathbf{u})_{z_k} = h_{z_k}\,(\mathbf{G} - \nabla p_{\mathrm{s}}/\rho_o)_{z_k}, \tag{12.44}$$

where

$$\begin{aligned}(h\,\mathbf{G})_{z_k} = &-(h\,\mathcal{M}\,\hat{\mathbf{z}}\wedge\mathbf{u}^\rho)_{z_k} - \nabla\cdot(h\,\mathbf{u}\,\mathbf{u}^\rho)_{z_k} \\ &-(h\,\nabla p_{\mathrm{b}})_{z_k}/\rho_o + (h\,\mathbf{F}^{(\mathbf{u})}_{(\mathrm{horz})})_{z_k} \\ &-[\,w\,\mathbf{u}^\rho - \kappa\,\mathbf{u}^\rho_{,z}\,]_{z_{k-1}} + [\,w\,\mathbf{u}^\rho - \kappa\,\mathbf{u}^\rho_{,z}\,]_{z_k}\end{aligned} \tag{12.45}$$

incorporates advection, the baroclinic pressure gradient, and friction. For the vertically integrated column dynamics, vertically integrated velocity

$$\mathbf{U} = \sum_k h_k\,\mathbf{u}_k \tag{12.46}$$

is time stepped, rather than the depth averaged velocity

$$\overline{\mathbf{u}}^z = \mathbf{U}/D, \tag{12.47}$$

where

$$\begin{aligned} D &= \sum_k h_k \\ &= H + \eta \end{aligned} \tag{12.48}$$

is the time-dependent total depth of ocean fluid at a particular horizontal position. The evolution of $\mathbf{U}$ takes the form

$$\begin{aligned} \partial_t \mathbf{U} &= \partial_t \sum_k (h\,\mathbf{u})_{z_k} \\ &= -f\,\hat{\mathbf{z}} \wedge \mathbf{U} - D\,\nabla p_s/\rho_o + \mathbf{G}, \end{aligned} \tag{12.49}$$

where equation (12.44) was used for $\partial_t (h\,\mathbf{u})_{z_k}$, and the vertically integrated forcing is given by

$$\mathbf{G} = \sum_k (h\,\mathbf{G})_{z_k}, \tag{12.50}$$

with $\mathbf{G}_{z_k}$ defined in equation (12.45). With $\mathbf{G} = 0$, the system reduces to the linear shallow water system with inertia-gravity wave solutions. As discussed in Sections 4.6.3 and 12.8.1, gravity waves have a speed $c_{\text{shallow}} = \sqrt{g\,H}$, which is about $100 - 200\,\text{m}\,\text{s}^{-1}$ in the deep ocean. These barotropic gravity waves determine the discrete time step to be used when temporally discretizing the depth integrated dynamics of fluid columns.

12.3.5 Baroclinic/barotropic split of the velocity

As with the approach of Bryan (1969), Semtner (1974), and Cox (1984), the velocity at an arbitrary depth level k and baroclinic time τ' is split into two components

$$\mathbf{u}_k(\tau') = B_{km}(\tau)\,\mathbf{u}_m(\tau') + [\delta_{km} - B_{km}(\tau)]\,\mathbf{u}_m(\tau'), \tag{12.51}$$

where δ_{km} is the Kronecker delta or unit tensor, and summation is implied over the repeated vertical level index $m \in [1, N_k]$. This equation is an identity valid for any baroclinic times τ' and τ and any operator $B_{km}(\tau)$. In Griffies et al. (2001), $\tau' = \tau + \Delta\tau$ was chosen since the baroclinic system was solved prior to the barotropic. However, it is more convenient to solve the barotropic system first, and $\tau' = \tau$, thus leading to*

$$\begin{aligned} \mathbf{u}_k(\tau) &= B_{km}(\tau)\,\mathbf{u}_m(\tau) + [\delta_{km} - B_{km}(\tau)]\,\mathbf{u}_m(\tau) \\ &\equiv \widehat{\mathbf{u}}_k(\tau) + \overline{\mathbf{u}}^z(\tau). \end{aligned} \tag{12.52}$$

The utility of this velocity field split relies on the form of the *baroclinicity* operator

$$B_{km}(\tau) = \delta_{km} - D(\tau)^{-1}\,h^{\text{u}}_m(\tau), \tag{12.53}$$

where

$$\begin{aligned} D(\tau) &= \sum_k h^{\text{u}}_k(\tau) \\ &= H + \eta^{\text{u}}(\tau) \end{aligned} \tag{12.54}$$

*The baroclinic velocity $\widehat{\mathbf{u}}_k(\tau)$ should not be confused with the effective two-dimensional horizontal transport velocity (equation (9.22) on page 195).

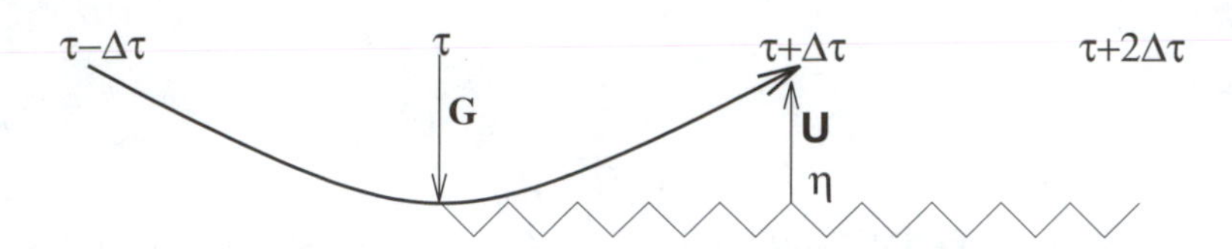

Figure 12.2 Schematic of a split-explicit time stepping scheme for the barotropic and baroclinic modes. A leap-frog scheme forms the basis for this method. Time increases to the right. Baroclinic time steps are denoted by $\tau - \Delta\tau, \tau, \tau + \Delta\tau$, and $\tau + 2\Delta\tau$. The curved line represents a baroclinic leap-frog time step, and the smaller barotropic time steps $N\,\Delta t = 2\,\Delta\tau$ are denoted by the zig-zag line. The vertically integrated forcing $\mathbf{G}(\tau)$ (equation (12.49)) computed at baroclinic time step τ represents the interaction between the barotropic and baroclinic motions. A leap-frog integration carries the surface height and vertically integrated velocity from τ to $\tau + 2\Delta\tau$ using N barotropic time steps of length Δt. During these barotropic time steps, the forcing $\mathbf{G}(\tau)$ and the ocean depth $D(\tau)$ are held fixed in time, along with the tracer, density, and surface fresh water. Time averaging the barotropic fields over the $N+1$ time steps (endpoints included) centers the vertically integrated velocity at the baroclinic time step $\tau + \Delta\tau$. A baroclinic leap-frog time step carries the surface height to $\tau + \Delta\tau$ using the convergence of the time averaged vertically integrated velocity taken from baroclinic time step τ.

is the ocean depth at baroclinic time τ over a column of velocity points, with H the time-independent resting ocean depth. In contrast to the rigid lid case discussed in Section 12.10, where all cells have time-independent thicknesses, the baroclinicity operator used here is based on the instantaneous distribution of cell thicknesses at time τ. However, just as in a rigid lid model, the baroclinicity operator eliminates the depth-independent part of a field and leaves an approximation to the baroclinic portion. Hence, $\widehat{\mathbf{u}}_k$ is an approximation to the baroclinic velocity field, and $\overline{\mathbf{u}}^z$ is an approximation to the barotropic.

If the baroclinic/barotropic split introduced in equation (12.52) is successful, the baroclinic velocity field $\widehat{\mathbf{u}}_k(\tau)$ evolves on a long time scale $\Delta\tau$ and the barotropic velocity $\overline{\mathbf{u}}^z(\tau)$ evolves on the short time scale

$$\Delta t = (2/N)\,\Delta\tau, \tag{12.55}$$

with N determined by the ratio of external to internal gravity wave speeds. The overall method therefore proceeds by separately updating $\widehat{\mathbf{u}}_k(\tau)$ and $\overline{\mathbf{u}}^z(\tau)$ by exploiting the time scale split. Upon doing so, the right-hand side of the identity (12.52) is specified, hence allowing for an update of the full velocity field. This approach defines the *splitting* between the fast and slow modes.

12.4 A LEAP-FROG ALGORITHM

The purpose of this section is to document a leap-frog method for updating both the barotropic and baroclinic velocities. Leap-frog methods have been used for decades, and so it is useful to present this material as a basic scheme from which alternatives are proposed. This method is critiqued in Section 12.5 with alternatives presented in Sections 12.6 and 12.7.

Figure 12.2 and its caption summarizes the algorithm. In the following, discrete baroclinic times and time steps are denoted by the Greek τ and $\Delta\tau$, respectively, whereas the smaller barotropic analogs use the Latin t and Δt.

12.4.1 Barotropic updates with a leap-frog

To start, assume knowledge of the velocity and tracer fields at baroclinic times τ and earlier. Then, one can update the surface height and vertically integrated velocity with a leap-frog scheme using the small barotropic time step Δt. For the non-Boussinesq model, mass conservation over a column of discrete fluid (see equation (11.27), page 228) leads to the time discretized surface height and transport equations

$$\frac{\rho_{z_1}(\tau)\,[\eta^{(\mathrm{b})}(\tau,t_{n+1}) - \eta^{(\mathrm{b})}(\tau,t_{n-1})]}{2\,\Delta t}$$
$$= -\rho_o\,\nabla\cdot\mathbf{U}^{(\mathrm{b})}(\tau,t_n) + \rho_o\,q_{\mathrm{w}}(\tau) - \sum_{k=1}^{N_k} h_k^{\mathrm{t}}(\tau)\,\partial_\tau\rho_{z_k}(\tau), \quad (12.56)$$

$$\frac{\mathbf{U}^{(\mathrm{b})}(\tau,t_{n+1}) - \mathbf{U}^{(\mathrm{b})}(\tau,t_{n-1})}{2\Delta t}$$
$$= -f\,\hat{\mathbf{z}}\wedge\mathbf{U}^{(\mathrm{b})}(\tau,t_n) - D(\tau)\,\nabla\,\tilde{p}_s^{(\mathrm{b})}(\tau,t_n) + \mathbf{G}(\tau). \quad (12.57)$$

For the Boussinesq fluid, the surface height evolution takes on the simpler form

$$\frac{\eta^{(\mathrm{b})}(\tau,t_{n+1}) - \eta^{(\mathrm{b})}(\tau,t_{n-1})}{2\,\Delta t} = -\nabla\cdot\mathbf{U}^{(\mathrm{b})}(\tau,t_n) + q_{\mathrm{w}}(\tau). \quad (12.58)$$

In these equations, a raised "(b)" denotes values of surface height and vertically integrated velocity updated with the small barotropic time steps. The τ time label on $\eta^{(\mathrm{b})}$ and $\mathbf{U}^{(\mathrm{b})}$ denotes the baroclinic time at which certain variables are held for the duration of the barotropic time stepping over a single cycle. The variables held fixed are the vertically integrated forcing $\mathbf{G}(\tau)$, the tracer and density fields, the fresh water flux $q_{\mathrm{w}}(\tau)$ and fresh water density $\rho_{\mathrm{w}}(\tau)$, thickness of a grid cell $h_k^{\mathrm{t}}(\tau)$, and total thickness of an ocean column $D(\tau)$. The time τ is also the time that sets the barotropic time steps via

$$t_n = \tau + n\,\Delta t \quad (12.59)$$

with n an integer. The density scaled surface pressure is evaluated via

$$\tilde{p}_s^{(\mathrm{b})}(\tau,t_n) = g\,\eta^{(\mathrm{b})}(\tau,t_n)\,\rho(\tau)_{k=1}/\rho_o. \quad (12.60)$$

To get started, assume the initial conditions

$$\begin{aligned}\eta^{(\mathrm{b})}(\tau,t_{n=-1}) &= \eta^{(\mathrm{b})}(\tau,t_{n=0})\\ &= \overline{\eta^{(\mathrm{b})}}(\tau),\end{aligned} \quad (12.61)$$

with

$$\overline{\eta^{(\mathrm{b})}}(\tau) = \frac{1}{N+1}\sum_{n=0}^{N}\eta^{(\mathrm{b})}(\tau-\Delta\tau,t_n) \quad (12.62)$$

the time averaged surface height taken from the previous barotropic cycle. Likewise,

$$\begin{aligned}\mathbf{U}^{(b)}(\tau, t_{n=-1}) &= \mathbf{U}^{(b)}(\tau, t_{n=0}) \\ &= \overline{\mathbf{U}^{(b)}}(\tau),\end{aligned} \tag{12.63}$$

with

$$\overline{\mathbf{U}^{(b)}}(\tau) = \frac{1}{N+1} \sum_{n=0}^{N} \mathbf{U}^{(b)}(\tau - \Delta\tau, t_n) \tag{12.64}$$

the time averaged vertically integrated transport. Notably, because of the time averaging, it has not been found necessary to introduce a Robert-Asselin time filter (Haltiner and Williams, 1980) in the barotropic portion of the integration. However, just as a Robert-Asselin time filter, the time averaging reduces the order of accuracy of the method from second to first. Additionally, because of the identification of the fields at $t_{n=-1}$ with those at $t_{n=0}$, the very first time step of a barotropic cycle must be a forward Euler step instead of a leap-frog step.

Upon reaching $t_{n=N} = \tau + 2\Delta\tau$, the vertically integrated velocity is time averaged to produce the updated vertically integrated velocity at baroclinic time $\tau + \Delta\tau$

$$\begin{aligned}\mathbf{U}(\tau + \Delta\tau) &\equiv \overline{\mathbf{U}^{(b)}}(\tau + \Delta\tau) \\ &= \frac{1}{N+1} \sum_{n=0}^{N} \mathbf{U}(\tau, t_n).\end{aligned} \tag{12.65}$$

The surface height at the new baroclinic time step is then determined via a baroclinic leap-frog using the following form

$$\begin{aligned}&\frac{\rho_{z_1}(\tau)\,[\eta(\tau + \Delta\tau) - \eta^{\mathrm{F}}(\tau - \Delta\tau)]}{2\,\Delta\tau} \\ &\qquad = -\rho_o \nabla \cdot \mathbf{U}(\tau) + \rho_o\, q_{\mathrm{w}}(\tau) - \sum_{k=1}^{N_k} h_k^{\mathrm{t}}(\tau)\, \partial_\tau \rho_{z_k}(\tau)\end{aligned} \tag{12.66}$$

for the non-Boussinesq fluid, and

$$\frac{\eta(\tau + \Delta\tau) - \eta^{\mathrm{F}}(\tau - \Delta\tau)}{2\,\Delta\tau} = -\nabla \cdot \mathbf{U}(\tau) + q_{\mathrm{w}}(\tau) \tag{12.67}$$

for the Boussinesq fluid. The use of this "big-leap-frog" scheme for the surface height ensures compatibility between the mass/volume budgets and the tracer budgets. More discussion of this point is provided in Sections 11.2 and 12.5.3.

In general, some form of time filter is needed to maintain integrity of the surface height field due to the leap-frog splitting mode (e.g., Haltiner and Williams, 1980; Durran, 1999) in equations (12.66) and (12.67). There are various approaches available, with the following method discussed in Griffies et al. (2001):

$$\eta^{\mathrm{F}}(\tau) = \overline{\eta^{(b)}}(\tau). \tag{12.68}$$

12.4.2 Updating the baroclinic velocity

By construction, evolution of the velocity field $\widehat{\mathbf{u}}_k(\tau)$ is unaffected by vertically independent forces, such as those from surface pressure gradients. Therefore, it is sufficient to update the "primed" velocity, which is affected by all terms in the the momentum equation (12.44) except the surface pressure gradient. Without the surface pressure gradient, the primed velocity can be updated using the longer baroclinic time step. If we use a leap-frog scheme for the time tendency, it is necessary to use the time filtered lagged velocity

$$\mathbf{u}_k^{\mathrm{R}}(\tau-\Delta\tau) = \mathbf{u}_k(\tau-\Delta\tau) + (\alpha/2)\,[\mathbf{u}_k(\tau) - 2\,\mathbf{u}_k(\tau-\Delta\tau) + \mathbf{u}_k^{\mathrm{R}}(\tau-2\Delta\tau)]. \quad (12.69)$$

Robert-Asselin filtering with $\alpha = 0.05$ has been found sufficient in applications to suppress splitting between the two leap-frog branches in the baroclinic system. It is also useful to dampen fast dynamics that may partially leak through the baroclinicity operator due to the generally imperfect separation between the slow and fast dynamics. We have more to say regarding the utility of such filters in Section 12.5.1, where it is noted that significant temporal noise can be present in regions where the solution has large temporal variability, such as the equator.

The baroclinic piece of the primed velocity $\mathbf{u}_k'(\tau+\Delta\tau)$ yields the baroclinic velocity

$$\widehat{\mathbf{u}}_k(\tau+\Delta\tau) = B_{km}(\tau+\Delta\tau)\,\mathbf{u}_m'(\tau+\Delta\tau), \quad (12.70)$$

To complete an update of the full velocity field, add the updated baroclinic velocity to the updated barotropic velocity

$$\mathbf{u}_k(\tau+\Delta\tau) = \widehat{\mathbf{u}}_k(\tau+\Delta\tau) + \mathbf{U}(\tau+\Delta\tau)/D(\tau+\Delta\tau). \quad (12.71)$$

Knowledge of the updated velocity, along with updated tracer $T(\tau+\Delta\tau)$, allows for construction of the vertically integrated forcing $\mathbf{G}(\tau+\Delta\tau)$, and hence movement onto the next time step. This completes a single split-explicit time step cycle for velocity using a leap-frog scheme.

12.5 DISCRETIZATION OF TIME TENDENCIES

How one discretizes the time tendency largely determines how other terms can be time discretized. The most popular method for ocean climate models is the leap-frog approach described in Section 12.4 (Griffies et al., 2000a). This method has served modelers for decades. Unfortunately, it has fundamental problems.

The purpose of this section is to discuss some issues which arise when considering how to time discretize the equations of an ocean model. In particular, problems with the use of a leap-frog time tendency are highlighted, with possible improvements proposed. The issues addressed here and in the subsequent sections are generic to the Boussinesq and non-Boussinesq equations. Hence, for simplicity of presentation, it is sufficient to focus on the Boussinesq equations for the remainder of this chapter.

12.5.1 Leap-frog noise

We start this section with a brief comment regarding some experience with temporal noise using a leap-frog scheme in global climate models based on the Modular

Ocean Model (Griffies et al., 2004). The use of a Robert-Asselin time filter has been found to be sufficient to maintain numerical stability. That is, the two leap-frog solution branches do not diverge. However, temporal noise can be nontrivial in certain regions. For the one-degree class of global ocean models, most temporal variability is at the equator. Off-equatorial solutions tend to be more slowly evolving due to geostrophy (Section 4.7.2). Leap-frog noise is therefore much more significant in equatorial regions, where noise can result in the strong zonal currents completely switching directions over the course of a single time step. This simulation behavior can occur over a broad range of time filtering strengths. It is very unphysical and highly unsatisfying. This result provides a strong reason to consider schemes that do not admit time splitting.

12.5.2 Thickness weighting

As discussed in Section 11.5.3, integrating the continuous tracer budget over the thickness of a discrete grid cell leads to the semi-discrete tracer equation

$$\partial_t(h\,C) = -\nabla_z \cdot (h\,\mathbf{F}) + F^z_k - F^z_{k-1} + h\,\mathcal{S}^{(C)} \tag{12.72}$$

where $\mathbf{F}$ here embodies both the advective and turbulent fluxes, and the discrete vertical label k is exposed just when needed. There are two possible starting points for time discretizing this equation. One can time step $h\,C$ as a product, which results in the *thickness weighted* approach. Alternatively, one can perform the product rule

$$h\,\partial_t\,C = -C\,\partial_t\,h - \nabla_z \cdot (h\,\mathbf{F}) + F^z_k - F^z_{k-1} + h\,\mathcal{S}^{(C)} \tag{12.73}$$

and separately time discretize $\partial_t\,C$ and $\partial_t\,h$ (Killworth et al., 1991; Dukowicz and Smith, 1994; Griffies et al., 2001). In this method, the tracer concentration budget has an extra time tendency due to undulations of the cell thickness. This term has the form of an added source, as it cannot be written as the convergence of a flux. Hence, it deviates from a finite volume formulation, whereby one aims to formulate finite difference tendencies in terms of flux convergences. This extra term also complicates the expression for a time discrete total tracer content in the model (Griffies et al., 2001). It is for these reasons that the thickness weighted approach is preferred in this book.

12.5.3 Compatible volume and tracer budgets

As mentioned in Section 11.2, the budget for tracer in a Boussinesq model reduces to that for volume in the case that the tracer is a space-time constant. This is a trivial statement in the space-time continuum, and it remains in practice a trivial constraint on how one discretizes in space. However, when discretizing time, it constrains the tracer and volume budgets to employ the same time stepping scheme. We use the term *compatible* to refer to discrete tracer and volume budgets that are so connected.

To help understand the importance of compatible tracer and volume budgets, consider the case of a discrete ocean model with tracer concentration the same uniform value throughout the model. Allow wind to blow on the surface, yet do not introduce any boundary fluxes of tracer. Wind forcing causes the surface height to evolve. When the volume and tracer budgets are compatible, evolution of the surface height will not lead to evolution of tracer concentration. That is, tracer concentration remains fixed, as it should since there are no boundary fluxes of tracer. If the budgets are incompatibly discretized, such as by using different time stepping schemes for the two budgets, then this experiment would result in nonzero tendencies for tracer concentration. This behavior is undesirable, since changes in tracer concentration lead to spurious changes in density, which lead to spurious currents, and further expose the coupled climate system to spurious feedbacks with the sea ice and atmosphere components. Hence, compatible budgets are critical.

As discussed by Griffies et al. (2001), compatible tracer and volume budgets are said to maintain a *locally conservative* tracer budget. As further discussed in that paper, local conservation does not necessarily lead to global conservation. Indeed, the two can be mutually incompatible in some time stepping schemes, such as a time filtered leap-frog scheme discussed in Section 12.4. We return to this point below in Section 12.5.4. Griffies et al. (2001) preferred local conservation over global, so that the above wind experiment is satisfied. Unfortunately, lack of global conservation can lead to unsatisfying trends in tracer content which can be unacceptable for many purposes, especially climate simulations. Hence, ocean climate simulations require both local and global tracer conservation.

12.5.4 Absence of global tracer conservation with leap-frog tendencies

Compatible three-time level leap-frog discretizations of the volume and thickness weighted tracer tendencies leads to

$$\partial_t h \approx \frac{h(\tau+\Delta\tau) - h(\tau-\Delta\tau)}{2\,\Delta\tau} \tag{12.74}$$

$$\partial_t (h\,C) \approx \frac{h(\tau+\Delta\tau)\,C(\tau+\Delta\tau) - h(\tau-\Delta\tau)\,C(\tau-\Delta\tau)}{2\,\Delta\tau}. \tag{12.75}$$

The analogous discretization of the tracer concentration equation (12.73) was considered by Griffies et al. (2001), whereby

$$h\,\partial_t C + C\,\partial_t h \approx \frac{h(\tau)\,[C(\tau+\Delta\tau) - C(\tau-\Delta\tau)]}{2\,\Delta\tau} + \frac{C(\tau)\,[h(\tau+\Delta\tau) - h(\tau-\Delta\tau)]}{2\,\Delta\tau}. \tag{12.76}$$

Although they have quite different appearances, both discretizations (12.75) and (12.76) are second order accurate in time. Hence, they are both legitimate leap-frog methods for discretizing the tracer and volume tendencies.

Use of either (12.75) or (12.76) requires time filtering on tracer concentration C and cell thickness h to suppress the unphysical leap-frog time splitting mode.* Time filtering reduces the order of accuracy from second order to first (Section 2.3.5 of Durran, 1999). There is an alternative approach that has been used by models inherited from Cox (1984). Here, one leap-frog branch is periodically removed via a two-time level Euler backward or forward time step. This approach, however,

*Recall that for a z-model, only the surface cell has a time-dependent thickness $h = \Delta z + \eta$. However, it is useful to retain some generality since models with generalized vertical coordinates can have interior cells with time-dependent thicknesses.

is unsatisfying since it introduces an unphysical period to the model dynamics. It also complicates model algorithms and diagnostics.

Time filtering borrows from the past and future to alter the present. This borrowing, when applied to the prognostic variables, breaks global conservation of fields such as tracer and volume. Mathematically, lack of conservation can be seen by exposing the filtering operator placed on the lagged variables,

$$h(\tau - \Delta\tau) \rightarrow h^{\mathrm{F}}(\tau - \Delta\tau) \tag{12.77}$$

$$C(\tau - \Delta\tau) \rightarrow C^{\mathrm{R}}(\tau - \Delta\tau). \tag{12.78}$$

Note that the time filter operator can generally be different for the thickness and tracer concentration. As in Griffies et al. (2001), thickness is often time filtered via time averaging over the many small barotropic time steps (equation (12.68)), whereas a Robert-Asselin filter is typically used for the tracer (equation (12.69)). Either filter suppresses time splitting, with the time average on the surface height more successful than the Robert-Asselin in also suppressing the B-grid spatial checkerboard mode afflicting the surface height field. We have more to say about the checkerboard mode in Section 12.9.

Rearranging the time tendency, with the filter operator exposed, leads to the time discrete thickness weighted tracer budget (see Section 11.5)

$$\begin{aligned} h(\tau+\Delta\tau)\, C(\tau+\Delta\tau) &= h^{\mathrm{F}}(\tau-\Delta\tau)\, C^{\mathrm{R}}(\tau-\Delta\tau) \\ &\quad - 2\,\Delta\tau\, \nabla_z \cdot [h(\tau)\, \mathbf{F}(\tau, \tau-\Delta\tau)] \\ &\quad + 2\,\Delta\tau\, [F^z_k(\tau, \tau-\Delta\tau) - F^z_{k-1}(\tau, \tau-\Delta\tau)] \\ &\quad + h(\tau)\, \mathcal{S}^{(C)}. \end{aligned} \tag{12.79}$$

The double argument on the flux components manifests the use of time step τ for the advective flux.* The $\tau - \Delta\tau$ time label is for dissipative fluxes (Section 12.8.4).

To illustrate problems with conservation, it is sufficient to assume zero tracer sources, zero tracer fluxes entering the domain through the solid boundaries, yet nonzero fluxes entering through the ocean surface. Global integration of the tracer budget then leads to

$$\begin{aligned} \int \mathrm{d}x\, \mathrm{d}y\, h(\tau+\Delta\tau)\, C(\tau+\Delta\tau) &= \int \mathrm{d}x\, \mathrm{d}y\, h^{\mathrm{F}}(\tau-\Delta\tau)\, C^{\mathrm{R}}(\tau-\Delta\tau) \\ &\quad - 2\,\Delta\tau \int \mathrm{d}x\, \mathrm{d}y\, F^z_{k=0}(\tau, \tau-\Delta\tau). \end{aligned} \tag{12.80}$$

In this equation, $F^z_{k=0}(\tau, \tau - \Delta\tau)$ represents the advective and turbulent fluxes which cross the ocean surface, with the advective fluxes associated with fresh water (i.e., precipitation, evaporation, and river runoff), and turbulent fluxes associated with such processes as sensible and latent heating. Absence of global tracer conservation is manifest, since the integrated tracer at $\tau + \Delta\tau$ is not equal to the integrated tracer at $\tau - \Delta\tau$ plus boundary contributions. Instead, the tracer content at $\tau - \Delta\tau$ is a filtered value, with

$$\int \mathrm{d}x\, \mathrm{d}y\, h^{\mathrm{F}}(\tau-\Delta\tau)\, C^{\mathrm{R}}(\tau-\Delta\tau) \neq \int \mathrm{d}x\, \mathrm{d}y\, h(\tau-\Delta\tau)\, C(\tau-\Delta\tau). \tag{12.81}$$

*With a leap-frog time tendency, centered difference advection schemes temporally discretize advection on the τ time step. Upwind biased schemes, however, discretize advection in time just like diffusion (Section 12.8.2), and so place the operator at the $\tau - \Delta\tau$ time step.

Furthermore, when using a Robert-Asselin filter, the filtered field at time $\tau - \Delta\tau$ is not known until time τ. Again, filtering borrows from the past and future, and it is not constrained to maintain global conservation. Even if the surface height were constant, as in a rigid lid model, inequality (12.81) would hold in the presence of nontrivial surface flux forcing. In summary, the result of time filtering with the leap-frog method is a scheme that does not conserve global tracer. Similar arguments hold for lack of volume conservation when time filtering the surface height.

The degree to which global tracer and volume are modified depends on details of the model, such as forcing, time steps, physical processes, etc. Hence, it is difficult to make rigorous statements regarding the level of the non-conservation, or whether the tracer content will exhibit a trend or random fluctuations. Instead, it is more fruitful to jettison the leap-frog method in favor of an alternative where tracer content is conserved.

12.5.5 Two-time level discretizations of the time tendency

A compatible two-time level discretization of the tendency for thickness and thickness weighted tracer leads to

$$\partial_t h \approx \frac{h(\tau + \Delta\tau) - h(\tau)}{\Delta\tau} \tag{12.82}$$

$$\partial_t (h\,C) \approx \frac{h(\tau + \Delta\tau)\,C(\tau + \Delta\tau) - h(\tau)\,C(\tau)}{\Delta\tau}. \tag{12.83}$$

Although naively first order accurate, the degree of temporal accuracy actually depends on the details of the way the forcing terms are time discretized. We return to this point in Section 12.6.2, where it is shown that time staggering allows for this update to be second order accurate.

The following points represent a summary of the motivation for using two-time level discretizations of the time tendency.

- No time steps are skipped and so there are no time splitting modes. Hence, there is no need to apply Robert-Asselin time filtering.

- So long as the form of the thickness and tracer budgets remain compatible upon splitting into baroclinic and barotropic modes, the two-time level schemes manifest both local and global tracer conservation.

- Since the continuum equations involve only first order temporal derivatives, only a single time level is formally necessary to begin a discrete integration. The leap-frog scheme, in contrast, requires two time levels to begin an integration. These two time levels double the size of restart files needed when stopping and restarting the model integration.

- Time step stability is enhanced relative to the leap-frog scheme for many processes when using a two-time level tendency. We detail these issues in Section 12.8.

12.5.6 Implicit physics with thickness weighted updates

Vertical physical processes are generally represented implicitly in time in order to maintain numerical stability in the presence of strong vertical mixing. It is straightforward to employ implicit methods because vertical processes are generally localized to a single vertical column, so the dimension of the matrix inversion is set just by the number of depth levels. These sorts of matrix inversions are modest in size and so can be readily handled in ocean models.

Details of how to numerically handle implicit vertical processes are straightforward when integrating the familiar tracer concentration and velocity equations. There are, however, some points that need to be kept in mind when updating the thickness weighted tracer concentration and thickness weighted velocity. We discuss these issues in this section.

12.5.6.1 At what time do we discretize the thickness?

Consider the time continuous thickness weighted tracer equation, focusing on vertical processes*

$$\partial_t (h\, C) = -h\, \partial_z F^z. \tag{12.84}$$

As discussed in Section 12.5.2, some models rearrange this equation into an update of the tracer concentration

$$\partial_t C = -(C/h)\, \partial_t h - \partial_z F^z \tag{12.85}$$

prior to time discretization. In this way, the vertical convergence, $-\partial_z F^z$, appears in the same form as in a rigid lid model.

In contrast to the rigid lid model, the vertical derivative operator ∂_z is based on time-dependent thicknesses. For models updating tracer concentration as in equation (12.85), ∂_z is temporally discretized using the vertical thickness at time τ. This time placement is necessitated by self-consistency, as the thickness in the source-like term $(C/h)\, \partial_t h$ is also evaluated at τ. Hence, the discrete time version of equation (12.85) is (using leap-frog for the tendency)

$$\frac{C(\tau+\Delta\tau) - C(\tau-\Delta\tau)}{2\,\Delta\tau} = -\frac{C(\tau)}{h(\tau)}\,\frac{h(\tau+\Delta\tau) - h(\tau-\Delta\tau)}{2\,\Delta\tau} + \frac{F^z_k - F^z_{k-1}}{h(\tau)}. \tag{12.86}$$

That is, the time discretized operator ∂_z is based on $h(\tau)$ regardless of when the vertical flux is evaluated.†

The thickness weighted equation has the continuous time result

$$\begin{aligned} \partial_t (h\, T) &= -h\, \partial_z F^z & (12.87)\\ &= -(F^z_{k-1} - F^z_k) & (12.88) \end{aligned}$$

where $h\, \partial_z$ is translated into a dimensionless vertical difference operator. This result is carried over to the discrete time. For a two-time level tendency, we thus have

$$\frac{h(\tau+\Delta\tau)\, C(\tau+\Delta\tau) - h(\tau)\, C(\tau)}{\Delta\tau} = F^z_k - F^z_{k-1}, \tag{12.89}$$

---

*Both tracer and velocity equations share the same numerical considerations for vertical processes. Hence, it is sufficient to focus on the tracer equation.

†The vertical flux may be evaluated at $\tau - \Delta\tau$ when using an explicit vertical diffusion scheme with leap-frog time tendencies; at time τ for vertical centered advection or vertical diffusion with two-time tendencies; or time $\tau + \Delta\tau$ for implicit vertical processes such as diffusion.

or for the three-time level leap-frog tendency

$$\frac{h(\tau+\Delta\tau)\,C(\tau+\Delta\tau) - h(\tau-\Delta\tau)\,C(\tau-\Delta\tau)}{2\,\Delta\tau} = F^z_k - F^z_{k-1}. \tag{12.90}$$

Rearrangement of the two-time level result leads to

$$C(\tau+\Delta\tau) = \frac{h(\tau)\,C(\tau)}{h(\tau+\Delta\tau)} + \Delta\tau\left(\frac{F^z_k - F^z_{k-1}}{h(\tau+\Delta\tau)}\right), \tag{12.91}$$

and the analogous leap-frog result is

$$C(\tau+\Delta\tau) = \frac{h(\tau-\Delta\tau)\,C(\tau-\Delta\tau)}{h(\tau+\Delta\tau)} + 2\,\Delta\tau\left(\frac{F^z_k - F^z_{k-1}}{h(\tau+\Delta\tau)}\right). \tag{12.92}$$

In both cases, the vertical thickness used to compute the discrete vertical convergence is centered at time $\tau+\Delta\tau$, instead of τ. Again, this result holds regardless of what time the vertical fluxes are evaluated.

In addition to determining the time to evalute the thickness for the vertical convergence operator, there is yet another thickness that is used to discretely compute the vertical turbulent fluxes

$$F^z = -\kappa\,\partial_z C. \tag{12.93}$$

When evaluating vertical mixing processes implicitly in time, strictly speaking each piece of this flux should be evaluated at time $\tau+\Delta\tau$. That is, the diffusivity κ, the vertical thickness appearing in ∂_z, and the tracer C should all be formally at $\tau+\Delta\tau$. However, doing so is not possible for modern turbulence schemes, which determine κ based on properties of the evolving fluid motion. Hence, in practice it is only the tracer concentration C that is evaluated at the future time $\tau+\Delta\tau$. The diffusivity and generally the thickness are evaluated at time τ.

12.5.6.2 Updating thickness when using implicit vertical physics

The surface height is updated to time $\tau+\Delta\tau$ along with the vertically integrated velocity $\mathbf{U}$. Updating $\mathbf{U}$ requires knowledge of the vertically integrated forcing of the full velocity equations, where all but the surface pressure gradient and Coriolis force are held constant over the course of the small barotropic time steps (Sections 12.4.1 and 12.7). Vertical friction is among the forcing terms contributing to the vertically integrated forcing. It is commonly implemented implicitly in time, as just discussed. Yet to compute implicit vertical friction using thickness weighted equations requires the updated thickness $h(\tau+\Delta\tau)$. Hence, this creates a problem with causality.

The problem is resolved by recognizing that all that is required from vertical friction during the update of the barotropic dynamics is its vertical integral. All terms appearing in the ocean interior cancel upon vertical integration, whether time discretizing friction explicitly or implicitly. The result is just the bottom and surface boundary forcings

$$-\int_{-H}^{\eta} \mathrm{d}z\,\partial_z F^z = \mathrm{smf} - \mathrm{bmf}, \tag{12.94}$$

where smf is the surface momentum flux, and bmf is the bottom momentum flux. These fluxes are generally evaluted at time τ, even when implementing vertical friction implicitly in time. Hence, the surface and bottom forces are known and so can be used to force the barotropic dynamics.*

12.6 A TIME STAGGERED ALGORITHM

The purpose of this section is to present a scheme which remedies the problems with the leap-frog based schemes discussed in Section 12.5. The alternative scheme can be interepreted as a time staggered method, where tracer and velocity are offset a half time step from one another. This approach is analogous to the spatially staggered methods commonly used in ocean models (Section 10.2).

12.6.1 Unstaggered time stepping

Before detailing a time staggered approach, it is useful to review the unstaggered leap-frog based scheme commonly used in ocean models. In this method, all variables are assumed to live at integer time steps.

A leap-frog update of the surface height is given by the big leap-frog step (equation (12.67))

$$\frac{\eta^{\tau+1} - \overline{\eta}^{\tau-1}}{2\,\Delta\tau} = -\nabla\cdot\mathbf{U}(\tau) + q_{w}(\tau) + \mathcal{S}^{(\eta)} \tag{12.95}$$

where $\mathcal{S}^{(\eta)}$ represents a possible source for volume in a fluid column. Compatibility is maintained with the tracer source so long as

$$\mathcal{S}^{(\eta)} = \mathcal{S}^{(C)}(C=1). \tag{12.96}$$

A shorthand notation was introduced in equation (12.95) whereby, for example,

$$\eta^{\tau+1} = \eta(\tau+\Delta\tau). \tag{12.97}$$

The bar on $\overline{\eta}^{\tau-1}$ symbolizes a time average taken over the barotropic cycle (Section 12.4.1).† Again, time averaging acts as a filter to suppress the leap-frog computational mode.

A compatible discretization of the thickness weighted tracer equation takes the form

$$\begin{aligned}\frac{h^{\tau+1}\,C^{\tau+1} - \overline{h}^{\tau-1}\,\overline{C}^{\tau-1}}{2\,\Delta\tau} =& -\nabla_{z}\cdot(h^{\tau}\,\mathbf{u}^{\tau}\,C^{\tau} + h^{\tau}\,\mathbf{F}^{\tau-1})\\ &+ [w^{\tau}\,C^{\tau} + F_{z}^{\tau+1}]_{k}\\ &- [w^{\tau}\,C^{\tau} + F_{z}^{\tau+1}]_{k-1}\\ &+ h^{\tau}\,\mathcal{S}^{(C)}.\end{aligned} \tag{12.98}$$

The lagged tracer concentration $\overline{C}^{\tau-1}$ has been filtered using a Robert-Asselin time filter to suppress the leap-frog mode. Vertical SGS tracer fluxes F_z are discretized at $\tau+\Delta\tau$, since vertical processes are generally included implicitly in time to maintain stability with large mixing coefficients. In contrast, horizontal SGS fluxes $\mathbf{F}$

*One may wish to evalute the bottom friction implicitly in time, for cases with large bottom drag and/or thin cells. Doing so requires further consideration.

†In equation (12.67), $\overline{\eta}$ is written η^{F}.

are discretized at time $\tau - \Delta\tau$. Updating thickness $h^{\tau+1}$ before thickness weighted tracer $h^{\tau+1} C^{\tau+1}$ allows for the tracer concentration to be evaluated

$$C^{\tau+1} = \frac{h^{\tau+1} C^{\tau+1}}{h^{\tau+1}}. \tag{12.99}$$

An unstaggered three-time level update of the horizontal velocity takes the form

$$\begin{aligned}\frac{h^{\tau+1}\,\mathbf{u}^{\tau+1} - \overline{h}^{\tau-1}\,\overline{\mathbf{u}}^{\tau-1}}{2\,\Delta\tau} = &- \mathcal{M}^{\tau}\,\hat{\mathbf{z}} \wedge h^{\tau}\,\mathbf{u}^{\tau} + (w^{\tau}\,\mathbf{u}^{\tau})_{k} - (w^{\tau}\,\mathbf{u}^{\tau})_{k-1} \\ &- \nabla_{z} \cdot (h^{\tau}\,\mathbf{u}^{\tau}\,\mathbf{u}^{\tau}) \\ &- h^{\tau}\,(f\,\hat{\mathbf{z}} \wedge \mathbf{u})_{\text{semi-implicit}} \\ &- h^{\tau}\,\nabla_{z}\,(p^{\tau}/\rho_{o}) \\ &+ h^{\tau}\,(\mathbf{F}^{\mathbf{u}})^{(\tau-1,\tau+1)}.\end{aligned} \tag{12.100}$$

The subscript "semi-implicit" refers to the semi-implicit treatment of the Coriolis force in which*

$$(f\,\hat{\mathbf{z}} \wedge \mathbf{u})_{\text{semi-implicit}} = f\,\hat{\mathbf{z}} \wedge (\alpha\,\mathbf{u}^{\tau} + \beta\,\mathbf{u}^{\tau+1}) \tag{12.101}$$

where $\alpha + \beta = 1$ (see Section 12.8.3). Second order accuracy is recovered by setting $\alpha = \beta = 1/2$. Bars on the $\tau - \Delta\tau$ variables in equation (12.100) refer to time filtered values. The superscript $(\tau - 1, \tau + 1)$ indicates that horizontal frictional dissipation is handled with a forward time step, and vertical frictional dissipation is handled with an implicit time step.

12.6.2 Volume and tracer updates with time staggering

Figure 12.3 illustrates a staggered temporal discretization of tracer and velocity. By convention, we define tracer, density, and pressure on half-integer time steps, whereas velocity is discretized on integer time steps. The surface height is discretized at both integer and half-integer time steps.

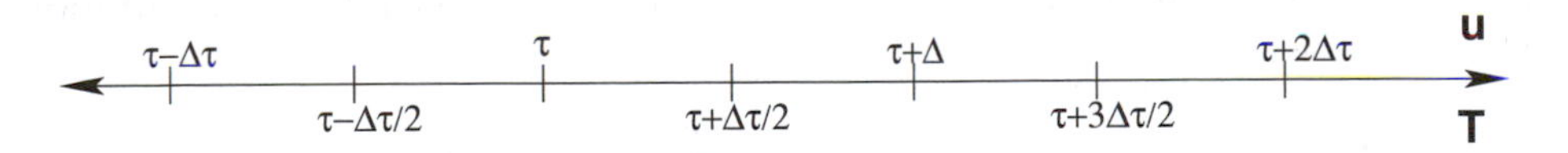

Figure 12.3 Schematic of the staggering used between tracer (on half-integer time steps) and velocity (on integer time steps), where the distance between integer and half-integer time steps is $\Delta\tau/2$. Along with tracer, the hydrostatic pressure p and density ρ are discretized at half-integer time steps. The surface height must be discretized at both integer and half-integer time steps in order to specify the thickness of the tracer and velocity cells.

*A semi-implicit treatment of the Coriolis force is available in B-grid models, where both components of the horizontal velocity are spatially coincident (see Figure 10.1). In contrast, C-grid models treat the Coriolis force explicitly since their velocity components are not spatially coincident. Note that most global models use a velocity time step that is sufficient to resolve inertial oscillations using an explicit treatment of the Coriolis force. However, when coupling to an ice model, it has been found that inertial instabilities can arise (Mike Winton, personal communication 2003), thus necessitating the use of semi-implicit Coriolis discretization.

Time staggering allows for a second order accurate discretization of the surface height onto a new half-integer time step

$$\frac{\eta^{\tau+1/2} - \eta^{\tau-1/2}}{\Delta\tau} = -\nabla \cdot \mathbf{U}(\tau) + q_{\mathrm{w}}(\tau) + \mathcal{S}^{(\eta)}. \tag{12.102}$$

This discrete time equation has the same form as the unstaggered leap-frog discretization (12.95). In particular, one can interpret equation (12.102) as one branch of an unstaggered leap-frog scheme whose time steps satisfy

$$2\Delta\tau_{\mathrm{leap}} = \Delta\tau_{\mathrm{staggered}}. \tag{12.103}$$

Conveniently, the surface height in equation (12.102) does not skip any of its half-integer time steps. Hence, there are no computational modes. Herein lies the basis for the staggered approach and the reason that it does not require time filtering. Notably, without time filtering, the second order accuracy of equation (12.102) is maintained.

A compatible staggered time discretization of thickness weighted tracer takes the form

$$\begin{aligned}\frac{h^{\tau+1/2}\, C^{\tau+1/2} - h^{\tau-1/2}\, C^{\tau-1/2}}{\Delta\tau} &= -\nabla_z \cdot (h\, \mathbf{u}^{\tau}\, C^{\tau-1/2} + h\, \mathbf{F}^{\tau-1/2}) \\ &\quad + [w^{\tau}\, C^{\tau-1/2} + F_z^{\tau+1/2}]_k \\ &\quad - [w^{\tau}\, C^{\tau-1/2} + F_z^{\tau+1/2}]_{k-1} \\ &\quad + h\, \mathcal{S}^{(\mathrm{C})}.\end{aligned} \tag{12.104}$$

Note that vertical SGS fluxes are discretized implicitly at $\tau + \Delta\tau/2$. It is assumed that velocity is known at time τ prior to updating the tracer to time $\tau + \Delta\tau/2$. Again, there is no leap-frog splitting mode since tracer does not skip any of its half-integer time steps, and because velocity is always evaluated at an integer time step. Hence, tracer does not require a time filter.*

As for the unstaggered method of Section 12.6.1, an update of thickness $h^{\tau+1/2}$ prior to thickness weighted tracer $h^{\tau+1/2}\, C^{\tau+1/2}$ allows for the tracer concentration to be evaluated

$$C^{\tau+1/2} = \frac{h^{\tau+1/2}\, C^{\tau+1/2}}{h^{\tau+1/2}}. \tag{12.105}$$

Because the surface height and thickness weighted tracer equations are compatible, and because there is no time averaging applied to the prognostic fields η and C, volume is globally conserved and tracer is conserved locally and globally.

Equation (12.104) has three thicknesses whose position in time have yet to be determined. One choice is to evaluate them all at $h^{\tau-1/2}$, which corresponds to the tracer time step. However, it is necessary to provide a consistent thickness of the velocity cell on an integer time step. We therefore prescribe a thickness weighted horizontal advection velocity according to

$$h\, \mathbf{u} = h^{\tau}\, \mathbf{u}^{\tau}. \tag{12.106}$$

*As discussed in Section 12.8.2, the natural method for implementing tracer advection with a forward time stepping scheme is via an upwind biased scheme. Centered differences are unstable.

This choice for the thickness weighted advection velocity must be maintained when computing the horizontal advection velocity acting to advect velocity (Section 12.6.3).

The remaining unspecified thicknesses multiplying the horizontal SGS flux and the tracer source are not constrained by compatibility to be at a particular time level. It is convenient to set

$$h\,\mathbf{F}^{\tau-1/2} = h^{\tau-1/2}\,\mathbf{F}^{\tau-1/2} \tag{12.107}$$

$$h\,\mathcal{S}^{(C)} = h^{\tau-1/2}\,\mathcal{S}^{(C)}. \tag{12.108}$$

With these choices, the tracer equation is updated according to

$$\begin{aligned}\frac{h^{\tau+1/2}\,C^{\tau+1/2} - h^{\tau-1/2}\,C^{\tau-1/2}}{\Delta\tau} &= -\nabla_z\cdot(h^{\tau}\,\mathbf{u}^{\tau}\,C^{\tau-1/2} + h^{\tau-1/2}\,\mathbf{F}^{\tau-1/2}) \\ &+ [w^{\tau}\,C^{\tau-1/2} + F_z^{\tau+1/2}]_k \\ &- [w^{\tau}\,C^{\tau-1/2} + F_z^{\tau+1/2}]_{k-1} \\ &+ h^{\tau-1/2}\,\mathcal{S}^{(C)}.\end{aligned} \tag{12.109}$$

12.6.3 Velocity updates with time staggering

The velocity equation requires some added consideration, both because of the splitting between barotropic and baroclinic modes, and because of time staggering. We now highlight the issues.

- Thickness of a grid cell must be mapped from the tracer cell to velocity cell. A technology for doing so on a B-grid is described in Griffies et al. (2004). Application of this technology is assumed when writing the velocity cell thickness in the following.

- Horizontal friction is stably time stepped with a forward time step. Hence, it is evaluated at time step τ when using a staggered scheme, and the velocity cell thickness weighting friction is at the same time. Vertical friction is updated implicitly and so is evaluated at $\tau + \Delta\tau$, with the cell thickness discretized according to the discussion in Section 12.5.6.

- As with the unstaggered time scheme discussed in Section 12.6.1, it is convenient to time step the Coriolis force either explicitly in time, thus placing it at time τ, or via a semi-implicit method for cases where added stability is needed to suppress numerically unstable inertial oscillations (Section 12.8.3).

- The hydrostatic pressure is naturally time discretized on the half-integer time steps used by the tracers, since hydrostatic pressure is proportional to vertically integrated density via

  $$p_{,z} = -\rho\,g. \tag{12.110}$$

  Hence, we use $p^{\tau+1/2}$ for the staggered scheme.

- The time tendency due to velocity self-advection is given by

  $$\partial_t\,(h\,\mathbf{u})_{\text{adv}} = [(w\,\mathbf{u})_k - (w\,\mathbf{u})_{k-1}] - \nabla_z\cdot(h\,\mathbf{u}\,\mathbf{u}) - \mathcal{M}\,\hat{\mathbf{z}}\wedge h\,\mathbf{u}. \tag{12.111}$$

The thickness weighted horizontal advection velocity $h\,\mathbf{u}$ is a remapped version of the same velocity used for advecting tracers (equation (12.106)). Advection is generally unstable with a forward time stepping scheme when both the advection velocity and advected field live at the same time step (e.g., Section 2.3 of Durran, 1999). This is the case for velocity self-advection using second order centered differences as in Bryan (1969). With this spatial discretization, we cannot time discretize velocity advection in a manner analogous to the advection of thickness weighted tracer. Another method must be considered.

The Adams-Bashforth scheme discussed in Durran (1999) provides an appealing approach.* This is a multi-time single-stage scheme. That is, the contribution from velocity advection over multiple time steps contributes to each update cycle. Either the second or third order accurate Adams-Bashforth schemes are attractive, with the third order scheme preferred in practice as it better suppresses computational modes. The Adams-Bashforth requires no more computation than the unstaggered leap-frog method, but it does require added memory in order to save the advection operator over multiple time steps. In practice, this is an acceptable price.

A summary of the velocity equation update is given by

$$\begin{aligned}\frac{h^{\tau+1}\,\mathbf{u}^{\tau+1} - h^{\tau}\,\mathbf{u}^{\tau}}{\Delta\tau} &= [(-\mathcal{M}\,\hat{\mathbf{z}}\wedge h\,\mathbf{u} + (w\,\mathbf{u})_k - (w\,\mathbf{u})_{k-1} - \nabla_z\cdot(h\,\mathbf{u}\,\mathbf{u})]_{\text{AB}}\\ &\quad - h^{\tau}\,[f\,\hat{\mathbf{z}}\wedge\mathbf{u}]_{\text{semi-implicit}}\\ &\quad - h^{\tau}\,\nabla_z\,(p^{\tau+1/2}/\rho_o)\\ &\quad + h^{\tau}\,(\mathbf{F}^{\mathbf{u}})^{(\tau,\tau+1)}.\end{aligned} \tag{12.112}$$

The subscript "semi-implicit" refers to the semi-implicit treatment of the Coriolis force also used in the unstaggered method of Section 12.6.1. The superscript $(\tau,\tau+1)$ indicates that horizontal frictional dissipation is handled with a forward time step and vertical frictional dissipation is handled with an implicit time step. The "AB" subscript refers to the third Adams-Bashforth treatment of velocity advection

$$(h\,\mathbf{A})_{\text{AB}} = c_1\,(h\,\mathbf{A})^{\tau} + c_2\,(h\,\mathbf{A})^{\tau-1} + c_3\,(h\,\mathbf{A})^{\tau-2}, \tag{12.113}$$

where $\mathbf{A}$ is the advection operator, and

$$(c_1, c_2, c_3) = (23/12, -16/12, 5/12) \tag{12.114}$$

are the coefficients needed to produce a stable third-order scheme.

12.7 BAROTROPIC UPDATES WITH A PREDICTOR-CORRECTOR

The gravity wave stability analysis in Section 12.8.1 indicates that two-time level discretizations of the time tendency lead to more stable schemes than the three-time level leap-frog approach. A two-time level method that has been found to

*The Adams-Bashforth scheme is also used by Campin et al. (2004).

be of use for the barotropic system in Hallberg (1997) and Griffies et al. (2004) is the following predictor-corrector scheme. This method represents a very useful alternative to the leap-frog scheme presented in Section 12.4.1. We document the predictor-corrector scheme here, assuming the baroclinic and tracer equations are updated using the staggered scheme from Section 12.6.

The first step "predicts" the surface height via

$$\frac{\eta^{(*)}(\tau, t_{n+1}) - \eta^{(b)}(\tau, t_n)}{\gamma\, \Delta t} = -\nabla \cdot \mathbf{U}^{(b)}(\tau, t_n) + q_w(\tau). \tag{12.115}$$

As in Section 12.4.1, a raised (b) denotes the surface height and vertically integrated velocity updated with the small barotropic time steps. The raised $(*)$ denotes an intermediate value of the surface height. This is the "predicted" value, to be later "corrected." The nondimensional parameter $0 \le \gamma$ acts to dissipate the small scales of motion (see Section 12.8). Setting $\gamma = 0$ recovers a second order accurate forward-backward scheme, in which case the predictor step (12.115) is eliminated. Larger values of γ reduce the order of accuracy, yet provide effective damping. As shown in Section 12.8, values of γ larger than $1/4$ can compromise the scheme's stability. The value $\gamma = 1/5$ has been found useful for many purposes.

The predicted surface height $\eta^{(*)}(\tau, t_{n+1})$ is used to compute the surface pressure via

$$\rho_o\, \tilde{p}_s^{(*)}(\tau, t_{n+1}) = g\, \eta^{(*)}(\tau, t_{n+1})\, \rho_{k=1}^{(\tau+1/2)} \tag{12.116}$$

where the applied pressure p_a has been dropped for brevity but can be trivially added. The surface pressure is used to update the vertically integrated velocity

$$\frac{\mathbf{U}^{(b)}(\tau, t_{n+1}) - \mathbf{U}^{(b)}(\tau, t_n)}{\Delta t} = \left[-f\, \hat{\mathbf{z}} \wedge \frac{\mathbf{U}^{(b)}(\tau, t_n) + \mathbf{U}^{(b)}(\tau, t_{n+1})}{2} - D(\tau)\, \nabla_z\, \tilde{p}_s^{(*)}(\tau, t_{n+1}) + \mathbf{G}(\tau) \right]. \tag{12.117}$$

For the vertically integrated transport, the Coriolis force is evaluated using the Crank-Nicholson semi-implicit time scheme in equation (12.117). Inverting provides an explicit update of the vertically integrated transport

$$U^{(b)}(\tau, t_{n+1}) = [1 + (f\, \Delta t/2)^2]^{-1}\, [U^{(\#)}(\tau, t_{n+1}) + (f\, \Delta t/2)\, V^{(\#)}(\tau, t_{n+1})] \tag{12.118}$$

$$V^{(b)}(\tau, t_{n+1}) = [1 + (f\, \Delta t/2)^2]^{-1}\, [V^{(\#)}(\tau, t_{n+1}) - (f\, \Delta t/2)\, U^{(\#)}(\tau, t_{n+1})] \tag{12.119}$$

where $\mathbf{U}^{(\#)}(\tau, t_{n+1})$ is updated just with the time-explicit tendencies

$$\frac{U^{(\#)}(\tau, t_{n+1}) - U^{(b)}(\tau, t_n)}{\Delta t} = (f/2)\, V^{(b)}(\tau, t_n) - D(\tau)\, \partial_x\, \tilde{p}_s^{(*)}(\tau, t_{n+1}) + G^x(\tau) \tag{12.120}$$

$$\frac{V^{(\#)}(\tau, t_{n+1}) - V^{(b)}(\tau, t_n)}{\Delta t} = -(f/2)\, U^{(b)}(\tau, t_n) - D(\tau)\, \partial_y\, \tilde{p}_s^{(*)}(\tau, t_{n+1}) + G^y(\tau). \tag{12.121}$$

The "corrector" part of the scheme steps the surface height using the updated transport

$$\frac{\eta^{(b)}(\tau,t_{n+1})-\eta^{(b)}(\tau,t_n)}{\Delta t} = -\nabla\cdot\mathbf{U}^{(b)}(\tau,t_{n+1})+q_w(\tau). \tag{12.122}$$

Note that $\eta^{(b)}(\tau,t_{n+1})$ is used rather than the predicted height $\eta^{(*)}(\tau,t_{n+1})$, since $\eta^{(*)}(\tau,t_{n+1})$ is computed with the altered time step $\gamma\,\Delta t$. Temporal dissipation is localized to the predictor portion of the time stepping, with the corrector part hidden from this dissipation. Because of the predictor step, the convergence of the vertically integrated transport is computed twice in the predictor-corrector scheme, thus increasing the cost relative to a forward-backward approach where $\gamma = 0$. The payoff is an extra parameter that can be used to tune the level of dissipation. Additionally, there is added stability towards representing gravity waves so that Δt can be longer than when using the leap-frog method (Section 12.8.1).

Let us detail how the barotropic steps accumulate over the course of a particular barotropic cycle. For this purpose, write out the first and second corrector steps (12.122) for the surface height

$$\frac{\eta^{(b)}(\tau,t_{n=1})-\eta^{(b)}(\tau,t_{n=0})}{\Delta t} = F^{(\eta)}(t_{n=0}) \tag{12.123}$$

$$\frac{\eta^{(b)}(\tau,t_{n=2})-\eta^{(b)}(\tau,t_{n=1})}{\Delta t} = F^{(\eta)}(t_{n=1}), \tag{12.124}$$

where the right-hand side of equation (12.122) is abbreviated as $F^{(\eta)}$. Adding these two equations leads to

$$\frac{\eta^{(b)}(\tau,t_{n=2})-\eta^{(b)}(\tau,t_{n=0})}{\Delta t} = F^{(\eta)}(t_{n=0})+F^{(\eta)}(t_{n=1}), \tag{12.125}$$

where the intermediate value $\eta^{(b)}(\tau,t_{n=1})$ has identically cancelled. This result easily generalizes, so that

$$\frac{\eta^{(b)}(\tau,t_{n=N})-\eta^{(b)}(\tau,t_{n=0})}{N\,\Delta t} = \frac{1}{N}\sum_{n=0}^{N-1}F^{(\eta)}(t_n). \tag{12.126}$$

This result does not hold for the leap-frog scheme presented in Section 12.4.1, since the intermediate values of the surface height do not generally cancel completely, as they do here for the predictor-corrector.

The only piece of the forcing $F^{(\eta)}(t_n)$ that changes during the barotropic cycle is the convergence of the vertically integrated velocity. The result (12.126) then suggests that the time averaged vertically integrated velocity should be given back to the baroclinic part of the model upon completion of the barotropic cycle. To have this velocity centered on the integer time step $\tau+\Delta\tau$, it is necessary to run the barotropic cycle to $\tau+2\,\Delta\tau$. Hence, upon reaching the last barotropic time step

$$t_{n=N} = \tau+2\Delta\tau, \tag{12.127}$$

the vertically integrated velocity is time averaged to produce the updated vertically integrated velocity at baroclinic time $\tau+\Delta\tau$

$$\mathbf{U}(\tau+\Delta\tau) \equiv \overline{\mathbf{U}^{(b)}}(\tau+\Delta\tau) \tag{12.128}$$

$$= \frac{1}{N}\sum_{n=0}^{N-1}\mathbf{U}^{(b)}(\tau,t_n). \tag{12.129}$$

The time average runs over $n \in [0, N-1]$ for the predictor-corrector, whereas we use $n \in [0, N]$ in the leap-frog case (12.64). In practice, the difference is trivial.

The surface height is needed at the integer time steps in order to specify the thickness of the velocity cells. There are two options for updating the surface height to time step $\tau + \Delta\tau$. First, we could take the instantaneous value from the barotropic portion of the cycle

$$\eta(\tau + \Delta\tau) \equiv \eta^{(b)}(\tau, t_{n=N/2}). \tag{12.130}$$

This approach has not been tried, since it likely leads to a meta-stable algorithm due to the absence of time averaging, depending on the predictor-corrector dissipation parameter γ. In contrast, extensive experience indicates that added stability is realized by using the time averaged surface height

$$\eta(\tau + \Delta\tau) = \frac{1}{N+1} \sum_{n=0}^{N} \eta^{(b)}(\tau, t_n). \tag{12.131}$$

Notably, tracer and volume conservation is not compromised by this specification since it is only used to define the surface height carried by the velocity cells.

12.8 STABILITY CONSIDERATIONS

The purpose of this section is to detail stability of various time stepping schemes used to temporally represent terms in the primitive equations.

12.8.1 Gravity waves

Linear gravity waves, either barotropic or baroclinic, are fluctuations whose representation in a numerical scheme is fundamental to the temporal stability. We focus here on barotropic gravity waves, but the ideas transfer to baroclinic waves.

To derive the equations for linear barotropic gravity waves, we consider the vertically integrated volume and velocity equations without the vertically integrated forcing $\mathbf{G}$, fresh water forcing q_w, a volume source, or a Coriolis force. We also assume the ocean depth D to be constant throughout the domain. These choices lead to the equations describing linear barotropic gravity waves in a flat bottom ocean

$$\partial_t \eta = -\nabla \cdot \mathbf{U} \tag{12.132}$$

$$\partial_t \mathbf{U} = -g\, D\, \nabla \eta. \tag{12.133}$$

Eliminating the transport $\mathbf{U}$ reveals the wave equation for surface height undulations

$$[\partial_{tt} - (c\,\nabla)^2]\, \eta = 0, \tag{12.134}$$

where

$$c = \sqrt{g\, D} \tag{12.135}$$

is the phase speed of the dispersionless gravity waves. Taking the divergence of the transport equation (12.133) leads to the same wave equation

$$[\partial_{tt} - (c\,\nabla)^2]\, \nabla \cdot \mathbf{U} = 0, \tag{12.136}$$

and so the divergence propagates without dispersion along with the surface height undulations. In contrast, the curl remains stationary, since

$$\partial_t (\nabla \wedge \mathbf{U}) = 0 \tag{12.137}$$

in a flat bottom ocean, thus leading to irrotational waves.

12.8.1.1 Staggered space and leap-frog time

Consider a two-dimensional discrete system where the surface height and velocity are placed at the same temporal grid point $t_n = n\,\Delta t$, with n an integer. Additionally, let the surface height be placed on the integer point $(x_i, y_j) = (i\,\Delta x, j\,\Delta y)$ in space, but with the velocity at $((i+1/2)\,\Delta x, (j+1/2)\,\Delta y)$ as on a B-grid. For simplicity, we assume that the grid spacings Δx and Δy are uniform in space.

Before starting the temporal stability analysis, we note some details about the discretization of gravity waves on the B-grid. For purposes of computing discrete horizontal derivatives, as in the divergence $\nabla \cdot \mathbf{U}$, we need to place the velocity at the appropriate face of the grid cell in order to have the divergence be centered at the tracer point. We do so via a spatial average operator whereby zonal and meridional velocity components are averaged to the zonal and meridional faces, respectively, of the tracer grid cell. This average then provides the B-grid with its version of the C-grid velocity components (see Figure 10.1). There must also be a spatial average to place horizontal pressure derivatives, naturally defined at the zonal and meridional faces of the tracer grid, onto the velocity point. This averaging process introduces a computational null mode to the B-grid gravity wave equations. This mode manifests as a checkerboard pattern in the surface height and velocity fields. We mention some numerical methods to suppress this mode in Section 12.9. For now, subtleties of spatial discretizations are ignored as our focus here is on temporal stability.

Given our assumption that velocity components for the present discussion refer to their C-grid versions, a leap-frog discretization of the linear gravity wave equations leads to

$$\frac{\eta^{n+1} - \eta^{n-1}}{2\,\Delta t} = -\left(\frac{U^n_{i+1/2} - U^n_{i-1/2}}{\Delta x}\right) - \left(\frac{V^n_{j+1/2} - V^n_{j-1/2}}{\Delta y}\right) \tag{12.138}$$

$$\frac{U^{n+1}_{i+1/2} - U^{n-1}_{i+1/2}}{2\,\Delta t} = -c^2 \left(\frac{\eta^n_{i+1} - \eta^n_i}{\Delta x}\right) \tag{12.139}$$

$$\frac{V^{n+1}_{j+1/2} - V^{n-1}_{j+1/2}}{2\,\Delta t} = -c^2 \left(\frac{\eta^n_{j+1} - \eta^n_j}{\Delta y}\right). \tag{12.140}$$

Note that if a discrete label is missing, it should be assumed that the field is at the central grid point. As the system is linear, it is sufficient to assume the space-time behavior takes the form of discrete modes

$$\eta^n_{i,j} = \eta_o\, e^{\iota\,(i\,\kappa_x\,\Delta x + j\,\kappa_y\,\Delta y - n\,\omega\,\Delta t)} \tag{12.141}$$

$$U^n_{i,j} = U_o\, e^{\iota\,(i\,\kappa_x\,\Delta x + j\,\kappa_y\,\Delta y - n\,\omega\,\Delta t)} \tag{12.142}$$

$$V^n_{i,j} = V_o\, e^{\iota\,(i\,\kappa_x\,\Delta x + j\,\kappa_y\,\Delta y - n\,\omega\,\Delta t)} \tag{12.143}$$

where we use

$$\iota = \sqrt{-1} \tag{12.144}$$

to distinguish from the discrete grid label i. The spatial wave numbers κ_x and κ_y have units inverse length, and the temporal frequency ω has units inverse time. As the equations model gravity waves without explicit dissipation, we expect ω to be real as occurs for oscillatory fluctuations. Plugging these modes into equations (12.138)–(12.140) leads to

$$\begin{aligned}\eta_o\,(e^{-\iota\,\omega\,\Delta t} - e^{\iota\,\omega\,\Delta t}) = &- U_o \left(\frac{2\,\Delta t}{\Delta x}\right)(e^{\iota\,\kappa_x\,\Delta x/2} - e^{-\iota\,\kappa_x\,\Delta x/2}) \\ &- V_o \left(\frac{2\,\Delta t}{\Delta y}\right)(e^{\iota\,\kappa_y\,\Delta y/2} - e^{-\iota\,\kappa_y\,\Delta y/2})\end{aligned} \tag{12.145}$$

$$U_o\,e^{\iota\,\kappa_x\,\Delta x/2}\,(e^{-\iota\,\omega\,\Delta t} - e^{\iota\,\omega\,\Delta t}) = -\eta_o \left(\frac{2\,c^2\,\Delta t}{\Delta x}\right)(e^{\iota\,\kappa_x\,\Delta x} - 1). \tag{12.146}$$

$$V_o\,e^{\iota\,\kappa_y\,\Delta y/2}\,(e^{-\iota\,\omega\,\Delta t} - e^{\iota\,\omega\,\Delta t}) = -\eta_o \left(\frac{2\,c^2\,\Delta t}{\Delta y}\right)(e^{\iota\,\kappa_y\,\Delta y} - 1). \tag{12.147}$$

Rearrangement yields

$$\eta_o\,\sin(\omega\,\Delta t) - U_o\,\frac{2\,\Delta t}{\Delta x}\,\sin(\kappa_x\,\Delta x/2) - V_o\,\frac{2\,\Delta t}{\Delta y}\,\sin(\kappa_y\,\Delta y/2) = 0 \tag{12.148}$$

$$\eta_o\,\frac{2\,c^2\,\Delta t}{\Delta x}\,\sin(\kappa_x\,\Delta x/2) - U_o\,\sin(\omega\,\Delta t) = 0 \tag{12.149}$$

$$\eta_o\,\frac{2\,c^2\,\Delta t}{\Delta y}\,\sin(\kappa_y\,\Delta y/2) - V_o\,\sin(\omega\,\Delta t) = 0. \tag{12.150}$$

Nontrivial amplitudes (η_o, U_o, V_o) are realized only when the determinant to this homogeneous linear system vanishes. This constraint gives the discrete dispersion relation

$$\sin(\omega\,\Delta t)\,\Big(\sin^2(\omega\,\Delta t) - (2\,\mu_x)^2\,\sin(\kappa_x\,\Delta x/2) - (2\,\mu_y)^2\,\sin(\kappa_y\,\Delta y/2)\Big) = 0, \tag{12.151}$$

with

$$\mu_x = c\,\Delta t/\Delta x \tag{12.152}$$

$$\mu_y = c\,\Delta t/\Delta y \tag{12.153}$$

the Courant numbers for the two spatial directions. The dispersion relation can be satisfied by a stationary wave, with $\omega = 0$. Stability for a nonstationary wave requires

$$\sin^2(\omega\,\Delta t) - (2\,\mu_x)^2\,\sin(\kappa_x\,\Delta x/2) - (2\,\mu_y)^2\,\sin(\kappa_y\,\Delta y/2) = 0. \tag{12.154}$$

Since we have assumed the frequency and wave numbers to be real, this relation can be satisfied in general only for cases where

$$\mu_x^2 + \mu_y^2 < 1/4. \tag{12.155}$$

As a constraint on the time step, this relation becomes

$$\Delta t < \frac{1}{2\,c}\,\Big((\Delta x)^{-2} + (\Delta y)^{-2}\Big)^{-1/2}. \tag{12.156}$$

In one space dimension, this constraint requires

$$\mu < 1/2, \tag{12.157}$$

or

$$\Delta t < \frac{\Delta x}{2\,c}. \tag{12.158}$$

The more restrictive constraint in two-dimensions arises since waves that propagate diagonal to a grid line have a smaller wave length (and so larger wave number) than waves along a grid line (see section 3.2.1 in Durran, 1999).

12.8.1.2 Staggered space and staggered time (equivalently a forward-backward)

The staggered scheme discussed in Section 12.6.2 provides an alternative to the leap-frog scheme. Using the staggered method for gravity waves leads to

$$\frac{\eta^{n+1/2} - \eta^{n-1/2}}{\Delta t} = -\left(\frac{U^n_{i+1/2} - U^n_{i-1/2}}{\Delta x}\right) - \left(\frac{V^n_{j+1/2} - V^n_{j-1/2}}{\Delta y}\right) \tag{12.159}$$

$$\frac{U^{n+1}_{i+1/2} - U^n_{i+1/2}}{\Delta t} = -c^2\left(\frac{\eta^{n+1/2}_{i+1} - \eta^{n+1/2}_{i}}{\Delta x}\right) \tag{12.160}$$

$$\frac{V^{n+1}_{j+1/2} - V^n_{j+1/2}}{\Delta t} = -c^2\left(\frac{\eta^{n+1/2}_{j+1} - \eta^{n+1/2}_{j}}{\Delta y}\right). \tag{12.161}$$

Although interpreted as a staggered temporal discretization, one can equally interpret this method as a forward-backward discretization. This is revealed by writing the equations for the barotropic system discussed in Section 12.7 with the dissipation parameter γ set to zero.

Substituting the modes (12.141)–(12.143) into equations (12.159)–(12.161), and setting the determinant to zero, leads to the discrete dispersion relation

$$\sin^2(\omega\,\Delta t) = \mu_x^2\,\sin(\kappa_x\,\Delta x/2) + \mu_y^2\,\sin(\kappa_y\,\Delta y/2). \tag{12.162}$$

This relation is similar to the leap-frog dispersion (12.154). However, the Courant numbers are multiplied by 2 with the leap-frog method. The absence of this factor leads to the less restrictive forward-backward stability constraint

$$\mu_x^2 + \mu_y^2 < 1, \tag{12.163}$$

where the leap-frog constraint (12.155) requires $\mu_x^2 + \mu_y^2 < 1/4$. As a constraint on the time step, this relation becomes

$$\Delta t < \frac{1}{c}\left((\Delta x)^{-2} + (\Delta y)^{-2}\right)^{-1/2}. \tag{12.164}$$

In one space dimension, this constraint requires

$$\mu < 1, \tag{12.165}$$

or

$$\Delta t < \frac{\Delta x}{c}. \tag{12.166}$$

Hence, it is possible to resolve gravity waves with a forward-backward, or staggered, scheme using twice the time step allowed by the leap-frog scheme (see Section 3.1.2 of Durran, 1999, for a similar discussion). This result certainly has important practical implications for the design of an ocean model. In particular, it provides motivation for using the predictor-corrector barotropic scheme presented in Section 12.7.

12.8.1.3 Staggered space and predictor-corrector time

To finish this discussion of gravity waves, we analyze stability properties of the predictor-corrector scheme discussed in Section 12.7. As just mentioned, the dissipationless predictor-corrector, which is a forward-backward scheme, can be interpreted as a time staggered scheme. The stability of this scheme to gravity waves is twice that for a leap-frog scheme. Our focus here is on establishing stability constraints on the dissipation parameter γ.

We first investigate the continuous space and discrete time system. This analysis serves to highlight the analog continuous equation to which the temporally discrete system corresponds. In two space dimensions, the predictor-corrector equations are

$$\frac{\eta^* - \eta^n}{\Delta t} = -\gamma \nabla \cdot \mathbf{U}^n \tag{12.167}$$

$$\frac{\mathbf{U}^{n+1} - \mathbf{U}^n}{\Delta t} = -c^2 \nabla \eta^* \tag{12.168}$$

$$\frac{\eta^{n+1} - \eta^n}{\Delta t} = -\nabla \cdot \mathbf{U}^{n+1}. \tag{12.169}$$

Eliminating the predicted surface height η^* leads to

$$\frac{\mathbf{U}^{n+1} - \mathbf{U}^n}{\Delta t} = -c^2 \nabla \eta^n + \gamma c^2 \Delta t \nabla \left[\nabla \cdot \mathbf{U}^n\right] \tag{12.170}$$

$$\frac{\eta^{n+1} - \eta^n}{\Delta t} = -\nabla \cdot \mathbf{U}^{n+1}. \tag{12.171}$$

To directly see how the surface height is effected, eliminate $\mathbf{U}$ to find

$$\frac{\eta^{n+1} - 2\,\eta^n + \eta^{n-1}}{(\Delta t)^2} = (c\,\nabla)^2 \eta^n + \gamma\,(c\,\nabla)^2 \left(\eta^n - \eta^{n-1}\right). \tag{12.172}$$

Now take the continuous time limit with $\gamma\,\Delta t$ constant. The time discrete equation then reduces to the dissipative wave equation

$$(\partial_{tt} - c^2 \nabla^2)\,\eta = (\gamma\,\Delta t)\,(c\,\nabla)^2\,\partial_t\,\eta. \tag{12.173}$$

A single spatial Fourier mode with wavenumber amplitude κ thus satisfies

$$\left(\mathrm{d}^2/\mathrm{d}t^2 + \gamma\,\Delta t\,(c\,\kappa)^2\,\mathrm{d}/\mathrm{d}t + (c\,\kappa)^2\right)\,\eta = 0. \tag{12.174}$$

This is the equation for a damped harmonic oscillator with inverse e-folding time $(1/2)\,\gamma\,\Delta t\,(c\,\kappa)^2$. In the presence of $\gamma > 0$, external gravity waves are selectively dissipated in regions where the surface height is changing in time, and where the spatial scales are small. Faster waves are damped more readily than slower waves. These properties is are useful to suppress the B-grid computational null mode discussed in Section 12.9.

Return now to the discrete space-time system. Our goal is to find a constraint on the dissipation parameter γ that allows for the system to maintain stability. Given the previous continous space-time analysis, we expect stable values of γ to depend on the space and time spacing as well as the gravity wave speed. The algebra is a bit tedious when $\gamma > 0$. Hence, focus on one space dimension.

The discrete space-time equations take the form

$$\frac{\eta_i^* - \eta_i^n}{\Delta t} = -\gamma \left(\frac{U_{i+1/2}^n - U_{i-1/2}^n}{\Delta x} \right) \tag{12.175}$$

$$\frac{U_{i+1/2}^{n+1} - U_{i+1/2}^n}{\Delta t} = -c^2 \left(\frac{\eta_{i+1}^* - \eta_i^*}{\Delta x} \right) \tag{12.176}$$

$$\frac{\eta_i^{n+1} - \eta_i^n}{\Delta t} = -\left(\frac{U_{i+1/2}^{n+1} - U_{i-1/2}^{n+1}}{\Delta x} \right). \tag{12.177}$$

Eliminating the predictor quantity η^* leads to

$$U_{i+1/2}^{n+1} - U_{i+1/2}^n = -\frac{c^2\,\Delta t}{\Delta x}\,(\eta_{i+1}^n - \eta_i^n) + \gamma\,\frac{c^2\,\Delta t}{\Delta x}\left(U_{i+3/2}^n - 2\,U_{i+1/2}^n - U_{i-1/2}^n\right) \tag{12.178}$$

$$\eta_i^{n+1} - \eta_i^n = -\frac{\Delta t}{\Delta x}\left(U_{i+1/2}^{n+1} - U_{i-1/2}^{n+1}\right). \tag{12.179}$$

Temporal dissipation results when $\gamma > 0$. This dissipation manifests as a complex frequency, which is written

$$\widetilde{\omega} = \omega + \iota\,\sigma. \tag{12.180}$$

The modal solutions are now written in the form

$$\eta_i^n = \eta_o\, e^{\iota\,(i\kappa\,\Delta x - n\,\omega\,\Delta t) + n\,\sigma\,\Delta t} \tag{12.181}$$

$$U_i^n = U_o\, e^{\iota\,(i\kappa\,\Delta x - n\,\omega\,\Delta t) + n\,\sigma\,\Delta t} \tag{12.182}$$

Substituting these solutions into the discrete system (12.178) and (12.179) leads to

$$U_o\left(e^{-\iota\,\omega\,\Delta t + \sigma\,\Delta t} - 1 + \gamma\,[2\,\mu\,\sin(\kappa\,\Delta x/2)]^2\right) + 2\,\iota\,\eta_o\,c\,\mu\,\sin(\kappa\,\Delta x/2) = 0 \tag{12.183}$$

$$2\,\iota\,U_o\,(\mu/c)\,\sin(\kappa\,\Delta x/2) + \eta_o\,(1 - e^{-\iota\,\omega\,\Delta t + \sigma\,\Delta t}) = 0. \tag{12.184}$$

Setting the determinant to zero yields the complex dispersion relation

$$[2\,\mu\,\sin(\kappa\,\Delta x/2)]^2\left(\gamma\,(e^{\iota\,\omega\,\Delta t - \sigma\,\Delta t} - 1) - 1\right) + \left(2 - e^{\iota\,\omega\,\Delta t - \sigma\,\Delta t} - e^{-\iota\,\omega\,\Delta t + \sigma\,\Delta t}\right) = 0. \tag{12.185}$$

Setting the imaginary part of this equation to zero gives the dispersion relation for the imaginary part of the frequency

$$e^{2\,\sigma\,\Delta t} = 1 - \gamma\,[2\,\mu\,\sin(\kappa\,\Delta x/2)]^2. \tag{12.186}$$

Satisfaction of this equation requires that the imaginary frequency be non-positive $\sigma \leq 0$. It also implies that σ vanishes when γ vanishes. Since the left-hand side of equation (12.186) is non-negative, equality holds only for cases where the right-hand side also is non-negative, which requires

$$\gamma < (2\,\mu)^{-2}. \tag{12.187}$$

Recall that for a stable forward-backward scheme, the Courant number μ is less than unity (equation (12.165)). For cases with maximum Courant number $\mu = 1$, the dissipation parameter must satisfy $\gamma < 1/4$. This value is hence a reasonably conservative setting for the dissipation parameter. Larger values stand the chance of leading to over-dissipation and so to an unstable scheme.

12.8.2 Advection

To detail the stability of advective transport

$$T_{,t} = -\partial_x\,(u\,T) - \partial_y\,(v\,T), \tag{12.188}$$

it is sufficient to consider a constant advection velocity. Note also that it is useful to separate the horizontal advection problem from the vertical for purposes of stability analysis. Much in this section follows Section 2.5 of Durran (1999).

12.8.2.1 Leap-frog time and centered space

A leap-frog discretization of the time tendency leads to the discrete advection equation

$$\frac{T^{n+1} - T^{n-1}}{2\,\Delta t} = -u\left(\frac{T^n_{i+1/2} - T^n_{i-1/2}}{\Delta x}\right) - v\left(\frac{T^n_{j+1/2} - T^n_{j-1/2}}{\Delta y}\right). \tag{12.189}$$

Advection schemes differ by how they evaluate the interface value of the tracer concentration. A traditional method is to use a second order centered discretization

$$2\,T^n_{i+1/2} = T^n_{i+1} + T^n_i \tag{12.190}$$

$$2\,T^n_{j+1/2} = T^n_{j+1} + T^n_j, \tag{12.191}$$

thus leading to

$$T^{n+1} = T^{n-1} + \mu_x\,(T^n_{i-1} - T^n_{i+1}) + \mu_y\,(T^n_{j-1} - T^n_{j+1}) \tag{12.192}$$

where

$$\mu_x = u\,\Delta t/\Delta x \tag{12.193}$$

$$\mu_y = v\,\Delta t/\Delta y \tag{12.194}$$

are the advective Courant numbers. If the tracer concentration takes the form of a linear oscillatory mode

$$T^n_{i,j} = T_o\,e^{\iota\,(i\,\kappa_x\,\Delta x + j\,\kappa_y\,\Delta y - n\,\omega\,\Delta t)}, \tag{12.195}$$

then the discrete advection equation leads to the dispersion relation

$$\sin(\omega\,\Delta t) = \mu_x\,\sin(\kappa_x\,\Delta x) + \mu_y\,\sin(\kappa_y\,\Delta y). \tag{12.196}$$

Stability is thus ensured if

$$|\mu_x + \mu_y| < 1. \tag{12.197}$$

As a constraint on the time step, this relation leads to

$$\Delta t < \left| \frac{u}{\Delta x} + \frac{v}{\Delta y} \right|^{-1}. \tag{12.198}$$

Assuming a one-dimensional stability analysis is sufficient for vertical advection, the vertical advection constraint is

$$\Delta t < \frac{\Delta z}{|w|}. \tag{12.199}$$

Time step stability in global models is often constrained by vertical advection in experiments with realistically strong wind forcing and refined vertical grid spacing. In particular, strong upwelling at the equator is often the source of time step instabilities. To test the suitability of a chosen time step prior to model integration, it is useful to assume a maximum advection velocity and test for stability based on a flow that reaches this maximum.

12.8.2.2 Leap-frog time and upwind space

With a positive zonal velocity component ($u > 0$), a first order upwind discretization of the zonal interface tracer concentration takes the form

$$T_{i-1/2} = T_{i-1} \tag{12.200}$$

$$T_{i+1/2} = T_i. \tag{12.201}$$

If the velocity component is negative, then

$$T_{i-1/2} = T_i \tag{12.202}$$

$$T_{i+1/2} = T_{i+1}. \tag{12.203}$$

Similar results hold for the meridional interface tracer. First order upwind advection is akin to diffusion, and as such it requires the use of a forward time difference (see Section 12.8.4). It is useful to prove this assertion, and for this purpose we consider the one-dimensional case where the upwind advection operator is centered in time and $u > 0$

$$T^{n+1} = T^{n-1} + 2\,\mu\,(T^n_{i-1} - T^n_i), \tag{12.204}$$

where $\mu = u\,\Delta t/\Delta x$. Now assume the discrete tracer takes the form

$$T^n_i = B^n\, e^{\iota\,(-n\,\omega\,\Delta t + i\kappa\,\Delta x)}, \tag{12.205}$$

which then leads to the complex dispersion relation

$$B\, e^{-\iota\,\omega\,\Delta t} - B^{-1}\, e^{\iota\,\omega\,\Delta t} = 2\,\mu\,(e^{-\iota\,\kappa\,\Delta x} - 1). \tag{12.206}$$

Rearranging the real and imaginary parts to this result leads to

$$B\,\sin(2\,\omega\,\Delta t) = 2\,\mu\,[\sin(\omega\,\Delta t - \kappa\,\Delta x) - \sin(\omega\,\Delta t)]. \tag{12.207}$$

The only choice that ensures $|B| < 1$ generally holds is the trivial case of $\mu = 0$. Otherwise, $|B|$ can be larger than unity, which signals an unstable scheme.

Now return to the two-dimensional case with a forward time step for the upwind advection operator

$$T^{n+1} = T^{n-1} + 2\,\mu_x\,(T^{n-1}_{i-1} - T^{n-1}_{i}) + 2\,\mu_y\,(T^{n-1}_{j-1} - T^{n-1}_{j}). \tag{12.208}$$

Substituting

$$T^{n}_{i,j} = B^{n}\,e^{\iota\,(-n\,\omega\,\Delta t + i\kappa_x\,\Delta x + j\,\kappa_y\,\Delta y)} \tag{12.209}$$

leads to

$$B^2\,e^{-\iota\,\omega\,\Delta t} = 1 + 2\,\mu_x\,(e^{-\iota\,\kappa_x\,\Delta x} - 1) + 2\,\mu_y\,(e^{-\iota\,\kappa_y\,\Delta y} - 1) \tag{12.210}$$

whose real and imaginary parts are

$$B^2\,\cos(2\,\omega\,\Delta t) = 1 + 2\,\mu_x\,(\cos(\kappa_x\,\Delta x) - 1) + 2\,\mu_y\,(\cos(\kappa_y\,\Delta y) - 1) \tag{12.211}$$

$$B^2\,\sin(2\,\omega\,\Delta t) = 2\,\mu_x\,\sin(\kappa_x\,\Delta x) + 2\,\mu_y\,\sin(\kappa_y\,\Delta y). \tag{12.212}$$

The next step requires a tremendous amount of algebra. We therefore restrict attention to one dimension, where squaring both equations and adding yields

$$B^4 = 1 - 4\,\mu\,(1 - 2\,\mu)\,[\,1 - \cos(\kappa\,\Delta x)\,]. \tag{12.213}$$

We ensure B is less than unity if

$$\mu < 1/2, \tag{12.214}$$

or as a time step constraint

$$\Delta t < \frac{\Delta z}{2\,|w|}, \tag{12.215}$$

where we focused on vertical advection as it typically provides the most stringent constraint in models of refined vertical resolution. The factor of 1/2 arises from the use of a leap-frog scheme for the time tendency.

12.8.2.3 Forward time and upwind space

An upwind implementation of advection, with a forward time step for the time tendency, leads to the discrete advection equation

$$T^{n+1} = T^{n} + \mu_x\,(T^{n}_{i-1} - T^{n}_{i}) + \mu_y\,(T^{n}_{j-1} - T^{n}_{j}). \tag{12.216}$$

Notice the absence of the factor of 2 present in the leap-frog upwind advection equation (12.208). The one-dimensional equation is therefore stable when

$$\mu < 1, \tag{12.217}$$

or

$$\Delta t < \frac{\Delta z}{|w|}. \tag{12.218}$$

Again we see how a two-time level implementation of the time tendency loosens the time step constraint.

Although first order upwind advection is overly diffusive for most practical applications, some form of upwind is typically part of advection schemes which provide a semblance of monotonicity (Durran, 1999). The upwind piece of the advection flux, whether it is first order or higher, must be implemented according to a forward time step. The present analysis indicates that a forward time scheme for the time tendency incurs a less stringent time step constraint than a leap-frog scheme.

12.8.2.4 Forward time and third order Adams-Bashforth

As discussed in Section 12.6.3, velocity advection can be stably represented using an Adams-Bashforth scheme when implementing centered spatial differences with a two-level time tendency scheme. Using second order centered differences for the fluxes, as done in Griffies et al. (2004) for velocity advection, leads to the one-dimensional version

$$\frac{T^{n+1} - T^{n}}{\Delta t} = -\frac{u}{\Delta x}\left(c_1\,(T^{n}_{i-1} - T^{n}_{i+1}) + c_2\,(T^{n-1}_{i-1} - T^{n-1}_{i+1}) + c_3\,(T^{n-2}_{i-1} - T^{n-2}_{i+1})\right), \tag{12.219}$$

where the constants (c_1, c_2, c_3) are given in equation (12.114). This scheme is stable so long as (see page 70 of Durran, 1999)

$$\Delta t < 0.724\,\frac{\Delta z}{|w|}, \tag{12.220}$$

where again we focused on the vertical advection process as it typically provides the most stringent constraint. This constraint is more restrictive than the second order centered constraint given by equation (12.199). However, experience indicates that this is not a problem for many cases in practice, where gravity waves and dissipation tend to provide the most stringent stability criteria.

12.8.3 Inertial oscillations

We follow Section 2.3.1 of Bryan (1989) and consider the stability of discretely realized inertial oscillations

$$u_{,t} - f\,v = 0 \tag{12.221}$$
$$v_{,t} + f\,u = 0, \tag{12.222}$$

where advection has been dropped. Introducing the complex horizontal velocity

$$\varpi = u + \iota v \tag{12.223}$$

leads to the oscillation equation

$$\varpi_{,t} = \iota f\,\varpi. \tag{12.224}$$

A leap-frog discretization takes the form

$$\varpi^{n+1} = \varpi^{n-1} + 2\,\iota\,\Delta t\, f\,\varpi^{n}. \tag{12.225}$$

Stability is maintained if

$$\Delta t < f^{-1}. \tag{12.226}$$

A forward in time discretization

$$\varpi^{n+1} = \varpi^{n} + \iota\,\Delta t\, f\,\varpi^{n} \tag{12.227}$$

is formally stable, yet admits growing modes that grow as $f\,\Delta t$ (see page 51 of Durran, 1999). A more favorable approach is to compute the Coriolis force semi-implicitly in time (equation (2.33) of Durran, 1999), whereby

$$\varpi^{n+1} = \varpi^{n} + \iota\,\Delta t\, f\,(\alpha\varpi^{n} + \beta\,\varpi^{n+1}) \tag{12.228}$$

with $\alpha + \beta = 1$. Defining the ratio

$$B \equiv \frac{\varpi^{n+1}}{\varpi^{n}} \tag{12.229}$$

leads to

$$|B|^2 = \frac{1 + (\alpha f \, \Delta t)^2}{1 + (\beta f \, \Delta t)^2}. \tag{12.230}$$

The second order accurate Crank-Nicolson scheme has $\alpha = \beta = 1/2$, and this scheme is neutrally stable.* When $\alpha < \beta$ the scheme is stable and damping, whereas $\alpha > \beta$ is unstable.

12.8.4 Diffusion

Diffusive processes impose a time step constraint associated with the level of mixing. In particular, given a time step and grid spacing, one cannot increase viscosity or diffusivity above a certain level. The level of mixing is determined by the *diffusive CFL contraint*, as discussed in this section.†

To detail the time step constraints, it is sufficient to consider a two-dimensional diffusion equation

$$T_{,t} = A\,(T_{,xx} + T_{,yy}) \tag{12.231}$$

where $A > 0$ is a constant diffusivity with units $\mathrm{m}^2\,\mathrm{s}^{-1}$. Consider a discretization of the time derivative using a leap-frog. It seems reasonable to center the diffusion operator in time to maximize order of accuracy used to represent diffusion. In this case we have

$$\frac{T^{n+1} - T^{n-1}}{2\,\Delta t} = \frac{A}{(\Delta x)^2}\left(T^n_{i+1} - 2T^n_i + T^n_{i-1}\right) + \frac{A}{(\Delta y)^2}\left(T^n_{j+1} - 2T^n_j + T^n_{j-1}\right). \tag{12.232}$$

Let modal solutions have the form

$$T^n_{i,j} = \mathcal{B}^n \, e^{\iota(-n\,\omega\,\Delta t + i\kappa_x\,\Delta x + j\,\kappa_y\,\Delta y)} \tag{12.233}$$

where ω is a real frequency, $\mathcal{B}$ is a real amplitude, and κ_x and κ_y are real wavenumbers. A stable scheme satisfies $|\mathcal{B}| < 1$. Substituting (12.233) into the discrete diffusion equation (12.232) leads to a complex dispersion relation. The only way to satisfy the imaginary part of the relation is for ω to vanish, which is expected since diffusion is a dissipative process rather than wave-like. The real part of the dispersion relation leads to

$$\mathcal{B}^2 + 8\,[\,\mu_x \sin^2(\kappa_x\,\Delta x/2) + \mu_y \sin^2(\kappa_y\,\Delta y/2)]\,\mathcal{B} - 1 = 0, \tag{12.234}$$

where

$$\mu_x = \frac{A\,\Delta t}{(\Delta x)^2} \tag{12.235}$$

$$\mu_y = \frac{A\,\Delta t}{(\Delta y)^2} \tag{12.236}$$

---

*This scheme has been found to be very useful for global climate models using the code of Griffies et al. (2004).

†Some of this material is also presented in Section 18.1.

are diffusive Courant numbers. There are two roots to equation (12.234). One of the roots is less than -1, regardless the value of the Courant numbers. This result indicates that the scheme is unstable. The problem, as we see next, is the time placement of the diffusion operator.

As discussed by Bryan (1989) and Durran (1999), stability for the diffusion equation is realized when discretizing the diffusion operator with a forward time step. Continuing to use a leap-frog for the time tendency then leads to

$$\begin{aligned}\frac{T^{n+1} - T^{n-1}}{2\,\Delta t} &= \frac{A}{(\Delta x)^2}\left(T_{i+1}^{n-1} - 2T_i^{n-1} + T_{i-1}^{n-1}\right)\\ &+ \frac{A}{(\Delta y)^2}\left(T_{j+1}^{n-1} - 2T_j^{n-1} + T_{j-1}^{n-1}\right).\end{aligned} \tag{12.237}$$

The real part of the dispersion relation yields the amplification equation

$$\mathcal{B}^{n+1} = \mathcal{B}^{n-1}\,[\,1 - 8\,\mu_x\,\sin^2(\kappa_x\,\Delta x/2) - 8\,\mu_y\,\sin^2(\kappa_y\,\Delta y/2)\,]. \tag{12.238}$$

Stability is satisfied if

$$-1 < 1 - 8\,\mu_x\,\sin^2(\kappa_x\,\Delta x/2) - 8\,\mu_y\,\sin^2(\kappa_y\,\Delta y/2) < 1, \tag{12.239}$$

which leads to the constraint

$$\mu_x + \mu_y < 1/4. \tag{12.240}$$

As a constraint on the time step, this relation becomes

$$\Delta t < \frac{1}{4\,A}\left((\Delta x)^{-2} + (\Delta y)^{-2}\right)^{-1/2}. \tag{12.241}$$

This constraint is analogous to that imposed on the leap-frog scheme when representing gravity waves (see equations (12.155) and (12.156)).

Now consider a forward time step for the tendency, as used in the staggered scheme in Section 12.6.2. In this case, the discrete diffusion equation is given by

$$\frac{T^{n+1} - T^{n}}{\Delta t} = \frac{A}{(\Delta x)^2}\left(T_{i+1}^{n} - 2T_i^{n} + T_{i-1}^{n}\right) + \frac{A}{(\Delta y)^2}\left(T_{j+1}^{n} - 2T_j^{n} + T_{j-1}^{n}\right). \tag{12.242}$$

A stability analysis leads to the constraint

$$\mu_x + \mu_y < 1, \tag{12.243}$$

which means the time step must satisfy

$$\Delta t < \frac{1}{A}\left((\Delta x)^{-2} + (\Delta y)^{-2}\right)^{-1/2}. \tag{12.244}$$

Just as for the case of gravity waves and for upwind advection, the two-time level discretization of the time tendency, as used in the staggered scheme, leads to a less stringent time step constraint than the leap-frog.

In summary, stabililty of diffusive processes requires a forward time step for the diffusion operator. If the time tendency is discretized with a leap-frog, then a forward time step for diffusion translates into an effective time step of $2\Delta\tau$. If the time tendency is discretized with a forward difference, then diffusion is time stepped with $\Delta\tau$. Therefore, if the diffusion operator is stable with a leap-frog time step of $2\Delta\tau$, it will also be stable with a forward time step of

$$\Delta\tau_{\text{forward}} = 2\Delta\tau_{\text{leap}}. \tag{12.245}$$

The same result holds for gravity waves and dissipative advection, as seen in the previous sections.

The $2\Delta\tau$ time step with the leap-frog tendency constrains the SGS coefficients (i.e., Laplacian or biharmonic diffusivities and viscosities) to be less than available when taking a $\Delta\tau$ time step with the two-level tendency. Although many modeling situations do not necessitate pushing the SGS mixing coefficients to their maximum numerically stable values, there are notable exceptions where the added constraint from the leap-frog can be onerous. The following describes some cases.

- For neutral physics implemented according to rotation, numerical stability in regions of steep neutral slopes requires a tapering to lower diffusivities (see Section 15.1). The use of two-time level schemes reduces the need to taper.
- By using $\Delta\tau$ rather than $2\,\Delta\tau$, the results from implicit mixing are more consistent with explicit vertical mixing, since accuracy is improved when taking smaller time steps.
- The time step difference between the horizontal and vertical flux components is smaller with the two-time level tendencies. (Recall that the horizontal flux components are discretized at τ for two-level tendencies and $\tau - \Delta\tau$ for leap-frog tendencies, whereas the vertical flux component is typically discretized implicitly at $\tau + \Delta\tau$).
- For models running with explicit momentum dissipation via a Laplacian and/or biharmonic operator, viscosity is numerically necessary to resolve western boundary currents (e.g., Griffies and Hallberg, 2000; Large et al., 2001; Smith and McWilliams, 2003). In some cases, such as near the equator, large viscosities may be warranted to ensure that currents are fully resolved.
- The coupling of sea ice to the ocean engenders the coupled system with a larger effective horizontal friction than deduced from the two separate systems.* This added level of friction, combined with relatively small grid cell dimensions commonly occurring in the Arctic, can initiate numerical instabilities that require either a reduction in time step, reduction in ocean Laplacian and/or biharmonic viscosity, and/or reduction in sea ice drag coefficient.

In summary, use of a two-time level discretization of the time tendency, a forward time step for the lateral SGS processes, and an implicit time scheme for the vertical processes, provides an extra level of stability and accuracy for the model equations relative to a leap-frog time tendency. This result has important practical implications for the design of an ocean model.

12.9 SMOOTHING THE SURFACE HEIGHT IN B-GRID MODELS

As discussed by Messinger (1973), Killworth et al. (1991), Pacanowski and Griffies (1999), and Griffies et al. (2001), there is a ubiquitous problem with B-grid models

*Mike Winton, personal communication 2003.

due to a null mode present when discretizing inviscid gravity waves. This mode manifests in the velocity field when using a relatively small viscosity. Additionally, it manifests in the surface height, especially in coarsely represented enclosed or semi-enclosed embayments where waves tend to resonate rather than to propagate. The pattern is stationary* and appears as a plus-minus pattern; i.e., as a checkerboard. As there is generally no dissipation in the surface height budget

$$\eta_{,t} = \nabla \cdot \mathbf{U} + q_{\mathrm{w}}, \tag{12.246}$$

suppression of the null mode requires some form of artificial dissipation.

Various methods have been described in the literature (e.g., Killworth et al., 1991; Griffies et al., 2001). The following filter has been found to be suitable for suppressing noise in the model of Griffies et al. (2004). It also maintains volume conservation and local and global tracer conservation

$$\partial_t (h\, T) = [\partial_t (h\, T)]_{\text{no filter}} + \delta_{k1}\, \nabla \cdot [A\, \nabla\, (h\, T)] \tag{12.247}$$

$$\partial_t\, \eta = [\partial_t\, \eta]_{\text{no filter}} + \nabla \cdot (A \nabla \eta), \tag{12.248}$$

where $A > 0$ is a diffusivity, and δ_{k1} vanishes for all but the surface grid cell at $k = 1$. For added scale selectivity, it is useful to employ a biharmonic operator instead of the Laplacian. Global conservation of volume and tracer is ensured by using no-flux conditions at the side boundaries. Note that a more conventional treatment of the tracer filter is given by the convergence of the thickness weighted flux $-h\, A\, \nabla\, T$. However, $\nabla \cdot (h\, A\, \nabla\, T)$ is not compatible with the filter applied to the surface height, and so this alternative approach is not acceptable.

12.10 RIGID LID STREAMFUNCTION METHOD

The rigid lid streamfunction method of Bryan (1969) may be used for Boussinesq ocean models. Over the years, this method has been remarkably versatile for z-models. Hence, it is useful to provide some comments on this method, even though it is largely obsolete.

12.10.1 Volume transport streamfunction

The basic assumption underlying the rigid lid method is that the time tendency of the ocean surface height is zero. Setting $\eta_{,t} = 0$ eliminates the fast barotropic gravity waves, and this is the key motivation for this assumption.

For the Boussinesq fluid, setting $\eta_{,t} = 0$ leads to a balance between the divergence of horizontal transport and surface fresh water forcing

$$\nabla \cdot \mathbf{U} = q_{\mathrm{w}}. \tag{12.249}$$

Although Huang (1993) described a rigid lid method realizing this balance, most rigid lid models set the surface fresh water flux to zero, thus giving

$$\nabla \cdot \mathbf{U} = 0. \tag{12.250}$$

Hence, the rigid lid Boussinesq model is fully volume conserving: all grid cells conserve volume and there are no fluxes of volume across the boundaries. Correspondingly, the vertical advective velocity (Section 11.3.2, page 225) vanishes at

*Hence the term "null," thus indicating it has a zero eigenvalue.

the surface. Herein lies the first major limiting aspect of the rigid lid streamfunction method: the effects of surface fresh water forcing must be incorporated into the model in an indirect and often unphysical manner. In particular, the effects on salinity are imposed via virtual salt fluxes across the ocean surface (for discussion, see Huang, 1993; Griffies et al., 2001). As mentioned in Section 11.5.4.2 (page 231), salt in the World Ocean is roughly constant, and in particular it is not transported across the ocean surface (except, for example, when interacting with sea ice). Hence, the virtual salt flux, which is quite real from the perspective of the rigid lid ocean model, is distinctly unphysical.

Since $\nabla \cdot \mathbf{U} = 0$, $\mathbf{U}$ can be specified by a single scalar *barotropic streamfunction*

$$\begin{aligned} \mathbf{U} &= H\,\overline{\mathbf{u}}^{z} \\ &= \nabla \wedge (\hat{\mathbf{z}}\psi) \\ &= -\hat{\mathbf{z}} \wedge \nabla \psi. \end{aligned} \tag{12.251}$$

Note that in this expression it is assumed that the depth of ocean fluid is equal to the depth of a resting fluid H. However, this assumption is not strictly true since $\eta_{,t} = 0$ does not imply $\eta = 0$. Indeed, there is a nonzero surface or *lid* pressure $p_{\text{lid}} = \rho_o\, g\, \eta$ which, although not needed for the streamfunction method, is computed via the *rigid lid surface pressure* method of Smith et al. (1992), Dukowicz et al. (1993), and Pinardi et al. (1995).

The barotropic streamfunction is specified to within a constant, and so only differences are physically relevant. In particular, Stokes's theorem implies that the vertically integrated advective transport between two points is given by

$$\begin{aligned} T_{a\rightarrow b} &= \int_a^b \mathrm{d}l\, \hat{\mathbf{n}} \cdot \int_{-H}^{0} \mathrm{d}z\, \mathbf{u} \\ &= \psi(b) - \psi(a), \end{aligned} \tag{12.252}$$

where $\mathrm{d}l$ is the line element along any path connecting the points a and b, and $\hat{\mathbf{n}}$ is a unit vector pointing perpendicular to the path in a rightward direction when facing the direction of integration. As written, $T_{a\rightarrow b}$ has units of volume per time, and so it represents a volume transport. Therefore, the difference in the barotropic streamfunction at two points represents the vertically integrated volume transport between the two points. It is for this reason that the barotropic streamfunction is sometimes called the *volume transport streamfunction*. Note that Bryan (1969) defined the barotropic streamfunction with an extra factor of the Boussinesq reference density ρ_o, such that his mass transport streamfunction has the dimensions of mass per time rather than volume per time. The difference is trivial for a Boussinesq fluid.

12.10.2 Time stepping the streamfunction

It is necessary to develop a prognostic equation for the streamfunction ψ in order to make use of ψ for time stepping the dynamics. This equation is formed by taking the curl of the momentum equation describing the dynamics of fluid columns (Section 4.6 on page 58), and so forming the barotropic vorticity equation. This equation involves an elliptic operator that must be inverted in order to specify the updated transport $\mathbf{U}(\tau + \Delta\tau)$ given the updated streamfunction $\psi(\tau + \Delta\tau)$. Inverting the elliptic operator is quite difficult when forcing the ocean with realistic time-dependent fluxes and using geometry with multiple islands and steep bottom topography such as occurs in the World Ocean. Herein lies the second major problem with the rigid lid streamfunction method: it leads to a difficult elliptic problem when running realistic ocean climate simulations. Furthermore, the elliptic problem involves specification of Dirichlet boundary conditions (the *island integrals*), and such nonlocal boundary conditions are very inefficient to impose on parallel computer architectures.

12.11 CHAPTER SUMMARY

Our discussions in this chapter focused on methods used to time step the ocean's hydrostatic primitive equations. There are many issues which arise when discretizing these equations in time, and this chapter aimed to expose the reader to some.

After a summary of the model equations in Secion 12.1, Section 12.2 presentated a discretization of the momentum equation in the vertical, as obtained by integrating the continuous equations over a discrete model grid cell. Care was given to the treatment of the surface boundary condition.

General strategies for time stepping the primitive equations were given in Section 12.3. Here, we introduced the notions of splitting the velocity into its slow and fast parts, commonly known, respectively, as the baroclinic or internal mode, and the barotropic or external mode. Section 12.4 then detailed the most common method to time step the ocean primitive equations, that being the method based on a leap-frog discretization of the time tendency. Section 12.5 then documented a number of problems with the leap-frog based schemes, and motivated the use of two-time level discretizations of the time tendency. Section 12.6 followed up that discussion with a time staggered scheme that remedies the problems with the leap-frog method. Section 12.7 then presented an alternative means to update the barotropic system via a predictor-corrector scheme. This scheme, as shown in Section 12.8, maintains some useful stability properties which motivate its use rather than the leap-frog scheme discussed in Section 12.4. Section 12.8 provided an extensive discussion of time step stability considerations for various terms in the primitive equations.

The final two sections are presented for completeness. Section 12.9 presents a brief discussion of a method to resolve a common problem with B-grid models, that being the presence of unacceptably large noise levels in the surface height. A filter was described which maintains volume and tracer conservation properties. Section 12.10 summarized the rigid lid method of Bryan (1969). This method has served the ocean climate modeling community for decades. However, its limitations associated with physical assumptions and computational inefficiencies have led most ocean climate modelers, regardless of their preferred vertical coordinate, to use one of the less restrictive free surface methods.

PART 4

Neutral Physics

Since the 1980's, ocean climate modelers using z-models have learned through experience how important closure terms are within the tracer equation. The original approach, whereby z-models employed a diffusion operator oriented according to geopotential surfaces, greatly compromised the simulation's physical integrity for climate purposes. The critical problem with the horizontally oriented operators is that they introduce unphysically large diapycnal mixing between simulated water masses. This problem was originally discussed by Veronis (1975), and hence is commonly known as the "Veronis effect."

Given the highly ideal fluid dynamics of the ocean interior, as revealed by ocean measurements discussed in Section 7.2.1 (page 157), most of the tracer transport in the interior is believed to occur along a locally referenced potential density direction, otherwise known as a *neutral direction*. Altering the orientation of the z-model's diffusion operator from horizontal-vertical to neutral-vertical brought the models more in line with the real ocean, and in turn greatly improved the simulation's realism.

In addition to a symmetric diffusive aspect to the subgrid scale (SGS) tracer transport, Chapter 9 noted that mesoscale eddies stir tracers in a reversible manner. This stirring corresponds to an antisymmetric component to the SGS tracer transport tensor. The combination of neutral diffusion and skew diffusion, and various other related schemes whose structure depend on neutral directions, are generically termed *neutral physics* in this book.

Adding a particular form of the *skew diffusive* component, as suggested by Gent and McWilliams (1990) and Gent et al. (1995) (referred to briefly at "GM" in this part of the book), again improved simulations in many regards. Notably, however, details of this component of the transport tensor remain part of an active research community, with many outstanding issues which will be raised in this part of the book. In addition, some results bring into question the utility of the GM approach.*
Nonetheless, most global ocean climate models are presently running with some form of neutral diffusion and neutral skew diffusion. This warrants a thorough examination of the methods used in the models, and this part of the book aims to provide such.

The technology for implementing neutral physics has become reasonably mature in z-models. Experience with these schemes has also matured, with nearly

*Section 11 of Griffies et al. (2000a) surveys model results. There are many indicating an improvement with neutral diffusion and GM. However, there are some that indicate a degradation (e.g., England and Holloway, 1998).

all z-model climate simulations today employing some form of neutral physics. Nonetheless, much remains to be improved. For example, the schemes can, in a manner analogous to numerical advection, exhibit nonmonotonic behavior under certain circumstances, with treatment of passive tracers a particularly difficult problem (Section 13.7). Additionally, it is arguable that the theoretical foundations for neutral physics operators remains to be fully articulated. Questions continually debated in the community include whether the schemes should be based on potential vorticity homogenization, available potential energy dissipation, nonequilibrium statistical physics principles, or otherwise. Finally, it remains an open question how to robustly handle, in theory and in practice, the interactions between mesoscale eddy closures and diabatic boundary layers.

The main focus in this part of the book is not so much to argue one way or another regarding some of the more controversial issues, although some prejudices will inevitably be revealed. Instead, the goal is to document the physical, mathematical, and numerical ideas comprising neutral physics as presently employed in z-models, and to highlight certain issues that remain to be resolved within the common practice. In particular, this discussion provides an opportunity to document some approaches implemented and working within the Modular Ocean Model described by Griffies et al. (2004). Quite generally, there are many details and choices to be made when implementing neutral physics schemes. Some of these issues are well known, yet others, many of which are quite important, tend to be overlooked, ignored, or hidden. Care is given to expose these issues in hopes that the reader will either come to understand what may appear to be an arbitrary or esoteric choice, or better yet be able to find fault with something and propose a better method.

Although the focus in this part of the book is on z-models, much is applicable to other models using a vertical coordinate whose surfaces are quasi-horizontal, such as pressure, z^*, and p^* (the latter two coordinates are defined in Section 11.9). Additionally, some ideas here find relevance, at least in principle, to the σ-models presented in Section 6.2.2. Unfortunately, the small slope approximation generally made for the interior of the z-model, whereby the neutral direction is assumed to slope no more than some 1% from constant z-surfaces, is not always appropriate for σ-models, since σ-surfaces can be sloped rather steeply relative to neutral directions (see Figure 6.5, page 150). Hence, more general discussions may be necessary when considering neutral physics for σ-models.

Chapter Thirteen

BASICS OF NEUTRAL PHYSICS

The purpose of this chapter is to introduce some of the physical and mathematical aspects of tracer transport associated with neutral physics in z-models. A conjecture is presented here regarding the general utility of neutral physics operators for a wide range of model resolutions, even those explicitly resolving the mesoscale. Given this conjecture, it is appropriate to work here within the context of the model equations derived in Chapter 8 resulting from an average over the small scales of motion.

13.1 CONCERNING THE UTILITY OF NEUTRAL PHYSICS

This chapter starts with a general discussion to motivate much of the presentations in this part of the book. The main idea is that neutral physics schemes are important for z-models using grids both coarser and finer than that needed to admit mesoscale eddies.

13.1.1 Small ocean interior mixing and spurious numerical mixing

As discussed in Chapters 7 and 9, the ocean is very close to an ideal fluid away from boundaries. This property means that tracer transport in the ocean interior occurs predominantly along neutral directions (Section 13.6), and the mass of fluid living between two isopycnals remains close to constant. Regardless of the ocean model mesh size relative to the mesoscale range, it is critical that numerical transport operators respect this property of the ocean.

When running coarse mesh models, there are important subgrid scale (SGS) fluxes that must be parameterized. Ocean climate model simulations are very sensitive to the form of these parameterizations. The work of Solomon (1971), Redi (1982), Olbers et al. (1985), McDougall and Church (1986), McDougall (1987a), Gent and McWilliams (1990), Gent et al. (1995), Griffies et al. (1998), Griffies (1998), and others stress the importance of neutral operators, whereby diffusion and skew diffusion are oriented according to the neutral directions. Although there remains work to be done in refining these operators, their use in coarse z-models has generally resulted in an improved physical integrity of the simulations relative to the older approaches based on horizontal diffusion (for a review, see Griffies et al., 2000a).

The extremely small levels of irreversible mixing within much of the ocean interior make it critical that ocean climate models respect these levels in both their explicit physical/numerical closure schemes, as well as the numerical realizations of all tracer transport operators such as advection. Unfortunately, as discussed in Section 7.5, and as shown by Roberts and Marshall (1998) and Griffies et al. (2000b), refining grid resolution so that mesoscale eddies are admitted can actually *exacerbate* problems with spurious mixing in z-models (and by implication, all other non-isopycnal based models).

Here are the basic causes of the problem. With an explicitly resolved quasi-geostrophic turbulent flow, tracer variance cascades down toward the grid scale. This variance must be numerically dissipated, or else the solution will degenerate into a noisy simulated sea. However, dissipation must be realized without incurring unphysically large levels of spurious diapycnal mixing. Unfortunately, as shown by Roberts and Marshall (1998), the highly scale-selective horizontally oriented biharmonic operators, which are quite useful for mopping up grid scale variance, also introduce unacceptably large levels of spurious mixing. Additionally, Griffies et al. (2000b) showed that truncation errors in common numerical advection schemes can stay relatively large even as the grid size is refined. The reason is that admitted scales of motion decrease upon grid refinement, *and* when the modeler reduces the level of applied numerical friction.* Because of the smaller scales admitted, truncation errors may increase when resolution is enhanced. Such truncation errors can lead to huge levels of spurious diapycnal mixing. Consequently, just as for the coarsely resolved models, refined models can be highly dependent on their explicit SGS transport operators as well as their numerical advection schemes. The bottom line is that a cavalier choice for how to dissipate the variance cascade is very risky.

13.1.2 A proposal and conjecture

What should be done? There are two basic approaches, with a combination perhaps the best. First, the numerical advection scheme must be good. Its dispersion errors should be small (e.g., second order centered differences induce too much dispersion), and its monotonicity properties must be handled in a high order manner (e.g., first order upwind limiters add too much spurious mixing). These points are detailed in Griffies et al. (2000b). Second, it is conjectured here that the *same* neutral operators employed at coarse resolution can provide a suitable *numerical closure* for fine resolutions, so long as they are run with suitably reduced diffusivities and perhaps with some modifications analogous to those used for the anisotropic friction schemes discussed in Chapter 17 (for a proposal, see Smith and Gent, 2004).

This conjecture is a variant of that proposed by Roberts and Marshall (1998), who advise the use of an adiabatic biharmonic operator. As argued in Section 14.5, biharmonic tracer operators should be used with some care, since they have the potential to introduce spurious extrema and anti-cabbeling. Instead, the familiar Laplacian based isotropic or anisotropic neutral operators should be sufficient. Therefore, it is conjectured that the basics of the neutral diffusion scheme of Redi (1982) and the skew diffusion scheme of Gent and McWilliams (1990) and Gent et al. (1995) are of primary importance for z-models at resolutions both above and below the mesoscale. This conjecture largely motivates the extensive attention

---

*Modelers nearly always reduce numerical friction when refining the grid, since the aim is to enhance the energetics of the flow and to (hopefully) reduce its dependence on generally *ad hoc* friction schemes. We have much more to say about model friction in Part 5 of this book.

paid to neutral physics in this book.

Importantly, the use of a "good" advection scheme and a robustly discretized neutral physics scheme *does not* guarantee that the levels of spurious mixing realized by a model will be below that of the real world mixing. However, it is hoped, and it remains a focus of research, that smart numerical forms of these transport operators can succeed. To remain far short of this goal will place into question the utility of z-models with refined eddying resolutions for ocean climate simulations.

Isopycnal models *must* run with some means to smooth potential density surfaces, or equivalently to smooth density interface height. Otherwise, the layers can and will develop negative thicknesses, which are unacceptable both physically and numerically. Since interface height smoothing is directly analogous to skew diffusion of potential density prescribed by Gent and McWilliams (1990) and Gent et al. (1995),* isopycnal models use some form of the closure proposed by Gent and McWilliams (1990), and have done so since their inception. When transporting tracers, they likewise diffuse tracers along isopycnal directions in a manner analogous to that proposed by Solomon (1971) and Redi (1982), either explicitly or via a dissipative advection scheme. Hence, the conjecture that neutral diffusion and skew diffusion should be used for both coarse and fine z-model resolutions is motivated by the ubiquitous practice in isopycnal models.

13.1.3 Comments on enhanced vertical viscosity

Rather than focusing on the tracer equation, one may wish to insert eddy-stirring effects into the momentum equation as an enhanced vertical friction (see, e.g., Greatbatch and Lamb, 1990; Greatbatch, 1998). The vertical friction approach induces no spurious diapycnal tracer mixing, by construction, since it only affects the momentum equation. However, the effects of vertical momentum friction on density are only available indirectly via geostrophy. Consequently, it is unclear whether vertical friction will prove sufficient to suppress grid scale power in the density field associated with dispersive advection schemes and/or a quasi-geostrophic cascade. Presently, no published studies document the ability of enhanced vertical friction to satisfy the needs of both physical closure and numerical closure in global ocean models.

Additionally, enhanced vertical friction only incorporates the skew diffusive piece of the transport tensor associated with Gent and McWilliams (1990). Neutral diffusion remains part of the tracer equation. As the combination of neutral diffusion plus skew diffusion within the tracer equation is often numerically simpler than the case of neutral diffusion alone (Section 14.2.2), in practice it may actually prove more convenient to keep the skew effects directly in the tracer equation. Again, these comments are speculative, and tests of the enhanced vertical friction method are warranted in realistic models to clearly determine its utility.

---

*Note that for flat bottom rigid lid models, interface height smoothing is equivalent to thickness diffusion. We discussed the mathematical equivalence in Section 9.5.4. However, in general it is crucial to note the difference (see, e.g., Holloway, 1997; Griffies et al., 2000a), with interface height smoothing the general method used in both isopycnal and z-models using the Gent and McWilliams (1990) scheme.

13.2 NOTATION AND SUMMARY OF SCALAR BUDGETS

The purpose of this section is to summarize some of the notation to be used in this part of the book, and recall the basic budgets of scalar properties of the fluid, such as mass and tracer.

13.2.1 Notation

The focus in this part of the book is on the equations describing scalar transport, along with their physical, mathematical, and numerical properties. Scalar fields remain invariant under coordinate transformations, with parcel mass and tracer concentration the scalar fields of concern here.

Scalars are the simplest of tensor fields. Consequently, one can safely dispense with many of the tools of general curvilinear tensor analysis needed to describe the evolution of momentum in a spherical geometry (Chapter 4), or to derive the form of the friction vector (Part 5). Instead, the use of tensors in this part of the book is limited to Cartesian tensors, and this usage is intermixed with common vector notation. Although not required, the use of covariant-contravariant index notation, needed in the case of curvilinear tensors, is maintained in this part. With this notation, indices are raised and lowered depending on whether the tensor is contravariant or covariant. Section 21.12 (page 481) provides a summary of these conventions. However, the reader need not penetrate that discussion for this part of the book. Rest assured that the use of tensor notation for scalar transport equations can be readily absorbed "on the fly."

The main piece of notation employed here is the comma notation, first introduced in Section 3.2.4 (page 29). With this notation, the partial derivative of a scalar field is written as

$$C_{,m} = \partial_m C. \tag{13.1}$$

This notation allows one to readily distinguish the mth element of a tensor field from the mth partial derivative. It also provides a very nice way to save ink.

13.2.2 Mass and tracer budgets

Mass and tracer budgets for fluid parcels were derived in Chapters 3, 4, and 5

$$\rho_{,t} + \rho_o \nabla \cdot \tilde{\mathbf{v}} = 0 \tag{13.2}$$

$$(\rho C)_{,t} + \rho_o \nabla \cdot (\tilde{\mathbf{v}} C) = 0, \tag{13.3}$$

where sources and molecular tracer diffusion are ignored for brevity, and $\rho_o \tilde{\mathbf{v}} = \rho \mathbf{v}$ is the linear momentum per volume of a fluid parcel. Compatibility between mass and tracer budgets manifests by having the tracer budget reduce to the mass budget in the case when the tracer field is everywhere constant.

Chapter 8 derived the ensemble mean primitive equations, with the ensemble mean mass and tracer budgets given by

$$\langle \rho \rangle_{,t} + \rho_o \nabla \cdot \langle \tilde{\mathbf{v}} \rangle = 0 \tag{13.4}$$

$$(\langle \rho \rangle \langle C \rangle^{\rho})_{,t} + \rho_o \nabla \cdot (\langle \tilde{\mathbf{v}} \rangle \langle C \rangle^{\rho}) = -\rho_o \nabla \cdot \langle \tilde{\mathbf{F}}_{\text{sgs}} \rangle, \tag{13.5}$$

where the SGS tracer flux is

$$\rho_o \tilde{\mathbf{F}}_{\text{sgs}} = \rho \, C'_{\rho} \, \mathbf{v}'_{\rho} \tag{13.6}$$

and interior sources of tracer are again omitted for brevity. To reach the averaged tracer budget required the introduction of the density weighted ensemble mean fields $\langle\rho\rangle\langle C\rangle^{\rho} = \langle\rho C\rangle$ and the fluctuation $C'_{\rho} = C - \langle C\rangle^{\rho}$ that satisfies $\langle\rho C'_{\rho}\rangle = 0$. Similar definitions apply to the density weighted velocity field.

For present concerns, the transport arising from $\rho C'_{\rho}\,\mathbf{v}'_{\rho}$ dominates that from molecular diffusion, so molecular diffusion is assumed to be incorporated into $\tilde{\mathbf{F}}_{\text{sgs}}$ without loss of generality. For those models not resolving mesoscale eddies, the piece of $\tilde{\mathbf{F}}_{\text{sgs}}$ associated with mesoscale eddies is interpreted in a manner introduced in Chapter 9 and further described by McIntosh and McDougall (1996), McDougall and McIntosh (2001), and Greatbatch and McDougall (2003). For those models that do resolve mesoscale eddies, the conjecture presented in Section 13.1 motivates the same form for the neutral tracer operators as used in the coarse models, yet with smaller diffusivities.

13.3 COMPATIBILITY IN THE MEAN FIELD BUDGETS

As noted above, compatibility between mass and tracer budgets for unaveraged fluid parcels manifests by having the tracer budget reduce to the mass budget in the case when the tracer field is everywhere constant. An analogous compatibility is useful for the mean fluid so that the mean fluid retains mass and tracer conservation regardless of the SGS fluxes. Doing so constrains the form of the SGS flux.

Compatibility for the ensemble averaged mass and tracer budgets is satisfied if $\nabla\cdot\langle\tilde{\mathbf{F}}_{\text{sgs}}\rangle = 0$ when $\langle C\rangle^{\rho}$ is everywhere constant. In the general case of a nontrivial mean tracer field, compatibility is maintained when the SGS tracer flux takes the form

$$\langle\tilde{\mathbf{F}}_{\text{sgs}}\rangle = \nabla\wedge\mathbf{B} - \mathbf{J}\cdot\nabla\langle C\rangle^{\rho}, \tag{13.7}$$

where $\mathbf{B}$ is an arbitrary vector field and $\mathbf{J}$ is a second order *tracer transport tensor*. In general, the transport tensor is a function of the fluid flow and tracer field. Terms of higher order in the tracer derivative also satisfy compatibility; they are dropped for simplicity.

Since the divergence of $\nabla\wedge\mathbf{B}$ vanishes, it does not contribute to the time tendency of the density weighted tracer. The freedom to add or subtract a curl to the tracer flux is exploited in the following. For present purposes it is dropped, thus bringing the mth component of the SGS flux to the form

$$\langle\tilde{F}^{m}_{\text{sgs}}\rangle = -J^{mn}\,\partial_n\langle C\rangle^{\rho}, \tag{13.8}$$

where J^{mn} are components to the tracer transport tensor $\mathbf{J}$, and the summation convention is assumed, in which case repeated indices are summed over their range.

Given these results, the mapping defined by Table 8.1 (page 187) leads to the mass and tracer equations to be discretized by the ocean model

$$\rho_{,t} + \rho_o\nabla\cdot\mathbf{v} = 0 \tag{13.9}$$

$$(\rho C)_{,t} + \rho_o\,\nabla\cdot(\mathbf{v}\,C) = -\rho_o\,\nabla\cdot\mathbf{F}, \tag{13.10}$$

where

$$F^{m} = -J^{mn}\,C_{,n} \tag{13.11}$$

is the model's version of the SGS tracer flux. Both the resolved and SGS tracer transport operators take the same mathematical form for the non-Boussinesq and Boussinesq fluids. The remainder of this chapter discusses mathematical and physical properties of this tracer flux.

13.4 THE SGS TRACER TRANSPORT TENSOR

This section establishes some general properties of the second order tracer transport tensor $\mathbf{J}$, decomposed as

$$\begin{aligned} J^{mn} &= (J^{mn} + J^{nm})/2 + (J^{mn} - J^{nm})/2 \\ &\equiv K^{mn} + A^{mn}, \end{aligned} \tag{13.12}$$

where K^{mn} is the symmetric part, and A^{mn} is the antisymmetric part. The tracer equation therefore takes the form

$$(\rho C)_{,t} + \rho_o \nabla \cdot (\mathbf{v}\, C) = \rho_o \nabla \cdot [(\mathbf{K} + \mathbf{A}) \cdot \nabla C]. \tag{13.13}$$

The transport properties of the symmetric and antisymmetric tensors are quite distinct.

13.4.1 Evolution of global tracer variance

We consider here some global properties of the tracer transport equation. For this purpose, examine the behavior of global tracer variance, defined as

$$\mathcal{V} = M^{-1} \int (\rho\, \mathrm{d}V)\, C^2 - \left(M^{-1} \int (\rho\, \mathrm{d}V)\, C \right)^2. \tag{13.14}$$

In this equation, $M = \int (\rho\, \mathrm{d}V)$ is the mass of seawater in the domain. Interest here regards how tracer variance evolves from the effects of SGS fluxes. Hence, it is sufficient to consider the case of an ocean with constant fluid mass and constant tracer mass. Tracer variance thus evolves according to the evolution of the first term in equation (13.14), which allows for the second term to be dropped,

$$\mathcal{V} = M^{-1} \int \mathrm{d}A \int_{-H}^{\eta} \mathrm{d}z\, \rho\, C^2. \tag{13.15}$$

To compute the time tendency $\mathcal{V}_{,t}$ requires the evolution equation

$$(\rho\, C^2)_{,t} + \rho_o \nabla \cdot (\mathbf{v}\, C^2) = -2\, \rho_o\, C\, \nabla \cdot \mathbf{F}, \tag{13.16}$$

where mass conservation was used to derive this result. The divergence theorem (Section 3.3) is also needed, along with the surface and solid earth kinematic boundary conditions

$$\hat{\mathbf{N}} \cdot \mathbf{v} = (\rho/\rho_o)\, \eta_{,t} - q_{\mathrm{w}} \qquad \text{at } z = \eta \tag{13.17}$$

$$\hat{\mathbf{N}} \cdot \mathbf{v} = 0 \qquad \text{at } z = -H, \tag{13.18}$$

with the orientation directions given by $\hat{\mathbf{N}} = \nabla(-\eta + z)$ at $z = \eta$ and $\hat{\mathbf{N}} = \nabla(H + z)$ at $z = -H$. The steps necessary are similar to those needed for the calculation of global energy budgets in Section 5.4.1 (page 105), and the result is

$$M\, \mathcal{V}_{,t} = \int_{z=\eta} \mathrm{d}A\, (\rho_{\mathrm{w}}\, q_{\mathrm{w}}\, C^2 - \rho_o\, C\, \mathbf{F} \cdot \hat{\mathbf{N}}) + 2\, \rho_o \int \mathrm{d}V\, \nabla C \cdot \mathbf{F}. \tag{13.19}$$

The boundary terms are not of concern here. Instead, focus now on the volume term.

13.4.2 Diffusive and skew diffusive tracer fluxes

In addition to the effects of boundary terms, tracer variance in equation (13.19) changes according to the global integral of

$$\begin{aligned} \nabla C \cdot \mathbf{F} &= -(A^{mn} + K^{mn})\, C_{,n}\, C_{,m} \\ &= -K^{mn}\, C_{,n}\, C_{,m} \end{aligned} \tag{13.20}$$

where $A^{mn}\, C_{,n}\, C_{,m} = 0$ due to the antisymmetry of A^{mn} and the symmetry of $C_{,n}\, C_{,m}$ (recall the summation convention). A particular form of the symmetric transport tensor of use in ocean modeling is one that is positive semidefinite,* for which

$$\nabla C \cdot \mathbf{K} \cdot \nabla C = C_{,m}\, K^{mn}\, C_{,n} \geq 0. \tag{13.21}$$

Transport tensors satisfying this property are known as *diffusion tensors*. The associated *diffusive* flux is directed down the gradient of the tracer

$$\nabla C \cdot \mathbf{F} \leq 0. \tag{13.22}$$

Such fluxes are dissipative, since within the ocean interior they reduce the value of $\int (\rho\, dV)\, C^{2n}$, with n a positive integer. This result is apparent for the tracer variance $n = 1$ by inspecting equation (13.19). It is derived for higher values of n via similar methods to those used here for $n = 1$.

Note that symmetric negative definite SGS tracer transport tensors are unstable, unless they are suitably balanced by a companion dissipation operator. Such operators are not common in the ocean climate literature, and only positive semidefinite symmetric transport tensors are considered in the following.

The antisymmetric transport tensor identically dropped out from the tendency for tracer variance. The reason is that its associated tracer flux $F^m = -A^{mn}\, C_{,n}$ is directed perpendicular to the tracer gradient

$$\begin{aligned} \mathbf{F} \cdot \nabla C &= -A^{mn}\, C_{,n}\, C_{,m} \\ &= 0. \end{aligned} \tag{13.23}$$

Tracer fluxes with this orientation are denoted *skew* fluxes in the following. Notably, skew fluxes do not alter the value of $\int (\rho\, dV)\, C^{2n}$. As such, they are akin to advection, which acts to stir the tracer without mixing.

13.4.3 Local dissipation of tracer extrema

Diffusion is the canonical dissipative process. The previous discussion illustrates what this means for the evolution of tracer variance. Now consider what it means locally for tracer extrema. At a tracer extremum, the tracer concentration satisfies $\nabla C = 0$. Inserting this property into the tracer equation (13.13) yields the evolution equation for an extremum

$$\rho\, C_{,t} = \rho_o\, K^{mn}\, C_{,mn}. \tag{13.24}$$

This result required $A^{mn}\, C_{,mn} = 0$, which arises from the antisymmetry of A^{mn} and the symmetry of $C_{,mn}$. Equation (13.24) says that tracer extrema are invisible to

---

**Positive semidefinite* means that there are no negative eigenvalues for the tensor. That is, all diffusivities are non-negative. Zero diffusivities are allowed, such as may occur in idealized studies with zero diapycnal diffusivity.

advection and skew diffusion, but not to diffusion. If the extremum is a minimum, the second derivative matrix $C_{,mn}$ is positive definite. Since K^{mn} is also positive semidefinite, the product $K^{mn}\, C_{,mn}$ is non-negative, thus creating a non-negative time tendency, which reduces the magnitude of the minimum. For example, if the minimum represents a cold region, then diffusion will warm this region and thus dissipate the minimum. The opposite holds for tracer maxima, in which case the time tendency is nonpositive, and so maxima are reduced. One thus concludes that tracer extrema are locally eroded away by diffusion. This property lends the intuition to diffusion as a local smoothing process.

13.5 ADVECTION AND SKEWSION

There are two complementary ways of interpreting the transport process associated with an antisymmetric tensor A^{mn}: via either the convergence of an advective flux, known as *advection*, or the convergence of a skew flux, called *skew diffusion* or *skewsion*. These issues were presented in Section 9.2 (page 191). Because of the importance of understanding these two perspectives in the formulation of neutral physics operators, and motivated by a desire to keep parts of this book effectively independent of one another, consider here an abbreviated version of the discussion given more completely in Section 9.2.

To start, consider an arbitrary three-dimensional divergence-free velocity field $\nabla \cdot \mathcal{U} = 0$. The divergence-free condition represents a diagnostic relation, or constraint, that reduces the functional degrees of freedom for the velocity field from three to two. This constraint can be satisfied identically by introducing a vector streamfunction

$$\mathcal{U} = \nabla \wedge \boldsymbol{\Upsilon}. \tag{13.25}$$

The vector streamfunction $\boldsymbol{\Upsilon}$ is not completely specified by this relation, since the equally valid streamfunction

$$\boldsymbol{\Upsilon}' = \boldsymbol{\Upsilon} + \nabla\lambda \tag{13.26}$$

corresponds to the same velocity field $\mathcal{U}$. The arbitrary scalar function λ is known as a *gauge function*, and the freedom to modify the vector streamfunction through the addition of λ is known as *gauge freedom*.

The *advective tracer flux*

$$\begin{aligned} \mathbf{F}_{(a)} &= \mathcal{U}\, C \\ &= (\nabla \wedge \boldsymbol{\Upsilon})\, C \end{aligned} \tag{13.27}$$

can be related to a skew flux

$$\begin{aligned} \mathbf{F}_{(s)} &= -\nabla C \wedge \boldsymbol{\Upsilon} \\ &= \mathbf{F}_{(a)} - \mathbf{F}_{(r)} \end{aligned} \tag{13.28}$$

through exploitation of the identity

$$\begin{aligned} \mathcal{U}\, C &= (\nabla \wedge \boldsymbol{\Upsilon})\, C \\ &= -\nabla C \wedge \boldsymbol{\Upsilon} + \nabla \wedge (C\, \boldsymbol{\Upsilon}). \end{aligned} \tag{13.29}$$

$\mathbf{F}_{(s)}$ qualifies as a skew flux since it is directed perpendicular to the tracer gradient $\mathbf{F}_{(s)} \cdot \nabla C = 0$. Since the *rotational flux*

$$\mathbf{F}_{(r)} = \nabla \wedge (C\,\boldsymbol{\Upsilon}) \tag{13.30}$$

has a vanishing divergence, the divergence of the skew and advective fluxes is identical

$$\nabla \cdot \mathbf{F}_{(s)} = \nabla \cdot \mathbf{F}_{(a)}. \tag{13.31}$$

Hence, if these fluxes enter into the evolution of a tracer, one may choose to use the skew flux or advective flux in describing the evolution. That is,

$$\begin{aligned}(\rho\, C)_{,t} + \rho_o\, \nabla \cdot (\mathbf{v}\, C) &= -\rho_o\, \nabla \cdot (\mathcal{U}\, C) \\ &= \rho_o\, \nabla \cdot (\nabla C \wedge \boldsymbol{\Upsilon}). \end{aligned} \tag{13.32}$$

Contact is made with results in Section 13.4.2 by introducing the antisymmetric transport tensor

$$\begin{aligned} F^m_{(s)} &= -\epsilon^{mnp}\, C_{,n}\, \Upsilon_p \\ &= -A^{mn}\, C_{,n} \end{aligned} \tag{13.33}$$

where A^{mn} represents a reorganization of the vector streamfunction

$$\begin{aligned} A^{mn} &= \epsilon^{mnp}\, \Upsilon_p \\ &= \begin{pmatrix} 0 & \Upsilon_3 & -\Upsilon_2 \\ -\Upsilon_3 & 0 & \Upsilon_1 \\ \Upsilon_2 & -\Upsilon_1 & 0 \end{pmatrix}, \end{aligned} \tag{13.34}$$

and ϵ^{mnp} is the totally antisymmetric Levi-Civita tensor defined in Section 20.12 (page 459). Gauge freedom is exploited often in this book by specifying a *vertical gauge* in which $\Upsilon_3 = 0$.

13.6 NEUTRAL TRACER FLUXES

In the formulation of neutral physics transport operators in a z-model, it is useful to understand both how neutral directions are determined and how the geometry of fluxes aligned according to these directions relate to the fixed $\hat{\mathbf{x}}, \hat{\mathbf{y}}, \hat{\mathbf{z}}$ directions. In particular, one is frequently concerned with rotation from the z-coordinate frame to neutral directions.

13.6.1 Specifying the neutral directions

As discussed in Section 7.2.1 (page 157), neutral directions represent directions in which the adiabatic displacement of a fluid parcel experiences no resistance from buoyancy forces. That is, neutral directions are parallel to lines of constant buoyancy.

Consider a fluid with a linear equation of state, in which buoyancy and density are linearly proportional. In this case, neutral directions are tangent to the density surface, and so are orthogonal to the local normal given by

$$\begin{aligned} \hat{\mathbf{n}} &= \hat{\boldsymbol{\rho}} \\ &= \frac{\nabla \rho}{|\nabla \rho|}. \end{aligned} \tag{13.35}$$

In a fluid with a nonlinear equation of state $\rho = \rho(\theta, s, p)$, such as seawater, pressure effects make it necessary to carefully specify how one computes density gradients to determine neutral directions. In particular, it is necessary to perform the gradient operation in a manner that removes pressure effects, just as done when checking for gravitational stability (e.g., Gill, 1982). That is, with pressure held fixed (i.e., locally referenced),

$$\begin{aligned} \nabla\rho &= \nabla\theta \left(\frac{\partial\rho}{\partial\theta}\right)_{s,p} + \nabla s \left(\frac{\partial\rho}{\partial s}\right)_{\theta,p} \\ &= \rho\,(-\alpha\,\nabla\theta + \beta\,\nabla s). \end{aligned} \tag{13.36}$$

Whenever neutral directions are computed, it is this derivative of density, again with pressure held fixed, which is needed. In these relations, θ is the potential temperature, s is the salinity, p is the pressure, and partial derivatives of density are taken with the specified variables held fixed. The functions

$$\alpha = -\left(\frac{\partial \ln \rho}{\partial\theta}\right)_{s,p} \tag{13.37}$$

and

$$\beta = \left(\frac{\partial \ln \rho}{\partial s}\right)_{\theta,p} \tag{13.38}$$

are thermal expansion and saline contraction coefficients. These functions can typically be determined analytically from an empirical expression for the equation of state. This is a useful means for computing these coefficients in a numerical model. Note that the local normal is perpendicular to potential temperature surfaces when salinity is uniform, regardless of the dependence on pressure.

As articulated by McDougall (1987a), neutral *surfaces* represent an accumulation of tangents to locally referenced potential density surfaces. However, McDougall and Jackett (1988) showed that due to pressure effects, neutral surfaces possess a helical topology. This nontrivial topology precludes neutral surfaces from being a useful means to partition the vertical direction in ocean models. Instead, it is possible only to unambiguously determine the locally defined neutral direction at each point in the ocean. An approximation to such directions is realized in z-models. The helical nature of neutral surfaces precludes one from defining a global ocean model based on neutral surfaces as defining the model's vertical coordinate.

13.6.2 Geometric relations for stably stratified density layers

Figure 13.1 shows a region on a potential density surface. At an arbitrary point along this surface, the unit normal direction $\hat{\mathbf{n}}$ and tangent direction $\hat{\mathbf{t}}$ are shown, along with the angle δ between the tangent and the horizontal axis, or equivalently between the normal and the vertical. The following considerations assume that the local tangent direction has some nonzero component in the horizontal. Such is the case for a stably stratified fluid, so long as the potential density surface is locally referenced to a point near to where the tangent is defined.

In the two-dimensional (x, z)-plane, an arbitrary vector aligned with the local tangent takes the form

$$\begin{aligned} \mathbf{F} &= |\mathbf{F}|\,\hat{\mathbf{t}} \\ &= F^{(x)}\,(\hat{\mathbf{x}} + S\,\hat{\mathbf{z}}), \end{aligned} \tag{13.39}$$

where

$$F^{(x)} = |\mathbf{F}| \cos\delta \tag{13.40}$$

is the horizontal component of the vector, and

$$\begin{aligned} S &= \tan\delta \\ &= z_{,x} \\ &= -\rho_{,x}/\rho_{,z} \end{aligned} \tag{13.41}$$

is the neutral direction slope with respect to the horizontal. In three dimensions, this slope takes the form

$$\begin{aligned} \mathbf{S} &= \nabla_\rho z \\ &= -\left(\frac{\nabla_z \rho}{\rho_{,z}}\right) \\ &= -\left(\frac{-\alpha\,\nabla_z\theta + \beta\,\nabla_z s}{-\alpha\,\theta_{,z} + \beta\,s_{,z}}\right), \end{aligned} \tag{13.42}$$

where ∇_ρ represents the horizontal gradient taken along the neutral direction, and ∇_z is the horizontal gradient along a constant z surface. The transformations between z-coordinates and isopycnal coordinates required for these steps are described in Section 6.5.3 (page 134).

Given these general results, a vector oriented parallel to the neutral direction at an arbitrary point in the ocean has horizontal $\mathbf{F}^{(h)}$ and vertical $F^{(z)}$ components related via

$$F^{(z)} = \mathbf{S} \cdot \mathbf{F}^{(h)}. \tag{13.43}$$

This result follows trivially, since $\mathbf{F} \cdot \nabla\rho = 0$. That is, $\mathbf{F}$ has zero projection in the $\hat{\boldsymbol{\rho}}$ direction, by construction.

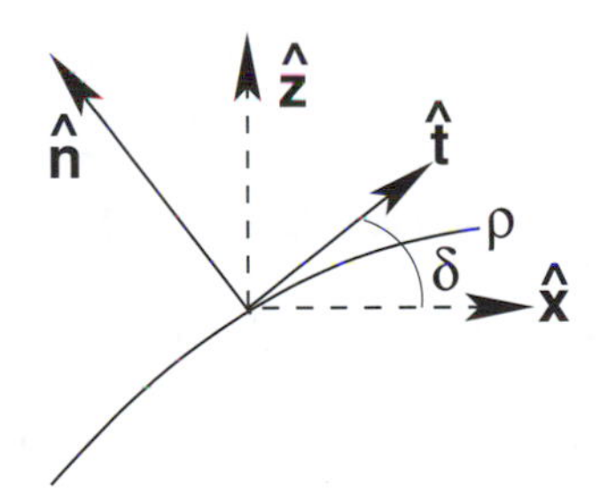

Figure 13.1 Geometry of a potential density surface ρ =const. At any point along the surface, a local normal $\hat{\mathbf{n}} = \hat{\boldsymbol{\rho}}$ and tangent $\hat{\mathbf{t}}$ can be defined, with δ being the angle between the tangent and the horizontal. Equivalently, it is the angle between the normal and the vertical. At the pressure where the potential density surface is defined, the local tangent defines the neutral direction.

13.6.3 Geometric relations for arbitrarily stratified density layers

The previous geometric analysis, restricted to stably stratified density layers, is sufficient for large-scale ocean modeling. Nonetheless, in order to understand the geometric basis for the Redi diffusion tensor (Redi, 1982), consider a more general approach where neutral directions can be arbitrarily oriented.

For this purpose, the symmetric diffusion tensor representing along and across neutral mixing can be written in the projection operator form

$$K^{mn} = A_I(\delta^{mn} - \hat{\rho}^m \hat{\rho}^n) + A_D \hat{\rho}^m \hat{\rho}^n, \tag{13.44}$$

where $\hat{\rho}^m$ are components to the local normal $\hat{\boldsymbol{\rho}}$ to the locally referenced potential density surface. The epineutral (along neutral) and dianeutral (across neutral) diffusivities A_I and A_D are non-negative and can in general be functions of space-time. Although not written such in her paper, this is indeed the Redi (1982) diffusion tensor, as noted by Olbers et al. (1985) and Griffies et al. (1998).

One may understand the Redi diffusion tensor in a geometric manner. For this purpose, note that the diffusive flux $F^m = -K^{mn} C_{,n}$ resulting from the diffusion tensor takes the form

$$\mathbf{F} = -A_I[\nabla C - \hat{\boldsymbol{\rho}}\,(\hat{\boldsymbol{\rho}} \cdot \nabla C)] - A_D\, \hat{\boldsymbol{\rho}}\,(\hat{\boldsymbol{\rho}} \cdot \nabla C). \tag{13.45}$$

The term

$$(\nabla C)_{\parallel} = \nabla C - \hat{\boldsymbol{\rho}}\,(\hat{\boldsymbol{\rho}} \cdot \nabla C) \tag{13.46}$$

is perpendicular to the local normal: $(\nabla C)_{\parallel} \cdot \hat{\boldsymbol{\rho}} = 0$. This vector therefore represents that part of the tracer gradient field that is oriented parallel to the neutral direction, that is, parallel to $\hat{\mathbf{t}}$. Consequently, $\delta^{mn} - \hat{\rho}^m \hat{\rho}^n$ is generally a projection operator that projects out that part of a vector in the direction parallel to the neutral direction. The second term,

$$(\nabla C)_{\perp} = \hat{\boldsymbol{\rho}}\,(\hat{\boldsymbol{\rho}} \cdot \nabla C), \tag{13.47}$$

is parallel to the local normal $\hat{\boldsymbol{\rho}}$, and so it represents that part of the tracer gradient field perpendicular to the neutral direction. $\hat{\rho}^m \hat{\rho}^n$ is therefore a projection operator that projects out that part of a vector in the $\hat{\boldsymbol{\rho}}$ direction.

13.7 CHAPTER SUMMARY AND A CAVEAT ON THE CONJECTURE

The main part of this chapter focused on kinematics associated with tracer transport via a second order transport tensor. To get started, Section 13.2 presented some notation and results from previous chapters, and then Section 13.3 made an argument for the general form of the SGS tracer flux. This argument, based on tracer and mass compatibility, motivated us to write the tracer flux in terms of a second order tensor. Section 13.4 then derived some of the basic properties of this tensor's symmetric, or diffusive, and antisymmetric, or skew diffusive, components. The properties are quite distinct, with diffusion acting to smooth extrema locally and reduce variance globally, whereas skew fluxes do nothing to extrema or variance. Instead, the convergence of skew fluxes stirs the tracer field, just as advection does. This discussion was followed by relating skew tracer fluxes to advective tracer fluxes in Section 13.5. The skew flux has a convergence which defines the skewsion operator, just as the convergence of advective fluxes leads to advection. As the fluxes differ only by a total curl, their convergences are identical, and so their effects on tracer evolution are the same. The chapter finished in

Section 13.6 by considering some basic geometric aspects of neutral tracer fluxes. Elementary trigonometry is sufficient to understand how to relate neutrally oriented fluxes to the horizontal and vertical directions.

A conjecture started this chapter. That is, neutral physics schemes will remain an essential aspect of SGS closures for z-models with resolutions both above and below the mesoscale. Indeed, most any model class, regardless of the vertical coordinate, should be using a neutral physics based closure for either physical or numerical reasons. This conjecture is based simply on the small levels of dianeutral mixing measured in the ocean. At coarse resolution, the schemes arguably take on a somewhat physical basis (even though there remains much debate regarding the physical interpretation of these schemes). At refined resolution, the schemes become mostly numerically based, as they provide a means to dissipate variance without introducing excessive transport across neutral directions.

For this conjecture to be realized, one must exercise care when implementing neutral physics schemes in z-models. The care presented in this part of the book is, unfortunately, not sufficient for all cases and so requires refinement. In particular, numerical truncation errors are difficult to control for passive tracer transport via neutral diffusion. The paper by Beckers et al. (1998) points out the numerical issues. The basic problem is that a projection of a neutrally oriented downgradient diffusive flux, onto the horizontal and vertical, generally results in upgradient flux components in at least one of the horizontal and/or vertical directions. Upgradient flux components are of no concern for stability, so long as the full three-dimensional flux vector is downgradient. However, when neutral diffusion is numerically realized on a finite grid, and when the neutral directions project strongly onto the vertical (i.e., steep neutral directions), truncation errors associated with the upgradient flux components can and will introduce nonmonotonic behavior distinctly out of character with the dissipative aspects of a diffusive process.

There is some help to be obtained by recognizing that, in many regions where the numerical neutral transport schemes (e.g., the one described in Chapter 16) have the most difficulty, physical processes such as enhanced diapycnal mixing help smooth out problems. However, problems still can arise. Relying on enhanced physical mixing may not always be sufficient. Therefore, the presence of nonmonotonic behavior suggests an implementation of flux limiters for neutral diffusion, much as done for advection schemes. Beckers et al. (2000) provides some suggestions. At present, these schemes have not been widely tested in the climate models. However, the coupling of ecosystem models into the circulation models promotes the issue to a higher level of priority. The problem is that if a biological tracer loses its positive-definiteness, some ecosystem models can exhibit very unstable and unrealistic behavior. Finding useful methods to remedy the nonmonotonicity of numerical neutral physics schemes remains an ongoing research topic.

Chapter Fourteen

NEUTRAL TRANSPORT OPERATORS

The purpose of this chapter is to detail mathematical and physical properties of the neutral diffusion scheme of Redi (1982) and the skew diffusion scheme of Gent and McWilliams (1990), Gent et al. (1995), and Griffies (1998). Some ideas are also introduced that are relevant for determining the value of the diffusivities based on properties of the flow regime. The chapter finishes with a discussion of biharmonic tracer operators. After summarizing some of those proposed in the literature, it is argued that such operators lack much utility for ocean climate modeling.

14.1 NEUTRAL DIFFUSION

From the standpoint of tracer transport via mesoscale eddies, a natural set of coordinates are those defined with respect to the neutral directions. These coordinates define what is termed here the *neutral frame*. Diffusive transport in this frame is assumed to be diagonal in the formulation of Redi (1982), with diffusion along the neutral directions on the order of 10^8 times greater than dianeutral diffusion.*

The simplest and most elegant way to derive the neutral diffusion tensor is through the projection operator form given by equation (13.44). The diffusion tensor is written down by inspection based on assumptions of the strength and orientation of the diffusive fluxes. Nonetheless, it is useful for many purposes to present the rotational approach used by Redi (1982) to derive this tensor. That is the purpose of this section. In addition, emphasis is given to a key balance between the neutral diffusive fluxes of active tracers. Maintaining this balance has proven to be crucial for realizing a numerically stable neutral diffusion scheme (Griffies et al., 1998).

14.1.1 Two bases of unit directions

As in Section 21.12 (page 481), it is often quite useful to use *physical tensor components* to represent tensors. Physical components each have the same dimensions. Furthermore, they are obtained by projecting a tensor onto a basis of dimensionless unit directions $\hat{\mathbf{e}}_m$, where $m = 1, 2, 3$ for the three-dimensional space considered here.

There are two sets of unit directions defining the coordinate frames of reference considered here. The first one, termed here the z-frame, is defined by the dimen-

*This number is based on measurements in the ocean interior, as discussed in Section 7.2.1 (page 157).

sionless unit directions

$$\hat{\mathbf{e}}_1 = \hat{\mathbf{x}} \tag{14.1}$$

$$\hat{\mathbf{e}}_2 = \hat{\mathbf{y}} \tag{14.2}$$

$$\hat{\mathbf{e}}_3 = \hat{\mathbf{z}}, \tag{14.3}$$

where $\hat{\mathbf{z}}$ is the usual vertical direction taken anti-parallel to gravity, and $\hat{\mathbf{x}}$ and $\hat{\mathbf{y}}$ represent generalized orthogonal horizontal directions. For example, with spherical coordinates, $\hat{\mathbf{x}}$ is in the longitudinal direction and $\hat{\mathbf{y}}$ is in the latitudinal direction. More general directions are available in most modern ocean climate models through the use of generalized orthogonal coordinates (see Sections 21.11 and 21.12).

The second reference frame is defined by the dimensionless unit directions

$$\hat{\mathbf{e}}_{\overline{1}} = \frac{\hat{\mathbf{z}} \wedge \nabla\rho}{|\hat{\mathbf{z}} \wedge \nabla\rho|} \tag{14.4}$$

$$\hat{\mathbf{e}}_{\overline{2}} = \hat{\mathbf{e}}_{\overline{3}} \wedge \hat{\mathbf{e}}_{\overline{1}}, \tag{14.5}$$

$$\hat{\mathbf{e}}_{\overline{3}} = \hat{\boldsymbol{\rho}} = \frac{\nabla\rho}{|\nabla\rho|}, \tag{14.6}$$

where $\nabla\rho$ is computed as in equation (13.36) (page 292) where pressure effects are removed. These three directions define the *neutral* frame. This frame is determined by the fluctuating geometry of a locally referenced isopycnal surface. A similar reference frame was discussed in Section 6.4 when discussing generalized vertical coordinates.

Both sets of basis directions define orthonormal reference frames. Hence, the transformation between the two frames is provided by a rotation matrix, which is an orthogonal matrix (unit determinant and inverse given by the transpose). This transformation can be interpreted in terms of Euler angles as discussed by Redi (1982).

As a tensor equation, the transformation between reference frames is written

$$\hat{\mathbf{e}}_{\overline{m}} = \Lambda^{m}{}_{\overline{m}}\, \hat{\mathbf{e}}_m, \tag{14.7}$$

where the summation convention is employed. The components of the transformation matrix are directional cosines determined via

$$\Lambda^{m}{}_{\overline{m}} = \hat{\mathbf{e}}^m \cdot \hat{\mathbf{e}}_{\overline{m}}, \tag{14.8}$$

where unit directions do not distinguish between raised and lowered indices. Substituting the explicit forms for the two bases, and organizing components of the transformation into a matrix-vector form, lead to

$$(\hat{\mathbf{e}}_{\overline{1}}\ \ \hat{\mathbf{e}}_{\overline{2}}\ \ \hat{\mathbf{e}}_{\overline{3}}) = (\hat{\mathbf{e}}_1\ \ \hat{\mathbf{e}}_2\ \ \hat{\mathbf{e}}_3) \begin{pmatrix} \frac{S_y}{S} & \frac{S_x}{S\sqrt{1+S^2}} & -\frac{S_x}{\sqrt{1+S^2}} \\ -\frac{S_x}{S} & \frac{S_y}{S\sqrt{1+S^2}} & -\frac{S_y}{\sqrt{1+S^2}} \\ 0 & \frac{S}{\sqrt{1+S^2}} & \frac{1}{\sqrt{1+S^2}} \end{pmatrix}, \tag{14.9}$$

where the neutral slope can be written in one of the following equivalant manners:

$$\begin{aligned} \mathbf{S} &= \nabla_\rho z \\ &= -\left(\frac{\nabla_z \rho}{\rho_{,z}}\right) \\ &= (S_x, S_y, 0). \end{aligned} \tag{14.10}$$

This slope vector was first introduced in Section 6.3.3 when discussing the slope of a surface relative to the horizontal. The slope vector has magnitude S.

14.1.2 Diffusion in the neutral frame

In the neutral frame, the gradient of a tracer is written

$$\nabla C = \hat{\mathbf{e}}_{\overline{1}} \partial_{\hat{\mathbf{e}}_{\overline{1}}} C + \hat{\mathbf{e}}_{\overline{2}} \partial_{\hat{\mathbf{e}}_{\overline{2}}} C + \hat{\mathbf{e}}_{\overline{3}} \partial_{\hat{\mathbf{e}}_{\overline{3}}} C, \tag{14.11}$$

where derivatives are taken along the three orthogonal directions. Following Redi (1982), downgradient diffusion in this frame is diagonally oriented, in which case the diffusion tensor takes the form

$$K^{\overline{mn}} = \begin{pmatrix} A_I & 0 & 0 \\ 0 & A_I & 0 \\ 0 & 0 & A_D \end{pmatrix}, \tag{14.12}$$

which can also be written in terms of projection operators

$$K^{\overline{mn}} = A_I(\delta^{\overline{mn}} - \hat{\rho}^{\overline{m}} \hat{\rho}^{\overline{n}}) + A_D \hat{\rho}^{\overline{m}} \hat{\rho}^{\overline{n}}, \tag{14.13}$$

as noted in Section 13.6.3. Note that in the neutral frame, the unit normal direction $\hat{\rho}$ has components $\hat{\rho} = (0, 0, 1)$. The diffusion operator is thus written

$$R(C) = \partial_{\hat{\mathbf{e}}_{\overline{1}}}(A_I \partial_{\hat{\mathbf{e}}_{\overline{1}}} C) + \partial_{\hat{\mathbf{e}}_{\overline{2}}}(A_I \partial_{\hat{\mathbf{e}}_{\overline{2}}} C) + \partial_{\hat{\mathbf{e}}_{\overline{3}}}(A_D \partial_{\hat{\mathbf{e}}_{\overline{3}}} C). \tag{14.14}$$

14.1.3 Diffusion in the z-frame

In order to describe the neutral diffusion process in the z-frame, components of the diffusion tensor $\mathbf{K}$ must be transformed. The simplest approach is to note that $\hat{\rho} = \nabla\rho/|\nabla\rho|$ in the z-frame, and to substitute this expression into the projection operator form of the tensor given by equation (14.13).

Another approach, employed by Redi (1982), is to use the rules of tensor analysis (Chapter 20) to transform the representation of the second order diffusion tensor from the neutral frame to the z-frame. In this way, the components $K^{\overline{mn}}$ of the diffusion tensor expressed in the neutral frame are transformed to the z-frame through

$$K^{mn} = \Lambda^m{}_{\overline{m}} K^{\overline{mn}} \Lambda^n{}_{\overline{n}}, \tag{14.15}$$

where the transformation matrix Λ has components given by equation (14.9).

Written as matrices with Λ having components $\Lambda^n{}_{\overline{n}}$, this transformation takes the form $K = \Lambda \overline{K} \Lambda^T$, where $\overline{K}$ has the diagonal form given in equation (14.12). Performing the matrix multiplication

$$K^{mn} = \begin{pmatrix} \frac{S_y}{S} & \frac{S_x}{S\sqrt{1+S^2}} & -\frac{S_x}{\sqrt{1+S^2}} \\ -\frac{S_x}{S} & \frac{S_y}{S\sqrt{1+S^2}} & -\frac{S_y}{\sqrt{1+S^2}} \\ 0 & \frac{S}{\sqrt{1+S^2}} & \frac{1}{\sqrt{1+S^2}} \end{pmatrix}$$

$$\times \begin{pmatrix} A_I & 0 & 0 \\ 0 & A_I & 0 \\ 0 & 0 & A_D \end{pmatrix} \begin{pmatrix} \frac{S_y}{S} & -\frac{S_x}{S} & 0 \\ \frac{S_x}{S\sqrt{1+S^2}} & \frac{S_y}{S\sqrt{1+S^2}} & \frac{S}{\sqrt{1+S^2}} \\ -\frac{S_x}{\sqrt{1+S^2}} & -\frac{S_y}{\sqrt{1+S^2}} & \frac{1}{\sqrt{1+S^2}} \end{pmatrix} \tag{14.16}$$

yields the diffusion tensor in the z-frame as written by Redi (1982)

$$K^{mn} = \frac{A_I}{(1+S^2)} \begin{pmatrix} 1+\frac{\rho_{,y}^2+\epsilon\rho_{,x}^2}{\rho_{,z}^2} & (\epsilon-1)\frac{\rho_{,x}\rho_{,y}}{\rho_{,z}^2} & (\epsilon-1)\frac{\rho_{,x}}{\rho_{,z}} \\ (\epsilon-1)\frac{\rho_{,x}\rho_{,y}}{\rho_{,z}^2} & 1+\frac{\rho_{,x}^2+\epsilon\rho_{,y}^2}{\rho_{,z}^2} & (\epsilon-1)\frac{\rho_{,y}}{\rho_{,z}} \\ (\epsilon-1)\frac{\rho_{,x}}{\rho_{,z}} & (\epsilon-1)\frac{\rho_{,y}}{\rho_{,z}} & \epsilon+S^2 \end{pmatrix}, \tag{14.17}$$

which can also be written

$$K^{mn} = \frac{A_I}{(1+S^2)} \begin{pmatrix} 1+S_y^2+\epsilon S_x^2 & (\epsilon-1)S_x S_y & (1-\epsilon)S_x \\ (\epsilon-1)S_x S_y & 1+S_x^2+\epsilon S_y^2 & (1-\epsilon)S_y \\ (1-\epsilon)S_x & (1-\epsilon)S_y & \epsilon+S^2 \end{pmatrix}. \tag{14.18}$$

This is the *Redi diffusion tensor.*

14.1.4 Small slope approximation

The bulk of processes leading to neutral diffusion processes, such as mixing along neutral directions by mesoscale eddying motions, are realized with neutral directions sloped no more than 5% from the horizontal. Hence, it is accurate, and quite efficient, to take the small angle or *small slope* approximation. This section details this approximation.

14.1.4.1 Small slope diffusion tensor

Gent and McWilliams (1990) first wrote down the following small slope diffusion tensor:

$$\begin{aligned} K^{mn}_{\text{small}} &= A_I \begin{pmatrix} 1 & 0 & -\frac{\rho_{,x}}{\rho_{,z}} \\ 0 & 1 & -\frac{\rho_{,y}}{\rho_{,z}} \\ -\frac{\rho_{,x}}{\rho_{,z}} & -\frac{\rho_{,y}}{\rho_{,z}} & \epsilon+S^2 \end{pmatrix} \\ &= A_I \begin{pmatrix} 1 & 0 & S_x \\ 0 & 1 & S_y \\ S_x & S_y & \epsilon+S^2 \end{pmatrix}. \end{aligned} \tag{14.19}$$

Note that the S^2 term was retained in the $(3,3)$ element. It is very small under the assumptions of the small slope approximation. However, it is potentially larger than the small ratio $\epsilon \approx 10^{-8}$. Cox (1987) originally retained the $(1,2) = (2,1)$ elements equal to $-S_x S_y$. In the small slope approximation, however, these terms are negligible. Additionally, and most crucially, by retaining these terms, the incorrect small angle tensor of Cox (1987) diffused buoyancy whereas the full tensor does not (see Section 14.1.6). Hence, it is important to use the physically consistent form of the small slope tensor which drops the $(1,2)$ and $(2,1)$ elements.

14.1.4.2 Small slope diffusive fluxes

The horizontal and vertical components to the small slope diffusive flux are given by

$$\mathbf{F}^{(h)} = -A_I\, \nabla_\rho C \tag{14.20}$$

$$F^{(z)} = -A_I\, \mathbf{S}\cdot\nabla_\rho C - A_D\, C_{,z} \tag{14.21}$$

where

$$\nabla_\rho = \nabla_z + \mathbf{S}\,\partial_z \tag{14.22}$$

is the neutral oriented gradient operator as derived in Section 6.5.3. The neutral portion of this flux satisfies $F^{(z)} = \mathbf{S} \cdot \mathbf{F}^{(h)}$. As discussed in Section 13.6.2, this relation is satisfied by an arbitrary vector oriented along the neutral direction in a region of stable stratification.

The dianeutral diffusive term reduces under the small slope approximation to a vertical downgradient diffusive flux. As seen in the following, this approximation is quite accurate since it neglects a term proportional to $A_D\, S^2$, which is very small.

14.1.4.3 Small slope tensor in the neutral frame

Given the small slope representation of the diffusion tensor in the z-frame, what is the representation of this tensor in the neutral frame? This tensor cannot be the diagonal form given in equation (14.12) since that form transformed into the full slope representation of equation (14.17). Using the full transformation back to the neutral frame, $K^{\overline{mn}}_{\text{small}} = \Lambda^{\overline{m}}{}_{m} K^{mn}_{\text{small}} \Lambda^{\overline{n}}{}_{n}$, or in matrix form

$$K^{\overline{mn}}_{\text{small}} = A_I \begin{pmatrix} \frac{S_y}{S} & -\frac{S_x}{S} & 0 \\ \frac{S_x}{S\sqrt{1+S^2}} & \frac{S_y}{S\sqrt{1+S^2}} & \frac{S}{\sqrt{1+S^2}} \\ -\frac{S_x}{\sqrt{1+S^2}} & -\frac{S_y}{\sqrt{1+S^2}} & \frac{1}{\sqrt{1+S^2}} \end{pmatrix} \tag{14.23}$$

$$\times \begin{pmatrix} 1 & 0 & (1-\epsilon)S_x \\ 0 & 1 & (1-\epsilon)S_y \\ (1-\epsilon)S_x & (1-\epsilon)S_y & \epsilon + S^2 \end{pmatrix} \tag{14.24}$$

$$\times \begin{pmatrix} \frac{S_y}{S} & \frac{S_x}{S\sqrt{1+S^2}} & -\frac{S_x}{\sqrt{1+S^2}} \\ -\frac{S_x}{S} & \frac{S_y}{S\sqrt{1+S^2}} & -\frac{S_y}{\sqrt{1+S^2}} \\ 0 & \frac{S}{\sqrt{1+S^2}} & \frac{1}{\sqrt{1+S^2}} \end{pmatrix} \tag{14.25}$$

yields the neutral frame representation of the small slope approximated tensor*

$$K^{\overline{mn}}_{\text{small}} = \begin{pmatrix} A_I & 0 & 0 \\ 0 & A_I(1+S^2) & 0 \\ 0 & 0 & A_D \end{pmatrix} + \frac{A_D S^2}{1+S^2} \begin{pmatrix} 0 & 0 & 0 \\ 0 & -1 & S \\ 0 & S & 1 \end{pmatrix}. \tag{14.26}$$

The small slope approximation thus adds a small amount of along neutral mixing (the $A_I\, S^2$ addition to the $(2,2)$ element) relative to the full neutral diffusion tensor. Additionally, there is a contribution proportional to $A_D S^2$. This term means that by approximating the dianeutral diffusive flux as a vertical diffusive flux, one is led to a difference on the order of $A_D\, S^2$, which is again trivially small. In conclusion, the small slope approximation leads to neutral and dianeutral diffusive fluxes that differ from the full slope by terms on the order $(A_I, A_D)\, S^2$, respectively, and so it is a very accurate approximation for purposes of ocean climate modeling.

*Note that the rotation need not transform the $\hat{\mathbf{x}} \leftrightarrow \hat{\mathbf{y}}$ directional symmetry present in the (x, y, z) form of the small slope mixing tensor into a $\hat{\mathbf{e}}_{\overline{1}} \leftrightarrow \hat{\mathbf{e}}_{\overline{2}}$ symmetry in the $(\hat{\mathbf{e}}_{\overline{1}}, \hat{\mathbf{e}}_{\overline{2}}, \hat{\mathbf{e}}_{\overline{3}})$ form. The (x, y) coordinate symmetry, however, is preserved.

14.1.5 Errors with horizontal diffusion in the ocean interior

Given the rotation formalism developed here, it is possible to consider errors made when using horizontal diffusion instead of neutral diffusion. For this purpose, let **I** represent the horizontal-vertical diffusion tensor, which is diagonal in the z-frame. Transforming this tensor to the neutral frame

$$I^{\overline{mn}} = \begin{pmatrix} \frac{S_y}{S} & -\frac{S_x}{S} & 0 \\ \frac{S_x}{S\sqrt{1+S^2}} & \frac{S_y}{S\sqrt{1+S^2}} & \frac{S}{\sqrt{1+S^2}} \\ -\frac{S_x}{\sqrt{1+S^2}} & -\frac{S_y}{\sqrt{1+S^2}} & \frac{1}{\sqrt{1+S^2}} \end{pmatrix} \begin{pmatrix} A_H & 0 & 0 \\ 0 & A_H & 0 \\ 0 & 0 & A_V \end{pmatrix} \times \begin{pmatrix} \frac{S_y}{S} & \frac{S_x}{S\sqrt{1+S^2}} & -\frac{S_x}{\sqrt{1+S^2}} \\ -\frac{S_x}{S} & \frac{S_y}{S\sqrt{1+S^2}} & -\frac{S_y}{\sqrt{1+S^2}} \\ 0 & \frac{S}{\sqrt{1+S^2}} & \frac{1}{\sqrt{1+S^2}} \end{pmatrix}, \tag{14.27}$$

yields

$$I^{\overline{mn}} = \frac{A_H}{1+S^2} \begin{pmatrix} 1+S^2 & 0 & 0 \\ 0 & 1+\tilde{\epsilon}S^2 & -S(1-\tilde{\epsilon}) \\ 0 & -S(1-\tilde{\epsilon}) & \tilde{\epsilon}+S^2 \end{pmatrix}, \tag{14.28}$$

where $\tilde{\epsilon} = A_V/A_H$ is the ratio of the vertical to horizontal diffusivities. Therefore, diffusion with **I** in the z-frame introduces first order in slope errors in the off-diagonal terms, whereas the diagonal terms contain second order in slope errors. The error in the $(3,3)$ component, however, is the most relevant as it represents an added and potentially huge source of dianeutral diffusion. For example, with the usual diffusivity ratio $\tilde{\epsilon} \approx 10^{-8}$, modest slopes larger than $\sqrt{10^{-8}} = 10^{-4}$ are sufficient to add spurious dianeutral mixing through the S^2 term, which is on the same order as $\tilde{\epsilon}$. Horizontal diffusion is thus not what happens in the ocean, at least to within the degree established by the measurements of a small level of dianeutral mixing.

14.1.6 Balance between active tracer neutral diffusive fluxes

Neutral diffusion does not affect the buoyancy field. That is, it does not diffuse locally referenced potential density. However, neutral diffusion does separately diffuse potential temperature and salinity along the neutral directions. As shown by equation (13.36) (page 292), neutral directions are themselves functions of the active tracers. Therefore, preserving the neutral directions while diffusing the active tracers requires a balance between the active tracer neutral diffusive fluxes.

14.1.6.1 Balance with the full diffusion tensor

The balance of active tracer neutral diffusion fluxes is a trivial consequence of the definition (13.44) of the neutral diffusion tensor, and the relation (13.36) for the local normal to the neutral direction. That is, with the neutral diffusive fluxes for potential temperature and salinity given by (recall equation (13.45))

$$\mathbf{F}(\theta) = -A_I(\nabla\theta - \hat{\boldsymbol{\rho}}(\hat{\boldsymbol{\rho}} \cdot \nabla\theta)) \tag{14.29}$$

$$\mathbf{F}(s) = -A_I(\nabla s - \hat{\boldsymbol{\rho}}(\hat{\boldsymbol{\rho}} \cdot \nabla s)), \tag{14.30}$$

one has

$$
\begin{aligned}
\rho(-\alpha\,\mathbf{F}(\theta) + \beta\,\mathbf{F}(s)) &= \left(\frac{\partial\rho}{\partial\theta}\right)_{s,p} \mathbf{F}(\theta) + \left(\frac{\partial\rho}{\partial s}\right)_{\theta,p} \mathbf{F}(s) \\
&= -A_I\,\rho_{,\theta}\,(\nabla\theta - \hat{\boldsymbol{\rho}}(\hat{\boldsymbol{\rho}}\cdot\nabla\theta)) - A_I\,\rho_{,s}\,(\nabla s - \hat{\boldsymbol{\rho}}(\hat{\boldsymbol{\rho}}\cdot\nabla s)) \\
&= -A_I\,(\rho_{,\theta}\,\nabla\theta + \rho_{,s}\,\nabla s) + A_I\,\hat{\boldsymbol{\rho}}\,(\hat{\boldsymbol{\rho}}\cdot(\rho_{,\theta}\,\nabla\theta + \rho_{,s}\,\nabla s)) \\
&= 0,
\end{aligned}
\tag{14.31}
$$

where substitution of equation (13.36) for $\hat{\boldsymbol{\rho}}$ led to the cancellation. Maintaining the balance

$$
\alpha\,\mathbf{F}(\theta) = \beta\,\mathbf{F}(s) \tag{14.32}
$$

on the lattice is crucial for realizing a stable discrete neutral diffusion operator (Griffies et al., 1998).

14.1.6.2 Balance with the small slope neutral diffusion tensor

Maintenance of the balance (14.32) is also afforded by small slope neutral diffusive fluxes. In this case, it is sufficient to show that the horizontal fluxes maintain a balance, since the vertical flux component is related to the horizontal through equation (13.43): $F^{(z)} = \mathbf{S}\cdot\mathbf{F}^{(h)}$. Equation (14.20) exhibits the horizontal neutral diffusive fluxes in the small slope limit

$$
\mathbf{F}^{(h)}(\theta) = -A_I\,\nabla_\rho\theta \tag{14.33}
$$

$$
\mathbf{F}^{(h)}(s) = -A_I\,\nabla_\rho s, \tag{14.34}
$$

where $\nabla_\rho = \nabla_z + \mathbf{S}\,\partial_z$ is the neutrally oriented gradient operator. Hence,

$$
\begin{aligned}
\rho_{,\theta}\,\mathbf{F}^{(h)}(\theta) + \rho_{,s}\,\mathbf{F}^{(h)}(s) &= -A_I\,(\rho_{,\theta}\,\nabla_\rho\theta + \rho_{,s}\,\nabla_\rho s) \\
&= -A_I\,(\rho_{,\theta}\,\nabla_z\theta + \rho_{,s}\,\nabla_z s + \mathbf{S}\,\rho_{,\theta}\,\partial_z\theta + \mathbf{S}\,\rho_{,s}\,\partial_z s) \\
&= 0,
\end{aligned}
\tag{14.35}
$$

where relation (13.42)

$$
\mathbf{S} = -\left(\frac{\rho_{,\theta}\,\nabla_z\theta + \rho_{,s}\,\nabla_z s}{\rho_{,\theta}\,\theta_{,z} + \rho_{,s}\,s_{,z}}\right) \tag{14.36}
$$

was used for the slope vector. Failure to maintain this balance proved to be the key reason that the Cox (1987) numerical realization of neutral diffusion was unstable (Griffies et al., 1998).

14.1.7 Cabbeling, thermobaricity, and halobaricity

When diffusing active tracers with a nonlinear equation of state, locally referenced potential density is affected by three nonflux forms of irreversible mixing known as *cabbeling*, *thermobaricity*, and *halobaricity* (McDougall, 1987b). This section presents the mathematical form of such mixing as associated with neutral diffusion. An alternative formulation of the following, in much more detail, can be found in McDougall (1991).

Under the effects of neutral diffusion, the material time tendency for locally referenced potential density is given by

$$\begin{aligned} \mathrm{d}\rho/\mathrm{d}t &= \rho(-\alpha\, \mathrm{d}\theta/\mathrm{d}t + \beta\, \mathrm{d}s/\mathrm{d}t) \\ &= \rho[\alpha\nabla\cdot\mathbf{F}_I(\theta) - \beta\nabla\cdot\mathbf{F}_I(s)] \\ &= -\nabla(\rho\alpha)\cdot\mathbf{F}_\mathbf{I}(\theta) + \nabla(\rho\beta)\cdot\mathbf{F}_I(s), \end{aligned} \tag{14.37}$$

where the active tracer balance $\alpha\mathbf{F}_I(\theta) = \beta\mathbf{F}_I(s)$ was used to reach the last equality. The forcing terms, which vanish for a linear equation of state and which cannot be written as the divergence of a flux, represent cabbeling, thermobaricity, and halobaricity. These processes provide irreversible, nondiffusive forms of mixing. Note for the special case of a single active tracer, neutral directions are aligned with the active tracer surfaces, thus providing for a zero neutral diffusive flux of the active tracer and therefore an absence of cabbeling, thermobaricity, and halobaricity.

It is useful to isolate the mathematical forms of the irreversible mixing, which can be done via the identities

$$\nabla s\cdot\mathbf{F}_I(\theta) = \nabla\theta\cdot\mathbf{F}_I(s) \tag{14.38}$$

$$\nabla(-\rho\alpha) = \rho_{,\theta\theta}\nabla\theta + \rho_{,\theta s}\nabla s + \rho_{,\theta\, p}\nabla p, \tag{14.39}$$

and likewise for $\nabla(\rho\beta)$. Note the presence of pressure gradients. These terms represent the effects of probing different pressure surfaces, and hence different potential density surfaces; that is, these are the thermobaric terms. Probing different pressure surfaces is necessary when computing the spatial gradients of the thermal and saline expansion coefficients. With these substitutions, the evolution of locally referenced potential density takes the form

$$\begin{aligned} \mathrm{d}\rho/\mathrm{d}t = {} & \rho_{,\theta\theta}\nabla\theta\cdot\mathbf{F}_I(\theta) + \rho_{,ss}\nabla s\cdot\mathbf{F}_I(s) \\ & + 2\rho_{,\theta s}\nabla s\cdot\mathbf{F}_I(\theta) + \nabla p\cdot(\rho_{,\theta\, p}\mathbf{F}_I(\theta) + \rho_{,s\, p}\mathbf{F}_I(s)). \end{aligned} \tag{14.40}$$

Some rearrangement then leads to

$$\begin{aligned} \mathrm{d}\rho/\mathrm{d}t = {} & \nabla\theta\cdot\mathbf{F}_I(\theta)\,[\rho_{,\theta\theta} - 2\rho_{,\theta s}(\rho_{,\theta}/\rho_{,s}) + \rho_{,ss}(\rho_{,\theta}/\rho_{,s})^2] \\ & + \nabla p\cdot(\rho_{,\theta\, p}\mathbf{F}_I(\theta) + \rho_{,s\, p}\mathbf{F}_I(s)). \end{aligned} \tag{14.41}$$

The first part of this expression can be written as the product of two quadradic forms by introducing a vector

$$\mathbf{W} = (1, \alpha/\beta) \tag{14.42}$$

and a metric

$$\rho_{,ab} = \partial_a\partial_b\rho, \tag{14.43}$$

where the labels a, b represent the two tracer fields θ, s. This definition renders

$$\mathrm{d}\rho/\mathrm{d}t = \nabla\theta\cdot\mathbf{F}_I(\theta)\, W^2 + (\rho_{,\theta\, p}\mathbf{F}_I(\theta) + \rho_{,s\, p}\mathbf{F}_I(s))\cdot\nabla p, \tag{14.44}$$

where the inner product

$$\begin{aligned} W^2 &= \mathbf{W}\cdot\mathbf{W} \\ &= \rho_{,ab}W^a\, W^b \\ &= \rho_{,\theta\theta} - 2\rho_{,s\theta}\,(\rho_{,\theta}/\rho_{,s}) + \rho_{,ss}\,(\rho_{,\theta}/\rho_{,s})^2 \\ &= -\alpha_{,\theta} - 2(\alpha/\beta)\alpha_{,s} + (\alpha/\beta)^2\beta_{,s} \end{aligned} \tag{14.45}$$

represents the squared length of the vector $\mathbf{W}$ on the curved potential density surface characterized locally by the metric $\rho_{,ab}$.

14.1.7.1 *Cabbeling*

The term

$$\text{cabbeling} = \nabla\theta \cdot \mathbf{F}_I(\theta)\, W^2 \tag{14.46}$$

represents the effects from cabbeling. As written here, cabbeling is seen to be the product of a term associated with the downgradient neutral diffusive flux of potential temperature ($\nabla\theta \cdot \mathbf{F}_I(\theta) \leq 0$), and a term associated with the local geometric properties intrinsic to the potential density surface ($\mathbf{W} \cdot \mathbf{W}$). In the ocean, the total or Gaussian curvature $\det(\rho_{,ab})(1 + \rho_{,\theta}^2 + \rho_{,s}^2)^{-1}$ is negative. This negative curvature renders $\mathbf{W} \cdot \mathbf{W} \leq 0$, which, when combined with downgradient neutral diffusion of potential temperature, always results in a nonnegative $\mathrm{d}\rho/\mathrm{d}t$ and a consequent downward dianeutral advection (McDougall, 1987b).

If the term $\nabla\theta \cdot \mathbf{F}_I(\theta)$ is sign-indefinite, then the contribution of cabbeling to the locally referenced potential density is also sign-indefinite. Sign-indefiniteness is not desirable for a subgrid scale process associated with mixing, since mixing in the ocean generally produces a sign-definite cabbeling term. Hence, one should avoid the use of continuum mixing operators that are not downgradient. As discussed in Section 14.5, biharmonic tracer operators are mixing operators that have sign-indefinite tracer fluxes. This is a key reason for eschewing these operators in realistic ocean climate simulations.

14.1.7.2 *Thermobaricity and halobaricity*

The terms proportional to the pressure gradient

$$\text{thermobaricity} + \text{halobaricity} = [\rho_{,\theta\,p}\mathbf{F}_I(\theta) + \rho_{,s\,p}\mathbf{F}_I(s)] \cdot \nabla p \tag{14.47}$$

represent the effects from thermobaricity and halobaricity. These terms depend on the pressure dependence of the equation of state for seawater. Note that the thermobaric term $\rho_{,\theta\,p}\mathbf{F}_I(\theta) \cdot \nabla p$ dominates the halobaric term $\rho_{,s\,p}\mathbf{F}_I(s) \cdot \nabla p$ (McDougall, 1987b). In contrast to cabbeling, thermobaricity and halobaricity do not provide a sign definite source for locally referenced potential density.

14.2 GENT-MCWILLIAMS STIRRING

The subgrid scale stirring suggested by Gent and McWilliams (1990) and Gent et al. (1995) (abbreviated "GM" in the following) is based on assuming that (i) mesoscale eddies locally provide an adiabatic sink of available potential energy (APE), and (ii) mesoscale eddies act on all tracers in the same fashion. Alternative proposals continue to be the subject of intense research (for a brief review with numerous references, see Section 11 of Griffies et al., 2000a). The discussion in this section does not question these assumptions. Instead, the purpose here is to develop some of basic properties of *GM stirring*. This is useful since this scheme is ubiquitous in ocean climate models.

14.2.1 The GM potential density skew flux

As discussed in Section 13.5, the GM scheme, as a parameterization of stirring, can be represented via either an advective or skew diffusive form. Griffies (1998) showed how the skew approach can be more convenient numerically than the advective approach. It is therefore useful to present both approaches in the following.

To get started, consider the case when GM skew fluxes act on potential density in an ideal Boussinesq ocean where the equation of state is a linear function of potential temperature, so there are no salinity or pressure effects on density. One may thus write the evolution of potential density in the form

$$\rho_{,t} = -\nabla \cdot (\mathbf{v}\,\rho + \mathbf{F}). \tag{14.48}$$

The first term on the right-hand side provides advective stirring of density via a divergence-free velocity. The second term is the convergence of a flux whose form is to be determined according to the assumptions of Gent and McWilliams (1990).

The arguments here are aimed at reducing the derivation of the GM skew flux to a geometric and local energetic argument. For this purpose, note that a flux aligned parallel to the potential density surfaces causes no mixing of potential density classes. Hence, the convergence of this flux results in an adiabatic operator. Note that a choice to orient the flux parallel to the potential density surface is a sufficient condition for producing an adiabatic transport operator. However, it is not necessary. There are other physically relevant adiabatic transport fluxes that have a nonzero cross-isopycnal component. The advective flux $\mathbf{v}\,\rho$ is an important example. Here, if advection is the only process taking place, $\rho_{,t} = -\mathbf{v} \cdot \nabla \rho$ leads to a nonstatic density so long as the velocity $\mathbf{v}$ has a nonzero projection onto the density surface's local normal direction.

Moving back to an assumption of a flux parallel to the potential density surface, geometric discussions in Section 13.6.2 (page 292) lead to the general form of this flux

$$\mathbf{F} = \mathbf{F}^{(h)} + \hat{\mathbf{z}}\,\mathbf{S} \cdot \mathbf{F}^{(h)}. \tag{14.49}$$

In this equation,

$$\mathbf{S} = -\left(\frac{\nabla_z \rho}{\rho_{,z}}\right) \tag{14.50}$$

is the slope of the potential density surface. Hence, aligning the flux along the potential density surface reduces the three unknown flux components down to two, as expected since the surface is two-dimensional. Additionally, because

$$\mathbf{F} \cdot \nabla \rho = 0, \tag{14.51}$$

$\mathbf{F}$ so aligned represents a skew flux of potential density.

Now consider the constraints on the flux imposed by assuming the operator locally and adiabatically reduces the available potential energy (APE) of the flow. A state of zero APE is one with no baroclinicity: isopycnals are flat. The gravitational potential energy of the adiabatic Boussinesq system is

$$P = g \int \mathrm{d}V\,\rho\,z. \tag{14.52}$$

The tendency associated with the unknown flux is given by

$$
\begin{aligned}
P_{,t} &= g \int \mathrm{d}V\, z\, \rho_{,t} \\
&= -g \int \mathrm{d}V\, (z\, \nabla \cdot \mathbf{F}) \\
&= -g \int \mathrm{d}V\, (z\, \partial_z F^{(z)}) \\
&= g \int \mathrm{d}V\, F^{(z)},
\end{aligned}
\tag{14.53}
$$

where $F^{(z)}$ is the vertical flux component. The neglect of surface and bottom effects is appropriate since the focus here is on baroclinic physics in the ocean interior isolated from solid earth and surface effects.* To provide a *local* APE sink requires

$$F^{(z)} \leq 0, \tag{14.54}$$

where zero occurs when the isopycnals are flat. It is sufficient to construct the vertical flux component using only the potential density field itself. For a stably stratified fluid in which $\rho_{,z} < 0$, the following form provides a local APE sink required by the GM assumptions:

$$F^{(z)} = \kappa\, S^2\, \rho_{,z} \tag{14.55}$$

where κ is a positive diffusivity setting the strength of the flux and S^2 is the squared isopycnal slope. Consequently, the horizontal flux component reduces to a downgradient diffusive flux

$$\mathbf{F}^{(h)} = -\kappa\, \nabla_z \rho. \tag{14.56}$$

In summary, the GM skew flux for potential density takes the form

$$\mathbf{F} = -\kappa\, (\nabla_z \rho - \hat{\mathbf{z}}\, S^2\, \rho_{,z}). \tag{14.57}$$

Notably, when breaking the three-dimensional GM skew flux into its horizontal and vertical components, the horizontal flux is downgradient whereas the vertical flux is upgradient. When properly realized numerically, there is no instability associated with the upgradient vertical flux, since it is the convergence of the full three-dimensional flux that creates the evolution of the potential density field, and the three-dimensional flux is parallel to potential density surfaces. However, numerical truncation errors can create some problems when discretizing upgradient flux components, as mentioned in Section 13.7.

The advective GM flux $\mathbf{v}^* \rho$ was presented in Gent et al. (1995). In their discussion, the divergence-free GM *eddy-induced* velocity takes the form

$$
\begin{aligned}
\mathbf{v}^* &= \nabla \wedge \boldsymbol{\Upsilon}_{\mathrm{gm}} \\
&= -\partial_z\, (\kappa\, \mathbf{S}) + \hat{\mathbf{z}}\, \nabla_z \cdot (\kappa\, \mathbf{S}),
\end{aligned}
\tag{14.58}
$$

with the vector streamfunction

$$\boldsymbol{\Upsilon}_{\mathrm{gm}} = \hat{\mathbf{z}} \wedge \kappa\, \mathbf{S}. \tag{14.59}$$

The advective form of the GM parameterization is generally more cumbersome to implement numerically than the skew flux form. The following section provides some motivation for this point, with Griffies (1998) presenting details and simulation examples.

*Boundary effects are in fact nontrivial and a focus of much research. Some issues are mentioned in Chapter 15.

14.2.2 GM skewsion and small slope neutral diffusion

Gent and McWilliams (1990) and Gent et al. (1995) assume that the same eddy-induced velocity $\mathbf{v}^*$ acts to transport all tracers, in addition to the model's resolved scale velocity field $\mathbf{v}$. In this case, Griffies (1998) showed that for an arbitrary tracer, the GM skew flux takes the form

$$\mathbf{F} = \kappa\,(\mathbf{S}\,C_{,z} - \hat{\mathbf{z}}\,\mathbf{S}\cdot\nabla_z C)\,. \tag{14.60}$$

Since $\mathbf{F}\cdot\nabla C = 0$, the GM skew flux is oriented parallel to surfaces of constant tracer, as expected from a skew flux.

In addition to stirring via GM skewsion, consider neutral diffusion where the small slope approximation is used. The combination of GM stirring and neutral diffusive mixing leads to the tracer flux

$$\mathbf{F} = -A_I\,\nabla_z C + (\kappa - A_I)\,\mathbf{S}\,C_{,z} - \hat{\mathbf{z}}\,[(A_I+\kappa)\,\mathbf{S}\cdot\nabla_z C + A_I\,S^2\,C_{,z}]. \tag{14.61}$$

This flux can be written in terms of the following neutral physics mixing tensor*

$$\begin{pmatrix} F^{(x)} \\ F^{(y)} \\ F^{(z)} \end{pmatrix} = \begin{pmatrix} A_I & 0 & (A_I-\kappa)\,S_x \\ 0 & A_I & (A_I-\kappa)\,S_y \\ (A_I+\kappa)\,S_x & (A_I+\kappa)\,S_y & A_I\,S^2 \end{pmatrix} \begin{pmatrix} C_{,x} \\ C_{,y} \\ C_{,z} \end{pmatrix}. \tag{14.62}$$

In ocean climate modeling, it is common to assume that the diffusivities associated with GM stirring and neutral diffusion are equal, at least within the ocean interior. In this special case, the combined tracer flux simplifies to

$$\mathbf{F} = -\kappa\,\nabla_z C - \hat{\mathbf{z}}\,\kappa\,(2\,\mathbf{S}\cdot\nabla_z C + S^2\,C_{,z}). \tag{14.63}$$

Notably, the 2×2 horizontal mixing tensor is diagonal. Hence, the horizontal tracer flux is the same as that which arises from downgradient horizontal tracer diffusion. The simplicity of the horizontal flux component reduces the total computational cost of these two schemes to less than the cost of either one alone. This is the key practical result of the Griffies (1998) paper. Additional numerical benefits arise since the discretized skew flux can be less prone to discretization errors, relative to the advective flux.

An additional point worth raising in regard to the skew versus advective form of GM stirring concerns the diagnosis of an effective horizontal/vertical diffusivity. In the special case of $A_I = \kappa$, this diagnosis becomes quite straightforward, since

$$\mathbf{F} = -\kappa\,\nabla_z C - K_{\text{eff}}\,C_{,z}. \tag{14.64}$$

That is, the effective horizontal diffusivity is simply κ, as mentioned in the previous paragraph. Although the effective vertical diffusivity K_{eff} is more complicated, it can be written in the reasonably elegant form

$$K_{\text{eff}} = \kappa\,\mathbf{S}\cdot(\mathbf{S} - 2\,\mathbf{S}_C), \tag{14.65}$$

where $\mathbf{S}_C$ defines the slope of tracer isolines

$$\mathbf{S}_C = -\left(\frac{\nabla_z C}{C_{,z}}\right). \tag{14.66}$$

*Note that equation (10.3.9) of Kantha and Clayson (2000a) is an incorrect form of this tensor.

As defined, K_{eff} can be either positive or negative, depending on the relative orientation of the neutral directions and tracer isolines. In particular, when the tracer is aligned with the neutral direction,

$$K_{\text{eff}} = -\kappa \, S^2, \tag{14.67}$$

which is the same negative effective diffusivity seen when considering GM skewsion acting on neutral density (i.e., equation (14.55)). Additionally, when the slope of the isopycnal is twice that of the tracer, then $K_{\text{eff}} = 0$, in which case the flux is purely downgradient in the horizontal direction.

14.3 SUMMARIZING THE NEUTRAL PHYSICS FLUXES

We consider here a brief summary of the tracer fluxes introduced in the previous sections. For all cases, the mixing operator $R(C)$ is given by the convergence of the flux

$$R(C) = -\nabla \cdot \mathbf{F}(C). \tag{14.68}$$

- The small slope neutral Laplacian diffusive flux for an arbitrary tracer is given by

$$\mathbf{F} = -A \, (\nabla_\rho + \hat{\mathbf{z}} \, \mathbf{S} \cdot \nabla_\rho) \, C, \tag{14.69}$$

  where

$$\begin{aligned} \mathbf{S} &= \nabla_\rho z \\ &= -\left(\frac{\nabla_z \rho}{\rho_{,z}} \right) \\ &= -\left(\frac{-\alpha \, \nabla_z \theta + \beta \, \nabla_z s}{-\alpha \, \theta_{,z} + \beta \, s_{,z}} \right) \end{aligned} \tag{14.70}$$

  is the neutral slope and

$$\nabla_\rho = \nabla_z + \mathbf{S} \, \partial_z \tag{14.71}$$

  is the neutral gradient operator. The neutral diffusive flux for locally referenced potential density vanishes:

$$\mathbf{F}(\rho) = 0. \tag{14.72}$$

  That is, neutral diffusion does not alter buoyancy, by definition.

- The GM skew flux for an arbitrary tracer is given by

$$\mathbf{F} = -\kappa \, (-\mathbf{S} \, \partial_z + \hat{\mathbf{z}} \, \mathbf{S} \cdot \nabla_z) \, C, \tag{14.73}$$

  and the skew flux for locally referenced potential density ρ is given by

$$\begin{aligned} \mathbf{F} &= -\kappa \, (\nabla_z + \hat{\mathbf{z}} \, \mathbf{S} \cdot \nabla_z) \, \rho \\ &= -\kappa \, (\nabla_z - \hat{\mathbf{z}} \, S^2 \, \partial_z) \, \rho. \end{aligned} \tag{14.74}$$

14.4 FLOW-DEPENDENT DIFFUSIVITIES

Most global ocean climate models run in the 1990's used neutral diffusive and GM skew diffusive mixing coefficients of equal value independent of the local flow properties: $A = \kappa$. A notable exception is when the diffusivities are tapered in regions of steep neutral direction slopes and the zero eddy flux condition is applied at the ocean surface (Chapter 15 discusses these topics). Recent studies, however, aim to provide a diffusivity within the ocean interior in terms of length and time scales set by properties of the model's resolved fields.

Currently, there are no strong theoretical arguments suggesting that it is best to use different diffusivities for the GM and neutral diffusion scheme, even when those diffusivities are functions of the flow. Indeed, Dukowicz and Smith (1997) provide arguments in favor of the same diffusivities. In practice, this choice has been made based on simplicity. However, some modelers, such as Wright (1997) and Gordon et al. (2000), have chosen to use unequal diffusivities. In their cases, a flow-dependent scheme prescribes the GM diffusivity, but a constant value is used for the neutral diffusivity. Little discussion is provided for the rationale. It is speculated that this choice simply produced the desired simulation qualities.

14.4.1 Horizontal and vertical dependent diffusivities

The papers by Held and Larichev (1996) and Visbeck et al. (1997) each propose methods for computing diffusivity in terms of a length and time scale. Properties of the resolved flow that are local in the horizontal but nonlocal in the vertical determine these length and time scales. Killworth (1997), following on a suggestion from Treguier et al. (1997), presents a theory based on linear baroclinic instability in which the diffusivity obtains both a horizontal and a vertical dependence. Notably, closures based on diffusing potential vorticity downgradient *must* incorporate a nontrivial vertical dependence to their diffusivities in order to maintain fundamental balances (e.g., Marshall and Shutts, 1981; Treguier et al., 1997; Killworth, 1997). Results from the high resolution channel model study by Treguier (1999) are consistent with Killworth's ideas. Smith and Vallis (2002) propose a diffusivity for potential vorticity dependent on both vertical and horizontal position, with their theory based on scalings from fully developed quasi-geostrophic turbulence.

Even though there are an increasing number of idealized numerical studies, there remains a dearth of large-scale coarse resolution studies to investigate the utility of flow-dependent diffusivities. In particular, there are no published tests of coarse global models using potential vorticity based closures. Instead, only closures based on the Gent and McWilliams (1990) ideas have been tested with nonconstant diffusivities. A notable account of nonconstant diffusivities run in a global model is that of Wright (1997). He compared a suite of constant coefficient experiments to a single run with the Visbeck et al. (1997) diffusivity. In these tests the flow dependence was given only to the GM diffusivity, with the neutral diffusion coefficient held constant. Tests were run with the Hadley Centre's 1.25° ocean model. Results were generally favorable toward the Visbeck et al. (1997) diffusivity, hence prompting the Hadley Centre to employ this scheme in their climate simulations (Gordon et al., 2000).

Unpublished studies in coarser simulations (e.g., coarser than 2°) suggest that nonconstant diffusivities may not be so advantageous at this and coarser resolutions.* The idea is that there needs to be a nontrivial structure to the resolved model flow for inhomogeneities in the diagnosed diffusivity to matter.

Additional research is ongoing to test the utility of vertically dependent diffusivities for large-scale modeling. However, there are no published studies exhibiting a significant improvement associated with the added complexity of a vertically dependent diffusivity. Notably, allowing for vertical dependence necessitates some added structure to the neutral tracer operators compared to the depth-independent form. At present most publicly supported ocean climate models only provide the framework for employing flow dependent diffusivities that are independent of depth. Hence, the remainder of this section focuses on depth independent diffusivities.

14.4.2 Length and time scales for Held and Larichev

Both Held and Larichev (1996) and Visbeck et al. (1997) develop diagnostic expressions for a time scale T and a length scale L according to properties of the model's resolved flow field. Thereafter, a diffusivity is computed via

$$\kappa = L^2/T. \tag{14.75}$$

The squared inverse time scale (squared growth rate) suggested by Held and Larichev (1996) is given by

$$T^{-2} = \frac{f^2}{D} \int_{-D_b}^{-D_t} \mathrm{Ri}^{-1}\, dz. \tag{14.76}$$

In this expression, f is the Coriolis parameter and Ri is the Richardson number determined by large-scale flow features. This growth rate corresponds to that of a growing Eady wave, and the time scale T is on the order of tens to hundreds of days. The integration depth range $D = D_b - D_t$ was suggested by Treguier et al. (1997) to be $D_t = 100$ m and $D_b = 2000$ m. This depth range is also taken in the Hadley Centre ocean model in which they implement the Visbeck et al. (1997) scheme. This range is intended to discount the unstratified mixed layer where the growth rate becomes unbounded.

The length scale suggested by Held and Larichev (1996) is related to the effective Rhines scale

$$L^{-1} = \beta_{\text{eff}}\, T. \tag{14.77}$$

The effective beta parameter β_{eff} incorporates both planetary vorticity gradients and gradients in the large-scale topography

$$\beta_{\text{eff}} = H |\nabla_z (f/H)|. \tag{14.78}$$

The resulting diffusivity is given by

$$\begin{aligned} \kappa &= L^2\, T^{-1} \\ &= (\beta_{\text{eff}}^2\, T^3)^{-1}. \end{aligned} \tag{14.79}$$

*See the report of the WOCE/CLIVAR Workshop on Ocean Modelling for Climate Studies, WOCE Report No. 165/99.

14.4.3 Length and time scales for Visbeck et al. (1997)

The time scale suggested by Visbeck et al. (1997) is given by

$$T^{-1} = \frac{f}{D} \int_{-D_b}^{-D_t} \mathrm{Ri}^{-1/2}\,\mathrm{d}z. \tag{14.80}$$

The square of this expression is not identical to the squared inverse time scale (14.76) from the Held and Larichev (1996) scheme. However, in both approaches the time scale is determined by scaling arguments. Hence, differences are not fundamentally significant and so can be absorbed into an overall tuning parameter.

The length scale L in the Visbeck et al. (1997) scheme is determined by the regional structure of the baroclinicity. This is the "width of the baroclinic zone," which is motivated by arguments from Green (1970). Details of how to compute this length scale are somewhat ambiguous for a three-dimensional flow. No algorithm was described in the Visbeck et al. (1997) paper as they considered only two-dimensional or zonally averaged flow. The Hadley Centre model of Gordon et al. (2000) employs a search algorithm described to the author by Malcolm Roberts (personal communication 1997). This scheme is documented in Pacanowski and Griffies (1999) and in Griffies et al. (2004).

14.4.4 Length scale given by the first baroclinic Rossby radius

Instead of the Rhines scale or width of the baroclinic zone, some prefer a length scale set according to the first baroclinic Rossby radius. Relevance of this length scale for eddy parameterization was suggested by theoretical studies of Stone (1972), satellite based data analysis by Stammer (1997), eddying ocean model studies by Bryan et al. (1999), and theoretical process studies of Smith and Vallis (2002), amongst others.

Outside an equatorial band of roughly $\pm 5°$, the mth-mode Rossby radius takes the form

$$\lambda_m = \frac{c_m}{|f|}, \tag{14.81}$$

where c_m is the phase speed of the mth-mode ($m \geq 1$) gravity wave in a non-rotating, continuously stratified flat bottom ocean (Gill, 1982). Within an equatorial band of $\pm 5°$, the Rossby radius is given by

$$\lambda_m = \left(\frac{c_m}{2\,\beta}\right)^{1/2}, \tag{14.82}$$

where β is the planetary beta-parameter. A useful approximate gravity wave phase speed can be obtained by the WKB form described in Chelton et al. (1998),

$$c_m \approx \frac{1}{m\,\pi} \int_{-H}^{0} N\,\mathrm{d}z, \tag{14.83}$$

where N is the buoyancy frequency. Notably, the vertical integral is over the full depth range. Unlike the Eady growth rate calculation, the vertical integration can be easily extended to unstratified regions since the integrand simply vanishes there. Even though the gravity wave speed feels the bottom, the study by Chelton et al. (1998) showed that the Rossby radius over the bulk of the World Ocean is largely dominated by the inverse Coriolis dependence (see their Figure 6).

14.4.5 Thermal wind Richardson number

Recall from Section 4.8.4 (page 70) that the Richardson number is given by the ratio of the squared buoyancy frequency to the squared vertical shear in the horizontal velocity

$$\mathrm{Ri} = \frac{N^2}{|\mathbf{u}_{,z}|^2}. \tag{14.84}$$

The eddy closure theories generally assume that the mesoscale eddy field is quasi-geostrophic. Consequently, the Richardson number should be based on vertical shears under thermal-wind balance with the buoyancy field (page 66)

$$\mathbf{u}_{,z} = -\left(\frac{g}{f\,\rho_o}\right)\hat{\mathbf{z}} \wedge \nabla \rho. \tag{14.85}$$

This assumption then leads to

$$|\mathbf{u}_{,z}| = \frac{N^2\,S}{|f|}, \tag{14.86}$$

where S^2 is the squared neutral direction slope vector and

$$\begin{aligned} N^2 &= -g\,\rho_{,z}/\rho_o \\ &= -(g/\rho_o)\,(\rho_{,\theta}\,\theta_{,z} + \rho_{,s}\,s_{,z}) \end{aligned} \tag{14.87}$$

is the squared buoyancy frequency based on the vertical gradient of locally referenced potential density. These results then lead to the large-scale Richardson number

$$\mathrm{Ri} = (f/N S)^2 \tag{14.88}$$

to be used for computing the diffusivities.

14.4.6 Rotational independence of the time scale

The thermal wind Richardson number (14.88) brings the squared inverse time scale written according to the Held and Larichev form (14.76) to

$$\begin{aligned} T^{-2} &= \frac{1}{D}\int_{-D_b}^{-D_t} N^2\,S^2\,\mathrm{d}z \\ &= \overline{(N\,S)^2}, \end{aligned} \tag{14.89}$$

where an overbar represents a vertically averaged quantity with the average over the chosen depth range. Likewise, the Visbeck et al. (1997) form of the inverse time scale (14.80) is

$$\begin{aligned} T^{-1} &= \frac{1}{D}\int_{-D_b}^{-D_t} N\,S\,\mathrm{d}z \\ &= \overline{N\,S}. \end{aligned} \tag{14.90}$$

As mentioned in Section 14.4.3, any difference between $(\overline{N\,S})^2$ and $\overline{(N\,S)^2}$ is not fundamentally significant and so can be ascribed to an overall dimensionless scaling factor.

Note how the explicit dependence on the Coriolis parameter f has canceled from the time scale. The source of this cancellation is the use of thermal wind balance for computing the Richardson number. Consequently, the time scale is an explicit function only of the vertically averaged horizontal and vertical stratification. Relatedly, the explicit cancellation of the Coriolis parameter allows for the time scale calculation to be naively applied globally, including at the equator where geostrophy is irrelevant.

The inverse time scale, or the growth rate, vanishes when the vertically integrated horizontal stratification vanishes, that is, when there is zero baroclinicity. Hence, the diffusivity vanishes when the neutral directions are flat, as one would expect for a configuration without available potential energy. The growth rate becomes large when the neutral direction slope becomes large, which is expected for regions of large available potential energy.

For testing the validity of a numerical implementation of this time scale, it is useful to estimate a maximum expected Eady growth rate. For this purpose, assume the buoyancy frequency to be on the order of f and the slopes to be a constant of $1/100$ throughout the column. This case leads to a maximum growth rate on the order of $T^{-1} = f/100$. Note that in unstratified regions the growth rate is actually unbounded. This behavior is undesirable numerically and signals a limitation of the quasi-geostrophic theory to extend into highly unstratified regions. Regularization methods proposed in Sections 14.4.7 and 14.4.8 aim to eliminate these singularities. Regardless of the details, the result of these regularization methods should be a growth rate no larger than the maximum estimate given here.

14.4.7 One means to provide a regularized diffusivity

In regions of zero vertical stratification, the Eady growth rate given in Section 14.4.6 becomes unbounded. It is therefore necessary to provide a regularization method for the numerical model. A method suggested by Treguier et al. (1997), and used in the Hadley Centre and Pacanowski and Griffies (1999) implementations of these schemes, is to restrict the vertical integration to regions where the vertical stratification is nontrivial. This approach, as mentioned in Section 14.4.2, leads to the somewhat *ad hoc* specification of two depths D_b and D_t. This section presents another method arguably just as valid. It aims to maintain consistency with the Rossby radius calculation of Chelton et al. (1998) discussed in Section 14.4.4, where the vertical integration is over the full ocean depth. However, the preferred approach should be based on model integrity, since the notions presented here are not rigorous.

To motivate a regularization method handling unstratified regions, write the first mode gravity wave phase speed in the form

$$\begin{aligned} \pi\, c_1 &= \int_{-H}^{0} N\, \mathrm{d}z \\ &= H\, \overline{N}. \end{aligned} \tag{14.91}$$

Hence, the first baroclinic Rossby radii for the equatorial and off-equatorial regions take the form

$$L_{<|5^\circ|} = \left(\frac{H \overline{N}}{2\,\pi\,\beta} \right)^{1/2} \tag{14.92}$$

$$L_{>|5^\circ|} = \frac{H \overline{N}}{\pi |f|}, \tag{14.93}$$

and the corresponding diffusivities $\kappa = L^2\, T^{-1}$ are

$$\kappa_{<|5^\circ|} = \left(\frac{H}{2\,\pi\,\beta} \right) (\overline{N})\, \overline{N\,S} \tag{14.94}$$

$$\kappa_{>|5^\circ|} = \left(\frac{H}{\pi\,f} \right)^2 (\overline{N})^2\, \overline{N\,S}, \tag{14.95}$$

where the inverse time scale (14.90) of Visbeck et al. (1997) was used. The numerical problem is that $\overline{N\,S}$ becomes infinite if any piece of a vertical column is unstratified. One means to resolve this problem is to write

$$\overline{N\,S} = \overline{\left(\frac{N^2\,S}{N} \right)} = \gamma \left(\frac{\overline{N^2\,S}}{\overline{N}} \right). \tag{14.96}$$

Now assume the dimensionless parameter γ is a constant, which leads to the regularized diffusivities

$$\kappa_{<|5^\circ|} = \gamma \left(\frac{H}{2\,\pi\,\beta} \right) \overline{N^2\,S} \tag{14.97}$$

$$\kappa_{>|5^\circ|} = \gamma \left(\frac{H}{\pi\,f} \right)^2 \overline{N}\, \overline{N^2\,S}. \tag{14.98}$$

Notably, the diffusivity within the equatorial band is only a function of the horizontal stratification via $\overline{N^2\,S}$, whereas the diffusivity outside the band is a function of both horizontal and vertical stratification via $\overline{N}\,\overline{N^2\,S}$. In practice, the dimensionless parameter γ can be used as a tuning parameter. Additionally, it is necessary to set a lower and upper limit on the diffusivity for purposes of maintaining numerical stability.

Regularization methods, such as this one and others to be presented in the following, remedy a numerical problem associated with an incomplete physical theory. In the present case, the regularization method numerically remedies a limitation of quasi-geostrophic theory to extend into vertically unstratified regions. One goal of the numericist designing regularization methods is to maintain numerical stability without egregiously compromising the physical integrity of the algorithm. In general, we counsel strong skepticism be given any regularization scheme, since they are often based more on numerical than on physical considerations.

A useful means of gauging the size of the tuning parameter is to consider a linear equation of state $\rho = \rho_o\,(1 - \alpha\theta)$ and a depth-independent neutral direction slope. In this case, the equatorial-band diffusivity becomes

$$\begin{aligned}
\kappa_{<|5^\circ|} &= \gamma \left(\frac{H}{2\,\pi\,\beta} \right) \overline{S\,N^2} \\
&= \gamma \left(\frac{S\,H}{2\,\pi\,\beta} \right) \overline{N^2} \\
&= \gamma \left(\frac{\alpha\, g\, S\, H}{2\,\pi\,\beta} \right) (\theta(z=0) - \theta(z=-H)).
\end{aligned} \tag{14.99}$$

With $\beta = 2.28 \times 10^{-11} (\mathrm{m\ s})^{-1}$ at the equator, $g = 9.8\,\mathrm{m\,s^{-2}}$, $\alpha = 2 \times 10^{-4} (^\circ\mathrm{K})^{-1}$, and $\theta(z=0) - \theta(z=-H) = 20^\circ\mathrm{K}$, then

$$\kappa_{<|5^\circ|} \approx S\,\gamma \left(3 \times 10^8\,\mathrm{m^2\,s^{-1}} \right). \tag{14.100}$$

If the depth averaged slope is $S = 10^{-3}$, then

$$\kappa_{<|5^\circ|} \approx \gamma \left(3 \times 10^5\,\mathrm{m^2\,s^{-1}} \right). \tag{14.101}$$

Based on experience with ocean models run with constant neutral physics diffusivities, where values are on the order $10^2 - 10^3\,\mathrm{m^2\,s^{-1}}$, one is motivated to set the tuning parameter to

$$\gamma = 0.01, \tag{14.102}$$

which is consistent with the dimensionless scaling parameter suggested by Visbeck et al. (1997). This choice then brings the diffusivity to

$$\kappa_{<|5^\circ|} \approx 3 \times 10^3\,\mathrm{m^2\,s^{-1}}. \tag{14.103}$$

14.4.8 Another regularization method

A cumbersome aspect of the flow-dependent diffusivities presented thus far is that they do not explicitly depend on the grid scale. Hence, there is no tendency for their values to decrease as the grid is refined. Indeed, for refined grids, the Eady growth rate increases since the model allows more vigorous mean flows with stronger gradients and slopes. One thus has the undesirable situation of larger flow-dependent diffusivities for refined grids.

A diffusivity decreasing upon grid refinement is useful since (1) refined grids allow for more active mean flows that one does not wish to overly diffuse, and (2) relatedly, with enough resolution, the model transitions to a regime where the parameterized quasi-geostrophic eddying flow becomes explicitly represented, thus motivating the reduction of the SGS operator's contribution to the time tendency in order not to double-count the effects of eddy transport. It is notable that frictional dissipation via the Smagorinsky viscosity scheme (Chapter 18) provides a case where the SGS operator naturally becomes less important as resolution is refined. This is the goal of the diffusivity proposed here.

14.4.8.1 Length scale

For purposes of quasi-geostrophic eddy parameterizations, coarseness of a model grid is based on how the grid size compares to the first baroclinic Rossby radius of deformation. If the grid is coarse, then a possible length scale for computing the diffusivity is given by the Rossby radius as discussed in Section 14.4.4. For refined models where the Rossby radius is larger than the grid, which is now quite common at least for models with refined tropical resolution, then use of the Rossby radius will produce an overly large diffusivity. Hence, it is proposed that the length scale be given by the smooth function of the Rossby radius and the grid scale

$$L = \frac{2\,\Delta s\,\lambda_1}{\Delta s + \lambda_1}, \tag{14.104}$$

where

$$\Delta s = \frac{2\,\Delta x\,\Delta y}{\Delta x + \Delta y}. \tag{14.105}$$

The ratios provide for a smoother length scale than the alternative usage of minimum functions. In particular, (a) when $\Delta s << \lambda_1$, then $L \approx 2\,\Delta s$; (b) when $\Delta s = \lambda_1$, then $L = \lambda_1$; and (c) when $\Delta s >> \lambda_1$, then $L \approx 2\,\lambda_1$.

14.4.8.2 Velocity scale

To remedy the numerical problem with an increasing Eady growth rate for refined grids, one may use the velocity scale

$$V = \frac{2\,V_{\mathrm{m}}\,V_{\mathrm{qg}}}{V_{\mathrm{m}} + V_{\mathrm{qg}}}. \tag{14.106}$$

The velocity scale V_{m} is a constant motivated from numerical considerations* and is a tuning parameter, with common values on the order

$$V_{\mathrm{m}} = 0.05\,\mathrm{m\,s^{-1}}. \tag{14.107}$$

The velocity scale

$$V_{\mathrm{qg}} = L\,T^{-1} \tag{14.108}$$

is motivated from the quasi-geostrophic theories using the Eady growth rate to compute T^{-1}, and the length scale is taken from equation (14.104). Instead of limiting the integration to be over an *a priori* specified depth range, as suggested by Treguier et al. (1997), one may prefer to integrate over the full depth range. Yet contributions from unstratified regions should be limited by the maximum slope allowed for numerical stability, which is typically around 0.01.

14.4.9 A comment about vertical depth range for the Eady calculation

The depth range $z \in [D_{\mathrm{b}}, D_{\mathrm{t}}]$ used to compute the Eady growth rate introduces some arbitrariness to the computation. However, by specifying $D_{\mathrm{t}} = 100$ m and $D_{\mathrm{b}} = 2000$ m, one explicitly recognizes that the expected regions where the quasi-geostrophic theories are to provide something sensible are limited to the open ocean, away from shallow shelves and the upper ocean mixed layer. Additionally, experience has shown that if the Eady growth rate is computed over the full depths, then the shallow regions tend to result in very large growth rates. These in turn lead to relatively large diffusivities that can overly smooth out boundary currents, such as those around the Labrador Sea. This problem is avoided by removing shallow regions from the integration, or by providing smaller weights for these regions than in the open ocean.

---

*Such a velocity scale was suggested by Eric Chassignet, personal communication 1999.

14.4.10 Two more possibilities for flow dependent diffusivities

The methods to compute the length scale can lead to very different values. For example, although switching to the equatorial Rossby radius in the tropics and so avoiding a singularity, the Rossby radius approach will generally yield a much larger length scale in the low latitudes than the baroclinic width approach. Many global modelers have thus tended to avoid the Rossby radius approach since the models commonly used today for climate studies do a reasonable job of explicitly resolving tropical transients. A larger value for the diffusivity in this region is therefore not desirable.

Recognizing that the length scale may be the least solid part of the diffusivity calculation, one may wish to consider two options while maintaining some flow dependene to the diffusivity. First, one can simply choose a global constant length scale.

Another approach is motivated by the discussion of regularization in Section 14.4.7. Within the equator, the length scale is proportional to the horizontal density gradient alone. That is, it is proportional to the baroclinicity. One may wish to consider this approach for the global ocean. Exposing some of the relevant dimensional length and time scales, this approach leads to

$$\kappa = \alpha \, \overline{|\nabla_z \rho|}^{z} \left(\frac{L^2 g}{\rho_o \, N_o} \right). \tag{14.109}$$

In this equation, α is a dimensionless tuning constant, L is a constant length scale set to roughly 50 or 100 km, N_o is a constant buoyancy frequency, with $N_o = 0.004\,\mathrm{s}^{-1}$ a representive value, and $\overline{|\nabla_z \rho|}^{z}$ is a depth average of the baroclinicity over the chosen depth range. To within a constant factor, a thermal wind balanced vertical shear (equation (14.85)) leads to the form*

$$\kappa = \alpha' \, \overline{|\mathbf{u}_{,z}|}^{z}, \tag{14.110}$$

where α' has dimensions of squared length.

14.5 BIHARMONIC OPERATORS

Biharmonic tracer operators dissipate tracer variance. As with the biharmonic friction operators discussed in Section 17.9 (page 402), they place more of their dissipation at the smaller scales than Laplacian operators do, thus allowing the larger scales to be less affected. It is for this reason that biharmonic operators are typically preferred for eddying simulations, where it is desirable to allow the flow to exhibit its natural tendency toward advectively dominant regimes, with associated fine scale structures and hydrodynamic instabilities.

Although true in general, the above comments about the need to employ biharmonic tracer dissipation may not be so critical in practice. For the purpose of enhancing the advectively dominant aspects of the flow, in many cases it is the use of biharmonic friction, not biharmonic tracer dissipation, that is most useful. Furthermore, as shown by Smith and McWilliams (2003) and Smith and Gent

*This is the form for the diffusivity originally suggested by Rong Zhang, personal communcation 2003.

(2004), generalizing the Laplacian form of both the momentum and tracer dissipation operators, so that they can be horizontally anisotropic, can enhance the flow energetics in a manner analogous to biharmonic operators.

There are two main reasons for preferring Laplacian operators. First, they are generally simpler to implement in an ocean model than their biharmonic relatives. Correspondingly, they are less expensive computationally. Second, neither biharmonic friction nor biharmonic tracer mixing can guarantee monotonic behavior, since biharmonic fluxes are not guaranteed to be downgradient. Lack of monotonicity is more problematic for tracers than momentum, since tracers generally have positive-definite properties in the real ocean. Additionally, for active tracers, sign-indefinite fluxes lead to sign-indefinite cabbeling, which is unphysical (Section 14.1.7). This result points to a fundamental problem with many of the biharmonic tracer operators discussed in this section. That is, they are not readily generalized to more than one active tracer.

Although the above discussion points out some fundamental reasons to avoid using biharmonic tracer operators, there are many eddying ocean models that employ these schemes. This section therefore provides a summary of the various operators that have been proposed, and presents some analysis of their properties. Points relevant for implementating these operators are also made. For simplicity, the discussion in this section is limited to Boussinesq fluids.

14.5.1 Tracer variance

As in Section 13.4.1 (page 288), consider the effects on tracer variance of the biharmonic operators. For Boussinesq fluids, tracer variance is given by

$$\mathcal{F} = V^{-1} \int \mathrm{d}V\, C^2 - V^{-1} \left(\int \mathrm{d}V\, C \right)^2, \tag{14.111}$$

where $V = \int \mathrm{d}V$ is the volume of seawater in the domain. In a conservative ocean system without sources and sinks, the total tracer mass $\int (\rho_o \, \mathrm{d}V)\, C$ and total volume are constant with time. However, the total squared tracer can change due to transport effects. Therefore, changes in tracer variance are equivalent to changes in the integrated squared tracer $\int \mathrm{d}V\, C^2$, and so for present purposes the two are considered the same.

14.5.2 Horizontal biharmonic mixing

Start with the horizontal biharmonic operator. This operator is an iteration on the horizontal Laplacian operator, and the horizontal biharmonic tracer flux is given by

$$\mathbf{F}_{\mathrm{B}} = \sqrt{B}\, \nabla_z L, \tag{14.112}$$

where

$$L = \nabla_z \cdot (\sqrt{B}\, \nabla_z C) \tag{14.113}$$

is the horizontal Laplacian operator acting on the tracer and $B > 0$ is a biharmonic mixing coefficient with units $(\text{length})^4\,(\text{time})^{-1}$. The convergence of the biharmonic flux $\mathbf{F}_{\mathrm{B}}$ yields the horizontal biharmonic mixing operator

$$R_{\mathrm{B}} = -\nabla \cdot \mathbf{F}_{\mathrm{B}}. \tag{14.114}$$

The biharmonic operator's effects on tracer variance can be seen via the following manipulations:

$$\begin{aligned}
\frac{1}{2}\partial_t\left(\int \mathrm{d}V\, C^2\right) &= -\int \mathrm{d}V\, C\, \nabla_z \cdot \mathbf{F}_\mathrm{B} \\
&= -\int \mathrm{d}V\, \nabla_z \cdot (C\, \mathbf{F}_\mathrm{B}) + \int \mathrm{d}V\, \nabla_z C \cdot \mathbf{F}_\mathrm{B} \\
&= \int \mathrm{d}V\, \sqrt{B}\, \nabla_z C \cdot \nabla_z L \\
&= \int \mathrm{d}V\, \nabla_z \cdot (\sqrt{B}\, L\, \nabla_z C) - \int \mathrm{d}V\, L^2 \\
&= -\int \mathrm{d}V\, L^2 \leq 0. \qquad (14.115)
\end{aligned}$$

To reach this result, it is necessary to assume that the horizontal Laplacian and biharmonic tracer fluxes satisfy no-flux side boundary conditions

$$\hat{\mathbf{n}} \cdot \sqrt{B}\, \nabla_z C = 0 \qquad (14.116)$$

$$\hat{\mathbf{n}} \cdot \mathbf{F}_\mathrm{B} = 0. \qquad (14.117)$$

Free surface boundary effects were also ignored. Note that even if the tracer gradients do not vanish at the boundaries, the flux can still vanish if the diffusivity is assumed to vanish there. Either way, these no-flux boundary conditions are easily ensured via masks placed on the numerical model's flux components.

Notably, without the square root on the biharmonic mixing coefficient, tracer variance is not generally dissipated when using a spatially dependent diffusivity, since it is no longer possible to form a perfect square. This point was discussed by Griffies and Hallberg (2000) in the context of biharmonic friction.

The orientation of the biharmonic flux relative to the tracer field

$$\mathbf{F}_\mathrm{B} \cdot \nabla_z C = \sqrt{B}\, \nabla_z L \cdot \nabla_z C, \qquad (14.118)$$

is not sign definite. Hence, the biharmonic flux is not generally down the tracer gradient. It therefore cannot be considered a downgradient diffusive flux, whereas the Laplacian analog is strictly downgradient. As discussed in Section 14.1.7, such sign-indefinite mixing flux leads to a sign-indefinite cabbeling term, which is unphysical. Additionally, biharmonic tracer mixing schemes can produce local extrema in the tracer field. Based on the discussion in Section 13.4.3 (page 289), such should not be surprising, since without a guaranteed downgradient flux, one cannot ensure that the flux always smooths the flow features. Merryfield and Holloway (2003) also discuss such problems in an appendix to their paper, and Di-Battista et al. (2001) provide a more rigorous discussion. So the ability to produce anti-cabbeling, and to increase tracer extrema, are the two main reasons for avoiding the use of any biharmonic tracer operator in ocean simulations.

14.5.3 Neutral biharmonic mixing

The Laplacian neutral diffusion of Solomon (1971) and Redi (1982) can in principle be extended into a neutral biharmonic mixing operator. This operator represents a direct analog of the horizontal biharmonic mixing operator. It also suffers from the same sign-indefinite nature of the local tracer flux, thus leading to unphysical cabbeling processes and the potential for creating local extrema.

As with horizontal biharmonic mixing, employ an iterative approach to construct the neutral biharmonic mixing operator. Hence, the mth component of the neutral biharmonic flux is given by

$$F_{\mathrm{B}}^{m} = K^{mn}\, L_{,n} \tag{14.119}$$

where

$$L = (K^{pq}\, C_{,q})_{,p} \tag{14.120}$$

is the Laplacian neutral diffusion operator. As a vector, the biharmonic flux is written

$$\mathbf{F}_{\mathrm{B}} = \sqrt{B}\,(\nabla_{\rho} + \hat{\mathbf{z}}\,\mathbf{S}\cdot\nabla_{\rho})\,L, \tag{14.121}$$

and the biharmonic operator is computed as the convergence of the neutral biharmonic flux

$$R_{\mathrm{B}} = -\nabla\cdot\mathbf{F}_{\mathrm{B}}. \tag{14.122}$$

As for horizontal biharmonic mixing in Section 14.5.2, it is best to apply the diffusivity symmetrically at each stage of the calculation with a square root of the full biharmonic diffusivity. This approach ensures variance reduction with nonconstant diffusivities, as seen via the following manipulations:

$$\begin{aligned}
\frac{1}{2}\,\partial_t\left(\int \mathrm{d}V\, C^2\right) &= -\int \mathrm{d}V\, C\,\nabla_z\cdot\mathbf{F}_{\mathrm{B}} \\
&= -\int \mathrm{d}V\,\nabla_z\cdot(C\,\mathbf{F}_{\mathrm{B}}) + \int \mathrm{d}V\,\nabla_z C\cdot\mathbf{F}_{\mathrm{B}} \\
&= \int \mathrm{d}V\, K^{mn}\, C_{,m}\, L_{,n} \\
&= \int \mathrm{d}V\,(L\,K^{mn}\, C_{,m})_{,n} - \int \mathrm{d}V\, L^2 \\
&= -\int \mathrm{d}V\, L^2 \le 0.
\end{aligned} \tag{14.123}$$

To reach this result, boundary forcing was dropped. Additionally, both the Laplacian and biharmonic fluxes were assumed to vanish at the solid earth boundaries

$$\hat{\mathbf{n}}_m\, K^{mn}\, C_{,n} = 0 \tag{14.124}$$

$$\hat{\mathbf{n}}_m\, K^{mn}\, L_{,n} = 0. \tag{14.125}$$

As for horizontal biharmonic mixing, these no-flux solid earth boundary conditions are easily established in the numerical model via masks. Furthermore, as for the Laplacian neutral diffusion case, the neutral biharmonic flux vanishes for tracers aligned with the neutral directions.

A numerical algorithm may proceed in a manner similar to the horizontal biharmonic operator. First, the full neutral Laplacian diffusion operator L is computed. For the second iteration, L is treated as the new "tracer" and the neutral biharmonic operator then constructed. Notably, the diagonal term in the vertical biharmonic flux contributes to the time evolution in an implicit manner, just as used when performing Laplacian neutral diffusion (see Section 16.8.3, page 373 for details).

14.5.4 The biharmonic stirring of Roberts and Marshall

Roberts and Marshall (1998) proposed the following divergence-free velocity to adiabatically dissipate grid scale variance in the density field

$$\begin{aligned} \mathbf{u}^* &= \nabla \wedge \boldsymbol{\Upsilon} \\ &= \partial_z \nabla_z^2 (B\,\mathbf{S}) - \hat{\mathbf{z}}\, \nabla_z \cdot \nabla_z^2 (B\,\mathbf{S}). \end{aligned} \tag{14.126}$$

In this equation, the vector streamfunction is given by

$$\boldsymbol{\Upsilon} = -\hat{\mathbf{z}} \wedge \nabla_z^2 (B\,\mathbf{S}) \tag{14.127}$$

with $B > 0$ a biharmonic mixing coefficient. The corresponding skew tracer flux is given by

$$\mathbf{F} = -C_{,z}\, \nabla_z^2\, (B\,\mathbf{S}) + \hat{\mathbf{z}}\, \nabla_z C \cdot \nabla_z^2\, (B\,\mathbf{S}). \tag{14.128}$$

Dropping the horizontal Laplacian ∇_z^2 and setting $B \rightarrow -\kappa$ recovers the Laplacian scheme of Gent et al. (1995) as discussed in Section 14.2.

Enhanced scale selectivity with the Roberts and Marshall (1998) operator is provided to the buoyancy field. Indeed, they show that for buoyancy, this operator is quite effective at dissipating grid scale variance while reducing spurious diapycnal mixing. For other tracers, the operator remains second order and so is not as scale selective as either the horizontal or neutral biharmonic mixing operators.

Roberts and Marshall (1998) termed their operator a "biharmonic GM" operator since it represents a straightforward generalization of the usual "Laplacian GM" operator. Yet, as they showed, and as shown in the following, their biharmonic operator does not generally dissipate potential energy, whereas the Laplacian GM operator does. Therefore, their operator is perhaps better considered another one of many possible *neutral biharmonic operators*.

14.5.4.1 *Effects on potential energy*

Consider how the Roberts and Marshall (1998) biharmonic operator affects the potential energy for the case when density is a linear function of potential temperature. A similar discussion was given in their paper, where they employed the advective flux formulation rather than the skew flux formulation used here. Focusing just on the biharmonic operator, the time tendency for gravitational potential energy is given by

$$\begin{aligned} P_{,t} &= g \int \mathrm{d}V\, z\, \rho_{,t} \\ &= -g \int \mathrm{d}V\, z\, \nabla \cdot \mathbf{F} \\ &= g \int \mathrm{d}V\, F^{(z)}, \end{aligned} \tag{14.129}$$

where boundary terms were dropped since concern here is with the effects that $g \int \mathrm{d}V\, F^{(z)}$ has on the evolution of potential energy. Introducing the vertical component to the skew flux leads to

$$\begin{aligned} \int \mathrm{d}V\, F^{(z)} &= -\int \mathrm{d}V\, B\, \rho_{,z}\, \mathbf{S} \cdot \nabla_z^2 \mathbf{S} \\ &= -\int \mathrm{d}V\, \nabla_z \cdot (B\, \rho_{,z}\, S_m\, \nabla_z S_m) + \int \mathrm{d}V\, \nabla_z (B\, \rho_{,z}\, S_m) \cdot \nabla_z S_m, \end{aligned} \tag{14.130}$$

where $m = 1, 2$ is summed. Again, drop boundary terms, which can be ensured by assuming that $B\, \rho_{,z}\, S_i$ vanishes next to the lateral ocean boundaries. This term may be set to zero via model masks, and it may formally be associated with a vanishing diffusivity at this boundary. The result is

$$\int dV\, F^{(z)} = \frac{1}{2} \int dV\, \nabla_z \left(B\, \rho_{,z}\right) \cdot \nabla_z S^2 + \int dV\, B\, \rho_{,z} \left(\partial_m \mathbf{S} \cdot \partial_m \mathbf{S}\right), \qquad (14.131)$$

where $S^2 = \mathbf{S} \cdot \mathbf{S}$, and $m = 1, 2$ is summed in the second term over the horizontal spatial dimensions. The second term is nonpositive in stably stratified fluids, for which $\rho_{,z} \leq 0$. It therefore represents a potential energy sink. The first term, however, is sign indefinite even when B is a constant. For the special case of $B\, \rho_{,z}$ independent of horizontal position, the first term vanishes. For example, the special density profile $\rho = \rho_o(z) + \rho_1 \cos(\mathbf{p} \cdot \mathbf{x})$, with $\mathbf{p}$ a horizontal wave vector, has $B\, \rho_{,z}$ horizontally constant if B is constant. In the slightly more general case of constant B and with $\rho_{,z} = \partial_z \rho_o + \partial_z \rho_1$, where $|\partial_z \rho_o| >> |\partial_z \rho_1|$, the first term in the expression for potential energy is nonzero but subdominant to the second term. So potential energy is again reduced in this case. It is unclear what happens for more general profiles.

It might be speculated that the inability to prove that the potential energy is generally reduced may indicate that the Roberts and Marshall (1998) operator is unstable. However, if numerically implemented according to the triad approach of Griffies et al. (1998), the discretized skew flux preserves density variance. As variance growth is typically associated with linearly unstable numerical schemes, any potential instability of the scheme will likely be nonlinear. However, the operator appears to be quite stable in a wide suite of both coarse and fine models (M. Roberts, personal communication 1998). Hence, it is unclear how important it is to provide a potential energy sink with the stirring operators.

14.5.4.2 Flux for a particular density profile

Recall from Section 14.2.1 that in the special case of a linear equation of state, the density skew flux from Gent and McWilliams (1990) takes the especially simple form

$$\mathbf{F}^{(h)}(\rho) = -\kappa\, \nabla_z \rho \qquad (14.132)$$

$$F^{(z)}(\rho) = \kappa\, S^2 \rho_{,z}. \qquad (14.133)$$

In general, the horizontal flux is directed down the density gradient, and the vertical component is upgradient. With the vertical density flux always upgradient, the Gent and McWilliams (1990) scheme decreases potential energy in a stably stratified fluid with nonzero baroclinicity.

As just seen, this potential energy dissipation property is not generally respected by the Roberts and Marshall (1998) operator. However, it is useful to highlight a case where these properties are shared, such as the case where density is given by

$$\rho = \rho_o(z) + \rho_1 \cos(\mathbf{p} \cdot \mathbf{x}), \qquad (14.134)$$

with $\rho_o(z)$ a stable vertical profile, ρ_1 an amplitude function, and $\mathbf{p} = (p_x, p_y, 0)$ a horizontal wave vector. The slope vector for this profile is given by

$$\mathbf{S} = \frac{\rho_1\, \mathbf{p}\, \sin(\mathbf{p} \cdot \mathbf{x})}{\rho_o'(z)}, \qquad (14.135)$$

and the horizontal Laplacian is

$$\nabla_z^2 \mathbf{S} = -p^2 \, \mathbf{S}, \tag{14.136}$$

where $p^2 = \mathbf{p} \cdot \mathbf{p}$. The biharmonic skew flux therefore takes the form

$$\mathbf{F}^{(h)} = (B\, p^2)\, \mathbf{S}\, C_{,z} \tag{14.137}$$

$$F^{(z)} = -(B\, p^2)\, \mathbf{S} \cdot \nabla_z C. \tag{14.138}$$

The biharmonic skew flux of density, linearly dependent on temperature, is given by

$$\mathbf{F}^{(h)}(\rho) = -(B\, p^2)\, \nabla_z \rho \tag{14.139}$$

$$F^{(z)}(\rho) = (B\, p^2)\, S^2\, \rho_{,z}. \tag{14.140}$$

Hence, for this profile, the horizontal biharmonic density skew flux components are down the horizontal density gradient, and the vertical skew flux component is up the vertical density gradient. Potential energy is therefore dissipated. The effective diffusivity is scale-dependent, with small scales, or large p^2, acted on with the largest effective diffusivity.

14.5.5 Gent's biharmonic stirring operator

As reported in Roberts and Marshall (1998), Peter Gent suggested the alternative velocity

$$\begin{aligned} \mathbf{u}^* &= \nabla \wedge \boldsymbol{\Upsilon} \\ &= -\partial_z\, \nabla_z M + \hat{\mathbf{z}}\, \nabla_z^2\, M, \end{aligned} \tag{14.141}$$

with vector streamfunction

$$\boldsymbol{\Upsilon} = \hat{\mathbf{z}} \wedge \nabla_z M \tag{14.142}$$

and

$$M = B \left(\frac{\nabla_z^2 \rho}{\rho_{,z}} \right). \tag{14.143}$$

The corresponding skew flux is given by

$$\mathbf{F} = C_{,z}\, \nabla_z\, M - \hat{\mathbf{z}}\, \nabla_z C \cdot \nabla_z\, M. \tag{14.144}$$

In contrast to the Roberts and Marshall (1998) operator, the Gent operator provides a sign definite sink of potential energy for any given density profile, as seen since the integrated vertical skew flux component is negative semidefinite in stably stratified water,

$$\begin{aligned} \int \mathrm{d}V\, F^{(z)} &= -\int \mathrm{d}V\, \nabla_z \left(B\, \frac{\nabla_z^2 \rho}{\rho_{,z}} \right) \cdot \nabla_z \rho \\ &= \int \mathrm{d}V\, \nabla_z \cdot (B\, \mathbf{S}\, \nabla_z^2 \rho) + \int \mathrm{d}V\, (B/\rho_{,z}) \left(\nabla_z^2 \rho \right)^2. \end{aligned} \tag{14.145}$$

The first term can be dropped when assuming $B\, \mathbf{S}\, \nabla_z^2 \rho$ vanishes at boundaries. The second term is nonpositive in stably stratified water, and so represents a potential energy sink for any density profile. Notably, the spectral representation of this operator when acting on the density profile $\rho = \rho_0(z) + \rho_1 \cos(\mathbf{k} \cdot \mathbf{x})$ is the same as the Roberts and Marshall (1998) operator given by equations (14.137) and (14.138).

The Gent biharmonic operator cannot be discretized using the proven technology developed by Griffies et al. (1998), Griffies (1998) (see chapters 15 and 16 for details). The reason is that the slope vector $\mathbf{S}$ is not a fundamental piece of the Gent operator, whereas it is fundamental for all other neutral physics schemes.

14.5.6 Neutral biharmonic filtering

The neutral biharmonic filtering operator is designed with the following points in mind.

1. Neutral biharmonic mixing does not act on buoyancy, except through effects due to the nonlinear equation of state. Hence, an alternative means for scale-selective buoyancy dissipation should be considered.
2. Horizontal biharmonic mixing provides an inexpensive, scale-selective dissipation mechanism.
3. However, the horizontal orientation of the horizontal biharmonic flux introduces too much diapycnal mixing, and so it is necessary to reorient the flux according to the neutral directions.

These points motivate the following tracer flux:

$$\mathbf{F} = \sqrt{B}\,(\nabla_z + \hat{\mathbf{z}}\,\mathbf{S} \cdot \nabla_z)\,L, \tag{14.146}$$

with

$$L = \nabla_z \cdot (\sqrt{B}\,\nabla_z C) \tag{14.147}$$

the horizontal Laplacian diffusion operator. The convergence of the horizontal flux components $\mathbf{F}^{(h)} = \sqrt{B}\,\nabla_z L$ yields the familiar horizontal biharmonic operator. Yet the vertical flux component $F^{(z)} = \mathbf{S} \cdot \mathbf{F}^{(h)}$ orients the three-dimensional flux vector parallel to the neutral direction (see Figure 13.1). That is,

$$\mathbf{F} \cdot \nabla \rho = 0, \tag{14.148}$$

which is analogous to the neutral diffusive fluxes. In contrast to the neutral diffusive fluxes, $\mathbf{F}(\rho) \neq 0$. Instead, $\mathbf{F}(\rho)$ is a skew flux and so has a corresponding advection velocity (see equation (14.152)).

When acting on the tracer profile

$$C = C_o(z) + C_1(z)\,\cos(\mathbf{p} \cdot \mathbf{x}) \tag{14.149}$$

with constant biharmonic coefficient B, the flux (14.146) takes the form

$$\mathbf{F} = -B\,p^2\,(\nabla_z + \hat{\mathbf{z}}\,\mathbf{S} \cdot \nabla_z)C. \tag{14.150}$$

The horizontal flux components are downgradient and scale-selective, yet the orientation of the vertical flux depends on the relative orientation of the tracer and the neutral slopes. When $C = \rho$,

$$\mathbf{F} = -B\,p^2\,(\nabla_z \rho - \hat{\mathbf{z}}\,S^2 \rho_{,z}). \tag{14.151}$$

Since the vertical component to the skew flux is upgradient, this profile has its potential energy dissipated. Note the slight generalization of this result compared to that of the Roberts and Marshall (1998) operator, where potential energy is dissipated only when $\rho_1(z) \rightarrow \rho_1$. Nonetheless, as with their operator, an arbitrary density profile is not guaranteed to have its potential energy dissipated.

It is useful to highlight the different perspective taken here as compared to Roberts and Marshall (1998) and Gent considered in Sections 14.5.4 and 14.5.5. Both of their schemes focused on the advective form of Gent et al. (1995), which led them to their higher order eddy-induced advection velocity. That is, they both proposed a stirring process. In contrast, the neutral biharmonic filter considered here only stirs density (assuming a linear equation of state), in which case one can define an advection velocity

$$\begin{aligned} \mathbf{u}^* &= \nabla \wedge \boldsymbol{\Upsilon} \\ &= -\partial_z \left(\sqrt{B}\, \frac{\nabla_z L}{\rho_{,z}} \right) + \hat{\mathbf{z}}\, \nabla_z \cdot \left(\sqrt{B}\, \frac{\nabla_z L}{\rho_{,z}} \right), \end{aligned} \tag{14.152}$$

where the vector streamfunction is

$$\boldsymbol{\Upsilon} = \hat{\mathbf{z}} \wedge \left(\sqrt{B}\, \frac{\nabla_z L}{\rho_{,z}} \right). \tag{14.153}$$

For other tracers, however, the operator produces mixing fluxes oriented along the neutral direction. Hence, there is no corresponding advection velocity.

Numerical discretization of the horizontal flux components can easily follow from the usual methods for horizontal biharmonic mixing. Notably, the horizontal fluxes are most simply applied for all values of the neutral direction slope. The vertical flux follows the triad approach described in Chapter 16, with tapering applied in steep sloped regions as described in Section 15.1.4 (page 332).

14.5.7 Biharmonic skew filtering operator

Another biharmonic operator, closely related to the neutral biharmonic filtering operator given in Section 14.5.6, is termed here a *biharmonic skew filtering operator*. The tracer flux for this operator is given by

$$\mathbf{F} = \sqrt{B}\, (\nabla_z + \hat{\mathbf{z}}\, \mathbf{S}(C) \cdot \nabla_z)\, L. \tag{14.154}$$

Importantly, the slope here is that of the particular tracer

$$\mathbf{S}(C) = -\left(\frac{\nabla_z C}{C_{,z}} \right). \tag{14.155}$$

In contrast, the neutral filter flux (14.146) is constructed with the neutral slope $\mathbf{S}(\rho)$ for all tracers. The two fluxes are the same in the special case of a linear equation of state with $\mathbf{S}(\rho) = \mathbf{S}(C)$. However, in general the fluxes are distinct.

When density is a function of both temperature and salinity, separately stirring temperature and salinity with the biharmonic skew filtering operator generally mixes density. Hence, biharmonic skew filtering is not useful for active tracers. Yet it may be a useful means to suppress noise in passive tracer fields. Unfortunately, the approach may become computationally burdensome when adding many passive tracers since the cost of computing successively more tracer slopes is nontrivial. Instead, one may be more motivated to consider the neutral biharmonic mixing discussed in Section 14.5.3 for passive tracers.

14.5.8 A note on iterative Laplacian skewsion processes

As noted in Section 14.5.2, the horizontal biharmonic flux is not generally oriented down the horizontal tracer gradient; likewise for the neutral biharmonic mixing operator. It is for this reason that these biharmonic operators are termed "mixing" operators rather than "diffusive" operators. Even so, they satisfy the usual global constraint of reducing tracer variance. Furthermore, the neutral biharmonic mixing operator remains true to the needs of fluxing tracers along the neutral directions. That is, the operator vanishes when a tracer is aligned with the neutral direction.

The three biharmonic operators acting on buoyancy each provide for an adiabatic stirring of buoyancy. However, none of them represents a straightforward iteration of a Laplacian stirring operator. There is good reason for this: iteration of a Laplacian skewsion process does not generally lead to a biharmonic skewsion process. Hence, there is ambiguity concerning the choice of an appropriate skewed biharmonic operator.

14.6 CHAPTER SUMMARY AND SOME CHALLENGES

This chapter focused on some mathematical properties of various neutral physics operators. Such issues prove important to those aiming to understand the effects on a simulation realized by these operators, and for those implementing these schemes in a numerical model. Methods were also presented for computing diffusivites used with these operators, with various flow-dependent approaches surveyed.

Section 14.1 started the chapter by focusing on the neutral diffusion tensor, both the full slope tensor of Redi (1982) and the small slope tensor of Gent and McWilliams (1990). The small slope tensor is commonly implemented in ocean climate models, whereas the full slope tensor has little justification. The reason is that in those regions where the small slope approximation is no longer valid, physical processes other than those associated with neutral physics are very active, such as boundary layer processes and atmospheric forcing.

Elements of the Gent and McWilliams (1990) and Gent et al. (1995) tracer stirring scheme were presented in Section 14.2. The skew perspective emphasized by Griffies (1998) complemented the advection velocity perspective of Gent et al. (1995). Section 14.3 briefly summarized the various fluxes that arise from the Laplacian neutral physics schemes. Section 14.5 provided an overview of biharmonic tracer operators. Their use is discouraged due to the possibilities for introducing tracer extrema and anti-cabbeling. Additionally, the neutral biharmonic operators can be quite expensive computationally.

Section 14.4 focused on various z-model implementations of flow-dependent diffusivities of potential use for the Laplacian neutral physics schemes. Such diffusivities are thought to be physically and numerically motivated for the following reasons. First, a single diffusivity for the World Ocean seems unreasonable given the many different flow regimes. Second, as model resolution is refined to the one-degree and finer scales, the explicitly resolved flow starts to exhibit a strengthening that was largely absent in the coarser models. That is, the flow becomes much less homogeneous, which again warrants a nonhomogeneous diffusivity.

The methods used to compute the diffusivity are many. However, climate models have so far only made use of two-dimensional diffusivities. The utility of three-dimensional dependence, as suggested by potential vorticity closure methods, remains an outstanding research question.

In the computation of the two-dimensional diffusivities, there is general agreement in the theories that the Eady growth rate should determine the time scale. However, there are a number of approaches to determine the length scale, and they do not necessarily agree. First, one may simply take a constant value and then let the spatial dependence of the diffusivity be given by the Eady growth rate. Second, the length scale may be taken proportional to the Rossby radius. Notably, the tropics are a region that present day global models tend to resolve reasonably well. Thus, one may wish to actually reduce the level of subgrid scale fluxes in this region to avoid double counting. With a larger Rossby radius in the tropics, its use to determine the length scale is not appropriate globally. A third approach sets the length scale via the baroclinic zone radius. This radius is not well defined, and tends to require the use of somewhat *ad hoc* search algorithms for their computation.

After working through and testing a number of different methods, including a "compromise" approach whereby the diffusivity is directly proportional to the baroclinicity, modelers may readily conclude that there are a lot of issues, some fundamental and some practical, that are not well understand about the closures. More tests, both idealized and realistic, need to be run in order to provide some guidance regarding what is an important detail versus what is irrelevant.

Finally, theories for mesoscale eddy mixing are predominantly based on semi-local ideas in that they are local in the horizontal but nonlocal vertically. This may not be sufficient for global models. Instead, horizontally and vertically nonlocal theories may need to be considered, perhaps inspired by approaches used for boundary layer processes (e.g., Large et al., 1994). Observational data may provide some insights, although this remains an incredibly tough problem due to the needs for high resolution data in all three spatial dimensions and in time.

Chapter Fifteen

NEUTRAL PHYSICS NEAR THE SURFACE BOUNDARY

To implement neutral physics schemes in an ocean climate model, one must address the question of what to do when neutral slopes steepen. This will inevitably happen in realistic simulations. Numerical stability requires some form of tapering toward a non-neutral oriented scheme, or at least the removal of the neutral processes. Are there physical arguments promoting one form of tapering versus another? How sensitive are the solutions to the methods? These questions, and related ones, are addressed in part within this chapter.

Although some issues are generic to both the surface and solid earth boundaries, discussions here focus on the surface boundary. The upper ocean is where more attention has been placed in the research. Critical elements of how the ocean interacts with the atmosphere and sea ice, as well as surface boundary layer processes, are at the forefront of the issues to be addressed when considering neutral physics in the ocean surface boundary layer.

Simulations are generally sensitive to details of how the neutral physics schemes are handled in the boundary layers. The discussion in this chapter presents material that is rapidly changing, as it represents an area of active research with many open questions. The reader should therefore consider this material with healthy skepticism.

15.1 LINEAR STABILITY FOR NEUTRAL DIFFUSION

There are various methods to preserve linear numerical stability of neutral diffusion in regions of steep neutral slopes. One method is to employ the full Redi (1982) diffusion tensor (Section 14.1). This tensor has finite values regardless of the value of the neutral slopes. Hence, its discretization generally requires no artificial numerical regularization, so long as a reasonable time step condition is maintained (for discussion, see Appendix C of Griffies et al., 1998). However, the full Redi tensor contains terms that are very small throughout the bulk of the ocean where slopes are less than $1/100$, and these terms are awkward and expensive to compute. Additionally, as argued in later sections in this chapter, there is little physical motivation to maintain the integrity of neutrally oriented diffusion in regions where the neutral slopes are steep. That is, there is little physical motivation to use the full Redi tensor since the small angle approximation (Section 14.1) is sufficient for all ocean climate related purposes using z-models.

Hence, the small slope neutral diffusion tensor is typically used in z-model simulations. This tensor, however, is not bounded as the neutral slopes increase, and so a regularization method is needed for simulations where the slopes can be vertical. Relatedly, when the slope of the neutral direction is large, it is possible for the neutral diffusive flux to project substantially onto the vertical direction. Because the vertical grid spacing is generally much smaller than the horizontal, care is needed to preserve numerical stability in regions of steep neutral slopes.

One approach for maintaining numerical stability is to solve the neutral diffusion operator implicitly, as is already done in most models for their vertical physics schemes. However, since neutral diffusion has lateral flux components, an implicit solution method involves the whole horizontal grid. Additionally, the possibility of infinite slopes, and correspondingly infinite fluxes, makes even an implicit scheme numerically sensitive. Such an approach is therefore not practical in a realistic climate model and so is not pursued. The remainder of this section presents other options. The discussion here represents in part an abbreviated form of that given in Appendix C of Griffies et al. (1998).

15.1.1 Linear stability analysis for Laplacian neutral diffusion

The linear stability constraints described in Section 18.1.2 (page 410) for unrotated diffusion are extended here to the case of rotated diffusion. It has been found sufficient in practice to assume that stability for rotated diffusion is ensured if stability holds for each of the three directions separately. Less stringent constraints may result from a more general analysis, but this is not pursued here.

The linear numerical constraint from the one-dimensional diffusion equation, as discussed in Sections 12.8.4 and 18.1.2.1 (see also Cox, 1987), indicates that an explicit numerical scheme with a $\Delta t_{\text{forward}}$ forward time step is stable if the grid Courant-Friedricks-Levy (CFL) number satisfies

$$\frac{|K^{mn}|\,\Delta t_{\text{forward}}}{\Delta x_m\,\Delta x_n} \leq \frac{1}{2}, \tag{15.1}$$

where K^{mn} are components of the diffusion tensor (14.19), and the inequality must hold for all combinations of $m,n = x,y,z$. The one-dimensional diffusion constraint has traditionally been used, rather than the two-dimensional constraint which leads to $1/4$ rather than $1/2$ on the right-hand side (Section 18.1.2.1). The one-dimensional constraint has been found sufficient in practice, and so is used in the following discussion.

For models that discretize the tracer time tendency $\partial_t C$ using a leap-frog scheme,

$$\Delta t_{\text{forward}} = 2\,\Delta t \tag{15.2}$$

is needed for numerical stability (Section 12.8.4), where Δt is the model's time step. In contrast, two-time level discretizations of the time tendency have

$$\Delta t_{\text{forward}} = \Delta t. \tag{15.3}$$

It is notable that the extra factor of 2 on the diffusive CFL constraint for the leap-frog tendency limits the value available for the product of the neutral diffusivity and the neutral slope to be one-half that available for models that use a two-time level scheme to discretize the time tendency. This can be an undesired constraint in cases where one aims to employ the neutral diffusion operator for steep slopes, refined vertical resolutions, and/or large neutral diffusivities.

Assuming a geophysically relevant vertical to horizontal aspect ratio for the grid ($\Delta z/\Delta x \leq 1/100$), the two-dimensional horizontal subsystem is stable when the diffusion equation in the horizontal is stable. In general, satisfying this stability constraint in the horizontal is trivial and so is not considered further. Solving the vertical K^{zz} diagonal piece implicitly, as done by Cox (1987), points to the K^{xz} and K^{yz} cross terms as setting the most restrictive constraint. From these terms, the small angle neutral diffusion equation is linearly stable when, for each grid cell,

$$|S_a| \leq \frac{\Delta a \, \Delta z}{2 \, A_I \, \Delta t_{\text{forward}}} \equiv \delta, \tag{15.4}$$

where Δa is either Δx or Δy. The parameter δ represents the maximum allowable slope that can be used before some prescription must be employed to ensure numerical stability. For many large-scale ocean model configurations, this slope check parameter is roughly $1/100$.

As discussed in Section 14.1.4, neutral slopes are typically smaller than $1/100$ for the interior ocean away from mixed layers and convective regions. For slopes larger than $1/100$, the motivation to remain "true" to the neutrally directed diffusion process remains less physically compelling. The reason is that in such large sloped regions the ocean admits many other leading processes, such as three-dimensional boundary layer turbulence which generally mixes across as well as along neutral directions. Hence, restricting neutral diffusion to act in an unregularized manner only for slopes less than $1/100$, or even $1/500$, is not physically restrictive. Importantly, details of the regularization schemes generally do matter, largely because they affect the upper ocean where boundary layer processes are crucial to establishing the ocean climate model's physical integrity. Hence, great care must be exercised, and it is furthermore important to test solution sensitivity to the various regularization approaches.

15.1.2 Slope clipping the neutral diffusive fluxes

The first approach to maintaining numerical stability is the *slope clipping* scheme of Cox (1987). Slope clipping limits the slope along which neutral diffusion occurs. For example, when the x-slope satisfies $|S_x| > S_{\text{clip}}$, with S_{clip} some maximum slope usually taken as a fraction of the grid parameter δ given in equation (15.4), then

$$\rho_{,z} \rightarrow (\rho_{,z})_{\text{clip}}, \tag{15.5}$$

where

$$S_{\text{clip}} = \frac{|\rho_{,x}|}{|\rho_{,z}|_{\text{clip}}}. \tag{15.6}$$

The same sort of clipping is applied independently to the y-component of the slope.

Limiting the slopes along which the fluxes are directed introduces significant false dianeutral fluxes in those regions where the actual slopes exceed the clipped slope. As mentioned in Section 14.1.4, one may expect increased cross-neutral fluxes in regions of steep slopes where there is the potential for enhanced energy release at all scales. Therefore, one may guess that slope clipping is a reasonable parameterization of these effects. However, the amount of uncontrolled fluxes can be egregiously large and unphysical.

An example should convince the reader that slope clipping is not appropriate. Consider the case where density is a linear function only of potential temperature and there are no vertical or meridional gradients. If maintaining a true neutral nature to the diffusive flux, each component of the temperature flux should vanish. However, as mentioned above, the clipped flux does not vanish. In particular, the off-diagonal piece of the vertical neutral flux of temperature becomes

$$F^{(z)}_{\mathrm{clip}} = A_I |\theta_{,x}| \, S_{\mathrm{clip}} > 0. \tag{15.7}$$

The diagonal piece $S^2 \theta_{,z}$ is handled via an implicit time stepping scheme. Focusing on the clipped off-diagonal term here, consider the typical case where $A_I = 10^3 \, \mathrm{m^2 \, s^{-1}}$, $S_{\mathrm{clip}} = 0.01$, $\theta_{,x} = 10^{-5} \, {}^\circ\mathrm{K\,m^{-1}}$. In this case, the clipped piece contributes to a vertical flux of potential temperature with magnitude $10^{-4} \, \mathrm{m} \, {}^\circ\mathrm{K\,s^{-1}}$. When multiplied by $\rho_o C_p \approx 4 \times 10^6 \, \mathrm{J\,m^{-3}} \, {}^\circ\mathrm{K^{-1}}$, the clipped flux contributes a heat flux of roughly 400 $\mathrm{W\,m^{-2}}$. This is a substantial heat flux. It has been seen to affect the solution integrity, especially in convective regions of models run with this scheme. This result motivates other, more physically satisfying, means to satisfy linear stability constraints in regions of steep neutral slopes.

15.1.3 Quadratic tapering the neutral diffusive fluxes

Two methods aim to maintain numerical stability via a rescaling or tapering of the neutral diffusivity. The diffusive flux remains oriented along the neutral direction, yet its magnitude is scaled smaller according to the steepness of the slope, once the slope reaches above some *a priori* set maximum slope.

The prescription of Gerdes et al. (1991) was the first method aiming to preserve the neutral orientation of the flux regardless of the slopes. It does so via a quadratic rescaling of the diffusivity to a smaller value. That is, if the slope satisfies $|S| > S_{\mathrm{max}}$, where S_{max} is the maximum slope, then the corresponding diffusivity is rescaled as

$$A_I \rightarrow A_I \left(\frac{S_{\mathrm{max}}}{S} \right)^2. \tag{15.8}$$

A quadratic rescaling is suggested by the quadratic term $1 + S^2$ in the full slope neutral diffusion tensor given in equation (14.18).

Notably, as for the full slope tensor, the Gerdes et al. (1991) taper does not remove all components of the neutral diffusion tensor as the neutral direction becomes vertical. Instead, it leaves the diagonal $(3,3)$ term saturated at the value

$$A_I \, S^2 \rightarrow A_I \, S^2_{\mathrm{max}}, \tag{15.9}$$

whereas all other components taper to zero. With maximum slope of $1/100$ in each of the two horizontal directions, and diffusivity of $A_I = 10^3 \, \mathrm{m^2 \, s^{-1}}$,

$$A_I \, S^2_{\mathrm{max}} = 0.2 \, \mathrm{m^2 \, s^{-1}}. \tag{15.10}$$

This is a large vertical diffusivity. In some cases, its effects will be much smaller than those from convective adjustment. However, this diffusivity is comparable to that diagnosed from some mixed layer schemes. This confusion of vertical mixing processes motivates another approach that allows for a clean separation.

15.1.4 Exponential tapering the neutral diffusive fluxes

Danabasoglu and McWilliams (1995) introduced an alternative rescaling in which the quadratic factor of Gerdes et al. (1991) is changed to a hyperbolic tangent

$$A_I \rightarrow A_I \, T_{\text{tanh}}, \tag{15.11}$$

with

$$T_{\text{tanh}} = \left(1 + \tanh\left(\frac{S_{\text{max}} - |S|}{S_{\text{d}}}\right)\right) \tag{15.12}$$

an exponentially rapid taper function where S_{d} determines the width of the transition region. Since the hyperbolic tangent rapidly moves between its limiting values, it is possible to always taper the diffusivity instead of introducing the "if-test" check necessary for the quadratic tapering. Such continuous tapering is familiar from numerical implementations of physical processes such as vertical mixing.

This form of tapering ensures that all components of the neutral mixing tensor are rapidly scaled to zero as the slope becomes larger than S_{max}. Notably, even the $(3,3)$ diagonal term is exponentially tapered to zero.

15.2 LINEAR STABILITY FOR GM STIRRING

Skew fluxes from Gent et al. (1995) remain numerically stable so long as the corresponding eddy-induced velocity components maintain the same CFL constraint as the Eulerian velocity. For illustrative purposes, assume the maximum CFL velocity is $1 \, \text{m} \, \text{s}^{-1}$. The scaling relation

$$u^* \sim \frac{\kappa \, S}{\Delta z}, \tag{15.13}$$

with $\kappa = 10^3 \, \text{m}^2 \, \text{s}^{-1}$ and $\Delta z = 10 \, \text{m}$, indicates that the slope S can be no larger than 0.01 in order for u^* to be less than $1 \, \text{m} \, \text{s}^{-1}$. As discussed in Section 15.1, this slope corresponds to that typically setting the largest slope allowable for stable small slope neutral diffusion. Hence, it is practical to provide a numerically stable GM scheme if the GM flux is computed as if the slope times diffusivity cannot get any larger than the maximum value

$$\kappa \mathbf{S} \rightarrow (\kappa \, \mathbf{S})_{\text{max}} \tag{15.14}$$

in the regions where $S > S_{\text{max}}$.

To maintain a nontrivial GM flux even when the neutral slopes are vertical, care must be exercised when introducing numerical artifices commonly used to eliminate computational overflows. With $\rho_{,z} < 0$ for stable stratification, a useful method for this purpose is to compute the neutral slope as

$$\begin{aligned} \mathbf{S} &= -\left(\frac{\nabla_z \rho}{\rho_{,z} - \epsilon}\right) \\ &= -\left(\frac{-\alpha \, \nabla_z \theta + \beta \, \nabla_z s}{-\alpha \, \theta_{,z} + \beta \, s_{,z} - \epsilon}\right). \end{aligned} \tag{15.15}$$

The small number $\epsilon > 0$ is placed in the denominator to prevent computational overflows when there is no vertical stratification. Crucially, its contribution is subtracted from the denominator in order to provide the correct sign of the slope when $\rho_{,z} \rightarrow 0^-$. The example given by Figure 15.1 illustrates this point. In this figure, the x-projection of the neutral slope is positive, $S_x > 0$. Notably, the slope becomes vertical when moving toward the easternmost isopycnal surface region since $\rho_{,z} \rightarrow 0^-$. The subtraction of ϵ in the denominator is necessary to have the computed slope go to $+\infty$. If one instead used $+\epsilon$ in the denominator, the computed slope would go to $-\infty$ and so yield unphysical and unstable behavior.

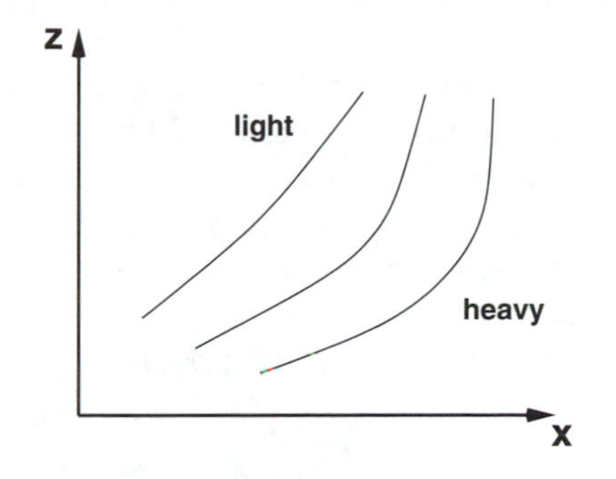

Figure 15.1 An example of steeply sloping isopycnals, where the slope in the x-direction $S_x = -\rho_{,x}/\rho_{,z}$ is large and positive, especially near the surface in the east. The numerical slope computation must maintain this positive sign to compute the correct sign for the GM fluxes.

15.3 NEUTRAL PHYSICS NEAR BOUNDARIES

What to do with the neutral physics schemes when entering boundary regions, which are often where steep neutral slopes occur, remains a research question. Should all neutral physics processes be turned off via some tapering scheme? Or should some of the parameterizations remain, such as GM? Should there be horizontal diffusion in the boundary layer, and if so, then what diffusivity should be used? Should neutral diffusivities remain the same as the skew diffusivities regardless of the neutral slope? How sensitive are simulations to the choices?

This section provides some discussion of these questions, with proposals for how to proceed. Most of the questions are physically based, although it is difficult to cleanly separate physical from numerical in steep neutral slope regions, since we know from the previous discussion in this chapter that numerical considerations are imposed in boundary regions. Given the research nature of these topics, healthy skepticism is suggested. Throughout the discussion, focus is placed on the surface boundary layer, although there is some comment on the bottom boundary layer.

15.3.1 General considerations

Surface and bottom boundary layers typically experience three-dimensional turbulent processes (for a review, see Large et al., 1994). Notably, the surface boundary layer is where strong diabatic processes associated with air-sea, ice-sea, and river-sea interactions occur. These physical considerations temper attempts to maintain the "purity" of numerical neutral diffusion in such regions, where indeed such purity is more difficult to maintain relative to the ocean interior. Additionally, Treguier et al. (1997) note that the lateral mixing effects from mesoscale eddies in the surface boundary layer should be parameterized via a *horizontal* diffusion, which is in fact diapycnal in regions where the neutral slopes are steeply sloped. These ideas suggest changing neutral diffusion in the surface boundary layer to horizontal diffusion.

Haine and Marshall (1998) point out that mesoscale eddies retain an adiabatic slumping effect in surface boundary layers. Similarly, Send and Marshall (1995) and Visbeck et al. (1996) cite similar effects in regions of deep convection. Maintaining nonzero GM in these regions may provide an effective parameterization of these processes. Hence, numerical methods for doing so should be considered.

These general considerations, and the discussion in this section, lead to the following proposals for how to handle neutral physics in the surface boundary region.

- For neutral diffusion, replace the small angle neutral flux components with horizontal diffusive flux components in steep neutral slope regions, and/or within some distance of boundaries.

- Maintain a nonzero GM skew flux in the surface boundary region where neutral slopes are large. Do so by linearly tapering the product $\kappa \mathbf{S}$ to zero over the depth of the boundary layer. This approach leads to a constant horizontal eddy-induced velocity throughout the boundary layer, and a vertical eddy-induced velocity that is linearly tapered to zero as the ocean surface is approached. Notably, one may wish to provide a vertical shear to the horizontal eddy-induced velocity. This remains a topic of research.

- In regions with shallow neutral slopes that are near the surface boundaries, taper the neutral flux components to parameterize the truncation of the vertical eddy transport due to the presence of a boundary.

- With nonzero GM fluxes in the boundary layer such as recommended above, it is essential to ensure that GM always acts on a stably stratified column. This may necessitate the use of a full convection scheme such as that from Rahmstorf (1993).

15.3.2 Surface boundary condition for fluxes from neutral physics

The tracer flux entering through the ocean surface arises from coupled processes, that is, air-sea, river-sea, or ice-sea interactions. The tracer flux entering through the ocean bottom is associated with geothermal and/or other geological effects. Both interfaces, however, have a zero tracer flux associated with neutral physics. This can be understood from quasi-Stokes transport ideas discussed in Section 9.3.5.1 (page 199). Analogously, one can understand these no-flux conditions as a statement of tracer conservation, whereby all boundary fluxes of tracer are associated with transport from or to other components in the climate system, and are represented by physical processes not part of the neutral physics schemes.

Figure 15.2 illustrates these points for a surface model grid cell. Notably, because there is always a zero vertical flux at the ocean surface from neutral physics processes, there is a nontrivial jump in the vertical flux across the surface grid cell. Hence, as the surface grid cell thickness is refined, the vertical divergence of the vertical neutral physics tracer flux generally increases across the cell.

This increased vertical gradient over the top cell announces the need for some careful tapering of the neutral flux as the surface is approached. This is important regardless of the neutral slope, but it is most apparent in simulations where neutral slopes are nearly vertical in the upper ocean, as within the boundary layer. In this case, tracer fluxes from neutral physical processes, from either neutral diffusion or GM skew diffusion, can be large and so can exhibit an especially large vertical gradient across the surface grid cell. In either case, a large surface-trapped divergence reflects a physically based need to taper the neutral processes toward zero as the boundary is approached. Consider the following reasons for doing so.

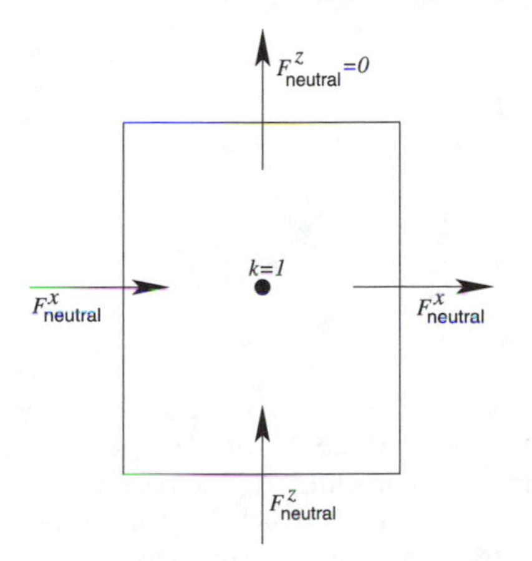

Figure 15.2 Details of how neutral physics is handled over the surface model grid cell. There is a zero neutral physics tracer flux crossing the top of this cell, yet there is generally a nonzero flux crossing the bottom. Horizontal fluxes cross the sides of the cell, as for interior cells away from boundaries. The inverse picture is realized for tracer cells next to the ocean bottom, where there is zero vertical tracer flux from neutral physics processes entering through the bottom face of the bottom cells.

15.3.3 A kinematic argument for tapering near a boundary

Recall from Section 9.3.6 that an approximation for the vertical displacement of a fluid parcel living on an undulating potential density surface is given by equation (9.55)

$$\xi = -\rho'/\partial_z \rho. \tag{15.16}$$

In this equation, ρ is the potential density field explicitly carried by the numerical model, and ρ' is an eddy fluctuation. Now assume* that the fluctuation ρ' is

*Thanks to Geoff Vallis (2002) for making this suggestion.

proportional to a mixing length times the horizontal density gradient $\nabla_z \rho$. For mesoscale eddying motions, Smith and Vallis (2002) argued that the first baroclinic Rossby radius λ_1 provides a useful mixing length. In this case, the vertical displacement takes the form

$$\begin{aligned} \xi &\sim -\lambda_1 \, |\nabla_z \rho / \partial_z \rho| \\ &= \lambda_1 |\mathbf{S}|, \end{aligned} \tag{15.17}$$

with $|\mathbf{S}|$ the magnitude of the neutral slope.

An analogous argument given in Appendix B of Large et al. (1997) and Section 8 of McDougall and McIntosh (2001) suggests that in the absence of a boundary, a typical neutrally directed displacement of a parcel is proportional to the first baroclinic Rossby radius λ_1. For typical neutral slopes less than $1/100$, this lateral displacement is roughly the same as the horizontal displacement, and the corresponding vertical displacement is given by $\xi \sim \lambda_1 \, |\mathbf{S}|$, which is the same result as derived above. Figure 15.3 illustrates this argument.

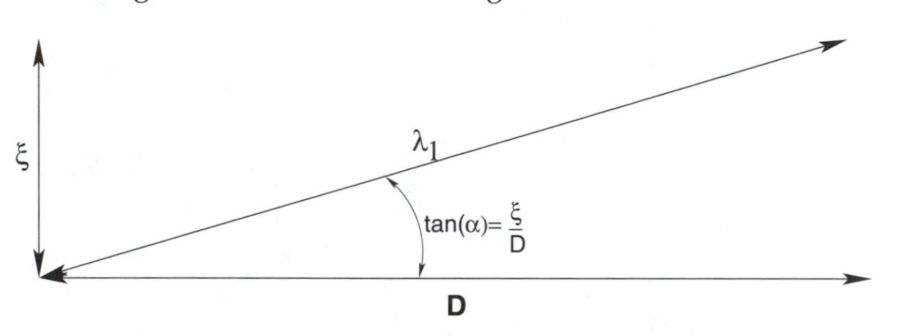

Figure 15.3 A fluid parcel within a mesoscale eddy undergoes both horizontal and vertical displacements. The scale of the displacement along neutral directions is on the order of the first baroclinic Rossby radius λ_1. Let the angle of the neutral direction α from the vertical be such that the neutral direction is shallow sloped: $|\mathbf{S}| = \tan\alpha << 1$. In this case, the corresponding vertical displacement is given by $\xi = D\,|\mathbf{S}| \approx \lambda_1 \, |\mathbf{S}|$.

When a fluid parcel nears a boundary, if the vertical displacement $\xi \sim \lambda_1 \, |\mathbf{S}|$ is greater than the distance to the boundary, the density surface associated with that eddy is truncated or clipped by the boundary (for discussions, see Treguier et al., 1997; Held and Schneider, 1999; McDougall and McIntosh, 2001). In turn, the quasi-Stokes transport associated with this eddy is truncated. Consequently, one is obliged to reflect this truncation in the parameterized neutral physics fluxes.

15.3.4 Summary and recommendations

There are two reasons for tapering neutral physics schemes: numerical and kinematic. Numerical constraints are discussed in Sections 15.1 and 15.2. They preclude unregularized neutral physical processes in regions of steep neutral slopes. Kinematic reasons for tapering near boundaries reflect the fact that undulating density surfaces can outcrop or incrop, at which point their effects on transport are truncated.

There are two general methods to taper neutral physical processes. First, one may taper toward zero all neutral physics processes as the boundary and/or steep neutral slope region is approached. This is the approach favored by Large et al. (1997) and McDougall and McIntosh (2001). Alternatively, one may taper to zero only those pieces of the neutral physics schemes that affect vertical transport. That is, one may continue to provide a horizontal downgradient diffusion and horizontal eddy-induced velocity transport. The vertical eddy-induced velocity must continue to be tapered toward zero as it reaches the boundary.

It is proposed here that it is appropriate to maintain a nontrivial parameterized horizontal advective and diffusive transport near boundaries and/or in steep neutral slope regions. Particular motivation comes from discussions in Treguier et al. (1997) and Held and Schneider (1999). They note that, although eddies within the surface boundary layer can no longer transport vertically across the boundary, when averaged over many realizations, their ensemble mean affects a horizontal transport. Additional arguments for maintaining a nontrivial horizontal eddy-induced transport in the surface boundary layer are provided by the studies of Send and Marshall (1995), Visbeck et al. (1996), and Haine and Marshall (1998).

15.3.4.1 Review of the mathematical form for the fluxes

For use in the following, recall from Section 14.2.2 that the horizontal and vertical flux components arising from small angle neutral diffusion, GM skewsion, plus vertical diffusion take the form

$$\begin{pmatrix} F^{(x)} \\ F^{(y)} \\ F^{(z)} \end{pmatrix} = \begin{pmatrix} A_I & 0 & (A_I - \kappa)\,S_x \\ 0 & A_I & (A_I - \kappa)\,S_y \\ (A_I + \kappa)\,S_x & (A_I + \kappa)\,S_y & A_D + A_I\,S^2 \end{pmatrix} \begin{pmatrix} C_{,x} \\ C_{,y} \\ C_{,z} \end{pmatrix}, \tag{15.18}$$

which leads to the untapered flux components

$$\mathbf{F}^{(h)} = -A_I\,\nabla_z C - (A_I - \kappa)\,\mathbf{S}\,C_{,z} \tag{15.19}$$

$$F^{(z)} = -(A_D + A_I\,S^2)\,C_{,z} - (A_I + \kappa)\,\mathbf{S}\cdot\nabla_z C. \tag{15.20}$$

In these equations, A_I is the neutral diffusivity, A_D is the vertical diffusivity, and κ is the GM diffusivity. The first terms in the respective flux components are termed "diagonal," whereas the second terms are "off-diagonal."

15.3.4.2 Suggestions for neutral diffusion

Here we propose a method for neutral diffusion. Skew diffusion will be discussed in Section 15.3.4.3. Following the discussion in Section 15.1, the recommended tapering method for steep neutral slope regions employs the exponential taper of Danabasoglu and McWilliams (1995). However, this taper should be employed selectively. For the horizontal flux components, the taper is applied only to the off-diagonal pieces of the neutral diffusive flux. No taper is applied to the horizontal diagonal diffusive terms, thus maintaining a nontrivial downgradient horizontal component to the flux, even in steep neutral slope regions. The diagonal vertical piece of the neutral diffusion operator is exponentially tapered, along with the off-diagonal pieces of the vertical flux, in order to remove all vertical flux contributions to neutral diffusion in steep slope regions. In this way, the diffusive flux components in steep neutral sloped regions reduce to those from horizontal-vertical diffusion,

$$|\mathbf{S}| >> S_{\max} \Rightarrow \begin{cases} \mathbf{F}^{(h)} &= -A_I \nabla_z C \\ F^{(z)} &= -A_D \, C_{,z}. \end{cases} \tag{15.21}$$

Figure 15.4 illustrates this approach. The integrity of the horizontal neutral diffusive flux components is "broken" in the steep sloped regions, thus compromising the neutral orientation of the flux. Yet, as argued in Section 15.3.1 and Treguier et al. (1997), neutral orientation of the diffusive fluxes is not necessarily relevant in such boundary layer regions. Notably, this approach does not lead to the numerical problems identified in Section 15.1.2 associated with slope clipping, since the difficulties there arose from a clipped form of the off-diagonal diffusion terms, which are here properly tapered to zero.

Additional tapering is suggested in the near-boundary regions according to the arguments in Section 15.3.3. That is, it is recommended that one taper in these regions even if the neutral slope is shallow. In particular, ensure that there is a finite vertical divergence across a surface or bottom boundary cell from neutral diffusive fluxes. Since this tapering is meant to continuously truncate the neutral processes as the boundary is reached, an exponential tapering is arguably too rapid. Instead, the nearly linear tapering suggested by Large et al. (1997) may be preferable. In this case, to handle the surface boundary, compute at each grid point the dimensionless ratio of depth of the grid point d to slope scaled Rossby radius

$$r = \frac{d}{\lambda_1 \, |\mathbf{S}|}, \tag{15.22}$$

and then compute the boundary tapered slope times diffusivity

$$A_I \, S \rightarrow (A_I \, S) \, T_{\text{sine}}, \tag{15.23}$$

where

$$T_{\text{sine}} = [1 + \sin \pi \, (r - 1/2)]/2 \tag{15.24}$$

defines a sine-taper function. Hence, in shallow sloped regions where $|\mathbf{S}| \leq S_{\max}$ and the depth is shallower than $\lambda_1 \, |\mathbf{S}|$, then the neutral flux components are truncated to

$$\begin{matrix} |\mathbf{S}| \leq S_{\max} \\ |z| < \lambda_1 \, |\mathbf{S}| \end{matrix} \Rightarrow \begin{cases} \mathbf{F}^{(h)} &= -A_I \nabla_z C - T_{\text{sine}} \, (A_I \, \mathbf{S}) \, C_{,z} \\ F^{(z)} &= -A_D \, C_{,z} - T_{\text{sine}} \, (A_I \, \mathbf{S}) \cdot (\mathbf{S} \, C_{,z} + \nabla_z C). \end{cases} \tag{15.25}$$

For simplicity in implementation, the first baroclinic Rossby radius is given by $\lambda_1 = c/f$, with $f = 2\,\Omega \, \sin\phi$ the Coriolis parameter. Large et al. (1997) suggested the value $c = 2\,\text{m}\,\text{s}^{-1}$ as an approximate first baroclinic gravity wave speed. To handle the singularity at the equator, Large et al. (1997) restrict λ_1 to the range $15\,\text{km} \leq \lambda_1 \leq 100\,\text{km}$. An example is illustrative. Let the slope over the upper ocean be the modest value $S = 1/1000$ and let $\lambda_1 = 100\,\text{km}$, thus yielding $\lambda_1 \, |\mathbf{S}| = 100\,\text{m}$. If the model grid cells are 10 m thick over this part of the ocean, then for the $k = 1$ grid cell, $r = 0.1$ and $(A_I \, S)_{k=1} \approx .025 \, (A_I \, S)$. For the $k = 9$ cell, $r = 0.9$ and $(A_I \, S)_{k=9} \approx .975 \, (A_I \, S)$.

In summary, the following tapering algorithms are proposed for neutral diffusion.

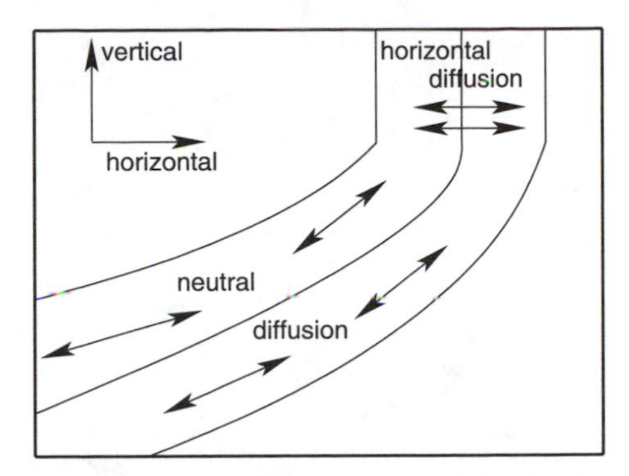

Figure 15.4 The recommended method for orienting the tracer fluxes from the neutral diffusion operator. In the ocean interior, where neutral slopes are modest, the neutral diffusion fluxes are oriented along neutral directions. In the regions where neutral slopes are steep, such as in the surface mixed layer, the diagonal horizontal pieces of the neutral diffusion operator are maintained, whereas the off-diagonal terms and vertical flux component are tapered to zero. Combined with the vertical diffusion operator, the resulting tracer diffusion is oriented in a horizontal-vertical manner in regions of steep neutral slopes.

- For all neutral slopes, regardless of the distance of the grid point from the ocean surface, the diagonal horizontal diffusive terms are maintained in an untapered manner.
- If the magnitude of the slope is greater than S_{max}, then the exponential taper (15.12) is applied to all off-diagonal terms of the neutral diffusion flux components, as well as the diagonal vertical term (the dianeutral diffusion term remains unaltered).
- If the magnitude of the slope is less than S_{max} yet the grid cell is shallower than $\lambda_1 |\mathbf{S}|$, than the sine taper (15.24) is applied to all except the horizontal diagonal terms.
- If the magnitude of the slope is less than S_{max} and the grid cell is deeper than $\lambda_1 |\mathbf{S}|$, then no tapering is applied.
- For the bottom, much remains to be studied. A reasonable and popular approach is to transition to an along-topography diffusion scheme such as that proposed by Beckmann and Döscher (1997) and Döscher and Beckmann (1999). More sophisticated bottom boundary layer schemes are at the cutting edge of current research, with comments and references provided in the Griffies et al. (2000a) review paper.

15.3.4.3 *Suggestions for GM skewsion*

As described by Send and Marshall (1995), Visbeck et al. (1996), and Haine and Marshall (1998), baroclinic eddies are active near regions of deep convection as

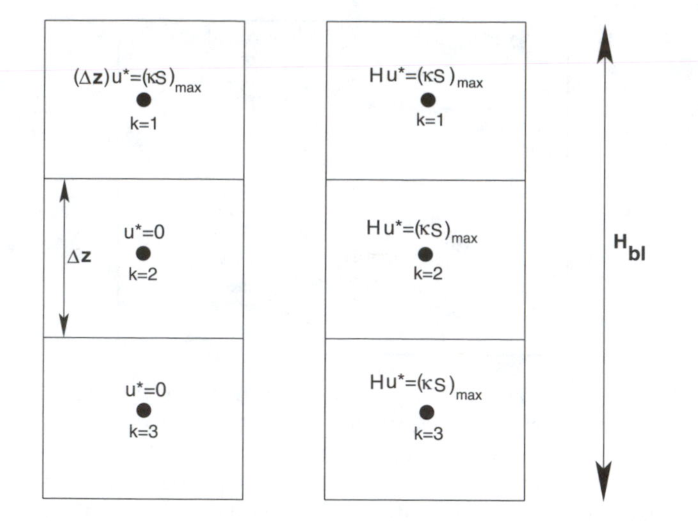

Figure 15.5 Horizontal eddy-induced velocity resulting from two different treatments of GM in the regions of steep neutral slopes. For simplicity, the vertical thickness of the cells is assumed to be the same. In both treatments, the depth integrated horizontal transport is $(\kappa S)_{\text{max}}$. The treatments differ in how they specify depths over which this transport occurs. In the left panel, the horizontal eddy induced velocity is nonzero only in the top cell, and so all the horizontal transport occurs in this one cell. This approach leads to an unbounded jump in the vertical shear of the eddy-induced velocity as the vertical grid resolution is refined. In the right panel, the same transport is evenly distributed throughout the boundary layer region. This option leads to a well-defined limit as the vertical grid spacing is refined, and provides for a constant horizontal eddy velocity throughout the boundary layer instead of a large velocity just in the surface cell.

well as more shallow surface boundary layers, both of which have steep neutral slopes. Consequently, mesoscale eddies provide a tendency to adiabatically flux density laterally and vertically as they extract the large store of available potential energy in the strongly baroclinic regions. One may thus choose to maintain a nontrivial GM flux in these regions to parameterize this process in coarse resolution models.

To motivate the form of this flux, focus here on issues critical to the surface boundary layer. In particular, consider a region of the upper ocean with vertical neutral directions, and ignore for now the near-boundary discussion in Section 15.3.3. If one sets the product of GM diffusivity and slope to its saturated value $(\kappa\,\mathbf{S})_{\text{max}}$ within a column of cells with steeply sloping neutral directions, as allowed by numerical stability (see Section 15.2), then there is zero horizontal eddy-induced velocity (equation (14.58)) since $\mathbf{u}^* = -\partial_z\,(\kappa\,\mathbf{S})_{\text{max}} = 0$. The vertical velocity component $w^* = \nabla_z \cdot (\kappa\,\mathbf{S})_{\text{max}}$ is vertically constant and generally nonzero.

However, since all vertical components of the neutral physics fluxes vanish at the ocean top (see Figure 15.2), as noted by Gent et al. (1995), the vertical eddy-induced velocity must transition from a vertically constant value at the bottom of the surface grid cell to zero at the grid cell's top. This rapid transition induces a large horizontal velocity $\mathbf{u}^*$ within the surface grid cell needed to satisfy continuity. As the thickness of the surface grid cell becomes smaller via grid cell refinement or free surface height undulations (see Figure 11.1 on page 224), the transition region becomes thinner and so induces a larger jump in eddy-induced velocity across the surface grid cell, ultimately becoming a delta function in the limit similar to that discussed by Bretherton (1966), Killworth (1997), and Killworth (2001). The left panel of Figure 15.5 illustrates this point.

A numerically robust approach capturing the horizontal eddy-induced transport allows the product $\kappa\,\mathbf{S}$ to linearly change from its saturated value $(\kappa\,\mathbf{S})_{max}$ at the base of the boundary layer, to zero at the top of the surface grid cell. This approach was suggested by Treguier et al. (1997) and Greatbatch and Li (2000). In this way, the horizontal eddy velocity $\mathbf{u}^*$ is vertically constant throughout the boundary layer with a magnitude inversely proportional to the boundary layer depth. The vertical velocity w^* also smoothly goes to zero at the surface.* The right panel of Figure 15.5 illustrates this choice. Mathematically, this choice leads to the GM flux in steeply sloped regions

$$|\mathbf{S}| >> S_{max} \Rightarrow \begin{cases} \mathbf{F}^{(h)} & = (\kappa\,\mathbf{S})_{\text{linear taper}}\, C_{,z} \\ F^{(z)} & = -(\kappa\,\mathbf{S})_{\text{linear taper}} \cdot \nabla_z C. \end{cases} \tag{15.26}$$

The subscript "linear taper" symbolizes linear tapering discussed above. Note that because of the maximum value of $(\kappa\,\mathbf{S})$ at the base of the boundary layer, it is here that restratification tendencies from GM are strongest. Finally, the vector character of $(\kappa\,\mathbf{S})_{\text{linear taper}}$ signifies that its sign is taken from the sign of the actual slope. A regularization method for computing slopes whose values approach infinity is given in Section 15.2. Figure 15.6 illustrates the tendency of the corresponding eddy-induced streamfunction.

Consider now the notions discussed in Section 15.3.3. For shallow sloped regions within a distance $\lambda_1\,|\mathbf{S}|$ of the ocean surface, yet outside the regions where the slopes are steeper than S_{max}, one may truncate the GM flux components according to

$$\begin{matrix} |\mathbf{S}| < S_{max} \\ |z| < \lambda_1\,|\mathbf{S}| \end{matrix} \Rightarrow \begin{cases} \mathbf{F}^{(h)} & = (\kappa\,\mathbf{S})\, T_{\text{sine}}\, C_{,z} \\ F^{(z)} & = -(\kappa\,\mathbf{S})\, T_{\text{sine}} \cdot \nabla_z C, \end{cases} \tag{15.27}$$

where T_{sine} is the sine-taper function defined by equation (15.24).

In summary, the following tapering algorithm for GM skewsion is proposed:

- If a grid point is within the surface boundary layer, defined as that region starting at the ocean surface where neutral slopes are steeper than S_{max}, then let $\kappa\,\mathbf{S}$ linearly move from its saturated value $(\kappa\,\mathbf{S})_{max}$ at the base of the boundary layer, to zero at the top of the surface grid cell. This approach ensures a vertically constant horizontal eddy-induced velocity within the boundary layer with a magnitude proportional to the depth of the boundary layer, and a vertical eddy-induced velocity that linearly decays to zero at the ocean surface.

*A method which leads to nonzero vertical shear for $\mathbf{u}^*$ within the boundary layer has been suggested by Young (1994). This method has yet to be documented in ocean climate models.

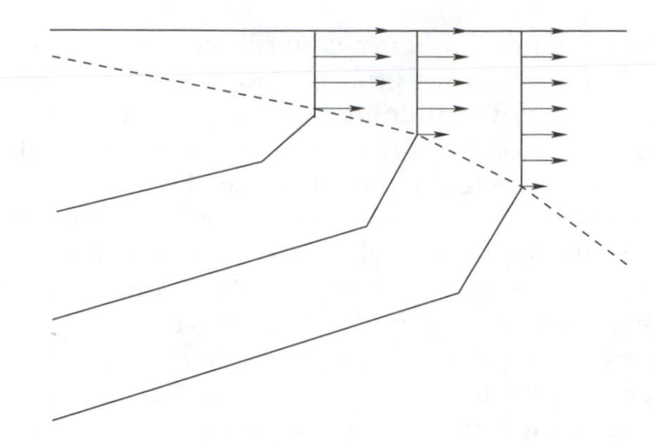

Figure 15.6 Schematic where isopycnals steepen as they enter the upper planetary boundary layer (region above the dashed line). In this layer, the horizontal GM flux is nonzero and oriented in a manner to maintain a nontrivial horizontal eddy-induced velocity (arrows), which moves light water over heavy water, thus slumping the isopycnals and reducing baroclinicity. This slumping effect is achieved via the nonzero vertical shear in the horizontal eddy-induced velocity present at the base of the mixed layer. As shown in Figure 15.5, the strength of the horizontal eddy-induced velocity is inversely proportional to the boundary layer depth. A more sophisticated parameterization may include vertical shear within the planetary boundary layer in a manner to increase the tendency to restratify. Note that the vertical component to the eddy-induced velocity, which vanishes at the ocean surface, is not shown here.

- If a grid point is outside the boundary layer, which means the slope is shallow ($|\mathbf{S}| < S_{\max}$), yet the grid point is shallower than the penetration depth $\lambda_1 |\mathbf{S}|$, then the sine-taper function (15.24) is applied to all pieces of the GM flux.
- If the slope at a grid point is shallow ($|\mathbf{S}| < S_{\max}$) and the depth is greater than $\lambda_1 |\mathbf{S}|$, then no tapering is applied.

By retaining a nonzero eddy-induced velocity for all neutral slopes, it is essential that there be no regions where the simulation realizes an imaginary buoyancy frequency N. Such regions would be acted upon by the GM scheme in a manner analogous to how it acts on any stably stratified baroclinic zone, that is, to flatten isopycnals. However, if the isopycnals are slightly bent over on themselves so that $N^2 < 0$, then GM will further increase the vertically unstable profile; that is, it will make N^2 even more negative. If this process is followed by enhanced vertical mixing from the mixed layer scheme, and this mixing again only partially removes the $N^2 < 0$ water parcels, then a rather unphysical interaction between the mixed layer physics and GM ensues, with the result being unphysically deep, and choppy, boundary layer solutions. Using a full convective adjustment scheme such as Rahmstorf (1993) ensures that no profile will have an imaginary buoyancy frequency, and this eliminates the problem.

Note that there have in the past been problems with convective adjustment schemes, which were largely overcome by employing a large vertical diffusivity time stepped implicitly. However, when combined with a large vertical diffusivity, the need to employ convective adjustment is reduced, and so, it is conjectured, are the problems encountered in models with only convective adjustment to stabilize the profiles.

15.4 CHAPTER SUMMARY AND CAVEATS

The issues of what to do with the mesoscale eddy parameterization schemes next to boundaries remains an active area of research. Therefore, nearly everything in this chapter, except perhaps the linear stability analysis presented in Sections 15.1 and 15.2, should be questioned. Sensitivity of the solutions to the choices can be large, since details often affect the interactions between the ocean and other component models. Historically, these issues were first addressed by those aiming to implement the neutral physics schemes in global climate models, with numerical considerations as the main guide. The early approaches were crude. With added modeling experience, and now some focus by the theoretical and process communities, there is hope that the schemes will become more rational and robust.

The basic question that we asked is what to do to the various pieces of the diffusion and skew diffusion tensor near boundaries, with focus here on the surface boundary. In general, it was proposed that one maintains horizontal downgradient diffusion within the mixed layer. The reason is that lateral stirring effects of mesoscale eddies can be significant here, and they stir across the vertical isopycnals in this region. Numerically, maintaining horizontal diffusion provides for a useful level of diapycnal mixing in a region where the monotonicity of the numerical neutral physics schemes can be stressed (see Section 13.7). For the Gent et al. (1995) scheme, it was proposed that one maintain a nonzero eddy-induced velocity within the mixed layer, with zero vertical shear. This is realized by linearly tapering the product of the diffusivity times neutral slope to zero as the surface is reached.* Maintaining both horizontal diffusion and a nonzero eddy velocity is nontraditional, with most modeling experience based on removing all pieces of the mixing tensor. Consequently, it is important to highlight two points to keep in mind when maintaining a nonzero eddy velocity within the mixed layer.

- The strength of the horizontal eddy-induced velocity in the mixed layer is directly proportional to the maximum slope S_{max}. For example, if the GM diffusivity is $10^3\,\mathrm{m^2\,s^{-1}}$, $S_{max} = 1/100$, and $\Delta z = 10$ m is the vertical grid spacing, then $|u^*| \sim 1\,\mathrm{m\,s^{-1}}$. If this velocity is uniformly distributed over the mixed layer, as proposed in this method, then for a 100 m deep mixed layer in a model with 100 km horizontal grid spacing, then 100 Sv of horizontal eddy-induced transport will arise from this prescription. This is a huge number. Hence, this approach should be implemented with some care. In particular, practical ways to modulate the induced transport are by reducing the diffusivity, as occurs in many parts of the ocean when using a flow-dependent scheme such as one of those discussed in Section 14.4, or by reducing S_{max}. The sensitivity of this approach to these parameters is of

*More generally, we taper the quasi-Stokes streamfunction to zero as the surface is reached. See Section 9.3.5.1 starting on page 199 for discussions of the quasi-Stokes streamfunction.

some concern, and so the details likely require plenty of tests.

- As discussed in Section 15.3.4.3, it is essential to maintain a full convective adjustment scheme when allowing for a nonzero eddy-induced velocity in regions of vertical neutral directions. Otherwise, an unphysical interaction between the mixed layer scheme and GM will ensue, with overly deep and choppy simulated mixed layers.

Much of the material in this chapter has taken the form of a handbook, with recommendations based on some personal experience. Indeed, this material was written for those actually implementing and testing neutral physics schemes in realistic models. It is hoped that by exposing the issues here, the reader can better understand what is the state-of-the-art in z-model implementations of these schemes.

Chapter Sixteen

FUNCTIONAL DISCRETIZATION OF NEUTRAL PHYSICS

The purpose of this chapter is to derive the discrete small angle neutral diffusion operator. This discretization is also of use for discrete Gent and McWilliams (1990) skewsion according to Griffies (1998). As all details focus on fields defined at the tracer point, the discussion is generic to z-models with arbitrary horizontal grid arrangements. The material in this chapter documents the algorithm discretized in the Modular Ocean Model of Griffies et al. (2004).

Before getting started, it is interesting to highlight a mathematical connection between this operator and the friction operator discussed in Chapter 19. Namely, both operators are self-adjoint elliptic operators. Hence, they correspond to the functional derivative of a sign-definite functional. Having a functional formulation provides for a "higher principle" that can be used when discretizing these operators. This principle is akin to an energy principle which is often used to guide the discretization of the inviscid part of the model equations. For the neutral physics and friction operators, the principle states that the neutral diffusion and friction operators should dissipate, respectively, variance and kinetic energy for all admitted flow features. Furthermore, the skew diffusion operator, which corresponds to a stirring process, should keep variance unchanged. This principle helps organize the many steps encountered when deriving the discrete operators, and thus it aids in a quest to remove arbitrary steps often made when casting the continuum equations on the lattice.

16.1 FOUNDATIONS FOR DISCRETE NEUTRAL PHYSICS

The purpose of this section is to set the foundations for discretization of the neutral diffusion and skew diffusion operators.

16.1.1 The diffusion dissipation functional

The diffusion operator, when acting on a passive tracer, is a linear self-adjoint operator. Consequently, it has an associated negative semidefinite functional (e.g., Courant and Hilbert, 1953, 1962). For example, the Laplacian operator $\nabla^2 C$ is identified with the functional derivative $\nabla^2 C = \delta\mathcal{F}/\delta C$, where

$$\mathcal{F} \equiv -(1/2)\int |\nabla C|^2 \mathrm{d}V \qquad (16.1)$$

is the associated functional. We now derive this result for a general diffusion tensor.

To motivate the introduction of the *diffusion dissipation functional*, recall the discussion in Section 13.4.1, where the global tracer variance

$$\mathcal{V} = M^{-1} \int (\rho \, \mathrm{d}V) \, C^2 - \left(M^{-1} \int (\rho \, \mathrm{d}V) \, C \right)^2 \tag{16.2}$$

was shown to evolve according to

$$M \, \mathcal{V}_{,t} = 2 \, \rho_o \int \mathrm{d}V \, \nabla C \cdot \mathbf{F} \tag{16.3}$$

in the case where the total ocean seawater mass $M = \int \rho \, \mathrm{d}V$ and total tracer content $\int T \, \rho \, \mathrm{d}V$ are held fixed by removing all boundary forcing.* Interest here is on downgradient diffusion. A diffusive flux can generally be written

$$F^m = -K^{mn} \, C_{,n} \tag{16.4}$$

with K^{mn} a symmetric and positive semidefinite diffusion tensor.† For a diffusion tensor, the integrand in equation (16.3) takes the form of a negative semidefinite quadratic form

$$\nabla C \cdot \mathbf{F} = -C_{,m} \, K^{mn} \, C_{,n} \leq 0. \tag{16.5}$$

Now define the *diffusion dissipation functional* as

$$\mathcal{F} = \int \mathrm{d}V \, \mathcal{L}, \tag{16.6}$$

where the nonpositive integrand is given by

$$\begin{aligned} 2 \, \mathcal{L} &= \mathbf{F} \cdot \nabla C \\ &= -C_{,m} \, K^{mn} \, C_{,n}. \end{aligned} \tag{16.7}$$

As defined, the relation between the dissipation functional and the evolution of global tracer variance is given by

$$\mathcal{F} = \left(\frac{M}{4 \, \rho_o} \right) \mathcal{V}_{,t}. \tag{16.8}$$

Hence, properties of a discrete realization of the functional $\mathcal{F}$ are shared by evolution of the discrete tracer variance.

16.1.2 Functional derivative of the dissipation functional

The diffusion operator is defined by

$$R(C) = -\nabla \cdot \mathbf{F}(C). \tag{16.9}$$

This operator is connected to the functional derivative of $\mathcal{F}$. To show this result, consider functional variations of the tracer field δC, thus leading to variations in the functional given by‡

$$\delta \mathcal{F} = \int \mathrm{d}\mathbf{x} \left(\frac{\delta \mathcal{L}}{\delta C} \, \delta C + \frac{\delta \mathcal{L}}{\delta C_{,m}} \, \delta C_{,m} \right). \tag{16.10}$$

---

*Equation (16.3) is the same as equation (13.19), but without the boundary terms.

†Recall the summation convention, where repeated indices are summed over their range, as well as the notation $C_{,n} = \partial_n \, C$, which is shorthand for partial derivative.

‡To keep the notation tidy, use $\mathrm{d}V = \mathrm{d}\mathbf{x}$ for the three-dimensional volume element.

Integration by parts leads to

$$\delta\mathcal{F} = \int \mathrm{d}\mathbf{x} \left[\frac{\delta\mathcal{L}}{\delta C}\, \delta C + \partial_m \left(\frac{\delta\mathcal{L}}{\delta C_{,m}}\, \delta C \right) - \partial_m \left(\frac{\delta\mathcal{L}}{\delta C_{,m}} \right) \delta C \right]. \tag{16.11}$$

The second term is a total derivative that integrates to a surface term. Concern for the moment is not with boundary forcing. Hence, it is sensible to assume that the tracer field satisfies *natural boundary conditions*, whereby

$$\hat{\mathbf{n}} \cdot \frac{\delta\mathcal{L}}{\delta\nabla C} = 0, \tag{16.12}$$

with $\hat{\mathbf{n}}$ an outward normal at boundaries. Doing so allows for the functional variation to be written

$$\delta\mathcal{F} = \int \mathrm{d}\mathbf{x} \left[\frac{\delta\mathcal{L}}{\delta C} - \partial_m \left(\frac{\delta\mathcal{L}}{\delta C_{,m}} \right) \right] \delta C. \tag{16.13}$$

Consequently, the functional derivative is given by

$$(\mathrm{d}\mathbf{y})^{-1}\, \frac{\delta\mathcal{F}}{\delta C(\mathbf{y})} = \frac{\delta\mathcal{L}}{\delta C} - \partial_m \left(\frac{\delta\mathcal{L}}{\delta C_{,m}} \right), \tag{16.14}$$

where $\mathrm{d}\mathbf{y}$ is the volume element at the field point $\mathbf{y}$. To reach the last step, it was necessary to use the identity

$$\frac{\delta C(\mathbf{x})}{\delta C(\mathbf{y})} = \mathrm{d}\mathbf{y}\, \delta(\mathbf{x} - \mathbf{y}), \tag{16.15}$$

where $\delta(\mathbf{x} - \mathbf{y})$ is the Dirac delta function satisfying

$$\int \mathrm{d}\mathbf{y}\, \delta(\mathbf{x} - \mathbf{y}) = 1, \tag{16.16}$$

so long as the integration range includes the singular point $\mathbf{x} = \mathbf{y}$. Note that, by definition, the delta function has physical dimensions of inverse volume L^{-3}.

Now reintroduce the specific form of the diffusion integrand

$$2\,\mathcal{L} = -C_{,m}\, K^{mn}\, C_{,n} \tag{16.17}$$

thus leading to

$$\begin{aligned} (\mathrm{d}\mathbf{y})^{-1}\, \frac{\delta\mathcal{F}}{\delta C(\mathbf{y})} &= -\partial_m \left(\frac{\delta\mathcal{L}}{\delta C_{,m}} \right) \\ &= -\partial_m \,(K^{mn}\, C_{,n}). \end{aligned} \tag{16.18}$$

The second equality identifies the diffusion operator, thus yielding the general result

$$(\mathrm{d}\mathbf{y})^{-1}\, \frac{\delta\mathcal{F}}{\delta C(\mathbf{y})} = R(C(\mathbf{y})). \tag{16.19}$$

Natural boundary conditions are maintained by tracer fluxes satisfying the usual no-normal flux boundary condition

$$\hat{\mathbf{n}} \cdot \mathbf{F} = 0, \tag{16.20}$$

with

$$\frac{\delta\mathcal{L}}{\delta\nabla C} = \mathbf{F}(C). \tag{16.21}$$

Thus completes the connection between the dissipation functional, the diffusion fluxes, and the diffusion operator.

16.1.3 General approach on the discrete lattice

On the discrete lattice, not every consistent* discrete diffusion operator corresponds to a negative semidefinite discrete functional. Therefore, a consistent numerical diffusion operator does not necessarily possess the dissipative properties of the continuum operator. For the Laplacian operator in an isotropic medium, it is trivial to produce a dissipative numerical operator without any special apparatus.† In the anisotropic case, such as neutral diffusion, it is nontrivial. Indeed, the discretization of neutral diffusion proposed by Cox (1987) is numerically consistent but not always dissipative. Griffies et al. (1998) provide examples. There is hence some motivation to consider an approach whereby the discrete version of the dissipation functional is derived, then the discrete functional derivative is taken in order to produce the discrete diffusive fluxes. This approach ensures that the discretized diffusion operator dissipates tracer variance.

16.1.4 Concerning neutral directions

Within the functional framework, provision is made for a discretization of the diffusive fluxes that are aligned according to a *self-consistent* approximation to the neutral directions. Self-consistency means that there is a zero neutral diffusive flux of locally referenced potential density. As shown in Section 14.1.6, a zero neutral flux of locally referenced potential density implies a balance between the neutral direction diffusive flux of potential temperature and salinity

$$\alpha \mathbf{F}(\theta) = \beta \mathbf{F}(s), \tag{16.22}$$

where

$$\alpha = -\left(\frac{\partial \ln \rho}{\partial \theta}\right)_{s,p} \tag{16.23}$$

and

$$\beta = \left(\frac{\partial \ln \rho}{\partial s}\right)_{\theta,p} \tag{16.24}$$

are the thermal expansion and saline contraction coefficients. In order to ensure this balance in the discrete equations, it is sufficient to compute the density gradients in terms of the active tracer gradients and the thermal and saline coefficients. Special care must be taken when choosing the reference points for evaluating these gradients, with details given in Section 16.5.2. Without a self-consistent discretization that guarantees a zero flux of locally referenced potential density, the neutral diffusion operator can produce grid noise even if it ensures that variance does not increase (Griffies et al., 1998).

16.1.5 How to formally handle active and passive tracers

When diffusing passive tracers, the neutral diffusion operator is linear since the diffusion tensor is independent of the passive tracer fields. However, when diffusing potential temperature and salinity, which are active tracers, the neutral diffusion operator is nonlinear since the diffusion tensor is itself a function of the

---

*Consistent here means that the discretization reduces to the correct continuum operator as the grid size goes to zero.

†This *5-point* operator is discussed in Section 19.2 when introducing discrete friction operators.

active tracers. Since the functional formalism assumes the diffusion operator to be a linear self-adjoint operator, one may question the validity of using this formalism for the active tracers.

However, there is no problem applying the formalism since the diffusion operator is structurally the same for both the active and passive tracers. What is done is to use the functional machinery assuming the tracer is passive. When it comes time to discretize the neutral directions, the understanding of how to properly align the slopes so that the diffusion operator is self-consistent (i.e., so that it does not flux locally referenced potential density along the neutral directions) is incorporated.

16.1.6 Dissipation functional for neutral diffusion

We now record the dissipative functional in the continuum for neutral diffusion for the full Redi (1982) diffusion tensor (Section 14.1.3) and the small slope approximation (Section 14.1.4). Note that only the small slope diffusion tensor is discretized in z-models, as the full slope tensor is more expensive and not justified physically (Chapter 15).

As discussed in Sections 13.6.3 and 14.1, it is possible to write the full Redi diffusion tensor in the projection operator form

$$\begin{aligned} K^{mn} &= A_I(\delta^{mn} - \hat{\rho}^m \hat{\rho}^n) + A_D \hat{\rho}^m \hat{\rho}^n \\ &= (A_I - A_D)(\delta^{mn} - \hat{\rho}^m \hat{\rho}^n) + A_D \delta^{mn}, \end{aligned} \tag{16.25}$$

with δ^{mn} the Kronecker delta and

$$\begin{aligned} \hat{\boldsymbol{\rho}} &= \frac{\nabla \rho}{|\nabla \rho|} \\ &= \frac{-\alpha \, \nabla \theta + \beta \, \nabla s}{|-\alpha \, \nabla \theta + \beta \, \nabla s|} \end{aligned} \tag{16.26}$$

the unit vector normal to the neutral direction. The diffusivities A_I and A_D are nonnegative and can in general be functions of space-time. Given this diffusion tensor, the dissipation functional for neutral diffusion is

$$2\,\mathcal{F} = -\int \mathrm{d}V \, (A_I - A_D) \frac{|\nabla \rho \wedge \nabla C|^2}{|\nabla \rho|^2} - \int \mathrm{d}V \, A_D |\nabla C|^2. \tag{16.27}$$

Hence, the diffusive flux is

$$\begin{aligned} \mathbf{F} &= \frac{\delta \mathcal{L}}{\delta \, \nabla C} \\ &= -(A_I - A_D)(\hat{\boldsymbol{\rho}} \wedge \nabla C) \wedge \hat{\boldsymbol{\rho}} - A_D \nabla C. \end{aligned} \tag{16.28}$$

The first term in the flux vanishes when the tracer isosurfaces are parallel to the neutral directions. For most oceanographic cases, the difference $A_I - A_D \approx A_I$ to many orders of accuracy. Note the presence of $A_D C_{,x}$ and $A_D C_{,y}$ in the horizontal flux components. In the case of very steep neutral directions, these terms become relevant.

The small slope approximation amounts to taking the limit of $|\rho_{,x}|, |\rho_{,y}| << |\rho_{,z}|$, and dropping terms of order $slope * (A_D/A_I)$, with *slope* the small neutral slope. The resulting functional is

$$\begin{aligned} 2\,\mathcal{F}^{\text{small}} &= -\int \mathrm{d}V\, A_I \frac{(C_{,x}\rho_{,z} - C_{,z}\rho_{,x})^2 + (C_{,y}\rho_{,z} - C_{,z}\rho_{,y})^2}{\rho_{,z}^2} - \int \mathrm{d}V\, A_D (C_{,z})^2 \\ &= -\int \mathrm{d}V\, A_I\, (\nabla_z C + \mathbf{S}\, C_{,z})^2 - \int \mathrm{d}V\, A_D (C_{,z})^2 \\ &= -\int \mathrm{d}V\, A_I\, (\nabla_\rho C)^2 - \int \mathrm{d}V\, A_D (C_{,z})^2 \end{aligned} \tag{16.29}$$

where $\mathbf{S}$ is the neutral slope vector. The corresponding diffusive flux is given by

$$\begin{aligned} \mathbf{F} &= \frac{\delta \mathcal{L}}{\delta \nabla C} \\ &= -\left(\frac{A_I}{\rho_{,z}}\right) (\nabla\rho \wedge \nabla C) \wedge \hat{\mathbf{z}} - \hat{\mathbf{z}}\, A_D\, C_{,z} \end{aligned} \tag{16.30}$$

which has horizontal component

$$\begin{aligned} \mathbf{F}^{(h)} &= -A_I\, (\nabla_z C + \mathbf{S} C_{,z}) \\ &= -A_I\, \nabla_\rho C, \end{aligned} \tag{16.31}$$

and vertical component

$$\begin{aligned} F^z &= \mathbf{S} \cdot \mathbf{F}^{(h)} - A_D C_{,z} \\ &= -A_I (\mathbf{S} \cdot \nabla_z C + S^2\, C_{,z}) - A_D C_{,z}. \end{aligned} \tag{16.32}$$

As for the full tensor, the A_I term in these expressions vanishes when a tracer isosurface parallels the neutral direction.

16.2 INTRODUCTION TO THE DISCRETIZATION

The remainder of this chapter focuses on the small angle neutral diffusion operator and the associated skew fluxes arising from Gent and McWilliams (1990) (GM in the following). Pacanowski and Griffies (1999) also presented a discretization of the full neutral diffusion tensor. For reasons stated earlier (Section 14.1.4), the small angle diffusion tensor is sufficient for ocean climate modeling purposes with z-models, and so discretization of the full tensor of Redi (1982) is not discussed here.

16.2.1 Summary and Caveats

As documented by Griffies et al. (1998), the discrete neutral diffusion operator described in this chapter satisfies the following properties.

1. It produces a zero horizontal flux of locally referenced potential density as embodied by the balance given in equation (14.32). The Cox (1987) scheme did not respect this property, and this problem led to the nonlinear numerical instability of that scheme.

2. It reduces tracer variance, and produces downgradient oriented tracer fluxes along neutral directions when considering a particular finite volume region of the discrete lattice (see Appendix D of Griffies et al., 1998).

3. It computes a second order approximation to the neutral slopes.

4. It requires zero background horizontal diffusion to remain numerically stable; the Cox (1987) scheme is unstable without this diffusion.

Some limitations of the discrete neutral diffusion operator include the following.

1. Because part of the vertical flux is computed explicitly in time and part implicitly, the balance between the vertical component of the neutral diffusive salinity and temperature fluxes (equation (14.32)) is slightly broken. This problem has proven to be minor in realistic settings, though it is useful to keep in mind. See Section 16.8.3 for more discussion.

2. Because the discretization guarantees downgradient oriented tracer fluxes only over regions larger than tracer cells, tracers within a cell can move outside their physically constrained range. Consequently, the scheme is not monotonic in general. This problem appears most pernicious for passive tracers in steep sloped regions. The temperature and salinity, being constrained by the need to produce a zero flux of locally referenced potential density, appear less problematic, though they too can move outside physical bounds. In order to deal with the pure neutral diffusion problem for all slopes, it appears necessary to employ a positive definite neutral diffusion scheme. Proposals are discussed by Beckers et al. (1998) and Beckers et al. (2000). However, as argued in Chapter 15 (see especially Section 15.3, page 333), it is physically justifiable to apply horizontal fluxes in regions where neutral slopes are steep. In this case, the numerical issues appearing in such regions become less problematic.

3. The discrete neutral diffusion scheme requires no horizontal background diffusion, in contrast to the case with the unstable Cox (1987) scheme. Unfortunately, removing horizontal diffusion exposes the simulations to problems with dispersive advection schemes. These problems are often apparent next to topography in regions where the tracer isolines are aligned with neutral directions, in which case the diffusive fluxes are weak. This "Peclet grid noise" problem is fundamental to dispersive advection schemes, and it prompts the use of horizontal diffusion in boundary layers. Another means for removing these problems is to use the sigma diffusion scheme of Beckmann and Döscher (1997) and Döscher and Beckmann (1999), which is a rudimentary scheme aiming to move dense water down slopes.

16.2.2 Notation and conventions

The presentation given here introduces grid labels on fields living on a discrete lattice. In particular, discrete tracer fields in this chapter are denoted by the symbol $T_{i,j,k}$, with i, j, k the discrete labels for points in the x, y, z directions, respectively. Note that for the remainder of this chapter alone, we switch from the letter C for tracer concentration to the letter T, for consistency with treatment of this subject in the literature (e.g., Griffies et al., 1998; Griffies, 1998; Pacanowski and Griffies, 1999). Other notation, such as the grid factors, is based on that used in the Modular

Ocean Model (MOM4) of Griffies et al. (2004). Indeed, the material in this chapter documents the algorithm discretized in MOM4.

Explicit sums and averages are used in most places in this chapter, rather than choosing among the more elegant alternative notations. A more explicit notation is chosen since it leads to less ambiguity for the reader. Furthermore, explicit formulae are ultimately required by the numericist when translating an algorithm into computer code. Experience has shown that having the algorithm written nearly in its "full glory" in LaTeX reduces the chance for bugs to penetrate the computer code. This is especially crucial for complicated schemes such as the one documented in this chapter. Hence, for pedogogical and "quality control" reasons, the explicit grid label notation is preferred.

16.2.3 General procedure for discretization

Motivated from the discussion in Section 16.1.1, discretization of the diffusion operator at a particular grid point is derived from the functional derivative of the discretized neutral diffusion functional $\mathcal{F}[T]$. On a discrete lattice, the functional derivative becomes a partial derivative

$$R(T)_{i,j,k} = \frac{1}{V_{T_{i,j,k}}} \frac{\partial \mathcal{F}[T_{i,j,k}]}{\partial T_{i,j,k}}, \tag{16.33}$$

where

$$V_{T_{i,j,k}} = dxt_{i,j}\, dyt_{i,j}\, dht_{i,j,k} \tag{16.34}$$

is the volume of the tracer cell (T-cell). This result corresponds to a discrete version of the continuum relation given by equation (16.19).

The general procedure for discretization is to identify those pieces of the discretized functional that contain contributions from the discretized tracer value $T_{i,j,k}$. These are the pieces possessing nonzero functional derivatives, and so contribute to the diffusion operator for this T-cell. An enumeration of these pieces depends on the particular discretization of the functional. Most models choose second order numerics. Higher order numerics are not motivated for an operator aiming to dissipate via mixing. One overriding principle used to guide this discretization is to recover the familiar 5-point discretization (Section 19.2.1) of the Laplacian in the case of zero neutral slopes, where neutral diffusion reduces to horizontal diffusion.

All details about the discretization are made at the level of the functional. These details include the particular grid choice and the choice for reference points to be used in approximating the neutral directions.

16.3 A ONE-DIMENSIONAL WARM-UP

In order to illustrate the general framework provided by the functional approach, consider the case of one-dimensional diffusion along the zonal direction. Neutral diffusion does not exist in one dimension, so this example cannot illustrate issues related to discretization of the neutral directions. The issues of partial vertical cells are also avoided here and will be addressed later.

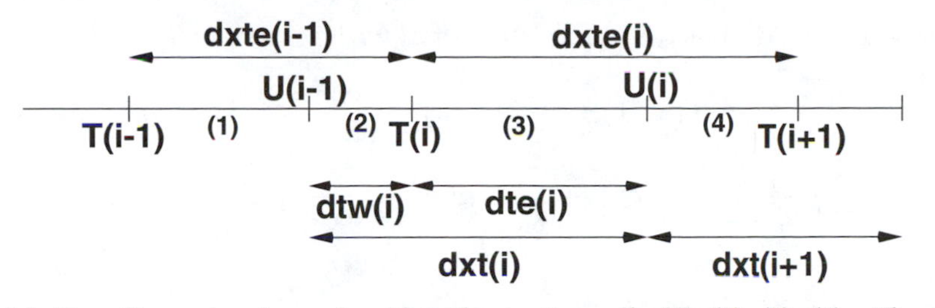

Figure 16.1 One-dimensional zonal grid with quarter cells (1), (2), (3), (4). The quarter cells are bounded by their adjacent velocity and tracer points: quarter cell (1) is bounded by T_{i-1} and U_{i-1}, quarter cell (2) by U_{i-1} and T_i, quarter cell (3) by T_i and U_i, and quarter cell (4) by U_i and T_{i+1}. The dimension of a tracer cell is dxt_i, velocity points live on the boundary of tracer cells, the distance between tracer points T_i and T_{i+1} is $dxte_i$, the distance from the tracer point T_i to the western boundary of the tracer cell is dtw_i, and the distance to the eastern boundary is dte_i.

Figure 16.1 shows the zonal grid. The grid is partitioned into four subcells called *quarter cells* in the following, with each possessing a generally different diffusivity $A(n)$ and length $V(n)$ (generalized to volume in the three-dimensional case). The quarter cells each contain a contribution from the central tracer point T_i upon discretizing the functional with second order numerics. That is, consider a second order discretization of the functional

$$\mathcal{F} = -\frac{1}{2}\sum_i \sum_n A(n)V(n)(\delta_x T(n))^2, \tag{16.35}$$

where the n sum is over the subcells relevant for each T-cell. The four terms containing a contribution from T_i are given by

$$\begin{aligned} 2\mathcal{F}[T_i] = -A(1)V(1)(\delta_x T_{i-1})^2 - A(2)V(2)(\delta_x T_{i-1})^2 \\ - A(3)V(3)(\delta_x T_i)^2 - A(4)V(4)(\delta_x T_i)^2. \end{aligned} \tag{16.36}$$

Note that the tracer derivative is assumed to be the same for the two subcells 1 and 2 and the two subcells 3 and 4. The "volumes" of the subcells are

$$V(1) = dte_{i-1} \tag{16.37}$$
$$V(2) = dtw_i \tag{16.38}$$
$$V(3) = dte_i \tag{16.39}$$
$$V(4) = dtw_{i+1} \tag{16.40}$$

and the finite difference derivatives are

$$\delta_x T_{i-1} = \frac{T_i - T_{i-1}}{dxte_{i-1}} \tag{16.41}$$

$$\delta_x T_i = \frac{T_{i+1} - T_i}{dxte_i}. \tag{16.42}$$

Taking the partial derivative of the functional with respect to T_i leads to

$$\begin{aligned} -2\frac{\partial \mathcal{F}}{\partial T_i} &= \frac{\delta_x T_{i-1}}{dxte_{i-1}}\,(A(1)\,V(1)+A(2)\,V(2)) - \frac{\delta_x T_i}{dxte_i}\,(A(3)\,V(3)+A(4)\,V(4)) \\ &= \frac{\delta_x T_{i-1}}{dxte_{i-1}}\,(A(1)\,dte_{i-1}+A(2)\,dtw_i) - \frac{\delta_x T_i}{dxte_i}\,(A(3)\,dte_i+A(4)\,dtw_{i+1})\,. \end{aligned} \tag{16.43}$$

Defining a distance weighted average diffusivity

$$A(1)\,dte_{i-1}+A(2)\,dtw_i = 2\left(\overline{dtx\,A}^{x}\right)_{i-1} \tag{16.44}$$

$$A(3)\,dte_i+A(4)\,dtw_{i+1} = 2\left(\overline{dtx\,A}^{x}\right)_{i} \tag{16.45}$$

leads to

$$\frac{\partial \mathcal{F}}{\partial T_i} = dxt_i\,\delta_x\left(\frac{\left(\overline{dtx\,A}^{x}\right)_{i-1}}{dxte_{i-1}}\,\delta_x T_{i-1}\right) \tag{16.46}$$

and to the diffusion operator

$$\begin{aligned} R(T)_i &= \frac{1}{dxt_i}\frac{\partial \mathcal{F}}{\partial T_i} \\ &= \delta_x\left(\frac{\left(\overline{dtx\,A}^{x}\right)_{i-1}}{dxte_{i-1}}\,\delta_x T_{i-1}\right). \end{aligned} \tag{16.47}$$

Note that for a constant grid with a constant diffusivity, this discretization reduces to the standard 3-point discrete one-dimensional Laplacian.

The main points to be taken from this example are the following:

1. The tracer gradient is assumed to be the same across the two adjacent subcells 1 & 2 and 3 & 4, respectively.
2. A grid weighted average operator naturally appeared.
3. A diffusion coefficient is used for each quarter tracer cell as opposed to a full tracer cell.
4. The assumption about tracer gradients across subcells is necessary to have the functional-based discretization reduce to the 3-point Laplacian when the grid and diffusivity are constant.

16.4 ELEMENTS OF THE DISCRETE DISSIPATION FUNCTIONAL

This section details the basic elements of the discrete neutral diffusion functional. It is here that issues are addressed regarding the three-dimensional lattice, including the bottom partial cells of Adcroft et al. (1997) and Pacanowski and Gnanadesikan (1998) commonly used in z-models to represent topography.

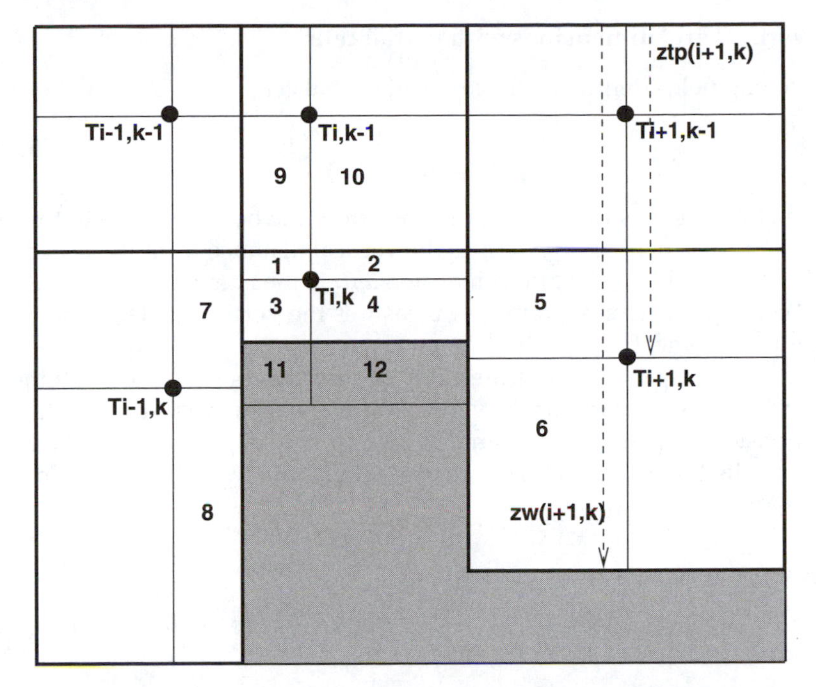

Figure 16.2 The x-z plane for the case with partial bottom cells. T-cells are surrounded by dark solid lines. Rock is shaded gray. Thin and thick solid lines define the quarter cells. The vertical position of a T-point within a T-cell maintains the same ratio for all the cells, whether full or partial. 12 quarter cells are shown, and these correspond to the cells where $T_{i,j,k}$ contributes to the functional. Note that quarter-cells 11 and 12 are in rock for this particular example. The vertical distances from the ocean surface to the bottom of the T-cells (zw) and to the T-cell center (ztp) are shown by lines with arrows. Note that the field $zw_{i,j,k}$ is actually one-dimensional in MOM4, but its full three-dimensional structure can be deduced from other information in the model.

16.4.1 Grid partitioning

Use of the small angle functional given by equation (16.29) motivates a discretization which projects separately onto the two planar slices x-z and y-z. Indeed, it is sufficient to work through details for the two-dimensional x-z plane. Generalization to the three-dimensional case is straightforward and will be made at the end. Figure 16.2 shows the projection for the x-z plane. Within a plane, the central T-cell is partitioned into 4 generally non-equal squares, with one corner being the T-point and the other corners being the corners of the T-cell. Shown here are cells next to the partial cell bottom. Note that discretization of the dianeutral term $A_D(T_{,z})^2$ can be done using the one-dimensional warm-up formalism just discussed, and so is ignored in the following.*

*Although providing a useful discretization of the one-dimensional diffusion equation, the functional approach from Section 16.3 is not commonly employed in z-models for vertical diffusion. The reason is that vertical diffusivities are often computed on the cell faces and are not considered equal across a quarter cell.

16.4.2 Vertical grid dimensions with partial cells

The following fields define the thickness of partial cells:

$$dht_{i,j,k} = zw_{i,j,k} - zw_{i,j,k-1} \tag{16.48}$$

$$dhwt_{i,j,k-1} = ztp_{i,j,k} - ztp_{i,j,k-1}. \tag{16.49}$$

The field $zw_{i,j,k}$ defines the vertical distance from the ocean surface to the bottom of T-cell $T_{i,j,k}$. The field $ztp_{i,j,k}$ defines the vertical distance from the ocean surface to the T-cell point $T_{i,j,k}$. The grid dimension $dht_{i,j,k}$ measures the vertical thickness of the T-cell $T_{i,j,k}$, whereas $dhwt_{i,j,k-1}$ measures the vertical distance between the grid points $T_{i,j,k}$ and $T_{i,j,k-1}$.

Inspection of Figure 16.2 indicates that in order to construct the volume of the quarter cells, it is necessary to take vertical differences such as $zw_{i,j,k} - ztp_{i,j,k}$ (vertical dimension of quarter cells 3 and 4), and $zt_{i,j,k} - zw_{i,k-1,j}$ (vertical dimension of quarter cells 1 and 2). It is therefore useful to define the following field

$$\delta^{(i,j,k1)}_{(i,j,k2)} = (-1)^{k1-k2}(zw_{i,j,k1} - zt_{i,j,k2}). \tag{16.50}$$

Explicitly, for example,

$$\delta^{i,j,k-1+kr}_{i,j,k} = \begin{cases} (zt_{i,j,k} - zw_{i,j,k-1}) & \text{if } kr = 0 \\ (zw_{i,j,k} - zt_{i,j,k}) & \text{if } kr = 1 \end{cases} \tag{16.51}$$

and

$$\delta^{i,j,k}_{i,j,k+kr} = \begin{cases} (zw_{i,j,k} - zt_{i,j,k}) & \text{if } kr = 0 \\ (zt_{i,j,k+1} - zw_{i,j,k}) & \text{if } kr = 1. \end{cases} \tag{16.52}$$

Note that in Pacanowski and Griffies (1999), an extra multiplicative factor of 2 was used to facilitate easier comparison with full bottom cell results given in Griffies et al. (1998). This factor cancels out later and so is not exposed here.

16.4.3 Horizontal tracer gradients and the minimum thickness rule

With partial bottom cells, tracers in horizontally adjacent grid cells generally live at different depths. Before taking horizontal gradients between these tracers, linear interpolation is used to approximate the deeper tracer as if it actually lived at the depth of the shallower tracer point. For example, when taking a horizontal derivative between points $T_{i,j,k}$ and $T_{i+1,j,k}$ shown in Figure 16.2, the value of $T_{i+1,j,k}$ is approximated by a linear interpolation to the same depth as $T_{i,j,k}$. The method and motivation for this interpolation is described in Section 26.2.3 of Pacanowski and Griffies (1999) as well as the partial cell paper of Pacanowski and Gnanadesikan (1998).

The interpolation generally is an interpolation to the *minimum* depth between two laterally adjacent tracer cells. Hence, it effectively reduces the vertical spacing of quarter cell 5 to that of quarter cell 2. A similar consideration applies between quarter cells 7 and 1, for which the vertical spacing for quarter cell 7 is effectively the same as the smaller vertical spacing in quarter cell 1. With this prescription for determining the horizontal gradients and grid cell volumes, the discrete neutral diffusion scheme is compatible with the horizontal diffusion scheme described in Section 26.2.3 of Pacanowski and Griffies (1999) as well as the partial cell paper of Pacanowski and Gnanadesikan (1998).

In general, the above *minimum thickness rule* implies that it is sufficient to only explicitly denote the vertical height of the quarter cells living in the eastern and northern halves, say, of a particular T-cell. Such is the case for full vertical cells as well, in which the vertical height is a function only of the vertical grid position. Therefore, the only difference between the volume specification for the partial cells and the full cells is that for the partial cells, it is necessary to introduce the minimum operation.

With the minimum thickness rule, the vertical spacing relevant for defining the x-z plane quarter cell volumes is written

$$\Delta_{(i,j,k2)}^{(i,j,k1)} = \min\left(\delta_{(i,j,k2)}^{(i,j,k1)}, \delta_{(i+1,j,k2)}^{(i+1,j,k1)}\right). \tag{16.53}$$

Likewise, the vertical spacing relevant for defining the y-z plane quarter cell volumes is written

$$\Delta_{(i,j,k2)}^{(i,j,k1)} = \min\left(\delta_{(i,j,k2)}^{(i,j,k1)}, \delta_{(i,j+1,k2)}^{(i,j+1,k1)}\right). \tag{16.54}$$

Explicitly, the vertical spacings for the 12 x-z plane quarter cells in Figure 16.2 are given by (dropping the j label for brevity)

$$\begin{aligned}
\delta(1) &= (zt_{i,k} - zw_{i,k-1}) = \delta_{i,k}^{i,k-1} \\
\delta(2) &= (zt_{i,k} - zw_{i,k-1}) = \delta_{i,k}^{i,k-1} \\
\delta(3) &= (zw_{i,k} - zt_{i,k}) = \delta_{i,k}^{i,k} \\
\delta(4) &= (zw_{i,k} - zt_{i,k}) = \delta_{i,k}^{i,k} \\
\delta(5) &= (zt_{i+1,k} - zw_{i+1,k-1}) = \delta_{i+1,k}^{i+1,k-1} \\
\delta(6) &= (zw_{i+1,k} - zt_{i+1,k}) = \delta_{i+1,k}^{i+1,k} \\
\delta(7) &= (zt_{i-1,k} - zw_{i-1,k-1}) = \delta_{i-1,k}^{i-1,k-1} \\
\delta(8) &= (zw_{i-1,k} - zt_{i-1,k}) = \delta_{i-1,k}^{i-1,k} \\
\delta(9) &= (zw_{i,k-1} - zt_{i,k-1}) = \delta_{i,k-1}^{i,k-1} \\
\delta(10) &= (zw_{i,k-1} - zt_{i,k-1}) = \delta_{i,k-1}^{i,k-1} \\
\delta(11) &= (zt_{i,k+1} - zw_{i,k}) = \delta_{i,k+1}^{i,k} \\
\delta(12) &= (zt_{i,k+1} - zw_{i,k}) = \delta_{i,k+1}^{i,k},
\end{aligned} \tag{16.55}$$

whereas the vertical spacing used to compute the volumes of these quarter cells is given by

$$\begin{aligned}
\Delta(1) &= \min\left(\delta_{i-1,k}^{i-1,k-1}, \delta_{i,k}^{i,k-1}\right) = \Delta_{(i-1,k)}^{(i-1,k-1)} \\
\Delta(2) &= \min\left(\delta_{i,k}^{i,k-1}, \delta_{i+1,k}^{i+1,k-1}\right) = \Delta_{(i,k)}^{(i,k-1)} \\
\Delta(3) &= \min\left(\delta_{i-1,k}^{i-1,k}, \delta_{i,k}^{i,k}\right) = \Delta_{(i-1,k)}^{(i-1,k)} \\
\Delta(4) &= \min\left(\delta_{i,k}^{i,k}, \delta_{i+1,k}^{i+1,k}\right) = \Delta_{(i,k)}^{(i,k)} \\
\Delta(5) &= \min\left(\delta_{i,k}^{i,k-1}, \delta_{i+1,k}^{i+1,k-1}\right) = \Delta_{(i,k)}^{(i,k-1)} \\
\Delta(6) &= \min\left(\delta_{i,k}^{i,k}, \delta_{i+1,k}^{i+1,k}\right) = \Delta_{(i,k)}^{(i,k)} \\
\Delta(7) &= \min\left(\delta_{i-1,k}^{i-1,k-1}, \delta_{i,k}^{i,k-1}\right) = \Delta_{(i-1,k)}^{(i-1,k-1)} \\
\Delta(8) &= \min\left(\delta_{i-1,k}^{i-1,k}, \delta_{i,k}^{i,k}\right) = \Delta_{(i-1,k)}^{(i-1,k)} \\
\Delta(9) &= \min\left(\delta_{i-1,k-1}^{i-1,k-1}, \delta_{i,k-1}^{i,k-1}\right) = \Delta_{(i-1,k-1)}^{(i-1,k-1)} \\
\Delta(10) &= \min\left(\delta_{i,k-1}^{i,k-1}, \delta_{i+1,k-1}^{i+1,k-1}\right) = \Delta_{(i,k-1)}^{(i,k-1)} \\
\Delta(11) &= \min\left(\delta_{i-1,k+1}^{i-1,k}, \delta_{i,k+1}^{i,k}\right) = \Delta_{(i-1,k+1)}^{(i-1,k)} \\
\Delta(12) &= \min\left(\delta_{i,k+1}^{i,k}, \delta_{i+1,k+1}^{i+1,k}\right) = \Delta_{(i,k+1)}^{(i,k)}.
\end{aligned} \tag{16.56}$$

16.4.4 Quarter cell volumes

In order to discretize the functional, it is necessary to construct the volumes $V(n)$ of the various subcells which partition the grid into the quarter cells shown in Figure 16.2. Referring to Figure 16.2 yields volumes for the 12 quarter cells

$$\begin{aligned}
V(1) &= dtw_{i,j}\, dyt_{i,j}\, \Delta_{(i-1,k)}^{(i-1,k-1)} \\
V(2) &= dte_{i,j}\, dyt_{i,j}\, \Delta_{(i,k)}^{(i,k-1)} \\
V(3) &= dtw_{i,j}\, dyt_{i,j}\, \Delta_{(i-1,k)}^{(i-1,k)} \\
V(4) &= dte_{i,j}\, dyt_{i,j}\, \Delta_{(i,k)}^{(i,k)} \\
V(5) &= dtw_{i+1,j}\, dyt_{i+1,j}\, \Delta_{(i,k)}^{(i,k-1)} \\
V(6) &= dtw_{i+1,j}\, dyt_{i+1,j}\, \Delta_{(i,k)}^{(i,k)} \\
V(7) &= dte_{i-1,j}\, dyt_{i-1,j}\, \Delta_{(i-1,k)}^{(i-1,k-1)} \\
V(8) &= dte_{i-1,j}\, dyt_{i-1,j}\, \Delta_{(i-1,k)}^{(i-1,k)} \\
V(9) &= dtw_{i,j}\, dyt_{i,j}\, \Delta_{(i-1,k-1)}^{(i-1,k-1)} \\
V(10) &= dte_{i,j}\, dyt_{i,j}\, \Delta_{(i,k-1)}^{(i,k-1)} \\
V(11) &= dtw_{i,j}\, dyt_{i,j}\, \Delta_{(i-1,k+1)}^{(i-1,k)} \\
V(12) &= dte_{i,j}\, dyt_{i,j}\, \Delta_{(i,k+1)}^{(i,k)}.
\end{aligned} \tag{16.57}$$

The j label has been omitted in the Δ expressions for purposes of brevity. The array $dtw_{i,j}$ represents the distance from the point $T_{i,j,k}$ to the western boundary of the tracer cell, whereas $dte_{i,j}$ is the distance to the eastern boundary. All grid distances are in meters, and have the appropriate non-Euclidean metric or stretching functions absorbed according to the discussion in Section 21.12.4 (page 485).

16.4.5 Tracer gradients within the quarter cells

Tracer and density gradients are required within the quarter cells. The second order accurate difference operators are given here. There are further details regarding the reference points to be used in determining the density gradient, and these reference point issues are discussed in Section 16.5.2. The finite difference operators approximating derivatives are written

$$\begin{aligned}
\delta_x T_{i,j,k} &= \frac{T_{i+1,k,j} - T_{i,j,k}}{dxte_{i,j}} \\
\delta_y T_{i,j,k} &= \frac{T_{i,j+1,k} - T_{i,j,k}}{dytn_{i,j}} \\
\delta_z T_{i,j,k} &= \frac{T_{i,j,k} - T_{i,j,k+1}}{dhwt_{i,j,k}}.
\end{aligned} \tag{16.58}$$

The array $dxte_{i,j}$ is the zonal distance between $T_{i+1,k,j}$ and $T_{i,j,k}$, $dytn_{i,j}$ is the meridional distance between $T_{i,j+1,k}$ and $T_{i,j,k}$, and $dhwt_{i,j,k}$ is the vertical distance between $T_{i,j,k}$ and $T_{i,j,k+1}$.

The difference operators are located at the T-cell faces and provide second order accurate approximations to the continuous derivatives. For partial cells, horizontal gradients are computed according to the minimum thickness rule and the associated linear interpolation described in Section 16.4.3. For brevity, such interpolation is assumed in the notation here, that is, each horizontal derivative next to the bottom is computed using this interpolation algorithm.

Suppressing the j-index for brevity leads to the tracer gradient in the 12 quarter cells

$$\begin{aligned}
\nabla T(1) &= \hat{\mathbf{x}}\,\delta_x T_{i-1,k} + \hat{\mathbf{y}}\,\delta_y T_{i,k} + \hat{\mathbf{z}}\,\delta_z T_{i,k-1} \\
\nabla T(2) &= \hat{\mathbf{x}}\,\delta_x T_{i,k} + \hat{\mathbf{y}}\,\delta_y T_{i,k} + \hat{\mathbf{z}}\,\delta_z T_{i,k-1} \\
\nabla T(3) &= \hat{\mathbf{x}}\,\delta_x T_{i-1,k} + \hat{\mathbf{y}}\,\delta_y T_{i,k} + \hat{\mathbf{z}}\,\delta_z T_{i,k} \\
\nabla T(4) &= \hat{\mathbf{x}}\,\delta_x T_{i,k} + \hat{\mathbf{y}}\,\delta_y T_{i,k} + \hat{\mathbf{z}}\,\delta_z T_{i,k} \\
\nabla T(5) &= \hat{\mathbf{x}}\,\delta_x T_{i,k} + \hat{\mathbf{y}}\,\delta_y T_{i+1,k} + \hat{\mathbf{z}}\,\delta_z T_{i+1,k-1} \\
\nabla T(6) &= \hat{\mathbf{x}}\,\delta_x T_{i,k} + \hat{\mathbf{y}}\,\delta_y T_{i+1,k} + \hat{\mathbf{z}}\,\delta_z T_{i+1,k} \\
\nabla T(7) &= \hat{\mathbf{x}}\,\delta_x T_{i-1,k} + \hat{\mathbf{y}}\,\delta_y T_{i-1,k} + \hat{\mathbf{z}}\,\delta_z T_{i-1,k-1} \\
\nabla T(8) &= \hat{\mathbf{x}}\,\delta_x T_{i-1,k} + \hat{\mathbf{y}}\,\delta_y T_{i-1,k} + \hat{\mathbf{z}}\,\delta_z T_{i-1,k} \\
\nabla T(9) &= \hat{\mathbf{x}}\,\delta_x T_{i-1,k-1} + \hat{\mathbf{y}}\,\delta_y T_{i,k-1} + \hat{\mathbf{z}}\,\delta_z T_{i,k-1} \\
\nabla T(10) &= \hat{\mathbf{x}}\,\delta_x T_{i,k-1} + \hat{\mathbf{y}}\,\delta_y T_{i,k-1} + \hat{\mathbf{z}}\,\delta_z T_{i,k-1} \\
\nabla T(11) &= \hat{\mathbf{x}}\,\delta_x T_{i-1,k+1} + \hat{\mathbf{y}}\,\delta_y T_{i,k+1} + \hat{\mathbf{z}}\,\delta_z T_{i,k} \\
\nabla T(12) &= \hat{\mathbf{x}}\,\delta_x T_{i,k+1} + \hat{\mathbf{y}}\,\delta_y T_{i,k+1} + \hat{\mathbf{z}}\,\delta_z T_{i,k}
\end{aligned} \tag{16.59}$$

and the functional derivative of these x-z plane gradients is given by

$$
\begin{aligned}
\frac{\partial \nabla T(1)}{\partial T_{i,j}} &= \frac{\hat{\mathbf{x}}}{dxte_{i-1,j}} - \frac{\hat{\mathbf{y}}}{dytn_{i,j}} - \frac{\hat{\mathbf{z}}}{dhwt_{i,j,k-1}} \\
\frac{\partial \nabla T(2)}{\partial T_{i,k}} &= -\frac{\hat{\mathbf{x}}}{dxte_{i,j}} - \frac{\hat{\mathbf{y}}}{dytn_{i,j}} - \frac{\hat{\mathbf{z}}}{dhwt_{i,k-1,j}} \\
\frac{\partial \nabla T(3)}{\partial T_{i,k}} &= \frac{\hat{\mathbf{x}}}{dxte_{i-1,j}} - \frac{\hat{\mathbf{y}}}{dytn_{i,j}} + \frac{\hat{\mathbf{z}}}{dhwt_{i,k,j}} \\
\frac{\partial \nabla T(4)}{\partial T_{i,k}} &= -\frac{\hat{\mathbf{x}}}{dxte_{i,j}} - \frac{\hat{\mathbf{y}}}{dytn_{i,j}} + \frac{\hat{\mathbf{z}}}{dhwt_{i,k,j}} \\
\frac{\partial \nabla T(5)}{\partial T_{i,k}} &= -\frac{\hat{\mathbf{x}}}{dxte_{i,j}} \\
\frac{\partial \nabla T(6)}{\partial T_{i,k}} &= -\frac{\hat{\mathbf{x}}}{dxte_{i,j}} \\
\frac{\partial \nabla T(7)}{\partial T_{i,k}} &= \frac{\hat{\mathbf{x}}}{dxte_{i-1,j}} \\
\frac{\partial \nabla T(8)}{\partial T_{i,k}} &= \frac{\hat{\mathbf{x}}}{dxte_{i-1,j}} \\
\frac{\partial \nabla T(9)}{\partial T_{i,k}} &= -\frac{\hat{\mathbf{z}}}{dhwt_{i,k-1,j}} \\
\frac{\partial \nabla T(10)}{\partial T_{i,k}} &= -\frac{\hat{\mathbf{z}}}{dhwt_{i,k-1,j}} \\
\frac{\partial \nabla T(11)}{\partial T_{i,k}} &= \frac{\hat{\mathbf{z}}}{dhwt_{i,k,j}} \\
\frac{\partial \nabla T(12)}{\partial T_{i,k}} &= \frac{\hat{\mathbf{z}}}{dhwt_{i,k,j}}.
\end{aligned}
\tag{16.60}
$$

16.4.6 Schematic form of the discretized functional

For each of the two vertical planes x-z and y-z, there are 12 components to the functional that contain contributions from the grid tracer value $T_{i,j,k}$. These 12 components correspond to the 12 quarter cells shown in Figure 16.2. For example, the x-z plane functional can be written (see equation (16.29) for the continuum version)

$$
\begin{aligned}
\mathcal{F}^{(x-z)} &= -\frac{1}{2} \sum_{i,k} \sum_{n=1}^{12} A(n)\, V(n)\, [\delta_x T(n) + S_{(x)}(n)\, \delta_z T(n)]^2 \\
&\equiv \sum_{i,k} L_{i,k}
\end{aligned}
\tag{16.61}
$$

where the discretized quadratic form

$$
\begin{aligned}
L_{i,k} &= -\frac{1}{2} \sum_{n=1}^{12} A(n)\, V(n)\, [\delta_x T(n) + S_{(x)}(n)\, \delta_z T(n)]^2 \\
&\equiv \sum_{n=1}^{12} L_{i,k}^{(n)}
\end{aligned}
\tag{16.62}
$$

consists of 12 nonpositive contributions, with $A(n)$ the nonnegative diffusion coefficient for each of the different quarter cells.

16.5 TRIAD STENCILS AND SOME MORE NOTATION

This section details methods used to compute the density gradients and the corresponding neutral directions. Along the way, the *triad* structure is described for elements comprising the discrete functional.

16.5.1 Tracer and density triads

Consider the component of the x-z plane functional arising from the first subcell (see Figure 16.2), which has the form

$$L_{i,k}^{(1)} = -\left(\frac{A(1)\,V(1)}{2}\right)\left(\delta_x T_{i-1,k} - \delta_z T_{i,k-1}\,\frac{\delta_x \rho_{i-1,k}}{\delta_z \rho_{i,k-1}}\right)^2. \tag{16.63}$$

This term has contributions from the tracer and density at the three grid points $T_{i-1,k}$, $T_{i,k}$, and $T_{i,k-1}$. These three points form a triangle, or *triad*, in the x-z plane. All pieces of the functional are discretized into these triad groups. That is, triads are the fundamental combinations of tracer and density points within the discrete functional. For each triad there is a unique quarter cell volume and diffusivity.

16.5.2 Reference points for computing density derivatives

Horizontal and vertical derivatives of density are required for computing neutral slopes (Section 13.6.1, page 291). Hence, the density derivatives appearing in the discrete neutral diffusion functional are discretized as finite differences of locally referenced potential density. A self-consistent choice for the reference point must be chosen. That is, one must choose a temperature, salinity, and pressure to reference the horizontal and vertical density derivatives. Given the prominence of the triad structure in the discrete functional, a useful reference point is the triad corner. Alternatives could be considered, but in practice they likely will lead to minimal differences from the triad corner. With the corner as the reference point, the triad forming $L_{i,k}^{(1)}$ uses the T-cell point $T_{i,j,k}$ as the reference. This choice is valid whether the triad is wholly in a full vertical cell region, or has a part within a partial cell. Other triads use similarly defined reference points.

With the triad corner as reference point, each of the three discrete densities appearing in a triad employ the *same* thermal and saline partial derivatives referenced to this single point. Therefore, the density derivatives appearing in $L_{i,k}^{(1)}$ are discretized as

$$\delta_x \rho_{i-1,j,k} = (\rho_{,\theta})_{i,j,k}\,\delta_x \theta_{i-1,j,k} + (\rho_{,s})_{i,j,k}\,\delta_x s_{i-1,j,k} \equiv \delta_x \rho_{i-1,j,k}^{(i,j,k)} \tag{16.64}$$

$$\delta_z \rho_{i,j,k-1} = (\rho_{,\theta})_{i,j,k}\,\delta_z \theta_{i,j,k-1} + (\rho_{,s})_{i,j,k}\,\delta_y s_{i,j,k-1} \equiv \delta_z \rho_{i,j,k-1}^{(i,j,k)}. \tag{16.65}$$

In these expressions, the thermal and saline partial derivatives are

$$(\rho_{,\theta})_{i,j,k} = (\partial\rho/\partial\theta)\left(\theta_{i,j,k}, s_{i,j,k}, p_{i,j,k}\right) \tag{16.66}$$

$$(\rho_{,s})_{i,j,k} = (\partial\rho/\partial s)\left(\theta_{i,j,k}, s_{i,j,k}, p_{i,j,k}\right), \tag{16.67}$$

where the arguments of the density partial derivatives indicate the value of the potential temperature, salinity, and pressure reference for use in their computation. The derivatives $\rho_{,\theta}$ and $\rho_{,s}$ are computed analytically and tabulated along with the equation of state. The superscripts introduced on the density gradients in equations (16.64) and (16.65) allow for a compact notation that exposes information about the reference point.

In an early derivation of the discretized diffusion operator, issues of reference points were ignored until the very end of the derivation. At that point, reasonable choices were made for choosing the reference points, which consisted of referencing on the various sides of the T-cells consistent with the location of the diffusive fluxes. However, it was soon realized that the numerical constraint of defining a dissipative diffusion operator was not satisfied by this choice. That is, such choices led to the possibility of a sign indefinite discrete neutral diffusion functional, and this allowed for the corresponding discrete neutral diffusion operator to increase variance. This point emphasizes the importance of addressing all discretization details when discretizing the functional in order to ensure a dissipative discrete neutral diffusion operator.

16.5.3 Notation for neutral direction slopes and diffusivities

The discretized neutral direction slopes appear throughout the functional, so it is useful to devise a notation to help reduce the clutter. For this purpose, let

$$Sx^{(i,k)}_{(i1,k1|i2,k2)} \equiv -\left(\frac{\delta_x \rho^{(i,k)}_{i1,k1}}{\delta_z \rho^{(i,k)}_{i2,k2}}\right). \tag{16.68}$$

Given this notation and the methods of regularization discussed in Section 15.3, it is also useful to introduce an associated diffusivity with notation

$$A(n) \rightarrow A^{(i,k)}_{(i1,k1|i2,k2)}. \tag{16.69}$$

The regularization of $A^{(i,k)}_{(i1,k1|i2,k2)}$ is determined by the value of $Sx^{(i,k)}_{(i1,k1|i2,k2)}$ according to one of the schemes discussed in Section 15.3.

16.5.4 Notation for the quarter cell dimensions

In the derivative of the x-z functional, write the horizontal area of quarter cells living in adjacent tracer cells as

$$(dtwe\, dyt)^{(ip)}_{i,j} = \begin{cases} dtw_{i+1,j}\, dyt_{i+1,j} & \text{if } ip = 1 \\ dte_{i,j}\, dyt_{i,j} & \text{if } ip = 0. \end{cases} \tag{16.70}$$

When considering the y-z functional, use the analogous notation

$$(dxt\, dtsn)^{(jq)}_{i,j} = \begin{cases} dxt_{i,j+1}\, dts_{i,j+1} & \text{if } jq = 1 \\ dxt_{i,j}\, dtn_{i,j} & \text{if } jq = 0. \end{cases} \tag{16.71}$$

It is also useful to introduce the following notation for the zonal and meridional dimensions of quarter cells within a tracer cell:

$$dtew_{i,j}^{(ip)} = \begin{cases} dte_{i,j} & \text{if } ip = 1 \\ dtw_{i,j} & \text{if } ip = 0 \end{cases} \tag{16.72}$$

$$dtns_{i,j}^{(jq)} = \begin{cases} dtn_{i,j} & \text{if } jq = 1 \\ dts_{i,j} & \text{if } jq = 0. \end{cases} \tag{16.73}$$

16.6 THE DISCRETE DIFFUSION OPERATOR

This section details the construction of the discrete diffusion operator as derived from the functional derivative of the discrete diffusion functional.

16.6.1 The discretized x-z dissipation functional

The different components of the neutral diffusion functional have been considered, and so it is now time to piece things together. It is sufficient to continue our focus on the x-z plane, as y-z results are symmetric. The x-z plane quadratic form contains contributions from the T-point $T_{i,j,k}$ in the following 12 terms (the j-index is suppressed)

$$\begin{aligned}
L_{i,k}^{(1)} &= -\frac{1}{2} V(1)\, A_{(i-1,k|i,k-1)}^{(i,k)} \left(\delta_x T_{i-1,k} + Sx_{(i-1,k|i,k-1)}^{(i,k)}\, \delta_z T_{i,k-1}\right)^2 \\
L_{i,k}^{(2)} &= -\frac{1}{2} V(2)\, A_{(i,k|i,k-1)}^{(i,k)} \left(\delta_x T_{i,k} + Sx_{(i,k|i,k-1)}^{(i,k)}\, \delta_z T_{i,k-1}\right)^2 \\
L_{i,k}^{(3)} &= -\frac{1}{2} V(3)\, A_{(i-1,k|i,k)}^{(i,k)} \left(\delta_x T_{i-1,k} + Sx_{(i-1,k|i,k)}^{(i,k)}\, \delta_z T_{i,k}\right)^2 \\
L_{i,k}^{(4)} &= -\frac{1}{2} V(4)\, A_{(i,k|i,k)}^{(i,k)} \left(\delta_x T_{i,k} + Sx_{(i,k|i,k)}^{(i,k)}\, \delta_z T_{i,k}\right)^2 \\
L_{i,k}^{(5)} &= -\frac{1}{2} V(5)\, A_{(i,k|i+1,k-1)}^{(i+1,k)} \left(\delta_x T_{i,k} + Sx_{(i,k|i+1,k-1)}^{(i+1,k)}\, \delta_z T_{i+1,k-1}\right)^2 \\
L_{i,k}^{(6)} &= -\frac{1}{2} V(6)\, A_{(i,k|i+1,k)}^{(i+1,k)} \left(\delta_x T_{i,k} + Sx_{(i,k|i+1,k)}^{(i+1,k)}\, \delta_z T_{i+1,k}\right)^2 \\
L_{i,k}^{(7)} &= -\frac{1}{2} V(7)\, A_{(i-1,k|i-1,k-1)}^{(i-1,k)} \left(\delta_x T_{i-1,k} + Sx_{(i-1,k|i-1,k-1)}^{(i-1,k)}\, \delta_z T_{i-1,k-1}\right)^2 \\
L_{i,k}^{(8)} &= -\frac{1}{2} V(8)\, A_{(i-1,k|i-1,k)}^{(i-1,k)} \left(\delta_x T_{i-1,k} + Sx_{(i-1,k|i-1,k)}^{(i-1,k)}\, \delta_z T_{i-1,k}\right)^2 \\
L_{i,k}^{(9)} &= -\frac{1}{2} V(9)\, A_{(i-1,k-1|i,k-1)}^{(i,k-1)} \left(\delta_x T_{i-1,k-1} + Sx_{(i-1,k-1|i,k-1)}^{(i,k-1)}\, \delta_z T_{i,k-1}\right)^2 \\
L_{i,k}^{(10)} &= -\frac{1}{2} V(10)\, A_{(i,k-1|i,k-1)}^{(i,k-1)} \left(\delta_x T_{i,k-1} + Sx_{(i,k-1|i,k-1)}^{(i,k-1)}\, \delta_z T_{i,k-1}\right)^2 \\
L_{i,k}^{(11)} &= -\frac{1}{2} V(11)\, A_{(i-1,k+1|i,k)}^{(i,k+1)} \left(\delta_x T_{i-1,k+1} + Sx_{(i-1,k+1|i,k)}^{(i,k+1)}\, \delta_z T_{i,k}\right)^2 \\
L_{i,k}^{(12)} &= -\frac{1}{2} V(12)\, A_{(i,k+1|i,k)}^{(i,k+1)} \left(\delta_x T_{i,k+1} + Sx_{(i,k+1|i,k)}^{(i,k+1)}\, \delta_z T_{i,k}\right)^2 .
\end{aligned}$$

16.6.2 Derivative of the x-z dissipation functional

To construct the diffusion operator, it is necessary to take the discrete functional derivative of $\mathcal{F}$ according to equation (16.33). When taking this derivative, all terms in $\mathcal{F}$ independent of $T_{i,j,k}$ drop out. Hence, it is only necessary to consider that part of $\mathcal{F}$ including contributions from $T_{i,j,k}$. This fact motivated the focus on just the 12 quarter cells shown in Figure 16.2. Taking the derivative of the 12 contributions to the functional in the x-z plane yields

$$\frac{\partial L^{(1)}_{i,k}}{\partial T_{i,j,k}} = -V(1)\, A^{(i,k)}_{(i-1,k|i,k-1)} \left(\delta_x T_{i-1,k} + Sx^{(i,k)}_{(i-1,k|i,k-1)}\, \delta_z T_{i,k-1}\right) \left(\frac{1}{dxte_{i-1,j}} - \frac{Sx^{(i,k)}_{(i-1,k|i,k-1)}}{dhwt_{i,k-1}}\right)$$

$$\frac{\partial L^{(2)}_{i,k}}{\partial T_{i,j,k}} = -V(2)\, A^{(i,k)}_{(i,k|i,k-1)} \left(\delta_x T_{i,k} + Sx^{(i,k)}_{(i,k|i,k-1)}\, \delta_z T_{i,k-1}\right) \left(-\frac{1}{dxte_{i,j}} - \frac{Sx^{(i,k)}_{(i,k|i,k-1)}}{dhwt_{i,k-1}}\right)$$

$$\frac{\partial L^{(3)}_{i,k}}{\partial T_{i,j,k}} = -V(3)\, A^{(i,k)}_{(i-1,k|i,k)} \left(\delta_x T_{i-1,k} + Sx^{(i,k)}_{(i-1,k|i,k)}\, \delta_z T_{i,k}\right) \left(\frac{1}{dxte_{i-1,j}} + \frac{Sx^{(i,k)}_{(i-1,k|i,k)}}{dhwt_{i,k}}\right)$$

$$\frac{\partial L^{(4)}_{i,k}}{\partial T_{i,j,k}} = -V(4)\, A^{(i,k)}_{(i,k|i,k)} \left(\delta_x T_{i,k} + Sx^{(i,k)}_{(i,k|i,k)}\, \delta_z T_{i,k}\right) \left(-\frac{1}{dxte_{i,j}} + \frac{Sx^{(i,k)}_{(i,k|i,k)}}{dhwt_{i,k}}\right)$$

$$\frac{\partial L^{(5)}_{i,k}}{\partial T_{i,j,k}} = -V(5)\, A^{(i+1,k)}_{(i,k|i+1,k-1)} \left(\delta_x T_{i,k} + Sx^{(i+1,k)}_{(i,k|i+1,k-1)}\, \delta_z T_{i+1,k-1}\right) \left(-\frac{1}{dxte_{i,j}}\right)$$

$$\frac{\partial L^{(6)}_{i,k}}{\partial T_{i,j,k}} = -V(6)\, A^{(i+1,k)}_{(i,k|i+1,k)} \left(\delta_x T_{i,k} + Sx^{(i+1,k)}_{(i,k|i+1,k)}\, \delta_z T_{i+1,k}\right) \left(-\frac{1}{dxte_{i,j}}\right)$$

$$\frac{\partial L^{(7)}_{i,k}}{\partial T_{i,j,k}} = -V(7)\, A^{(i-1,k)}_{(i-1,k|i-1,k-1)} \left(\delta_x T_{i-1,k} + Sx^{(i-1,k)}_{(i-1,k|i-1,k-1)}\, \delta_z T_{i-1,k-1}\right) \left(\frac{1}{dxte_{i-1,j}}\right)$$

$$\frac{\partial L^{(8)}_{i,k}}{\partial T_{i,j,k}} = -V(8)\, A^{(i-1,k)}_{(i-1,k|i-1,k)} \left(\delta_x T_{i-1,k} + Sx^{(i-1,k)}_{(i-1,k|i-1,k)}\, \delta_z T_{i-1,k}\right) \left(\frac{1}{dxte_{i-1,j}}\right)$$

$$\frac{\partial L^{(9)}_{i,k}}{\partial T_{i,j,k}} = -V(9)\, A^{(i,k-1)}_{(i-1,k-1|i,k-1)} \left(\delta_x T_{i-1,k-1} + Sx^{(i,k-1)}_{(i-1,k-1|i,k-1)}\, \delta_z T_{i,k-1}\right) \left(-\frac{Sx^{(i,k-1)}_{(i-1,k-1|i,k-1)}}{dhwt_{i,k-1}}\right)$$

$$\frac{\partial L^{(10)}_{i,k}}{\partial T_{i,j,k}} = -V(10)\, A^{(i,k-1)}_{(i,k-1|i,k-1)} \left(\delta_x T_{i,k-1} + Sx^{(i,k-1)}_{(i,k-1|i,k-1)}\, \delta_z T_{i,k-1}\right) \left(-\frac{Sx^{(i,k-1)}_{(i,k-1|i,k-1)}}{dhwt_{i,k-1}}\right)$$

$$\frac{\partial L^{(11)}_{i,k}}{\partial T_{i,j,k}} = -V(11)\, A^{(i,k+1)}_{(i-1,k+1|i,k)} \left(\delta_x T_{i-1,k+1} + Sx^{(i,k+1)}_{(i-1,k+1|i,k)}\, \delta_z T_{i,k}\right) \left(\frac{Sx^{(i,k+1)}_{(i-1,k+1|i,k)}}{dhwt_{i,k}}\right)$$

$$\frac{\partial L^{(12)}_{i,k}}{\partial T_{i,j,k}} = -V(12)\, A^{(i,k+1)}_{(i,k+1|i,k)} \left(\delta_x T_{i,k+1} + Sx^{(i,k+1)}_{(i,k+1|i,k)}\, \delta_z T_{i,k}\right) \left(\frac{Sx^{(i,k+1)}_{(i,k+1|i,k)}}{dhwt_{i,k}}\right).$$

To organize the results more compactly, we rearrange terms to identify components to the discrete diffusive flux. For this purpose, it is useful to consider the three terms from quarter cells $1+7+9$, with the expressions for volumes given in

Section 16.4.4:

$$\frac{1}{\partial T_{i,j,k}}\left(L_{i,k}^{(1+7+9)}\right) = -\left(\frac{\Delta_{(i-1,k)}^{(i-1,k-1)}}{dxte_{i-1,j}}\right)$$
$$\times\left(A_{(i-1,k|i,k-1)}^{(i,k)} dtw_{i,j}\, dyt_{i,j}\,(\delta_x T_{i-1,k} + Sx_{(i-1,k|i,k-1)}^{(i,k)}\,\delta_z T_{i,k-1})\right.$$
$$\left. + A_{(i-1,k|i-1,k-1)}^{(i-1,k)} dte_{i-1,j}\, dyt_{i-1,j}\,(\delta_x T_{i-1,k} + Sx_{(i-1,k|i-1,k-1)}^{(i-1,k)}\,\delta_z T_{i-1,k-1})\right)$$
$$+\left(\frac{dtw_{i,j}\, dyt_{i,j}}{dhwt_{i,k-1}}\right)$$
$$\times\left(\Delta_{(i-1,k)}^{(i-1,k-1)}\,(A\ Sx)_{(i-1,k|i,k-1)}^{(i,k)}(\delta_x T_{i-1,k} + Sx_{(i-1,k|i,k-1)}^{(i,k)}\,\delta_z T_{i,k-1})\right.$$
$$\left. + \Delta_{(i-1,k-1)}^{(i-1,k-1)}\,(A\ Sx)_{(i-1,k-1|i,k-1)}^{(i,k-1)}(\delta_x T_{i-1,k-1} + Sx_{(i-1,k-1|i,k-1)}^{(i,k-1)}\,\delta_z T_{i,k-1})\right)$$

To clean up this expression, introduce notation for the quarter cell area from Section 16.5.4 to yield

$$\frac{1}{\partial T_{i,j,k}}\left(L_{i,k}^{(1+7+9)}\right) = -\left(\frac{\Delta_{(i-1,k)}^{(i-1,k-1)}}{dxte_{i-1,j}}\right)\sum_{ip=0}^{1}(dtwe\, dyt)_{(i-1,j)}^{(ip)}\, A_{(i-1,k|i-1+ip,k-1)}^{(i-1+ip,k)}$$
$$\times\left(\delta_x T_{i-1,k} + Sx_{(i-1,k|i-1+ip,k-1)}^{(i-1+ip,k)}\,\delta_z T_{i-1+ip,k-1}\right)$$
$$+\left(\frac{dtw_{i,j}\, dyt_{i,j}}{dhwt_{i,k-1}}\right)\sum_{kr=0}^{1}\Delta_{(i-1,k-1+kr)}^{(i-1,k-1)}\,(A\ Sx)_{(i-1,k-1+kr|i,k-1)}^{(i,k-1+kr)}$$
$$\times\left(\delta_x T_{i-1,k-1+kr} + Sx_{(i-1,k-1+kr|i,k-1)}^{(i,k-1+kr)}\,\delta_z T_{i,k-1}\right). \tag{16.74}$$

Results for the other three combinations of three quarter cells in the x-z plane follow similarly. Thus,

$$\frac{1}{\partial T_{i,j,k}}\left(L_{i,k}^{(2+5+10)}\right) = \left(\frac{\Delta_{(i,k)}^{(i,k-1)}}{dxte_{i,j}}\right)\sum_{ip=0}^{1}(dtwe\, dyt)_{i,j}^{(ip)}\, A_{(i,k|i+ip,k-1)}^{(i+ip,k)}$$
$$\times\left(\delta_x T_{i,k} + Sx_{(i,k|i+ip,k-1)}^{(i+ip,k)}\,\delta_z T_{i+ip,k-1}\right)$$
$$+\left(\frac{dte_{i,j}\, dyt_{i,j}}{dhwt_{i,k-1}}\right)\sum_{kr=0}^{1}\Delta_{(i,k-1+kr)}^{(i,k-1)}\,(A\ Sx)_{(i,k-1+kr|i,k-1)}^{(i,k-1+kr)}$$
$$\times\left(\delta_x T_{i,k-1+kr} + Sx_{(i,k-1+kr|i,k-1)}^{(i,k-1+kr)}\,\delta_z T_{i,k-1}\right) \tag{16.75}$$

$$\frac{1}{\partial T_{i,j,k}}\left(L_{i,k}^{(3+8+11)}\right) = -\left(\frac{\Delta_{(i-1,k)}^{(i-1,k)}}{dxte_{i-1,j}}\right)\sum_{ip=0}^{1}(dtwe\,dyt)_{i-1,j}^{(ip)}\,A_{(i-1,k|i-1+ip,k)}^{(i-1+ip,k)}$$

$$\times\left(\delta_x T_{i-1,k} + Sx_{(i-1,k|i-1+ip,k)}^{(i-1+ip,k)}\,\delta_z T_{i-1+ip,k}\right)$$

$$-\left(\frac{dtw_{i,j}\,dyt_{i,j}}{dhwt_{i,k}}\right)\sum_{kr=0}^{1}\Delta_{(i-1,k+kr)}^{(i-1,k)}\,(A\;Sx)_{(i-1,k+kr|i,k)}^{(i,k+kr)}$$

$$\times\left(\delta_x T_{i-1,k+kr} + Sx_{(i-1,k+kr|i,k)}^{(i,k+kr)}\,\delta_z T_{i,k}\right) \quad (16.76)$$

and finally

$$\frac{1}{\partial T_{i,j,k}}\left(L_{i,k}^{(4+6+12)}\right) = \left(\frac{\Delta_{(i,k)}^{(i,k)}}{dxte_{i,j}}\right)\sum_{ip=0}^{1}(dtwe\,dyt)_{i,j}^{(ip)}\,A_{(i,k|i+ip,k)}^{(i+ip,k)}$$

$$\times\left(\delta_x T_{i,k} + \delta_z T_{i+ip,k}\,Sx_{(i,k|i+ip,k)}^{(i+ip,k)}\right)$$

$$-\left(\frac{dte_{i,j}\,dyt_{i,j}}{dhwt_{i,k}}\right)\sum_{kr=0}^{1}\Delta_{(i,k+kr)}^{(i,k)}\,(A\;Sx)_{(i,k+kr|i,k)}^{(i,k+kr)}$$

$$\times\left(\delta_x T_{i,k+kr} + Sx_{(i,k+kr|i,k)}^{(i,k+kr)}\,\delta_z T_{i,k}\right). \quad (16.77)$$

Bringing the 12 pieces together leads to the functional derivative

$$\frac{\partial \mathcal{F}^{(x-z)}}{\partial T_{i,j,k}} = \left(\frac{1}{dxte_{i,j}}\right)\sum_{kr=0}^{1}\Delta_{(i,k)}^{(i,k-1+kr)}\sum_{ip=0}^{1}(dtwe\,dyt)_{i,j}^{(ip)}\,A_{(i,k|i+ip,k-1+kr)}^{(i+ip,k)}$$

$$\times\left(\delta_x T_{i,k} + Sx_{(i,k|i+ip,k-1+kr)}^{(i+ip,k)}\,\delta_z T_{i+ip,k-1+kr}\right)$$

$$-\left(\frac{1}{dxte_{i-1,j}}\right)\sum_{kr=0}^{1}\Delta_{(i-1,k)}^{(i-1,k-1+kr)}\sum_{ip=0}^{1}(dtwe\,dyt)_{i-1,j}^{(ip)}\,A_{(i-1,k|i-1+ip,k-1+kr)}^{(i-1+ip,k)}$$

$$\times\left(\delta_x T_{i-1,k} + Sx_{(i-1,k|i-1+ip,k-1+kr)}^{(i-1+ip,k)}\,\delta_z T_{i-1+ip,k-1+kr}\right)$$

$$+\left(\frac{dyt_{i,j}}{dhwt_{i,k-1}}\right)\sum_{ip=0}^{1}dtwe_{i,j}^{(ip)}\sum_{kr=0}^{1}\Delta_{(i-1+ip,k-1+kr)}^{(i-1+ip,k-1)}\,(A\;Sx)_{(i-1+ip,k-1+kr|i,k-1)}^{(i,k-1+kr)}$$

$$\times\left(\delta_x T_{i-1+ip,k-1+kr} + Sx_{(i-1+ip,k-1+kr|i,k-1)}^{(i,k-1+kr)}\,\delta_z T_{i,k-1}\right)$$

$$-\left(\frac{dyt_{i,j}}{dhwt_{i,k}}\right)\sum_{ip=0}^{1}dtwe_{i,j}^{(ip)}\sum_{kr=0}^{1}\Delta_{(i-1+ip,k+kr)}^{(i-1+ip,k)}\,(A\;Sx)_{(i-1+ip,k+kr|i,k)}^{(i,k+kr)}$$

$$\times\left(\delta_x T_{i-1+ip,k+kr} + Sx_{(i-1+ip,k+kr|i,k)}^{(i,k+kr)}\,\delta_z T_{i,k}\right). \quad (16.78)$$

16.6.3 Discretized diffusion operator

Based on manipulations in the previous subsection, and generalizations to three-dimensional geometry, we identify the discretized neutral diffusion operator

$$
\begin{aligned}
R[T_{i,j,k}] &= \frac{1}{V_{T_{i,j,k}}} \frac{\partial \mathcal{F}[T_{i,j,k}]}{\partial T_{i,j,k}} \\
&= \left(\frac{1}{dyt_{i,j}\, dht_{i,j,k}}\right) \delta_x \left[\frac{1}{dxte_{i-1,j}} \sum_{kr=0}^{1} \Delta_{(i-1,k)}^{(i-1,k-1+kr)} \sum_{ip=0}^{1} (dtwe\, dyt)_{i-1,j}^{(ip)} \right. \\
&\qquad \left. \times A_{(i-1,k|i-1+ip,k-1+kr)}^{(i-1+ip,k)} \left(\delta_x T_{i-1,k} + Sx_{(i-1,k|i-1+ip,k-1+kr)}^{(i-1+ip,k)} \, \delta_z T_{i-1+ip,k-1+kr} \right) \right] \\
&+ \left(\frac{1}{dxt_{i,j}\, dht_{i,j,k}}\right) \delta_y \left[\frac{1}{dytn_{i,j-1}} \sum_{kr=0}^{1} \Delta_{(j-1,k)}^{(j-1,k-1+kr)} \sum_{jq=0}^{1} (dxt\, dtwe)_{i,j-1}^{(jq)} \right. \\
&\qquad \left. \times A_{(j-1,k|j-1+jq,k-1+kr)}^{(j-1+jq,k)} \left(\delta_y T_{j-1,k} + Sy_{(j-1,k|j-1+jq,k-1+kr)}^{(j-1+jq,k)} \, \delta_z T_{j-1+jq,k-1+kr} \right) \right] \\
&+ \left(\frac{1}{dxt_{i,j}}\right) \delta_z \left[\frac{1}{dhwt_{i,k-1}} \sum_{ip=0}^{1} (dtew)_{i,j}^{(ip)} \sum_{kr=0}^{1} \Delta_{(i-1+ip,k-1+kr)}^{(i-1+ip,k-1)} \right. \\
&\qquad \left. \times (A\, Sx)_{(i-1+ip,k-1+kr|i,k-1)}^{(i,k-1+kr)} \left(\delta_x T_{i-1+ip,k-1+kr} + Sx_{(i-1+ip,k-1+kr|i,k-1)}^{(i,k-1+kr)} \, \delta_z T_{i,k-1} \right) \right] \\
&+ \left(\frac{1}{dyt_{i,j}}\right) \delta_z \left[\frac{1}{dhwt_{i,k-1}} \sum_{jq=0}^{1} (dtns)_{i,j}^{(jq)} \sum_{kr=0}^{1} \Delta_{(j-1+jq,k-1+kr)}^{(j-1+jq,k-1)} \right. \\
&\qquad \left. \times (A\, Sy)_{(j-1+jq,k-1+kr|j,k-1)}^{(j,k-1+kr)} \left(\delta_y T_{j-1+jq,k-1+kr} + Sy_{(j-1+jq,k-1+kr|j,k-1)}^{(j,k-1+kr)} \, \delta_z T_{j,k-1} \right) \right].
\end{aligned}
\tag{16.79}
$$

16.7 DIFFUSIVE FLUX COMPONENTS

Although equation (16.79) provides the desired discretized diffusion operator, it is useful to identify the three diffusive flux components since ocean models are typically coded in terms of flux components defined at the tracer cell faces. Implementing no-flux boundary conditions is also easier when identifying fluxes.

16.7.1 Continuum considerations

Given generalized horizontal coordinates and partial cells, some care is required to define flux components. For this purpose, recall that in the continuum the diffusion operator is

$$R(T) = -\nabla \cdot \mathbf{F}, \tag{16.80}$$

where

$$\mathbf{F} = \hat{\mathbf{x}}\, F^{(x)} + \hat{\mathbf{y}}\, F^{(y)} + \hat{\mathbf{z}}\, F^{(z)} \tag{16.81}$$

is the flux vector expressed in terms of physical vector components.* In generalized orthogonal horizontal coordinates, Section 21.11.4 notes that the divergence of a vector is written

$$\nabla \cdot \mathbf{F} = (\mathrm{d}y)^{-1} (\mathrm{d}y\, F^{(x)})_{,x} + (\mathrm{d}x)^{-1} (\mathrm{d}x\, F^{(y)})_{,y} + F^{(z)}_{,z}, \tag{16.82}$$

*Section 21.12.3 on page 485 discusses physical tensor components.

where the squared line element is given by

$$\begin{aligned}(ds)^2 &= (h_1\, d\xi^1)^2 + (h_2\, d\xi^2)^2 + (\mathrm{d}z)^2 \\ &= (\mathrm{d}x)^2 + (\mathrm{d}y)^2 + (\mathrm{d}z)^2,\end{aligned} \tag{16.83}$$

with h_1 and h_2 the stretching functions and (ξ^1, ξ^2) the generalized orthogonal coordinates. For spherical longitude-latitude coordinates $(\xi^1, \xi^2) = (\lambda, \phi)$, the stretching functions are $(h_1, h_2) = (R, R\cos\phi)$ where R is the earth's effective radius.*

When the vertical grid spacing is a function of horizontal and vertical position, as for the case with partial bottom cells, the divergence takes the more general form

$$\begin{aligned}\nabla \cdot \mathbf{F} &= (dy\,\mathrm{d}z)^{-1}\,(dy\,\mathrm{d}z\,F^{(x)})_{,x} + (dx\,\mathrm{d}z)^{-1}\,(dx\,\mathrm{d}z\,F^{(y)})_{,y} + F^{(z)}_{,z} \\ &\equiv (dy\,\mathrm{d}z)^{-1}\,(F^{(x)}_{\text{area}})_{,x} + (dx\,\mathrm{d}z)^{-1}\,(F^{(y)}_{\text{area}})_{,y} + F^{(z)}_{,z}.\end{aligned} \tag{16.84}$$

The area weighted flux components $F^{(x)}_{\text{area}}$ and $F^{(y)}_{\text{area}}$ have units

$$[\mathbf{F}^{(h)}_{\text{area}}] = (L^2 \times \text{tracer concentration} \times L/t). \tag{16.85}$$

16.7.2 Discrete fluxes

Given the above continuum notions, one can identify the discrete versions of the diffusive flux. The flux components are identified directly from equations (16.79) and (16.84):

$$R[T]_{i,j,k} = -\left(\frac{\delta_x F^{(x)}_{i-1,j,k}}{dyt_{i,j}\, dht_{i,j,k}} + \frac{\delta_y F^{(y)}_{i,j-1,k}}{dxt_{i,j}\, dht_{i,j,k}} + \delta_z F^{(z)}_{i,k-1,j} \right), \tag{16.86}$$

where the subscript "area" was dropped from the horizontal fluxes for brevity. The area weighted zonal flux component defined at the east face of T-cell $T_{i,j,k}$ is

$$\begin{aligned}F^{(x)}_{i,j,k} = &- \frac{1}{dxte_{i,j}} \sum_{kr=0}^{1} \Delta^{(i,k-1+kr)}_{(i,k)} \sum_{ip=0}^{1} (dtwe\, dyt)^{(ip)}_{i,j}\, A^{(i+ip,k)}_{(i,k|i+ip,k-1+kr)} \\ &\times \left(\delta_x T_{i,k} + Sx^{(i+ip,k)}_{(i,k|i+ip,k-1+kr)}\, \delta_z T_{i+ip,k-1+kr} \right).\end{aligned} \tag{16.87}$$

The area weighted meridional flux component defined at the north face of T-cell $T_{i,j,k}$ is

$$\begin{aligned}F^{(y)}_{i,j,k} = &- \frac{1}{dytn_{i,j}} \sum_{kr=0}^{1} \Delta^{(j,k-1+kr)}_{(j,k)} \sum_{jq=0}^{1} (dxt\, dtsn)^{(jq)}_{i,j}\, A^{(j+jq,k)}_{(j,k|j+jq,k-1+kr)} \\ &\times \left(\delta_y T_{j,k} + Sy^{(j+jq,k)}_{(j,k|j+jq,k-1+kr)}\, \delta_z T_{j+jq,k-1+kr} \right).\end{aligned} \tag{16.88}$$

---

*For more on generalized orthogonal coordinates, see Section or 21.12.2 (page 484) for a summary, or the discussions in Chapters 20 and 21 for a complete treatment.

The vertical flux component defined at the bottom face of T-cell $T_{i,j,k}$ is

$$
\begin{aligned}
F^{(z)}_{i,j,k} = &- \left(\frac{1}{dxt_{i,j}\, dhwt_{i,j,k}} \right) \sum_{ip=0}^{1} dtew^{(ip)}_{i,j} \sum_{kr=0}^{1} \Delta^{(i-1+ip,k)}_{(i-1+ip,k+kr)} (A\, Sx)^{(i,k+kr)}_{(i-1+ip,k+kr|i,k)} \\
&\times \left(\delta_x T_{i-1+ip,k+kr} + Sx^{(i,k+kr)}_{(i-1+ip,k+kr|i,k-1)}\, \delta_z T_{i,k} \right) \\
&- \left(\frac{1}{dyt_{i,j}\, dhwt_{i,j,k}} \right) \sum_{jq=0}^{1} dtns^{(jq)}_{i,j} \sum_{kr=0}^{1} \Delta^{(j-1+jq,k)}_{(j-1+jq,k+kr)} (A\, Sy)^{(j,k+kr)}_{(j-1+jq,k+kr|j,k)} \\
&\times \left(\delta_y T_{j-1+jq,k+kr} + Sy^{(j,k+kr)}_{(j-1+jq,k+kr|j,k)}\, \delta_z T_{j,k} \right).
\end{aligned} \tag{16.89}
$$

16.7.3 Identifying diagonal components to the diffusion tensor

The off-diagonal terms of the small angle diffusion tensor (14.19) are not separable from the tracers. However, the diagonal terms are separable. It is useful to explicitly identify the nonnegative diagonal elements since they can be computed once per time step and used repeatedly for all tracers.

From the area weighted zonal flux component (16.87), one can identify the area weighted diagonal tensor component

$$
K^{11}_{i,j,k} = \frac{1}{dxte_{i,j}} \sum_{kr=0}^{1} \Delta^{(i,k-1+kr)}_{(i,k)} \sum_{ip=0}^{1} (dtwe\, dyt)^{(ip)}_{i,j} A^{(i+ip,k)}_{(i,k|i+ip,k-1+kr)} \tag{16.90}
$$

which then leads to the area weighted zonal flux component

$$
\begin{aligned}
F^{x}_{i,j,k} = &- K^{11}_{i,j,k}\, \delta_x T_{i,j,k} \\
&- \frac{1}{dxte_{i,j}} \sum_{kr=0}^{1} \Delta^{(i,k-1+kr)}_{(i,k)} \sum_{ip=0}^{1} (dtwe\, dyt)^{(ip)}_{i,j} \,(Sx\, A)^{(i+ip,k)}_{(i,k|i+ip,k-1+kr)}\, \delta_z T_{i+ip,k-1+kr}.
\end{aligned} \tag{16.91}
$$

Likewise, the area weighted meridional flux component (16.88) leads to the area weighted diagonal tensor component

$$
K^{22}_{i,j,k} = \frac{1}{dytn_{i,j}} \sum_{kr=0}^{1} \Delta^{(j,k-1+kr)}_{(j,k)} \sum_{jq=0}^{1} (dxt\, dtsn)^{(jq)}_{i,j} A^{(j+jq,k)}_{(j,k|j+jq,k-1+kr)} \tag{16.92}
$$

which then renders the area weighted meridional flux component

$$
\begin{aligned}
F^{y}_{i,j,k} = &- K^{22}_{i,j,k}\, \delta_y T_{i,j,k} \\
&- \frac{1}{dytn_{i,j}} \sum_{kr=0}^{1} \Delta^{(j,k-1+kr)}_{(j,k)} \sum_{jq=0}^{1} (dxt\, dtsn)^{(jq)}_{i,j} \,(Sy\, A)^{(j+jq,k)}_{(j,k|j+jq,k-1+kr)}\, \delta_z T_{j+jq,k-1+kr}.
\end{aligned} \tag{16.93}
$$

From the vertical flux component (16.89), one has

$$
\begin{aligned}
dhwt_{i,j,k}\, K^{33}_{i,j,k} = &\frac{1}{dxt_{i,j}} \sum_{ip=0}^{1} (dtew)^{(ip)}_{i,j} \sum_{kr=0}^{1} \Delta^{(i-1+ip,k)}_{(i-1+ip,k+kr)} ((Sx)^2\, A)^{(i,k+kr)}_{(i-1+ip,k+kr|i,k)} \\
&+ \frac{1}{dyt_{i,j}} \sum_{jq=0}^{1} (dtns)^{(jq)}_{i,j} \sum_{kr=0}^{1} \Delta^{(j-1+jq,k)}_{(j-1+jq,k+kr)} ((Sy)^2 A)^{(j,k+kr)}_{(j-1+jq,k+kr|j,k)}
\end{aligned} \tag{16.94}
$$

which then leads to the vertical flux component

$$
\begin{aligned}
dhwt_{i,j,k}\, F^{z}_{i,j,k} = & - dhwt_{i,j,k}\, K^{33}_{i,j,k}\, \delta_z T_{i,j,k} \\
& - \frac{1}{dxt_{i,j}} \sum_{ip=0}^{1} (dtew)^{(ip)}_{i,j} \sum_{kr=0}^{1} \Delta^{(i-1+ip,k)}_{(i-1+ip,k+kr)} (A\, Sx)^{(i,k+kr)}_{(i-1+ip,k+kr|i,k)}\, \delta_x T_{i-1+ip,k+kr} \\
& - \frac{1}{dyt_{i,j}} \sum_{jq=0}^{1} (dtns)^{(jq)}_{i,j} \sum_{kr=0}^{1} \Delta^{(j-1+jq,k)}_{(j-1+jq,k+kr)} (A\, Sy)^{(j,k+kr)}_{(j-1+jq,k+kr|j,k)}\, \delta_y T_{j-1+jq,k+kr}.
\end{aligned}
\tag{16.95}
$$

For later discussion given in Section 16.8.3, refer to the $K^{33}\,\delta_z T_{i,j,k}$ term as the *diagonal* portion of the vertical flux component, and the remaining terms represent the *off-diagonal* portion.

For diagnostic purposes, it is often useful to map the diagonal tensor components K^{11}, K^{22}, K^{33}. As defined above, K^{11} and K^{22} represent area weighted versions of their continuum analogs. However, it is relevant to map a field with dimensions of diffusivity in order to compare with the input neutral diffusivity A_I. For this purpose, one may wish to map

$$
\frac{K^{11}_{i,j,k}}{dyt_{i,j}\, dht_{i,j,k}} \tag{16.96}
$$

$$
\frac{K^{22}_{i,j,k}}{dxt_{i,j}\, dht_{i,j,k}} \tag{16.97}
$$

whose dimensions are those of a diffusivity. On a uniform grid in Cartesian space using full vertical cells and neutral slopes less than the critical slope, these fields reduce to the value of the neutral diffusivity A_I. In a spherical geometry using spherical coordinates, full vertical cells, and slopes less than the critical slope, K^{22} is generally less than A_I, whereas K^{11} equals A_I. In a spherical geometry with generalized orthogonal coordinates and/or next to a partial cell bottom, the reduction generally does not occur for either field. This lack of reduction to the input diffusivity is not a problem with the diffusion algorithm, rather it is a limitation of the diagnostic.

16.7.4 Stencils for small angle flux components

A figure is useful to garner insight into the stencil deduced from the functional discretization. Figure 16.3 provides such a stencil for the zonal component to the small angle neutral flux $F^{x}_{i,k}$. Each triad is weighted by the smallest vertical distance consistent with the minimum thickness rule discussed in Section 16.4.3. A similar stencil holds for the meridional component of the neutral diffusive flux.

For the vertical diffusive flux component (Figure 16.4), a set of four triads are used, each of which is rotated by 90 degrees relative to the triads shown for the zonal component to the flux. The weighting for each triad is a combination of the zonal grid spacing and a ratio of the minimum thickness rule spacing, normalized by the relevant vertical T-cell distance. As a result of the rotation, the Δ weighting for each of the four triads is generally different.

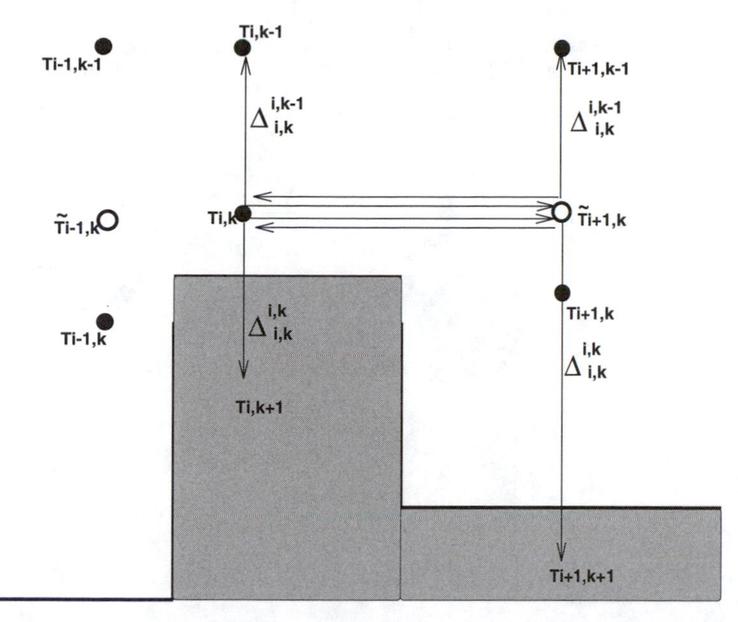

Figure 16.3 Stencil for computing the zonal component to the neutral diffusive flux $F^x_{i,k}$ (equation (16.87)). This flux component is located at the east face of T-cell $T_{i,k}$. Shown are four density triads, each weighted by the relevant vertical spacing factor Δ. For the two triads extending upwards, the weighting is $\Delta^{(i,k-1)}_{(i,k)}$, whereas the two triads extending downwards are weighted by $\Delta^{(i,k)}_{(i,k)}$. The tracer points with the open circles denote tracer values which are determined through linear interpolation. For the two triads extending into rock (shaded regions), they are zeroed out in the code since they should not contribute to the diffusion operator.

16.8 FURTHER ISSUES OF NUMERICAL IMPLEMENTATION

This section presents some general comments and details regarding the implementation of the neutral diffusion scheme.

16.8.1 Gent-McWilliams skew flux plus neutral diffusive flux

Discretization of the small angle neutral diffusion tensor also provides a discretization for the GM skew diffusion process. All that is needed is to make appropriate changes to the diffusivities according to the discussions in Chapter 15.

16.8.2 Discretization of the GM velocity for diagnostics

It is sometimes useful to compute the eddy-induced velocity (equation 14.58)) for diagnostic purposes. In early realizations of the Gent and McWilliams (1990) scheme, such diagnostics were used to compute the meridional-vertical transport streamfunction. However, a simpler and more direct method is available as de-

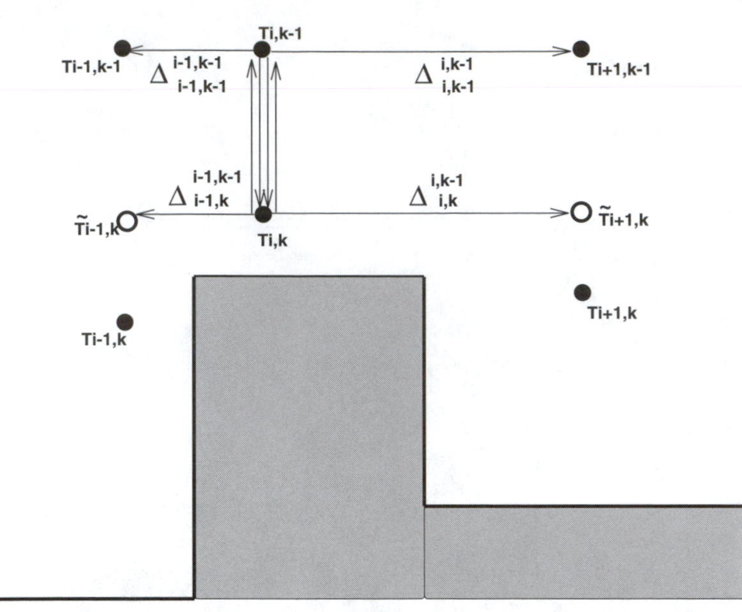

Figure 16.4 Stencil for computing the vertical component neutral diffusive flux $F^{z}_{i,k-1}$ (equation (16.89)), which is located at the bottom face of T-cell $T_{i,k-1}$. Shown are four density triads, each weighted by the relevant vertical spacing factor Δ. Note the rotation of the triads relative to those used to construct the zonal flux in Figure 16.3. Each of the four triads in this figure employ a generally different vertical weighting factor Δ. As in Figure 16.3, tracer points with the open circles denote tracer values which are determined through linear interpolation.

scribed in Section 9.5.3 (page 210).

The following discussion illustrates one choice for discretizing the GM eddy-induced velocity. Additionally, it highlights the ambiguity inherent in the discretization, thus further motivating the skew flux approach. That is, a discretization of GM skewsion relies on the functional formalism of Section 16.1. In contrast, there is no general principle to guide discretization of GM advection. More than one approach has been tried in the literature (see, e.g., Danabasoglu and McWilliams, 1995; Weaver and Eby, 1997). The form chosen here uses unbroken triad-based neutral slopes. In particular, averaging operators are not applied to the numerator and denominator of the discretized form of $\nabla_{z}\rho/\rho_{,z}$. Use of unbroken triads means that the neutral slopes computed for the diffusion calculation can be employed for diagnosis of the eddy-induced velocity.

For computing the zonal eddy-induced velocity, defined at the zonal face of the tracer cell $T_{i,j,k}$, we take the difference between diffusivity weighted slopes averaged so that they live at the bottom-zonal corners of a tracer cell. Thus, define the diffusivity weighted slope

$$\begin{aligned}4\,(\kappa\,Sx)_{i,k} = {}& \kappa_{i,k}\,Sx^{(i,k)}_{(i,k|i,k)} + \kappa_{i+1,k}\,Sx^{(i+1,k)}_{(i,k|i+1,k)} \\ & + \kappa_{i,k+1}\,Sx^{(i,k+1)}_{(i,k+1|i,k)} + \kappa_{i+1,k+1}\,Sx^{(i+1,k+1)}_{(i,k+1|i+1,k)}, \qquad (16.98)\end{aligned}$$

which represents a zonal and vertical average of the four diffusivity weighted triad slopes surrounding the bottom-zonal corner of the tracer cell $T_{i,k-1}$. Likewise, define the diffusivity weighted meridional slope

$$4\,(\kappa\,Sy)_{j,k} = \kappa_{j,k}\,Sy^{(j,k)}_{(j,k|j,k)} + \kappa_{j+1,k}\,Sy^{(j+1,k)}_{(j,k|j+1,k)} + \kappa_{j,k+1}\,Sy^{(j,k+1)}_{(j,k+1|j,k)} + \kappa_{j+1,k+1}\,Sy^{(j+1,k+1)}_{(j,k+1|j+1,k)}. \quad (16.99)$$

Given these slopes, the zonal eddy-induced advection velocity, defined at the zonal face of the tracer cell $T_{i,j,k}$, is given by

$$u^{*}_{i,k} = \frac{(\kappa\,Sx)_{i,k-1} - (\kappa\,Sx)_{i,k}}{dht_{i,k}}, \quad (16.100)$$

and the meridional eddy-induced advection velocity, defined at the meridional face of the tracer cell $T_{i,j,k}$, is given by

$$v^{*}_{j,k} = \frac{(\kappa\,Sy)_{j,k-1} - (\kappa\,Sy)_{j,k}}{dht_{j,k}}. \quad (16.101)$$

The vertical component w^{*} of the eddy-induced advection velocity is obtained by vertically integrating the convergence of the horizontal eddy-advection velocities, as done to compute the resolved scale vertical advection velocity.

16.8.3 Computing the vertical flux convergence

The vertical flux convergence $-\delta_z F^{(z)}$ presents a special case numerically due to the small grid spacing in the vertical relative to the horizontal. As noted by Cox (1987), it is numerically efficient to split the off-diagonal portions of $F^{(z)}$ from the diagonal K^{33} portion. The K^{33} term presents the most severe limitation on the time step, and so it is useful to time step this term implicitly, along with the usual vertical diffusion term. That is, it is common to have the following portion of the diffusion equation time stepped implicitly:

$$\partial_t T = \partial_z\,[(A_D + K^{33})\,T_{,z}]. \quad (16.102)$$

The off-diagonal terms in $F^{(z)}$ present a less severe time step constraint, and so they are time stepped explicitly. Indeed, it is not feasible to time step the off-diagonal components implicitly since to do so would involve inversion of a matrix the size of the model.

The above approach is useful for maintaining numerical stability with long time steps and/or refined vertical grid spacing. However, because of the time split of the vertical flux component $F^{(z)}$, it is generally not possible to rigorously maintain the balance within the vertical flux component of active tracers according to the discussion in Section 14.1.6 (page 301). That is, the continuum relation

$$\alpha\,F^{(z)}(\theta) = \beta\,F^{(z)}(s) \quad (16.103)$$

is not fully satisfied by the discretized flux, so long as the K^{33} piece is time stepped implicitly. As discussed in Griffies et al. (1998), this limitation of the scheme has not proven to be a problem in realistic simulations. Even so, it has been found useful to reduce the imbalance in the vertical flux component by splitting K^{33} into explicit and implicit pieces

$$K^{33} = K^{33}_{\text{explicit}} + K^{33}_{\text{implicit}}. \tag{16.104}$$

The explicit piece is set according to the linear stability analysis given in Section 15.1, with

$$K^{33}_{\text{explicit}} \leq (\Delta z)^2/(2\,\Delta t_{\text{forward}}). \tag{16.105}$$

The time step $\Delta t_{\text{forward}}$ is twice the tracer time step when using a leap-frog time tendency, and equals the tracer time step with a forward time tendency (Section 12.8.4). In either case, the explicit piece is combined with the rest of the vertical flux for explicit time stepping, whereas the implicit piece is split off for implicit time stepping. When $K^{33} \leq (\Delta z)^2/(2\,\Delta t_{\text{forward}})$, then $K^{33}_{\text{implicit}} = 0$, thus keeping the calculation completely explicit.

16.9 CHAPTER SUMMARY

Section 16.1 started the chapter with a theoretical basis for the functional discretization used for the neutral diffusion operator. This approach provides a facility to organize the many details of the discretization within a general principle: the discrete diffusion operator should reduce the global tracer variance.

Section 16.2 began a detailed discussion of the discretization. Outlined here was the overal plan of attack, based on guidance from the continuum functional methodology in Section 16.1. Before moving onto the neutral diffusion operator, Section 16.3 paused to present a one-dimensional warm-up to introduce the functional methods in practice. Then Section 16.4 moved onto the full problem at hand by summarizing the various terms needed to construct the discrete dissipation functional. Section 16.5 described the triad structure of the scheme, whereby groups of three tracer points are singled out as fundamental elements in the discrete dissipation functional. Section 16.6 presented the discrete diffusion operator in its glory. It turns out to be quite beautiful. Even so, it is useful to express this operator in terms of its fundamental diffusive flux components, and that work is presented in Section 16.7. Finally, Section 16.8 outlined some further issues of concern for implementation in a computer model.

As useful as this discretization has proven for z-models, it remains imperfect in ways mentioned earlier in this part of the book. In particular, as described in Section 13.7, there is no guarantee that the discrete operator is monotonic. That is, tracers can, and sometimes do, go outside their physical bounds. In some cases, it is necessary in models to take a "strong-arm" approach and simply revert the neutral physics operator to horizontal diffusion in regions where the tracer is moving well beyond what is sensible. This approach is related to flux limiter methods common in advection schemes. However, it is naive at best, since it only operates when the tracer goes outside some globally defined range. More sophisticated schemes should be considered, like those described by Beckers et al. (1998) and Beckers et al. (2000).

Although an imperfect scheme, the notions introduced in this chapter, many of which are fundamental to any neutral physics discretization, may play a role in shaping future schemes. At the least, the discussion hopefully provides a pedagogical treatment of a method commonly used in z-models run for global climate simulations.

PART 5

Horizontal Friction

Frictional dissipation of kinetic energy in the ocean is tiny (Section 5.3.2). Yet kinetic energy dissipation in ocean climate models is huge. Why the distinction? Ocean models *require friction for numerical reasons*. These reasons are of no concern to the real ocean. One way to understand the numerical need is that the Reynolds number for traditional advection schemes should be kept less than unity, with a length scale given by the model grid scale used to set the Reynolds number. With grid scales of many kilometers rather than the millimeters corresponding to the ocean's Kolmogorov scale, numerical viscosity must be increased accordingly.

Linear momentum is not conserved by fluid parcels. Pressure effects preclude this. Hence, the subgrid scale (SGS) transport of linear momentum is not considered in the same way as a passive tracer (see Chapter 7), where material tracer conservation prompts an analog to Brownian motion (Section 7.1.3). That is, frictional dissipation, realized via diffusion of linear momentum, does not have the same basis as passive tracer diffusion. Gent and McWilliams (1996) address some of these points in the context of closing the momentum equation for mesoscale eddies. They conclude, nonetheless, that one may still wish to close the horizontal momentum equation in a diffusive manner. Others conclude the same when focusing on vertical momentum exchange, such as within the mixed layer and below, as well as lateral momentum mixing along the equator (Richards and Edwards, 2003).

In general, there remains a large degree of uncertainty regarding the details of horizontal momentum closure. For the most part, ocean modelers employ lateral friction as a numerical closure. Many techniques (or more frankly, "tricks") have been tried, such as high order filters, variations on the Laplacian and biharmonic operators (as described in this part of the book), or dissipative momentum transport operators (an approach gaining some popularity with atmospheric modelers). As a numerical closure, the aim is to reduce friction to low levels while suppressing numerical instabilities and/or noise. Hence, friction is *tuned*. Experience with global models reveals a nontrivial sensitivity to details of friction. For example, friction affects transport through frictionally controlled passages, modifies the Gulf Stream separation point (Chassignet and Garraffo, 2001), strongly influences the equatorial undercurrent strength (Large et al., 2001), and changes poleward heat transport by altering transport in boundary currents and gyres. In practice, friction tuning can be one of the least satisfying aspects of ocean climate model construction.

Given the need to apply friction for numerical reasons, is it necessary to damp toward a state of zero kinetic energy? Certainly that is what molecular viscosity does, but the ocean models are so far from resolving the Kolmogorov scale that analogies to molecular friction are unwarranted. Nonetheless, damping toward rest is the common approach, and it is the one explicitly described in this part of the book. There are, however, arguments that suggest dissipation should instead damp the solution toward a nonzero flow state. Holloway (1992) proposes a practical means to realize this approach. It represents a minor extension to the common implementation of ocean model friction. Hence, the discussions here can be readily generalized to Holloway's approach. More generally, his work raises some fundamental questions that fall outside the scope of this book and remain well within the realm of research.

The formulation of friction to be described in this part of the book follows from standard methods in continuum mechanics. Many of the mathematical tools developed in Part 6 are useful here. In particular, these tools help to maintain some basic requirements of symmetry on the sphere, whose absence would disrespect the continuum system's symmetries. However, the sophistication of the mathematics involved in formulating the friction operator should not lead the reader to assume that the material here is physically rigorous. Instead, it is best regarded as fancy engineering with the practical goal, again, of maintaining small levels of friction without incurring unacceptable levels of grid noise.

Chapter Seventeen

HORIZONTAL FRICTION IN MODELS

The purpose of this chapter is to describe the fundamentals of horizontal momentum friction used in many ocean climate models. We make use of an array of techniques and ideas common from elasticity theory. This requires the use of tensor analysis presented in Part 6. However, the reader should not be misled into thinking that the stress-strain relations inspired from elasticity theory have a direct physical connection to the ocean. They may, but this remains unproven. Instead, our aim is to systematically derive friction operators whose mathematical form manifests symmetries of the sphere, and whose realizations satisfy the numerical closure needs of global simulations.

17.1 BOUSSINESQ AND NON-BOUSSINESQ FRICTION

The focus in this chapter is on establishing friction within the context of the unaveraged equations of motion derived in Chapters 3 and 4. This is strictly a misrepresentation of the unaveraged equations, since they employ dissipation operators due to the tiny molecular friction, not the huge friction necessary for the ocean climate models at issue here. Nonetheless, this approach is taken, largely for purposes of tidiness. When transferring the present results to model fields, one should follow the mapping given by Table 8.1 (page 187). In particular, $\mathbf{v} \rightarrow \mathbf{v}^{\rho}$ for the non-Boussinesq fluid. Notably, this mapping means that it is the velocity $\mathbf{v}^{\rho} = \mathbf{v}\,(\rho_o/\rho)$ that is used to construct the friction operator in the non-Boussinesq model, *not* the velocity $\mathbf{v}$. For the Boussinesq model, there is no distinction. In addition, the friction operator in the model represents a discrete version of the averaged momentum flux

$$\langle \tilde{\mathbf{F}}^{(\mathbf{u})}_{\text{sgs}} \rangle \rightarrow \mathbf{F}^{(\mathbf{u})}_{\text{model}}. \tag{17.1}$$

Hence, when considering model implementations, factors of density reduce to their Boussinesq value ρ_o, which simplifies the numerical treatment in the non-Boussinesq case.

17.2 INTRODUCTION AND GENERAL FRAMEWORK

The framework used here for deriving the form of the momentum friction operator is the following.

- Fluid motion preserves the hydrostatic balance.
- Horizontal kinetic energy, which dominates the vertical kinetic energy in hydrostatic flows, is dissipated by friction.

- Friction does not introduce interior sources or sinks of angular momentum. Hence, friction vanishes when the fluid undergoes a solid body rotation. On a plane, friction also vanishes under uniform translations.

- Stresses associated with friction in the interior of the fluid are directly proportional to the local strain acting on fluid parcels. This assumption follows the generalized Hooke's Law approach commonly used in elasticity theory (e.g., Aris, 1962; Landau and Lifshitz, 1986; Smagorinsky, 1993).

It is common to additionally assume that frictional stresses maintain transverse isotropy in which gravity picks out the only special direction (e.g., Smagorinsky, 1993; Pacanowski and Griffies, 1999; Griffies and Hallberg, 2000). However, Smith and McWilliams (2003) suggest that anisotropic transverse stresses are more appropriate. The freedom afforded from transverse anisotropy allows one, in particular, to prescribe a friction force that is a function of the flow direction and/or an *a priori* specified direction. This has been found to be of use for ocean climate modeling, especially when aiming to increase the strength of simulated equatorial undercurrents (Large et al., 2001).

For simplicity, the analysis in much of this chapter uses Cartesian coordinates and Cartesian tensors. However, we aim to keep results coordinate invariant and so follow the general index placement necessary with curvilinear tensors.* This notation helps in Section 17.7, which details the friction operator appropriate for a generalized orthogonal coordinate ocean model.

17.3 PROPERTIES OF THE STRESS TENSOR

As seen in Section 4.3.2 (page 48), forces acting on an element of a continuous medium can be classified into two categories. *External or body* forces act throughout the medium, with forces from gravity (including tides), Coriolis, and electromagnetism common examples. *Internal or contact* forces act on an element of volume through its bounding surface, with forces from pressure or other stresses common examples. The balance between all forces with acceleration leads, through Newton's second law, to the equations of motion. Furthermore, if all torques acting on a fluid parcel arise from macroscopic forces, which is the case for Newtonian fluids such as seawater, then fluid parcels conserve angular momentum, and so should ocean model solutions. Notably, however, the presence of meridional boundaries precludes angular momentum conservation in the World Ocean, except in regions of the Southern Ocean devoid of meridional barriers.†

As introduced in Section 4.3.4 (page 49), the stresses acting within a continuous medium can be organized into a second order stress tensor with generally 3×3 independent elements. The divergence of these stresses gives rise to the internal forces acting in the medium. A proper account of the angular momentum budget implies that the stress tensor is symmetric, which brings the number of independent stress elements down to six.

---

*See section Section 21.1 (page 466) and 21.12 (page 481) for discussion of curvilinear tensors.

†More is said regarding angular momentum in Sections 4.10 and 4.11.

17.3.1 The strain tensor

Consider two infinitesimally close fluid parcels with material coordinates ζ^a and $\zeta^a + \mathrm{d}\zeta^a$, where $a = 1, 2, 3$. The components

$$(u^1, u^2, u^3) = \frac{\mathrm{d}}{\mathrm{d}t}(x^1, x^2, x^3) \tag{17.2}$$

of the velocity for these two parcels differ by the increment

$$\begin{aligned} \mathrm{d}u^m &= \frac{\partial u^m}{\partial \zeta^a}\,\mathrm{d}\zeta^a \\ &= \frac{\partial u^m}{\partial \zeta^a}\frac{\partial \zeta^a}{\partial x^n}\,\mathrm{d}x^n \\ &= \frac{\partial u^m}{\partial x^n}\,\mathrm{d}x^n \\ &= u^m_{,n}\,\mathrm{d}x^n, \end{aligned} \tag{17.3}$$

where the summation convention is used in which repeated indices are summed over their range. The velocity derivatives

$$u^m_{,n} = \frac{\partial u^m}{\partial x^n} \tag{17.4}$$

form the components to a second order tensor. In order to attach physical significance to this tensor, it is useful to separately consider its symmetric and antisymmetric components,* which are written

$$u_{m,n} = \Omega_{mn} + e_{mn}, \tag{17.5}$$

where

$$2\,\Omega_{mn} = u_{m,n} - u_{n,m} \tag{17.6}$$

$$2\,e_{mn} = u_{m,n} + u_{n,m} \tag{17.7}$$

and $u_m = g_{mn}\,u^n$ are the covariant components to the velocity vector. Note that the components to the metric tensor g_{mn} are equal to the Kronecker symbol δ_{mn} when working with Cartesian coordinates.† In curvilinear coordinates, the partial derivatives appearing in Ω_{mn} and e_{mn} generalize to covariant derivatives. This generalization is considered in Section 17.7.1, where the comma notation for partial derivative introduced in equation (17.4) is replaced by a semicolon for covariant derivative.‡

The antisymmetric piece of the velocity derivative tensor is related to the vorticity through

$$\begin{aligned} 2\,\omega^m &= -\epsilon^{mnp}\,\Omega_{np} \\ &= -\epsilon^{mnp}\,u_{n,p} \end{aligned} \tag{17.8}$$

where ϵ^{mnp} is the Levi-Civita symbol (Section 20.12, page 459). In Cartesian vector notation, this result takes the form

$$\boldsymbol{\omega} = \frac{1}{2}\,\nabla \wedge \mathbf{u}. \tag{17.9}$$

---

*A similar decomposition is considered in Section 13.4 (page 288) which discusses subgrid scale tracer transport tensors.

†More general metric tensors are described in Chapters 6, 20, and 21. See in particular the summary discussion in Section 21.12 (page 481).

‡Chapter 21 discusses covariant derivatives.

Standard results from fluid mechanics establish the connection between vorticity and rigid body rotation of a fluid parcel.* If the motion is completely rigid, which means that it consists of a translation plus a rotation, then the symmetric part of the velocity derivative tensor vanishes. Consequently, the symmetric tensor e_{mn} is called the *deformation* or *rate of strain* tensor since it represents deviations from rigid body motion.

To provide a further interpretation of the strain tensor, consider the squared arc-distance $(\mathrm{d}s)^2 = \mathrm{d}s^2$ between two infinitesimally close material parcels of fluid,

$$\begin{aligned} \mathrm{d}s^2 &= g_{mn}\,\mathrm{d}x^m\,\mathrm{d}x^n \\ &= g_{mn}\,\frac{\partial x^m}{\partial \zeta^a}\frac{\partial x^n}{\partial \zeta^b}\,\mathrm{d}\zeta^a\,\mathrm{d}\zeta^b. \end{aligned} \tag{17.10}$$

The material time derivative of this distance is given by

$$\frac{\mathrm{d}(\mathrm{d}s^2)}{\mathrm{d}t} = g_{mn}\left(\frac{\partial u^m}{\partial \zeta^a}\frac{\partial x^n}{\partial \zeta^b} + \frac{\partial x^m}{\partial \zeta^a}\frac{\partial u^n}{\partial \zeta^b}\right)\mathrm{d}\zeta^a\,\mathrm{d}\zeta^b, \tag{17.11}$$

where $\mathrm{d}\zeta^a/\mathrm{d}t = 0$ since ζ^a are material coordinates. Use of the chain rule in the forms

$$\frac{\partial u^m}{\partial \zeta^a}\,\mathrm{d}\zeta^a = u^m_{\ ,n}\,\mathrm{d}x^n \tag{17.12}$$

$$\frac{\partial x^m}{\partial \zeta^a}\,\mathrm{d}\zeta^a = \mathrm{d}x^m \tag{17.13}$$

yields

$$\frac{\mathrm{d}(\mathrm{d}s^2)}{\mathrm{d}t} = 2\,e_{mn}\,\mathrm{d}x^m\,\mathrm{d}x^n, \tag{17.14}$$

or equivalently

$$\frac{1}{\mathrm{d}s}\frac{\mathrm{d}(\mathrm{d}s)}{\mathrm{d}t} = e_{mn}\,\frac{\mathrm{d}x^m}{\mathrm{d}s}\frac{\mathrm{d}x^n}{\mathrm{d}s}. \tag{17.15}$$

Now $\mathrm{d}x^m/\mathrm{d}s$ is a component of the unit vector pointing from one fluid parcel to the other. Hence, equation (17.15) says that the rate of change of the infinitesimal distance separating the two parcels, as a fraction of the distance, is related to the relative position of the parcels through the strain tensor. Section 4.42 of Aris (1962), among others, provides further elaboration.

17.3.2 Relating strain to stress

The divergence of the stress tensor yields the internal forces acting on a fluid parcel. Stress is related to strain, where strain arises from the kinematics of parcel deformations. In elasticity theory, the relation between stress and strain is typically assumed to follow some form of Hooke's law. In its simplest form, this "law" linearly relates the stress to the strain.† In fluid dynamics, it is common

---

*See also Section 4.10.2 (page 76) for a discussion of solid body motion of a particle on a rotating sphere.

†A similar relation between dynamics and kinematics forms the basis of general relativity, where stresses associated with matter/energy are directly proportional to space-time curvature.

to also assume a stress-strain relation in the form of Hooke's law. Details of this relation often depend on properties of the fluid as well as the flow state. Such dependencies can generally make the fluid's stress-strain relation nonlinear.

When the fluid is static and exposed only to buoyancy forces, it remains in hydrostatic balance, and the only form of stress on a fluid parcel is due to the pressure. Hence, the stress tensor for such a state takes the form

$$T^{mn} = -p\,\delta^{mn}, \tag{17.16}$$

where p is the pressure and δ^{mn} is the Kronecker delta.*

When the fluid undergoes deformations, further stresses bring the stress tensor to the more general form

$$T^{mn} = -p\,\delta^{mn} + \tilde{\tau}^{mn}. \tag{17.17}$$

For purposes of formulating friction to be used in an ocean model, the divergence of $\tilde{\tau}^{mn}$ is associated here with dissipative stresses in the fluid, thus motivating the name *frictional stress tensor*. For a Newtonian fluid, the frictional stress tensor can be written

$$\tilde{\tau}^{mn} = \rho\, C^{mnpq}\, e_{pq}. \tag{17.18}$$

In general, this relation between stress and strain is of the form of Hooke's law, where the components C^{mnpq} of the fourth order kinematic *viscosity tensor* can depend on the state of the fluid. Assuming this form for the internal stresses, the essential problem with subgrid scale parameterization of momentum fluxes reduces to determining appropriate forms for C^{mnpq}.

17.3.3 Angular momentum and symmetry of the stress tensor

As mentioned in Section 17.3.1, the symmetric rate of strain tensor e_{mn} vanishes for motion consisting of rigid rotation plus uniform translation. In such cases, the generalized Hooke's law (17.18) says that the stress tensor T^{mn} reduces to $-p\,\delta^{mn}$ since $\tilde{\tau}^{mn}$ vanishes. The purpose of this section is to provide some further details regarding these ideas and their connection to conservation of angular momentum. Some of these ideas were described in Section 4.11 (page 80).

As seen in Section 4.3.6 (page 51), Newton's second law is given by

$$\rho\,\frac{\mathrm{d}u^m}{\mathrm{d}t} = \rho\, f^m + T^{mn}_{,n} \tag{17.19}$$

where ρ is the mass density, and f^i are components to external or body forces such as those arising from gravity and the Coriolis force. $T^{mn}_{,n}$ is the divergence of the stress tensor, where T^{mn} is written in the form (17.17) which incorporates the pressure. A component of the angular momentum for a fluid parcel is given by

$$L_m = \epsilon_{mnp}\, x^n\, u^p\, \rho\, \mathrm{d}V, \tag{17.20}$$

where $\rho\,\mathrm{d}V$ is the mass of the infinitesimal parcel. The material time derivative of the parcel's angular momentum is given by†

$$\frac{\mathrm{d}L_m}{\mathrm{d}t} = \epsilon_{mnp}\, x^n\, \frac{\mathrm{d}u^p}{\mathrm{d}t}\, \rho\, \mathrm{d}V, \tag{17.21}$$

*Chapter 1 of Salmon (1998) provides some comments on the implicit identification of this mechanical pressure with thermodynamic pressure.

†For a Boussinesq fluid, ρ appears as the constant ρ_o, and $\mathrm{d}(\mathrm{d}V)/\mathrm{d}t = 0$ then follows from volume conservation.

where $\mathrm{d}(\rho\,\mathrm{d}V)/\mathrm{d}t = 0$ follows from mass conservation, and $\epsilon_{mnp}\,u^n\,u^p = 0$. Substituting Newton's law into this expression leads to

$$\frac{\mathrm{d}L_m}{\mathrm{d}t} = \epsilon_{mnp}\,(x^n\,\rho\,f^p + x^n\,T^{pq}_{,q})\,\mathrm{d}V. \tag{17.22}$$

The first term accounts for torques placed on the parcel from external forces. The second term arises from torques on the fluid from internal stresses. To further interpret the second term, consider the budget for total angular momentum of the fluid, which is obtained by integrating over the fluid volume

$$\int \frac{\mathrm{d}L_m}{\mathrm{d}t} = \int \epsilon_{mnp}\,(x^n\,\rho\,f^p + x^n\,T^{pq}_{,q})\,\mathrm{d}V. \tag{17.23}$$

Now integrate by parts on the stress tensor term to find

$$\int \epsilon_{mnp}\,x^n\,T^{pq}_{,q}\,\mathrm{d}V = \int \epsilon_{mnp}\,[\partial_q\,(x^n\,T^{pq}) - T^{pn}]\,\mathrm{d}V. \tag{17.24}$$

The first term integrates to a boundary contribution, which is nonvanishing for cases in which there are torques arising from boundary stresses. The second term is a volume contribution and it picks out the term $\epsilon_{mnp}\,\tilde{\tau}^{pn}$, since $\epsilon_{mnp}\,\delta^{pn} = 0$. For Newtonian fluids, such as ocean water, the internal torques are balanced and so there is no net contribution to angular momentum from internal stresses. This case can be ensured if the frictional stress tensor is symmetric,

$$\tilde{\tau}^{mn} = \tilde{\tau}^{nm}, \tag{17.25}$$

thus giving $\epsilon_{mnp}\,\tilde{\tau}^{pn} = 0$. This result provides the reason for considering only symmetric stress tensors in the following.

17.3.4 Quasi-hydrostatic approximation

As pointed out by Smith and McWilliams (2003), there is some ambiguity regarding use of the quasi-hydrostatic approximation insofar as it reduces the form of the stress tensor. We follow here their pragmatic approach whereby it is noted that the aim is to derive friction for a fluid satisfying the hydrostatic approximation, where friction is applied explicitly only to the horizontal momentum. Friction due to deformations of the vertical velocity is ignored, since the vertical momentum equation is reduced to hydrostatic balance. This forms the *quasi-hydrostatic* approximation, where the "quasi" label signifies that the fluid is not static, although the hydrostatic balance is maintained.

These considerations suggest splitting the frictional stress tensor into two subtensors

$$\tilde{\tau}^{mn} = \tau^{mn}_{\text{vert}} + \tau^{mn}. \tag{17.26}$$

The first part of the stress tensor, τ^{mn}_{vert}, is assumed to arise only from vertical deformations of the horizontal velocity field. It can be written as

$$\begin{aligned} \tau^{mn}_{\text{vert}} &= 2\,\rho\,\kappa\,\delta^{n3}\,e^{m3} \\ &\approx \rho\,\kappa\,\delta^{n3}\,\partial_3\,u^m, \end{aligned} \tag{17.27}$$

with κ a non-negative viscosity and $m = 1, 2$. The second tensor, τ^{mn}, embodies the stress arising from horizontal deformations of the horizontal velocity field.

Consequently, the labels $m, n = 1, 2$ range only over the transverse coordinates, with τ^{mn} set to zero if one of the indices equals 3. It is symmetric, according to the requirements of angular momentum conservation discussed in Section 17.3.3. Hence, the general form of the frictional stress tensor is given by

$$\tilde{\tau}^{mn} = \begin{pmatrix} \tau^{11} & \tau^{12} & \rho\,\kappa\,u_{,z} \\ \tau^{12} & \tau^{22} & \rho\,\kappa\,v_{,z} \\ \rho\,\kappa\,u_{,z} & \rho\,\kappa\,v_{,z} & 0 \end{pmatrix}. \tag{17.28}$$

Note that $\tilde{\tau}^{33} = 0$ is consistent with use of the hydrostatic approximation, which reduces the vertical momentum equation to the inviscid hydrostatic balance.

17.3.5 Trace-free frictional stress

The frictional stress tensor under consideration here is a deviatoric stress tensor (e.g., Smagorinsky, 1993; Salmon, 1998), which has a zero trace

$$\begin{aligned} \delta_{mp}\,\tilde{\tau}^{mp} &= \tilde{\tau}^{1}_{1} + \tilde{\tau}^{2}_{2} + \tilde{\tau}^{3}_{3} \\ &= 0. \end{aligned} \tag{17.29}$$

One index was lowered on the stress tensor components in order to yield a coordinate invariant expression for the trace. Pressure incorporates the trace part of the general stress tensor, as seen in equation (17.17). As mentioned in Section 17.3.4, the quasi-hydrostatic approximation allows one to ignore stresses due to deformations in the vertical velocity field, and so $\tilde{\tau}^{33} = 0$ without loss of generality. The trace-free and symmetric conditions then lead to the transverse stress tensor

$$\tau^{mn} = \begin{pmatrix} \tau^{11} & \tau^{12} \\ \tau^{12} & -\tau^{11} \end{pmatrix}. \tag{17.30}$$

Specification of this tensor, and the corresponding friction vector, provides the focus for the remainder of this chapter.

17.3.6 Forces acting on a horizontal plane

Recall from Section 4.3.5 (page 50) that the internal force acting on a surface is given by the area integrated stress acting on that surface. With stress due to pressure and friction, the force is

$$\mathbf{F}_{(\text{internal})} = \int (\tilde{\boldsymbol{\tau}} \cdot \hat{\mathbf{n}} - p\,\hat{\mathbf{n}})\,\mathrm{d}A, \tag{17.31}$$

where $\tilde{\boldsymbol{\tau}} \cdot \hat{\mathbf{n}}$ is shorthand for $\tilde{\tau}^{ab}\,n_b$ and the integral is taken over the bounding surface of the domain whose outward normal is $\hat{\mathbf{n}}$. Consider an example of this force for the case of internal forces acting within the fluid on an imaginary horizontal plane. Figure 17.1 illustrates such a situation, with a vertically sheared zonal velocity field and an imaginary horizontal plane inserted in the middle of the flow.

The pressure force acting on the top side of this horizontal plane, with outward normal $\hat{\mathbf{n}} = \hat{\mathbf{z}}$, is

$$\mathbf{F}^{(\text{top})}_{(\text{pressure})} = -\hat{\mathbf{z}} \int p\,\mathrm{d}A, \tag{17.32}$$

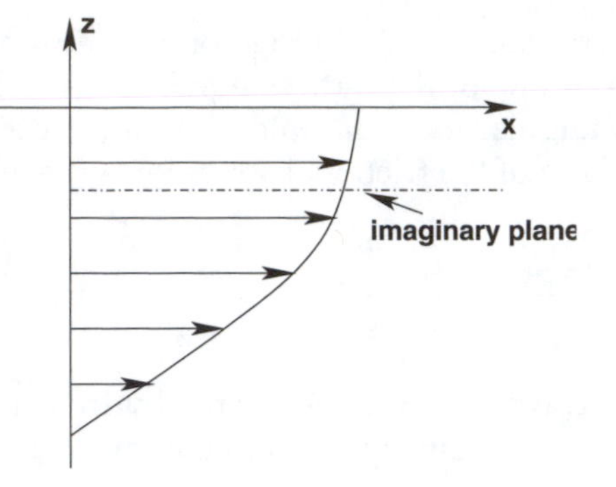

Figure 17.1 Schematic of a horizontal velocity field whose strength increases toward the ocean surface. Drawn is an imaginary plane which experiences forces from the viscous fluid.

where $\mathrm{d}A = \mathrm{d}x\,\mathrm{d}y$ is the horizontal area element for the plane. The pressure force acting on the bottom side of the plane, with outward normal $\hat{\mathbf{n}} = -\hat{\mathbf{z}}$, has the opposite sign,

$$\mathbf{F}^{(\text{bottom})}_{(\text{pressure})} = \hat{\mathbf{z}} \int p\,\mathrm{d}A, \tag{17.33}$$

thus expressing the compressive nature of the pressure force as it acts antiparallel to the surface's outward normal.

For the frictional force on the top of the plane, $\widetilde{\boldsymbol{\tau}} \cdot \hat{\mathbf{n}} = \widetilde{\tau}^{m3}$ since $\hat{\mathbf{n}} = \hat{\mathbf{z}} = (0,0,1)$, thus leading to

$$F^{m}_{(\text{friction})} = \int \widetilde{\tau}^{m3}\,\mathrm{d}A. \tag{17.34}$$

In words this equation makes clear that $\widetilde{\tau}^{m3}$ is the frictional stress in the mth direction acting on the top of the horizontal surface with outward normal $\hat{\mathbf{n}} = (0,0,1)$. For the quasi-hydrostatic frictional stress tensor (17.28), this frictional force acts within the horizontal plane, and thus takes the form

$$\mathbf{F}^{(\text{top})}_{(\text{friction})} = \int (\rho\,\kappa\,\mathbf{u}_{,z})\,\mathrm{d}A. \tag{17.35}$$

For the case with a zonal current increasing toward the ocean surface, so that $u_{,z} > 0$ as in Figure 17.1, zonal friction on top of the plane imparts a positive zonal force, whereas on the bottom of the plane

$$\mathbf{F}^{(\text{bottom})}_{(\text{friction})} = -\int (\rho\,\kappa\,\mathbf{u}_{,z})\,\mathrm{d}A. \tag{17.36}$$

Consider a special case with $\rho\,\kappa$ vertically constant and a linear shear so that $u_{,z}$ is constant. The viscous force on top of the infinitesimally thin plane then balances that below. If there is curvature to the shear, as in Figure 17.1, viscous forces will not balance, thus leading to stress divergence and so a net force within the plane.

17.3.7 Concerning the sign convention for stress and strain

This is an opportune point to comment on the sign convention used to define stress and strain. Recall that the stress tensor components $\widetilde{\tau}^{mn}$ are forces per unit area acting in the mth direction on a plane whose outward normal is in the nth direction. The rate of strain tensor components e_{mn} are likewise defined. The convention to orient stress and strain components according to the outward normal to a surface could just as well be modified to orient the components according to the inward normal. No signs would change in the stress-strain relation given by the Hooke's law expression (17.18). Given this sign, and given the concern with viscous stresses that act to dissipate kinetic energy in the flow, one is led to the Hooke's law expression (17.18) with stress and strain related by a positive sign.

To pursue these comments a bit further, revisit the expression for stress acting on a horizontal plane in the vertically sheared fluid of Figure 17.1: $\widetilde{\tau}^{m3} = \pm\rho\,\kappa\,u^{m}_{,z}$, with the positive sign for the upper side of the plane, and the negative for the lower side. Again, this relation yields the mth component of viscous stress acting on a plane whose outward normal is in the $\pm\hat{\mathbf{z}}$ direction, respectively. This relation for stress can be compared to those for the vertical flux of horizontal momentum arising from viscous forces. For the vertically sheared example of Figure 17.1, where shear has a nontrivial curvature, viscous forces slow the near surface fluid and speed the deeper fluid. That is, momentum fluxes due to viscosity are downgradient and take the form $f^{m}_{(\text{momentum})} = -\rho\,\kappa\,u^{m}_{,z}$. Hence, the process whereby viscosity leads to a net frictional force acting on a point in the fluid can be thought of equally in two ways: (1) diverging viscous stresses or (2) converging downgradient momentum fluxes.

17.4 PROPERTIES OF THE VISCOSITY TENSOR

For three-dimensional systems, the viscosity tensor C^{mnpq} generally contains 81 degrees of freedom ($81 = 3 \times 3 \times 3 \times 3$). However, the properties just described, and others to follow, greatly reduce this number. Focusing on τ^{mn} for $m, n = 1, 2$ allows one to start with $16 = 2 \times 2 \times 2 \times 2$ degrees of freedom. Hence, in the following discussion all labels are assumed to run over just the transverse coordinates $1, 2$.

17.4.1 Hooke's law

The Hooke's law form for stress

$$\tau^{mn} = \rho\, C^{mnpq}\, e_{pq} \tag{17.37}$$

indicates that the only relevant forms of the viscosity tensor are those satisfying

$$C^{mnpq} = C^{mnqp} \tag{17.38}$$

since the rate of strain tensor e_{pq} is symmetric. This constraint reduces the degrees of freedom in the viscosity tensor to $2 \times 2 \times 3 = 12$. The 3 arises from the $2 + 1 = 3$ degrees of freedom in symmetric 2×2 matrices. The general viscosity tensor

$$C^{mnpq} = \begin{pmatrix} C^{1111} & C^{1112} & C^{1121} & C^{1122} \\ C^{1211} & C^{1212} & C^{1221} & C^{1222} \\ C^{2111} & C^{2112} & C^{2121} & C^{2122} \\ C^{2211} & C^{2212} & C^{2221} & C^{2222} \end{pmatrix} \tag{17.39}$$

can therefore be written in the restricted form

$$C^{mnpq} = \begin{pmatrix} C^{1111} & C^{1112} & C^{1112} & C^{1122} \\ C^{1211} & C^{1212} & C^{1212} & C^{1222} \\ C^{2111} & C^{2112} & C^{2112} & C^{2122} \\ C^{2211} & C^{2212} & C^{2212} & C^{2222} \end{pmatrix}. \tag{17.40}$$

17.4.2 Angular momentum

Assuming a symmetric stress tensor brings about the following symmetry on the viscosity tensor:

$$C^{mnpq} = C^{nmpq}. \tag{17.41}$$

Consequently, the degrees of freedom become $3 \times 3 = 9$, and the viscosity tensor takes the form

$$C^{mnpq} = \begin{pmatrix} C^{1111} & C^{1112} & C^{1112} & C^{1122} \\ C^{1211} & C^{1212} & C^{1212} & C^{1222} \\ C^{1211} & C^{1212} & C^{1212} & C^{1222} \\ C^{2211} & C^{2212} & C^{2212} & C^{2222} \end{pmatrix}. \tag{17.42}$$

17.4.3 Kinetic energy dissipation

The budget for horizontal kinetic energy per volume of a fluid parcel is given by

$$\begin{aligned} \frac{\rho}{2} \frac{\mathrm{d}(\delta_{mn}\, u^m\, u^n)}{\mathrm{d}t} &= \rho\, \delta_{mn}\, u^m\, f^n + \delta_{mn}\, u^m\, T^{np}_{\ ,p} \\ &= \rho\, u_m\, f^m + (u_n\, T^{np})_{,p} - u_{n,p}\, T^{np} \\ &= \rho\, u_m\, f^m + (u_n\, T^{np})_{,p} - e_{np}\, T^{np} \\ &= \rho\, u_m\, f^m + (u_n\, T^{np})_{,p} + p\, u^n_{\ ,n} - e_{np}\, \tau^{np}, \end{aligned} \tag{17.43}$$

where Cartesian coordinates are used for simplicity.* The first term on the right-hand side arises from work done by external forces. The second term, when integrated over the fluid domain, accounts for work done at boundaries by the stresses. The third term arises from pressure work against changes in the parcel's volume. This term vanishes for a volume conserving fluid. The fourth term is present throughout the fluid domain, and it can be written

$$e_{mn}\, \tau^{mn} = \rho\, e_{mn}\, C^{mnpq}\, e_{pq}. \tag{17.44}$$

In general, this term is sign-indefinite. However, if one insists that the frictional stress tensor manifests dissipative friction at each point in the fluid, then

$$e_{mn}\, C^{mnpq}\, e_{pq} \geq 0. \tag{17.45}$$

Since the rate of strain tensor e_{mn} is symmetric, this constraint is satisfied if

$$C^{mnpq} = C^{pqmn}. \tag{17.46}$$

---

*See Section 5.3.2, page 97 for the general case.

This constraint brings the number of degrees of freedom in the viscosity tensor down to $6 = 3 + 2 + 1$, which is the number of degrees of freedom in a 3×3 symmetric matrix. The resulting viscosity tensor takes the form

$$C^{mnpq} = \begin{pmatrix} C^{1111} & C^{1112} & C^{1112} & C^{1122} \\ C^{1112} & C^{1212} & C^{1212} & C^{1222} \\ C^{1112} & C^{1212} & C^{1212} & C^{1222} \\ C^{1122} & C^{1222} & C^{1222} & C^{2222} \end{pmatrix}. \tag{17.47}$$

17.4.4 Trace-free transverse stress

As discussed in Section 17.3.5, the transverse stress tensor is trace-free. Hence,

$$\begin{aligned} \tau^{11} + \tau^{22} &= \left(C^{1111} + C^{1122}\right) e_{11} + 2\left(C^{1112} + C^{2212}\right) e_{12} + \left(C^{1122} + C^{2222}\right) e_{22} \\ &= 0. \end{aligned} \tag{17.48}$$

Given the general independence of the three strain tensor components, this constraint is satisfied for all cases only when the three terms individually vanish,

$$C^{1122} = -C^{1111} \tag{17.49}$$

$$C^{2212} = -C^{1112} \tag{17.50}$$

$$C^{1122} = -C^{2222}. \tag{17.51}$$

There are now three degrees of freedom in the viscosity tensor:

$$C^{mnpq} = \begin{pmatrix} \alpha & \gamma & \gamma & -\alpha \\ \gamma & \delta & \delta & -\gamma \\ \gamma & \delta & \delta & -\gamma \\ -\alpha & -\gamma & -\gamma & \alpha \end{pmatrix}, \tag{17.52}$$

where

$$C^{1111} = \alpha \tag{17.53}$$

$$C^{1112} = \gamma \tag{17.54}$$

$$C^{1212} = \delta. \tag{17.55}$$

introduces a shorthand notation for the three viscosities.

17.5 TRANSVERSE ISOTROPY

Gravity breaks three-dimensional isotropy down to transverse isotropy about the local vertical direction $\hat{\mathbf{z}}$. Before considering the more general anisotropic case in Section 17.6, consider here the isotropic case.

17.5.1 Viscosity tensor

Transverse, or axial, isotropy means two things. First, the physical system remains invariant under arbitrary rotations about the $\hat{\mathbf{z}}$ direction. Second, the physical system remains invariant under the transformation $z \rightarrow -z$, and $x \rightarrow y, y \rightarrow x$, which is a transformation between two right-handed coordinate systems, with the vertical pointing up and down, respectively. Since friction is directly proportional to viscosity, and since the physical system is directly affected by friction, transverse isotropy for the physical system means that the viscosity tensor must also manifest this symmetry.

The transformation matrix for rotation about the vertical takes the form

$$\Lambda^{\overline{m}}_{m} = \begin{pmatrix} \cos\varphi & \sin\varphi & 0 \\ -\sin\varphi & \cos\varphi & 0 \\ 0 & 0 & 1 \end{pmatrix}, \tag{17.56}$$

and the transformation matrix between right-handed coordinate systems takes the form

$$\Lambda^{\overline{m}}_{m} = \begin{pmatrix} 0 & 1 & 0 \\ 1 & 0 & 0 \\ 0 & 0 & -1 \end{pmatrix}. \tag{17.57}$$

Under an arbitrary transformation, the fourth order viscosity tensor transforms as*

$$C^{\overline{m}\,\overline{n}\,\overline{p}\,\overline{q}} = \Lambda^{\overline{m}}_{m}\,\Lambda^{\overline{n}}_{n}\,\Lambda^{\overline{p}}_{p}\,\Lambda^{\overline{q}}_{q}\,C^{mnpq}. \tag{17.58}$$

Note that restricting attention to the two-dimensional transverse subspace allows us to focus on the two-dimensional portion of these transformation matrices. In general, transverse isotropy imposes the constraint on the viscosity tensor

$$\begin{aligned} C^{\overline{m}\,\overline{n}\,\overline{p}\,\overline{q}} &= \Lambda^{\overline{m}}_{m}\,\Lambda^{\overline{n}}_{n}\,\Lambda^{\overline{p}}_{p}\,\Lambda^{\overline{q}}_{q}\,C^{mnpq} \\ &\equiv C^{mnpq}, \end{aligned} \tag{17.59}$$

where $\Lambda^{\overline{m}}_{m}$ is one of the given transformation matrices.

Determining the relations between the elements of C^{mnpq} requires enumeration of the possibilities. For example, with a rotation angle of $\pi/2$ about $\hat{\mathbf{z}}$, rotational symmetry implies

$$\begin{aligned} C^{\overline{1}\,\overline{2}\,\overline{2}\,\overline{2}} &= -C^{2111} \\ &\equiv C^{1222}, \end{aligned} \tag{17.60}$$

which is a constraint also imposed by the trace-free condition discussed in Section 17.4.4. However, the transformation between two right-handed coordinate systems implies

$$C^{1222} = C^{2111}. \tag{17.61}$$

These two results are satisfied only if

$$\begin{aligned} C^{1222} &= C^{2111} \\ &= 0 \\ &\Rightarrow \gamma = 0. \end{aligned} \tag{17.62}$$

*Chapters 6, 20, and 21 provide discussions of such transformation rules, and Section 21.12 (page 481) provides a summary of certain salient points.

Likewise, insisting on isotropy when rotating by $\pi/4$ implies

$$C^{1111} = (C^{1111} + C^{1122} + 2\,C^{1212})/2, \tag{17.63}$$

or

$$\begin{aligned} C^{1212} &= (C^{1111} - C^{1122})/2 \\ &\Rightarrow \delta = \alpha. \end{aligned} \tag{17.64}$$

Consequently, the transverse isotropic viscosity tensor takes the simple form

$$C^{mnpq} = \alpha \begin{pmatrix} 1 & 0 & 0 & -1 \\ 0 & 1 & 1 & 0 \\ 0 & 1 & 1 & 0 \\ -1 & 0 & 0 & 1 \end{pmatrix}, \tag{17.65}$$

thus manifesting only a single viscous degree of freedom.

17.5.2 Stress tensor

The transverse stress tensor satisfying transverse isotropy is given by

$$\begin{aligned} \tau^{mn} &= \rho\,A \begin{pmatrix} (e_{11} - e_{22}) & 2\,e_{12} \\ 2\,e_{12} & (e_{22} - e_{11}) \end{pmatrix} \\ &= \rho\,A\,(2\,e^{mn} - g^{mn}\,e^{p}_{p}), \end{aligned} \tag{17.66}$$

where $\alpha = A$ as this is the common notation for viscosity in the isotropic case (e.g., Pacanowski and Griffies, 1999; Griffies and Hallberg, 2000). Notice how the stress is a function only of the two combinations of the rate of strain $e_{11} - e_{22}$ and $2\,e_{12}$, instead of the three components e_{11}, e_{22}, e_{12}. For this reason, it is common to define the tension and shearing rate of strain, also known as the *deformation rates*, as

$$\begin{aligned} e_{\mathrm{T}} &= e_{11} - e_{22} \\ &= u_{1,1} - u_{2,2} \end{aligned} \tag{17.67}$$

$$\begin{aligned} e_{\mathrm{S}} &= 2\,e_{12} \\ &= u_{1,2} + u_{2,1}. \end{aligned} \tag{17.68}$$

Just like the rate of strain tensor, the deformation rates have dimensions of inverse time.

The second expression in equation (17.66) is written in a covariant form, with g^{mn} the components to the inverse metric tensor (Section 21.12 on page 481). This expression also introduced the trace of the strain tensor

$$\begin{aligned} g_{mn}\,e^{mn} &= e^{n}_{n} \\ &= e^{1}_{1} + e^{2}_{2}, \end{aligned} \tag{17.69}$$

which is a scalar and hence invariant under transformations of coordinates. The covariant form of the stress tensor (17.66) is valid for all coordinate systems. General orthogonal coordinates are considered in Section 17.7.

17.5.3 The friction vector

The friction vector is given by the covariant divergence of the frictional stress tensor. In Cartesian coordinates, the covariant divergence is computed via the usual partial derivative operator

$$\rho F^m = \tilde{\tau}^{mn}_{,n}. \tag{17.70}$$

Performing the divergence using the transverse stress tensor in equation (17.66), as part of the full frictional stress tensor $\tilde{\tau}^{mn}$ given by (17.28), leads to the friction components

$$\rho F^1 = \nabla_z \cdot (\rho A \nabla_z u^1) + \hat{\mathbf{z}} \cdot \nabla_z u^2 \wedge \nabla_z (\rho A) + [\rho \kappa (u^1_{,z})]_{,z} \tag{17.71}$$

$$\rho F^2 = \nabla_z \cdot (\rho A \nabla_z u^2) - \hat{\mathbf{z}} \cdot \nabla_z u^1 \wedge \nabla_z (\rho A) + [\rho \kappa (u^2_{,z})]_{,z} \tag{17.72}$$

$$\rho F^3 = 0. \tag{17.73}$$

In these expressions, the horizontal divergence operator $\nabla_z = (\partial_1, \partial_2, 0)$ was introduced, $z = \xi^3$ is the vertical coordinate, and A is the viscosity associated with horizontal deformations. Note that when making the quasi-hydrostatic approximation, the vertical friction F^3 is ignored so that the vertical momentum equation reduces to the inviscid hydrostatic equation. The extra cross-product terms appearing in the transverse friction vanish when using a constant viscosity. Their importance when using a spatially nonconstant viscosity is highlighted in the next section.

17.5.4 The case of nonconstant viscosity

It is common for global ocean modelers to employ a nonconstant viscosity for various numerical reasons. As emphasized by Wajsowicz (1993), implementations of the corresponding friction vector often ignore the importance of formulating friction as the divergence of a symmetric stress tensor. Namely, what is sometimes done is to take the friction appropriate for a constant viscosity Boussinesq fluid

$$F^1_{\text{const}} = \nabla_z \cdot (A \nabla_z u^1) + [\kappa (u^1_{,z})]_{,z} \tag{17.74}$$

$$F^2_{\text{const}} = \nabla_z \cdot (A \nabla_z u^2), + [\kappa (u^2_{,z})]_{,z} \tag{17.75}$$

and allow A to be nonconstant. That is, the cross-product terms in equations (17.71) and (17.72) are dropped. Focusing on the two-dimensional transverse subspace, dropping the cross-product terms amounts to employing the nonsymmetric stress tensor

$$\tau^{mn}_{\text{NS}} = \rho_o A \begin{pmatrix} u^1_{,1} & u^1_{,2} \\ u^2_{,1} & u^2_{,2} \end{pmatrix}. \tag{17.76}$$

It is easy to show that the chosen friction dissipates kinetic energy since it is written as a Laplacian. However, for a fluid in uniform rotation

$$\mathbf{u} = \mathbf{\Omega} \wedge \mathbf{x}, \tag{17.77}$$

where $\mathbf{u} = (u^1, u^2, 0)$, $\mathbf{x} = (x^1, x^2, 0)$, and $\mathbf{\Omega}$ is spatially constant, the horizontal friction vector takes the form

$$\mathbf{F}^{(h)} = -\nabla \wedge (A \mathbf{\Omega}), \tag{17.78}$$

and it vanishes only when A is a constant. Consequently, by using friction derived from a nonsymmetric stress tensor and with a nonconstant viscosity, a uniformly rotating fluid feels a nonzero stress. Conversely, such a stress tensor can introduce uniform rotation; i.e., it can act as an internal source or sink of angular momentum. Unless one has a physical reason for doing so, such viscosity-dependent sources of angular momentum should be avoided.

17.6 TRANSVERSE ANISOTROPY

Smith and McWilliams (2003) argued that the orientation of friction should be determined according to the velocity direction. For example, in a zonal current, such as that along the equatorial region, friction arising from shears in the direction of the current ($u_{,x}$) should be preferentially larger than friction from shears perpendicular to the current ($u_{,y}$). They presented some compelling arguments that suggest this anisotropy is relevant for a physically based closure of the momentum equation. More pragmatically, and consistent with the perspective in this chapter, both Smith and McWilliams (2003) and Large et al. (2001) noted that numerical constraints allow for anisotropy, and exploiting this freedom enhances the simulation's physical integrity.

17.6.1 Viscosity tensor

Following Smith and McWilliams (2003), start with a viscosity tensor of the form

$$\begin{aligned} C^{mnpq} = & P\,\delta^{mn}\,\delta^{pq} + Q\,(\delta^{mp}\,\delta^{nq} + \delta^{mq}\,\delta^{np}) + R\,(\delta^{mn}\,s^{pq} + s^{mn}\,\delta^{pq}) \\ & + S\,(\delta^{mp}\,s^{nq} + \delta^{mq}\,s^{np} + s^{mp}\,\delta^{nq} + s^{mq}\,\delta^{np}) + T\,s^{mn}\,s^{pq}, \end{aligned} \tag{17.79}$$

where P, Q, R, S, T are viscosities and s^{mn} is a symmetric tensor defined as an outer product

$$s^{mn} = \hat{s}^m\,\hat{s}^n. \tag{17.80}$$

Smith and McWilliams (2003) suggest orienting viscosity according to the instantaneous velocity field, in which case

$$\hat{\mathbf{s}} = \frac{\mathbf{u}}{|\mathbf{u}|}. \tag{17.81}$$

In contrast, Large et al. (2001) suggest orienting the viscosity according to the local grid lines. For example, with an eastward orientation using spherical coordinates,

$$\hat{\mathbf{s}} = (1, 0). \tag{17.82}$$

This orientation has been found useful to increase the strength of the predominantly zonal equatorial currents in the Pacific (Large et al., 2001). In contrast, for time-dependent eddying flows, the dynamic orientation of equation (17.81) may spread friction over a broader domain in the time mean, and so lead to larger effective friction than for the case with a static orientation. In general, tests are necessary to determine what is best for the particular application.

Regardless of the choice for the *orientation tensor* s^{mn}, assume that its components satisfy the two identities

$$s^{11} + s^{22} = 1 \tag{17.83}$$

$$s^{11}\,s^{22} = s^{12}\,s^{12}. \tag{17.84}$$

These identities are used to simplify the form of the stress tensor in Section 17.6.2. Additionally, C^{mnpq} as written in equation (17.79) satisfies the trace-free property from Section 17.4 if

$$2\,(P+Q)+R=0 \tag{17.85}$$
$$R+2\,S+T/2=0. \tag{17.86}$$

These relations are assumed in the following.

17.6.2 Stress tensor

The transverse stress tensor $\tau^{mn}=\rho\,C^{mnpq}\,e_{pq}$ is given by

$$\tau^{mn}=\rho\begin{pmatrix} C^{1111}\,(e_{11}-e_{22})+2\,C^{1112}\,e_{12} & C^{1112}\,(e_{11}-e_{22})+2\,C^{1212}\,e_{12} \\ C^{1112}\,(e_{11}-e_{22})+2\,C^{1212}\,e_{12} & -C^{1111}\,(e_{11}-e_{22})-2\,C^{1112}\,e_{12} \end{pmatrix} \tag{17.87}$$

where the viscosity tensor components are

$$\begin{aligned} C^{1111} &= (P+2\,Q)+2\,(R+2\,S+(T/2)\,s^{11})\,s^{11} \\ &= (P+2\,Q)-T\,s^{11}\,s^{22} \end{aligned} \tag{17.88}$$

$$\begin{aligned} C^{1112} &= (R+2\,S+T\,s^{11})\,s^{12} \\ &= (T/2)\,s^{12}\,(s^{11}-s^{22}) \end{aligned} \tag{17.89}$$

$$\begin{aligned} C^{1212} &= Q+S\,(s^{11}+s^{22})+T\,s^{12}\,s^{12} \\ &= (P+2\,Q-T/4)+T\,s^{12}\,s^{12}. \end{aligned} \tag{17.90}$$

Since the combination of viscosities $P+2\,Q$ appears throughout the stress tensor, there are only two viscous degrees of freedom instead of the three present in Section 17.4.4. The choice for the tensor s^{mn} as the outer product $s^{mn}=\hat{s}^m\,\hat{s}^n$ eliminates one of the three degrees of freedom.

Aiming to maintain compatibility with the notation of Smith and McWilliams (2003), introduce the new viscosities

$$A=P+2\,Q \tag{17.91}$$
$$D=T/2 \tag{17.92}$$

thus leading to the stress tensor components

$$\rho^{-1}\,\tau^{11}=(A-2\,D\,s^{11}\,s^{22})\,(e_{11}-e_{22})+2\,D\,s^{12}\,(s^{11}-s^{22})\,e_{12} \tag{17.93}$$
$$\rho^{-1}\,\tau^{12}=D\,s^{12}\,(s^{11}-s^{22})\,(e_{11}-e_{22})+2\,(A-D/2+2\,D\,s^{12}\,s^{12})\,e_{12} \tag{17.94}$$
$$\tau^{22}=-\tau^{11}. \tag{17.95}$$

Using $s^{11}+s^{22}=1$ and $s^{11}\,s^{22}=s^{12}\,s^{12}$ gives

$$\rho^{-1}\,\tau^{mn}=A\begin{pmatrix} e_{\mathrm{T}} & e_{\mathrm{S}} \\ e_{\mathrm{S}} & -e_{\mathrm{T}} \end{pmatrix}+D\,\Delta\begin{pmatrix} 2\,s^{12} & -(s^{11}-s^{22}) \\ -(s^{11}-s^{22}) & -2\,s^{12} \end{pmatrix} \tag{17.96}$$

where $e_{\mathrm{T}}=e_{11}-e_{22}$ and $e_{\mathrm{S}}=2\,e_{12}$ are the deformation rates introduced in Section 17.5.2, and

$$2\,\Delta=e_{\mathrm{S}}\,(s^{11}-s^{22})-2\,e_{\mathrm{T}}\,s^{12}. \tag{17.97}$$

Note that the isotropic results from Section 17.5 are recovered upon setting $D = 0$.

To help understand the nature of the new anisotropic contribution to the stress tensor,

$$\tau^{mn}_{\text{aniso}} = \rho\, D\, \Delta \begin{pmatrix} 2\, s^{12} & -(s^{11} - s^{22}) \\ -(s^{11} - s^{22}) & -2\, s^{12} \end{pmatrix}, \tag{17.98}$$

it is useful to consider both its orientation and how it affects kinetic energy evolution. First, write the orientation vector as

$$\hat{\mathbf{s}} = (\cos\theta, \sin\theta) \tag{17.99}$$

to give

$$\tau^{mn}_{\text{aniso}} = \rho\, D\, \Delta \begin{pmatrix} \sin 2\theta & -\cos 2\theta \\ -\cos 2\theta & -\sin 2\theta \end{pmatrix} \tag{17.100}$$

and

$$2\, \Delta = e_{\text{S}} \cos 2\theta - e_{\text{T}} \sin 2\theta. \tag{17.101}$$

In the special case of zonal orientation (17.82), $\theta = 0$, whereas for the general flow-dependent case (17.81), $\mathbf{s} = \mathbf{u}/|\mathbf{u}|$ is oriented according to the horizontal velocity vector. The anisotropic stress tensor is proportional to a rotation matrix

$$R^{mn}(\hat{\mathbf{u}}) = \begin{pmatrix} \sin 2\theta & -\cos 2\theta \\ -\cos 2\theta & -\sin 2\theta \end{pmatrix}, \tag{17.102}$$

and the corresponding friction is oriented according to this rotation matrix.*

The contribution from friction to kinetic energy evolution (Section 17.4.3) takes the form

$$-e_{mn}\, \tau^{mn} = -\rho\, A\, (e_{\text{T}}^2 + e_{\text{S}}^2) + 2\, \rho\, D\, \Delta^2. \tag{17.103}$$

Hence, the anisotropic stress reduces the magnitude of frictional dissipation of kinetic energy. Because Δ is proportional to the scalar dissipation term $e_{mn}\, \tau^{mn}$ when $A = 0$, Δ is also a scalar and so takes the same value regardless of the coordinate system. Taking the special case $\hat{\mathbf{s}} = (1, 0)$ and $2\, \Delta = e_{\text{S}}$ leads to

$$-e_{mn}\, \tau^{mn} = -\rho\, A\, (e_{\text{T}}^2 + e_{\text{S}}^2) + \rho\, (D/2)\, e_{\text{S}}^2 \qquad \text{for } \hat{\mathbf{s}} = (1, 0). \tag{17.104}$$

Setting the anisotropic viscosity according to

$$D < 2\, A \tag{17.105}$$

ensures kinetic energy is dissipated by friction.

The above considerations help to determine a covariant form of the stress tensor. Summarizing the previous results, first note that the isotropic contribution takes the covariant form $\rho\, A\, (2\, e^{mn} - g^{mn}\, e^{p}_{p})$, as shown by equation (17.66). Second, the multiplier Δ is proportional to the scalar $e_{mn}\, \tau^{mn}$ when $A = 0$, and so is itself a scalar. Third, the rotation matrix $R(\hat{\mathbf{s}})$ is determined by an orientation vector, such as the local flow direction, which is coordinate invariant. Hence, a covariant form of the stress tensor is given by

$$\tau^{mn} = \rho\, A\, (2\, e^{mn} - g^{mn}\, e^{p}_{p}) + \rho\, D\, \Delta\, R^{mn}(\hat{\mathbf{s}}). \tag{17.106}$$

This result is used in Section 17.7 when deriving the friction given by the covariant divergence of the stress tensor.

---

*A rotation matrix is characterized by a unit determinant and an inverse equal to its transpose.

17.7 GENERALIZED ORTHOGONAL COORDINATES

The tensor analysis of Chapters 20 and 21, as summarized in Section 21.12 (page 481), is now used to derive the Laplacian frictional operator in generalized orthogonal coordinates. This section can be easily skipped for those uninterested in general tensor analysis.

17.7.1 Deformation rates

As noted in Section 17.5.2 and 17.6.2, the frictional stress tensor is a function of two independent combinations of rate of strain tensor elements. To determine the forms of these *deformation rates* appropriate for general coordinates, reconsider the local effects on kinetic energy of the isotropic frictional stress tensor

$$\begin{aligned} e_{mn}\,\tau^{mn}_{\mathrm{iso}} &= \rho\,A\,(2\,e^{mn}\,e_{mn} - g^{mn}\,e_{mn}\,e^{p}_{p}) \\ &= \rho\,A\,[(e^{1}_{1} - e^{2}_{2})^2 + 4\,e^{1}_{2}\,e^{2}_{1}]. \end{aligned} \tag{17.107}$$

This result motivates the deformation due to tension

$$\begin{aligned} e_{\mathrm{T}} &= e^{1}_{1} - e^{2}_{2} \\ &= e^{x}_{x} - e^{y}_{y}, \end{aligned} \tag{17.108}$$

and shears

$$\begin{aligned} e_{\mathrm{S}} &= 2\,(h_1/h_2)\,e^{1}_{2} \\ &= 2\,(h_2/h_1)\,e^{2}_{1} \\ &= 2\,e^{x}_{y} \\ &= 2\,e^{y}_{x}, \end{aligned} \tag{17.109}$$

so that

$$-e_{mn}\,\tau^{mn} = -\rho\,A\,[(e_{\mathrm{T}})^2 + (e_{\mathrm{S}})^2] + \rho\,(D/2)\,(e_{\mathrm{S}}\cos 2\theta - e_{\mathrm{T}}\sin 2\theta)^2. \tag{17.110}$$

Note that the Cartesian symbols x and y label physical components to tensors.* These deformation rates reduce to the forms given in Section 17.5.2 when using Cartesian coordinates on a flat manifold.

To express the deformation rates in terms of the velocity fields, note that the strain tensor components are generally defined in terms of covariant derivatives of the velocity field (e.g., Aris, 1962). From the manipulations presented in Section 17.10.2, one is led to the expressions valid for orthogonal transverse coordinates

$$e_{\mathrm{T}} = (\mathrm{d}y)\,(u/\mathrm{d}y)_{,x} - (\mathrm{d}x)\,(v/\mathrm{d}x)_{,y} \tag{17.111}$$

$$e_{\mathrm{S}} = (\mathrm{d}x)\,(u/\mathrm{d}x)_{,y} + (\mathrm{d}y)\,(v/\mathrm{d}y)_{,x}. \tag{17.112}$$

These forms reduce to the Cartesian case of Section 17.3.1 when the generalized orthogonal grid increments $(\mathrm{d}x, \mathrm{d}y) = (h_1\,\mathrm{d}\xi^1, h_2\,\mathrm{d}\xi^2)$ can be removed from the partial derivatives, as in the case of Cartesian coordinates on a flat space.

---

*As discussed in Section 21.11.1 (page 476), physical components of tensors have dimensions that are physically meaningful, whereas tensorial components of tensors can have arbitrary dimensions depending on the choice of coordinates. Physical components do not transform in the convenient manner that tensorial components do. Hence, it is best to work with the more abstract tensorial components for mathematical manipulations. When reaching a point where one aims to clarify the dimensions and physical content of a mathematical statement, and to clean up expressions prior to discretization, then it is relevant to introduce physical tensor components.

17.7.2 Stress tensor components

We now express the covariant form of the stress tensor (Section 17.6.2)

$$\tau^{mn} = \rho\, A\,(2\, e^{mn} - g^{mn}\, e^{p}_{p}) + \rho\, D\, \Delta\, R^{mn}(\hat{\mathbf{s}}) \tag{17.113}$$

in terms of the deformation rates. The $(1,1)$ element is given by

$$\begin{aligned} \rho^{-1}\, \tau^{11} &= A\,(2\, e^{11} - g^{11}\, e^{1}_{1} - g^{11}\, e^{2}_{2}) + D\, \Delta\, R^{11} \\ &= g^{11}\, [A\,(e^{1}_{1} - e^{2}_{2}) + D\, \Delta\, R^{1}_{1}], \end{aligned} \tag{17.114}$$

thus leading to the physical component of the stress tensor

$$\begin{aligned} \tau^{xx} &= h_1\, h_1\, \tau^{11} \\ &= \rho\,(A\, e_{\mathrm{T}} + D\, \Delta\, R^{x}_{x}), \end{aligned} \tag{17.115}$$

where the Cartesian symbols x and y label physical components to tensors. Likewise,

$$\begin{aligned} \tau^{yy} &= h_2\, h_2\, \tau^{22} \\ &= \rho\,(-A\, e_{\mathrm{T}} + D\, \Delta\, R^{y}_{y}). \end{aligned} \tag{17.116}$$

For the off-diagonal term,

$$\begin{aligned} \rho^{-1}\, \tau^{12} &= 2\, A\, e^{12} + D\, \Delta\, R^{12} \\ &= 2\,(h_1\, h_2)^{-1}\, A\, e^{x}_{y} + (h_2\, h_2)^{-1}\, D\, \Delta\, R^{1}_{2}, \end{aligned} \tag{17.117}$$

thus leading to

$$\tau^{xy} = \rho\,(A\, e_{\mathrm{S}} + D\, \Delta\, R^{x}_{y}). \tag{17.118}$$

In summary, the physical component version of the stress tensor is given by

$$\begin{aligned} \tau^{(m)(n)} &= \begin{pmatrix} \tau^{xx} & \tau^{xy} \\ \tau^{xy} & -\tau^{xx} \end{pmatrix} \\ &= \rho \begin{pmatrix} (A\, e_{\mathrm{T}} + D\, \Delta\, R^{x}_{x}) & (A\, e_{\mathrm{S}} + D\, \Delta\, R^{x}_{y}) \\ (A\, e_{\mathrm{S}} + D\, \Delta\, R^{y}_{x}) & (-A\, e_{\mathrm{T}} + D\, \Delta\, R^{y}_{y}) \end{pmatrix} \end{aligned} \tag{17.119}$$

and the rotation matrix

$$R^{(m)}_{(n)} = \begin{pmatrix} \sin 2\theta & -\cos 2\theta \\ -\cos 2\theta & -\sin 2\theta \end{pmatrix}. \tag{17.120}$$

The parenthesis notation is used to distinguish physical components of tensors, which use the Cartesian labels x and y, from the more abstract tensorial components, which use the numerals 1 and 2 (see Section 21.11.1 and footnote following equation (17.110)). Note that when using the flow-dependent orientation as in Smith and McWilliams (2003), the orientation of the physical velocity components

$$\begin{aligned} \mathbf{u} &= (u, v) \\ &= (h_1\, u^1, h_2\, u^2) \end{aligned} \tag{17.121}$$

serves to orient the anisotropic portion of the stress tensor according to equation (17.80), rather than the dimensionally ambiguous tensorial components (u^1, u^2). In spherical coordinates with the shallow ocean approximation (Section 2.4.2),

$$(u, v) = R\,(\cos\phi\, \mathrm{d}\lambda/\mathrm{d}t, \mathrm{d}\phi/\mathrm{d}t) \tag{17.122}$$

are the familiar zonal and meridional velocity components, with $\mathrm{d}/\mathrm{d}t$ the material time derivative.

17.7.3 The friction vector

Friction arising from horizontal and vertical deformations is given by the covariant divergence of the frictional stress tensor. From Section 21.3 (page 470), one can write the covariant divergence as

$$\begin{aligned}\rho F^m &= \tilde{\tau}^{mn}_{;n} \\ &= \tilde{\tau}^{mn}_{,n} + \Gamma^m_{nc}\,\tilde{\tau}^{nc} + \Gamma^n_{nc}\,\tilde{\tau}^{mc},\end{aligned} \tag{17.123}$$

where $\tilde{\tau}^{mn}$ is the frictional stress tensor, whose transverse portion is given by τ^{mn} (Section 17.3.4). For implementation in an ocean model, it is necessary to massage equation (17.123). Details are given in Section 17.10.3. The final results are the physical friction components

$$\rho F^x = (dy)^{-2}\,[(dy)^2\,\tau^{xx}]_{,x} + (dx)^{-2}\,[(dx)^2\,\tau^{xy}]_{,y} + (\rho\,\kappa\,u_{,z})_{,z} \tag{17.124}$$

$$\rho F^y = (dx)^{-2}\,[(dx)^2\,\tau^{yy}]_{,y} + (dy)^{-2}\,[(dy)^2\,\tau^{xy}]_{,x} + (\rho\,\kappa\,v_{,z})_{,z}, \tag{17.125}$$

where

$$\tau^{xx} = -\tau^{yy}. \tag{17.126}$$

17.8 DISSIPATION FUNCTIONAL

The purpose of this section is to introduce the dissipation functional corresponding to the friction operator. This functional is fundamental to the method used to discretize friction in Chapter 19. The formalism is analogous to how the tracer variance is related to the diffusion operator (Section 16.1). A discussion of the dissipation functional was given by Griffies and Hallberg (2000) and Smith and McWilliams (2003).

17.8.1 Motivation for the functional approach

There are many ways to discretize an operator. One approach is to take the continuum form and provide reasonable second order accurate discretizations, with averaging used whenever needed to place quantities in their proper grid position. This was the common approach taken for friction in the past, with the continuum formulas of Bryan (1969) and Wajsowicz (1993) or Smagorinsky (1963, 1993) as the starting point. This direct approach is not always satisfying since it may produce a discrete operator with undesirable properties. An example of the problems with naive discretizations is described by Griffies et al. (1998) for the neutral diffusion operator.

After having some success with the reformulation of neutral diffusion, in which the diffusion operator was shown to be equivalent to the functional derivative of the tracer variance, it was decided to attempt a similar approach for friction. As shown in Section 17.8.2, the friction operator is equivalent to the functional derivative of the local kinetic energy dissipation. Both of these results follow from the self-adjointness of the diffusion and friction operators. Through this connection, the approach is to first discretize the appropriate functional. Thereafter, the functional derivative of the discrete functional is used to derive the discretized friction operator.

It is notable that discretization of the friction operator on a B-grid via first discretizing the functional with second order numerics leads to the use of velocity triads. These triads result because on the B-grid, both velocity components are at a single point. For second order discretization of the neutral diffusion operator, density triads are fundamental (Griffies et al., 1998). Triads emerge for diffusion since all quantities of interest live on the tracer grid.

17.8.2 Defining the functional

As noted in Section 17.5.2 and 17.6.2, the effects on local kinetic energy dissipation from horizontal deformations in the fluid take the form

$$-e_{mn}\,\tau^{mn} = -\rho\,A\,(e_{\mathrm{T}}^2 + e_{\mathrm{S}}^2) + 2\,\rho\,D\,\Delta^2, \tag{17.127}$$

where

$$2\,\Delta = e_{\mathrm{S}}\,\cos 2\theta - e_{\mathrm{T}}\,\sin 2\theta. \tag{17.128}$$

The angle θ sets the orientation of friction. The first term on the right-hand side of equation (17.127) is associated with transverse isotropic stresses, whereas the second term arises from anisotropic stresses. Ensuring a sign definite dissipation of kinetic energy requires (equation (17.105))

$$D < 2\,A. \tag{17.129}$$

Integration over the full ocean domain leads to the positive semidefinite *dissipation functional*

$$\begin{aligned} \mathcal{S} &\equiv \int \mathrm{d}V\, e_{mn}\,\tau^{mn} \\ &= \int \rho\,\mathrm{d}V\,[A\,(e_{\mathrm{T}}^2 + e_{\mathrm{S}}^2) - 2\,D\,\Delta^2], \end{aligned} \tag{17.130}$$

where

$$\mathrm{d}V = \sqrt{\mathcal{G}}\,\mathrm{d}\xi^1\,\mathrm{d}\xi^2\,\mathrm{d}z \tag{17.131}$$

is the invariant volume element written using generalized horizontal coordinates,* and $\sqrt{\mathcal{G}} = h_1\,h_2\,h_3$ is the square root of the metric determinant. With the shallow ocean approximation (Section 3.2), the vertical stretching function $h_3 = 1$.

As shown in this section, the functional derivative $\delta\mathcal{S}/\delta u^a$ is proportional to the friction $g_{ab}\,F^b$. The connection between a dissipation functional and the friction operator is afforded through the self-adjointness of the friction operator. This result leads to a numerical discretization of friction that is ensured to dissipate kinetic energy on the discrete lattice. The same approach for the tracer diffusion operator is detailed in Section 16.1.1.

Two assumptions are needed to make the friction operator self-adjoint: (1) The viscosity is functionally independent of the velocity field. (2) The flow satisfies *natural boundary conditions*, of which no-slip is one. The Smagorinsky viscosity does not satisfy the first assumption. Nonetheless, the functional approach leads to a discretization inside of which one can employ the Smagorinsky viscosity. A similar assumption was used to discretize the neutral diffusion operator when diffusing active tracers, where in that case the operator is a nonlinear function of the active tracers (Section 16.1.5).

---

*See Section 20.11, page 457 for a discussion of invariant volume elements.

17.8.3 Performing the functional derivative

This section presents the functional derivative, thus providing details to the above statements. For this purpose, write the dissipation functional as

$$\mathcal{S} = \int \mathrm{d}V\, \mathcal{L}, \tag{17.132}$$

with

$$\begin{aligned} \mathcal{L} &= e_{mn}\, \tau^{mn} \\ &= \rho\, A\, (e_{\mathrm{T}}^2 + e_{\mathrm{S}}^2) - 2\, \rho\, D\, \Delta^2. \end{aligned} \tag{17.133}$$

Now consider a variation in the tensorial components to the velocity field

$$u^a \to u^a + \delta u^a, \tag{17.134}$$

with the underlying space-time geometry held fixed; then the dissipation functional $\mathcal{S}$ changes according to

$$\delta\, \mathcal{S} = \int \mathrm{d}V\, \delta \mathcal{L}. \tag{17.135}$$

Since $\mathcal{L} = \mathcal{L}[u^a, u^a_{,b}]$ is a functional of the velocity and its partial derivatives, its variation leads to

$$\begin{aligned} \delta\, \mathcal{S} &= \int \sqrt{\mathcal{G}}\, \mathrm{d}\xi^1\, \mathrm{d}\xi^2\, \mathrm{d}z \left[\frac{\delta\, \mathcal{L}}{\delta u^a}\, \delta u^a + \frac{\delta\, \mathcal{L}}{\delta u^a_{,b}}\, \delta u^a_{,b} \right] \\ &= \int \mathrm{d}\xi^1\, \mathrm{d}\xi^2\, \mathrm{d}z \left[\sqrt{\mathcal{G}}\, \frac{\delta\, \mathcal{L}}{\delta u^a}\, \delta u^a + \partial_b \left(\sqrt{\mathcal{G}}\, \frac{\delta\, \mathcal{L}}{\delta u^a_{,b}}\, \delta u^a \right) - \partial_b \left(\sqrt{\mathcal{G}}\, \frac{\delta\, \mathcal{L}}{\delta u^a_{,b}} \right) \delta u^a \right], \end{aligned} \tag{17.136}$$

where an integration by parts has been performed. The total derivative reduces to a surface term, which vanishes when either

$$\delta u^a = 0 \tag{17.137}$$

on all solid boundaries, or

$$\hat{n}_b\, (\delta\, \mathcal{L} / \delta u^a_{,b})\, \delta u^a = 0, \tag{17.138}$$

where $\hat{n}_b$ are components to the outward normal at the boundaries. These two conditions define natural boundary conditions for the operator. If velocity, and hence its variation, satisfies the no-slip condition, then $\delta u^a = 0$ on all boundaries and the total derivative can be dropped. More general boundary conditions can be derived from the second type of natural boundary condition.

With natural boundary conditions, variation of the dissipation functional takes the form

$$\delta\, \mathcal{S} = \int \sqrt{\mathcal{G}}\, \mathrm{d}\xi^1\, \mathrm{d}\xi^2\, \mathrm{d}z \left[\frac{\delta\, \mathcal{L}}{\delta u^a} - \mathcal{G}^{-1/2}\, \partial_b \left(\sqrt{\mathcal{G}}\, \frac{\delta\, \mathcal{L}}{\delta u^a_{,b}} \right) \right] \delta u^a, \tag{17.139}$$

thus leading to the functional derivative

$$[\mathrm{d}V(\mathbf{y})]^{-1}\, \frac{\delta\, \mathcal{S}}{\delta u^a} = \frac{\delta\, \mathcal{L}}{\delta u^a} - \mathcal{G}^{-1/2}\, \partial_b \left(\sqrt{\mathcal{G}}\, \frac{\delta\, \mathcal{L}}{\delta u^a_{,b}} \right), \tag{17.140}$$

where $\mathrm{d}V(\mathbf{y})$ is the volume element evaluated at the field point $\mathbf{y}$. To reach this result, it was necessary to use the identity

$$\frac{\delta u^a(\mathbf{x})}{\delta u^b(\mathbf{y})} = \mathrm{d}V(\mathbf{y})\, \delta^a_b\, \delta(\mathbf{x}-\mathbf{y}), \tag{17.141}$$

where $\delta(\mathbf{x}-\mathbf{y})$ is the Dirac delta function. The delta function has physical dimensions of inverse volume L^{-3}. Hence,

$$\int \mathrm{d}V(\mathbf{y})\, \delta(\mathbf{x}-\mathbf{y}) = 1, \tag{17.142}$$

so long as integration is over a domain containing the singular point $\mathbf{x} = \mathbf{y}$; otherwise, the integral vanishes. The derivation of equation (17.140) should be familiar to those having studied the Euler-Lagrange equations used in classical mechanics (e.g., Marion and Thornton, 1988).

17.8.4 The connection to friction

Now that the functional derivative of the dissipation $\mathcal{S}$ has been computed, it remains to determine its connection to friction. For this purpose, write a general variation of $\mathcal{L}$ as

$$\begin{aligned}\delta\mathcal{L} &= 2\,\rho\,[\delta e_{\mathrm{T}}\,(A\, e_{\mathrm{T}} + D\,\Delta\,\sin 2\theta) + \delta e_{\mathrm{S}}\,(A\, e_{\mathrm{S}} - D\,\Delta\,\cos 2\theta)] \\ &= 2\,(\tau^{xx}\,\delta e_{\mathrm{T}} + \tau^{xy}\,\delta e_{\mathrm{S}}),\end{aligned} \tag{17.143}$$

where the second step introduced expressions for the stress tensor components given in Section 17.7.2. Now recall the discussion in Section 17.10.2, where it was shown that the horizontal tension can be written

$$e_{\mathrm{T}} = u^1{}_{,1} - u^2{}_{,2} + u^m\,\partial_m\,\ln(h_1/h_2) \tag{17.144}$$

and the horizontal shearing strain is

$$e_{\mathrm{S}} = \frac{h_1}{h_2}\,u^1{}_{,2} + \frac{h_2}{h_1}\,u^2{}_{,1}. \tag{17.145}$$

Consequently, the deformation rates have functional derivatives given by

$$\frac{\delta e_{\mathrm{T}}}{\delta u^a} = \delta^m_a\,\partial_m\,\ln(h_1/h_2) \tag{17.146}$$

$$\frac{\delta e_{\mathrm{T}}}{\delta u^a{}_{,b}} = \delta^b_1\,\delta^1_a - \delta^b_2\,\delta^2_a \tag{17.147}$$

$$\frac{\delta e_{\mathrm{S}}}{\delta u^a} = 0 \tag{17.148}$$

$$\frac{\delta e_{\mathrm{S}}}{\delta u^a{}_{,b}} = (h_1/h_2)\,\delta^b_2\,\delta^1_a + (h_2/h_1)\,\delta^b_1\,\delta^2_a \tag{17.149}$$

thus leading to

$$\rho^{-1}\,\frac{\delta\mathcal{L}}{\delta u^a} = 2\,\tau^{xx}\,\delta^m_a\,\partial_m\,\ln(h_1/h_2) \tag{17.150}$$

$$\rho^{-1}\,\frac{\delta\mathcal{L}}{\delta u^a_{,b}} = 2\,\tau^{xx}\,(\delta^b_1\,\delta^1_a - \delta^b_2\,\delta^2_a) + 2\,\tau^{xy}\left(\frac{h_1}{h_2}\,\delta^1_a\,\delta^b_2 + \frac{h_2}{h_1}\,\delta^2_a\,\delta^b_1\right). \tag{17.151}$$

For $a = 1$, these results yield

$$
\begin{aligned}
-\frac{1}{2\,\mathrm{d}V}\frac{\delta \mathcal{S}}{\delta u^1} &= \tau^{xx}\,\partial_1\,\ln(h_1/h_2) + \frac{1}{h_1\,h_2}\,\partial_b\left(h_1\,h_2\,\delta^b_1\,\tau^{xx} + (h_1/h_2)\,\delta^b_2\,\tau^{xy}\right) \\
&= h_1\left((\mathrm{d}y)^{-2}\,[(\mathrm{d}y)^2\,\tau^{xx}]_{,x} + (\mathrm{d}x)^{-2}\,[(\mathrm{d}x)^2\,\tau^{xy}]_{,y}\right) \\
&= \rho\,h_1\,F^x. \qquad (17.152)
\end{aligned}
$$

The intermediate step utilized manipulations described in Section 17.10, and the last step identified the friction vector (sans the piece arising from vertical shears) given in Section 17.7.3. Hence, in general,

$$
-\frac{1}{2\,\mathrm{d}V}\frac{\delta \mathcal{S}}{\delta u^{(a)}} = \rho\,F^{(a)}, \qquad (17.153)
$$

where the parentheses indicate the physical tensor components. This result establishes the desired connection between the functional derivative of the dissipation functional and the friction operator. The discrete analog of this result is employed in Section 19.3 to derive a discrete version of the friction operator.

17.9 BIHARMONIC FRICTION

The previous derivations were all concerned with second order, or *Laplacian*, friction. It is often useful to consider another method of dissipating momentum through use of fourth order, or *biharmonic*, friction. As reviewed by Griffies and Hallberg (2000), biharmonic friction acts more strongly on the small scales than Laplacian friction, and less strongly on the large scales. Each property is desirable, especially when aiming to realize some form of a quasi-geostrophic turbulent cascade in which enstrophy cascades to the small scales and energy to the large scales. Biharmonic friction is therefore commonly used in eddying ocean simulations.

The goal of this section is to derive the appropriate form of the biharmonic operator that dissipates kinetic energy yet does not introduce spurious sources of angular momentum. As with the previous derivations, some work is necessary in order to realize these properties on a sphere with a generally nonconstant viscosity.

Recall that the quasi-hydrostatic approximation allowed for the separation of the transverse or horizontal subspace from the vertical subspace. Consequently, friction arising from the vertical gradient of the vertical strain, $\partial_z\,(\kappa\mathbf{u}_{,z})$, was isolated from gradients in the horizontal deformations. As a result, the following focuses solely on deriving the biharmonic operator for shears arising in the two-dimensional transverse subspace. Hence, all labels in this section only run over the transverse space $m, n, p = 1, 2$.

The utility of an anisotropic biharmonic friction operator has not been established. It is trivial to implement an anisotropic biharmonic friction, once the formalism for the corresponding Laplacian operator is established. For brevity, the following focuses on isotropic biharmonic friction.

17.9.1 General formulation

The general formulation of biharmonic friction is a straightforward extension of the Laplacian friction given in the previous sections. What is done is to iterate twice on the Laplacian approach. More precisely, the components F_B^m of the biharmonic friction vector are derived from the covariant divergence*

$$\rho\, F_B^m = \Theta^{mn}_{;n} \tag{17.154}$$

where

$$\Theta^{mn} = -\rho\, B\,(2\, E^{mn} - g^{mn}\, E^p_p), \tag{17.155}$$

$B > 0$ has units of $L^2\, t^{-1/2}$. As shown in the next subsection, use of this "square root" biharmonic viscosity is prompted by the desire to dissipate kinetic energy. This detail matters for cases with a nonconstant viscosity. Θ^{mn} has the same form as the stress tensor used for the isotropic Laplacian friction discussed in Section 17.7.2, except with a minus sign. Components of the symmetric "strain" tensor are given by

$$E_{mn} = (F_{m;n} + F_{n;m})/2. \tag{17.156}$$

F^m are components to the friction vector determined through the covariant divergence of the Laplacian frictional stress tensor

$$\begin{aligned} \rho\, F^m &= \tau^{mn}_{;n} \\ &= [B\, \rho\, (2\, e^{mn} - g^{mn}\, e^p_p)]_{;n} \end{aligned} \tag{17.157}$$

as derived in the previous sections, where the only difference is that the viscosity used for computing the stress tensor τ^{mn} is now set to B, and the dimensions on F^m are $L\, t^{-3/2}$.

This approach ensures that the biharmonic friction is derived from the divergence of a symmetric tensor Θ^{mn}, hence ensuring a proper angular momentum budget. Additionally, the computational work necessary to compute the Laplacian friction is directly employed for the biharmonic friction. Finally, as shown in the next subsection, this form for biharmonic friction also dissipates kinetic energy.

17.9.2 Effects on kinetic energy

The manipulations necessary to show that the biharmonic friction dissipates kinetic energy are analogous to those used for Laplacian friction in Section 17.7. As with that discussion, the relevant contribution from horizontal biharmonic friction is given by $\mathrm{d}V\, u_m\, \Theta^{mn}_{;n}$. Assuming either no-slip or no-normal Θ stress at the boundaries brings this expression to the form

$$\begin{aligned} \int \mathrm{d}V\, u_m\, \Theta^{mn}_{;n} &= -\int \mathrm{d}V\, \Theta^{mn}\, e_{mn} \\ &= \int \mathrm{d}V\, \rho\, B\, [2\, E^{mn}\, e_{mn} - e^n_n\, E^m_m]. \end{aligned} \tag{17.158}$$

For the product of traces, one has

$$\begin{aligned} \int \mathrm{d}V\, \rho\, B\, e^n_n\, E^m_m &= \int \mathrm{d}V\, \rho\, B\, e^n_n\, F^m_{;m} \\ &= \int \mathrm{d}V\, [(\rho\, B\, e^n_n\, F^m)_{;m} - (\rho\, B\, e^n_n)_{;m}\, F^m]. \end{aligned} \tag{17.159}$$

*Recall that all labels in this section run over $m, n, p = 1, 2$.

The first term reduces to a boundary contribution, which vanishes if the Laplacian friction F^m vanishes on the boundaries. For the contraction of the two strain tensors, one has

$$\begin{aligned} 2\int \mathrm{d}V\,\rho\,B\,e^{mn}\,E_{mn} &= 2\int \mathrm{d}V\,\rho\,B\,e^{mn}\,F_{m\,;n} \\ &= -2\int \mathrm{d}V\,F_m\,(\rho\,B\,e^{mn})_{\,;n} \end{aligned} \tag{17.160}$$

where again the boundary term $(\rho\,B\,e^{mn}\,F_m)_{;n}$ was assumed to vanish. Combining the two contributions leads to

$$\begin{aligned} \int \mathrm{d}V\,u_m\,\Theta^{mn}_{;n} &= -\int \mathrm{d}V\,F_m\,[2\,B\,\rho\,e^{mn} - g^{mn}\,B\,\rho\,e^k_k]_{\,;n} \\ &= -\int \mathrm{d}V\,\rho\,F_m\,F^m, \end{aligned} \tag{17.161}$$

which is nonpositive, thus ensuring that kinetic energy is dissipated by biharmonic friction in the ocean interior.

If the viscosity B is distributed nonsymmetrically, then kinetic energy is guaranteed to be dissipated by the biharmonic friction only for the special case of constant viscosity. That is, in Cartesian coordinates, the operator $\nabla_z \cdot B\,\nabla_z\,(\nabla_z \cdot B\nabla_z\psi)$ is dissipative for all $B > 0$, whereas $\nabla_z \cdot B^2\,\nabla_z\,(\nabla_z^2\psi)$ or $\nabla_z^2\,(\nabla_z \cdot B^2\nabla_z\psi)$ can be shown to be dissipative only for constant B.

17.10 SOME MATHEMATICAL DETAILS

The purpose of this section is to present some of the steps in the derivation of the deformation rates and Laplacian friction operator. The material may be easily skipped for those uninterested in mathematical details.

17.10.1 Results from Chapters 20 and 21

The infinitesimal distance between two points using orthogonal coordinates on the sphere is given by

$$\begin{aligned} (\mathrm{d}s)^2 &= g_{mn}\,\mathrm{d}\xi^m\,\mathrm{d}\xi^n \\ &= (h_1\,\mathrm{d}\xi^1)^2 + (h_2\,\mathrm{d}\xi^2)^2 + (h_3\,\mathrm{d}\xi^3)^2 \\ &= (\mathrm{d}x)^2 + (\mathrm{d}y)^2 + (\mathrm{d}z)^2, \end{aligned} \tag{17.162}$$

where physical displacements are given by

$$(\mathrm{d}x, \mathrm{d}y, \mathrm{d}z) = (h_1\,\mathrm{d}\xi^1, h_2\,\mathrm{d}\xi^2, h_3\,\mathrm{d}\xi^3) \tag{17.163}$$

and dimensions are length. With z-coordinates we take $h_3 = 1$. Furthermore, the shallow ocean approximation (Section 3.2) sets the transverse stretching functions h_1, h_2 to be functions only of the transverse coordinates (ξ^1, ξ^2). In the special case of geographical spherical coordinates, $(\xi^1, \xi^2) = (\lambda, \phi)$ are the dimensionless longitude and latitude, and $(h_1, h_2) = R\,(\cos\phi, 1)$ are the stretching functions with R the earth's radius.

The determinant of the metric tensor g_{mn} is positive definite. Using orthogonal coordinates, the determinant is given by

$$\begin{aligned} \mathcal{G} &= g_{11}\, g_{22}\, g_{33} \\ &= (h_1\, h_2\, h_3)^2, \end{aligned} \tag{17.164}$$

where again $h_3 = 1$. The metric determinant appears in certain differential formulas, such as the invariant volume element (Section 20.11).

Physical components to the transverse partial derivative operator are needed in the following:

$$(\partial_x, \partial_y) = (h_1^{-1}\, \partial_1, h_2^{-1}\, \partial_2), \tag{17.165}$$

as well as the physical velocity components

$$(u, v) = (h_1\, u^1, h_2\, u^2). \tag{17.166}$$

In spherical coordinates

$$(u, v) = R\,(\cos\phi\, \mathrm{d}\lambda/\mathrm{d}t, \mathrm{d}\phi/\mathrm{d}t) \tag{17.167}$$

are the familiar zonal and meridional velocity components, with $\mathrm{d}/\mathrm{d}t$ the material time derivative. Likewise,

$$(\partial_x, \partial_y) = ((R\, \cos\phi)^{-1}\, \partial_\lambda, (R)^{-1}\, \partial_\phi) \tag{17.168}$$

are the familiar partial derivative operators. A full discussion of physical tensor components using orthogonal coordinates is provided in Section 21.12.3. See also the summary discussion in Section 21.12.

17.10.2 Details of the deformation rate derivation

The methods described in Chapters 20 and 21, as summarized above, lead to

$$\begin{aligned} e_{\mathrm{T}} &= e^1_{\ 1} - e^2_{\ 2} \\ &= u^1_{\ ;1} - u^2_{\ ;2} \\ &= u^1_{\ ,1} - u^2_{\ ,2} + u^m\,(\Gamma^1_{1m} - \Gamma^2_{2m}) \\ &= u^1_{\ ,1} - u^2_{\ ,2} + \frac{1}{2}\, u^m\, \partial_m\, \ln(g_{11}/g_{22}) \\ &= (h_2/h_1)\, \Big(u^1\,(h_1/h_2)\Big)_{,1} - (h_1/h_2)\, \Big(u^2\,(h_2/h_1)\Big)_{,2} \\ &= h_2\,(u/h_2)_{,x} - h_1\,(v/h_1)_{,y} \\ &= (\mathrm{d}y)\,(u/\mathrm{d}y)_{,x} - (\mathrm{d}x)\,(v/\mathrm{d}x)_{,y} \end{aligned} \tag{17.169}$$

where the stretching functions h_1 and h_2 were eliminated in the last step via the properties articulated in Section 21.11.2. Additionally, Γ^p_{mn} are the Christoffel symbols defined in Section 21.2.2. They can be written in terms of the metric tensor as (Section 21.4)

$$\Gamma^p_{mn} = \frac{1}{2}\, g^{pq}(g_{qm,n} + g_{qn,m} - g_{mn,q}). \tag{17.170}$$

Proceeding in a similar manner leads to

$$
\begin{aligned}
(h_1/h_2)\, e_S &= 2\, e^2_1 \\
&= u^2_{;1} + g^{22}\, g_{11}\, u^1_{;2} \\
&= u^2_{,1} + \Gamma^2_{1m}\, u^m + g^{22}\, g_{11}\, (u^1_{,2} + \Gamma^1_{2m}\, u^m) \\
&= u^2_{,1} + \frac{1}{2}\, g^{2d}\, (g_{1d,m} + g_{md,1} - g_{1m,d})\, u^m + g^{22}\, g_{11}\, u^1_{,2} \\
&\quad + \frac{1}{2}\, g^{22}\, g_{11}\, g^{1d}\, (g_{2d,m} + g_{md,2} - g_{2m,d})\, u^m \\
&= g^{22}\, (g_{11}\, u^1_{,2} + g_{22}\, u^2_{,1}),
\end{aligned}
\tag{17.171}
$$

which brings the horizontal shearing strain to

$$
\begin{aligned}
e_S &= h_1\, (u/h_1)_{,y} + h_2\, (v/h_2)_{,x} \\
&= (\mathrm{d}x)\, (u/\mathrm{d}x)_{,y} + (\mathrm{d}y)\, (v/\mathrm{d}y)_{,x}.
\end{aligned}
\tag{17.172}
$$

17.10.3 Details of the friction derivation

To deduce the form of the friction vector appropriate for an ocean model, it is necessary to massage equation (17.123). For this purpose, start from the equivalent expression

$$
\rho\, F^m = (\sqrt{\mathcal{G}})^{-1} \left(\sqrt{\mathcal{G}}\, \tau^{mn}\right)_{,n} + \Gamma^m_{ab}\, \tau^{ab},
\tag{17.173}
$$

which is valid for any metric. Its derivation is given in Section 21.6. We now exploit the diagonal form of the metric tensor. First, the contraction $\Gamma^n_{ab}\, \tau^{ab}$ is given by

$$
\begin{aligned}
\Gamma^m_{ab}\, \tau^{ab} &= \frac{1}{2}\, g^{md}\, (g_{ad,b} + g_{bd,a} - g_{ab,d})\, \tau^{ab} \\
&= g^{mm}\, (g_{am,b} - \frac{1}{2}\, g_{ab,m})\, \tau^{ab} \\
&= g^{mm}\, g_{mm,b}\, \tau^{mb} - \frac{1}{2}\, g^{mm}\, g_{ab,m}\, \tau^{ab},
\end{aligned}
\tag{17.174}
$$

where there is no sum on the m label. Plugging this result into equation (17.173) for the friction vector yields

$$
\rho\, F^m = (g_{mm}\, \sqrt{\mathcal{G}})^{-1} \left(g_{mm}\, \sqrt{\mathcal{G}}\, \tau^{mn}\right)_{,n} - \frac{1}{2}\, g^{mm}\, g_{ab,m}\, \tau^{ab}.
\tag{17.175}
$$

Now recall that the transverse stress tensor considered here has a zero trace (Section 17.3.5), which in general coordinates is given by the invariant expression

$$
\begin{aligned}
g_{mn}\, \tau^{mn} &= \tau^m_m \\
&= \tau^1_1 + \tau^2_2 \\
&= 0.
\end{aligned}
\tag{17.176}
$$

Consequently, the generalized friction component with $m = 1$ is given by

$$
\begin{aligned}
\rho F^1 &= (g_{11}\sqrt{\mathcal{G}})^{-1}\left(g_{11}\sqrt{\mathcal{G}}\,\tau^{1n}\right)_{,n} - \frac{1}{2} g^{11} g_{ab,1}\,\tau^{ab} \\
&= \tau^{1n}_{,n} + \tau^{1n}\,\partial_n \ln(g_{11}\sqrt{\mathcal{G}}) - \tau^{11}\,\partial_1 \ln\sqrt{g_{11}} - \frac{1}{2}\tau^{22} g^{11} g_{22,1} \\
&= \tau^{11}_{,1} + \tau^{11}\,\partial_1 \ln(g_{11}\sqrt{\mathcal{G}}) + (g_{11}\sqrt{\mathcal{G}})^{-1}(g_{11}\sqrt{\mathcal{G}}\,\tau^{12})_{,2} - \tau^{11}\,\partial_1 \ln\sqrt{g_{11}} + \tau^{11}\,\partial_1 \ln\sqrt{g_{22}} \\
&= \tau^{11}_{,1} + \tau^{11}\,\partial_1 \ln\mathcal{G} + (g_{11}\sqrt{\mathcal{G}})^{-1}(g_{11}\sqrt{\mathcal{G}}\,\tau^{12})_{,2} \\
&= \mathcal{G}^{-1}(\mathcal{G}\,\tau^{11})_{,1} + (g_{11}\sqrt{\mathcal{G}})^{-1}(g_{11}\sqrt{\mathcal{G}}\,\tau^{12})_{,2} \\
&= [h_1\,(\mathrm{d}y)^2]^{-1}\,[(\mathrm{d}y)^2\,h_1\,h_1\,\tau^{11}]_{,x} + [h_1\,(\mathrm{d}x)^2]^{-1}\,[(\mathrm{d}x)^2\,h_1\,h_2\,\tau^{12}]_{,y}.
\end{aligned}
$$

Multiplying by $h_1 = \sqrt{g_{11}}$ determines the physical component to the generalized zonal friction

$$\rho F^x = (\mathrm{d}y)^{-2}\,[(\mathrm{d}y)^2\,\tau^{xx}]_{,x} + (\mathrm{d}x)^{-2}\,[(\mathrm{d}x)^2\,\tau^{xy}]_{,y} + (\rho\,\kappa\,u_{,z})_{,z} \tag{17.177}$$

where the contribution due to vertical strains was reintroduced, and the physical components to the friction and stress tensor are given by

$$F^x = h_1\,F^1 \tag{17.178}$$

$$\tau^{xx} = h_1\,h_1\,\tau^{11} \tag{17.179}$$

$$\tau^{xy} = h_1\,h_2\,\tau^{12}. \tag{17.180}$$

Letting $x \leftrightarrow y$ leads to the generalized meridional friction component

$$\rho F^y = (\mathrm{d}x)^{-2}\,[(\mathrm{d}x)^2\,\tau^{yy}]_{,y} + (\mathrm{d}y)^{-2}\,[(\mathrm{d}y)^2\,\tau^{xy}]_{,x} + (\rho\,\kappa\,v_{,z})_{,z}. \tag{17.181}$$

17.11 CHAPTER SUMMARY

This chapter presented a rationalization of friction as commonly used in global ocean climate models. The approach followed that commonly applied in other areas of continuum mechanics, most notably elasticity theory. Although based on a physical theory, the formulation may actually have little direct relevance to a proper subgrid scale closure of the momentum budget in large-scale ocean models. Indeed, it likely does not, since its simplest form leads to momentum diffusion, which is likely an inappropriate closure for linear momentum at scales larger than meters. Nonetheless, this form of friction is nearly ubiquitous in ocean climate models, as it provides a useful numerical closure. For this reason alone it is important to carefully formalize the mathematical form of the operators, and that was the general aim of this chapter.

Section 17.2 started the main discussion by presenting the physics underlying the friction operator. Section 17.3 developed some properties of the frictional stress tensor, such as symmetry and zero trace. Section 17.4 followed up with discussion of the fourth order viscosity tensor, which relates stress to strain via a generalization of Hooke's law. An explicit form for friction was computed in Section 17.5 for the special case of horizontally isotropic frictional stress. This is the common manner whereby frictional stress is applied in ocean climate models. Section 17.6 followed up with the anisotropic approach, which has shown some promise for ocean climate simulations, especially when aiming to increase the strength of the simulated equatorial currents in coarse resolution models. Section 17.7 then generalized the Cartesian discussion to generalized orthogonal horizontal coordinates. This generalization is required for application in the global models. To set the stage for discretization discussed in Chapter 19, Section 17.8 presented a discussion of the dissipation functional. This functional is related, via a functional derivative, to the friction operator. Section 17.9 then gave a brief derivation of the form that biharmonic friction occurs using the formalism of this chapter. Biharmonic friction is commonly used in refined resolution models, where increasing the energetics of the eddies is a primary goal. Finally, Section 17.10 presented some mathematical details that are needed to fill in the steps for some of the derivations in the earlier sections.

Chapter Eighteen

Choosing the Horizontal Viscosity

The purpose of this chapter is to discuss how to choose the viscosity used in horizontal friction schemes. For simplicity, we consider only the Boussinesq expressions here, though the issues discussed are general. There are many ideas presented here for how to choose the viscosity. None work for all cases. The modeler should therefore consider them as proposals, not necessarily recommendations, some of which are worth testing for one's chosen model configuration.

18.1 STABILITY AND RESOLUTION CONSIDERATIONS

The purpose of this section is to review some of the basic numerical considerations that constrain values for the mixing coefficient used with Laplacian and biharmonic mixing operators.

18.1.1 Balance between advection and diffusion

The representation of advection in a numerical model is nontrivial. In practice, most concern has gone into improving the representation of tracer advection. Advection of momentum is still commonly discretized with the centered difference scheme used by Bryan (1969), although some global modelers have tested alternatives, such as Webb et al. (1998b) (for further discussion, see Griffies et al., 2000a). Centered advection of momentum provides for an energetically consistent scheme (see Bryan, 1969), which in turn eliminates the nonlinear instability of Phillips (1959).

The purpose of this section is to summarize numerical stability issues arising when using centered differences for advection. Most notably, it will be shown that a nonzero level of mixing is needed to eliminate a computational mode. The arguments are taken from Appendix B of Bryan et al. (1975).

Consider the steady state balance between one-dimensional advection and diffusion

$$U\,\psi_{,x} = A\,\psi_{,xx} \tag{18.1}$$

where U is a constant advection velocity, ψ is any field that is advected, such as a component of velocity or any tracer field, and A is a constant Laplacian viscosity or diffusivity. Using centered differences in space on a constant grid of size Δ, the finite difference counterpart to this equation takes the form

$$(R-2)\,\psi_{i+1} + 4\,\psi_i - (R+2)\,\psi_{i-1} = 0, \tag{18.2}$$

where

$$R = \frac{U\,\Delta}{A}. \tag{18.3}$$

When ψ is one of the velocity components, R is the grid Reynolds number, where the qualifier "grid" is used since the length scale is the grid scale. When ψ is a tracer, R is the grid Peclet number. In either case, R measures the ratio of advection to diffusion.

A constant solves the finite difference equation (18.2), as does a power

$$\psi_s = C\,\xi^s, \tag{18.4}$$

where C is a constant and the s superscript on the right-hand side represents an integer power. Substituting this function into the finite difference equation yields the quadratic equation

$$(R-2)\,\xi^2 + 4\,\xi - (R+2) = 0. \tag{18.5}$$

The two real roots to this equation are

$$\xi = 1 \tag{18.6}$$

$$\xi = \frac{2+R}{2-R}. \tag{18.7}$$

The first root is a constant as already mentioned. The second root is more relevant. With $\psi_s = C\,\xi^s$, if ξ^s is negative, then ψ_s oscillates in space with a wave length given by the grid size. Such behavior is not physical and so should be avoided. Ensuring that $R < 2$ or

$$\frac{U\,\Delta}{A} < 2 \tag{18.8}$$

keeps the second root positive and so eliminates the unphysical behavior. Therefore, with centered difference discretization of advection, it is necessary to also include a nonzero amount of viscosity or diffusivity sufficient to maintain this inequality.

For the case of tracers, various discretizations other than centered differences can either reduce the need to include diffusivity or introduce diffusivity in regions where the Peclet number does not satisfy the above constraint. Experience has shown that if the Reynolds/Peclet number constraint is not satisfied, the solution has a tendency to degrade slowly via the introduction of grid scale noise.

18.1.2 Linear stability of the diffusion equation

Convergence of the meridians with spherical coordinates makes it possible that a horizontal mixing coefficient appropriate for the mid-latitudes is too large in the high latitudes. More generally, anisotropy in the grid spacing introduces the need to consider spatial variations in the mixing coefficient. This section discusses the linear numerical constraints, where it is sufficient to consider the situation in Cartesian coordinates.

18.1.2.1 Laplacian mixing

Consider two-dimensional Laplacian mixing in Cartesian coordinates*,

$$\psi_{,t} = A\,(\psi_{,xx} + \psi_{,yy}). \tag{18.9}$$

---

*Some of the following material was presented in Section 12.8.4. It is presented here for completeness.

With a uniform grid, the discrete form of this equation is given by

$$\psi_{i,j}^{n+1} = (1 - 2\sigma_x - 2\sigma_y)\, \psi_{i,j}^{n-1} + \sigma_x\, (\psi_{i+1,j}^{n-1} + \psi_{i-1,j}^{n-1}) + \sigma_y\, (\psi_{i,j+1}^{n-1} + \psi_{i,j-1}^{n-1}), \quad (18.10)$$

where

$$\sigma_x = A \left(\frac{2\,\Delta t}{(\Delta x)^2} \right) \quad (18.11)$$

$$\sigma_y = A \left(\frac{2\,\Delta t}{(\Delta y)^2} \right) \quad (18.12)$$

and n labels the discrete time step. In time, this equation has the form of a *forward discretization* with a time step $2\Delta t$. As discussed in Section 12.8.4, this is the appropriate time stepping for models using a leap-frog for the time tendency. If instead we used the time staggered method discussed in Section 12.6, the $2\Delta t$ factor would become Δt.

A von Neumann stability analysis is sufficient for the present purposes. This method determines the constraints necessary for numerical stability of an arbitrary grid wave taking the form

$$\psi_{i,j}^{n} = \mathcal{B}^n\, e^{i\,\mu\, x_i}\, e^{i\,\nu\, y_j}, \quad (18.13)$$

where $x_i = i\,\Delta x$, $y_j = j\,\Delta y$, and μ and ν are arbitrary wavenumbers of dimension inverse length. With this wave ansätz, the finite difference form of the diffusion equation becomes

$$\mathcal{B}^{n+1} = \Omega\, \mathcal{B}^{n-1}, \quad (18.14)$$

where the amplification factor is

$$\Omega = 1 - 4\,\sigma_x\, \sin^2(\mu\, x_i/2) - 4\,\sigma_y\, \sin^2(\nu\, y_j/2). \quad (18.15)$$

To ensure stability, $-1 < \Omega < 1$ is required. So long as σ_x and σ_y are positive, $\Omega < 1$ is trivial to satisfy. The opposite inequality requires

$$\sigma_x\, \sin^2(\mu\, x_i/2) + \sigma_y\, \sin^2(\nu\, y_j/2) < 1/2. \quad (18.16)$$

The most conservative form of this constraint occurs when the sine terms are unity, in which case one finds

$$A < \frac{1}{2\,(2\,\Delta t)} \left((\Delta x)^{-2} + (\Delta y)^{-2} \right)^{-1}. \quad (18.17)$$

For one spatial dimension, this constraint implies

$$A^{\text{one-dim}} < \frac{(\Delta s)^2}{2\,(2\,\Delta t)}, \quad (18.18)$$

where again $2\Delta t$ is the leap-frog time step.

Two dimensions, again being conservative and so choosing $\Delta x = \Delta y = \Delta s$, have the constraint

$$A^{\text{two-dim}} < \frac{(\Delta s)^2}{4\,(2\,\Delta t)}. \quad (18.19)$$

A final conservative approximation is to take Δs as the minimum of Δx and Δy within this equation.

The constraints discussed here can be likened to the CFL constraint placed on the time step due to wave propagation. That is, a diffusive or viscous flux represents a transfer of information across the grid. If this transfer occurs too fast relative to the model's time step, then the numerical scheme will go unstable. Correspondingly, in one dimension the constraint is less strong by a factor of 1/2. The increased restriction placed on the two-dimensional problem is similar to the CFL constraint on waves, as discussed in Section 5-6-7 of Haltiner and Williams (1980). Namely, the effective size of the smallest grid wave is reduced through the addition of an extra dimension, and so the constraint is stronger.

In the above, use of $2\,\Delta t$ for the time step has been highlighted. Again, this is needed in models using a leap-frog to discretize the time tendency. The factor of 2 imposes an added constraint on the dissipation beyond that arising when using a two-time level scheme for the time tendency. We provide further comment on this point in Section 12.8.4. As with the CFL condition, experience has shown that if the linear diffusion equation constraint is not satisfied, then a numerical model soon becomes wildly unstable. The instability can be removed by reducing either the time step or the mixing coefficient.

18.1.2.2 Biharmonic mixing

The same type of analysis for the two-dimensional Cartesian biharmonic equation

$$\psi_{,t} = -B\,(\psi_{,xxxx} + 2\,\psi_{,xxyy} + \psi_{,yyyy}), \tag{18.20}$$

where $B > 0$ is the biharmonic viscosity, leads to the biharmonic amplification factor

$$\Omega = 1 - 16\,[\sqrt{\sigma_x}\,\sin^2(\mu\,\Delta x/2) + \sqrt{\sigma_y}\,\sin^2(\nu\,\Delta y/2)]^2, \tag{18.21}$$

where now

$$\sigma_x = B\left(\frac{2\,\Delta t}{(\Delta x)^4}\right) \tag{18.22}$$

$$\sigma_y = B\left(\frac{2\,\Delta t}{(\Delta y)^4}\right). \tag{18.23}$$

Restricting $-1 < \Omega < 1$ implies

$$[\sqrt{\sigma_x}\,\sin^2(\mu\,\Delta x/2) + \sqrt{\sigma_y}\,\sin^2(\nu\,\Delta y/2)]^2 < 1/8. \tag{18.24}$$

This constraint is satisfied if the following more conservative constraint is satisfied:

$$(\sqrt{\sigma_x} + \sqrt{\sigma_y})^2 < 1/8, \tag{18.25}$$

or

$$B < \frac{1}{16\,\Delta t}\left((\Delta x)^{-2} + (\Delta y)^{-2}\right)^{-2}. \tag{18.26}$$

The most conservative constraint is to set

$$B < \frac{(\Delta s)^4}{64\,\Delta t} \tag{18.27}$$

where Δs is the minimum grid spacing. Again, for a two-level discretization of the time tendency, the stability constraint is weaker by a factor of two.

18.1.3 Western boundary currents

Models with meridional boundaries have boundary currents. The Munk layer (Munk, 1950; Gill, 1982) is relevant for determining the width of the boundary layer in ocean models. As discussed by Bryan et al. (1975), the model must resolve this layer with at least one grid point (optimally more than one grid point) in order to maintain numerical stability. Under-resolution of the Munk layer shows up most visibly in the vertically integrated velocity field (i.e., the barotropic streamfunction when using the rigid lid, or the free surface height with the free surface). In addition, the work of Griffies et al. (2000b) emphasized the importance of having at least two grid points in the Munk layer in order to minimize the level of spurious diapycnal mixing associated with tracer advection. This issue is of most concern for non-isopycnal models where spurious mixing can readily corrupt climate simulations. We discuss these points more extensively in Section 13.7.

With N grid points within the Munk layer, the Laplacian viscosity must satisfy

$$A > \beta(N\,\Delta s\sqrt{3}/\pi)^3, \qquad (18.28)$$

where $\beta = \partial_y f = 2.28 \times 10^{-11}(\mathrm{m\,s})^{-1}\cos\phi$ is the meridional gradient of the planetary vorticity. Hence, for Δs in meters

$$A(\mathrm{m^2\,s^{-1}}) > 3.82 \times 10^{-12}(N\,\Delta s)^3\cos\phi. \qquad (18.29)$$

For example, with $\Delta s = 100 \times 10^3$ m at the equator, having one grid cell within the Munk boundary layer requires a viscosity of $3.82 \times 10^3\,\mathrm{m^2\,s^{-1}}$, whereas with $\Delta s = 10 \times 10^3$ m, the viscosity must be larger than only $3.82\,\mathrm{m^2\,s^{-1}}$.

18.1.4 B-grid computational modes associated with gravity waves

As discussed by Killworth et al. (1991), discretization of gravity waves on a B-grid can admit a stationary grid scale checkerboard pattern (for atmospheric discussions, see Messinger, 1973; Janjić, 1977). This pattern is associated with an unsuppressed grid splitting that can be initiated through grid scale forcing, such as topography. That is, the discrete equations admit a *checkerboard null mode*.* This null mode is seen quite frequently in the surface height field, since this field is typically updated without dissipation. Methods to smooth the surface height are discussed in the above references, as well as Pacanowski and Griffies (1999) and Griffies et al. (2001). Section 12.9 presents a new approach which has been found useful in global climate simulations using the code in Griffies et al. (2004).

In addition to noise in the surface height, the computational mode typically is seen when running with very low levels of viscosity. This mode manifests most clearly as zig-zag or checkerboard velocity vectors. For example, zig-zag patterns can be visible in quiescent regions of a model simulation when using a Smagorinsky viscosity (Section 18.3) with no background viscosity.

Although the model may remain numerically stable, enhanced power at the grid scale arises from accepting too much unresolved structure at the model's smallest scale. There is no justification for considering such structure physically relevant, since it is quite unresolved. Instead, it is a numerical artifact. The extent to which

*The term "null" is used here since an eigenvalue decomposition will reveal that this mode has zero eigenvalue, and so is stationary.

this power influences the larger scales depends on the situation. One egregious problem occurs over long integration periods when noisy velocity fields enhance the levels of spurious mixing of tracers (Griffies et al., 2000b).

18.1.5 Considerations for viscosity on the sphere

Problems associated with the convergence of the meridians in the Arctic have been resolved with models using alternative horizontal coordinates, such as one of the tripolar grids designed by Murray (1996). Nonetheless, it is relevant to present here some considerations that arise using viscosity with grids whose spacing can change over the sphere. The case of spherical coordinates on the sphere is a useful example to illustrate these points.

On a sphere using spherical coordinates, grid spacing in the zonal direction changes according to $\cos\phi$. From many numerical and physical perspectives, it is useful to employ square grids in which the latitudinal resolution is kept abreast with the converging meridians

$$\Delta\phi = \cos\phi\,\Delta\lambda. \tag{18.30}$$

In this way, a grid cell is roughly square with squares becoming smaller as one moves poleward.

Consistent with the desire to employ square grids, it might be useful to prescribe a momentum friction that damps a particular grid scale anomaly with the same time scale regardless of the position on the sphere. As shown in Section 18.2, the damping times for a constant viscosity used for Laplacian and biharmonic friction in one Cartesian dimension is given by

$$\tau_{\text{lap}} = (\Delta/2)^2/A \tag{18.31}$$

$$\tau_{\text{bih}} = (\Delta/2)^4/B, \tag{18.32}$$

where Δ is the grid spacing, and A and B are the Laplacian and biharmonic viscosities, respectively. Preserving the damping time as Δ changes on the sphere suggests letting A have a $\cos^2\phi$ dependence and B have a $\cos^4\phi$ dependence.

Besides providing for a constant damping time, a latitudinally dependent friction can be prescribed that relieves the time step constraint given in Section 18.1.2 that ensues when employing a constant viscosity over the extent of the sphere. That constraint becomes more restrictive on the size of $A\,\Delta t$ when moving toward the poles, or any region of refined resolution. Again, letting A have a $\cos^2\phi$ dependence relieves this constraint. Similar stability considerations with biharmonic friction leads to a biharmonic coefficient with $\cos^4\phi$ dependence.

The above considerations neglect the lower bound considerations given in Sections 18.1.1 and 18.1.3. Notably, if viscosity gets too small, the flow becomes numerically unstable and/or noisy. Therefore, as a compromise, instead of a $\cos^2\phi$ dependence, the Laplacian viscosity is typically given a $\cos\phi$ dependence

$$A = A_o\,\cos\phi, \tag{18.33}$$

where A_o has no latitudinal dependence. This form for the viscosity is furthermore not carried all the way to the pole. In practice, the $\cos\phi$ dependence appears sufficient to alleviate the time step restrictions arising from friction. For biharmonic friction, the analogous viscosity is given by

$$B = B_o\,\cos^3\phi, \tag{18.34}$$

where B_o has no latitudinal dependence.

18.1.6 Summary of stability considerations for viscosity

In summary, there are three main constraints placed on the viscosity. The Reynolds number and Munk boundary layer constraints provide a lower bound on viscosity, whereas the linear diffusion equation constraint provides an upper bound. In general, modelers hope to reduce the viscosity to its lowest value consistent with these constraints. The reason is that it allows the flow to become more advectively dominant, which typically leads to more realistic simulations. Unfortunately, that effort is often difficult to achieve, largely due to the Reynolds number and Munk constraints.

As proposed in Section 18.3, the Smagorinsky scheme provides a means of satisfying each of the above three constraints with only one adjustable dimensionless constant. An alternative, which provides two horizontal viscous degrees of freedom, is the anisotropic scheme proposed by Large et al. (2001) and Smith and McWilliams (2003) (see Sections 17.6 and 18.5).

18.2 COMPARING LAPLACIAN AND BIHARMONIC MIXING

To understand the differences and similarities between Laplacian and biharmonic dissipation, it is sufficient to consider the following linear evolution equations in one Cartesian space dimension (Semtner and Mintz, 1977):

$$\psi_{,t} = A\,\psi_{,xx} \tag{18.35}$$

$$\psi_{,t} = -B\,\psi_{,xxxx} \tag{18.36}$$

where A is the Laplacian mixing coefficient and B is the biharmonic mixing coefficient. Discretizing the Laplacian operator on a Cartesian grid of constant size Δ yields

$$\begin{aligned}\psi_{,t} &= (A/\Delta^2)\,(\psi_{i+1} - 2\psi_i + \psi_{i-1}) \\ &= (A/\Delta^2)\left(D^{1/2} - D^{-1/2}\right)^2 \psi_i,\end{aligned} \tag{18.37}$$

where

$$D^m \psi_i = \psi_{i+m} \tag{18.38}$$

is a linear shift operator. A similar discretization of the biharmonic operator leads to

$$\begin{aligned}\psi_{,t} &= -(B/\Delta^4)\,(\psi_{i+2} - 4\psi_{i+1} + 6\psi_i - 4\psi_{i-1} + \psi_{i-2}) \\ &= -(B/\Delta^4)\left(D^{1/2} - D^{-1/2}\right)^4 \psi_i.\end{aligned} \tag{18.39}$$

To garner a sense of how the dissipation compares, consider a monochromatic grid wave

$$\psi = c(t)\,e^{i k x_i}, \tag{18.40}$$

where k is a wavenumber and $x_i = i\,\Delta$. For such a wave,

$$(D^{1/2} - D^{-1/2})\,e^{i k x_i} = 2\,i\,\sin(k\,\Delta/2)\,e^{i k x_i} \tag{18.41}$$

$$(D^{1/2} - D^{-1/2})^2\,e^{i k x_i} = -4\,\sin^2(k\,\Delta/2)\,e^{i k x_i}. \tag{18.42}$$

$$(D^{1/2} - D^{-1/2})^4\,e^{i k x_i} = 16\,\sin^4(k\,\Delta/2)\,e^{i k x_i}. \tag{18.43}$$

As such, the continuous time evolution equations for this wave under the two forms of dissipation are given by

$$\frac{\mathrm{d} \ln c(t)}{\mathrm{d}t} = -A \left(\frac{\sin(k\,\Delta/2)}{(\Delta/2)} \right)^2 \tag{18.44}$$

$$\frac{\mathrm{d} \ln c(t)}{\mathrm{d}t} = -B \left(\frac{\sin(k\,\Delta/2)}{(\Delta/2)} \right)^4 . \tag{18.45}$$

The solution to these equations is an exponential damping $c(t) = c(0)\, e^{-t/\tau}$, where the inverse damping times are

$$\tau^{-1}_{\text{lap}}(k) = A \left(\frac{\sin(k\,\Delta/2)}{\Delta/2} \right)^2 \tag{18.46}$$

$$\tau^{-1}_{\text{bih}}(k) = B \left(\frac{\sin(k\,\Delta/2)}{\Delta/2} \right)^4 . \tag{18.47}$$

The damping times are equal whenever

$$A = B \left(\frac{\sin(k\,\Delta/2)}{\Delta/2} \right)^2 . \tag{18.48}$$

For example, the smallest grid wave that can live on a discrete grid has size

$$\lambda_{\text{smallest}} = 2\Delta, \tag{18.49}$$

which means the wavenumber for this wave is

$$k_{\text{largest}} = \pi/\Delta. \tag{18.50}$$

In this case, $\sin(k\Delta/2) = 1$, and so the damping times for this wave are given by

$$\tau_{\text{lap}} = (\Delta/2)^2/A \tag{18.51}$$

$$\tau_{\text{bih}} = (\Delta/2)^4/B. \tag{18.52}$$

This is the strongest damping available from either form of dissipation; waves of smaller wavenumber, or longer wavelength, are less damped and so have longer damping times τ. For the $k = \pi/\Delta$ grid wave, if the dissipation coefficients satisfy

$$B = (\Delta/2)^2 A, \tag{18.53}$$

then the damping times are the same. For a typical mid-latitude eddy permitting model, let $\Delta = 0.25 \times \cos(\pi/6) \times 111 \times 1000 = 2.4 \times 10^4$ m. A biharmonic coefficient $B = 10^{11}\,\text{m}^4\,\text{s}^{-1}$ leads to a damping time of 2.4 days for the smallest grid scale waves. Relation (18.53) says that if $A = 7 \times 10^2\,\text{m}^2\,\text{s}^{-1}$, the Laplacian dissipation leads to the same damping time.

18.3 SMAGORINSKY VISCOSITY

As considered in hydrostatic ocean models, the Smagorinsky scheme provides expressions for horizontal momentum viscosity. The viscosities are flow-dependent and hence are nonlinear. The ideas and history of the method are nicely summarized in Smagorinsky (1993). See also the paper by Griffies and Hallberg (2000) for a discussion focusing on ocean climate modeling.

18.3.1 General ideas and motivation

Smagorinsky (1963, 1993) proposed that the effective Laplacian viscosity due to unresolved scales should be proportional to the resolved horizontal deformation rate times the squared grid spacing. This scheme is a physically plausible parameterization of the effects of three-dimensional isotropic turbulence, and it has found much use in large-eddy simulations (e.g., see Galperin and Orszag, 1993). For large-scale geophysical fluid simulations, however, it has little physical justification since the unresolved scales are dominated by quasi-two-dimensional geostrophic turbulence. For this reason, Leith (1968, 1996) proposed an alternative approach based on two-dimensional turbulence. Leith's viscosity is proportional to the horizontal gradient of the relative vorticity times the cubed grid spacing. This approach has found some use in atmospheric models (e.g., Boer and Shepherd, 1983), but it is not commonly used in ocean models.

The Smagorinsky scheme has found notable use in large-scale ocean models for pragmatic reasons (e.g., Blumberg and Mellor, 1987; Rosati and Miyakoda, 1988; Bleck and Boudra, 1981; Bleck et al., 1992; Griffies and Hallberg, 2000). First, it is more convenient to compute than the Leith (1968, 1996) viscosity due to the smaller required grid stencil, and because the deformation rate used to compute the Smagorinsky viscosity is also needed to compute the stress tensor. Second, as with the Leith scheme, the Smagorinsky viscosity tailors the local dissipation to both the local flow state and the local grid resolution with only a single, nondimensional adjustable parameter. If this parameter is properly chosen, the resulting viscosity ensures that the flow respects the relevant numerical stability properties, even when simulating multiple flow and grid regimes such as occur in realistic ocean simulations.

18.3.2 The dimensionless Smagorinsky scaling parameter

In the Smagorinsky scheme, one introduces an adjustable dimensionless parameter which in practice is a tuning parameter. It acts to set the overall magnitude of the viscosity. In an attempt to bring the definition of this parameter into line with that used by others, it is useful to present the following discussion, based largely on that given in Section 1.9 of Smagorinsky (1993).

Recall that kinetic energy is locally dissipated by isotropic transverse frictional stresses according to

$$-e_{mn}\,\tau^{mn} = A\,\rho\,(e_{\mathrm{T}}^2 + e_{\mathrm{S}}^2) \tag{18.54}$$

where e_{T} and e_{S} are the transverse deformation rates due to transverse strains (Section 17.7.1). Smagorinsky chose his viscosity according to

$$A = (\Upsilon/k_{\mathrm{m}})^2\,|E| \tag{18.55}$$

where

$$E^2 = e_{\mathrm{T}}^2 + e_{\mathrm{S}}^2 \tag{18.56}$$

is the squared total deformation rate whose units are squared inverse time. As written, the viscosity depends in a nonlinear manner on the flow, and so is often termed a *nonlinear viscosity*. The wavenumber k_{m} will be specified below. Υ is a dimensionless adjustable parameter. For certain flow regimes, it has been predicted by theories and determined by experiments. Originally, Smagorinsky (1963) was motivated to set $\Upsilon = 0.4$, which is the value expected from wall boundary layer turbulence with 0.4 the von Karman constant. However, this value for Υ is not necessarily relevant for an ocean climate model, where grid lengths less than a kilometer are rarely achieved. Consequently, there is little theoretical justification for choosing one particular value of Υ, and the modeler should consider tuning this parameter.

Notably, as emphasized by Griffies and Hallberg (2000), one practical advantage of the Smagorinsky approach is that the value for Υ is often appropriate across a broad range of grid sizes. That is, if simulating a flow with the Smagorinsky scheme, changing to a higher resolution may typically be done without changing Υ, so long as similar flow regimes are present. This is not the case in general when tuning the dimensionful viscosity. Hence, the Smagorinsky approach helps to reduce the burden on retuning the model upon changing resolution.

In the Modular Ocean Model (Griffies et al., 2004), k_{m} is defined to be the largest resolvable wavenumber local to the grid cell of interest. That is, for a given horizontal grid scale Δs, $2\Delta s$ represents the smallest explicitly resolvable wave length. The corresponding largest resolvable wavenumber is

$$\begin{aligned} k_{\mathrm{m}} &= (2\pi)/(2\Delta s) \\ &= \pi/\Delta s. \end{aligned} \tag{18.57}$$

Hence, the viscosity is determined by

$$A = (\Upsilon \Delta s/\pi)^2 |E|. \tag{18.58}$$

This result differs by a factor of π from that employed by Smagorinsky (1963).

It is convenient to choose the grid spacing as

$$\Delta s \equiv \frac{2\,\Delta x\,\Delta y}{\Delta x + \Delta y}. \tag{18.59}$$

This choice for Δs has continuous derivatives, in contrast to alternative choice of $\Delta s = \min(\Delta x, \Delta y)$.

Note that when using the isotropic form of the stress tensor discussed in Section 17.5, it is important to employ a single transverse viscosity, not a separate one for each of the two horizontal directions as suggested by Rosati and Miyakoda (1988). The reason is that if two viscosities are introduced, then that would generally break the trace-free and symmetric properties of the stress tensor discussed in Section 17.3. A general way to consistently introduce two horizontal viscosities is described in Section 17.6, which follows the approach of Smith and McWilliams (2003). Issues for choosing these two viscosities are presented in Section 18.5.

18.3.3 Choosing the scaling coefficient

There are at least three things to consider when deciding what value to use for the scaling coefficient Υ. First, the viscosity used in a centered differenced momentum advection scheme must be large enough so that the cell Reynolds number satisfies the advection-diffusion constraint established in Section 18.1.1,

$$\begin{aligned} R &= U\Delta s/A \\ &< 2 \\ &\Rightarrow A > U\Delta s/2, \end{aligned} \tag{18.60}$$

where U is a velocity scale and Δs the corresponding grid spacing. In order to get a rough value for Υ based on this constraint, let $|E| \approx U/\Delta s$, which means that the Smagorinsky viscosity is roughly $A \approx (\Upsilon/\pi)^2\, U\, \Delta s$. Forcing $R < 2$ implies that

$$\Upsilon > \pi/\sqrt{2} \approx 2.2. \tag{18.61}$$

Actually, this estimate provides an upper bound on the viscosity realized with the scheme, since $\delta U \approx U$ only in regions of strong horizontal shear. In other regions, the nonlinear viscosity is significantly smaller.

The second consideration involves the western boundary layer considered in Section 18.1.3. Again, with N grid points within the Munk layer and grid spacing Δs in meters, the viscosity must satisfy

$$A(\mathrm{m}^2\,\mathrm{s}^{-1}) > 3.82 \times 10^{-12}(N\,\Delta s)^3 \cos\phi. \tag{18.62}$$

In a 100×10^3 m resolution model, typical velocities are $0.1\ \mathrm{m\,s}^{-1}$, and so the maximum Smagorinsky nonlinear viscosities are on the order of $\Upsilon^2 \times 10^3\ \mathrm{m}^2\,\mathrm{s}^{-1}$. With $\Upsilon = 2.2$, the model might be safely resolving the Munk layer since the regions where the layer is formed are also where there are strong shears. A larger value of Υ, however, might be necessary, and so tests are appropriate.

The Munk boundary layer considerations can be computed prior to any model computations since they do not involve flow properties. If the result of this calculation is that the needed Munk viscosity far exceeds a reasonable estimate based on a chosen Υ, such as the 2.2 based on the cell Reynolds number, then there are two choices that can be made. The first is to run with a time-independent background viscosity set large enough to satisfy the Munk boundary layer condition. If this choice is made, there is little reason to employ the Smagorinsky scheme since the computed viscosities are dominated by the background. The second choice is to increase Υ so that the nonlinear viscosity satisfies the Munk boundary layer condition, at least for the cells within the boundary layer.

For constant viscosity, the linear stability constraints from Section 18.1.2 become especially important in regions of small grid spacing. For spherical grids, such regions typically are in the high latitudes where meridians converge. Because the Smagorinsky scheme has the proper grid stretching factors, it naturally produces small viscosities in the high latitudes. That is, Υ generally need not have any latitudinal dependence. This property is quite convenient for global models and any model with variable resolution.

18.3.4 What about tracers?

For the dissipation of tracer variance through horizontal diffusion, Smagorinsky (1963) assumed a unit value for the horizontal turbulent Prandtl number.* That is, the horizontal viscosity and diffusivity are set equal. Such an approach has been followed by some ocean modelers who use the Smagorinsky scheme (e.g., Blumberg and Mellor, 1987).

Since geostrophic turbulence is the dominant mixing/stirring process in the large-scale ocean, most modelers interested in the long-term properties of ocean tracers have not used the Smagorinsky algorithm for computing diffusivities for

*The Prandtl number is the ratio of the viscosity to diffusivity.

horizontal tracer diffusion, isopycnal tracer diffusion, or thickness diffusion. Instead, alternative theories based on quasi-geostrophic turbulence are more common. Such theories tend to enhance diffusivities in regions of strong vertical thermal wind shears, not horizontal shears as in the Smagorinsky method. These ideas are discussed in Section 14.4.

A further shortcoming of the Smagorinsky approach for horizontal tracer diffusion is that the horizontal Smagorinsky diffusivities tend to be largest near boundaries, where the shears are largest. Hence, spurious upwelling at the side boundaries caused by large horizontal diffusivities, a process known as the *Veronis effect* (Veronis, 1973, 1975), can actually be enhanced with Smagorinsky tracer diffusivities when used in geopotentially aligned horizontal diffusion operators.

For the above reasons, the Smagorinsky scheme is of use in ocean climate models *only* for use in determining the horizontal viscosity for dissipating kinetic energy through friction. It is not of use for determining tracer diffusivities at the grid resolutions of interest for ocean climate models.

18.3.5 Biharmonic Smagorinsky friction

What about using the Smagorinsky scheme for biharmonic viscosity? Recall that linear stability of the diffusion equation using a leap-frog time tendency suggests that the Smagorinsky viscosity should satisfy

$$A < \frac{(\Delta s)^2}{4\,\Delta t}, \tag{18.63}$$

where $\Delta s \equiv (2\,\Delta x\,\Delta y)/(\Delta x + \Delta y)$. A biharmonic viscosity should likewise satisfy

$$B < \frac{(\Delta s)^4}{32\,\Delta t}. \tag{18.64}$$

These results suggest setting the Smagorinsky biharmonic viscosity according to

$$B = A\,(\Delta s)^2/8, \tag{18.65}$$

where A is the Laplacian Smagorinsky viscosity. This relation also holds for two-time level discretizations of the time tendency. Overall, use of the Smagorinsky viscosity in a biharmonic operator has the advantages of the Laplacian Smagorinsky scheme while providing the enhanced scale selectivity of a biharmonic operator (Griffies and Hallberg, 2000).

18.4 BACKGROUND VISCOSITY

The Smagorinsky viscosity suffers from one problem necessitating the addition of a nontrivial background viscosity. The problem is that in regions where the flow is weak, the Smagorinsky viscosity can be quite small. When such regions are next to boundaries, the Munk boundary layer can go unresolved. In this case, the simulated velocity field becomes quite noisy, especially in B-grid models which contain a null mode (Section 18.1.4). Increasing the nondimensional Smagorinsky parameter Υ reduces the noise, but at the cost of over-dissipating other regions where the Smagorinsky viscosity is sufficient.

To alleviate this problem,* consider a grid scale-dependent viscosity to set a minimum allowable viscosity. The idea is to let the minimum Laplacian viscosity be

$$A_{\text{min}} = U_{\text{A}}\, \Delta s, \tag{18.66}$$

where U_{A} is a constant velocity scale and

$$\Delta s = \frac{2\,\Delta x\,\Delta y}{\Delta x + \Delta y}. \tag{18.67}$$

When allowing for multiple regimes of flow, such as with a more highly resolved equatorial region, it may be necessary to provide an *a priori* spatial dependence to U. For biharmonic viscosity,

$$B_{\text{min}} = U_{\text{B}}\,(\Delta s)^3, \tag{18.68}$$

where U_{B} is another velocity scale.

When comparing experiments run with Laplacian versus biharmonic operators, it is often useful to use grid scale-dependent viscosities that damp grid scale waves at the same rate. As seen in Section 18.2, if $B = (\Delta/2)^2 A$ then the biharmonic operator damps a grid wave at the same rate as the corresponding Laplacian operator. If each viscosity is set at its minimum given above, then

$$B_{\text{min}} = U_{\text{B}}\,(\Delta s)^3 \tag{18.69}$$

$$= (\Delta s)^2 A_{\text{min}}/4 \tag{18.70}$$

$$= (\Delta s)^3\, U_{\text{A}}/4, \tag{18.71}$$

thus leading to the correspondence between the velocity scales

$$U_{\text{B}} = U_{\text{A}}/4. \tag{18.72}$$

18.5 VISCOSITIES FOR ANISOTROPIC FRICTION

If we continue to consider horizontal friction used in ocean climate models as a numerical closure, the goal for choosing friction is to satisfy the numerical constraints of Section 18.1 without overly dissipating the simulation. The anisotropic friction operator considered in Section 17.6 affords one extra degree of freedom when specifying the transverse friction. This extra degree of freedom has been shown to be useful to reduce model dissipation, while maintaining numerical integrity by dissipating noise. This section presents some guidelines for choosing the isotropic viscosity A and anisotropic viscosity D. Similar considerations were discussed by Smith and McWilliams (2003).

There are two main numerical constraints from Section 18.1 that guide how one chooses the viscosities. First, frictional boundary layers must be resolved. Doing so requires enough friction be applied in the cross-boundary direction. Second, friction must be large enough to satisfy the grid Reynolds number constraint. Doing so requires enough friction be applied in the along-stream direction.

---

*Eric Chassignet from the University of Miami suggested this approach to the author in 1999.

To see how these requirements translate into setting the viscosities, consider the stress and friction associated with flow aligned with the coordinate directions. For this purpose, recall the stress tensor is generally given by (Section 17.7.2 on page 397)

$$\tau^{(m)(n)} = \rho \begin{pmatrix} (A\, e_{\mathrm{T}} + D\,\Delta\,\sin 2\theta) & (A\, e_{\mathrm{S}} - D\,\Delta\,\cos 2\theta) \\ (A\, e_{\mathrm{S}} - D\,\Delta\,\cos 2\theta) & (-A\, e_{\mathrm{T}} - D\,\Delta\,\sin 2\theta) \end{pmatrix} \tag{18.73}$$

where the deformation rates are (Section 17.7.1)

$$e_{\mathrm{T}} = (\mathrm{d}y)\,(u/\mathrm{d}y)_{,x} - (\mathrm{d}x)\,(v/\mathrm{d}x)_{,y} \tag{18.74}$$

$$e_{\mathrm{S}} = (\mathrm{d}x)\,(u/\mathrm{d}x)_{,y} + (\mathrm{d}y)\,(v/\mathrm{d}y)_{,x} \tag{18.75}$$

the anisotropic term is (Section 17.6.2)

$$2\,\Delta = e_{\mathrm{S}}\,\cos 2\theta - e_{\mathrm{T}}\,\sin 2\theta, \tag{18.76}$$

and θ sets the orientation of the friction. With the flow-dependent orientation of Smith and McWilliams (2003), θ is given by

$$\mathbf{u} = |\mathbf{u}|\,(\cos\theta, \sin\theta). \tag{18.77}$$

In the fixed method of Large et al. (2001), θ is set by the grid lines, such as a zonal direction in the equatorial region. Consider friction oriented so that $\theta = \pi/2$, which occurs for the flow-dependent case when currents are moving in the generalized meridional direction; then $2\,\Delta = -e_{\mathrm{S}}$ and the stress tensor is

$$\tau^{(m)(n)} = \rho \begin{pmatrix} A\, e_{\mathrm{T}} & (A - D/2)\, e_{\mathrm{S}} \\ (A - D/2)\, e_{\mathrm{S}} & -A\, e_{\mathrm{T}} \end{pmatrix}. \tag{18.78}$$

For simplicity, assume Cartesian coordinates on a plane with a Boussinesq fluid and constant viscosities. Doing so leads to the friction vector

$$F^{x} = (A + D/2)\, u_{,xx} + (A - D/2)\, u_{,yy} - (D/2)\, \partial_x \nabla \cdot \mathbf{u} \tag{18.79}$$

$$F^{y} = (A + D/2)\, v_{,yy} + (A - D/2)\, v_{,xx} - (D/2)\, \partial_y \nabla \cdot \mathbf{u}. \tag{18.80}$$

With a small horizontal divergence $\nabla \cdot \mathbf{u}$, the last term in these expressions is subdominant to the first and second terms. For meridional flow away from boundaries, the along-stream viscosity

$$A_{\text{along}} = A + D/2 \tag{18.81}$$

must satisfy the grid Reynolds number constraint. For meridional flow aligned with a meridional boundary, the cross-stream viscosity

$$A_{\text{cross}} = A - D/2 \tag{18.82}$$

must satisfy the Munk boundary layer constraint. Finally, recall from Section 17.6.2 that $D < 2\,A$ is required to ensure that kinetic energy is dissipated by the friction operator (see equation (17.105)), which means that the cross-stream viscosity can never go negative.

18.6 CHAPTER SUMMARY

This chapter aimed to provide the reader with some of the basic considerations related to choosing horizontal viscosity for an ocean climate model. The choices are many, with no clear best approach for all situations. Instead, the modeler is encouraged to test various methods, with the simplest useful one normally preferred. Again, this methodology is based on the common notion that horizontal friction is needed for reasons of numerical closure. Physical closure ideas are speculative for the horizontal momentum equation, with no general consensus as to what is appropriate. Indeed, some even suggest that the common forms used in the models are sufficient for parameterizing some aspects of subgrid scale momentum transfer, whereas others argue they are completely wrong.

Tuning horizontal friction in global ocean climate models is a nontrivial task requiring a great deal of patience and persistence. There are many aspects of the flow that are affected by friction, such as boundary currents, transport through straights and through-flows, equatorial currents, gyre structures, heat transport, and so on. The array of ideas introduced in this chapter, from the stability arguments which provide lower and upper bounds on the viscosity, to the newer ideas of anisotropic operators, should be considered as mere proposals. General recommendations are difficult to make, since there are numerous factors that determine the suitability of a particular scheme. Nonetheless, the global modeler is wise to digest some of the basic ideas here, since with the present state-of-the-art in ocean climate models, they form a central role in what is used to develop a simulation that looks like the real ocean.

Chapter Nineteen

FUNCTIONAL DISCRETIZATION OF FRICTION

The purpose of this chapter is to formulate a discrete version of the horizontal momentum friction operators derived in Chapter 17. The basis for this discretization is the functional approach described in Section 17.8. This chapter will appeal to those interested in implementing friction in either a B-grid or C-grid ocean model. Notably, the formulation on the C-grid is far simpler, as mentioned by Griffies and Hallberg (2000). Hence, the bulk of this chapter focuses on details for the B-grid.

19.1 COMMENTS ON NOTATION

As with the discrete neutral physics operator in Chapter 16, the presentation here introduces grid labels on fields living on a discrete lattice. Explicit sums and averages are also used in most places rather than choosing among the more elegant alternatives. This more explicit notation adds clarity which is ultimately required by the numericist translating an algorithm into computer code. Experience has shown that having the algorithm written nearly in its "full glory" in LaTeX reduces the chances for bugs to penetrate the computer code.

Chapter 17 established the form of the horizontal friction operator within the context of the unaveraged equations of motion derived in Chapters 3 and 4. This is strictly speaking a misrepresentation of the unaveraged equations, since they employ dissipation operators due to molecular friction, not the subgrid scale (SGS) friction of focus here. Nonetheless, we took this approach for purposes of tidiness. To transfer results from Chapter 17 to model fields, it is necessary to follow the mapping given by Table 8.1. Only the non-Boussinesq case is of concern, whereby $\mathbf{v} \rightarrow \mathbf{v}^{\rho}$ for the non-Boussinesq fluid. Notably, this mapping means that it is the velocity $\mathbf{v}^{\rho} = \mathbf{v}\,(\rho_o/\rho)$ that is used to construct the friction operator in the non-Boussinesq model, *not* the velocity $\mathbf{v}$. In addition, the friction operator in the model represents a discrete version of the SGS momentum flux

$$\langle \tilde{\mathbf{F}}^{(\mathbf{u})}_{\text{sgs}} \rangle \rightarrow \mathbf{F}^{(\mathbf{u})}_{\text{model}}. \tag{19.1}$$

For z-model implementations, factors of density reduce to the Boussinesq reference value ρ_o, which simplifies the numerical treatment in the non-Boussinesq case.

In the present chapter, in order to maintain a minimum set of labels, we are *not* careful to abide by the above notation for the velocity. Hence, as emphasized by Greatbatch et al. (2001), it is important to recognize that when constructing friction within a non-Boussinesq z-model, one should use $\mathbf{v}^{\rho}$ instead of $\mathbf{v}$ to ensure that kinetic energy is dissipated.

The presentation given here introduces grid labels on fields living on a discrete

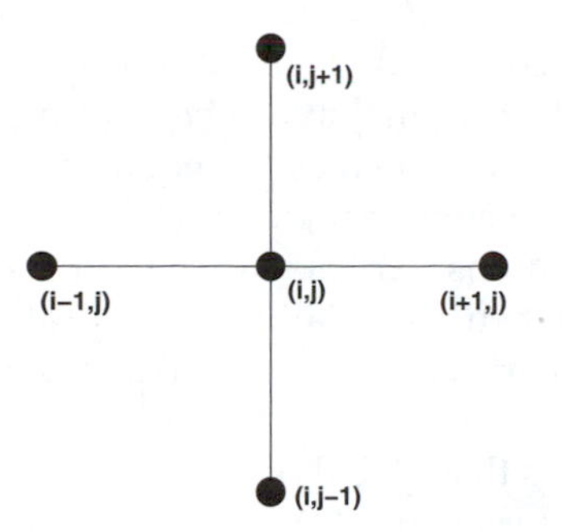

Figure 19.1 Stencil for the 5-point Laplacian operator commonly used for models on the plane.

lattice. In particular, i, j, k denote the discrete labels for points in the $x, y, -z$ directions, respectively. Other notation, such as the grid factors, is based on that used in version 4 of the Modular Ocean Model (MOM4) of Griffies et al. (2004). Indeed, the material in this chapter documents the algorithm discretized in MOM4.

19.2 SUMMARY OF THE VARIOUS FORMULATIONS

This section provides a summary of common formulations of friction used in ocean climate models.

19.2.1 5-point Laplacian operator

On a horizontal plane with second order numerics, the discrete Laplacian operator forms a 5-point stencil. That is, with uniform grid spacing Δ in the two directions,

$$\Delta^2 (\psi_{,xx} + \psi_{,yy}) \approx (\psi_{i+1,j} - 2\,\psi_{i,j} + \psi_{i-1,j}) + (\psi_{i,j+1} - 2\,\psi_{i,j} + \psi_{i,j-1}), \quad (19.2)$$

which incorporates the 5 points shown in Figure 19.1. This stencil acts to smooth a discrete field regardless of the configuration, since the 5-point operator has no spurious computational modes.* It is for this reason that many ocean models employ a Laplacian operator to dissipate momentum (and tracer) in order to satisfy various numerical stability requirements.

19.2.2 Friction on the sphere

On a sphere and in the presence of nonhomogeneous grids, boundaries, and various flow regimes, a Laplacian operator using a constant viscosity is often suboptimal in the sense that it may over-dissipate the simulated flow. The paper by Griffies and Hallberg (2000) motivates consideration of the Smagorinsky viscosity (Section 18.3) in either a second order (Laplacian) or fourth order (biharmonic) friction operator. The work of Large et al. (2001) and Smith and McWilliams (2003)

*In this context, a *computational mode* is a field configuration, for example a grid wave, that is undamped by the discretized operator.

note that by removing the commonly made assumption of horizontal isotropy, one garners an extra degree of freedom that can be readily exploited for increasing the flow strength in critical regions, such as the equatorial undercurrent, yet without incurring noise in boundary current regions. In general, these approaches require some additional considerations to preserve angular momentum on a sphere, and these warrant some care when formulating the operator beyond that of the familiar Cartesian form of the Laplacian operator. These issues were fully discussed in Chapter 17.

There are two basic ways that global ocean models handle the added complications in the friction operator arising from sphericity and nonconstant viscosities. First, there are the formulas of Bryan (1969), Wajsowicz (1993), Smith et al. (1995), and Murray and Reason (2002) appropriate for horizontally isotropic friction. In this approach, friction is written as a "Laplacian plus frictional metric" form. The Laplacian piece is the spherical coordinate form of a Laplacian operator acting on a component of the velocity, as if the velocity component was a scalar field. The additional frictional metric piece accounts for the fact that velocity is a vector, not a scalar. The friction metric term arises for two general reasons: (1) the sphericity of the earth (Section 4.11), and (2) nonconstant viscosities (Section 17.5.4). Without the frictional metric term, the corresponding stress tensor looses its symmetry. A nonsymmetric stress tensor can generally introduce spurious torques, thus compromising angular momentum budgets. Bryan (1969) used a friction metric appropriate for constant viscosities, Wajsowicz (1993) showed how it is modified for use with nonconstant viscosities, and Smith et al. (1995) and Murray and Reason (2002) generalized these results to arbitrary orthogonal coordinates.

The frictional metric term is cumbersome to discretize, especially with generalized orthogonal coordinates. Additionally, its discretization on a B-grid admits computational modes. Hence, the frictional metric is sometimes ignored, as in Cox (1984). If maintaining the metric term, the presence of the scalar Laplacian appears to render the computational modes harmless. Nonetheless, the formulation is not satisfying since ideally no part of the friction operator should admit computational modes. An alternative formulation is therefore desired.

A second way to discretize the friction operator is to apply the stress tensor formalism of Smagorinsky (1963, 1993) in the process of discretizing the Smagorinsky nonlinear viscosity. This was the approach of Rosati and Miyakoda (1988). In the continuum, this approach is equivalent to Wajsowicz (1993) (see Section 19.4 for details). Yet there are differences on the discrete lattice. Furthermore, the tensor formalism is considerably more concise, and the ability to generalize to arbitrary orthogonal curvilinear coordinates is more elegant. It is for these reasons that this method is preferred here. This approach was originally proposed by Griffies and Hallberg (2000) and used by Smith and McWilliams (2003).

19.3 HORIZONTAL FRICTION DISCRETIZATION

The purpose of this section is to discretize the horizontal friction operator on a B-grid. Much here follows the functional approach used for the neutral diffusion discretization detailed in Chapter 16. However, the present discussion is simpler because there is need to only discretize a two-dimensional (x-y plane) operator, and because there are no subtleties associated with neutral directions. The approach here is therefore a straightforward application of the functional formalism described in Section 17.8.

19.3.1 Concerning a C-grid discretization

We start this section by noting that the issues of discretizing friction on the C-grid are significantly simpler than on the B-grid. As mentioned in Section 10.2, the C-grid is ideally designed for discretizing transport operators, such as advection, diffusion, and friction. Details for how to construct the friction operator on the C-grid are described in Griffies and Hallberg (2000). Focus in the following is given to the more extensive details needed for the B-grid.

19.3.2 Discrete dissipation functional for the B-grid

The discrete dissipation functional takes the form

$$\begin{aligned}\mathcal{S} &= \sum_{i,j} \sum_{n=1}^{12} V(n) \left[A(n) \left(e_{\mathrm{T}}^2(n) + e_{\mathrm{S}}^2(n)\right) - 2\, D(n)\, \Delta^2(n)\right] \\ &\equiv \sum_{i,j} \mathcal{S}_{i,j}.\end{aligned} \tag{19.3}$$

Figure 19.2 illustrates the nearest neighbor stencil used for discretizing the dissipation functional using second order numerics. The summation $n = 1, 12$ arises from the 12 subcells to which the velocity point $U_{i,j}$ contributes when discretizing the functional on a B-grid. $V(n)$ are the volumes of each of the subcells, $A(n)$ are the viscosities, and $e_{\mathrm{T}}(n)$, $e_{\mathrm{S}}(n)$, and $\Delta(n)$ are the corresponding tensions, strains, and anisotropic contributions, respectively.

The physical component of the friction vector acting at the velocity cell $U_{i,j,k}$ is given by the discrete functional derivative (see Section 17.8.4)

$$\begin{aligned}-F_{i,j}^{(b)} &= \frac{1}{2\, V_{i,j}^{U}} \frac{\partial \mathcal{S}_{i,j}}{\partial (u^{(b)})_{i,j}} \\ &= \frac{1}{V_{i,j}^{U}} \sum_{n=1}^{12} V(n) \\ &\quad \times \left(A(n)\, e_{\mathrm{T}}(n) \frac{\partial e_{\mathrm{T}}(n)}{\partial (u^{(b)})_{i,j}} + A(n)\, e_{\mathrm{S}}(n) \frac{\partial e_{\mathrm{S}}(n)}{\partial (u^{(b)})_{i,j}} - 2\, D(n)\, \Delta(n) \frac{\partial \Delta(n)}{\partial (u^{(b)})_{i,j}} \right) \\ &= \frac{1}{\rho_o\, V_{i,j}^{U}} \sum_{n=1}^{12} V(n) \left(\frac{\partial e_{\mathrm{T}}(n)}{\partial (u^{(b)})_{i,j}}\, \tau^{xx}(n) + \frac{\partial e_{\mathrm{S}}(n)}{\partial (u^{(b)})_{i,j}}\, \tau^{xy}(n) \right).\end{aligned} \tag{19.4}$$

In this equation, $b = 1, 2$ labels the generalized zonal and meridional directions, the discrete depth label k was dropped since all points are at the same level,

$$\left((u^{(1)})_{i,j}, (u^{(2)})_{i,j} \right) = (u_{i,j}, v_{i,j}) \tag{19.5}$$

are physical components of the velocity vector,

$$V_{i,j,k}^{U} = dxu_{i,j}\, dyu_{i,j}\, dhu_{i,j,k} \tag{19.6}$$

is the velocity cell volume,

$$\tau^{xx}(n) = \rho_o \left[A(n)\, e_T(n) + D(n)\, \Delta(n)\, \sin 2\theta\right] \tag{19.7}$$

$$\tau^{xy}(n) = \rho_o \left[A(n)\, e_S(n) - D(n)\, \Delta(n)\, \cos 2\theta\right] \tag{19.8}$$

are the stress tensor components, and the orientation angle θ determines the orientation of the friction according to the discussion in Section 17.6.1. The remainder of this section is devoted to performing the discrete functional derivatives and then manipulating the results into a tidy expression for the discrete friction operator.

19.3.3 Grid cell distances and subcell volumes

Enumerating the 12 quarter cell volumes constitutes an important part of the discretization. Assume knowledge of the distance along each of the four sides of a velocity and tracer cell, as well as the distance from the velocity and tracer points to the sides of their respective cells. Although nonuniform grids with general coordinates do not allow an exact (i.e., analytically exact) area calculation, the horizontal area of the quarter cells is well approximated using this information. Even so, for the purpose of discretizing friction, there is no reason to take pains to compute a very accurate quarter cell area. Indeed, the aim is to realize the dissipative property of friction with minimal computational expense. Hence, make the simplification that the volume of a quarter cell is equal to one-quarter the volume of the corresponding velocity cell. In summary, we employ the following grid information (refer to Figure 19.2):

- The longitudinal spacing, in meters, between $U_{i,j}$ and $U_{i+1,j}$ is $dxue_{i,j}$. The latitudinal spacing, in meters, between $U_{i,j}$ and $U_{i,j+1}$ is $dyun_{i,j}$.

- The volume of a quarter cell is taken equal to one-quarter the volume of the velocity cell where the quarter cell lives. For example, quarter cells 1, 2, 3, and 4 live inside velocity cell $U_{i,j,k}$ and so have volume

$$(1/4)\, V^{U}_{i,j,k} = (1/4)\, dxu_{i,j}\, dyu_{i,j}\, dhu_{i,j,k}, \tag{19.9}$$

whereas quarter cells 5 and 6 live inside velocity cell $U_{i+1,j,k}$ and so have volume

$$(1/4)\, V^{U}_{i+1,j,k} = (1/4)\, dxu_{i+1,j}\, dyu_{i+1,j}\, dhu_{i+1,j,k}. \tag{19.10}$$

In summary, the volumes of the 12 quarter cells are taken to be

$$\begin{aligned} V(1) &= V(2) = V(3) = V(4) = (1/4)\, V^{U}_{i,j,k} \\ V(5) &= V(6) = (1/4)\, V^{U}_{i+1,j,k} \\ V(7) &= V(8) = (1/4)\, V^{U}_{i-1,j,k} \\ V(9) &= V(10) = (1/4)\, V^{U}_{i,j+1,k} \\ V(11) &= V(12) = (1/4)\, V^{U}_{i,j-1,k}. \end{aligned} \tag{19.11}$$

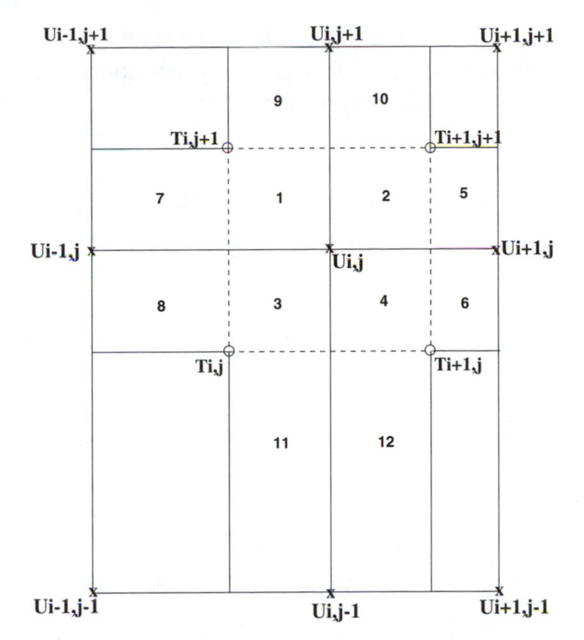

Figure 19.2 Stencil for the discrete horizontal friction functional. The 12 quarter cells each contain contributions to the functional from the central velocity point $U_{i,j}$. The tracer points $T_{i,j}$ are oriented according to the B-grid (see Section 10.2).

19.3.4 Derivative operators

Second order finite difference approximations to the derivative operators are given by

$$\delta_x u_{i,j} = \frac{u_{i+1,j} - u_{i,j}}{dxue_{i,j}} \tag{19.12}$$

$$\delta_y u_{i,j} = \frac{u_{i,j+1} - u_{i,j}}{dyun_{i,j}}. \tag{19.13}$$

The distances $dxue_{i,j}$ and $dyun_{i,j}$ represent the zonal and meridional distances between the velocity points. All metric stretching factors are absorbed into these grid distances according to the discussion in Section 21.12.4. Second order accuracy is realized since the derivative operators live at intermediate points, that is, at the cell faces.

19.3.5 Tension and strain for the subcells

A decision must be made regarding the form of tension and strain to discretize. The goal here is to work with a discretization involving the least amount of grid factors in order to minimize computational expense. The first forms of tension and strain are more symmetric,

$$e_{\mathrm{T}} = \mathrm{d}y\,(u/\mathrm{d}y)_{,x} - \mathrm{d}x\,(v/\mathrm{d}x)_{,y} \tag{19.14}$$

$$e_{\mathrm{S}} = \mathrm{d}x\,(u/\mathrm{d}x)_{,y} + \mathrm{d}y\,(v/\mathrm{d}y)_{,x} \tag{19.15}$$

and lead to a compact form for the discretized friction operator (see below), directly analogous to the continuum form given by equations (17.124) and (17.125). The second forms isolate the Cartesian expression

$$e_{\mathrm{T}} = u_{,x} - v_{,y} + v\,\partial_y \ln \mathrm{d}x - u\,\partial_x \ln \mathrm{d}y \tag{19.16}$$

$$e_{\mathrm{S}} = u_{,y} + v_{,x} - u\,\partial_y \ln \mathrm{d}x - v\,\partial_x \ln \mathrm{d}y. \tag{19.17}$$

The second forms require less computation, since the metric terms

$$(M_{\mathrm{T}})_{i,j} = -u_{i,j}\,(\partial_x \ln \mathrm{d}y)_{i,j} + v_{i,j}\,(\partial_y \ln \mathrm{d}x)_{i,j} \tag{19.18}$$

$$(M_{\mathrm{S}})_{i,j} = -u_{i,j}\,(\partial_y \ln \mathrm{d}x)_{i,j} - v_{i,j}\,(\partial_x \ln \mathrm{d}y)_{i,j} \tag{19.19}$$

are common to each of the four surrounding triads, and the grid factors

$$dh1dy_{i,j} = (\partial_y \ln \mathrm{d}x)_{i,j} \tag{19.20}$$

$$dh2dx_{i,j} = (\partial_x \ln \mathrm{d}y)_{i,j} \tag{19.21}$$

can be computed at the start of the integration. Hence, we choose the first form of the deformation rates to develop the discrete friction, and the second form to evaluate the deformation rates within the discrete friction. For purposes of completeness, both discrete forms of the deformation rates are displayed in the following.

The first form for tension in the 12 subcells is

$$e_{\mathrm{T}}(1) = dyue_{i-1,j}\,\delta_x\,(u_{i-1,j}/dyu_{i-1,j}) - dxun_{i,j}\,\delta_y\,(v_{i,j}/dxu_{i,j}) \tag{19.22}$$

$$e_{\mathrm{T}}(2) = dyue_{i,j}\,\delta_x\,(u_{i,j}/dyu_{i,j}) - dxun_{i,j}\,\delta_y\,(v_{i,j}/dxu_{i,j}) \tag{19.23}$$

$$e_{\mathrm{T}}(3) = dyue_{i-1,j}\,\delta_x\,(u_{i-1,j}/dyu_{i-1,j}) - dxun_{i,j-1}\,\delta_y\,(v_{i,j-1}/dxu_{i,j-1}) \tag{19.24}$$

$$e_{\mathrm{T}}(4) = dyue_{i,j}\,\delta_x\,(u_{i,j}/dyu_{i,j}) - dxun_{i,j-1}\,\delta_y\,(v_{i,j-1}/dxu_{i,j-1}) \tag{19.25}$$

$$e_{\mathrm{T}}(5) = dyue_{i,j}\,\delta_x\,(u_{i,j}/dyu_{i,j}) - dxun_{i+1,j}\,\delta_y\,(v_{i+1,j}/dxu_{i+1,j}) \tag{19.26}$$

$$e_{\mathrm{T}}(6) = dyue_{i,j}\,\delta_x\,(u_{i,j}/dyu_{i,j}) - dxun_{i+1,j-1}\,\delta_y\,(v_{i+1,j-1}/dxu_{i+1,j-1}) \tag{19.27}$$

$$e_{\mathrm{T}}(7) = dyue_{i-1,j}\,\delta_x\,(u_{i-1,j}/dyu_{i-1,j}) - dxun_{i-1,j}\,\delta_y\,(v_{i-1,j}/dxu_{i-1,j}) \tag{19.28}$$

$$e_{\mathrm{T}}(8) = dyue_{i-1,j}\,\delta_x\,(u_{i-1,j}/dyu_{i-1,j}) - dxun_{i-1,j-1}\,\delta_y\,(v_{i-1,j-1}/dxu_{i-1,j-1}) \tag{19.29}$$

$$e_{\mathrm{T}}(9) = dyue_{i-1,j+1}\,\delta_x\,(u_{i-1,j+1}/dyu_{i-1,j+1}) - dxun_{i,j}\,\delta_y\,(v_{i,j}/dxu_{i,j}) \tag{19.30}$$

$$e_{\mathrm{T}}(10) = dyue_{i,j+1}\,\delta_x\,(u_{i,j+1}/dyu_{i,j+1}) - dxun_{i,j}\,\delta_y\,(v_{i,j}/dxu_{i,j}) \tag{19.31}$$

$$e_{\mathrm{T}}(11) = dyue_{i-1,j-1}\,\delta_x\,(u_{i-1,j-1}/dyu_{i-1,j-1}) - dxun_{i,j-1}\,\delta_y\,(v_{i,j-1}/dxu_{i,j-1}) \tag{19.32}$$

$$e_{\mathrm{T}}(12) = dyue_{i,j-1}\,\delta_x\,(u_{i,j-1}/dyu_{i,j-1}) - dxun_{i,j-1}\,\delta_y\,(v_{i,j-1}/dxu_{i,j-1}) \tag{19.33}$$

and the first form for strain is

$$e_S(1) = dxun_{i,j}\, \delta_y\, (u_{i,j}/dxu_{i,j}) + dyue_{i-1,j}\, \delta_x\, (v_{i-1,j}/dyu_{i-1,j}) \tag{19.34}$$
$$e_S(2) = dxun_{i,j}\, \delta_y\, (u_{i,j}/dxu_{i,j}) + dyue_{i,j}\, \delta_x\, (v_{i,j}/dyu_{i,j}) \tag{19.35}$$
$$e_S(3) = dxun_{i,j-1}\, \delta_y\, (u_{i,j-1}/dxu_{i,j-1}) + dyue_{i-1,j}\, \delta_x\, (v_{i-1,j}/dyu_{i-1,j}) \tag{19.36}$$
$$e_S(4) = dxun_{i,j-1}\, \delta_y\, (u_{i,j-1}/dxu_{i,j-1}) + dyue_{i,j}\, \delta_x\, (v_{i,j}/dyu_{i,j}) \tag{19.37}$$
$$e_S(5) = dxun_{i+1,j}\, \delta_y\, (u_{i+1,j}/dxu_{i+1,j}) + dyue_{i,j}\, \delta_x\, (v_{i,j}/dyu_{i,j}) \tag{19.38}$$
$$e_S(6) = dxun_{i+1,j-1}\, \delta_y\, (u_{i+1,j-1}/dxu_{i+1,j-1}) + dyue_{i,j}\, \delta_x\, (v_{i,j}/dyu_{i,j}) \tag{19.39}$$
$$e_S(7) = dxun_{i-1,j}\, \delta_y\, (u_{i-1,j}/dxu_{i-1,j}) + dyue_{i-1,j}\, \delta_x\, (v_{i-1,j}/dyu_{i-1,j}) \tag{19.40}$$
$$e_S(8) = dxun_{i-1,j-1}\, \delta_y\, (u_{i-1,j-1}/dxu_{i-1,j-1}) + dyue_{i-1,j}\, \delta_x\, (v_{i-1,j}/dyu_{i-1,j}) \tag{19.41}$$
$$e_S(9) = dxun_{i,j}\, \delta_y\, (u_{i,j}/dxu_{i,j}) + dyue_{i-1,j+1}\, \delta_x\, (v_{i-1,j+1}/dyu_{i-1,j+1}) \tag{19.42}$$
$$e_S(10) = dxun_{i,j}\, \delta_y\, (u_{i,j}/dxu_{i,j}) + dyue_{i,j+1}\, \delta_x\, (v_{i,j+1}/dyu_{i,j+1}) \tag{19.43}$$
$$e_S(11) = dxun_{i,j-1}\, \delta_y\, (u_{i,j-1}/dxu_{i,j-1}) + dyue_{i-1,j-1}\, \delta_x\, (v_{i-1,j-1}/dyu_{i-1,j-1}) \tag{19.44}$$
$$e_S(12) = dxun_{i,j-1}\, \delta_y\, (u_{i,j-1}/dxu_{i,j-1}) + dyue_{i,j-1}\, \delta_x\, (v_{i,j-1}/dyu_{i,j-1}). \tag{19.45}$$

The second forms for tension and strain are

$$e_T(1) = \delta_x\, u_{i-1,j} - \delta_y\, v_{i,j} + (M_T)_{i,j} \qquad e_S(1) = \delta_y\, u_{i,j} + \delta_x\, v_{i-1,j} + (M_S)_{i,j} \tag{19.46}$$
$$e_T(2) = \delta_x\, u_{i,j} - \delta_y\, v_{i,j} + (M_T)_{i,j} \qquad e_S(2) = \delta_y\, u_{i,j} + \delta_x\, v_{i,j} + (M_S)_{i,j} \tag{19.47}$$
$$e_T(3) = \delta_x\, u_{i-1,j} - \delta_y\, v_{i,j-1} + (M_T)_{i,j} \qquad e_S(3) = \delta_y\, u_{i,j-1} + \delta_x\, v_{i-1,j} + (M_S)_{i,j} \tag{19.48}$$
$$e_T(4) = \delta_x\, u_{i,j} - \delta_y\, v_{i,j-1} + (M_T)_{i,j} \qquad e_S(4) = \delta_y\, u_{i,j-1} + \delta_x\, v_{i,j} + (M_S)_{i,j} \tag{19.49}$$
$$e_T(5) = \delta_x\, u_{i,j} - \delta_y\, v_{i+1,j} + (M_T)_{i+1,j} \qquad e_S(5) = \delta_y\, u_{i+1,j} + \delta_x\, v_{i,j} + (M_S)_{i+1,j} \tag{19.50}$$
$$e_T(6) = \delta_x\, u_{i,j} - \delta_y\, v_{i+1,j-1} + (M_T)_{i+1,j} \qquad e_S(6) = \delta_y\, u_{i+1,j-1} + \delta_x\, v_{i,j} + (M_S)_{i+1,j} \tag{19.51}$$
$$e_T(7) = \delta_x\, u_{i-1,j} - \delta_y\, v_{i-1,j} + (M_T)_{i-1,j} \qquad e_S(7) = \delta_y\, u_{i-1,j} + \delta_x\, v_{i-1,j} + (M_S)_{i-1,j} \tag{19.52}$$
$$e_T(8) = \delta_x\, u_{i-1,j} - \delta_y\, v_{i-1,j-1} + (M_T)_{i-1,j} \qquad e_S(8) = \delta_y\, u_{i-1,j-1} + \delta_x\, v_{i-1,j} + (M_S)_{i-1,j} \tag{19.53}$$
$$e_T(9) = \delta_x\, u_{i-1,j+1} - \delta_y\, v_{i,j} + (M_T)_{i,j+1} \qquad e_S(9) = \delta_y\, u_{i,j} + \delta_x\, v_{i-1,j+1} + (M_S)_{i,j+1} \tag{19.54}$$
$$e_T(10) = \delta_x\, u_{i,j+1} - \delta_y\, v_{i,j} + (M_T)_{i,j+1} \qquad e_S(10) = \delta_y\, u_{i,j} + \delta_x\, v_{i,j+1} + (M_S)_{i,j+1} \tag{19.55}$$
$$e_T(11) = \delta_x\, u_{i-1,j-1} - \delta_y\, v_{i,j-1} + (M_T)_{i,j-1} \qquad e_S(11) = \delta_y\, u_{i,j-1} + \delta_x\, v_{i-1,j-1} + (M_S)_{i,j-1} \tag{19.56}$$
$$e_T(12) = \delta_x\, u_{i,j-1} - \delta_y\, v_{i,j-1} + (M_T)_{i,j-1} \qquad e_S(12) = \delta_y\, u_{i,j-1} + \delta_x\, v_{i,j-1} + (M_S)_{i,j-1}. \tag{19.57}$$

19.3.6 Functional derivative of e_T and e_S

Choosing to work with the discretized first form of tension (equation (19.14)) leads to the functional derivative of the tension within the central four subcells

$$\frac{\partial e_T(1)}{\partial (u^{(b)})_{i,j}} = \frac{dyue_{i-1,j}}{dxue_{i-1,j}}\, \frac{\delta_b^1}{dyu_{i,j}} + \frac{dxun_{i,j}}{dyun_{i,j}}\, \frac{\delta_b^2}{dxu_{i,j}} \tag{19.58}$$

$$\frac{\partial e_T(2)}{\partial (u^{(b)})_{i,j}} = -\frac{dyue_{i,j}}{dxue_{i,j}}\, \frac{\delta_b^1}{dyu_{i,j}} + \frac{dxun_{i,j}}{dyun_{i,j}}\, \frac{\delta_b^2}{dxu_{i,j}} \tag{19.59}$$

$$\frac{\partial e_T(3)}{\partial (u^{(b)})_{i,j}} = \frac{dyue_{i-1,j}}{dxue_{i-1,j}}\, \frac{\delta_b^1}{dyu_{i,j}} - \frac{dxun_{i,j-1}}{dyun_{i,j-1}}\, \frac{\delta_b^2}{dxu_{i,j}} \tag{19.60}$$

$$\frac{\partial e_T(4)}{\partial (u^{(b)})_{i,j}} = -\frac{dyue_{i,j}}{dxue_{i,j}}\, \frac{\delta_b^1}{dyu_{i,j}} - \frac{dxun_{i,j-1}}{dyun_{i,j-1}}\, \frac{\delta_b^2}{dxu_{i,j}} \tag{19.61}$$

and to the strain within the same cells

$$\frac{\partial e_{\mathrm{S}}(1)}{\partial (u^{(b)})_{i,j}} = -\frac{dxun_{i,j}}{dyun_{i,j}}\frac{\delta_b^1}{dxu_{i,j}} + \frac{dyue_{i-1,j}}{dxue_{i-1,j}}\frac{\delta_b^2}{dyu_{i,j}} \tag{19.62}$$

$$\frac{\partial e_{\mathrm{S}}(2)}{\partial (u^{(b)})_{i,j}} = -\frac{dxun_{i,j}}{dyun_{i,j}}\frac{\delta_b^1}{dxu_{i,j}} - \frac{dyue_{i,j}}{dxue_{i,j}}\frac{\delta_b^2}{dyu_{i,j}} \tag{19.63}$$

$$\frac{\partial e_{\mathrm{S}}(3)}{\partial (u^{(b)})_{i,j}} = \frac{dxun_{i,j-1}}{dyun_{i,j-1}}\frac{\delta_b^1}{dxu_{i,j}} + \frac{dyue_{i-1,j}}{dxue_{i-1,j}}\frac{\delta_b^2}{dyu_{i,j}} \tag{19.64}$$

$$\frac{\partial e_{\mathrm{S}}(4)}{\partial (u^{(b)})_{i,j}} = \frac{dxun_{i,j-1}}{dyun_{i,j-1}}\frac{\delta_b^1}{dxu_{i,j}} - \frac{dyue_{i,j}}{dxue_{i,j}}\frac{\delta_b^2}{dyu_{i,j}}. \tag{19.65}$$

The functional derivatives for tension and strain in the other eight cells are given by

$$\frac{\partial e_{\mathrm{T}}(5)}{\partial (u^{(b)})_{i,j}} = -\frac{dyue_{i,j}}{dxue_{i,j}}\frac{\delta_b^1}{dyu_{i,j}} \qquad \frac{\partial e_{\mathrm{S}}(5)}{\partial (u^{(b)})_{i,j}} = -\frac{dyue_{i,j}}{dxue_{i,j}}\frac{\delta_b^2}{dyu_{i,j}} \tag{19.66}$$

$$\frac{\partial e_{\mathrm{T}}(6)}{\partial (u^{(b)})_{i,j}} = -\frac{dyue_{i,j}}{dxue_{i,j}}\frac{\delta_b^1}{dyu_{i,j}} \qquad \frac{\partial e_{\mathrm{S}}(6)}{\partial (u^{(b)})_{i,j}} = -\frac{dyue_{i,j}}{dxue_{i,j}}\frac{\delta_b^2}{dyu_{i,j}} \tag{19.67}$$

$$\frac{\partial e_{\mathrm{T}}(7)}{\partial (u^{(b)})_{i,j}} = \frac{dyue_{i-1,j}}{dxue_{i-1,j}}\frac{\delta_b^1}{dyu_{i,j}} \qquad \frac{\partial e_{\mathrm{S}}(7)}{\partial (u^{(b)})_{i,j}} = \frac{dyue_{i-1,j}}{dxue_{i-1,j}}\frac{\delta_b^2}{dyu_{i,j}} \tag{19.68}$$

$$\frac{\partial e_{\mathrm{T}}(8)}{\partial (u^{(b)})_{i,j}} = \frac{dyue_{i-1,j}}{dxue_{i-1,j}}\frac{\delta_b^1}{dyu_{i,j}} \qquad \frac{\partial e_{\mathrm{S}}(8)}{\partial (u^{(b)})_{i,j}} = \frac{dyue_{i-1,j}}{dxue_{i-1,j}}\frac{\delta_b^2}{dyu_{i,j}} \tag{19.69}$$

$$\frac{\partial e_{\mathrm{T}}(9)}{\partial (u^{(b)})_{i,j}} = \frac{dxun_{i,j}}{dyun_{i,j}}\frac{\delta_b^2}{dxu_{i,j}} \qquad \frac{\partial e_{\mathrm{S}}(9)}{\partial (u^{(b)})_{i,j}} = -\frac{dxun_{i,j}}{dyun_{i,j}}\frac{\delta_b^1}{dxu_{i,j}} \tag{19.70}$$

$$\frac{\partial e_{\mathrm{T}}(10)}{\partial (u^{(b)})_{i,j}} = \frac{dxun_{i,j}}{dyun_{i,j}}\frac{\delta_b^2}{dxu_{i,j}} \qquad \frac{\partial e_{\mathrm{S}}(10)}{\partial (u^{(b)})_{i,j}} = -\frac{dxun_{i,j}}{dyun_{i,j}}\frac{\delta_b^1}{dxu_{i,j}} \tag{19.71}$$

$$\frac{\partial e_{\mathrm{T}}(11)}{\partial (u^{(b)})_{i,j}} = -\frac{dxun_{i,j-1}}{dyun_{i,j-1}}\frac{\delta_b^2}{dxu_{i,j}} \qquad \frac{\partial e_{\mathrm{S}}(11)}{\partial (u^{(b)})_{i,j}} = \frac{dxun_{i,j-1}}{dyun_{i,j-1}}\frac{\delta_b^1}{dxu_{i,j}} \tag{19.72}$$

$$\frac{\partial e_{\mathrm{T}}(12)}{\partial (u^{(b)})_{i,j}} = -\frac{dxun_{i,j-1}}{dyun_{i,j-1}}\frac{\delta_b^2}{dxu_{i,j}}. \qquad \frac{\partial e_{\mathrm{S}}(12)}{\partial (u^{(b)})_{i,j}} = \frac{dxun_{i,j-1}}{dyun_{i,j-1}}\frac{\delta_b^1}{dxu_{i,j}}. \tag{19.73}$$

19.3.7 Tidy form for the discretized friction

Focus first on the zonal friction with $b = 1$. Using the volumes enumerated in Section 19.3.3, and dropping the vertical label k since it is the same for all terms, yields for the tension

$$\begin{aligned} & -\sum_{n=1}^{12} V(n)\, \tau^{xx}(n)\, \frac{\partial e_{\mathrm{T}}(n)}{\partial (u^{(1)})_{i,j}} \\ &= \left(\frac{dyue_{i,j}}{4\, dxue_{i,j}\, dyu_{i,j}}\right)\left([\tau^{xx}(5) + \tau^{xx}(6)]\, V^{U}_{i+1,j} + [\tau^{xx}(2) + \tau^{xx}(4)]\, V^{U}_{i,j}\right) \\ &\quad - \left(\frac{dyue_{i-1,j}}{4\, dxue_{i-1,j}\, dyu_{i-1,j}}\right)\left([\tau^{xx}(1) + \tau^{xx}(3)]\, V^{U}_{i,j} + [\tau^{xx}(7) + \tau^{xx}(8)]\, V^{U}_{i-1,j}\right). \end{aligned} \tag{19.74}$$

Figure 19.2 indicates that a velocity point $U_{i,j}$ is associated with four triads, each of which is used to construct a tension and strain along with a viscosity. This arrangement motivates the following notation (see Figure 19.3):

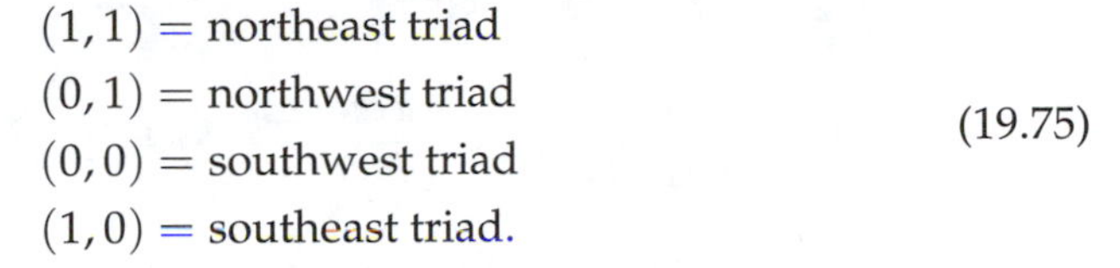

$$\begin{aligned}(1,1) &= \text{northeast triad}\\ (0,1) &= \text{northwest triad}\\ (0,0) &= \text{southwest triad}\\ (1,0) &= \text{southeast triad.}\end{aligned} \tag{19.75}$$

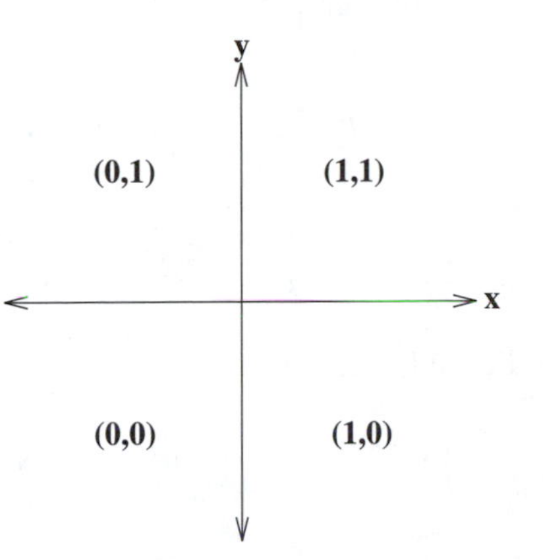

Figure 19.3 Notation for the quadrants surrounding a velocity point.

With this notation,

$$-\sum_{n=1}^{12} V(n)\,\tau^{xx}(n)\,\frac{\partial e_{\mathrm{T}}(n)}{\partial(u^{(1)})_{i,j}} = \left(\frac{dyue_{i,j}}{4\,dxue_{i,j}\,dyu_{i,j}}\right)\sum_{ip=0}^{1} V^{U}_{i+ip,j}\sum_{jq=0}^{1}(\tau^{xx})^{(1-ip,jq)}_{(i+ip,j)} - \left(\frac{dyue_{i-1,j}}{4\,dxue_{i-1,j}\,dyu_{i,j}}\right)\sum_{ip=0}^{1} V^{U}_{i-1+ip,j}\sum_{jq=0}^{1}(\tau^{xx})^{(1-ip,jq)}_{(i-1+ip,j)}. \tag{19.76}$$

The two terms on the right-hand side are centered on the east and west faces, respectively, of the $U_{i,j}$ velocity cell. Consequently, introduce the finite difference derivative operator to yield

$$-\sum_{n=1}^{12} V(n)\,\tau^{xx}(n)\,\frac{\partial e_{\mathrm{T}}(n)}{\partial(u^{(1)})_{i,j}} = \left(\frac{dxu_{i,j}}{4\,dyu_{i,j}}\right)\delta_x\left[\frac{dyue_{i-1,j}}{dxue_{i-1,j}}\sum_{ip=0}^{1} V^{U}_{i-1+ip,j}\sum_{jq=0}^{1}(\tau^{xx})^{(1-ip,jq)}_{(i-1+ip,j)}\right]. \tag{19.77}$$

Dividing by the velocity cell volume $V^{U}_{i,j} = dxu_{i,j}\,dyu_{i,j}\,dhu_{i,j}$ leads to

$$-\left(\frac{1}{V^{U}_{i,j}}\right)\sum_{n=1}^{12} V(n)\,\tau^{xx}(n)\,\frac{\partial e_{\mathrm{T}}(n)}{\partial(u^{(1)})_{i,j}} = \left(\frac{1}{4\,dhu_{i,j}\,(dyu_{i,j})^2}\right)\delta_x\left[\frac{dyue_{i-1,j}}{dxue_{i-1,j}}\sum_{ip=0}^{1} V^{U}_{i-1+ip,j}\sum_{jq=0}^{1}(\tau^{xx})^{(1-ip,jq)}_{(i-1+ip,j)}\right]. \tag{19.78}$$

Similar manipulations apply for the strain terms, thus yielding

$$-\left(\frac{1}{V^{U}_{i,j}}\right)\sum_{n=1}^{12} V(n)\,\tau^{xy}(n)\,\frac{\partial e_{S}(n)}{\partial(u^{(1)})_{i,j}}$$
$$= \left(\frac{1}{4\,dhu_{i,j}\,(dxu_{i,j})^{2}}\right)\delta_{y}\left[\frac{dxun_{i,j-1}}{dyun_{i,j-1}}\sum_{jq=0}^{1} V^{U}_{i,j-1+jq}\sum_{ip=0}^{1}(\tau^{xy})^{(ip,1-jq)}_{(i,j-1+jq)}\right]. \quad (19.79)$$

Bringing the two pieces together leads to the zonal friction acting at the velocity cell $U_{i,j}$

$$\rho_{o}\,F^{(x)}_{i,j} = \left(\frac{1}{4\,dhu_{i,j}\,(dyu_{i,j})^{2}}\right)\delta_{x}\left[\frac{dyue_{i-1,j}}{dxue_{i-1,j}}\sum_{ip=0}^{1} V^{U}_{i-1+ip,j}\sum_{jq=0}^{1}(\tau^{xx})^{(1-ip,jq)}_{(i-1+ip,j)}\right]$$
$$+ \left(\frac{1}{4\,dhu_{i,j}\,(dxu_{i,j})^{2}}\right)\delta_{y}\left[\frac{dxun_{i,j-1}}{dyun_{i,j-1}}\sum_{jq=0}^{1} V^{U}_{i,j-1+jq}\sum_{ip=0}^{1}(\tau^{xy})^{(ip,1-jq)}_{(i,j-1+jq)}\right]. \quad (19.80)$$

By inspection, the meridional friction is given by

$$\rho_{o}\,F^{(y)}_{i,j} = \left(\frac{1}{4\,dhu_{i,j}\,(dyu_{i,j})^{2}}\right)\delta_{x}\left[\frac{dyue_{i-1,j}}{dxue_{i-1,j}}\sum_{ip=0}^{1} V^{U}_{i-1+ip,j}\sum_{jq=0}^{1}(\tau^{xy})^{(1-ip,jq)}_{(i-1+ip,j)}\right]$$
$$- \left(\frac{1}{4\,dhu_{i,j}\,(dxu_{i,j})^{2}}\right)\delta_{y}\left[\frac{dxun_{i,j-1}}{dyun_{i,j-1}}\sum_{jq=0}^{1} V^{U}_{i,j-1+jq}\sum_{ip=0}^{1}(\tau^{xx})^{(ip,1-jq)}_{(i,j-1+jq)}\right]. \quad (19.81)$$

Comparison with the continuum friction components given by equations (17.124) and (17.125) indicates that the discretization is consistent, that is, the discrete friction reduces to the continuum friction as the grid size goes to zero.

19.3.8 Tension and strain in the quadrants

There are four tensions and strains corresponding to the four triads surrounding each velocity point. Referring to Figure 19.3, assuming the central point is $U_{i,j}$, discretize the tensions and strains starting from the second form of the continuous tension and strain (equations (19.16) and (19.17)) to find

$$(e_{T})_{i,j,(0,1)} = \delta_{x}\,u_{i-1,j} - \delta_{y}\,v_{i,j} + (M_{T})_{i,j} \quad (19.82)$$
$$(e_{T})_{i,j,(1,1)} = \delta_{x}\,u_{i,j} - \delta_{y}\,v_{i,j} + (M_{T})_{i,j} \quad (19.83)$$
$$(e_{T})_{i,j,(0,0)} = \delta_{x}u_{i-1,j} - \delta_{y}\,v_{i,j-1} + (M_{T})_{i,j} \quad (19.84)$$
$$(e_{T})_{i,j,(1,0)} = \delta_{x}\,u_{i,j} - \delta_{y}\,v_{i,j-1} + (M_{T})_{i,j} \quad (19.85)$$
$$(e_{S})_{i,j,(0,1)} = \delta_{y}\,u_{i,j} + \delta_{x}\,v_{i-1,j} + (M_{S})_{i,j} \quad (19.86)$$
$$(e_{S})_{i,j,(1,1)} = \delta_{y}\,u_{i,j} + \delta_{x}\,v_{i,j} + (M_{S})_{i,j} \quad (19.87)$$
$$(e_{S})_{i,j,(0,0)} = \delta_{y}\,u_{i,j-1} + \delta_{x}\,v_{i-1,j} + (M_{S})_{i,j} \quad (19.88)$$
$$(e_{S})_{i,j,(1,0)} = \delta_{y}\,u_{i,j-1} + \delta_{x}\,v_{i,j} + (M_{S})_{i,j}. \quad (19.89)$$

In general, the four tensions can be written

$$(e_{T})_{i,j,(ip,jq)} = \delta_{x}\,u_{i+ip-1,j} - \delta_{y}\,v_{i,j+jq-1} + (M_{T})_{i,j} \quad (19.90)$$

and the four strains can be written

$$(e_S)_{i,j,(ip,jq)} = \delta_y\, u_{i,j+jq-1} + \delta_x\, v_{i+ip-1,j} + (M_S)_{i,j} \tag{19.91}$$

where $ip = 0, 1$ and $jq = 0, 1$. Notably, the metric terms

$$(M_T)_{i,j} = -u_{i,j}\,(\partial_x \ln dy)_{i,j} + v_{i,j}\,(\partial_y \ln dx)_{i,j} \tag{19.92}$$

$$(M_S)_{i,j} = -u_{i,j}\,(\partial_y \ln dx)_{i,j} - v_{i,j}\,(\partial_x \ln dy)_{i,j} \tag{19.93}$$

are common to the four triads, and so only need be computed once per velocity point. Generally, the four tensions and strains are computed in the model and are then used to compute the friction operator. When the Smagorinsky viscosity scheme (Section 18.3) is enabled, they are used to compute the Smagorinsky viscosity as well.

19.3.9 Comments

Note a few points here related to the proposed discretization.

- The tension and strain for ocean points next to land contain a contribution from a velocity living at the land-sea interface. This velocity, due to the no-slip condition used in B-grid ocean models, has a zero value. In order to provide a full accounting of the generally strong shears next to no-slip walls present in the B-grid ocean models, it is important to include such contributions rather than masking them out.

- In the special case of a uniform Cartesian grid, a constant isotropic viscosity, and a zero anisotropic viscosity, the functionally derived discrete friction operator reduces to the 5-point discrete Laplacian discussed in Section 19.2.1.

- Practical experience has revealed problems with discretization for bottom grid cells when these cells are thin partial cells (e.g., see Figure 6.2 on page 125) that are surrounded by thicker partial cells. The problem is that contributions from surrounding thick cells are then normalized by the thin dhu of the central cell. To alleviate this problem, it is effective to use the traditional 5-point Laplacian operator for computing friction in the bottom grid cells.

19.3.10 Discretized Smagorinsky viscosity

The nonlinear Smagorinsky viscosity coefficient is determined in terms of the deformation rates e_T and e_S as well as the grid spacing. Since e_T and e_S involve terms with derivatives in both horizontal directions, an averaging must be performed to place them at a common grid position.

Pacanowski et al. (1991) defined both deformation rates at the north face of the U-cell. This is the natural position for the meridional derivative terms. However, to get the zonal derivative terms defined at the north face, it is necessary to average over the four zonal derivatives surrounding the north face. The problem with such "4 point" averages on the B-grid is that they can introduce computational modes. Computational modes are not always problematical if there are other processes that can suppress the growth of the modes. The problem with the computational modes in the Smagorinsky scheme is that they allow nontrivial field configurations yielding a zero deformation rate. Hence, they produce a zero Smagorinsky viscosity and so are not dissipated. Furthermore, these modes represent grid scale waves, which are the waves an ideal implementation of the Smagorinsky scheme should dissipate the most. Therefore, it is not acceptable to allow these modes in the discretized Smagorinsky scheme. Another approach is necessary.

The functional discretization described in this chapter eliminates the computational modes. For each velocity triad, there is a corresponding Smagorinsky viscosity. In particular, referring to the deformation rates defined in equation (19.89) yields the corresponding Smagorinsky diffusivities

$$A_{i,j,(ip,jq)} = (\Upsilon \Delta s/\pi)^2 |E|_{i,j,(ip,jq)} \tag{19.94}$$

where $ip = 0, 1$, $jq = 0, 1$ are the triad labels, and

$$[E_{i,j,(ip,jq)}]^2 = [(e_{\mathrm{T}})_{i,j,(ip,jq)}]^2 + [(e_{\mathrm{S}})_{i,j,(ip,jq)}]^2 \tag{19.95}$$

is the discrete total deformation rate. As mentioned earlier, one advantage of the functional approach over the "Laplacian plus metric" approach (see Section 19.4) is the exploitation of the deformation rates for computing both the Smagorinsky viscosity and the friction operator.

19.4 LAPLACIAN PLUS METRIC FORM OF ISOTROPIC FRICTION

The purpose of this section is to illustrate the mathematical equivalence, in the continuum, between the "Laplacian plus metric" form of friction and the stress tensor approach. For this purpose, we focus only on isotropic friction using spherical coordinates.

19.4.1 Spherical form of Laplacian friction

In spherical coordinates, the metric takes the form

$$\begin{aligned} g_{ij} &= \mathrm{diag}\,(g_{11}, g_{22}, g_{33}) \\ &= \mathrm{diag}\,(R^2 \cos^2 \phi, R^2, 1) \end{aligned} \tag{19.96}$$

and horizontal increments are

$$(\mathrm{d}x, \mathrm{d}y, \mathrm{d}z) = (R \cos\phi \,\mathrm{d}\lambda, R\,\mathrm{d}\phi, \mathrm{d}z). \tag{19.97}$$

Consequently, it is only g_{11} that has nonzero spatial dependence. The friction then can be written

$$F^x = (A\, e_{\mathrm{T}})_{,x} + \cos^{-2}\phi\,(A\, \cos^2\phi\, e_{\mathrm{S}})_{,y} + (\kappa\, u_{,z})_{,z} \tag{19.98}$$

$$F^y = (A\, e_{\mathrm{S}})_{,x} - \cos^{-2}\phi\,(A\, \cos^2\phi\, e_{\mathrm{T}})_{,y} + (\kappa\, v_{,z})_{,z} \tag{19.99}$$

where the deformation rates are

$$\begin{aligned} e_{\mathrm{T}} &= u_{,x} - v_{,y} - (v/R)\, \tan\phi \\ &= (R\cos\phi)^{-1}\,(u_{,\lambda} - v_{,\phi}\, \cos\phi - v\, \sin\phi) \end{aligned} \tag{19.100}$$

$$\begin{aligned} e_{\mathrm{S}} &= v_{,x} + \cos\phi\,(u/\cos\phi)_{,y} \\ &= (R\cos\phi)^{-1}\,(v_{,\lambda} + u_{,\phi}\, \cos\phi + u\, \sin\phi). \end{aligned} \tag{19.101}$$

The terms associated with horizontal deformations

$$F^u = (R\cos\phi)^{-1}\,(A\,e_T)_{,\lambda} + (R\cos^2\phi)^{-1}\,(A\cos^2\phi\,e_S)_{,\phi} \tag{19.102}$$

$$F^v = (R\cos\phi)^{-1}\,(A\,e_S)_{,\lambda} - (R\cos^2\phi)^{-1}\,(A\cos^2\phi\,e_T)_{,\phi} \tag{19.103}$$

can be massaged into the form presented by Bryan (1969) and Wajsowicz (1993). That is the purpose of the remainder of this section.

19.4.2 Zonal friction

The lateral friction acting on the zonal velocity takes the expanded form

$$\begin{aligned}
R\,F^u =& \frac{1}{\cos\phi}\,(e_T\,A_{,\lambda} + A\,\partial_\lambda e_T) \\
&+ \frac{1}{\cos^2\phi}\,(e_S\,A_{,\phi}\cos^2\phi + A\,\partial_\phi e_S\cos^2\phi - 2\,A\,e_S\cos\phi\sin\phi) \\
=& \frac{1}{\cos\phi}\,(e_T\,A_{,\lambda} + e_S\,A_{,\phi}\cos\phi) \\
&- \frac{2A}{R\cos\phi}(v_{,\lambda}\tan\phi + u_{,\phi}\sin\phi + u\sin\phi\tan\phi) \\
&+ \frac{A}{R\cos\phi}(u_{,\lambda\lambda}\sec\phi + u_{,\phi\phi}\cos\phi + u_{,\phi}\sin\phi + u\sec\phi) \\
=& \frac{1}{\cos\phi}\,(e_T\,A_{,\lambda} + e_S\,A_{,\phi}\cos\phi) \\
&+ \frac{A}{R}\left(u_{,\lambda\lambda}\sec^2\phi + u_{,\phi\phi} - u_{,\phi}\tan\phi + u\,(\sec^2\phi - 2\tan^2\phi) - 2v_{,\lambda}\sec^2\phi\sin\phi\right) \\
=& \frac{1}{\cos\phi}\,(e_T\,A_{,\lambda} + e_S\,A_{,\phi}\cos\phi) \\
&+ \frac{A}{R}\left(u_{,\lambda\lambda}\sec^2\phi + \sec\phi\,(u_{,\phi}\cos\phi)_{,\phi} + u(1-\tan^2\phi) - 2v_{,\lambda}\sec^2\phi\sin\phi\right),
\end{aligned} \tag{19.104}$$

which renders

$$\begin{aligned}
F^u =& A\left(\nabla_z^2\,u + \frac{u(1-\tan^2\phi)}{R^2} - \frac{2v_{,\lambda}\sin\phi}{R^2\cos^2\phi}\right) \\
&+ \frac{1}{R\cos\phi}\,(e_T\,A_{,\lambda} + e_S\,A_{,\phi}\cos\phi).
\end{aligned} \tag{19.105}$$

The second bracketed term in this expression arises from the spatial dependence of the viscosity coefficient, and should be present for any nonconstant viscosity coefficient model. These nonconstant viscosity coefficient terms amount to those identified by Wajsowicz (1993).

It is useful to perform one final step in order to bring the nonconstant viscosity coefficient inside the Laplacian. For this purpose, the Laplacian and nonconstant viscosity coefficient terms are expanded to yield

$$
\begin{aligned}
F^u =& \frac{A}{R^2\cos^2\phi} u_{,\lambda\lambda} + \frac{A}{R^2\cos\phi}(u_{,\phi}\cos\phi)_{,\phi} \\
&+ \frac{A\,u\,(1-\tan^2\phi)}{R^2} - \frac{2\,A\,v_{,\lambda}\sin\phi}{R^2\cos^2\phi} \\
&+ \frac{A_{\lambda}}{R^2\cos^2\phi}(u_{,\lambda} - v_{,\phi}\cos\phi - v\sin\phi) \\
&+ \frac{A_{,\phi}}{R^2\cos\phi}(v_{,\lambda} + u_{,\phi}\cos\phi + u\sin\phi) \\
=& \frac{1}{R^2\cos^2\phi}(A\,u_{,\lambda\lambda} + A_{,\lambda}\,u_{,\lambda}) \\
&+ \frac{1}{R^2\cos\phi}(A\,u_{,\phi}\cos\phi)_{,\phi} - A_{,\phi}\,u_{,\phi}R^{-2} \\
&+ \frac{A\,u\,(1-\tan^2\phi)}{R^2} - \frac{2\,A\,v_{,\lambda}\sin\phi}{R^2\cos^2\phi} \\
&- \frac{A_{,\lambda}}{R^2\cos^2\phi}(v_{,\phi}\cos\phi + v\sin\phi) \\
&+ R^{-2}A_{\phi}(v_{,\lambda}\sec\phi + u_{,\phi} + u\tan\phi) \\
=& \nabla_z\cdot(A\,\nabla_z u) + \frac{A\,u\,(1-\tan^2\phi)}{R^2} - \frac{2\,A\,v_{,\lambda}\sin\phi}{R^2\cos^2\phi} \\
&- \frac{A_{,\lambda}}{R^2\cos\phi}(v_{,\phi} + v\tan\phi) + \frac{A_{,\phi}}{R^2\cos\phi}(v_{,\lambda} + u\sin\phi).
\end{aligned}
\tag{19.106}
$$

This expression can be written

$$
\nabla_z\cdot(A\,\nabla_z u) + \text{old\_metric}^u + \text{new\_metric}^u, \tag{19.107}
$$

where

$$
\nabla_z\cdot(A\,\nabla_z u) = \frac{1}{R^2\cos^2\phi}(Au_{,\lambda})_{,\lambda} + \frac{1}{R^2\cos\phi}(A\,u_{,\phi}\cos\phi)_{,\phi} \tag{19.108}
$$

is the horizontal Laplacian with the generally nonconstant viscosity coefficient inserted. Note that this Laplacian is acting on the zonal velocity component u as a scalar field. The term

$$
\text{old\_metric}^u = \frac{A\,u\,(1-\tan^2\phi)}{R^2} - \frac{2\,A\,v_{,\lambda}\sin\phi}{R^2\cos^2\phi} \tag{19.109}
$$

is the metric term employed for constant horizontal viscosity coefficient (Bryan, 1969), and

$$
\text{new\_metric}^u = -\frac{A_{,\lambda}}{R^2\cos\phi}(v_{,\phi} + v\tan\phi) + \frac{A_{,\phi}}{R^2\cos\phi}(v_{,\lambda} + u\sin\phi) \tag{19.110}
$$

is the metric term arising from spatial dependence in the viscosity coefficient (Wajsowicz, 1993).

19.4.3 Meridional friction

Repeating the exercise just performed for the zonal friction yields the following lines of algebra for the meridional friction

$$
\begin{aligned}
R\,F^{v} =& \frac{1}{\cos\phi}\,(e_{S}\,A_{,\lambda} + A\,\partial_{\lambda}\,e_{S}) \\
& - \frac{1}{\cos^{2}\phi}\,(e_{T}\,A_{,\phi}\cos^{2}\phi + A\,\partial_{\phi}e_{T}\cos^{2}\phi - 2\,A\,e_{T}\cos\phi\sin\phi) \\
=& \frac{1}{\cos\phi}\,(e_{S}\,A_{,\lambda} - e_{T}\,A_{,\phi}\cos\phi) \\
& + \frac{2A}{R\cos\phi}(u_{,\lambda}\tan\phi - v_{,\phi}\tan\phi\cos\phi - v\tan\phi\sin\phi) \\
& + \frac{A}{R\cos\phi}\,(v_{,\lambda\lambda}\sec\phi + v_{,\phi\phi}\cos\phi + v_{,\phi}\sin\phi + v\sec\phi) \\
=& \frac{1}{\cos\phi}\,(e_{S}\,A_{,\lambda} - e_{T}\,A_{,\phi}\cos\phi) \\
& + \frac{A}{R}\left(v_{,\lambda\lambda}\sec^{2}\phi + v_{,\phi\phi} - v_{,\phi}\tan\phi + v(\sec^{2}\phi - 2\tan^{2}\phi) + 2u_{,\lambda}\sec^{2}\phi\sin\phi\right) \\
=& \frac{1}{\cos\phi}\,(e_{S}\,A_{,\lambda} - e_{T}\,A_{,\phi}\cos\phi) \\
& + \frac{A}{R}\left(v_{,\lambda\lambda}\sec^{2}\phi + \sec\phi\,(v_{,\phi}\cos\phi)_{,\phi} + v(1 - \tan^{2}\phi) + 2u_{,\lambda}\sec^{2}\phi\sin\phi\right),
\end{aligned}
\tag{19.111}
$$

which yields

$$
\begin{aligned}
F^{v} =& A\left(\nabla_{z}^{2}\,v + \frac{v(1-\tan^{2}\phi)}{R^{2}} + \frac{2u_{,\lambda}\sin\phi}{R^{2}\cos^{2}\phi}\right) \\
& + \frac{1}{R\cos\phi}\,(e_{S}\,A_{,\lambda} - e_{T}\,A_{,\phi}\cos\phi).
\end{aligned}
\tag{19.112}
$$

Again, it is useful to expand this friction one more step in order to bring the viscosity coefficient inside the Laplacian. This manipulation yields

$$
F^{v} = \nabla_{z}\cdot(A\,\nabla_{z}v) + \text{old\_metric}^{v} + \text{new\_metric}^{v}, \tag{19.113}
$$

where

$$
\text{old\_metric}^{v} = \frac{A\,v\,(1-\tan^{2}\phi)}{R^{2}} + \frac{2\,A\,u_{,\lambda}\sin\phi}{R^{2}\cos^{2}\phi} \tag{19.114}
$$

is the metric employed for constant horizontal viscosity coefficient, and

$$
\text{new\_metric}^{v} = \frac{A_{,\lambda}}{R^{2}\cos\phi}(u_{,\phi} + u\tan\phi) + \frac{A_{,\phi}}{R^{2}\cos\phi}(-u_{,\lambda} + v\sin\phi). \tag{19.115}
$$

19.5 CHAPTER SUMMARY

This chapter presented a derivation of the discrete horizontal friction operator of use in a generalized horizontal coordinate ocean model. Methods followed from the functional approach described in Section 17.8. It was noted at the start that a discretization on a C-grid is simpler than on the B-grid, as the C-grid is more friendly to the discretization of transport operators such as advection, diffusion, and friction. Hence, attention here focused on the B-grid, with the bulk of the discussion given in Section 19.3. Afterward, a derivation was presented that connects the continuous form of horizontal friction commonly used in models, the "Laplacian plus metric" form, to the form arising from taking the divergence of the stress tensor. The forms are typically distinct on the lattice. Preference was given here to the stress tensor formulation, since it leads to a straightforward application to generalized horizontal grids. Additionally, it computes deformation rates, which are themselves of use when choosing the Smagorinsky nonlinear viscosity scheme.

It is hoped that the reader of this chapter, as well as Chapter 16, has garnered an appreciation for the utility of the functional methodology for discretizing elliptic dissipation operators. Notably, it provides the numericist with a basis for choosing a particular discretization that results in an operator satisfying desired global properties, in this case dissipation of kinetic energy. The use of such principles facilitates the derivation of discrete algorithms that maintain properties analogous to the continuum equations from which they are based.

PART 6

Tensor Analysis

This part of the book aims to familiarize the reader with basic tools of curvilinear tensor analysis. An understanding of tensor analysis is necessary to formulate the equations of rotating fluid dynamics in a spherical geometry. The presentation is geared toward those with a solid foundation in undergraduate vector calculus. No exposure to Cartesian or general tensors is assumed. Since much attention in ocean model development during the past few years has been placed on both general horizontal and vertical coordinates, it is useful to have a presentation of tensor analysis that focuses on the needs of ocean models. A goal for this part of the book is to at least partially fill this niche.

Chapter Twenty

ELEMENTARY TENSOR ANALYSIS

The purpose of this chapter is to introduce elementary notions of tensor analysis fundamental to the mathematical formulation of geophysical fluid mechanics on a non-Euclidean space, such as occurs in a spherical geometry. Our treatment is pedagogical in nature, with reference given to issues arising in ocean model formulations.

20.1 INTRODUCTION

Physics is independent of coordinates. A coordinate-independent framework is at the heart of theoretical physics. However, the deduction of physical laws can at times be initiated through working within the framework of a particular set of coordinates. For example, it is often convenient to use Cartesian coordinates when describing physics on a flat manifold. Additionally, the implementation of physical laws for numerical computation *requires* a chosen set of coordinates. Hence, those working in geophysical fluid dynamics (GFD), and in particular in an area of numerical GFD such as ocean modeling, may find it useful to be familiar with various sets of coordinates and the transformations of physical laws from one set of coordinates to another.

Differential geometry is an area of mathematics that provides systematic tools facilitating a description of physics occuring on arbitrary smooth manifolds. In particular, it helps to expose the underlying physical essence of a mathematical statement, without being encumbered with coordinate-dependent details. Differential geometry has its roots in classical vector analysis and calculus. Thus, many of its results are familiar to those who have studied elementary calculus. Over time, the language and notation of differential geometry has been refined to the point where indices are nearly absent in modern treatments. This *index free* approach is not taken here, since to do so would exclude a large number of readers from geophysical fluid dynamics backgrounds. Rather, the treatment employs tensorial notation, with a sprinkling of the modern language thrown in for culture. Correspondingly, the purpose of this chapter and the next is to expose the reader to elements of calculus on manifolds, also known as *tensor analysis*, with particular attention placed on applications to GFD. In the process, many formulae proving useful in this book are derived and listed.

As with any mathematical subject, there is a fair amount of notation and convention to absorb. Indeed, the rules and gymnastics involved can appear quite tedious. However, the rules are systematic and have sound motivation. A goal of this chapter and the next is to expose some of the reasoning leading to the formalism, and so to reduce the tedium and intimidation typically associated with tensor analysis.

Those readers interested in somewhat more complete treatments of differential geometry, with physical applications in mind, are encouraged to read one or more of the following books.

- Aris (1962): *Vectors, Tensors and the Basic Equations of Fluid Mechanics*. This book presents the basic ideas of tensor analysis within the context of fluid dynamics. The treatment is general and the notation is clean. Aris's approach is used throughout this book, and it is highly recommended.
- Dutton (1976): *The Ceaseless Wind*, and Pielke (2002): *Mesoscale Meteorological Modeling*. These books on atmospheric dynamics and modeling provide succinct discussions of tensor analysis which parallel much of the material in this book.
- Schutz (1980): *Geometrical Methods of Mathematical Physics* and Schutz (1985): *A First Course in General Relativity*. The first book provides a discussion of mathematical ideas central to much of modern physics, with some application to fluid mechanics. The second book is a very readable presentation of introductory general relativity. Much of the approach taken in the relativity book is followed here, and it provides another of the central references for this chapter. Additionally, note that nearly all introductory general relativity books include a discussion of tensor analysis in a manner consistent with that given here.
- Flanders (1989): *Differential Forms with Applications to the Physical Sciences*. This book, first published in 1963, is an early work aiming toward the use of index-free notation in physical theories. Flanders provides some applications to fluid mechanics.
- Morse and Feshbach (1953): *Methods of Theoretical Physics*. This two-volume book is a classic. It is listed here since it articulates an older and more cumbersome formalism not employed in the present book. Nevertheless, certain authors still employ Morse and Feshbach as their primary reference for tensor analysis. Hence, it might be useful for some to become familiar with their notation.

20.2 SOME PRACTICAL MOTIVATION

It is useful to begin with some practical motivation for learning the methods of tensor analysis for use in GFD.

20.2.1 Generalized horizontal coordinates

When studying geophysical fluids moving in a spherical geometry, the horizontal coordinates have traditionally been the spherical latitude-longitude coordinates (λ, ϕ). The infinitesimal distance along the two orthogonal coordinate axes is given by

$$(\mathrm{d}x, \mathrm{d}y) = r\,(\cos\phi\,\mathrm{d}\lambda, \mathrm{d}\phi), \tag{20.1}$$

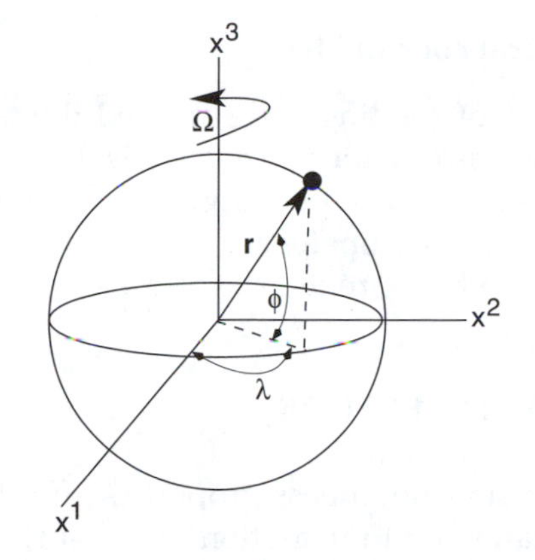

Figure 20.1 The position vector for a point in 3D Euclidean space can be represented in terms of many sets of coordinates, such as the Cartesian coordinates (x^1, x^2, x^3) or the spherical coordinates (λ, ϕ, r). In a geophysical context, the angular coordinate $0 \leq \lambda \leq 2\pi$ is the longitude, with positive values measured eastward from a meridian passing through Greenwich, England. The angular coordinate ϕ is the latitude, with values $\phi = 0$ at the equator and $\phi = \pi/2(-\pi/2)$ at the north (south) poles. The radial distance r is measured here with respect to the center of the sphere. The coordinate transformation between Cartesian and spherical is given by $(x^1, x^2, x^3) = r\,(\cos\phi\cos\lambda, \cos\phi\sin\lambda, \sin\phi)$.

where r is the radial position. Specifically, the angular coordinate ϕ is latitude, which increases northward and is zero at the equator. Likewise, the angular coordinate λ is longitude, which increases eastward with zero defined at an arbitrary longitude (e.g., Greenwich, England); (see Figure 20.1). Spherical coordinates are locally orthogonal, thus greatly simplifying the representation of the governing equations of motion.

Generalizations of spherical coordinates are common in global ocean models. The central reason is to avoid the spherical coordinate singularity at the North Pole. Generalized horizontal coordinates can be designed, after a bit of mathematical analysis, so that the polar singularity is moved over land instead of being in the ocean. As the Arctic circulation is quite important for global climate simulations, it is necessary to avoid coordinate artifacts, such as a polar singularity, in these models. Additionally, practical limitations of the time step available when the meridians converge greatly hinder the utility of spherical coordinates.

In the generalizations of spherical coordinates, ocean modelers have predominantly chosen not to sacrifice the simplifications inherent in orthogonal curvilinear coordinates. Section 21.11 presents many of the mathematical formulae needed to derive the fluid equations of motion with generalized orthogonal coordinates. Once the fluid equations are written in this form, one of the many orthogonal coordinate choices (e.g., Murray, 1996) is available to the modeler in order to remove the polar singularity from the ocean domain. The papers by Smith et al. (1995), Murray (1996), and Madec and Imbard (1996) provide discussions of ocean modeling in generalized horizontal coordinates.

20.2.2 Generalized vertical coordinates

GFD consists of the study of rotating and stratified fluids. To describe stratified fluids, it is often useful to employ specialized vertical coordinates that are distinct from the usual vertical distance from the bottom topography. The mathematical issues of generalized vertical coordinates are discussed in Chapter 6. Many aspects of tensor analysis described here prove to be of use.

20.3 COORDINATES AND VECTORS

A surface possessing certain smoothness properties affords the usual rules of calculus, such as differentiation and integration. Such surfaces are known as *differential manifolds*. Importantly, these manifolds are locally Euclidean, which means that they possess flat tangent planes. These are the sorts of surfaces of interest in this book. One familiar two-dimensional example is the surface of a sphere S^2. Another is three-dimensional Euclidean space $\mathcal{R}^3$. A third is a constant density surface, commonly known as an *isopycnal* surface, in a stratified fluid. An isopycnal surface results from taking a suitable space-time or ensemble average to smooth out microscale fluctuations (see Chapter 7). For present purposes, it simply represents a smooth undulating two-dimensional surface in the ocean.

To specify the position of a point on the manifold, an observer living on the manifold may decide to set up a grid in some systematic fashion using a set of coordinates. Let the observer decide to call these coordinates ξ^a, where the index a can take on integer values up to M, the dimension of the manifold.* For example, a person living in $\mathcal{R}^3$ may decide to use Cartesian coordinates or spherical coordinates, depending on the application. Figure 20.1 illustrates their relation. In contrast, a person confined to living on the surface of a sphere most likely will decide to use the spherical angles (λ, ϕ), thus setting up the familiar latitude and longitude lines on S^2. Such is the case for inhabitants of the earth. A person confined to living on a general undulating surface, such as an isopycnal, may decide to use some general curvilinear coordinates. The analytical relation between these general coordinates and the more familiar Cartesian coordinates might be quite complicated, assuming such a relation even exists.

At any point on a differential manifold, the directions in which the coordinates increase define special directions. For an M-dimensional manifold, M directions form a basis for representing all vectors originating from this point. An arbitrary *first order tensor* can be written as the linear combination

$$\begin{aligned} \boldsymbol{\xi} &= \sum_{a=1}^{M} \xi^a \, \vec{e}_a \\ &\equiv \xi^a \, \vec{e}_a, \end{aligned} \tag{20.2}$$

where $\vec{e}_a$ are a set of M linearly independent vectors forming a basis for the manifold. A first order tensor represented in terms of a basis of vectors is often called a *vector* for brevity. Note the use of an arrow to denote a basis vector and boldface to denote a general tensor. The second expression in equation (20.2) defines the Einstein summation convention, for which repeated indices are summed when one

*This index should not be confused with ξ raised to power a.

index is raised and the other lowered. The choice to lower the index on the basis vectors $\vec{e}_a$ is motivated by this very useful convention, whose power will become apparent in subsequent discussions. To summarize, it is said that $\boldsymbol{\xi}$ is a first order tensor that has a *representation* given by the coordinates ξ^a in terms of the basis vectors $\vec{e}_a$.

Using the above notation, the components of a basis vector, as written in terms of another basis, take the form

$$\vec{e}_a = \Lambda^{\bar{b}}{}_a \, \vec{e}_{\bar{b}}, \tag{20.3}$$

where $\Lambda^{\bar{b}}{}_a$ is a number representing the components of the basis vector $\vec{e}_a$ as written in terms of another basis $\vec{e}_{\bar{b}}$. That is, $\Lambda^{\bar{b}}{}_a$ are the components to the transformation matrix between the bases. We will have more to say about transformations in the following.

If the bases are identical, then $\vec{e}_{\bar{a}} = \vec{e}_a$. To formalize this trivial result, introduce components of the identity transformation. The components of this transformation are given by the Kronecker symbol $\Lambda^{\bar{b}}{}_a = \delta^{\bar{b}}{}_a$, where

$$\delta^{\bar{b}}{}_a = \begin{cases} 1 \text{ if } a = \bar{b}, \\ 0 \text{ if } a \neq \bar{b}. \end{cases} \tag{20.4}$$

The Kronecker symbol forms an invariant representation of the second order *unit* tensor. It is invariant since it takes the same numerical values when written in any coordinate basis. The issues of tensorial order are discussed later in this chapter.

Some examples are useful. First, in the Euclidean space $\mathcal{R}^3$, the following set of vectors, based on the spherical coordinates, are suitable as a basis for all points except the poles:

$$\vec{e}_1 = r \cos\phi \, \hat{\boldsymbol{\lambda}} \tag{20.5}$$

$$\vec{e}_2 = r \, \hat{\boldsymbol{\phi}} \tag{20.6}$$

$$\vec{e}_3 = \hat{\mathbf{r}}. \tag{20.7}$$

The unit directions, denoted by hats, point in the longitudinal (eastward) direction $\hat{\boldsymbol{\lambda}}$, latitudinal (northward) direction $\hat{\boldsymbol{\phi}}$, and radial direction $\hat{\mathbf{r}}$. At the poles, $\phi = \pm\pi/2$. Since $\vec{e}_1$ vanishes at the pole, the spherical basis becomes singular there. Another set of basis vectors is the familiar Cartesian unit vectors

$$\vec{e}_1 = \hat{\mathbf{x}} \tag{20.8}$$

$$\vec{e}_2 = \hat{\mathbf{y}} \tag{20.9}$$

$$\vec{e}_3 = \hat{\mathbf{z}}. \tag{20.10}$$

These vectors are nonsingular everywhere.

Both the spherical and Cartesian basis vectors form mutually orthogonal sets when using the familiar Euclidean metric. Furthermore, the Cartesian set is normalized under the Euclidean metric, so that

$$\vec{e}_a \cdot \vec{e}_b = \delta_{ab}, \tag{20.11}$$

where δ_{ab} is yet another representation of the unit tensor to be discussed in the next section. The spherical basis is not normalized, however, and the motivation for not normalizing this basis will become apparent later.

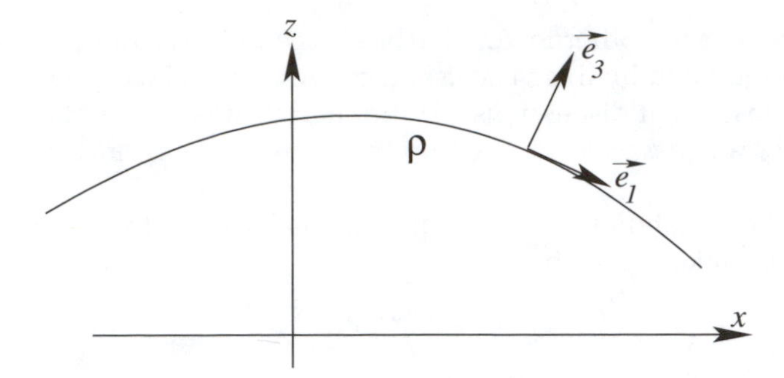

Figure 20.2 Cross-section of an isopycnal surface ρ =const in the ocean, shown here with a local set of orthogonal unit vectors at an arbitrary point.

For describing a general surface, such as an isopycnal surface, one may use the orthonormal basis vectors

$$\vec{e}_1 = \frac{\hat{\mathbf{y}} \wedge \nabla \rho}{|\hat{\mathbf{y}} \wedge \nabla \rho|} \tag{20.12}$$

$$\vec{e}_2 = \vec{e}_3 \wedge \vec{e}_1 \tag{20.13}$$

$$\vec{e}_3 = \hat{\boldsymbol{\rho}}, \tag{20.14}$$

where the wedge symbol $\wedge$ represents the usual Cartesian vector cross product. The unit normal

$$\hat{\boldsymbol{\rho}} = \frac{\nabla \rho}{|\nabla \rho|} \tag{20.15}$$

points outward to the surfaces of constant density $\rho(x, y, z, t)$. The gradient has the following representation in terms of the Cartesian basis vectors

$$\nabla \rho = \hat{\mathbf{x}}\, \rho_{,x} + \hat{\mathbf{y}}\, \rho_{,y} + \hat{\mathbf{z}}\, \rho_{,z}, \tag{20.16}$$

where

$$\rho_{,a} = \partial_a \, \rho \tag{20.17}$$

is a useful notation further motivated in later discussions. Note that when the isopycnal surface has an outward normal $\hat{\boldsymbol{\rho}}$ parallel to the vertical, then the basis vectors reduce to

$$\vec{e}_1 \rightarrow \hat{\mathbf{x}} \tag{20.18}$$

$$\vec{e}_3 \rightarrow \hat{\mathbf{y}} \tag{20.19}$$

$$\vec{e}_3 \rightarrow \hat{\mathbf{z}}. \tag{20.20}$$

That is, the surface is locally flat, or in the horizontal plane. In the general case, each of the vectors changes whenever the isopycnal surface evolves. Figure 20.2 illustrates a two-dimensional cross-section of an isopycnal surface.

20.4 THE METRIC AND COORDINATE TRANSFORMATIONS

Upon specifying the position of a point with respect to some origin, it is of interest to determine how far this point is from some other point. Since the points are assumed to live on a smooth manifold, it is sufficient to consider the distance between two infinitesimally close points, and then use integration to find the distance between finitely displaced points. Let the coordinates for the two points be given by ξ^a and $\xi^a + \mathrm{d}\xi^a$, and let $\mathrm{d}\boldsymbol{\xi} = \mathrm{d}\xi^a\,\vec{e}_a$ be the infinitesimal vector pointing from one point to the other, where $\vec{e}_a$ are the basis vectors. If these points live in the flat Euclidean $\mathcal{R}^3$ space, and the coordinates ξ^a are the usual Cartesian coordinates, then the squared distance between them is given by the Euclidean norm

$$\begin{aligned} \mathrm{d}s^2 &= \mathrm{d}\boldsymbol{\xi} \cdot \mathrm{d}\boldsymbol{\xi} \\ &= \vec{e}_a \cdot \vec{e}_b\,\mathrm{d}\xi^a\,\mathrm{d}\xi^b \\ &= \delta_{ab}\,\mathrm{d}\xi^a\,\mathrm{d}\xi^b. \end{aligned} \tag{20.21}$$

In this expression, $(\mathrm{d}s)^2 \equiv \mathrm{d}s^2$ is the conventional way to write the infinitesimal squared length. The Kronecker symbol δ_{ab} is symmetric

$$\delta_{ab} = \delta_{ba}, \tag{20.22}$$

vanishes when $a \neq b$, and is unity when $a = b$. As defined by equation (20.21), δ_{ab} forms the Cartesian components of the *metric tensor* for Euclidean space. Note that the properties of δ_{ab} are also shared by $\delta^{\overline{a}}_{\ b}$ as defined by equation (20.4). These very simple properties of the Kronecker symbol, or the unit tensor, make much of the formalism of general tensor analysis unnecessary when applied to Euclidean space using Cartesian coordinates. Yet when working on non-Euclidean spaces, such as a sphere or an arbitrary surface, tensor analysis provides a very transparent and elegant framework.

As seen by equation (20.21), the representation of the metric tensor for Euclidean space using Cartesian coordiantes is δ_{ab}. To determine the representation in another set of coordinates, consider the transformation from Cartesian coordinates ξ^a to arbitrary coordinates $\xi^{\overline{a}}$. Use of the chain rule leads to the equivalent expression for the squared infinitesimal length,

$$\begin{aligned} \mathrm{d}s^2 &= \delta_{ab}\left(\frac{\partial \xi^a}{\partial \xi^{\overline{a}}}\frac{\partial \xi^b}{\partial \xi^{\overline{b}}}\right)\mathrm{d}\xi^{\overline{a}}\,\mathrm{d}\xi^{\overline{b}} \\ &= \delta_{ab}\Lambda^{a}_{\ \overline{a}}\,\Lambda^{b}_{\ \overline{b}}\,\mathrm{d}\xi^{\overline{a}}\,\mathrm{d}\xi^{\overline{b}} \\ &\equiv g_{\overline{a}\overline{b}}\,\mathrm{d}\xi^{\overline{a}}\,\mathrm{d}\xi^{\overline{b}}, \end{aligned} \tag{20.23}$$

where

$$g_{\overline{a}\overline{b}} = \delta_{ab}\Lambda^{a}_{\ \overline{a}}\,\Lambda^{b}_{\ \overline{b}} \tag{20.24}$$

define the components to the metric tensor as represented by the new set of coordinates $\xi^{\overline{a}}$. Also introduced in these expressions are elements to the transformation matrix

$$\Lambda^{a}_{\ \overline{a}} = \frac{\partial \xi^a}{\partial \xi^{\overline{a}}}. \tag{20.25}$$

This matrix, whose rows are denoted by a and columns by $\overline{a}$, is nonsingular for well-defined coordinate transformations. Its determinant, called the *Jacobian of the transformation*, is nonvanishing and single signed for such cases.

Once the metric is determined, the distance along a curve between two finitely separated points on the manifold is given by the integration

$$L = \int \sqrt{\mathrm{d}s^2}$$
$$= \int_{\varphi_1}^{\varphi_2} \left| g_{ab} \frac{\mathrm{d}\xi^a}{\mathrm{d}\varphi} \frac{\mathrm{d}\xi^b}{\mathrm{d}\varphi} \right|^{1/2} \mathrm{d}\varphi, \tag{20.26}$$

where φ is a parameter specifying the curve, and $\varphi_{1,2}$ are the curve's endpoints. For example, if the curve is a path on a space-time manifold traversed by a particle, φ would be the time measured by a clock carried along by the particle (the *proper time*) (Schutz, 1985).

Before proceeding with the general discussion, it is useful to establish some explicit results with the two most familiar sets of coordinates used in GFD: Cartesian and spherical.

20.4.1 Transformation between Cartesian and spherical

Consider the transformation between Cartesian and spherical coordinates (see Figure 20.1)

$$(\xi^1, \xi^2, \xi^3) = (x^1, x^2, x^3) \tag{20.27}$$
$$(\xi^{\overline{1}}, \xi^{\overline{2}}, \xi^{\overline{3}}) = (\lambda, \phi, r). \tag{20.28}$$

These coordinates are related by the expressions

$$x^1 = r \cos\phi \cos\lambda \tag{20.29}$$
$$x^2 = r \cos\phi \sin\lambda \tag{20.30}$$
$$x^3 = r \sin\phi. \tag{20.31}$$

The transformation matrix is given by

$$\Lambda^a{}_{\overline{a}} = \begin{pmatrix} -r\cos\phi\sin\lambda & -r\sin\phi\cos\lambda & \cos\phi\cos\lambda \\ r\cos\phi\cos\lambda & -r\sin\phi\sin\lambda & \cos\phi\sin\lambda \\ 0 & r\cos\phi & \sin\phi \end{pmatrix}, \tag{20.32}$$

and the determinant of the transformation is

$$\det(\Lambda^a{}_{\overline{a}}) = r^2 \cos\phi. \tag{20.33}$$

This transformation matrix can be used, for example, to determine the transformation of the partial derivative operator

$$\partial_{\overline{a}} = \Lambda^a{}_{\overline{a}} \partial_a. \tag{20.34}$$

Organized as matrix-vector multiplication, this result takes the form

$$(\partial_\lambda, \partial_\phi, \partial_r) = (\partial_1, \partial_2, \partial_3) \begin{pmatrix} -r\cos\phi\sin\lambda & -r\sin\phi\cos\lambda & \cos\phi\cos\lambda \\ r\cos\phi\cos\lambda & -r\sin\phi\sin\lambda & \cos\phi\sin\lambda \\ 0 & r\cos\phi & \sin\phi \end{pmatrix}, \tag{20.35}$$

where the partial derivatives do not act on the components of the transformation matrix.

20.4.2 Metric for S^2

Consider the two-sphere S^2 with radius R. Use the angles $(\xi^1, \xi^2) = (\lambda, \phi)$ as the intrinsic coordinates, and the vectors $\vec{e}_1 = R\cos\phi\,\hat{\boldsymbol{\lambda}}\,, \vec{e}_2 = R\,\hat{\boldsymbol{\phi}}$ as the corresponding basis vectors. Embedding S^2 in the Euclidean space $\mathcal{R}^3$ recovers the geometric situation for classical GFD in which the sphere represents an idealization of the Earth embedded in the three-dimensional Euclidean space of Newtonian physics. Embeddings of objects in more complicated space-times are studied in general relativity.

The metric for 3D Euclidean space is

$$g_{ab} = \delta_{ab} = \begin{pmatrix} 1 & 0 & 0 \\ 0 & 1 & 0 \\ 0 & 0 & 1 \end{pmatrix}, \tag{20.36}$$

where the labels a, b run from $1, 2, 3$. It will also be of some use to introduce labels α, β which run from $1, 2$. These Greek labels are used for coordinates $(\xi^1, \xi^2) = (\lambda, \phi)$ intrinsic to the sphere.

Consider two very close points living on S^2, and let the Cartesian coordinates for these points differ by $\mathrm{d}x^a$. Since the sphere is assumed to be embedded in $\mathcal{R}^3$, the squared distance between the two points is determined through the Euclidean norm

$$\mathrm{d}s^2 = \delta_{ab}\,\mathrm{d}x^a\,\mathrm{d}x^b. \tag{20.37}$$

Now use the fact that the coordinates x^a for points on the sphere are functions of the sphere's intrinsic coordinates ξ^α. The chain rule then brings the squared distance to the form

$$\mathrm{d}s^2 = \delta_{ab}\left(\frac{\partial x^a}{\partial \xi^\alpha}\frac{\partial x^b}{\partial \xi^\beta}\right)\mathrm{d}\xi^\alpha\,\mathrm{d}\xi^\beta. \tag{20.38}$$

Consequently, one can identify the metric for a sphere S^2 embedded in $\mathcal{R}^3$,

$$\begin{aligned} g_{\alpha\beta} &= \delta_{ab}\left(\frac{\partial x^a}{\partial \xi^\alpha}\frac{\partial x^b}{\partial \xi^\beta}\right) \\ &= \mathrm{diag}(R^2\cos^2\phi, R^2) \\ &= \vec{e}_\alpha \cdot \vec{e}_\beta. \end{aligned} \tag{20.39}$$

The first and second columns of the transformation matrix (20.32) were used to deduce this result. The last expression provides an elegant practical means to derive the metric for an arbitrary surface embedded in Euclidean space.

20.5 TRANSFORMATIONS OF A VECTOR

In differential geometry, a fundamental question one asks about an object is how it changes under coordinate transformations. The transformation properties provide valuable mathematical and physical information about the object. It turns out that many objects transform under ways familiar from elementary calculus, and they have been assigned the name *tensors*. The transformation properties determine the tensor's *order* or *rank*.

The simplest tensor is a *scalar*. This is a rank zero tensor, or equivalently a tensor of zeroth order. The defining property of a scalar is that it has the same coordinate representation in any set of coordinates. In other words, scalars remain unchanged or invariant under coordinate transformations. Ocean tracers are examples of scalar fields (Section 5.1.1 on page 87).

As mentioned in Section 20.3, a first order tensor represented in terms of a basis of vectors is often called a *vector* for brevity. Vectors are objects of physical interest whose coordinate representations are affected by coordinate transformations. Here we determine how a vector transforms under changes in coordinates, and hence determine its tensorial attributes. First, consider the representation of an arbitrary vector in terms of a particular set of basis vectors

$$\mathbf{F} = F^{a}\, \vec{e}_{a}. \tag{20.40}$$

Now consider a representation of the same vector in terms of another set of basis vectors

$$\mathbf{F} = F^{\overline{a}}\, \vec{e}_{\overline{a}}. \tag{20.41}$$

Note that the tensor $\mathbf{F}$ did not change, only its coordinate *representations* F^{a} and $F^{\overline{a}}$ changed. The tensor $\mathbf{F}$ is therefore abstracted from any particular coordinate representation. This property is desired for objects with physical significance that transcends coordinate representations. Its representation, however, depends on the arbitrary choice of coordinates which assign numerical values to its components according to these coordinates.

In general, the two sets of basis vectors differ by a transformation, thus allowing the second basis to be written in terms of the first through the relation

$$\mathbf{F} = F^{\overline{a}}\, \Lambda^{a}{}_{\overline{a}} \vec{e}_{a}, \tag{20.42}$$

where $\Lambda^{a}{}_{\overline{a}}$ are components to the transformation matrix relating the two sets of basis vectors. Equating the right-hand sides of equations (20.40) and (20.42) implies that the two coordinate representations of $\mathbf{F}$ are related by

$$F^{a} = \Lambda^{a}{}_{\overline{a}}\, F^{\overline{a}}. \tag{20.43}$$

Similar arguments yield

$$F^{\overline{a}} = \Lambda^{\overline{a}}{}_{a}\, F^{a}, \tag{20.44}$$

where $\Lambda^{\overline{a}}{}_{a}$ are components to the inverse transformation Λ^{-1}. In the older tensor analysis literature, these transformation rules are used to define a *first order contravariant tensor*. These ideas and manipulations form the core of practical tensor analysis, and so will be used throughout the following.

20.6 ONE-FORMS

As discussed in Section 20.4, the metric tensor determines how to take the scalar product between two vectors. The result is the squared distance between the two vectors. More abstractly, the metric tensor takes two vectors as arguments, and yields the scalar distance

$$\mathbf{g}(\vec{F}, \vec{E}) = [\text{distance}(\vec{F}, \vec{E})]^{2}, \tag{20.45}$$

where it is convenient in this section to use the arrow to denote a first order tensor expressed in terms of its vector basis. When supplied with only one vector, the metric produces an object known as a *one-form*

$$\widetilde{F} \equiv \mathbf{g}(\vec{F}, \quad), \tag{20.46}$$

where the tilde label is used to distinguish a one-form from a vector. In turn, a one-form takes a vector as argument and produces a scalar

$$\widetilde{F}(\vec{E}) = \mathbf{g}(\vec{F}, \vec{E}). \tag{20.47}$$

These definitions are symmetric, so that a vector acts on a one-form and also produces a scalar

$$\vec{E}(\widetilde{F}) = \mathbf{g}(\vec{E}, \vec{F}). \tag{20.48}$$

Since the metric tensor is symmetric, the ordering of its vector arguments is irrelevant, and so the result of a one-form acting on a vector is the same as that vector acting on the one-form

$$\widetilde{F}(\vec{E}) = \vec{E}(\widetilde{F}). \tag{20.49}$$

Just as for vectors, there are special one-forms defining a basis for all one-forms. Hence, any one-form can be written as a linear sum of the basis one-forms. For any basis of vectors, there is a corresponding basis of one-forms related through

$$\vec{e}_a(\widetilde{e}^b) = \delta^b_{\ a}. \tag{20.50}$$

For example, with the basis of vectors for the surface of the sphere of radius R,

$$\vec{e}_1 = R \cos\phi \, \hat{\boldsymbol{\lambda}} \tag{20.51}$$

$$\vec{e}_2 = R \, \hat{\boldsymbol{\phi}} \,, \tag{20.52}$$

the corresponding basis of one-forms is

$$\widetilde{e}^1 = (R \cos\phi)^{-1} \, \hat{\boldsymbol{\lambda}} \tag{20.53}$$

$$\widetilde{e}^2 = R^{-1} \, \hat{\boldsymbol{\phi}} \,. \tag{20.54}$$

Using the basis of one-forms, an arbitrary one-form is written as the linear sum

$$\widetilde{F} = F_a \, \widetilde{e}^a, \tag{20.55}$$

where F_a is the coordinate representation of the one-form $\widetilde{F}$. Using the orthonormality relation between the one-form and vector basis, it is possible to determine the coordinate components to a one-form through letting the one-form "eat" one of the basis vectors

$$\begin{aligned} \widetilde{F}(\vec{e}_b) &= F_a \, \widetilde{e}^a(\vec{e}_b) \\ &= F_a \, \delta^a_{\ b} \\ &= F_b. \end{aligned} \tag{20.56}$$

20.7 MAPPING BETWEEN VECTORS AND ONE-FORMS

To each one-form, there is a corresponding vector related through the metric tensor. This result is easy to see through examining again the expression for the one-form components

$$\begin{aligned} F_a &= \widetilde{F}(\vec{e}_a) \\ &= \mathbf{g}(\vec{F}, \vec{e}_a) \\ &= \mathbf{g}(F^b \vec{e}_b, \vec{e}_a) \\ &= F^b \mathbf{g}(\vec{e}_b, \vec{e}_a) \\ &= F^b g_{ba}. \end{aligned} \tag{20.57}$$

The ability to take the component F^b outside the argument of the metric tensor follows since F^b is simply a number. In short, the component of a one-form F_a is related to the component of a vector F^b through the relation

$$F_a = g_{ab} F^b. \tag{20.58}$$

As a result, one finds that a one-form is a tensor of type $\binom{0}{1}$, and the metric tensor maps one-forms to vectors. The previous arguments can be used to show that the inverse metric provides for the relation

$$F^a = g^{ab} F_b, \tag{20.59}$$

where g^{ab} are components of the inverse metric $\mathbf{g}^{-1}$. It is in this manner that a one-form and its corresponding vector are said to maintain a *duality* relationship, which is a generalization of the duality maintained between row vectors and column vectors in linear algebra.

These results establish the general manner in which to transform the representation of a tensor. Namely, the metric and its inverse transform the type or rank of a tensor through acting on the tensor's labels. For example, the inner product of two vectors can be written in either of the equivalent forms

$$\begin{aligned} \mathbf{g}(\vec{F}, \vec{E}) &= \widetilde{F}(\vec{E}) \\ &= F_a E^a \\ &= g_{ab} F^b E^a \\ &= g^{ab} F_a E_b. \end{aligned} \tag{20.60}$$

It is through this transformation that one sees why it is often irrelevant whether a tensor is considered in its one-form or vector representation, since they are easily mapped to one another via the metric tensor. The fundamental notion is the order of the tensor, where both one-forms and vectors are first order tensors. Further examples are encountered in the following.

20.8 TRANSFORMATION OF A ONE-FORM

The position of indices on the one-form components are dual to the indices on the vector components. One therefore suspects that the transformation of a one-form occurs through the components to the inverse transformation used for vectors. Indeed, this result is easy to establish using the transformation properties of the metric, the mapping property between vectors and one-forms, and the transformation properties of a vector:

$$\begin{aligned} F_{\bar{a}} &= g_{\bar{a}\bar{b}} F^{\bar{b}} \\ &= g_{\bar{a}\bar{b}} \Lambda^{\bar{b}}{}_{b} F^{b} \\ &= \Lambda^{a}{}_{\bar{a}} \Lambda^{c}{}_{\bar{b}} g_{ac} \Lambda^{\bar{b}}{}_{b} F^{b} \\ &= \Lambda^{a}{}_{\bar{a}} \delta^{c}_{b} g_{ac} F^{b} \\ &= \Lambda^{a}{}_{\bar{a}} F_{a}. \end{aligned} \tag{20.61}$$

To establish this result required the relation

$$\Lambda^{c}{}_{\bar{b}} \Lambda^{\bar{b}}{}_{b} = \delta^{c}_{b} \tag{20.62}$$

which follows since $\Lambda^{\bar{b}}{}_{b}$ are components to the matrix Λ^{-1} and $\Lambda^{b}{}_{\bar{b}}$ are components to the matrix Λ.

20.9 ARBITRARY TENSORS AND THEIR TRANSFORMATIONS

The previous formalism provides the framework for determining how any tensor transforms under a coordinate change. For example, consider a *second order* tensor

$$\mathbf{T} = T^{ab} \vec{e}_a \vec{e}_b. \tag{20.63}$$

The numbers T^{ab} form the components to $\mathbf{T}$ as expanded in terms of the basis vectors $\vec{e}_a$. Considering $\mathbf{T}$ as an abstract object allows one to also consider the representations

$$\mathbf{T} = T^{a}_{b} \vec{e}_a \vec{e}^{b}, \tag{20.64}$$

or

$$\mathbf{T} = T_{ab} \vec{e}^{a} \vec{e}^{b}. \tag{20.65}$$

In classical tensor analysis, T^{ab} is termed the second order contravariant representation of $\mathbf{T}$; T^{a}_{b} is the first order contravariant and first order covariant representation; and T_{ab} is the second order covariant representation. An easy way to remember the language is the mnemonic "co-lo," which associates covariant with lowered indices. Modern tensor analysis avoids the use of these terms in favor of denoting the *type* of tensor as symbolized by $\binom{M}{N}$. T^{ab} is termed the $\binom{2}{0}$ representation; T^{a}_{b} is the $\binom{1}{1}$ representation; and T_{ab} is the $\binom{0}{2}$ representation. Both languages are employed in the following, as it is useful to be familiar with these terms when reading the literature.

The transformation properties of the different representations of $\mathbf{T}$ are straightforward to determine. First, consider the transformation to another basis set. Recall that vectors and one-forms transform as

$$\vec{e}_a = \Lambda^{\bar{a}}{}_{a} \vec{e}_{\bar{a}} \tag{20.66}$$

$$\vec{e}^{a} = \Lambda^{a}_{\bar{a}} \vec{e}^{\bar{a}}. \tag{20.67}$$

Consequently,

$$\begin{aligned}\mathbf{T} &= T^{ab}\,\vec{e}_a\,\vec{e}_b \\ &= T^{ab}\,\Lambda^{\bar{a}}_{a}\,\Lambda^{\bar{b}}_{b}\,\vec{e}_{\bar{a}}\,\vec{e}_{\bar{b}} \\ &= T^{\bar{a}\bar{b}}\,\vec{e}_{\bar{a}}\,\vec{e}_{\bar{b}},\end{aligned} \tag{20.68}$$

where

$$T^{\bar{a}\bar{b}} = T^{ab}\,\Lambda^{\bar{a}}_{a}\,\Lambda^{\bar{b}}_{b}. \tag{20.69}$$

Similarly,

$$T^{\bar{a}}_{\bar{b}} = T^{a}_{b}\,\Lambda^{\bar{a}}_{a}\,\Lambda^{b}_{\bar{b}} \tag{20.70}$$

and

$$T_{\bar{a}\bar{b}} = T_{ab}\,\Lambda^{a}_{\bar{a}}\,\Lambda^{b}_{\bar{b}}. \tag{20.71}$$

Recall that in Section 20.4, it was determined that the metric transforms under a change in coordinates as

$$g_{\bar{a}\bar{b}} = \Lambda^{a}{}_{\bar{a}}\,\Lambda^{b}{}_{\bar{b}}\,g_{ab}, \tag{20.72}$$

where this relation was established in the special case of $g_{ab} = \delta_{ab}$. In general, it is simple to show that any two representations of the metric are related by equation (20.72). Hence, the metric transforms as a second order tensor, justifying its appellation as the *metric tensor*.

20.10 TENSORIAL PROPERTIES OF THE GRADIENT OPERATOR

The partial derivative operator

$$\partial_a = \frac{\partial}{\partial \xi^a} \tag{20.73}$$

naturally carries its index "downstairs." One might therefore suspect that it forms the first order covariant components to the gradient tensor ∇. The chain rule

$$\begin{aligned}\partial_{\bar{a}} &= \frac{\partial}{\partial \xi^{\bar{a}}} \\ &= \frac{\partial \xi^a}{\partial \xi^{\bar{a}}}\,\frac{\partial}{\partial \xi^a} \\ &= \Lambda^{a}{}_{\bar{a}}\,\partial_a\end{aligned} \tag{20.74}$$

establishes that, indeed, the partial derivative operator transforms as a first order covariant tensor, thus establishing that the gradient operator

$$\nabla = \tilde{e}^a\,\partial_a \tag{20.75}$$

is a first order tensor naturally represented as a one-form. Raising the label on the derivative gives

$$\partial^a = g^{ab}\,\partial_b, \tag{20.76}$$

which form the components to the contravariant or vector representation of the gradient operator

$$\nabla = \vec{e}_a\,\partial^a. \tag{20.77}$$

20.11 THE INVARIANT VOLUME ELEMENT

The volume element for Euclidean space, as written with Cartesian coordinates, takes the form

$$dV = dx^1\, dx^2\, dx^3. \tag{20.78}$$

Volume elements arise, for example, when integrating over a finite volume, or for setting the volume of an infinitesimal fluid parcel. It is useful to generalize this result to arbitrary coordinates on an arbitrary manifold.

20.11.1 Derivation of the invariant volume element

To derive an expression for the volume element for any manifold using any coordinates, begin with the familiar rules from elementary calculus relating the differential elements $d\xi^1\, d\xi^2\, d\xi^3$ and $d\xi^{\bar{1}}\, d\xi^{\bar{2}}\, d\xi^{\bar{3}}$ in two sets of coordinates

$$\begin{aligned} d\xi^1\, d\xi^2\, d\xi^3 &= \left(\frac{\partial(\xi^1, \xi^2, \xi^3)}{\partial(\xi^{\bar{1}}, \xi^{\bar{2}}, \xi^{\bar{3}})} \right) d\xi^{\bar{1}}\, d\xi^{\bar{2}}\, d\xi^{\bar{3}} \\ &= \det(\Lambda^{a}{}_{\bar{a}})\, d\xi^{\bar{1}}\, d\xi^{\bar{2}}\, d\xi^{\bar{3}}, \end{aligned} \tag{20.79}$$

where $\det(\Lambda^{a}{}_{\bar{a}})$ is the Jacobian, or determinant of the transformation matrix. The transformation is well defined so long as the Jacobian does not vanish. Note that labels on the transformation matrix are maintained inside the determinant symbol to indicate between which sets of coordinates this tranformation matrix is referred. It is thus possible to explicitly note the matching of indices on both sides of this equation. This notation is especially useful when more than one transformation is considered.

For certain cases, the Jacobian is tedious to compute. In these cases, the computation can be simplified by using the following relation between the Jacobian and the determinant of the metric. For this purpose, recall that the transformation of the metric (Section 20.9) is given by

$$g_{\bar{a}\bar{b}} = \Lambda^{a}{}_{\bar{a}} \Lambda^{b}{}_{\bar{b}}\, g_{ab}. \tag{20.80}$$

Written as a matrix equation, this result takes the form

$$\bar{\mathbf{g}} = \Lambda^T\, \mathbf{g}\, \Lambda \tag{20.81}$$

where Λ^T is the transposed matrix. This equation is valid upon taking determinants of both sides. Now recall two properties of the determinant:

1. $\det(A\,B) = \det(A)\,\det(B)$, for any two matrices A and B. This result is proven in various linear algebra texts, such as page 203 of Noble and Daniel (1977).

2. $\det(\Lambda^T) = \det(\Lambda)$. This result is trivial to prove.

Consequently, exposing the labels on Λ and the metric gives

$$\det(\Lambda^{a}{}_{\bar{a}}) = \left(\frac{\det(g_{\bar{a}\bar{b}})}{\det(g_{ab})} \right)^{1/2}, \tag{20.82}$$

where it has been assumed that the manifold has a positive definite metric, which is true for surfaces, such as the sphere, considered in this book. The determininat of the metric appears quite often in the following, and so it is useful to introduce the notation

$$\mathcal{G} = \det(g_{ab}) \tag{20.83}$$

$$\overline{\mathcal{G}} = \det(g_{\bar{a}\bar{b}}). \tag{20.84}$$

Note that most books use the symbol g rather than $\mathcal{G}$, which for our purposes is not useful since g is the acceleration of gravity at the earth's surface (Section 4.2.3). In summary, one now has the very useful relation

$$\sqrt{\mathcal{G}}\, d\xi^1\, d\xi^2\, d\xi^3 = \sqrt{\overline{\mathcal{G}}}\, d\xi^{\bar{1}}\, d\xi^{\bar{2}}\, d\xi^{\bar{3}}. \tag{20.85}$$

This expression leads one to conclude that the expression

$$dV \equiv \sqrt{\mathcal{G}}\, d\xi^1\, d\xi^2\, d\xi^3 \tag{20.86}$$

is invariant under coordinate transformations; that is, it is a scalar. Hence, it forms the appropriate generalization to arbitrary coordinates of the differential volume element.

20.11.2 Invariant volume element for some common coordinates

It is useful to consider some examples of how the volume element appears when writing it in various coordinates. First, as noted above, the volume element for three-dimensional Euclidean space using Cartesian coordinates (ξ^1, ξ^2, ξ^3) is

$$dV = dx^1\, dx^2\, dx^3. \tag{20.87}$$

Spherical coordinates, $(\xi^1, \xi^2, \xi^3) = (\lambda, \phi, r)$ (Figure 20.1), with metric $g_{ab} = \mathrm{diag}(r^2 \cos^2\phi, r^2, 1)$, have a volume element

$$\begin{aligned} dV &= \mathcal{G}^{1/2}\, dr\, d\lambda\, d\phi \\ &= r^2 \cos\phi\, dr\, d\lambda\, d\phi. \end{aligned} \tag{20.88}$$

The equality

$$dx^1\, dx^2\, dx^3 = r^2 \cos\phi\, dr\, d\lambda \tag{20.89}$$

is familiar from elementary calculus.

For an arbitrary nonsingular and orthogonal set of coordinates, defined such that the line element takes the diagonal form

$$ds^2 = (h_1\, d\xi^1)^2 + (h_2\, d\xi^2)^2 + (h_3\, d\xi^3)^2, \tag{20.90}$$

the invariant volume element is

$$\begin{aligned} dV &= \mathcal{G}^{1/2}\, d\xi^1\, d\xi^2\, d\xi^3 \\ &= h_1\, h_2\, h_3\, d\xi^1\, d\xi^2\, d\xi^3. \end{aligned} \tag{20.91}$$

These coordinates are called *generalized orthogonal curvilinear coordinates*. They correspond to the orthogonal set of basis vectors

$$\vec{e}_a = h_a\, \hat{\mathbf{e}}_{(a)}, \tag{20.92}$$

where no summation is implied and $\hat{\mathbf{e}}_{(a)}$ are the dimensionless unit directions.* Hence, the functions h_a, which generally depend on space-time, determine the local "stretching" in the three orthogonal unit directions $\hat{\mathbf{e}}_{(a)}$ for use in measuring distance. A common choice in global models is to define the horizontal stretching functions h_1 and h_2 so that the North Pole lives over land rather than ocean (e.g., Murray, 1996; Madec and Imbard, 1996; Smith et al., 1995; Griffies et al., 2004). In this way, the ocean domain does not contain the annoying coordinate singularity possessed by spherical coordinates at the North Pole. Much more is said about orthogonal coordinates in Section 21.11.

Finally, consider the case of isopycnal coordinates (x, y, ρ). Section 6.5.2 derived the transformation matrix between Cartesian and isopycnal coordinates. The Jacobian of that transformation is the specific thickness $z_{,\rho}$, which gives the expression for the volume element

$$dV = z_{,\rho}\, dx^1\, dx^2\, d\rho. \tag{20.93}$$

This expression can also be found through the formula $dV = \mathcal{G}^{1/2}\, dx^1\, dx^2\, d\rho$, where $\mathcal{G}$ is the determinant of the metric tensor for three-dimensional Euclidean space as written using isopycnal coordinates. Using the transformation matrix $\Lambda^a_{\bar{a}}$ from Section 6.5.2, and the tensor transformation rules from Section 20.9, it is straightforward to derive the expression for the metric (see Section 6.6)

$$\begin{aligned} g_{\bar{a}\bar{b}} &= \Lambda^a_{\bar{a}}\, \Lambda^b_{\bar{b}} \delta_{ab} \\ &= \begin{pmatrix} 1+z_{,x}^2 & z_{,x}\, z_{,y} & z_{,x}\, z_{,\rho} \\ z_{,x}\, z_{,y} & 1+z_{,y}^2 & z_{,y}\, z_{,\rho} \\ z_{,x}\, z_{,\rho} & z_{,y}\, z_{,\rho} & z_{,\rho}^2 \end{pmatrix}, \end{aligned} \tag{20.94}$$

where δ_{ab} is the metric for Euclidean space written in terms of Cartesian coordinates. A simple calculation shows that indeed $\mathcal{G}^{1/2} = z_{,\rho}$.

20.12 DETERMINANTS AND THE LEVI-CIVITA SYMBOL

Consider a transformation matrix $\Lambda^a{}_{\bar{a}}$ between two sets of coordinates, and let the manifold be two-dimensional. As already mentioned, a well-defined transformation has a nonzero determinant

$$\det(\Lambda^a{}_{\bar{a}}) = \Lambda^1{}_{\bar{1}}\, \Lambda^2{}_{\bar{2}} - \Lambda^1{}_{\bar{2}}\, \Lambda^2{}_{\bar{1}}. \tag{20.95}$$

Determinants, especially those associated with transformations, are ubiquitous in tensor analysis. It turns out that for many manipulations it is useful to introduce the permutation or *Levi-Civita symbol*, which brings the determinant to the form

$$\det(\Lambda^a{}_{\bar{a}}) = \epsilon_{ab}\, \Lambda^a{}_{\bar{1}}\, \Lambda^b{}_{\bar{2}}, \tag{20.96}$$

where

$$\epsilon_{ab} = \begin{cases} 0 & \text{if any two labels are the same,} \\ 1 & \text{if } a, b \text{ is an even permutation of } 1, 2, \\ -1 & \text{if } a, b \text{ is an odd permutation of } 1, 2. \end{cases} \tag{20.97}$$

*As explained in Section 21.11 on page 474, the index on the unit directions is enclosed in parentheses to advertise the fact that this label is not tensorial; i.e., the unit directions do not transform as tensors. Rather, the functions h_a carry the tensorial properties of the basis vectors $\vec{e}_a$.

This symbol is defined to have numerically the same values whether the labels are raised or lowered: $\epsilon^{ab} = \epsilon_{ab}$. As seen, the Levi-Civita symbol is totally antisymmetric on its two labels. It has generalizations to any number of dimensions, each of which adds one more label to the symbol, and one more number added to the permutation string. For example, in three dimensions,

$$\epsilon_{abc} = \begin{cases} 0 & \text{if any two labels are the same,} \\ 1 & \text{if } a, b, c \text{ is an even permutation of } 1, 2, 3, \\ -1 & \text{if } a, b, c \text{ is an odd permutation of } 1, 2, 3. \end{cases} \tag{20.98}$$

Through its connection to the determinant, the Levi-Civita symbol is also useful for writing the curl of two vectors living in $\mathcal{R}^3$. For example, in Cartesian coordinates,

$$(\nabla \wedge \vec{A})^a = \epsilon^{abc} A_{c,b}. \tag{20.99}$$

To determine how the Levi-Civita symbol transforms, return to two dimensions, where things are simple to verify explicity. The central formula is

$$\epsilon_{ab} \Lambda^a{}_{\bar{a}} \Lambda^b{}_{\bar{b}} = \epsilon_{\bar{a}\bar{b}} \det(\Lambda^a{}_{\bar{a}}), \tag{20.100}$$

which follows directly from the definition of the determinant. This expression is valid so long as the transformed Levi-Civita symbol $\epsilon_{\bar{a}\bar{b}}$ is numerically identical to ϵ_{ab}.

The relation (20.100) indicates that the Levi-Civita symbol *does not* transform as the components to a second order covariant tensor, unless the determinant of the transformation is unity. Unit determinants occur for special transformations, such rotations and identity transformations. However, they are not unity in general.

One can define a *covariant Levi-Civita symbol* by considering

$$\varepsilon_{ab} = \mathcal{G}^{1/2} \epsilon_{ab}. \tag{20.101}$$

This scaling with the metric determinant is identical to that which was used to produce a coordinate invariant volume element in Section 20.11. These factors are indeed related, and in more complete discussions (e.g., Schutz, 1980), the connection is made explicit. For present purposes, it is sufficient to note that ε_{ab} transforms as

$$\begin{aligned} \Lambda^a{}_{\bar{a}} \Lambda^b{}_{\bar{b}} \varepsilon_{ab} &= \Lambda^a{}_{\bar{a}} \Lambda^b{}_{\bar{b}} \mathcal{G}^{1/2} \epsilon_{ab} \\ &= \mathcal{G}^{1/2} \epsilon_{\bar{a}\bar{b}} \det(\Lambda^a{}_{\bar{a}}) \\ &= \overline{\mathcal{G}}^{1/2} \epsilon_{\bar{a}\bar{b}} \\ &= \varepsilon_{\bar{a}\bar{b}}, \end{aligned} \tag{20.102}$$

where equations (20.82) and (20.100) were employed. Therefore, ε_{ab} transforms as components to a second order covariant tensor, without the extra determinant factors floating around. Likewise,

$$\varepsilon^{ab} = \mathcal{G}^{-1/2} \epsilon^{ab} \tag{20.103}$$

transforms as the components to a second order contravariant tensor. In subsequent developments, ε_{ab} is referred to as the *covariant* Levi-Civita tensor.

Each of these expressions derived in two dimensions are valid in any number of dimensions, as can be shown through induction. As hinted at above, one useful application of the third order tensor ε_{abc} is in defining the covariant curl operation (Section 21.8), which generalizes the usual curl from Cartesian vector analysis to curved manifolds.

20.13 SURFACES EMBEDDED IN EUCLIDEAN SPACE

In Chapter 6, we discussed some of the basic properties of surfaces through study of the metric for an isopycnal surface (see in particular Section 6.6). This section systematizes and extends that discussion.

20.13.1 Induced metric tensor for a surface

Consider the line element in three-dimensional Euclidean space

$$ds^2 = g_{ab}\,d\xi^a\,d\xi^b. \tag{20.104}$$

The components of the metric tensor g_{ab} are those for Euclidean space as written in terms of an arbitrary coordinate system $\xi^a = (\xi^1, \xi^2, \xi^3)$. If the line element lives on a two-dimensional surface, then the chain rule gives

$$d\xi^a = \left(\frac{\partial \xi^a}{\partial w^\alpha}\right) dw^\alpha, \tag{20.105}$$

where $w^\alpha = (w^1, w^2)$ are coordinates intrinsic to the two-dimensional surface, and $\xi^a = \xi^a(w^\alpha)$ for points on the surface. In this section, Greek labels refer to coordinates intrinsic to the surface, and Latin labels refer to the Euclidean space. Equation (20.105) provides a transformation between infinitesimal displacements dw^α on the surface, and the corresponding displacements $d\xi^a$ in the space tangent to the surface. The numbers

$$\Lambda^a_{\ \alpha} = \left(\frac{\partial \xi^a}{\partial w^\alpha}\right) \tag{20.106}$$

form components to the transformation matrix which transforms between the surface and the embedding Euclidean space. It follows that the line element for points on the surface takes the form

$$\begin{aligned} ds^2 &= g_{ab}\,\Lambda^a_{\ \alpha}\,\Lambda^b_{\ \beta}\,dw^\alpha\,dw^\beta \\ &\equiv G_{\alpha\beta}\,dw^\alpha\,dw^\beta, \end{aligned} \tag{20.107}$$

where $G_{\alpha\beta}$ are components to the metric tensor for the two-dimensional surface. This metric is said to be *induced* by the embedding of the surface in the three-dimensional Euclidean space. As an aside, note that if the same surface is embedded in some other space, such as a non-Euclidean space, then the induced metric would generally differ.

The above results were used when deriving the metric for an isopycnal surface in Section 6.6.1 (see also Section 20.11.2). For that case, the intrinsic coordinates are the usual (x, y) Cartesian coordinates of the point on the surface $\rho(x, y, z) = \rho_{\text{const}}$. The resulting surface metric $G_{\alpha\beta}$ is given by the nondiagonal 2×2 matrix in equation (6.56).

20.13.2 The invariant surface area element

The derivation of an invariant volume element given in Section 20.11 is general to any space dimension. For a two-dimensional space, such as a surface embedded in Euclidean space, the invariant "volume" or area element takes the form

$$dA = \sqrt{G}\,dw^1\,dw^2, \tag{20.108}$$

where $G = \det(G_{\alpha\beta})$ is the determinant of the surface metric tensor. This is the appropriate form of the surface area element for use in forming surface integrals. For example, if one wishes to integrate a field, such as temperature, over a particular isopycnal surface $\rho = \rho_1$ using the isopycnal coordinates (x, y, ρ), then the metric given by equation (6.56) indicates that the surface integral should be of the form

$$\int_{\rho=\rho_1} T(x, y, \rho)\, \mathrm{d}A_{(\rho)} = \int_{\rho=\rho_1} T(x, y, \rho) \left(1 + |\nabla_\rho z|^2\right)^{1/2} \mathrm{d}x\, \mathrm{d}y. \tag{20.109}$$

20.13.3 Surface and space components of a tensor

Recall from Section 20.5 that a vector field $\vec{F}$ living in Euclidean space can be represented in terms of a basis of vectors that span Euclidean space

$$\vec{F} = F^a \, \vec{e}_a. \tag{20.110}$$

The components F^a are sometimes called the *space* components, as they are the components of $\vec{F}$ relative to the Euclidean space basis $\vec{e}_a$. For points on some surface, the chain rule transformation (20.105) allows for $\vec{F}$ also to be written in the form

$$\begin{aligned} \vec{F} &= F^a \, \vec{e}_a \\ &= \Lambda^a_{\ \alpha} \, F^\alpha \, \vec{e}_a \\ &\equiv F^\alpha \, \vec{e}_\alpha, \end{aligned} \tag{20.111}$$

where

$$F^\alpha = \Lambda^\alpha_{\ a} \, F^a \tag{20.112}$$

defines the *surface* components of $\vec{F}$, and

$$\vec{e}_\alpha = \Lambda^a_{\ \alpha} \, \vec{e}_a \tag{20.113}$$

defines the surface basis vectors. As seen in the next subsection, the surface components F^α are the projection of the space components F^a onto the surface; that is, they are tangent to the surface.

The covariant surface components F_α of the tensor are related to the contravariant components through the surface metric tensor

$$\begin{aligned} F_\alpha &= \Lambda^a_{\ \alpha} \, F_a \\ &= \Lambda^a_{\ \alpha} \, g_{ab} \, F^b \\ &= \Lambda^a_{\ \alpha} \, \Lambda^b_{\ \beta} \, g_{ab} \, F^\beta \\ &= G_{\alpha\beta} \, F^\beta. \end{aligned} \tag{20.114}$$

Consequently, the surface metric $G_{\alpha\beta}$ maps between surface vectors and surface one-forms, just as the space metric g_{ab} maps between space vectors and space one-forms.

20.13.4 Normals and tangents

Recall from Section 20.12 that the generalized form of a cross product between two vectors is conveniently written as

$$A_a = \varepsilon_{abc}\, B^b\, C^c, \tag{20.115}$$

where $\varepsilon_{abc} = \mathcal{G}^{1/2}\,\epsilon_{abc}$ is the covariant form of the Levi-Civita symbol. A_a form the covariant components to a one-form $\widetilde{A} = A_a\,\widetilde{e}^a$. It is easy to show that $\widetilde{A}$ is orthogonal to either of the vectors $\vec{B} = B^b\,\vec{e}_b$ or $\vec{C} = C^c\,\vec{e}_c$. For example,

$$\begin{aligned} \widetilde{A}(\vec{B}) &= A_a\, B^a \\ &= \varepsilon_{abc}\, B^b\, C^c\, B^a \\ &= 0, \end{aligned} \tag{20.116}$$

which follows from the antisymmetry of ε_{abc}.

If the two vectors $\vec{B}$ and $\vec{C}$ are not colinear, they define a two-dimensional surface that is tangent to these two vectors at each point in space. In this case, one can write the cross-product as

$$\begin{aligned} A_a &= \varepsilon_{abc}\, B^b\, C^c \\ &= \varepsilon_{abc}\, \Lambda^b{}_\beta\, \Lambda^c{}_\gamma\, B^\beta\, C^\gamma. \end{aligned} \tag{20.117}$$

The factor $\varepsilon_{abc}\,\Lambda^b{}_\beta\,\Lambda^c{}_\gamma$ is antisymmetric under interchange of the surface tensor labels, viz.,

$$\begin{aligned} \varepsilon_{abc}\, \Lambda^b{}_\beta\, \Lambda^c{}_\gamma &= \varepsilon_{acb}\, \Lambda^c{}_\beta\, \Lambda^b{}_\gamma \\ &= -\varepsilon_{abc}\, \Lambda^b{}_\gamma\, \Lambda^c{}_\beta. \end{aligned} \tag{20.118}$$

Hence, one can introduce the covariant tensor N_a through

$$N_a\, \varepsilon_{\beta\gamma} = \varepsilon_{abc}\, \Lambda^b{}_\beta\, \Lambda^c{}_\gamma, \tag{20.119}$$

bringing the cross-product to the form

$$A_a = N_a\, \varepsilon_{\beta\gamma}\, B^\beta\, C^\gamma. \tag{20.120}$$

In this expression, $\varepsilon_{\beta\gamma} = \sqrt{G}\,\epsilon_{\beta\gamma}$ is the covariant Levi-Civita tensor on the surface. With $\varepsilon_{\beta\gamma}\,\varepsilon^{\beta\gamma} = 2$, it is possible to write

$$N_a = \frac{1}{2}\,\varepsilon_{abc}\,\varepsilon^{\beta\gamma}\,\Lambda^b{}_\beta\,\Lambda^c{}_\gamma. \tag{20.121}$$

By construction, the one-form $\widetilde{N} = N_a\,\widetilde{e}^a$ is orthogonal to all spatial vectors tangent to the surface. To see this property, note that

$$\begin{aligned} \widetilde{N}(\vec{B}) &= N_a\, B^a \\ &= N_a\, \Lambda^a{}_\mu\, B^\mu \\ &= \frac{1}{2}\,\varepsilon_{abc}\,\varepsilon^{\beta\gamma}\,\Lambda^b{}_\beta\,\Lambda^c{}_\gamma\,\Lambda^a{}_\mu\, B^\mu \\ &= 0, \end{aligned} \tag{20.122}$$

which follows because μ must equal either β or γ. Therefore, the surface components of an arbitrary one-form $F_\alpha = \Lambda^a{}_\alpha F_a$ can be thought of as the projection of the space components F_a onto the surface. If $F_a = \overline{F}_a + c\, N_a$, then

$$\begin{aligned} F_\alpha &= \Lambda^a{}_\alpha (\overline{F}_a + c\, N_a) \\ &= \Lambda^a{}_\alpha \overline{F}_a, \end{aligned} \tag{20.123}$$

which shows that the component normal to the surface has been destroyed. In this way, $F_\alpha = \Lambda^a{}_\alpha F_a$ represents the components of the one-form $\widetilde{F}$ which are tangent to the surface. With $c = N^b F_b$, the projection of an arbitrary covariant space tensor onto the surface is given by

$$\overline{F}_a = F_a - (N^b F_b)\, N_a. \tag{20.124}$$

In addition to describing a surface in terms of two noncolinear vectors, one often describes a surface in terms of the envelope of constant values of a scalar function of space. For example, the algebraic equation

$$\rho(\xi^a) = \rho_{\text{const}} \tag{20.125}$$

defines isopycnal surfaces. Following the ideas from Section 20.3, which are based on traditional vector analysis,

$$\begin{aligned} N_a &= \frac{\rho_{,a}}{|\rho_{,b}\, \rho^{,b}|^{1/2}} \\ &= \frac{\rho_{,a}}{|g^{bc}\, \rho_{,b}\, \rho_{,c}|^{1/2}} \end{aligned} \tag{20.126}$$

provides the components to a outward normal one-form for the surface. In this expression, g^{bc} are components to the inverse metric for the embedding space. The previous formalism follows with this definition of the normal one-form. In particular, the projection of an arbitrary one-form onto the surface is given by

$$\overline{F}_a = F_a - \frac{F^b\, \rho_{,b}\, \rho_{,a}}{g^{bc}\, \rho_{,b}\rho_{,c}}. \tag{20.127}$$

These are the components of $\widetilde{F}$ which are tangent to the surface.

20.14 CHAPTER SUMMARY

This chapter presented a set of results from calculus on curved manifolds. The presentation was hopefully found to be accessible for those familiar with elementary calculus as well as partial differential equations. Some attention was given to examples relevant for ocean modeling.

After providing recommendations for texts on this subject in Section 20.1, and then some practical motivation in Section 20.2, we started the main part of the chapter in Section 20.3 by considering how one goes about describing the position of a point on a surface. This discussion introduced the notions of coordinates and vectors. Once possessing a way to mark points and directions, one is posed with the question of how to measure the distance between points. This question led to the metric tensor in Section 20.4. This is a second order symmetric tensor which is very fundamental to calculus on curved manifolds.

One of the more useful aspects of tensor analysis is that it provides a formalism to transform equations from one set of coordinates into another. To start considering how to perform such transformations, Section 20.5 considered issues arising when working with a vector. Sections 20.6 and 20.8 considered the analogous issues for a one-form, where a one-form is an object dual to a vector. Section 20.7 highlighted the property of a metric as the map between one-forms and vectors. In this way, the metric provides a means to transform the representations of an arbitrary tensor, such as shown in Section 20.9. The transformation properties of an object define the order of a tensor. Section 20.10 considered how to classify the gradient operator. The formalism identified it as a first order tensor, with a typical representation as a one-form. However, use of the metric tensor allows it to be transformed into a vector.

Volume or area elements are needed for various purposes in calculus to measure regions. Section 20.11 formalized the definition of a volume element to arbitrary coordinates. Doing so allows one to access an invariant form of the volume element which is applicable for any set of coordinates. In discussions of volume, and determinants as well, it turns out that the totally antisymmetric tensor, known as the Levi-Civita tensor, proves quite fundamental. Section 20.12 provides some understanding of the utility of this tensor. Section 20.13 finished the chapter with a formal discussion of calculus on surfaces embedded in Euclidean space, such as an isopycnal surface.

Chapter Twenty-One

CALCULUS ON CURVED MANIFOLDS

The purpose of this chapter is to derive many of the formulae of vector calculus with the underlying manifold curved. Our focus is aimed toward applying these results to the equations for fluid mechanics on an arbitrary smooth surface.

21.1 FUNDAMENTAL CHARACTER OF TENSOR EQUATIONS

Differential physical laws are most useful when written so that they remain form invariant under arbitrary coordinate transformations. In this way, the essence of the laws can be revealed, rather than being confused with potentially coordinate-dependent artifacts. The laws then realize the postulate that physics does not care about coordinates. This fundamental postulate, which often goes by the name *covariance*, is at the root of modern physics.

Although physics does not care about coordinates, for many practical situations it is often convenient, if not necessary, to work in a specific coordinate system that is suited to the geometry. For example, when working in Euclidean space it is simplest for many purposes to use the familiar Cartesian coordinate system. If the physical system exhibits a symmetry, such as spherical symmetry, then this may motivate the use of a particular set of coordinates that reflect this symmetry. Finally, for numerical modeling, it is necessary to choose a particular set of coordinates in which to discretize the continuous equations.

After deriving a physical law in one set of coordinates, it is of interest to establish the form of the law in another set of coordinates. How does the physical law, typically represented as a differential equation, transform into other coordinates? Does one have to rederive the law from first principles working with the new set of coordinates? Fortunately, so long as the equations are written in a proper tensorial form, in which they exemplify covariance, then the equations are form invariant and so the derivation need not be repeated in each set of coordinates. In practice, an equation exemplifies covariance if all the tensor indices are properly matched on both sides of the equation and each derivative is covariant. The elegance allowed by this property is the key reason that tensor analysis is ubiquitous in theoretical physics.

To motivate these ideas, note that the partial derivative of a field appears in most differential physical laws. For example, the divergence of velocity appears in the equation for mass conservation in a fluid. In Cartesian coordinates, this divergence takes the form

$$\begin{aligned} \nabla \cdot (v^a \, \vec{e}_a) &= \tilde{e}^a_m \, \partial_a \, (\vec{e}^m_b \, v^b) \\ &= v^a_{,a}. \end{aligned} \tag{21.1}$$

The intermediate expression is a rather cumbersome way of writing what is more often seen in the more concise final form. Yet the final form only arises since the basis vectors $\vec{e}_a$ are each constants in space-time. In more general coordinates, such as spherical coordinates, the basis vectors are not constants, and so the partial derivative operator picks up terms proportional to the derivative of the basis vectors. The result is a more complicated divergence. Additionally, it possesses clumsy terms depending on the nature of the chosen coordinates. Hence, the divergence as defined using partial derivatives does not remain form invariant under coordinate transformations. The notion of a covariant divergence described in Section 21.2 accounts for these extra terms so that the equations remain form invariant.

We now summarize some of the properties characterizing covariance (taken after page 153 of Schutz, 1985):

1. Certain manipulations of tensor components are called *permissible tensor operations* because they produce components of new tensors. The following are permissible operations:
   (a) Multiplication of a tensor by a scalar field produces a new tensor of the same type.
   (b) Addition of components of two tensors of the same type gives components of a new tensor of the same type. In particular, only tensors of the same type can be equal.
   (c) Multiplication of components of two tensors of arbitrary type gives components of a new tensor whose type is given by the sum of the types for the individual tensors. This operation is called the *outer product*.
   (d) Covariant differentiation (discussed in Section 21.2) of the components of a tensor of type $\binom{M}{N}$ gives the components to a tensor of type $\binom{M}{N+1}$.
   (e) Contraction on a pair of indices of the components of a tensor of type $\binom{M}{N}$ gives the components of a tensor of type $\binom{M-1}{N-1}$. For example, contraction of the components $T^a_{\ b}$ of the second order tensor $\mathbf{T}$ produces the scalar $V = T^a_{\ a}$. Note that contraction is defined only between an upper and lower index.

2. If two tensors of the same type have equal components in a given basis, then they have equal components in all bases. Hence, they are identical. For example, if the components of the third rank tensors $\mathbf{A}$ and $\mathbf{B}$ satisfy $A^{ab}_c = B^{ab}_c$ in one set of coordinates, for all possible combinations of the indices a, b, c, then this equality holds in all sets of coordinates and so the tensors are identical. In particular, if all components of a tensor vanish in one set of coordinates, they vanish in all coordinates. This property is easy to prove through use of the transformation properties of tensors established in Section 20.9.

3. It follows from the previous properties that if an equation consists of tensors combined only by the permissible tensor operations, and if the equation is true in one basis, then it is true in any basis. If the equations involve covariant derivatives, then the equations remain form invariant under changes in coordinates.

21.2 COVARIANT DIFFERENTIATION

Differentiation of tensors on curved manifolds is a logical extension of the usual differentiation on flat manifolds. The name given to such operations is *covariant differentiation*. As mentioned earlier, covariant differentiation of a tensor of type $\binom{M}{N}$ gives the components to a tensor of type $\binom{M}{N+1}$. This result is exemplified below. In determining the form of the covariant derivatives of tensorial quantities, it is necessary to be aware that the basis vectors, basis one-forms, and coordinate representation of the tensor can each have nonzero partial derivatives.

21.2.1 The gradient one-form

Recall that a scalar function V can be thought of as a tensor of type $\binom{0}{0}$. As such, it maintains the same numerical value in any coordinate representation since it needs no basis vectors to represent it. When taking its partial derivative

$$\partial_a V = V_{,a} \tag{21.2}$$

the result is a type $\binom{0}{1}$ tensor, as follows since

$$\partial_{\bar{a}} V = \Lambda^a{}_{\bar{a}} V_{,a} \tag{21.3}$$

defines the transformation rule of a type $\binom{0}{1}$ tensor. This result could be anticipated from the results in Section 20.10 where the transformation rules of the partial derivative were established. Written as a one-form, the partial derivatives of a scalar are components to the *gradient one-form*

$$\nabla V \equiv (\tilde{e}^a \, \partial_a) \, V, \tag{21.4}$$

where $\tilde{e}^a$ define a basis of one-forms.

21.2.2 Covariant derivative of a vector

Consider a vector field living on the manifold $\vec{F}(\xi^a)$. The partial derivative of this vector is given by

$$\partial_b \vec{F} = \partial_b (F^a \, \vec{e}_a). \tag{21.5}$$

To determine the coordinate components of the right-hand side, account must be taken of the generally nontrivial dependence of the basis vectors on the coordinates ξ^a. As a result,

$$\begin{aligned} \partial_b \vec{F} &= (\partial_b \, F^a) \, \vec{e}_a + F^a \, \partial_b \, \vec{e}_a \\ &= F^a{}_{,b} \, \vec{e}_a + F^a \, \Gamma^c_{ba} \, \vec{e}_c \\ &= \left(F^a{}_{,b} + F^c \, \Gamma^a_{bc} \right) \vec{e}_a, \end{aligned} \tag{21.6}$$

where

$$\partial_b \, \vec{e}_a = \Gamma^c_{ba} \, \vec{e}_c \tag{21.7}$$

defines the Christoffel symbols Γ^c_{ba}. The Christoffel symbols are the coordinate representations of the partial derivatives of the basis vectors. If the manifold is the flat Euclidean space, and the basis vectors are the Cartesian unit vectors, then all of the Christoffel symbols vanish. However, for a Euclidean space described by non-Cartesian basis vectors, the Christoffel symbols do not vanish. This result indicates that the Christoffel symbols are *not* components to a tensor, since a tensor vanishing in one set of coordinates vanishes for all sets of coordinates (Section 21.1).

Contracting $\partial_b \vec{F}$ with the basis one-form $\tilde{e}^b$ yields the $\binom{1}{1}$ tensor

$$\tilde{e}^b\, \partial_b \vec{F} = \tilde{e}^b \left(F^a{}_{,b} + F^c\, \Gamma^a_{bc} \right) \vec{e}_a. \tag{21.8}$$

It is clear that this is a $\binom{1}{1}$ tensor because of the explicit presence of the basis one-form and basis vector. This tensor is known as the covariant derivative of the vector $\vec{F}$. It is typically written in one of the following manners

$$\tilde{e}^b\, \partial_b \vec{F} = (\nabla_b \vec{F})^a\, \vec{e}_a \tag{21.9}$$
$$= F^a{}_{;b}\, \tilde{e}^b\, \vec{e}_a$$
$$= (F^a{}_{,b} + \Gamma^a_{bc}\, F^c)\, \tilde{e}^b\, \vec{e}_a. \tag{21.10}$$

The semicolon notation is used in the following. It generalizes the comma notation used for the partial derivative. When all of the Christoffel symbols vanish, the components of the covariant derivative reduce to those of the partial derivative. Additionally, the results from Section 21.2.1 show that the covariant derivative of a scalar is identical to the partial derivative of that scalar,

$$V_{;a} = V_{,a}. \tag{21.11}$$

This result is obvious, since the scalar field requires no basis vectors for its representation, and so there are no resulting Christoffel symbols.

21.2.3 Covariant derivative of a one-form

The results for the covariant derivative of a scalar and vector provide tools sufficient for determining the covariant derivatives of any arbitrary tensor. For example, the covariant derivative of a one-form can be found by considering the result of taking the covariant derivative on the inner product of the one-form and a vector

$$(E_a\, F^a)_{;b} = (E_a\, F^a)_{,b} \tag{21.12}$$

where equation (21.11) was used since $E_a\, F^a$ is a scalar. Expanding the partial derivative yields

$$(E_a\, F^a)_{;b} = F^a\, \partial_b\, E_a + E_a\, \partial_b\, F^a$$
$$= F^a\, \partial_b\, E_a + E_a (F^a{}_{;b} - \Gamma^a_{bc} F^c)$$
$$= F^a (\partial_b\, E_a - \Gamma^c_{ba} E_c) + E_a\, F^a{}_{;b}$$
$$\equiv F^a\, E_{a;b} + E_a\, F^a{}_{;b}. \tag{21.13}$$

The last equality defines the components to the covariant derivative of the one form

$$E_{a;b} = \partial_b\, E_a - \Gamma^c_{ba} E_c, \tag{21.14}$$

which leads to

$$\partial_b \tilde{E} = E_{a;b}\, \tilde{e}^a. \tag{21.15}$$

These results establish that the covariant derivative of a one-form results in a tensor of type $\binom{0}{2}$.

21.2.4 Covariant derivative of the metric

As discussed in Section 21.1, one of the central reasons for employing tensor analysis in physics is that when the tensorial properties of an expression are established in one coordinate system, the same expression is also valid in an arbitrary coordinate system. Notably, when written in Cartesian coordinates, the covariant derivative of the metric for Euclidean space vanishes,

$$g_{ab;c} = 0, \tag{21.16}$$

simply because the metric is the unit tensor δ_{ab}, and all the Christoffel symbols vanish. Previous results established the tensorial nature of the covariant derivative. Hence, $g_{ab;c} = 0$ is a valid result for *all* coordinates. This result is often called the *metricity* condition. It represents a self-consistency condition required for the manifolds considered in this book.

21.3 COVARIANT DERIVATIVE OF A SECOND ORDER TENSOR

We are concerned in this book with manifolds that have zero *torsion*. For these purposes, a zero torsion manifold makes the Christoffel symbol symmetric on its lower two labels. Given this property, compute the covariant derivative of an arbitrary tensor on torsionless manifolds. To illustrate the general ideas, consider the covariant derivative of a second order tensor. To start, note that the covariant derivative of the following scalar quantity is

$$\begin{aligned}
(T_{ab}\, F^a\, F^b)_{;c} &= (T_{ab}\, F^a\, F^b)_{,c} \\
&= T_{ab,c}\, F^a\, F^b + T_{ab}\, (F^a\, F^b)_{,c} \\
&= T_{ab,c}\, F^a\, F^b + T_{ab}\, (F^a_{,c}\, F^b + F^a\, F^b_{,c}) \\
&= T_{ab,c}\, F^a\, F^b + T_{ab}\, [(F^a_{;c} - \Gamma^a_{cd}\, F^d)\, F^b + F^a\, (F^b_{;c} - \Gamma^b_{cd}\, F^d)] \\
&= F^a\, F^b\, (T_{ab,c} - T_{db}\, \Gamma^d_{ca} - T_{ad}\, \Gamma^d_{cd}) + T_{ab}\, (F^a\, F^b)_{;c} \\
&= T_{ab\,;c}\, (F^a\, F^b) + T_{ab}\, (F^a\, F^b)_{;c}
\end{aligned} \tag{21.17}$$

where

$$T_{ab\,;c} = T_{ab,c} - \Gamma^d_{ca}\, T_{db} - \Gamma^d_{cb}\, T_{ad} \tag{21.18}$$

defines the components to the covariant derivative of the second order tensor T_{ab}. To reach this result required use of the symmetry $\Gamma^a_{bc} = \Gamma^a_{cb}$. This covariant derivative results in a tensor of type $\binom{0}{3}$. Similar considerations yield the expressions

$$T^{ab}_{;c} = T^{ab}_{,c} + \Gamma^a_{cd}\, T^{db} + \Gamma^b_{cd}\, T^{ad}, \tag{21.19}$$

which results in a tensor of type $\binom{2}{1}$, as well as

$$T^a_{b\,;c} = T^a_{b\,,c} + \Gamma^a_{cd}\, T^d_b - \Gamma^d_{bc}\, T^a_d, \tag{21.20}$$

which results in a tensor of type $\binom{1}{2}$.

21.4 CHRISTOFFEL SYMBOLS IN TERMS OF THE METRIC

The vanishing covariant derivative of the metric, equation (21.16), combined with equation (21.18) leads to the identity

$$\begin{aligned} 0 &= g_{ab;c} \\ &= g_{ab,c} - \Gamma^d_{ca}\, g_{db} - \Gamma^d_{cb}\, g_{ad} \end{aligned} \tag{21.21}$$

which can be solved for the Christoffel symbols

$$\Gamma^c_{ab} = \frac{1}{2} g^{cd}(g_{da,b} + g_{db,a} - g_{ab,d}). \tag{21.22}$$

This expression exhibits the symmetry property of the lower two indices on the Christoffel symbols

$$\Gamma^c_{ab} = \Gamma^c_{ba}. \tag{21.23}$$

21.5 COVARIANT DIVERGENCE OF A VECTOR

There are many places in fluid mechanics where one needs to compute the covariant divergence of a vector. For example, the covariant divergence of the velocity field is proportional to the time rate of change of the density following a fluid parcel. The covariant convergence of a tracer flux contributes to the time tendency of the tracer. In general, the covariant divergence of the components to a vector results in a scalar

$$F^a_{;a} = F^a_{,a} + \Gamma^a_{ab} F^b. \tag{21.24}$$

Expression (21.22) for the Christoffel symbols yields for the contraction

$$\begin{aligned} \Gamma^a_{ab} &= \frac{1}{2} g^{ad}(g_{da,b} + g_{db,a} - g_{ab,d}) \\ &= \frac{1}{2} g^{ad} g_{ad,b} \end{aligned} \tag{21.25}$$

where symmetry of the metric tensor and its inverse was used.

The above expression for the covariant divergence can be put in a more convenient form. For this purpose, employ the following relation, which follows for any symmetric positive definite matrix such as the metric tensor,

$$\begin{aligned} \det(A) &= e^{\ln \det(A)} \\ &= e^{\ln(\Pi_i \Lambda_i)} \\ &= e^{\Sigma_i \ln \Lambda_i} \\ &= e^{\mathrm{Tr}(\ln A)}. \end{aligned} \tag{21.26}$$

The first equality is a simple identity. The second equality relates the determinant of a matrix to the product of its eigenvalues. The third equality is a simple identity. The fourth equality relates the sum of the eigenvalues of a matrix to the trace of this matrix. Each of these identities is trivial to verify using a set of coordinates in which the matrix is diagonal. For any symmetric and positive definite matrix, such a set of coordinates always exists. This result gives

$$\begin{aligned}\partial_c \ln \det(A) &= \partial_c[\mathrm{Tr}(\ln A)] \\ &= \mathrm{Tr}(\partial_c \ln A) \\ &= \mathrm{Tr}(A^{-1}\partial_c A).\end{aligned} \tag{21.27}$$

With A now set equal to the metric and $\mathcal{G} = \det(g_{ab})$, this result yields

$$\partial_c \ln \mathcal{G} = g^{ab}\, g_{ab,c} \tag{21.28}$$

which in turn yields for the contracted Christoffel symbol

$$\Gamma^a_{ac} = \partial_c \ln \sqrt{\mathcal{G}}. \tag{21.29}$$

This result brings the covariant divergence of a vector to the form

$$\begin{aligned}F^a_{;a} &= F^a_{,a} + F^a\, \partial_a \ln \sqrt{\mathcal{G}} \\ &= \frac{1}{\sqrt{\mathcal{G}}} \left(\sqrt{\mathcal{G}}\, F^a\right)_{,a}.\end{aligned} \tag{21.30}$$

This is a very convenient result since it requires the use only of partial derivatives in the coordinate system chosen. All coordinate dependent properties are captured by $\sqrt{\mathcal{G}}$, which is trivial to compute once the metric tensor is specified.

21.6 COVARIANT DIVERGENCE OF A SECOND ORDER TENSOR

In many applications, it is necessary to compute the covariant divergence of the components to a second order tensor T^{ab}. An important example includes the covariant divergence of the symmetric stress tensor, which typically parameterizes subgrid scale transfer of momentum (Chapter 17). The results from Section 21.3 give

$$T^{ab}_{;b} = T^{ab}_{,b} + \Gamma^a_{bd}\, T^{db} + \Gamma^b_{bd}\, T^{ad}. \tag{21.31}$$

To simplify this expression, it is useful to split the tensor components T^{ab} into symmetric and antisymmetric parts

$$S^{ab} = (T^{ab} + T^{ba})/2 \tag{21.32}$$

$$A^{ab} = (T^{ab} - T^{ba})/2. \tag{21.33}$$

Using equation (21.29), one finds the covariant divergence of the symmetric components to the tensor to be

$$S^{ab}_{;b} = \frac{1}{\sqrt{\mathcal{G}}} \left(\sqrt{\mathcal{G}}\, S^{ab}\right)_{,b} + \Gamma^a_{bd}\, S^{bd}. \tag{21.34}$$

The term $\Gamma^a_{bd}\, S^{bd}$ does not generally vanish since both Γ^a_{bd} and S^{bd} are symmetric under interchange of b, d. For the antisymmetric components, however, the analogous term does vanish, hence leading to the covariant divergence of an antisymmetric tensor

$$\begin{aligned}A^{ab}_{;b} &= A^{ab}_{,b} + \Gamma^b_{bd}\, A^{ad} \\ &= A^{ab}_{,b} + (\ln \sqrt{\mathcal{G}})_{,b}\, A^{ab} \\ &= \frac{1}{\sqrt{\mathcal{G}}} \left(\sqrt{\mathcal{G}}\, A^{ab}\right)_{,b}.\end{aligned} \tag{21.35}$$

This relation is analogous to the covariant divergence of a vector given by equation (21.30). In particular, the components

$$F^a \equiv A^{ab}_{\ ;b} \tag{21.36}$$

are those of a divergence-free vector field, since

$$\begin{aligned} F^a_{\ ;a} &= A^{ab}_{\ ;b;a} \\ &= 0. \end{aligned} \tag{21.37}$$

This result follows from antisymmetry of the components A^{ab} under interchange of a, b and symmetry of the double covariant derivative under the same interchange.

21.7 COVARIANT LAPLACIAN OF A SCALAR

Equation (21.11) showed that the covariant derivative of a scalar is the same as the partial derivative $V_{;a} = V_{,a}$. Recall also that $V_{,a}$ form the components to the gradient one-form $d\tilde{V} \equiv V_{,a}\, \tilde{e}^a$ (equation (21.4)). Raising the index on these components yields the components to a vector $g^{ab}\, V_{,a}$. Taking the covariant divergence of these components then yields a scalar field, which defines the covariant Laplacian of the scalar function V

$$(g^{ab}\, V_{,a})_{;b} = \frac{1}{\sqrt{\mathcal{G}}} \left(\sqrt{\mathcal{G}}\, g^{ab}\, V_{,a} \right)_{,b} \tag{21.38}$$

where the relation (21.30) for the covariant divergence of a vector was employed.

21.8 COVARIANT CURL OF A VECTOR

The scaled Levi-Civita symbol $\varepsilon_{ab} = \mathcal{G}^{1/2}\, \epsilon_{ab}$ from Section 20.12 is useful for generalizing the curl operation from Cartesian coordinates in Euclidean space to arbitrary coordinates on a curved manifold. Consequently, let

$$(\operatorname{curl} \vec{F}) = \varepsilon^{abc}\, F_{c;b}\, \vec{e}_a \tag{21.39}$$

define the covariant curl vector. To simplify the curl, recall the expression (21.14) for the covariant derivative $F_{c;b} = F_{c,b} - \Gamma^a_{cb}\, F_a$. Conveniently, the contraction $\varepsilon^{abc}\, \Gamma^a_{cb}$ vanishes identically since $\varepsilon^{abc} = -\varepsilon^{acb}$, whereas $\Gamma^a_{cb} = \Gamma^a_{bc}$. Hence, one is left with the general expression for the covariant curl, which involves just partial derivatives

$$(\operatorname{curl} \vec{F}) = \varepsilon^{abc}\, F_{c,b}\, \vec{e}_a. \tag{21.40}$$

21.9 COVARIANT LAPLACIAN OF A VECTOR

Recall from Section 21.2.2 that the covariant derivative of a vector is given by

$$\partial_a \vec{F} = F^b_{\ ;a}\, \vec{e}_b, \tag{21.41}$$

where $F^b_{;a} = F^b_{,a} + \Gamma^b_{ad} F^d$ is the covariant derivative of the components to the vector. The covariant Laplacian of a vector is given by

$$\begin{aligned}
\partial^a \partial_a \vec{F} &= \partial^a \left(F^b_{;a} \vec{e}_b \right) \\
&= g^{ac} \partial_c \left(F^b_{;a} \vec{e}_b \right) \\
&= g^{ac} \left(F^d_{;a,c} + \Gamma^d_{cb} F^b_{;a} \right) \vec{e}_d \\
&= g^{ac} F^d_{;a;c} \vec{e}_d. \qquad (21.42)
\end{aligned}$$

In general, the components $g^{ac} F^d_{;a;c}$ of the Laplacian are tedious to compute. Appendix 2 of Batchelor (1967) provides useful expressions for various orthogonal coordinates.

21.10 INTEGRAL THEOREMS

The integral theorems from Cartesian vector analysis transform in a straightforward manner to arbitrary coordinates in arbitrary smooth spaces. The easiest way to prove the theorems in general is to invoke the ideas from Section 21.1, in which the integral theorems are written in a tensorially correct manner and then partial derivatives are changed to covariant derivatives. The divergence theorem provides a useful example of the approach, and Section 3.3 provided a discussion. In summary, the volume integral of the divergence of a vector is transformed to the surface integral

$$\begin{aligned}
\int \mathrm{d}V \, F^a_{;a} &= \int \mathrm{d}^3\xi \, \sqrt{\mathcal{G}} \, \frac{1}{\sqrt{\mathcal{G}}} \left(\sqrt{\mathcal{G}} \, F^a \right)_{,a} \\
&= \int \mathrm{d}A_{(\hat{\mathbf{n}})} \, F^a \, \hat{\mathbf{n}}_a. \qquad (21.43)
\end{aligned}$$

Results from Section 20.11 were used to write the invariant volume element. Additionally, the covariant divergence was written as given in Section 21.5, $\hat{\mathbf{n}}_a$ are components to the outward normal one-form $\hat{\mathbf{n}}$ for the boundary surface, $\mathrm{d}A_{(\hat{\mathbf{n}})} = \mathrm{d}^2 w \, \sqrt{\mathcal{G}}$ represents the invariant surface area element (Section 20.13.2), and the Cartesian form of Gauss's law has been invoked. Other integral theorems are generalized in a similar manner.

21.11 ORTHOGONAL CURVILINEAR COORDINATES

The case of orthogonal curvilinear coordinates occurs frequently in applications. In particular, many ocean models are written using orthogonal curvilinear coordinates for the horizontal directions. Consequently, it is useful to specialize some of the previous discussions to this case. We focus here on the case occuring in z-models, with the vertical also orthogonal to the horizontal. However, the discussion can be generalized simply to the case with generalized vertical coordinates (Chapter 6), since the focus here is on the horizontal directions. Many of the following results can be found, among other places, at the end of chapter 1 in Morse and Feshbach (1953), various sections in Aris (1962), and Appendix 2 of Batchelor (1967).

Orthogonal coordinates are characterized by a diagonal metric tensor. Hence, the most general form of the metric written in orthogonal coordinates takes the form

$$g_{ab} = \text{diag}\left((h_1)^2, (h_2)^2, (h_3)^2\right). \tag{21.44}$$

This metric is associated with the infinitesimal line element

$$\begin{aligned} ds^2 &= g_{ab}\, d\xi^a\, d\xi^b \\ &= (h_1\, d\xi^1)^2 + (h_2\, d\xi^2)^2 + (h_3\, d\xi^3)^2. \end{aligned} \tag{21.45}$$

It follows that the components to the inverse metric are given by

$$g^{ab} = \text{diag}\left((h_1)^{-2}, (h_2)^{-2}, (h_3)^{-2}\right). \tag{21.46}$$

For some purposes, the single label on the metric functions h_a can be thought of as a single covariant label since the squares of the h_a yield components to the second order covariant metric tensor. With the metric as defined, the square root of its determinant is given by

$$\sqrt{\mathcal{G}} = h_1\, h_2\, h_3. \tag{21.47}$$

Using orthogonal coordinates, the basis vectors take the form

$$\vec{e}_1 = h_1\, \hat{\mathbf{e}}_{(1)} \tag{21.48}$$

$$\vec{e}_2 = h_2\, \hat{\mathbf{e}}_{(2)} \tag{21.49}$$

$$\vec{e}_3 = h_3\, \hat{\mathbf{e}}_{(3)}, \tag{21.50}$$

where $\hat{\mathbf{e}}_{(a)}$ are dimensionless orthonormal *unit directions*. The index on the unit directions is enclosed in parentheses to advertise the fact that this label is not tensorial; i.e., the unit directions do not transform as tensors. Rather, the functions h_a carry the tensorial properties of the basis vectors $\vec{e}_a$. Note that the basis vectors are determined by writing the metric tensor in the form

$$g_{ab} = \vec{e}_a \cdot \vec{e}_b. \tag{21.51}$$

The corresponding basis one-forms are

$$\tilde{e}^1 = (h_1)^{-1}\, \hat{\mathbf{e}}^{(1)} \tag{21.52}$$

$$\tilde{e}^2 = (h_2)^{-1}\, \hat{\mathbf{e}}^{(2)} \tag{21.53}$$

$$\tilde{e}^3 = (h_3)^{-1}\, \hat{\mathbf{e}}^{(3)}, \tag{21.54}$$

where the unit directions are identical whether their indices are raised or lowered: $\hat{\mathbf{e}}^{(a)} = \hat{\mathbf{e}}_{(a)}$.

Spherical coordinates (λ, ϕ, r) on Euclidean $\mathcal{R}^3$ provides the canonical example of a nontrivial set of orthogonal curvilinear coordinates

$$h_1 = r \cos\phi \tag{21.55}$$
$$h_2 = r \tag{21.56}$$
$$h_3 = 1 \tag{21.57}$$
$$\sqrt{\mathcal{G}} = r^2 \cos\phi \tag{21.58}$$
$$\vec{e}_1 = r \cos\phi \, \hat{\boldsymbol{\lambda}} \tag{21.59}$$
$$\vec{e}_2 = r \, \hat{\boldsymbol{\phi}} \tag{21.60}$$
$$\vec{e}_3 = \hat{\mathbf{r}} \tag{21.61}$$
$$\vec{e}^1 = (r \cos\phi)^{-1} \, \hat{\boldsymbol{\lambda}} \tag{21.62}$$
$$\vec{e}^2 = r^{-1} \, \hat{\boldsymbol{\phi}} \tag{21.63}$$
$$\vec{e}^3 = \hat{\mathbf{r}} \tag{21.64}$$

where the unit directions point in the radial, eastward, and northward directions, respectively (see Figure 20.1). With Cartesian coordinates for $\mathcal{R}^3$, all the h_a are trivially set to unity.

21.11.1 Physical components of tensors

In terms of a locally orthogonal basis of unit directions, an arbitrary vector takes the form

$$\begin{aligned}\vec{F} &= F^a \, \vec{e}_a \\ &= (F^a \, h_a) \, \hat{\mathbf{e}}_{(a)} \\ &\equiv \mathcal{F}^{(a)} \, \hat{\mathbf{e}}_{(a)}.\end{aligned} \tag{21.65}$$

Note that there is no summation performed on the $F^a \, h_a$ product. Since the unit directions $\hat{\mathbf{e}}_{(a)}$ are dimensionless, the terms $\mathcal{F}^{(a)}$ each have the same physical dimensions. Consequently, $\mathcal{F}^{(a)}$ are termed the *physical components* to the vector $\vec{F}$ (see, e.g., Section 7.41 of Aris, 1962). The physical components, due to their dimensional consistency, are often the forms for which tensorial quantities appear in physical theories. They are also the forms most likely to appear in numerical models. As with the unit vectors, the index on the physical components is enclosed in parentheses since physical components transform in a manner distinct from tensor components

$$\begin{aligned}\mathcal{F}^{(\overline{a})} &= F^{\overline{a}} \, h_{\overline{a}} \\ &= \Lambda^{\overline{a}}{}_{a} \, F^a \, h_{\overline{a}} \\ &= \frac{h_{\overline{a}}}{h_a} \, \Lambda^{\overline{a}}{}_{a} \, \mathcal{F}^{(a)}.\end{aligned} \tag{21.66}$$

The $h_{\overline{a}}/h_a$ factor distinguishes the transformation properties of physical components from true tensor components. Instead of the parentheses, it is useful to use the Cartesian symbols x, y, z to denote physical components

$$\mathcal{F}^{(a)} = (\mathcal{F}^x, \mathcal{F}^y, \mathcal{F}^z). \tag{21.67}$$

Expressed in terms of the orthogonal coordinates, an arbitrary one-form is written

$$\begin{aligned}\tilde{F} &= F_a \,\tilde{e}^a \\ &= (F_a/h_a)\,\hat{\mathbf{e}}^{(a)} \\ &= \mathcal{F}_{(a)}\,\hat{\mathbf{e}}^{(a)}.\end{aligned} \tag{21.68}$$

An important one-form operator is the gradient operator, to be considered in Section 21.11.3. It is useful to introduce the physical components to this operator via the use of physical components to the partial derivative operator

$$\begin{aligned}(\partial_x, \partial_y, \partial_z) &= (\partial_{(1)}, \partial_{(2)}, \partial_{(3)}) \\ &= ((h_1)^{-1}\,\partial_1, (h_2)^{-1}\,\partial_2, (h_3)^{-1}\,\partial_3).\end{aligned} \tag{21.69}$$

As for vectors, the fields $\mathcal{F}_{(a)}$ each have the same physical dimensions, and they are called the physical components to the one-form $\tilde{F}$. As with the physical components to the vectors, the physical components to the one-forms do not transform as tensors.

The above results for a vector and one-form generalize to an arbitrary tensor. For example, if one expresses a second order tensor in terms of its contravariant components, these components take the following form when using orthogonal coordinates.

$$\begin{aligned}\mathbf{T} &= T^{ab}\,\vec{e}_a\,\vec{e}_b \\ &= (h_a\,T^{ab}\,h_b)\,\hat{\mathbf{e}}_{(a)}\,\hat{\mathbf{e}}_{(b)} \\ &= \mathcal{T}^{(ab)}\,\hat{\mathbf{e}}_{(a)}\,\hat{\mathbf{e}}_{(b)}.\end{aligned} \tag{21.70}$$

Without any implied sums, the components

$$\mathcal{T}^{(ab)} = h_a\,T^{ab}\,h_b \tag{21.71}$$

are identified as the physical components of the contravariant tensor T^{ab}. Each of the physical components $\mathcal{T}^{(ab)}$ has the same dimensions. Likewise, the same second order tensor written in terms of its mixed second order components can be expressed in terms of the orthogonal coordinates as

$$\begin{aligned}\mathbf{T} &= T^a_b\,\vec{e}_a\,\tilde{e}^b \\ &= (h_a/h_b)\,T^a_b\,\hat{\mathbf{e}}_{(a)}\,\hat{\mathbf{e}}^{(b)} \\ &= \mathcal{T}^{(a)}_{(b)}\,\hat{\mathbf{e}}_{(a)}\,\hat{\mathbf{e}}^{(b)},\end{aligned} \tag{21.72}$$

where

$$\mathcal{T}^{(a)}_{(b)} = (h_a/h_b)\,T^a_b \tag{21.73}$$

are the physical components, each of which has the same dimensions, and again where there is no implied sum.

21.11.2 Eliminating the stretching functions

Orthogonal coordinates afford an added feature that is quite useful to exploit when discretizing the equations for an ocean model. Since the generalized coordinates are orthogonal, a coordinate increment in one direction is independent of the other direction. For example,

$$\partial_1 (\mathrm{d}\xi^2) = \partial_2 (\mathrm{d}\xi^1) = 0. \tag{21.74}$$

In particular, when computing the horizontal rates of deformation for a fluid parcel in Section 17.7.1, we made use of the following result for orthogonal coordinates

$$\begin{aligned} h_2 (u/h_2)_{,x} &= h_2 \, \mathrm{d}\xi^2 \, (u/h_2 \, \mathrm{d}\xi^2)_{,x} \\ &= \mathrm{d}y \, (u/\mathrm{d}y)_{,x}. \end{aligned} \tag{21.75}$$

This and similar results for other terms in the equations of motion provide the opportunity to write all formulae in terms of the physical displacements, and so to eliminate stretching functions h_1 and h_2. This property is convenient since the physical displacements are ultimately what are of physical interest, and what are of concern when discretizing the equations in an ocean model.

21.11.3 Gradient one-form

In orthogonal coordinates, the gradient one-form acting on a scalar is given by

$$\begin{aligned} \nabla V &= (\tilde{e}^a \, \partial_a) \, V \\ &= (\hat{\mathbf{e}}^{(a)} \, \partial_{(a)}) \, V. \end{aligned} \tag{21.76}$$

The second equality introduces the physical components to the gradient one-form operator

$$\nabla = \hat{\mathbf{e}}^{(a)} \, \partial_{(a)}, \tag{21.77}$$

with $\hat{\mathbf{e}}^{(a)}$ unit directions, and the physical components to the partial derivatives are

$$\partial_{(a)} = (h_{(a)})^{-1} \, \partial_a, \tag{21.78}$$

with no implied sum on the right-hand side. Often the physical components of the partial derivative are written $\partial_{(a)} = (\partial_x, \partial_y, \partial_z)$. Although suggestive, this notation should not lead one to consider the space to be flat. Indeed, the nonflat nature of the sphere manifests itself in the nonvanishing commutator

$$\begin{aligned} \left[\partial_x, \partial_y\right] &= \partial_x \, \partial_y - \partial_y \, \partial_x \\ &= (\partial_x \ln \mathrm{d}y) \, \partial_y - (\partial_y \ln \mathrm{d}x) \, \partial_x. \end{aligned} \tag{21.79}$$

It is useful to touch base with spherical coordinates, where the gradient of a scalar is

$$\begin{aligned} \nabla V &= V_{,\lambda} \, (R \cos\phi)^{-1} \, \hat{\boldsymbol{\lambda}} + V_{,\phi} \, R^{-1} \, \hat{\boldsymbol{\phi}} + \hat{\mathbf{z}} \, V_{,z} \\ &= V_{,x} \, \hat{\boldsymbol{\lambda}} + V_{,y} \, \hat{\boldsymbol{\phi}} + \hat{\mathbf{z}} \, V_{,z} \end{aligned} \tag{21.80}$$

and the gradient operator is

$$\nabla = \hat{\boldsymbol{\lambda}} \, (R \cos\phi)^{-1} \, \partial_\lambda + \hat{\boldsymbol{\phi}} \, R^{-1} \, \partial_\phi + \hat{\mathbf{z}} \, \partial_z. \tag{21.81}$$

21.11.4 Covariant divergence of a vector

From Section 21.5, the expression for the covariant divergence of a vector

$$F^a_{;a} = \frac{1}{\sqrt{\mathcal{G}}} \left(\sqrt{\mathcal{G}}\, F^a\right)_{,a} \tag{21.82}$$

takes the form in orthogonal coordinates

$$\begin{aligned} F^a_{;a} &= \frac{1}{h_1\,h_2\,h_3} \,(h_1\,h_2\,h_3\,F^a)_{,a} \\ &= \frac{1}{h_1\,h_2\,h_3} \left\{ \left(h_2\,h_3\,(h_1\,F^1)\right)_{,1} + \left(h_1\,h_3\,(h_2\,F^2)\right)_{,2} + \left(h_1\,h_2\,(h_3\,F^3)\right)_{,3} \right\} \\ &= \frac{1}{h_1\,h_2\,h_3} \left\{ \left(h_2\,h_3\,\mathcal{F}^{(1)}\right)_{,1} + \left(h_1\,h_3\,\mathcal{F}^{(2)}\right)_{,2} + \left(h_1\,h_2\,\mathcal{F}^{(3)}\right)_{,3} \right\}. \end{aligned} \tag{21.83}$$

Eliminating the stretching functions in favor of the physical displacements, according to the properties noted in Section 21.11.2, leads to

$$\begin{aligned} F^a_{;a} =& \nabla \cdot \mathbf{F} \\ =& (\mathrm{d}y\,\mathrm{d}z)^{-1}\,(\mathrm{d}y\,\mathrm{d}z\,\mathcal{F}^{(x)})_{,x} + (\mathrm{d}x\,\mathrm{d}z)^{-1}\,(\mathrm{d}x\,\mathrm{d}z\,\mathcal{F}^{(y)})_{,y} \\ &+ (\mathrm{d}x\,\mathrm{d}y)^{-1}\,(\mathrm{d}x\,\mathrm{d}y\,\mathcal{F}^{(z)})_{,z} \end{aligned} \tag{21.84}$$

where the first equality introduced the common notation used to represent the divergence of a vector. For spherical coordinates (λ, ϕ, r), with

$$(h_1, h_2, h_3) = (r \cos\phi, r, 1), \tag{21.85}$$

the divergence becomes

$$\nabla \cdot \mathbf{F} = \frac{1}{r^2\,\cos\phi} \left((r^2\,\cos\phi\,F^{\lambda})_{,\lambda} + (r^2\,\cos\phi\,F^{\phi})_{,\phi} + (r^2\,\cos\phi\,F^{r})_{,r} \right) \tag{21.86}$$

$$= (r\,\cos\phi)^{-1} \left((\mathcal{F}^{\lambda})_{,\lambda} + (\mathcal{F}^{\phi}\,\cos\phi)_{,\phi} \right) + r^{-2}\,(r^2\,\mathcal{F}^{r})_{,r}. \tag{21.87}$$

Note that with general vertical coordinates (Chapter 6), or even with partial cells (Adcroft et al., 1997; Pacanowski and Gnanadesikan, 1998) used in many z-coordinate ocean models (Section 6.2.1), the vertical distance $\mathrm{d}z$ is a function of both the vertical and horizontal position. Hence, $\mathrm{d}z$ cannot be removed from the horizontal partial derivative in equation (21.84). This point is brought out in Section 16.7 when discussing the divergence of tracer fluxes.

21.11.5 Covariant divergence of an antisymmetric tensor

From Section 21.6, the covariant divergence of an antisymmetric tensor is given by

$$\begin{aligned} A^{ab}_{;b} &= \frac{1}{\sqrt{\mathcal{G}}} \left(\sqrt{\mathcal{G}}\,A^{ab}\right)_{,b} \\ &= \frac{1}{h_1\,h_2\,h_3} \left(h_1\,h_2\,h_3\,A^{ab}\right)_{,b}. \end{aligned} \tag{21.88}$$

In particular,

$$\begin{aligned} F^1 &\equiv A^{1b}_{;b} \\ &= \frac{1}{h_1\,h_2\,h_3} \left[(h_1\,h_2\,h_3\,A^{12})_{,2} + (h_1\,h_2\,h_3\,A^{13})_{,3} \right] \\ &= \frac{1}{h_1\,h_2\,h_3} \left[(h_3\,\mathcal{A}^{(12)})_{,2} + (h_2\,\mathcal{A}^{(13)})_{,3} \right], \end{aligned} \tag{21.89}$$

and likewise

$$F^2 \equiv A^{2b}_{;b} = \frac{1}{h_1\, h_2\, h_3}\left[(h_3\,\mathcal{A}^{(21)})_{,1} + (h_1\,\mathcal{A}^{(23)})_{,3}\right] \tag{21.90}$$

$$F^3 \equiv A^{3b}_{;b} = \frac{1}{h_1\, h_2\, h_3}\left[(h_2\,\mathcal{A}^{(31)})_{,1} + (h_1\,\mathcal{A}^{(32)})_{,2}\right]. \tag{21.91}$$

The physical components of the vector field F^a are therefore given by

$$\mathcal{F}^{(1)} = h_1\, F^1 = \frac{1}{h_2\, h_3}\left[(h_3\,\mathcal{A}^{(12)})_{,2} + (h_2\,\mathcal{A}^{(13)})_{,3}\right] \tag{21.92}$$

$$\mathcal{F}^{(2)} = h_2\, F^2 = \frac{1}{h_1\, h_3}\left[(h_3\,\mathcal{A}^{(21)})_{,1} + (h_1\,\mathcal{A}^{(23)})_{,3}\right] \tag{21.93}$$

$$\mathcal{F}^{(3)} = h_3\, F^3 = \frac{1}{h_1\, h_2}\left[(h_2\,\mathcal{A}^{(31)})_{,1} + (h_1\,\mathcal{A}^{(32)})_{,2}\right]. \tag{21.94}$$

Eliminating the stretching functions leads to

$$\mathcal{F}^{(x)} = (\mathrm{d}z)^{-1}\,(\mathrm{d}z\,\mathcal{A}^{(12)})_{,y} + (\mathrm{d}y)^{-1}\,(\mathrm{d}y\,\mathcal{A}^{(13)})_{,z} \tag{21.95}$$

$$\mathcal{F}^{(y)} = (\mathrm{d}z)^{-1}\,(\mathrm{d}z\,\mathcal{A}^{(21)})_{,x} + (\mathrm{d}x)^{-1}\,(\mathrm{d}x\,\mathcal{A}^{(23)})_{,z} \tag{21.96}$$

$$\mathcal{F}^{(z)} = (\mathrm{d}y)^{-1}\,(\mathrm{d}z\,\mathcal{A}^{(31)})_{,x} + (\mathrm{d}x)^{-1}\,(\mathrm{d}x\,\mathcal{A}^{(32)})_{,y}. \tag{21.97}$$

In terms of spherical coordinates (λ, ϕ, r), the physical components take the form

$$\mathcal{F}^{(\lambda)} = \frac{1}{r}\left[(r\,\mathcal{A}^{(\lambda r)})_{,r} + (\mathcal{A}^{(\lambda\phi)})_{,\phi}\right] \tag{21.98}$$

$$\mathcal{F}^{(\phi)} = \frac{1}{r\,\cos\phi}\left[(r\,\cos\phi\,\mathcal{A}^{(\phi r)})_{,r} + (\mathcal{A}^{(\phi\lambda)})_{,\phi}\right] \tag{21.99}$$

$$\mathcal{F}^{(r)} = \frac{1}{r^2\,\cos\phi}\left[(r\,\mathcal{A}^{(r\lambda)})_{,\lambda} + (r\,\cos\phi\,\mathcal{A}^{(r\phi)})_{,\phi}\right]. \tag{21.100}$$

21.11.6 Covariant Laplacian of a scalar

In orthogonal coordinates, the covariant Laplacian of a scalar, derived in Section 21.7, takes the form

$$\frac{1}{\sqrt{\mathcal{G}}}\left(\sqrt{\mathcal{G}}\,g^{ab}\,V_{,a}\right)_{,b} = \frac{1}{h_1\, h_2\, h_3}\left\{\left(\frac{h_2\, h_3}{h_1}\,V_{,1}\right)_{,1} + \left(\frac{h_1\, h_3}{h_2}\,V_{,2}\right)_{,2} + \left(\frac{h_1\, h_2}{h_3}\,V_{,3}\right)_{,3}\right\}. \tag{21.101}$$

Eliminating the stretching functions leads to

$$\nabla^2 V = (\mathrm{d}y\,\mathrm{d}z)^{-1}\,(\mathrm{d}y\,\mathrm{d}z\,V_{,x})_{,x} + (\mathrm{d}x\,\mathrm{d}z)^{-1}\,(\mathrm{d}x\,\mathrm{d}z\,V_{,y})_{,y} + (\mathrm{d}x\,\mathrm{d}y)^{-1}\,(\mathrm{d}x\,\mathrm{d}y\,V_{,z})_{,z} \tag{21.102}$$

where $\nabla^2 V$ is the familiar notation for the Laplacian of a scalar. In spherical coordinates

$$\nabla^2 V = \frac{1}{r^2}(r^2\,V_{,r})_{,r} + \frac{1}{r^2\,\cos^2\phi}\,(V_{,\lambda\lambda} + (V_{,\phi}\,\cos\phi)_{,\phi}) + V_{,zz}. \tag{21.103}$$

21.11.7 Covariant curl of a vector

From Section 21.8, the covariant curl takes the form

$$\begin{aligned}\nabla \wedge \vec{F} &= \operatorname{curl} \vec{F} \\ &= \varepsilon^{abc}\, F_{c,b}\, \vec{e}_a,\end{aligned} \tag{21.104}$$

where the first expression is that commonly used in vector analysis. With orthogonal coordinates, the curl becomes

$$\begin{aligned}\operatorname{curl} \vec{F} &= \mathcal{G}^{-1/2}\, \epsilon^{abc}\, F_{c,b}\, \vec{e}_a \\ &= \frac{h_a\, \hat{e}_{(a)}}{h_1\, h_2\, h_3} \epsilon^{abc} \left(h_c\, \mathcal{F}_{(c)}\right)_{,b} \\ &= \frac{1}{h_1\, h_2\, h_3} \begin{vmatrix} \hat{\mathbf{e}}_{(1)}\, h_1 & \hat{\mathbf{e}}_{(2)}\, h_2 & \hat{\mathbf{e}}_{(3)}\, h_3 \\ \partial_1 & \partial_2 & \partial_3 \\ h_1\, \mathcal{F}_{(1)} & h_2\, \mathcal{F}_{(2)} & h_3\, \mathcal{F}_{(3)} \end{vmatrix},\end{aligned} \tag{21.105}$$

where the last expression is written in a determinantal form. Eliminating the stretching functions leads to

$$\begin{aligned}\operatorname{curl} \vec{F} =& \hat{\mathbf{e}}_{(1)} \left[(\mathrm{d}z)^{-1} (\mathrm{d}z\, \mathcal{F}_{(3)})_{,y} - (\mathrm{d}y)^{-1} (\mathrm{d}y\, \mathcal{F}_{(2)})_{,z} \right] \\ &+ \hat{\mathbf{e}}_{(2)} \left[(\mathrm{d}x)^{-1} (\mathrm{d}x\, \mathcal{F}_{(1)})_{,z} - (\mathrm{d}z)^{-1} (\mathrm{d}z\, \mathcal{F}_{(3)})_{,x} \right] \\ &+ \hat{\mathbf{e}}_{(3)} \left[(\mathrm{d}y)^{-1} (\mathrm{d}y\, \mathcal{F}_{(2)})_{,x} - (\mathrm{d}x)^{-1} (\mathrm{d}x\, \mathcal{F}_{(1)})_{,y} \right].\end{aligned} \tag{21.106}$$

21.12 SUMMARY OF CURVILINEAR TENSOR ANALYSIS

The purpose of this section is to summarize the main results of Chapters 20 and 21 that are of use throughout this book.

21.12.1 Rules for tensor analysis on manifolds

The following tensor rules are sufficient for applications in this book.

- **Conservation of indices**: Lower and upper tensor indices balance across equal signs.
- **Einstein summation convention**: Repeated indices are summed, unless otherwise noted.
- **Metric tensor**: The metric tensor provides a means to measure the distance between two points on a manifold:

  $$(\mathrm{d}s)^2 = g_{mn}\, \mathrm{d}\xi^m\, \mathrm{d}\xi^n. \tag{21.107}$$

  In this expression, $(\mathrm{d}s)^2$, often written $\mathrm{d}s^2$, is the squared infinitesimal arc-length between the points, ξ^m is a general coordinate for a point, and $m = 1, 2, 3$ labels the coordinate.

The metric for spherical coordinates on a sphere is diagonal. With coordinates $(\xi^1, \xi^2, \xi^3) = (\lambda, \phi, r)$, where λ is longitude and ϕ latitude, the metric is

$$\begin{aligned} g_{mn} &= \mathrm{diag}(g_{11}, g_{22}, g_{33}) \\ &= \mathrm{diag}((r\cos\phi)^2, r^2, 1). \end{aligned} \tag{21.108}$$

The inverse metric components g^{mn} are also needed, and they are given by

$$g^{mn} = \mathrm{diag}((r\ \cos\phi)^{-2}, r^{-2}, 1). \tag{21.109}$$

In Cartesian coordinates, the metric tensor is given by

$$\begin{aligned} g_{mn} &= \delta_{mn} \\ &= \delta^{mn} \\ &= \delta^m_n, \end{aligned} \tag{21.110}$$

where δ is the unit or Kronecker delta tensor. There is no distinction between raised and lowered indices in Cartesian coordinates, hence the ability to jettison the conservation of indices rule when working with Cartesian tensors. For curvilinear coordinates, however, conservation of indices is necessary.

- **Covariant and contravariant**: A lower label is often termed "covariant" and an upper label "contravariant." The mnemonic "co-lo" assists in remembering the terminology.

  Covariant and contravariant tensors can be considered dual, where the transformation is through the metric tensor. For example, the covariant components to the velocity vector u_m are related to the contravariant components through

$$u_m = g_{mn} u^n. \tag{21.111}$$

  Some examples are useful. In Cartesian coordinates, the velocity vector takes the familiar form

$$\begin{aligned} (u^1, u^2, u^3) &= (u_1, u_2, u_3) \\ &= \left(\frac{\mathrm{d}x}{\mathrm{d}t}, \frac{\mathrm{d}y}{\mathrm{d}t}, \frac{\mathrm{d}z}{\mathrm{d}t} \right), \end{aligned} \tag{21.112}$$

  where again there is no distinction between covariant and contravariant for Cartesian tensors. In spherical coordinates, however, the contravariant velocity components are

$$(u^1, u^2, u^3) = \left(\frac{\mathrm{d}\lambda}{\mathrm{d}t}, \frac{\mathrm{d}\phi}{\mathrm{d}t}, \frac{\mathrm{d}r}{\mathrm{d}t} \right), \tag{21.113}$$

  whereas the covariant components $u_m = g_{mn}\, u^n$ are

$$(u_1, u_2, u_3) = \left((r\ \cos\phi)^2 \frac{\mathrm{d}\lambda}{\mathrm{d}t}, r^2 \frac{\mathrm{d}\phi}{\mathrm{d}t}, \frac{\mathrm{d}r}{\mathrm{d}t} \right). \tag{21.114}$$

- **Notation**: As the above indicates, for curvilinear tensor analysis the difference between a raised and lowered label is important. Additionally, in order to avoid confusion, partial derivatives are denoted with a comma
$$u_{m,n} = \frac{\partial u_m}{\partial \xi^n}. \tag{21.115}$$
The alternative, more commonly used in GFD literature, omits the comma. Without the comma notation, it is unclear whether the subscript represents the component to a tensor or a partial derivative acting on a field.

- **Covariant derivative**: In order to account for nonconstant basis vectors on a curved manifold, it is necessary to generalize partial derivatives to covariant derivatives. In particular, the strain tensor (Section 17.3.1) has components
$$2\, e_{mn} = u_{m;n} + u_{n;m} \tag{21.116}$$
where the comma has been generalized to a semicolon.

For a "torsionless" manifold, such as a sphere or a smooth surface in the ocean, each component of the metric tensor has a vanishing covariant derivative
$$g_{mn;p} = 0. \tag{21.117}$$
This is a trivial property for Cartesian coordinates on a plane, in which case the metric is the constant unit tensor and the covariant derivative a partial derivative
$$\begin{aligned}\delta_{mn;p} &= \delta_{mn,p} \\ &= 0.\end{aligned} \tag{21.118}$$
However, for curvilinear coordinates $g_{mn;p} = 0$ is quite useful. For example, it provides for the convenient relation
$$\begin{aligned}u_{m;n} &= (g_{mp}\, u^p)_{;n} \\ &= g_{mp}\, u^p_{;n}.\end{aligned} \tag{21.119}$$
This result brings the strain tensor components to the form
$$2\, e_{mn} = g_{mp}\, u^p_{;n} + g_{np}\, u^p_{;m}. \tag{21.120}$$
In general, the covariant derivative of a vector on a torsionless manifold is given by
$$u^p_{;n} = u^p_{,n} + \Gamma^p_{mn}\, u^n \tag{21.121}$$
where Γ^p_{mn} are components to the Christoffel symbol
$$\Gamma^p_{mn} = \frac{1}{2}\, g^{pq}\, (g_{qm,n} + g_{qn,m} - g_{mn,q}). \tag{21.122}$$
The Christoffel symbol components form the expansion coefficients of the partial derivative of the basis vectors
$$\vec{e}_{a,b} = \Gamma^m_{ab}\, \vec{e}_m. \tag{21.123}$$
That is, the Christoffel symbol accounts for the nonzero changes in the basis vectors on a curved manifold. Note that it is symmetric on the lower two labels:
$$\Gamma^p_{mn} = \Gamma^p_{nm}, \tag{21.124}$$
which is the defining property of torsionless manifolds.

- **Transformation rules**: Under a coordinate transformation

$$\xi^{\overline{m}} = \xi^{\overline{m}}(\xi^m), \tag{21.125}$$

tensors transform as, for example,

$$e_{\overline{m}\overline{n}} = \Lambda^m_{\overline{m}} \Lambda^n_{\overline{n}} e_{mn}, \tag{21.126}$$

where the transformation matrix is given by the partial derivatives

$$\Lambda^m_{\overline{m}} = \frac{\partial \xi^m}{\partial \xi^{\overline{m}}}. \tag{21.127}$$

Sometimes it is useful to write the transformation matrix in traditional matrix form. The convention is that the index which is placed a bit closer to the Λ denotes the row (m in $\Lambda^m_{\overline{m}}$), and the one pushed away a bit is the column ($\overline{m}$ in $\Lambda^m_{\overline{m}}$). The inverse transformation of a tensor takes the form

$$e_{mn} = \Lambda^{\overline{m}}_m \Lambda^{\overline{n}}_n e_{\overline{m}\overline{n}}, \tag{21.128}$$

where the inverse transformation matrix has components given by

$$\Lambda^{\overline{m}}_m = \frac{\partial \xi^{\overline{m}}}{\partial \xi^m}. \tag{21.129}$$

Transformations of tensors with arbitrary rank generalize with an extra factor of the transformation matrix corresponding to each tensor label. An example of how the fourth order viscosity tensor transforms is given in Section 17.5.1.

21.12.2 Orthogonal coordinates

The metric tensor for orthogonal coordinates is diagonal,

$$g_{mn} = \mathrm{diag}(g_{11}, g_{22}, g_{33}), \tag{21.130}$$

where the components $g_{mn} = g_{mn}(t, \xi^1, \xi^2, \xi^3)$ are generally functions of space-time. The infinitesimal arc length measuring the distance between any two closely spaced points is therefore given by the diagonal quadratic-form

$$\begin{aligned}(\mathrm{d}s)^2 &= g_{11}(\mathrm{d}\xi^1)^2 + g_{22}(\mathrm{d}\xi^2)^2 + g_{33}(\mathrm{d}\xi^3)^2 \\ &= (h_1\, \mathrm{d}\xi^1)^2 + (h_2\, \mathrm{d}\xi^2)^2 + (h_3\, \mathrm{d}\xi^3)^2,\end{aligned} \tag{21.131}$$

where the positive metric or *stretching* functions (with no implied sum)

$$h_m = \sqrt{g_{mm}} \tag{21.132}$$

are often useful. Additionally, the relation between covariant and contravariant components of a tensor is given through a single multiplication. For example,

$$\begin{aligned}u_m &= g_{mn}\, u^n \\ &= g_{mm}\, u^m\end{aligned} \tag{21.133}$$

relates the covariant velocity components u_m to the contravariant components u^m. Importantly, there is no sum on the m label in the last expression.

21.12.3 Physical tensor components with orthogonal coordinates

In many applications, it is useful to introduce the *physical components* of a tensor.* For example, the velocity field using spherical coordinates is often written

$$\begin{aligned}(u, v, w) &= \left(\sqrt{g_{\lambda\lambda}}\, u^{\lambda}, \sqrt{g_{\phi\phi}}\, u^{\phi}, \sqrt{g_{rr}}\, u^{r}\right) \\ &= \left(r \cos\phi \frac{\mathrm{d}\lambda}{\mathrm{d}t}, r \frac{\mathrm{d}\phi}{\mathrm{d}t}, \frac{\mathrm{d}r}{\mathrm{d}t}\right).\end{aligned} \tag{21.134}$$

Additionally, the infinitesimal displacements along the coordinate directions on the sphere are given by

$$(\mathrm{d}x, \mathrm{d}y, \mathrm{d}z) = ((r \cos\phi)\, \mathrm{d}\lambda, r\, \mathrm{d}\phi, \mathrm{d}r)\,. \tag{21.135}$$

More generally, for any orthogonal coordinate system, the physical components of the displacement are written

$$(\mathrm{d}x, \mathrm{d}y, \mathrm{d}z) = (h_1\, \mathrm{d}\xi^1, h_2\, \mathrm{d}\xi^2, h_3\, \mathrm{d}\xi^3). \tag{21.136}$$

Likewise, the physical components of the velocity are

$$(u, v, w) = (h_1\, u^1, h_2\, u^2, h_3\, u^3). \tag{21.137}$$

Consequently, for example,

$$\begin{aligned}u^1_{,1} &= \sqrt{g_{11}}\, (u/\sqrt{g_{11}})_{,x} \\ &= h_1\, (u/h_1)_{,x}.\end{aligned} \tag{21.138}$$

Note that although the Cartesian notation x, y is used for convenience, the coordinates are generally curvilinear.

The physical components of a tensor each have the same dimensions. For example, the physical displacement components have dimensions of length, and the physical velocity components have dimensions of length per time. This property makes physical tensor components very useful in practice. However, the physical components are *not* components to a true tensor, since the tensorial transformation rules are corrupted by the square root of the metric. In particular, the physical components of the partial derivative operator do not generally commute

$$\begin{aligned}\left[\partial_x, \partial_y\right] &= \partial_x\, \partial_y - \partial_y\, \partial_x \\ &= (\partial_x \ln \mathrm{d}y)\, \partial_y - (\partial_y \ln \mathrm{d}x)\, \partial_x.\end{aligned} \tag{21.139}$$

This result vanishes only when the horizontal geometry is flat instead of curved, and we use Cartesian coordinates. Hence, as a rule of thumb it is best to perform mathematical manipulations with the tensorial components, and only after establishing the final result should the physical components be introduced into an expression.

21.12.4 Eliminating the stretching functions

Orthogonal coordinates afford an added feature that is quite useful to exploit when discretizing the equations. When the coordinates are orthogonal, a coordinate increment in one direction is independent of the other direction. For example, with orthogonal horizontal coordinates,

$$\partial_1\, (\mathrm{d}\xi^2) = \partial_2\, (\mathrm{d}\xi^1) = 0. \tag{21.140}$$

---

*Physical tensor components are treated in Section 7.4 of Aris (1962), Section of 4.8 Weinberg (1972), and Section 21.11.1 of this book.

This property allows for all formulae in an ocean model to be written in terms of the physical displacements $\mathrm{d}x$ and $\mathrm{d}y$, and so eliminate horizontal stretching functions h_1 and h_2. This property is convenient since the physical displacements are ultimately what are relevant for the ocean model.

PART 7
Epilogue

Chapter Twenty-Two

SOME CLOSING COMMENTS AND CHALLENGES

Ocean models represent a compendium of ocean understanding. Hence, the integrity of the model simulation is directly related to how well we understand the ocean. At their fundamental level, models rely on the laws of classical continuum physics. We therefore spent Part 1 of this book describing these laws as applied to the ocean. These equations are reasonably well established. There are few debates about whether they have validity for describing motions from the millimeter scale up to the planetary scale. This represents one of the great achievements of classical physics in describing natural phenomena.

Scientific debate ensues when trying to prescribe both a rational and a *practical* description of the ocean. By *practical* is meant a description that explicitly acknowledges an ignorance of processes occurring at space-time scales beneath the resolution of either measuring devices or computer model grid. Such subgrid scale (SGS) parameterization represents a fundamental problem in computational fluid dynamics, and generally in all of physics. An acceptance that one cannot measure all space-time scales, nor simulate them, motivates the derivation of various approximate or averaged descriptions of the fluid motion. These equations form the basis for the equations of an ocean model.

We touched on these matters in Part 2 of this book. There are whole monographs devoted to these subjects in the fluid mechanics literature. The goals in this book were relatively modest by comparison. However, it is hoped that the reader has come away from this discussion recognizing two basic points. First, an averaged description of the ocean depends on the reference frame where the "averaged observer" is situated. Both the scales of motion that fall within the subgrid scale and the coordinates chosen to describe the fluid determine details of the averaged description. Second, SGS closure is nontrivial. It is likely that the issues of SGS closure will remain a key aspect of ocean model development in years to come, even as model resolution is refined.

The equations in an ocean model represent a discrete realization of the continuum equations. Discretization breaks the continuum symmetries, and doing so introduces the possibility of spurious unphysical solutions. Such *computational* modes are the bane of numerical algorithm developers. They must be eliminated, or at least controlled, or else the simulated ocean may provide a poor rendition of the real ocean. In Part 3 we introduced the *finite volume* method for discretizing the continuum equations. Chapters 16 and 19 followed on this discussion with details of how to discretize some generalized Laplacian operators of use for SGS closure in z-models. The subject of how to bring the discrete equations in line with the continuum pervaded this discussion.

The problems of SGS parameterization are fundamental to ocean models. Indeed, the physical integrity of simulations is quite dependent on prescriptions for the SGS. This is very unsatisfying, since there are huge uncertainties in how best to parameterize the SGS. Much is being learned with improved observations and process studies. However, much remains in doubt and confused.

Parts 4 and 5 of this book presented issues of SGS lateral exchanges in both the tracer equation and the linear momentum equation. There is no lack of debate regarding how best to parameterize these processes in ocean climate models. Related to the physical uncertainties are numerical constraints that necessitate the use of dissipation to maintain numerical stability. The manner in which such *numerical closures* are implemented in the models often coincides with the form suggested from physical arguments. This could be simply a useful coincidence. Or it could be a confusion, where ambiguities arise when confusing the numerical closure for physical closure, and vice versa.

A fair amount of technology was used in Parts 4 and 5. This was motivated in order to provide a sound representation in the numerical models of the various closures. Cavalier numerical implementations have been found to be insufficient for purposes of ensuring that the models are performing in a manner consistent with the ideas underlying the closures. The mathematical and numerical sophistication employed should not be mistaken for sanctioning the physical rigor of the operators. Instead, many stages of the devolopment revealed uncertainties which need to be remedied before one can provide robust prescriptions for the models.

Although some relatively fancy numerical techniques were presented in Chapters 16 and 19, the methods documented in this book are rather common relative to others employed in the computational fluid dynamics literature. Indeed, ocean models have traditionally used relatively old-fashioned techniques. There are two reasons for this. First, building an ocean climate model is a nontrivial task. To alter the fundamental methodology is significant. Hence, there is a lot of inertia associated with a "working" model. Modelers must balance the desire to try new algorithms with the interests of using the model to study the ocean circulation. Second, there have actually been attempts to use more sophisticated techniques in ocean modeling. The book by Haidvogel and Beckmann (1999) provides a summary of different methods, such as spectral methods and finite element or unstructured grids. Each of these techniques presents the ocean modeler with possibilities for removing limitations in the traditional finite difference methods established in the 1950's-1960's. However, the needs of ocean climate modeling are tough to meet. There has yet to be an indication that fundamentally new approaches, beyond those available with the best of finite difference methods, will significantly penetrate the ocean climate community within a decade. Perhaps afterward.

A major effort in ocean model design today is aimed at providing an algorithmic framework that allows various vertical coordinates, time stepping schemes, physical parameterizations, etc., to be readily incorporated into a single modular set of code primitives. Out of this discussion has arisen the notion of a common *model repository*, or *model environment*. These ideas have indeed generated some excitement amongst modelers. As idealistically envisioned, a model environment allows for various model developers to access common pieces of code to build a model of their choosing. There is tremendous power to be gained by attracting developers to the same environment. It reduces the overhead required for code maintenance, since many common elements are shared. It also encourages scientists to interact much more closely. Notably, a model environment should not be confused with a top-down dictate to homogenize model algorithms. Instead, it aims to provide a cleaner and clearer way to compare and contrast various ideas and methods, and to evolve the model classes into more robust tools.

The evolution of model development into model environments is perhaps a natural result of ocean climate modeling having reached a level of adolescence. For example, ocean models are being used increasingly now for predictions, and this requires a tremendous level of overhead beyond that of the older research models. They are also a key part of coupled earth system models whose aim is to understand climate variability and climate change. Furthermore, the community has perhaps reached a stage in ocean model development where the limits of each individual model class have been reached. To make progress appears to require generalizing the vertical coordinate. This argument was presented in the review paper by Griffies et al. (2000a), and it represents a growing consensus in the community, although there remains no consensus regarding the details of what is the best choice for generalized coordinates.

The difficulty of building a sound algorithmic base to allow generalized vertical coordinates is nontrivial. It requires a great deal of collaboration, discussion, and debate within the community. Indeed, whether a model environment can cleanly incorporate all possible algorithms remains a topic of debate (e.g., Adcroft and Hallberg, 2004). Regardless, moving toward a common environment, at least as well as possible, will arguably accelerate model evolution. Doing so will hopefully facilitate the maturation of ocean climate modeling to a point that it becomes a robust scientific endeavor. This remains a profound challenge to present and future ocean modelers.

Bibliography

Adcroft, A., and R. W. Hallberg, 2004: On methods for solving the oceanic equations of motion in generalized vertical coordinates. *Ocean Modelling*, in press.

Adcroft, A., C. Hill, and J. Marshall, 1997: Representation of topography by shaved cells in a height coordinate ocean model. *Monthly Weather Review*, 125, 2293–2315.

Andrews, D. G., and M. E. McIntyre, 1978: An exact theory of nonlinear waves on a Lagrangian-mean flow. *Journal of Fluid Mechanics*, 89, 609–646.

Apel, J. R., 1987: *Principles of Ocean Physics*. International Geophysics Series, Vol. 38, Academic Press.

Arakawa, A., 1966: Computational design for long-term numerical integration of the equations of fluid motion: Two-dimensional incompressible flow. Part 1. *Journal of Computational Physics*, 1, 119–143.

Arakawa, A., and V. R. Lamb, 1977: The UCLA general circulation model. J. Chang, Ed., *Methods in Computational Physics: General Circulation Models of the Atmosphere*, Vol. 17, Academic Press, 174–265.

Aris, R., 1962: *Vectors, Tensors and the Basic Equations of Fluid Mechanics*. Dover Publishing.

Bacon, S., and N. P. Fofonoff, 1996: Oceanic heat flux calculation. *Journal of Atmospheric and Oceanic Technology*, 13, 1327–1329.

Batchelor, G. K., 1967: *An Introduction to Fluid Dynamics*. Cambridge University Press.

Becker, E., 2003: Frictional heating in global climate models. *Monthly Weather Review*, 131, 508–520.

Beckers, J. M., H. Burchard, J.-M. Campin, E. Deleersnijder, and P. P. Mathieu, 1998: Another reason why simple discretizations of rotated diffusion operators cause problems in ocean models: Comments on isoneutral diffusion in a *z*-coordinate ocean model. *Journal of Physical Oceanography*, 28, 1552–1559.

Beckers, J. M., H. Burchard, E. Deleersnijder, and P. P. Mathieu, 2000: Numerical discretization of rotated diffusion operators in ocean models. *Monthly Weather Review*, 128, 2711–2733.

Beckmann, A., 1998: The representation of bottom boundary layer processes in numerical ocean circulation models. E. P. Chassignet and J. Verron, Eds., *Ocean Modeling and Parameterization*, NATO ASI Mathematical and Physical Sciences Series, Vol. 516, Kluwer, 135–154.

Beckmann, A., and R. Döscher, 1997: A method for improved representation of dense water spreading over topography in geopotential-coordinate models. *Journal of Physical Oceanography*, 27, 581–591.

Black, T. L., 1994: The new NMC mesoscale eta model: description and forecast examples. *Weather and Forecasting*, 9, 265–278.

Bleck, R., 1978: Finite difference equations in generalized vertical coordinates. Part I: Total energy conservation. *Contributions to Atmospheric Physics*, 51, 360–372.

Bleck, R., 1998: Ocean modeling in isopycnic coordinates. E. P. Chassignet and J. Verron, Eds., *Ocean Modeling and Parameterization*, NATO ASI Mathematical and Physical Sciences Series, Vol. 516, Kluwer, 423–448.

Bleck, R., 2002: An oceanic general circulation model frame in hybrid isopycnic-cartesian coordinates. *Ocean Modelling*, 4, 55–88.

Bleck, R., and D. B. Boudra, 1981: Initial testing of a numerical ocean circulation model using a hybrid- (quasi-isopycnic) vertical coordinate. *Journal of Physical Oceanography*, 11, 755–770.

Bleck, R., C. Rooth, D. Hu, and L. T. Smith, 1992: Salinity-driven thermocline transients in a wind and thermohaline forced isopycnic coordinate model of the North Atlantic. *Journal of Physical Oceanography*, 22, 1486–1505.

Blumberg, A. F., and G. L. Mellor, 1987: A description of a three-dimensional coastal ocean circulation model. N. Heaps, Ed., *Three-Dimensional Coastal Ocean Models*, Coastal and Estuarine Series, Vol. 4, American Geophysical Union.

Boer, G. J., and T. G. Shepherd, 1983: Large-scale two-dimensional turbulence in the atmosphere. *Journal of Atmospheric Sciences*, 40, 164–184.

Bokhove, O., 2000: On hydrostatic flows in isentropic coordinates. *Journal of Fluid Mechanics*, 402, 291–310.

Bretherton, F. P., 1966: Critical layer instability in baroclinic flows. *Quarterly Journal of the Royal Meteorological Society*, 92, 325–334.

Bryan, K., 1969: A numerical method for the study of the circulation of the world ocean. *Journal of Computational Physics*, 4, 347–376.

Bryan, K., 1989: The design of numerical models of the ocean circulation. D. L. Anderson and J. Willebrand, Eds., *Oceanic Circulation Models: Combining Data and Dynamics*, NATO ASI Series. Series C, Vol. 284, Kluwer Academic Publishers, 465–511.

Bryan, K., and M. D. Cox, 1972: An approximate equation of state for numerical models of the ocean circulation. *Journal of Physical Oceanography*, 4, 510–514.

Bryan, K., J. K. Dukowicz, and R. D. Smith, 1999: On the mixing coefficient in the parameterization of bolus velocity. *Journal of Physical Oceanography*, 29, 2442–2456.

Bryan, K., and L. J. Lewis, 1979: A water mass model of the world ocean. *Journal of Geophysical Research*, 84, 2503–2517.

Bryan, K., S. Manabe, and R. C. Pacanowski, 1975: A global ocean-atmosphere climate model. Part II. The oceanic circulation. *Journal of Physical Oceanography*, 5, 30–46.

Callen, H. B., 1985: *Thermodynamics and an Introduction to Thermostatics*. John Wiley and Sons.

Campin, J.-M., A. Adcroft, C. Hill, and J. Marshall, 2004: Conservation of properties in a free-surface model. *Ocean Modelling*, 6, 221–244.

Chaikin, P. M., and T. C. Lubensky, 1995: *Principles of Condensed Matter Physics*. Cambridge University Press.

Chandrasekhar, S., 1961: *Hydrodynamic and Hydromagnetic Stability*. Dover Publications.

Chassignet, E. P., and Z. Garraffo, 2001: Viscosity parameterization and the Gulf Stream separation. P. Müller and D. Henderson, Eds., *From Stirring to Mixing in a Stratified Ocean*, Proceedings of the 12th 'Aha Huliko'a Hawaiian Winter Workshop, University of Hawaii at Manoa, 37–41.

Chassignet, E. P., and J. Verron, 1998: *Ocean Modeling and Parameterization*. NATO ASI Mathematical and Physical Sciences Series, Vol. 516, Kluwer Academic Publishers.

Chelton, D. B., R. A. DeSzoeke, M. G. Schlax, K. E. Naggar, and N. Siwertz, 1998: Geographical variability of the first baroclinic rossby radius of deformation. *Journal of Physical Oceanography*, 28, 433–460.

Courant, R., and D. Hilbert, 1953: *Methods of Mathematical Physics Volume I*. Wiley-Interscience.

Courant, R., and D. Hilbert, 1962: *Methods of Mathematical Physics Volume II: Partial Differential Equations*. Wiley-Interscience.

Cox, M. D., 1984: *A Primitive Equation, 3-Dimensional Model of the Ocean*. NOAA/Geophysical Fluid Dynamics Laboratory.

Cox, M. D., 1987: Isopycnal diffusion in a *z*-coordinate ocean model. *Ocean Modelling*, 74, 1–5.

Cushman-Roisin, B., 1994: *Introduction to Geophysical Fluid Dynamics*. Prentice-Hall.

Danabasoglu, G., and J. C. McWilliams, 1995: Sensitivity of the global ocean circulation to parameterizations of mesoscale tracer transports. *Journal of Climate*, 8, 2967–2987.

Davis, R. E., 1994a: Diapycnal mixing in the ocean: equations for large-scale budgets. *Journal of Physical Oceanography*, 24, 777–800.

Davis, R. E., 1994b: Diapycnal mixing in the ocean: the Osborn-Cox model. *Journal of Physical Oceanography*, 24, 2560–2576.

DeGroot, S. R., and P. Mazur, 1984: *Non-Equilibrium Thermodynamics*. Dover Publications.

DeSzoeke, R. A., and A. F. Bennett, 1993: Microstructure fluxes across density surfaces. *Journal of Physical Oceanography*, 23, 2254–2264.

DeSzoeke, R. A., and R. M. Samelson, 2002: The duality between the Boussinesq and non-Boussinesq hydrostatic equations of motion. *Journal of Physical Oceanography*, 32, 2194–2203.

Dewar, W. K., Y. Hsueh, T. J. McDougall, and D. Yuan, 1998: Calculation of pressure in ocean simulations. *Journal of Physical Oceanography*, 28, 577–588.

DiBattista, M. T., A. J. Majda, and M. J. Grote, 2001: Meta-stability of equilibrium statistical structures for prototype geophysical flows with damping and driving. *Physica D*, 151, 271–304.

Döscher, R., and A. Beckmann, 1999: Effects of a bottom boundary layer parameterization in a coarse-resolution model of the North Atlantic Ocean. *Journal of Atmospheric and Oceanic Technology*, 17, 698–707.

Dukowicz, J. K., and R. D. Smith, 1994: Implicit free-surface method for the Bryan-Cox-Semtner ocean model. *Journal of Geophysical Research*, 99, 7991–8014.

Dukowicz, J. K., and R. D. Smith, 1997: Stochastic theory of compressible turbulent fluid transport. *Physics of Fluids*, 9, 3523–3529.

Dukowicz, J. K., R. D. Smith, and R. C. Malone, 1993: A reformulation and implementation of the Bryan-Cox-Semtner ocean model on the connection machine. *Journal of Atmospheric and Oceanic Technology*, 10, 195–208.

Durran, D. R., 1999: *Numerical Methods for Wave Equations in Geophysical Fluid Dynamics*. Springer Verlag.

Dutton, J. A., 1976: *The Caseless Wind: An Introduction to the Theory of Atmospheric Motion*. McGraw-Hill.

Eckart, C., 1948: An analysis of the stirring and mixing processes in incompressible fluids. *Journal of Marine Research*, 7, 265–275.

England, M. H., and G. Holloway, 1998: Simulations of CFC content and water-mass age in the deep north atlantic. *Journal of Geophysical Research*, 103, 15 885–15 902.

Ezer, T., H. Arango, and A. F. Shchepetkin, 2002: Developments in terrain-following ocean models: Intercomparisons of numerical aspects. *Ocean Modelling*, 4, 249–267.

Favre, A., 1965: Équations des gaz turbulents compressibles. Parts I and II. *Journel de Méchanic*, 4, 361–421.

Feistel, R., 1993: Equilibrium thermodynamics of seawater revisited. *Progress in Oceanography*, 31, 101–179.

Feistel, R., 2003: A new extended Gibbs thermodynamic potential of seawater. *Progress in Oceanography*, 58, 43–114.

Feistel, R., and E. Hagen, 1995: On the Gibbs thermodynamic potential of seawater. *Progress in Oceanography*, 36, 249–327.

Ferrari, R., and A. R. Plumb, 2003: Residual circulation in the ocean. P. Müller and C. Garrett, Eds., *Near-Boundary Processes and Their Parameterization*, Proceedings of the 13th 'Aha Huliko'a Hawaiian Winter Workshop, University of Hawaii at Manoa, 219–228.

Fetter, A. L., and J. D. Walecka, 1980: *Theoretical Mechanics of Particles and Continua*. McGraw-Hill Book Company.

Flanders, H., 1989: *Differential Forms with Applications to the Physical Sciences*. Dover Publications.

Fofonoff, N. P., 1962: Physical properties of seawater. M. N. Hill, Ed., *The Sea*, Vol. 1, Wiley-Interscience, 3–30.

Fox-Kemper, B., R. Ferrari, and J. Pedlosky, 2003: A note on the indeterminacy of rotationaal and divergent eddy fluxes. *Journal of Physical Oceanography*, 33, 478–483.

Galperin, B., and S. A. Orszag, 1993: *Large Eddy Simulation of Complex Engineering and Geophysical Flows*. Cambridge University Press.

Gent, P. R., and J. C. McWilliams, 1990: Isopycnal mixing in ocean circulation models. *Journal of Physical Oceanography*, 20, 150–155.

Gent, P. R., and J. C. McWilliams, 1996: Eliassen-Palm fluxes and the momentum equations in non-eddy-resolving ocean circulation models. *Journal of Physical Oceanography*, 26, 2539–2546.

Gent, P. R., J. Willebrand, T. J. McDougall, and J. C. McWilliams, 1995: Parameterizing eddy-induced tracer transports in ocean circulation models. *Journal of Physical Oceanography*, 25, 463–474.

Gerdes, R., 1993: A primitive equation ocean circulation model using a general vertical coordinate transformation. 1. Description and testing of the model. *Journal of Geophysical Research*, 98, 14 683–14 701.

Gerdes, R., C. Köberle, and J. Willebrand, 1991: The influence of numerical advection schemes on the results of ocean general circulation models. *Climate Dynamics*, 5, 211–226.

Gill, A., 1982: *Atmosphere-Ocean Dynamics*. International Geophysics Series, Vol. 30, Academic Press.

Gordon, C., C. Cooper, C. A. Senior, H. Banks, T. C. J. J. M. Gregory, J. F. B. Mitchell, and R. A. Wood, 2000: The simulation of SST, sea ice extents and ocean heat transports in a version of the Hadley Centre coupled model without flux adjustments. *Climate Dynamics*, 16, 147–168.

Greatbatch, R. J., 1994: A note on the representation of steric sea level in models that conserve volume rather than mass. *Journal of Geophysical Research*, 99, 12767–12771.

Greatbatch, R. J., 1998: Exploring the relationship between eddy-induced transport velocity, vertical momentum transfer, and the isopycnal flux of potential vorticity. *Journal of Physical Oceanography*, 28, 422–432.

Greatbatch, R. J., and K. G. Lamb, 1990: On parameterizing vertical mixing of momentum in non-eddy resolving ocean models. *Journal of Physical Oceanography*, 20, 1634–1637.

Greatbatch, R. J., and G. Li, 2000: Alongslope mean flow and an associated upslope bolus flux of tracer in a parameterization of mesoscale turbulence. *Deep-Sea Research*, 47, 709–735.

Greatbatch, R. J., Y. Lu, and Y. Cai, 2001: Relaxing the Boussinesq approximation in ocean circulation models. *Journal of Atmospheric and Oceanic Technology*, 18, 1911–1923.

Greatbatch, R. J., and T. J. McDougall, 2003: The non-Boussinesq temporal-residual-mean. *Journal of Physical Oceanography*, 33, 1231–1239.

Greatbatch, R. J., and G. L. Mellor, 1999: An overview of coastal ocean models. C. Mooers, Ed., *Coastal Ocean Prediction*, Coastal and Estuarine Studies, Vol. 56, American Geophysical Union, 31–57.

Green, J. A., 1970: Transfer properties of the large-scale eddies and the general circulation of the atmosphere. *Quarterly Journal of the Royal Meteorological Society*, 96, 157–185.

Gregg, M. C., 1984: Entropy generation in the ocean by small-scale mixing. *Journal of Physical Oceanography*, 14, 688–711.

Gregg, M. C., 1987: Diapycnal mixing in the thermocline: a review. *Journal of Geophysical Research*, 92, 5249–5286.

Griffies, S. M., 1998: The Gent-McWilliams skew-flux. *Journal of Physical Oceanography*, 28, 831–841.

Griffies, S. M., and Coauthors, 2000a: Developments in ocean climate modelling. *Ocean Modelling*, 2, 123–192.

Griffies, S. M., and K. Bryan, 1997: Predictability of the North Atlantic multidecadal climate variability. *Science*, 275, 181–184.

Griffies, S. M., A. Gnanadesikan, R. C. Pacanowski, V. Larichev, J. K. Dukowicz, and R. D. Smith, 1998: Isoneutral diffusion in a *z*-coordinate ocean model. *Journal of Physical Oceanography*, 28, 805–830.

Griffies, S. M., and R. W. Hallberg, 2000: Biharmonic friction with a Smagorinsky viscosity for use in large-scale eddy-permitting ocean models. *Monthly Weather Review*, 128, 2935–2946.

Griffies, S. M., M. J. Harrison, R. C. Pacanowski, and A. Rosati, 2004: *A Technical Guide to MOM4*. NOAA/Geophysical Fluid Dynamics Laboratory.

Griffies, S. M., R. C. Pacanowski, and R. W. Hallberg, 2000b: Spurious diapycnal mixing associated with advection in a *z*-coordinate ocean model. *Monthly Weather Review*, 128, 538–564.

Griffies, S. M., R. C. Pacanowski, R. M. Schmidt, and V. Balaji, 2001: Tracer conservation with an explicit free surface method for *z*-coordinate ocean models. *Monthly Weather Review*, 129, 1081–1098.

Haidvogel, D. B., and A. Beckmann, 1999: *Numerical Ocean Circulation Modeling*. Imperial College Press.

Haine, T. W. N., and J. Marshall, 1998: Gravitational, symmetric, and baroclinic instability of the ocean mixed layer. *Journal of Physical Oceanography*, 28, 634–658.

Hallberg, R. W., 1997: Stable split time stepping schemes for large-scale ocean modeling. *Journal of Computational Physics*, 135, 54–65.

Hallberg, R. W., 2000: Time integration of diapycnal diffusion and Richardson number-dependent mixing in isopycnal coordinate ocean models. *Monthly Weather Review*, 128, 1402–1419.

Hallberg, R. W., 2003a: The suitability of large-scale ocean models for adapting parameterizations of boundary mixing and a description of a refined bulk mixed layer model. P. Müller and C. Garrett, Eds., *Near-Boundary Processes and Their Parameterization*, Proceedings of the 13th 'Aha Huliko'a Hawaiian Winter Workshop, University of Hawaii at Manoa, 187–203.

Hallberg, R. W., 2003b: Thermobaric instability in Lagrangian vertical coordinate ocean models. *Ocean Modelling*.

Haltiner, G. T., and R. T. Williams, 1980: *Numerical Prediction and Dynamic Meteorology*. John Wiley and Sons.

Haynes, P. H., and M. E. McIntyre, 1987: On the evolution of vorticity and potential vorticity in the presence of diabatic heating and frictional or other forces. *Journal of Atmospheric Sciences*, 44, 828–841.

Haynes, P. H., and M. E. McIntyre, 1990: On the conservation and impermeability theorems for potential vorticity. *Journal of Atmospheric Sciences*, 47, 2021–2031.

Held, I. M., and V. D. Larichev, 1996: A scaling theory for horizontally homogeneous baroclinically unstable flow on a beta plane. *Journal of Atmospheric Sciences*, 53, 946–952.

Held, I. M., and T. Schneider, 1999: The surface branch of the zonally averaged mass transport circulation in the troposphere. *Journal of Atmospheric Sciences*, 56, 1688–1697.

Hesselberg, T., 1926: Die Gesetze der ausgeglichenen atmosphaerischen Bewegungen. *Beiträgeder Physik der freien Atmosphere*, 12, 141–160.

Hinze, J. O., 1975: *Turbulence*. McGraw-Hill Publishers.

Holland, W. R., and P. B. Rhines, 1980: An example of eddy-induced ocean circulation. *Journal of Physical Oceanography*, 10, 1010–1031.

Holloway, G., 1989: Subgridscale representation. D. L. Anderson and J. Willebrand, Eds., *Oceanic Circulation Models: Combining Data and Dynamics*, NATO ASI Series. Series C, Vol. 284, Kluwer Academic Publishers, 513–593.

Holloway, G., 1992: Representing topographic stress for large-scale ocean models. *Journal of Physical Oceanography*, 22, 1033–1046.

Holloway, G., 1997: Eddy transport of thickness and momentum in layer and level models. *Journal of Physical Oceanography*, 27, 1153–1157.

Holloway, G., 1999: Moments of probable seas: statistical dynamics of Planet Ocean. *Physica D*, 133, 199–214.

Holloway, G., and P. Rhines, 1991: Angular momenta of modeled ocean gyres. *Journal of Geophysical Research*, 27, 843–846.

Holton, J. R., 1992: *An Introduction to Dynamic Meteorology*. Academic Press.

Houghton, J., Y. Ding, D. Griggs, M. Noguer, P. van der Linden, X. Dai, K. Maskell, and C. Johnson, 2001: *Climate Change 2001: The Scientific Basis*. Cambridge University Press.

Huang, K., 1987: *Statistical Mechanics*. John Wiley and Sons.

Huang, R. X., 1993: Real freshwater flux as a natural boundary condition for the salinity balance and thermohaline circulation forced by evaporation and precipitation. *Journal of Physical Oceanography*, 23, 2428–2446.

Huang, R. X., X. Jin, and X. Zhang, 2001: An oceanic general circulation model in pressure coordinates. *Advances in Atmospheric Physics*, 18, 1–22.

Hughes, C. W., C. Wunsch, and V. Zlotnicki, 2000: Satellite peers through the oceans from space. *EOS*, 81, 68.

Jackett, D. R., T. J. McDougall, R. Feistel, D. G. Wright, and S. M. Griffies, 2004: Updated algorithms for density, potential temperature, conservative temperature, and freezing temperature of seawater. *Journal of Atmospheric and Oceanic Technology*, submitted.

Jackson, J. D., 1975: *Classical Electrodynamics*. John Wiley and Sons.

Janjić, Z. I., 1977: Pressure gradient force and advection scheme used for forecasting with steep and small scale topography. *Contributions in Atmospheric Physics*, 50, 186–199.

Jaynes, E. T., 1957: Information theory and statistical mechanics. *Physical Review*, 106, 620–630.

Kantha, L. H., and C. A. Clayson, 2000a: *Numerical Models of Oceans and Oceanic Processes*. Academic Press.

Kantha, L. H., and C. A. Clayson, 2000b: *Small Scale Processes in Geophysical Fluid Flows*. Academic Press.

Killworth, P. D., 1997: On the parameterization of eddy transfer Part I: Theory. *Journal of Marine Research*, 55, 1171–1197.

Killworth, P. D., 2001: Boundary conditions on the quasi-Stokes velocities in parameterizations. *Journal of Physical Oceanography*, 31, 1132–1155.

Killworth, P. D., D. Stainforth, D. J. Webb, and S. M. Paterson, 1991: The development of a free-surface Bryan-Cox-Semtner ocean mode. *Journal of Physical Oceanography*, 21, 1333–1348.

Kraichnan, R. H., and D. Montgomery, 1980: Two-dimensional turbulence. *Reports on Progress in Physics*, 43, 547–619.

Kunze, E., and T. B. Sanford, 1996: Abyssal mixing: where it is not. *Journal of Physical Oceanography*, 26, 2286–2296.

Kushner, P. J., and I. M. Held, 1999: Potential vorticity thickness fluxes and wave-mean flow interaction. *Journal of Atmospheric Sciences*, 56, 948–958.

Landau, L. D., and E. M. Lifshitz, 1976: *Mechanics*. Pergamon Press.

Landau, L. D., and E. M. Lifshitz, 1986: *Theory of Elasticity*. Pergamon Press.

Landau, L. D., and E. M. Lifshitz, 1987: *Fluid Mechanics*. Pergamon Press.

Large, W. G., 1998: Modeling and parameterizing the ocean planetary boundary layer. E. P. Chassignet and J. Verron, Eds., *Ocean Modeling and Parameterization*, NATO ASI Mathematical and Physical Sciences Series, Vol. 516, Kluwer, 81–120.

Large, W. G., G. Danabasoglu, S. C. Doney, and J. C. McWilliams, 1997: Sensitivity to surface forcing and boundary layer mixing in a global ocean model: annual-mean climatology. *Journal of Physical Oceanography*, 27, 2418–2447.

Large, W. G., G. Danabasoglu, J. C. McWilliams, P. R. Gent, and F. O. Bryan, 2001: Equatorial circulation of a global ocean climate model with anisotropic horizontal viscosity. *Journal of Physical Oceanography*, 31, 518–536.

Large, W. G., J. C. McWilliams, and S. C. Doney, 1994: Oceanic vertical mixing: A review and a model with a nonlocal boundary layer parameterization. *Reviews of Geophysics*, 32, 363–403.

Lau, N.-C., and J. M. Wallace, 1979: On the distribution of horizontal transports by transient eddies in the north hemisphere wintertime circulation. *Journal of Atmospheric Sciences*, 36, 1844–1861.

Ledwell, J. R., A. J. Watson, and C. S. Law, 1993: Evidence for slow mixing across the pycnocline from an open-ocean tracer-release experiment. *Nature*, 364, 701–703.

Leith, C. E., 1968: Diffusion approximation for two-dimensional turbulence. *Physics of Fluids*, 10, 1409–1416.

Leith, C. E., 1996: Stochastic models of chaotic systems. *Physica D*, 98, 481–491.

Lin, S. J., 1997: A finite volume integration method for computing pressure gradient force in general vertical coordinates. *Quarterly Journal of the Royal Meteorological Society*, 123, 1749–1762.

Losch, M., A. Adcroft, and J.-M. Campin, 2004: How sensitive are coarse general circulation models to fundamental approximations in the equations of motion? *Journal of Physical Oceanography*, 34, 306–319.

Lu, Y., 2001: Including non-Boussinesq effects in Boussinesq ocean circulation models. *Journal of Physical Oceanography*, 31, 1616–1622.

Madec, G., and M. Imbard, 1996: A global ocean mesh to overcome the North Pole singularity. *CD*, 12, 381–388.

Marion, J. B., and S. T. Thornton, 1988: *Classical Dynamics of Particles and Systems*. Harcourt Brace Jovanovich.

Marshall, J., A. Adcroft, J.-M. Campin, and C. Hill, 2003: Atmosphere-ocean modeling exploiting fluid isomorphisms. *Journal of Physical Oceanography*. In press.

Marshall, J., C. Hill, L. Perelman, and A. Adcroft, 1997: Hydrostatic, quasi-hydrostatic, and nonhydrostatic ocean modeling. *Journal of Geophysical Research*, 102, 5733–5752.

Marshall, J., and F. Schott, 1999: Open-ocean convection: observations, theory, and models. *Reviews of Geophysics*, 37, 1–64.

Marshall, J., and G. Shutts, 1981: A note on rotational and divergent eddy fluxes. *Journal of Physical Oceanography*, 11, 1677–1680.

McDougall, T. J., 1987a: Neutral surfaces. *Journal of Physical Oceanography*, 17, 1950–1967.

McDougall, T. J., 1987b: Thermobaricity, cabbeling, and water-mass conversion. *Journal of Geophysical Research*, 92, 5448–5464.

McDougall, T. J., 1991: Parameterizing mixing in inverse models. P. Müller and D. Henderson, Eds., *Dynamics of Oceanic Internal Gravity Waves*, Proceedings of the 6th 'Aha Huliko'a Hawaiian Winter Workshop, University of Hawaii at Manoa, 355–386.

McDougall, T. J., 1995: The influence of ocean mixing on the absolute velocity vector. *Journal of Physical Oceanography*, 25, 705–725.

McDougall, T. J., 2003: Potential enthalpy: a conservative oceanic variable for evaluating heat content and heat fluxes. *Journal of Physical Oceanography*, 33, 945–963.

McDougall, T. J., and J. A. Church, 1986: Pitfalls with numerical representations of isopycnal and diapycnal mixing. *Journal of Physical Oceanography*, 16, 196–199.

McDougall, T. J., and R. Feistel, 2003: What causes the adiabatic lapse rate. *Deep-Sea Research*, 50, 1523–1535.

McDougall, T. J., and C. J. R. Garrett, 1992: Scalar conservation equations in a turbulent ocean. *Deep-Sea Research*, 11/12, 1953–1966.

McDougall, T. J., R. J. Greatbatch, and Y. Lu, 2003a: On conservation equations in oceanography: How accurate are Boussinesq ocean models? *Journal of Physical Oceanography*, 32, 1574–1584.

McDougall, T. J., and D. R. Jackett, 1988: On the helical nature of neutral trajectories in the ocean. *Progress in Oceanography*, 20, 153–183.

McDougall, T. J., D. R. Jackett, D. G. Wright, and R. Feistel, 2003b: Accurate and computationally efficient algorithms for potential temperature and density of seawater. *Journal of Atmospheric and Oceanic Technology*, 20, 730–741.

McDougall, T. J., and P. C. McIntosh, 2001: The temporal-residual-mean velocity. Part II: isopycnal interpretation and the tracer and momentum equations. *Journal of Physical Oceanography*, 31, 1222–1246.

McIntosh, P. C., and T. J. McDougall, 1996: Isopycnal averaging and the residual mean circulation. *Journal of Physical Oceanography*, 26, 1655–1660.

Mellor, G. L., 1996: *Introduction to Physical Oceanography*. American Institute of Physics.

Mellor, G. L., and T. Ezer, 1995: Sea level variations induced by heating and cooling: an evaluation of the Boussinesq approximation in ocean models. *Journal of Geophysical Research*, 100, 20 565–20 577.

Mellor, G. L., S. Häkkinen, T. Ezer, and R. Patchen, 2002: A generalization of a sigma coordinate ocean model and an intercomparison of model vertical grids. N. Pinardi and J. Woods, Eds., *Ocean Forecasting: Conceptual Basis and Applications*, Springer, 55–72.

Merryfield, W. J., and G. Holloway, 2003: Application of an accurate advection algorithm to sea-ice modelling. *Ocean Modelling*, 5, 1–15.

Messinger, F., 1973: A method for construction of second-order accurate difference schemes permitting no false two-grid-interval waves in the height field. *Tellus*, 25, 444–457.

Middleton, J. F., and J. W. Loder, 1989: Skew fluxes in polarized wave fields. *Journal of Physical Oceanography*, 19, 68–76.

Moritz, H., 2000: Geodetic reference system 1980. *Journal of Geodesy*, 74, 128–133.

Morse, P. M., and H. Feshbach, 1953: *Methods of Theoretical Physics Part I and II*. McGraw-Hill Book Company.

Müller, P., 1995: Ertel's potential vorticity theorem in physical oceanography. *Reviews of Geophysics*, 33, 67–97.

Munk, W., and C. Wunsch, 1998: Abyssal recipes II: Energetics of tidal and wind mixing. *Deep-Sea Research*, 45, 1977–2010.

Munk, W. H., 1950: On the wind-driven ocean circulation. *Journal of Meteorology*, 7, 79–93.

Murray, R. J., 1996: Explicit generation of orthogonal grids for ocean models. *Journal of Computational Physics*, 126, 251–273.

Murray, R. J., and C. J. C. Reason, 2002: A curvilinear version of the Bryan-Cox ocean model. *Journal of Computational Physics*, 171, 1–46.

Noble, B., and J. W. Daniel, 1977: *Applied Linear Algebra*. Prentice-Hall.

Nurser, A. G., and M.-M. Lee, 2003a: Isopycnal averaging at constant height. Part I: The formulation and a case study, preprint.

Nurser, A. G., and M.-M. Lee, 2003b: Isopycnal averaging at constant height. Part II: Relating to the residual streamfunction in eulerian space, preprint.

O'Brien, J. J., 1986: *Advanced Physical Oceanographic Numerical Modelling*. D. Reidel Publishing Company.

Olbers, D. J., M. Wenzel, and J. Willebrand, 1985: The inference of North Atlantic circulation patterns from climatological hydrographic data. *Reviews of Geophysics*, 23, 313–356.

Osborn, T. R., 1980: Estimates of the local rate of vertical diffusion from dissipation measurements. *Journal of Physical Oceanography*, 10, 83–89.

Osborn, T. R., and C. S. Cox, 1972: Oceanic fine structure. *Geophysical Fluid Dynamics*, 3, 321–345.

Pacanowski, R. C., 1995: *MOM2 Documentation, User's Guide, and Reference Manual*. NOAA/Geophysical Fluid Dynamics Laboratory.

Pacanowski, R. C., K. Dixon, and A. Rosati, 1991: *The GFDL Modular Ocean Model User Guide*. NOAA/Geophysical Fluid Dynamics Laboratory.

Pacanowski, R. C., and A. Gnanadesikan, 1998: Transient response in a z-level ocean model that resolves topography with partial-cells. *Monthly Weather Review*, 126, 3248–3270.

Pacanowski, R. C., and S. M. Griffies, 1999: *The MOM3 Manual*. NOAA/Geophysical Fluid Dynamics Laboratory.

Panton, R. L., 1996: *Incompressible Flow*. 2nd ed., John Wiley and Sons.

Pedlosky, J., 1987: *Geophysical Fluid Dynamics*. 2nd ed., Springer-Verlag.

Pedlosky, J., 1996: *Ocean Circulation Theory*. Springer-Verlag.

Pedlosky, J., 2003: *Waves in the Ocean and Atmosphere: Introduction to Wave Dynamics*. Springer-Verlag.

Peixoto, J. P., and A. H. Oort, 1992: *Physics of Climate*. American Institute of Physics.

Philander, S. G., 1998: *Is the Temperature Rising? The Uncertain Science of Global Warming*. Princeton University Press.

Phillips, N. A., 1957: A coordinate system having some special advantages for numerical forecasting. *Journal of Meteorology*, 14, 184–185.

Phillips, N. A., 1959: An example of a non-linear computational instability. B. Bolin, Ed., *The Atmosphere and the Sea in Motion: Rossby Memorial Volume*, Rockefeller Institute Press, 501–504.

Pickard, G. L., and W. J. Emery, 1990: *Descriptive Physical Oceanography*. 5th ed., Pergamon Press.

Pielke, R., 2002: *Mesoscale Meteorological Modeling*. International Geophysics Series, Academic Press.

Pinardi, N., A. Rosati, and R. C. Pacanowski, 1995: The sea surface pressure formulation of rigid lid models. Implications for altimetric data assimilation studies. *Journal of Marine Systems*, 6, 109–119.

Plumb, R. A., 1979: Eddy fluxes of conserved quantities by small-amplitude waves. *Journal of Atmospheric Sciences*, 36, 1699–1704.

Plumb, R. A., and J. D. Mahlman, 1987: The zonally averaged transport characteristics of the GFDL general circulation/transport model. *Journal of Atmospheric Sciences*, 44, 298–327.

Polzin, K. L., K. Speer, J. M. Toole, and R. W. Schmitt, 1996: Intense mixing of Antarctic bottom water in the equatorial Atlantic. *Nature*, 380, 54–57.

Polzin, K. L., J. M. Toole, J. R. Ledwell, and R. W. Schmitt, 1997: Spatial variability of turbulent mixing in the abyssal ocean. *Science*, 276, 93–96.

Price, J. F., and M. O. Baringer, 1994: Outflows and deep water production by marginal seas. *Progress in Oceanography*, 33, 161–200.

Rahmstorf, S., 1993: A fast and complete convection scheme for ocean models. *Ocean Modelling*, 101, 9–11.

Randall, D. A., 2000: *General Circulation Model Development*. Academic Press.

Redi, M. H., 1982: Oceanic isopycnal mixing by coordinate rotation. *Journal of Physical Oceanography*, 12, 1154–1158.

Reichl, L. E., 1987: *A Modern Course in Statistical Physics*. John Wiley and Sons.

Reif, F., 1965: *Fundamentals of Statistical and Thermal Physics*. McGraw-Hill.

Rhines, P. B., 1982: Basic dynamics of the large-scale geostrophic circulation. *WHOI 1982 Summer Study Program*, Woods Hole Oceanographic Institute.

Rhines, P. B., 1986: Vorticity dynamics of the oceanic general circulation. *Annual Review of Fluid Mechanics*, 18, 433–497.

Rhines, P. B., and W. R. Young, 1982: Homogenization of potential vorticity in planetary gyres. *Journal of Fluid Mechanics*, 122, 347–367.

Richards, K. J., and N. R. Edwards, 2003: Lateral mixing in the equatorial Pacific: The importance of inertial instability. *Geophysical Research Letters*, 30, 1888.

Roberts, M. J., and D. Marshall, 1998: Do we require adiabatic dissipation schemes in eddy-resolving ocean models? *Journal of Physical Oceanography*, 28, 2050–2063.

Roberts, M. J., and D. Marshall, 2000: On the validity of downgradient eddy closures in ocean models. *Journal of Geophysical Research*, 105, 28 613–28 628.

Rosati, A., and K. Miyakoda, 1988: A general circulation model for upper ocean simulation. *Journal of Physical Oceanography*, 18, 1601–1626.

Saffman, P. G., 1992: *Vortex Dynamics*. Cambridge University Press.

Salmon, R., 1988: Hamiltonian fluid mechanics. *Annual Review of Fluid Mechanics*, 20, 225–256.

Salmon, R., 1998: *Lectures on Geophysical Fluid Dynamics*. Oxford University Press.

Schmitt, R., 1998: Double-diffusive convection. E. P. Chassignet and J. Verron, Eds., *Ocean Modeling and Parameterization*, NATO ASI Mathematical and Physical Sciences Series, Vol. 516, Kluwer, 215–234.

Schmitt, R. W., 1994: Double diffusion in oceanography. *Annual Review of Fluid Mechanics*, 26, 255–285.

Schneider, T., I. M. Held, and S. T. Garner, 2003: Boundary effects in potential vorticity dynamics. *Journal of Atmospheric Sciences*, 60, 1024–1040.

Schutz, B. F., 1980: *Geometrical Methods of Mathematical Physics*. Cambridge University Press.

Schutz, B. F., 1985: *A First Course in General Relativity*. Cambridge University Press.

Semtner, A. J., 1974: An oceanic general circulation model with bottom topography. *Numerical Simulation of Weather and Climate*, Technical Report No. 9, UCLA Department of Meteorology.

Semtner, A. J., and Y. Mintz, 1977: Numerical simulation of the Gulf Stream and mid-ocean eddies. *Journal of Physical Oceanography*, 7, 208–230.

Send, U., and R. Käse, 1998: Parameterization of processes in deep convection regimes. E. P. Chassignet and J. Verron, Eds., *Ocean Modeling and Parameterization*, NATO ASI Mathematical and Physical Sciences Series, Vol. 516, Kluwer, 191–214.

Send, U., and J. Marshall, 1995: Integral effects of deep convection. *Journal of Physical Oceanography*, 25, 855–872.

Siedler, G., J. Gould, and J. Church, 2001: *Ocean Circulation and Climate: Observing and Modelling the Global Ocean*. Academic Press.

Smagorinsky, J., 1963: General circulation experiments with the primitive equations: I. The basic experiment. *Monthly Weather Review*, 91, 99–164.

Smagorinsky, J., 1993: Some historical remarks on the use of nonlinear viscosities. B. Galperin and S. A. Orszag, Eds., *Large Eddy Simulation of Complex Engineering and Geophysical Flows*, Cambridge University Press, 3–36.

Smith, K. S., and G. K. Vallis, 2002: The scales and equilibration of midocean eddies: Forced-dissipative flow. *Journal of Physical Oceanography*, 32, 1699–1721.

Smith, R. D., 1999: The primitive equations in the stochastic theory of adiabatic stratified turbulence. *Journal of Physical Oceanography*, 29, 1865–1880.

Smith, R. D., J. K. Dukowicz, and R. C. Malone, 1992: Parallel ocean general circulation modeling. *Physica D*, 60, 38–61.

Smith, R. D., and P. R. Gent, 2004: Anisotropic GM parameterization for ocean models. *Journal of Physical Oceanography*, in press.

Smith, R. D., S. Kortas, and B. Meltz, 1995: Curvilinear coordinates for global ocean models. *Los Alamos preprint*, LA-UR-95-1146.

Smith, R. D., and J. C. McWilliams, 2003: Anisotropic horizonal viscosity for ocean models. *Ocean Modelling*, 5, 129–156.

Solomon, H., 1971: On the representation of isentropic mixing in ocean models. *Journal of Physical Oceanography*, 1, 233–234.

Stacey, M. W., S. Pond, and Z. P. Nowak, 1995: A numerical model of the circulation in Knight Inlet, British Columbia, Canada. *Journal of Physical Oceanography*, 25, 1037–1062.

Stammer, D., 1997: Global characteristics of ocean variability estimated from regional TOPEX/POSEIDON altimeter measurements. *Journal of Physical Oceanography*, 27, 1743–1769.

Starr, V. P., 1945: A quasi-Lagrangian system of hydrodynamical equations. *Journal of Meteorology*, 2, 227–237.

Stone, P., 1972: A simplified radiative-dynamical model for the static stability of rotating atmospheres. *Journal of Atmospheric Sciences*, 29, 405–418.

Sun, S., R. Bleck, C. Rooth, J. Dukowicz, E. Chassignet, and P. D. Killworth, 1999: Inclusion of thermobaricity in isopycnic-coordinate ocean models. *Journal of Physical Oceanography*, 29, 2719–2729.

Tanden, A., and C. J. R. Garrett, 1996: On a recent parameterization of mesoscale eddies. *Journal of Physical Oceanography*, 26, 406–411.

Tomczak, M., and J. S. Godfrey, 1994: *Regional Oceanography: An Introduction*. Pergamon Press.

Toole, J., 1998: Turbulent mixing in the ocean. E. P. Chassignet and J. Verron, Eds., *Ocean Modeling and Parameterization*, NATO ASI Mathematical and Physical Sciences Series, Vol. 516, Kluwer Academic Publishers, 171–190.

Toole, J. M., and T. J. McDougall, 2001: Mixing and stirring in the ocean interior. G. Siedler, J. Gould, and J. Church, Eds., *Ocean Circulation and Climate: Observing and Modelling the Global Ocean*, Academic Press, 337–356.

Toole, J. M., K. L. Polzin, and R. W. Schmitt, 1994: Estimates of diapycnal mixing in the abyssal ocean. *Science*, 264, 1120–1123.

Toole, J. M., R. W. Schmitt, and K. L. Polzin, 1997: Near-boundary mixing above the flanks of a mid-latitude seamount. *Journal of Geophysical Research*, 102, 947–959.

Treguier, A. M., 1999: Evaluating eddy mixing coefficients from eddy-resolving ocean models: A case study. *Journal of Marine Research*, 57, 89–108.

Treguier, A. M., I. M. Held, and V. D. Larichev, 1997: On the parameterization of quasi-geostrophic eddies in primitive equation ocean models. *Journal of Physical Oceanography*, 27, 567–580.

Trenberth, K. E., 1992: Climate system modeling. K. E. Trenberth, Ed., *Climate System Modeling*, Cambridge University Press.

Veronis, G., 1973: Large scale ocean circulation. *Advances in Applied Mechanics*, 13, 2–92.

Veronis, G., 1975: The role of models in tracer studies. *Numerical Models of Ocean Circulation*, National Academy of Sciences.

Visbeck, M., J. C. Marshall, T. Haine, and M. Spall, 1997: Specification of eddy transfer coefficients in coarse resolution ocean circulation models. *Journal of Physical Oceanography*, 27, 381–402.

Visbeck, M., J. C. Marshall, and H. Jones, 1996: On the dynamics of convective "chimneys" in the ocean. *Journal of Physical Oceanography*, 26, 1721–1734.

Wajsowicz, R. C., 1993: A consistent formulation of the anisotropic stress tensor for use in models of the large-scale ocean circulation. *Journal of Computational Physics*, 105, 333–338.

Washington, W. M., and C. L. Parkinson, 1986: *An Introduction to Three-Dimensional Climate Modeling*. University Science Books.

Weaver, A. J., and M. Eby, 1997: On the numerical implementation of advection schemes for use in conjunction with various mixing parameterizations in the GFDL ocean model. *Journal of Physical Oceanography*, 27, 369–377.

Webb, D. J., A. C. Coward, B. A. de Cuevas, and C. S. Gwilliam, 1998a: *The first main run of the OCCAM global ocean model*. Internal Document No. 34, Southampton Oceanography Centre.

Webb, D. J., B. A. de Cuevas, and C. S. Richmond, 1998b: Improved advection schemes for ocean models. *Journal of Atmospheric and Oceanic Technology*, 15, 1171–1187.

Weinberg, S., 1972: *Gravitation and Cosmology*. John Wiley and Sons.

Winton, M., R. W. Hallberg, and A. Gnanadesikan, 1998: Simulation of density-driven frictional downslope flow in z-coordinate ocean models. *Journal of Physical Oceanography*, 28, 2163–2174.

Wright, D. K., 1997: A new eddy mixing parameterization in an ocean general circulation model. *International WOCE Newsletter*, 26, 27–29.

Young, W. R., 1994: The subinertial mixed layer approximation. *Journal of Physical Oceanography*, 24, 1812–1826.

Index